auto mechanics fundamentals

HOW and WHY of the
Design, Construction and
Operation of Automotive Units

MARTIN W. STOCKEL

Industrial Education
Consultant

South Holland, Illinois
THE GOODHEART-WILLCOX COMPANY, INC.
Publishers

Library of Congress Cataloging in Publication Data

Stockel, Martin Wall,
    Auto mechanics fundamentals.

    Bibliography: p.
    1. Automobiles--Design and construction.
    2. Automobile engineering.   I.   Title.
TL240.S84   1974   629.2'3   74—13985
ISBN 0—87006—183—6

INTRODUCTION

It is the aim of AUTO MECHANICS FUNDAMENTALS to provide the student with a thorough understanding of the Design, Construction and Operation of Automotive Units.

This text has many unusual features.

Each unit is approached by starting with basic theory, then parts are added until the unit is complete. By following this procedure, the function of each unit is explained, and its relationship to the complete car is made clear.

A build-it-yourself approach is used in a number of areas. The student "builds" on paper, the units being described. This stimulates student interest, and facilitates a good understanding of the basic principles that are so essential in learning the HOW and WHY of auto mechanics. Math, physics, chemistry, electricity, magnetism and hydraulics are covered right in the text, WHERE they apply.

Hundreds of illustrations were drawn especially for this text. Important areas are featured in these drawings, and many are exaggerated to place emphasis on parts being discussed. Nonrelated, unimportant features are minimized, or completely removed. This assists greatly in getting points across.

Unfamiliar words that may stump the student are defined immediately following the words. In addition, the Dictionary of Terms in the back of the book defines more than 1100 terms with which the student should be familiar, including hot rod references. Every effort has been made to retain the student's keen interest.

Safety Precautions and Safe Working Practices are stressed where jobs discussed involve the possibility of serious accident. In this way, safety instruction becomes meaningful. The student will better understand and appreciate the value of Safety.

A clear and concise summary is provided at the close of each chapter. Also provided are thought provoking questions which test the student's comprehension and retention, and determine the extent of his grasp of the overall picture presented in the chapter.

Most chapters include a brief "shop talk" aimed at pointing up the value of proper attitudes, good work and study habits, and the relationship between these traits and job success.

An entire chapter is devoted to exploring Career Opportunities, and the ways and means of obtaining additional training in auto mechanics. Advantages of graduating, compared to the plight of the "drop-out," are pointed out. An attempt is made to develop pride in the trade, and an awareness of the importance of the professional auto mechanic.

AUTO MECHANICS FUNDAMENTALS provides instruction as recommended by the Standards for Automotive Service instruction in Schools, prepared by the Automotive Industry-Vocational Educational Conference.

This text is intended for use in an Auto Mechanics Fundamentals Course in High School, Trade School or College. It also can be used successfully in more advanced auto shop classes to review and emphasize theoretical aspects.

Martin W. Stockel

CONTENTS

Chapter 1

BUILDING AN ENGINE

WHAT IS AN ENGINE?

An engine is a related group of parts assembled in a specific order. In operation, it is designed to convert the energy given off by burning fuel into a useful form.

There are many parts in a modern engine, each one being essential to the engine operation. For the time being, however, we may think of an engine as a device that allows us to pour fuel into one end and get power from the other. Fig. 1-1.

INTERNAL COMBUSTION GASOLINE ENGINE

Internal combustion means burning within. Gasoline indicates what is being burned. Engine refers to the device in which the gasoline is burned.

BUILDING A PAPER ENGINE

An excellent way to learn about engines is to build one--on paper. Make believe that the engine has not yet been invented. YOU will be the inventor. YOU will solve the problems involved, step by step.

WHAT TO USE FOR FUEL

If you are going to convert fuel into useful energy, you will need something that will ignite (burn) easily. It should burn cleanly, be reasonably inexpensive, and should produce sufficient power with good burning characteristics. It must be available in quantity, should be safe to use and easy to transport.

How about dynamite? NO, it is expensive and burns violently. In short, it would blow up the engine. Fig. 1-2.

Kerosene? NO, it is too hard to ignite and does not burn cleanly.

Gasoline? Now you have found a fuel which will serve your purpose.

WHAT IS GASOLINE?

Gasoline is a product obtained by refining crude oil. Fig. 1-4. Basically, the crude oil (obtained from oil wells in the earth) is treated in various ways to produce gasoline. Gasoline

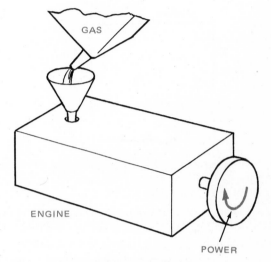

Fig. 1-1. Converting fuel into energy.

is only one of the many items. produced from crude oil, (petroleum).

Gasoline used in engines is a complex mixture of basic fuels and special additives. Since gasoline is a mixture of carbon and hydrogen atoms, it is termed a HYDROCARBON. Fig. 1-3.

Gasoline is available in various qualities. Less expensive

Fig. 1-2. It will explode.

7

gasoline is suitable for some engines, while high compression engines usually require the more expensive fuel. All gasoline qualities are given an OCTANE rating; the higher the octane number, the better the quality. The octane rating indicates how well the gasoline will resist detonation, or too rapid burning, in the engine cylinders.

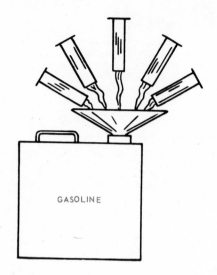

Fig. 1-3. Gasoline is a complex mixture.

Many factors enter into the quality of any gasoline. It must pass exhaustive tests, both in the laboratory and in actual use. Basically, gasoline must burn cleanly, ignite readily and resist vapor lock (turning too quickly from a liquid to a vapor from exposure to excessive heat). It should contain a minimum amount of harmful ingredients and be free from detonation.

WARNING * THE FOLLOWING STEPS, SHOWING HOW GASOLINE IS PREPARED FOR USE IN THE ENGINE, ARE FOR PURPOSES OF ILLUSTRATION ONLY AND SHOULD NOT BE TRIED BY THE STUDENT. GASOLINE IS DANGEROUS — TREAT IT WITH CARE AND RESPECT.

PREPARING THE FUEL

As you know, gasoline burns readily. However, to get the most power from this fuel and, in fact, to get it to run an engine, special treatment is required.

If you were to place a small amount of gasoline in a jar and drop a match in it, it would burn. Fig. 1-5.

Such burning is fine to produce heat, but it will not give us

Fig. 1-5. Gasoline burns.

an explosive burning effect which we need to run an engine. How can we get the gas to burn fast enough to produce an explosive force?

If you will examine Fig. 1-5, you will notice that the gasoline is burning on the top side. Why is it not burning along the sides and bottom of the container? In order to burn, gasoline must combine with oxygen in the air. The sides and bottom are not exposed to air so they do not burn.

BREAKING UP GASOLINE

For purposes of illustration, imagine having gasoline in the same amount and shape, but with no container, left. Fig. 1-6. It will then burn on all sides. This will increase the burning speed. However, it still will not burn quickly enough for use in an engine.

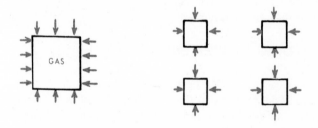

Fig. 1-6. Breaking gasoline into smaller particles exposes more surface to the air, increasing the burning rate.

What to do? The answer: break the gasoline into smaller particles. Notice that as you divide it into smaller particles, right, Fig. 1-6, you expose more and more edges to the air. If you ignite it now, it will burn considerably faster.

If you break up gasoline into very tiny particles, burning is fierce. Rapid burning produces a tremendous amount of heat which, in turn, causes a rapid and powerful expansion. The burning gasoline is now giving off ENERGY in the form of HEAT. Fig. 1-7.

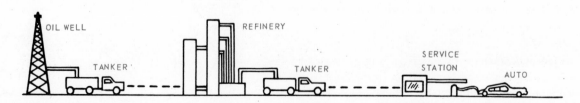

Fig. 1-4. Before reaching your auto, crude oil requires much handling.

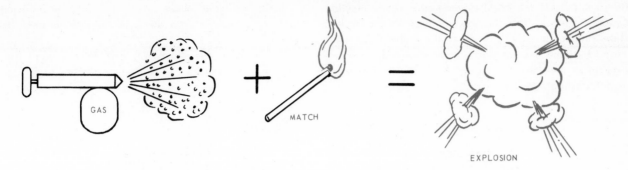

Fig. 1-7. When gas particles are small enough they will explode.

You now have the necessary basic force with which to do work. How do you harness it and make it work for you?

TRAPPING THE EXPLOSION

If you were to take a sturdy metal container, spray a gasoline and air mixture into it, place a lid over the top and then light it, the resulting explosion would blow the lid high into the air. Fig. 1-8.

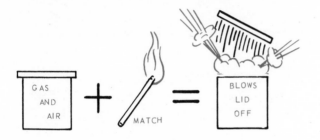

Fig. 1-8. Blowing the lid off.

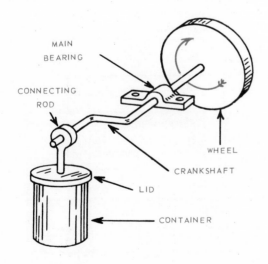

Fig. 1-9. A simple engine.

This is an example of using gasoline to do work. In this case the work is blowing the lid into the air. It is obvious that a flying lid will not push a car but, think a minute, and see if the flying lid does not suggest some way of converting the power into useful motion.

A SIMPLE ENGINE

Make the same setup, only this time hook the lid to a simple crankshaft by means of a connecting rod. Then place a wheel on the other end of the crankshaft. Fig. 1-9.

Now if you explode the mixture, the lid (as it blows up) will give the crankshaft a sharp upward push and cause the wheel to spin. You have built a very simple engine; not practical, but it is pointing the way!

WHAT IS WRONG WITH IT?

Many things. Let us discuss them one at a time. The lid would fly up; as the wheel spun, it would be forced down. It can come down in any position, but to work, it must be over the container.

Instead of putting the lid over the container, cut it so that it just slips inside. Make the container longer so the lid can push the crank to the top of its travel, and still not get out of the container. Fig. 1-10.

If you were to bolt the container and shaft bearing so they could not change position, you would then have an engine that would spin the wheel every time you fired a fuel mixture.

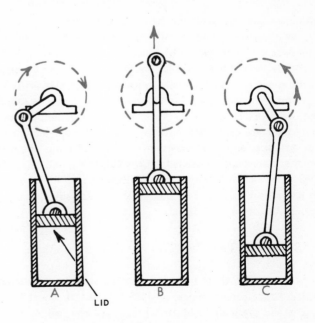

Fig. 1-10. Lid placed inside container forms a simple engine.

In order to cause the wheel to spin in the proper direction, you would have to fire the mixture with the crank in a position similar to A, Fig. 1-10. If the crank is in the B position, the lid could not fly up without pushing the crank bearing up or the container down. If it were fired with the crank in C position, the wheel would spin backwards.

Make the block heavy to give it strength to withstand the fuel explosions. Bring it down to support the main bearing. Fig. 1-11.

By bringing the block out and down around the main bearing, you now have a strong unit. Notice in B, Fig. 1-11, that the lower block end that forms a case around the crank,

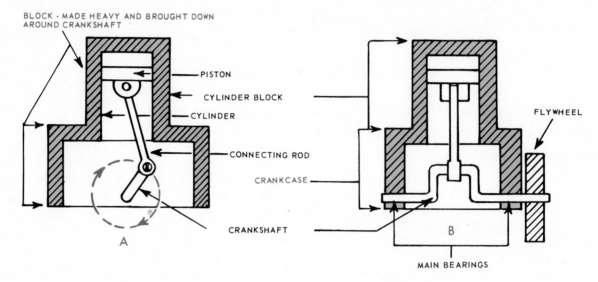

Fig. 1-11. Front, side views showing how engine is inverted, block strengthened and brought down and around crankshaft. This forms the crankcase.

It is important that the mixture be fired when the crank is in the proper position. By studying Fig. 1-10, you can see that the crankshaft changes the RECIPROCATING (up and down) motion of the lid into ROTARY (round and round) motion.

NAME THE PARTS

At this time, it is well to name the parts you have developed. By doing so, you will learn what to call parts in a gasoline engine that serve the same general purpose.

The container would be called the BLOCK. The hole in the container, or block, would be called the CYLINDER. The lid would be termed the PISTON. The shaft, with a section bent in the shape of a crank, would be called the CRANKSHAFT. The rod that connects the crankshaft to the piston, is called the CONNECTING ROD. The bearings that support the crankshaft are called MAIN BEARINGS. The connecting rod has an upper and lower bearing. The lower bearing, (the one on the crankshaft), is called a CONNECTING ROD BEARING. The wheel is called the FLYWHEEL.

Refer back to Fig. 1-9, and see if you can substitute the correct names for those listed.

FASTENING THE PARTS

Since you are going to fasten the parts so the main bearing and cylinder cannot move, it would be wise to invert your engine and place the cylinder on top. Just why, will soon be obvious.

allows you to have two main bearings. This lower block end is called a CRANKCASE.

LENGTHEN THE PISTON

Now make the piston longer. This stops it from tipping sideways in the cylinder. In order to avoid a piston that is too heavy, it can be made hollow as shown in Fig. 1-12.

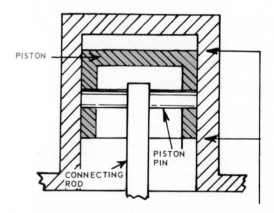

Fig. 1-12. Piston is lengthened to avoid tipping.

Tipping is illustrated in Fig. 1-13.

If the piston is to travel straight up and down, and the connecting rod is to swing back and forth in order to follow

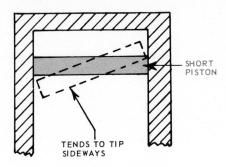

Fig. 1-13. Effect of short piston.

the crank, it is obvious that the connecting rod must be able to move where it is fastened to the piston. Fig. 1-14.

Now, drill a hole through the upper end of the connecting rod; also drill a hole throuh the piston. Line up the two and

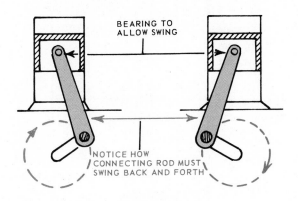

Fig. 1-14. Rod must swing.

pass a pin through them. This pin is called a PISTON or WRIST PIN. It is secured in various ways (more on this later). Fig. 1-15.

GETTING FUEL INTO THE ENGINE

You have probably noticed no way has been provided to get fuel into the upper cylinder area of your assembled engine. Your next problem is to develop a system to admit fuel, and to exhaust (blow out) the fuel after it is burned.

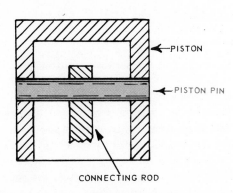

Fig. 1-15. Pin secures rod to piston.

One opening in the upper cylinder area (called the COMBUSTION CHAMBER) will not be adequate. You cannot very well admit fuel and exhaust the burned gas out the same opening. You need TWO openings.

REMOVABLE CYLINDER HEAD

Redesign your cylinder block and make the top removable. You will call this removable head a CYLINDER HEAD. It will be fastened in place with bolts. Fig. 1-16.

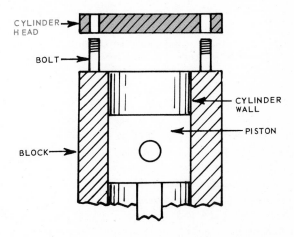

Fig. 1-16. Cylinder head. Top of cylinder has been removed.

FUEL INTAKE AND EXHAUST PASSAGES

Making your cylinder head removable has not yet solved all of the problems. You cannot take it off, and put it on, each time the engine fires.

Now make a block of much thicker metal and make two holes or passages like those shown in Fig. 1-17. This will give you a passage to take in fuel mixture, and one to exhaust it. These passages are called VALVE PORTS.

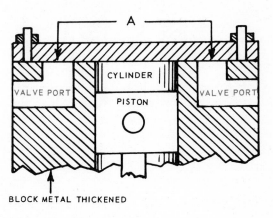

Fig. 1-17. Valve ports.

You will notice at A, Fig. 1-17, that the ports do not connect to the cylinder. To make a connection, you must make the cylinder head thicker and cut an opening in it that will straddle the intake and exhaust ports, plus the cylinder. Fig. 1-18.

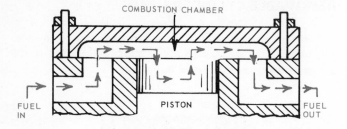

Fig. 1-18. Note opening in head allowing fuel mixture to enter and leave cylinder. This opening is called COMBUSTION CHAMBER.

VALVES

The next logical step is to provide some type of device to open and close the ports. If the ports are left open, and fuel explodes in the combustion chamber, it will blow out the openings and fail to push the piston down.

This port control device (valve) will have to be arranged so that it can be opened and closed when desired. Now arrange a valve in each opening. This may be done as shown in Fig. 1-19.

The valves may be held in a straight up and down position by a hole bored in the block metal at A, Fig. 1-19. This hole is called a VALVE GUIDE because it guides the valve up and down in a straight line.

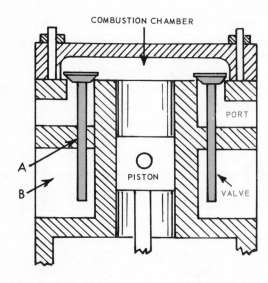

Fig. 1-19. Valves installed in ports.

Note that another opening, B, beneath each valve guide, is cut into the block. You will use this opening to install a spring, spring washer and keeper. The spring is necessary to keep the valve tightly closed against its seat. When the valve is opened, the spring will close it again. Arrange your spring, spring washer, and keepers, as shown in Fig. 1-20.

The valves provide a satisfactory method of opening and closing the passages.

FOUR-STROKE CYCLE

At this point you have an intake valve to allow fuel to enter the cylinder and an exhaust valve to let the burned fuel out. The next problem is how to get fuel into the cylinder and out when burned.

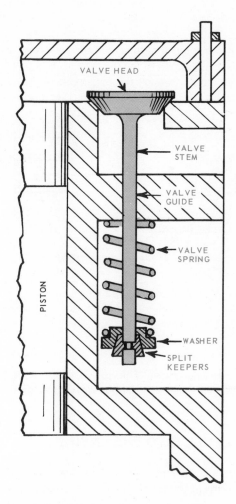

Fig. 1-20. Installation of valve, spring, washer, etc.

A VACUUM IS THE CLUE

You understand that the air in which we live presses on all things. Fig. 1-21. This pressure is approximately 15 lbs. per square inch at sea level.

Were we to draw all of the air out of a container, we would form a vacuum. A vacuum is unnatural and atmospheric pressure will do all it can to get into the low pressure area. If there is the slightest leak in the container, air will seep in until the pressure is the same on the inside as it is on the outside.

A vacuum, then, is any area in which the air pressure is lower than that of the prevailing or outside atmospheric pressure.

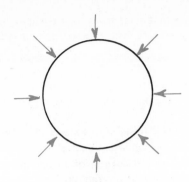

Fig. 1-21. Air presses on all things with a pressure of 15 lbs. per square inch.

YOU ALREADY HAVE A VACUUM PUMP

Fig. 1-22 illustrates a simple vacuum pump, or a cylinder into which a snug fitting piston is placed. You will see in A, Fig. 1-22, that the piston is against the end of the cylinder. Obviously there is no air between these two surfaces.

When you pull the piston through the cylinder, as in B, Fig. 1-22, you will have a large area between the face of the piston, and the end of the cylinder. If there was no air between them before, and if the piston fits snugly against the cylinder, there still could be no air between them. This gives you a large area in which there is no air, or in other words, a vacuum.

If you were to drill a hole into the closed end of the cylinder, C, Fig. 1-22, air would immediately rush in and fill the cylinder. Any material, such as a fuel mixture, that happened to be in the surrounding air would be drawn into the cylinder.

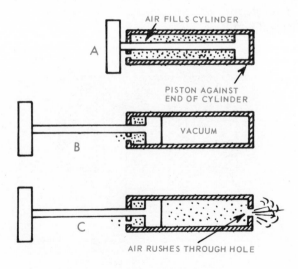

Fig. 1-22. Simple vacuum pump.

WHERE IS YOUR VACUUM PUMP?

If the cylinder and piston, in Fig. 1-22, forms a vacuum, the cylinder and piston in your engine will do the same. If the piston is at the top of the cylinder (with both valves closed)

and you turned the crankshaft, the piston will be drawn down into the cylinder. This forms a strong vacuum in the cylinder. If you now open the intake valve, the air will rush into the cylinder. This is called the INTAKE STROKE.

STROKE NO. 1 — THE INTAKE STROKE

The first stroke in your engine is the intake. Instead of opening the intake valve after you have drawn the piston down, you will find it better to open the intake valve as the piston starts down. This allows the air to draw fuel in all the time the piston is moving down. If you wait until the piston is down before opening the valve, the piston will be starting up before the cylinder can be filled with air. Fig. 1-23.

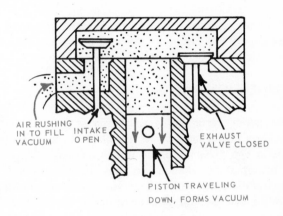

Fig. 1-23. Intake stroke.

Remember: The intake stroke starts with the piston at the top of the cylinder, intake valve open and exhaust valve closed, and it stops with the piston at the bottom of its travel. This requires one-half turn of the crankshaft.

STROKE NO. 2 — THE COMPRESSION STROKE

You have discovered that the smaller the particles of gasoline, mixed in air, the more powerful the explosion.

As the crankshaft continues to move, the piston is forced up through the cylinder. If you keep both valves closed, the fuel mixture will be squeezed, or compressed, as the piston reaches the top. This is called the compression stroke. Fig. 1-24. It, too, requires one-half turn of the crankshaft.

The compression stroke serves several purposes. First, it tends to break up the fuel into even smaller particles. This happens due to the sudden swirling and churning of the mixture as it is compressed.

When engine fuel mixture is subjected to a sudden sharp compression force, its temperature rises. This increase in temperature makes the mixture easier to ignite, and causes it to explode with greater force.

As the piston reaches the top of its travel on the compression stroke, it has returned to the proper position to be pushed back down by the explosion.

Remember: The compression stroke starts with the piston

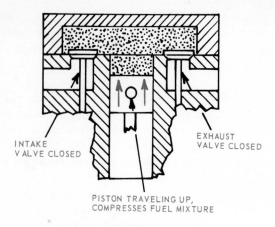

INTAKE VALVE CLOSED

EXHAUST VALVE CLOSED

PISTON TRAVELING UP, COMPRESSES FUEL MIXTURE

Fig. 1-24. Compression stroke.

at the bottom of the cylinder while both valves are closed, and it stops with the piston at the top of the cylinder. This requires an additional one-half turn of the crankshaft.

COMPRESSION RATIO

The amount your engine will compress the fuel mixture depends on how small a space the mixture is squeezed into. Notice in Fig. 1-25, the piston in A has traveled down 6 in. from the top of the cylinder. This is the intake stroke. In B the piston, on the compression stroke, has traveled up to within 1 in. of the cylinder top. It is obvious that 6 in. of cylinder volume have been squeezed into 1 in. of cylinder volume. This gives you a ratio of 6 to 1. This is termed the COMPRESSION RATIO.

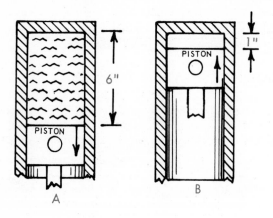

Fig. 1-25. Compression ratio.

STROKE NO. 3 — FIRING OR POWER STROKE

As the piston reaches the top of the compression stroke, the mixture is broken into tiny particles, and heated up. When ignited, it will explode with great force.

This is the right time to explode the mixture. A spark is provided inside the combustion chamber by means of a spark plug. The spark produced at the plug is formed by the ignition system. This will be discussed later.

Just imagine that a hot spark has beeen provided in the fuel mixture. The mixture will explode and, in turn, force the piston down through the cylinder. This gives the crankshaft a quick and forceful push.

Both valves must be kept closed during the firing stroke or the pressure of the burning fuel will squirt out the valve ports. Fig. 1-26.

Remember: The firing stroke starts with the piston at the top of the cylinder while both valves are closed, and it stops with the piston at the bottom of the cylinder. This requires another one-half turn of the crankshaft.

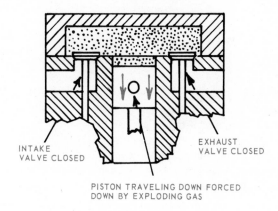

INTAKE VALVE CLOSED

EXHAUST VALVE CLOSED

PISTON TRAVELING DOWN FORCED DOWN BY EXPLODING GAS

Fig. 1-26. Firing stroke.

STROKE NO. 4 — THE EXHAUST STROKE

When the piston reaches the bottom of the firing stroke the exhaust valve opens. The spinning crankshaft forces the piston up through the cylinder blowing burned gases out of the cylinder. Fig. 1-27.

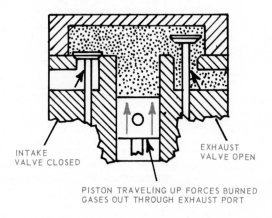

INTAKE VALVE CLOSED

EXHAUST VALVE OPEN

PISTON TRAVELING UP FORCES BURNED GASES OUT THROUGH EXHAUST PORT

Fig. 1-27. Exhaust stroke.

Remember: The exhaust stroke starts with the piston at the bottom of the cylinder while the exhaust valve is open and intake valve is closed. It stops with the piston at the top of the cylinder. This requires one more one-half turn of the crankshaft.

COMPLETED CYCLE

If you will count the number of one-half turns in the intake, compression, firing and exhaust strokes, you will find you have four one-half turns. This gives you two complete turns (called revolutions) of the crankshaft. Fig. 1-28.

While the crankshaft is turning around twice, it is receiving power only during one-half turn, or one-fourth of the time.

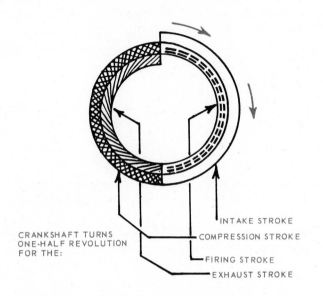

CRANKSHAFT TURNS
ONE-HALF REVOLUTION
FOR THE:

INTAKE STROKE
COMPRESSION STROKE
FIRING STROKE
EXHAUST STROKE

Fig. 1-28. Four strokes make two revolutions of the crankshaft.

CYCLE REPEATED

As soon as the piston reaches the top of the exhaust stroke, it starts down on another intake, compression, firing and exhaust cycle. This is repeated time after time. Each complete cycle consists of four strokes of the piston, hence the name FOUR-STROKE CYCLE engine. There is another type, called a two-stroke cycle, that will be covered later.

WHAT OPENS AND CLOSES THE VALVES?

You have seen that the intake valve must be opened for the intake stroke. Both valves remain closed during the compression and firing strokes. The exhaust valve opens during the exhaust stroke. You must now design a device to open and close the valves.

Start by machining a round shaft that will lie beneath the valve stem, at some distance. This shaft is supported in bearings placed in the front and the rear of the crankcase. See Fig. 1-29.

CAMSHAFT OPENS VALVES

Each shaft will have a bump, called a cam lobe, machined as a part of the shaft. This shaft, with the cam lobe, is called a CAMSHAFT. Fig. 1-30.

The distance the valve will be raised, how long it will stay open, how fast it opens and closes, can all be controlled by the

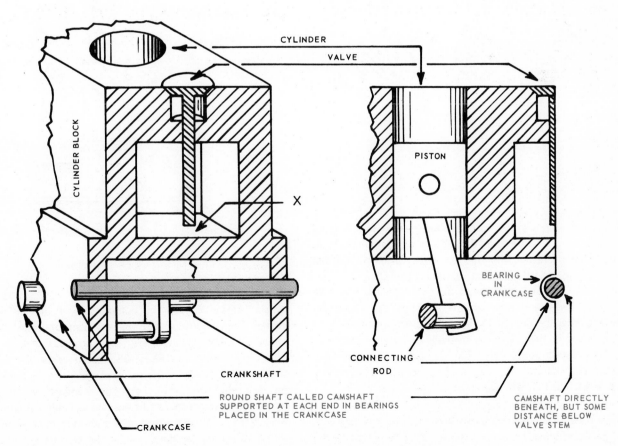

CYLINDER

VALVE

CYLINDER BLOCK

X

PISTON

BEARING
IN
CRANKCASE

CONNECTING
ROD

CRANKSHAFT

ROUND SHAFT CALLED CAMSHAFT
SUPPORTED AT EACH END IN BEARINGS
PLACED IN THE CRANKCASE

CAMSHAFT DIRECTLY
BENEATH, BUT SOME
DISTANCE BELOW
VALVE STEM

CRANKCASE

Fig. 1-29. Engine section showing how camshaft is positioned.

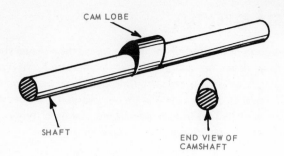

Fig. 1-30. Camshaft.

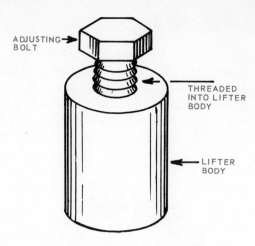

Fig. 1-32. Valve lifter.

height and shape of the lobe. The lobe must, of course, be located directly beneath the valve stem. Fig. 1-31.

As you will see later, it is impractical to have the cam lobe contact the end of the valve stem itself. You have placed the camshaft some distance below the end of the valve stem. When you turn the camshaft, the lobes will not even touch the valve stem.

CAMSHAFT SPEED

You have developed a method of opening and closing the valves. The next problem is how to turn the camshafts, and at WHAT SPEED.

Now stop a minute and think — each valve must be open for one stroke only. The intake is open during the intake, and

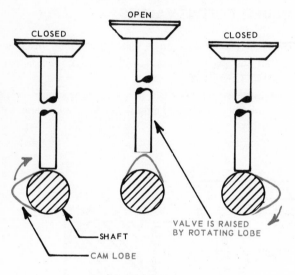

Fig. 1-31. How valve is opened.

VALVE LIFTER

Your next step is to construct a round, bar-like unit called a cam follower or VALVE LIFTER. This valve lifter will have an adjusting bolt screwed into the top so the distance between the valve stem and lifter may be adjusted. Fig. 1-32.

The lifter is installed between the cam lobe and the valve stem. The lower end rides on the lobe and the upper end almost touches the valve stem. The lifter slides up and down in a hole bored in the crankcase metal that separates the valve stem from the camshaft. Fig. 1-33.

THE VALVE TRAIN

You have now built the essential parts of the valve system. The parts, in their proper positions, are referred to as the VALVE TRAIN. Fig. 1-34.

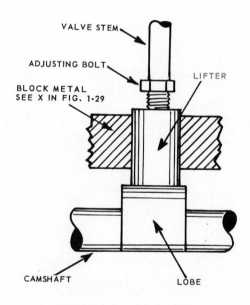

Fig. 1-33. Valve lifter installation.

remains closed during the compression, firing and exhaust strokes. This would indicate that the cam lobe must turn fast enough to raise the valve every fourth stroke. Fig. 1-35.

You can see, from Fig. 1-35, that it takes one complete revolution of the cam lobe for every four strokes. Remembering back, four strokes of the piston required two revolutions of the crankshaft. You can say then, that for every two revolutions of the crankshaft, the camshaft must turn once. If you are speaking of the speed of the camshaft, you can say that the camshaft must turn at one-half the crankshaft speed.

TURNING THE CAMSHAFT

If the crankshaft is turning, and the camshaft must always turn at one-half crankshaft speed, it seems logical to use the spinning crankshaft to turn the camshaft.

One very simple way to drive the camshaft would be by means of gears, one fastened on the end of the crankshaft and the other on the end of the camshaft.

If you put a small gear on the crankshaft and a larger gear with twice as many teeth on the camshaft, the crankshaft will turn the camshaft. It will also turn it at exactly one-half crankshaft speed. Fig. 1-36.

Fig. 1-37 shows a front view of the crankshaft gear and both camshaft gears as they would appear on the engine. In your engine, the crankshaft gear will turn clockwise (to the right). The timing or camshaft gears will then be driven counterclockwise (to the left).

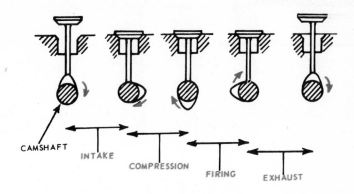

Fig. 1-35. Intake valve is opened once every fourth stroke. Each stroke turns camshaft one-quarter turn.

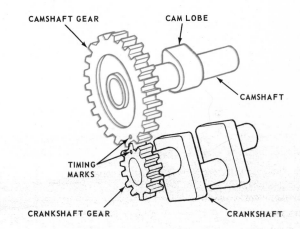

Fig. 1-36. Crankshaft turns camshaft at one-half crankshaft speed. Drive is through gears.

TIMING THE VALVES

You have now developed a method of not only turning the camshaft, but also at the correct speed. The next question to be answered is how will you get the valves to open at the proper time. This is called VALVE TIMING.

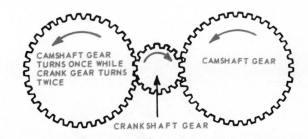

Fig. 1-37. Front view. Crankshaft gear driving both camshaft gears.

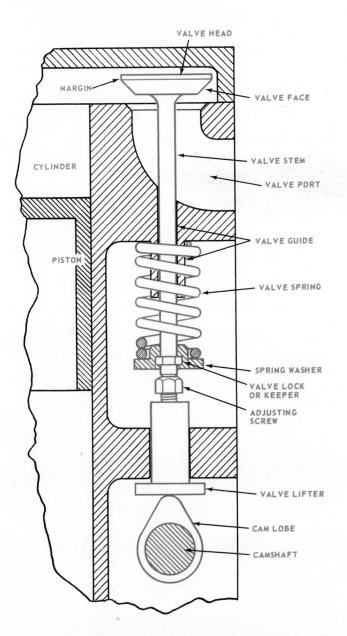

Fig. 1-34. Valve train. Note that the cam lobe has forced the valve open against the pressure of the valve spring.

Start with intake stroke. As you have discovered, the intake valve must open as the piston starts down in the cylinder. Place your piston at top dead center (TDC — at the topmost point of piston travel), Fig. 1-38.

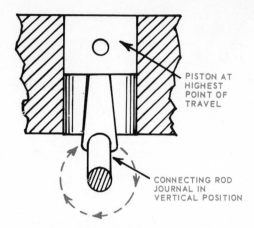

Fig. 1-38. Piston at top dead center (TDC).

Insert the intake camshaft, then turn it in a counterclockwise direction until the heel of the cam lobe contacts the lifter. The timing gear should be meshed with the crankshaft gear. Fig. 1-39.

MARK THE GEARS

You should punch a mark on the crankshaft gear, and one on the camshaft gear. These are called TIMING MARKS. See Fig. 1-40.

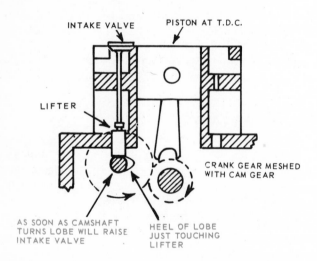

Fig. 1-39. Position of cam lobe at the start of the intake stroke.

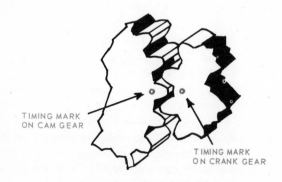

Fig. 1-40. Timing gears meshed and marked.

If the camshafts are removed, they may be reinstalled by merely lining up the timing marks.

Were you to crank the engine, the crankshaft would pull the piston down on the intake stroke. It would also rotate the camshaft, causing the cam lobe to turn and raise and lower the valve. The crankshaft would make one-half turn, the camshaft one-quarter turn during this stroke. Fig. 1-41.

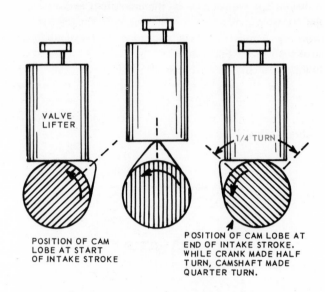

Fig. 1-41. Camshaft makes one-quarter turn during each stroke.

INTAKE REMAINS CLOSED

As the crank continues to turn, it will push the piston up on compression, down on firing, and up on exhaust. During these three strokes, the intake cam lobe continues to turn. When the piston reaches top dead center (TDC) on the exhaust stroke, the heel of the cam lobe will again be touching the intake lifter.

As your next stroke will be the intake, the piston is ready to start down and the cam lobe is ready to raise the lifter to provide proper timing. Fig. 1-42.

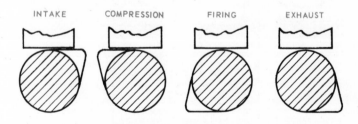

Fig. 1-42. Position of lobe at start of each stroke.

EXHAUST TIMED THE SAME WAY

The only difference between the intake and exhaust timing is that you place the piston on bottom dead center (BDC). Turn the exhaust camshaft until the lobe contacts the lifter, then mesh and mark the gears. Use two marks on the crankshaft gear and two on the camshaft gear for the exhaust camshaft. This prevents aligning with the wrong marks.

THE FLYWHEEL

When you crank an engine, it turns through all the necessary strokes. The only time the crankshaft receives power is during the firing stroke. After the firing stroke is completed, the crankshaft must continue to turn to exhaust the burned gases, take in fresh fuel and then compress it. One power stroke is not enough to keep the crankshaft turning during the four required strokes. Fig. 1-43.

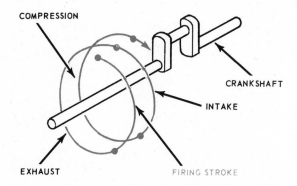

Fig. 1-43. After the firing stroke, the crankshaft must coast through the next three strokes.

You will recall that one end of the crankshaft has a timing gear attached. If you will fasten a fairly large, heavy wheel, called a FLYWHEEL, to the other end, the engine will run successfully.

The flywheel, which is caused to spin by the firing stroke, will continue to spin because it is heavy. In other words, the inertia built up within the flywheel will cause it to keep turning. As it is attached to the crankshaft, it will cause the shaft to continue to turn. The shaft will now spin long enough to reach the next firing stroke. Fig. 1-44.

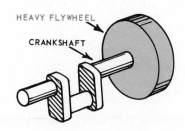

Fig. 1-44. Flywheel inertia helps engine to run.

ENGINE WILL RUN MORE SMOOTHLY

You can see that every time the engine fires, the crank receives a hard push. This will speed it up. As the power stroke is finished, the shaft coasts. This slows it down. The alternating fast and slow speed would make for a very rough running engine.

With the heavy flywheel in place, the firing stroke cannot increase the crankshaft speed so quickly since it must also speed up the flywheel. When the crankshaft is coasting, it cannot slow down so fast because the flywheel wants to keep spinning. This will cause the crankshaft speed to become more constant and will, in turn, give a smooth running engine.

BASIC ENGINE COMPLETED

You have now completed a basic engine. Fig. 1-45. With the addition of fuel, spark and oil, it will run. This is a one cylinder engine, and its uses would be confined to such small tasks as powering lawnmowers, scooters, generators, carts, etc.

There are many modifications and improvements that can be made to your simple engine. These will be discussed in following chapters.

SUMMARY

How an engine works:

The piston is drawn down by the crankshaft. As it starts down, the camshaft opens the intake valve and a gasoline and air mixture is drawn into the cylinder.

When the piston reaches the bottom of its stroke, the intake valve closes, and the crankshaft forces the piston back up through the cylinder. This compresses the air-fuel mixture in the combustion chamber.

As the piston nears the top of the compression stroke, the air-fuel mixture is ignited by a spark from the spark plug. This explodes the mixture, and the pressure of the rapidly expanding gas drives the piston down through the cylinder. Both valves are closed during this firing stroke.

After reaching the bottom of the firing stroke, the exhaust valve opens, and the spinning crankshaft forces the piston up through the cylinder once again. This time all the burned gases are driven, or exhausted, from the cylinder and combustion chamber.

When the piston reaches the top of the exhaust stroke, the exhaust valve closes and the intake opens. The piston is drawn down on another intake stroke.

Once the cylinder has fired, the heavy flywheel will keep the crankshaft spinning long enough to go through the exhaust, intake and compression strokes. It will then receive power from another firing stroke. This cycle is repeated over and over, and thus the engine will run on its own power.

DO NOT BE CONFUSED

Regardless of the size, number of cylinders, and horsepower a four-stroke cycle engine may have, it will contain the same basic parts that do the same job, in the same way, as the small engine you have just built. The parts may be arranged in different ways and their shape changed, but you will not find ANY different BASIC parts, just more of them.

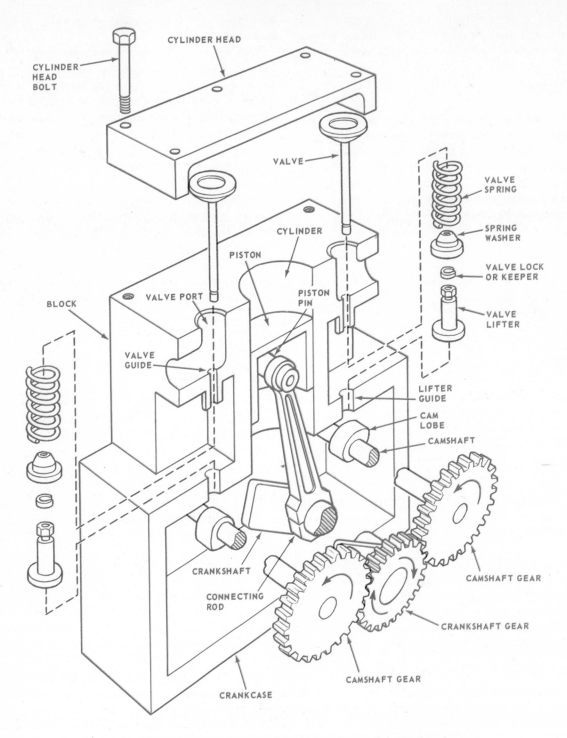

Fig. 1-45. The various parts of the basic engine you have developed.

A MOST IMPORTANT CHAPTER

The chapter you have just completed is most important. To properly understand the theory and operation of the internal combustion engine, it is essential that you COMPLETELY UNDERSTAND ALL THE MATERIAL PRESENTED IN THIS CHAPTER.

To help you check your grasp of the chapter, answer the following questions. When you CAN answer all of the questions correctly, you are on your way.

REVIEW QUESTIONS – CHAPTER 1

Write on separate sheet of paper. DO NOT write in this book.

1. What is an engine?
2. Gasoline is obtained from _____ _____ .
3. Gasoline is a mixture of _____ and _____ atoms.
4. List four features of a good gasoline.

5. In order to burn, gasoline must have _____.
6. What must be done to gasoline to speed up the burning rate?
7. Gasoline, when it reaches the cylinder, must be _____ into tiny _____, and mixed with _____.
8. When the gasoline explodes in the cylinder, what movable part of the engine first receives the force?
9. The connecting rod must have a bearing at _____.
10. What is a vacuum?
11. How is a vacuum developed in an engine?
12. What do you need a vacuum for?
13. What is the purpose of valve ports?
14. What is placed in each port to open and close the port?
15. List the strokes, in order, in a four-stroke cycle engine.
16. The position of each valve during the various strokes is important. Give the position of each valve for each of the strokes in question 15.

17. What opens and closes the valves?
18. The camshaft is turned by the _____ through means of _____.
19. What is the speed of the camshaft in relation to that of the crankshaft?
20. Give two uses for the flywheel.
21. Fig. 1-45 shows all of the parts of your engine. Examine each part. List the part names and use of each of the parts.

STORY:

Do not be like the eager bear hunter who went charging through the woods. When he encountered a bear, he found he had forgotten to load his gun!

MORAL:

There are many BEARS in this book. This chapter contains a lot of ammunition — LOAD UP!

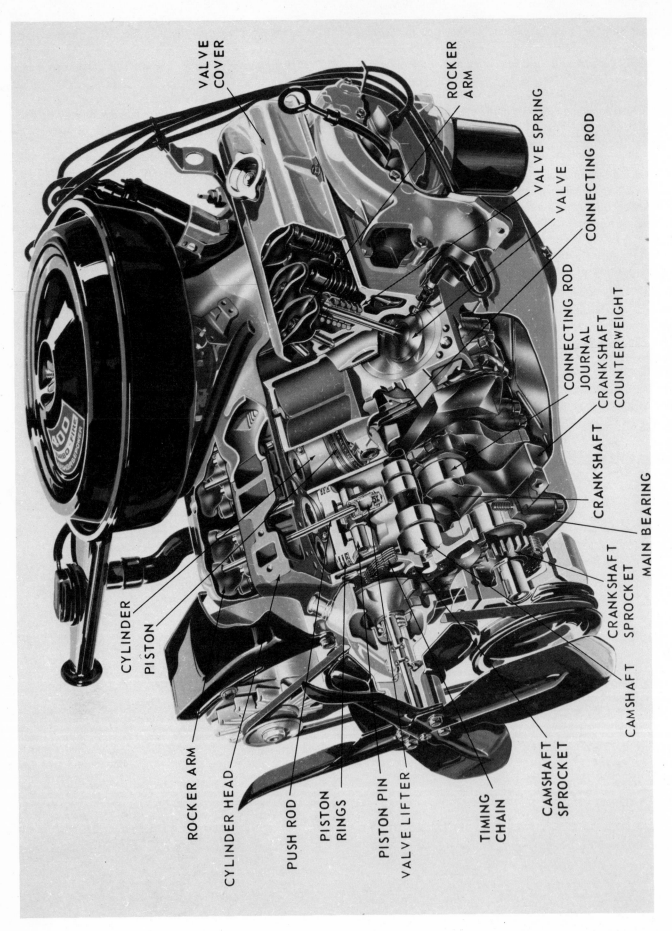

VALVE COVER

ROCKER ARM

VALVE SPRING

VALVE

CONNECTING ROD

CONNECTING ROD JOURNAL

CRANKSHAFT COUNTERWEIGHT

CRANKSHAFT

MAIN BEARING

CRANKSHAFT SPROCKET

CAMSHAFT

CYLINDER

PISTON

ROCKER ARM

CYLINDER HEAD

PUSH ROD

PISTON RINGS

PISTON PIN

VALVE LIFTER

TIMING CHAIN

CAMSHAFT SPROCKET

Cutaway view of a typical V-8 engine. (Chevrolet)

Chapter 2

DESIGN, CONSTRUCTION, APPLICATION OF ENGINE COMPONENTS

You have already designed and built an engine, on paper. To give you a clear idea of how the engine works, it was necessary to move from step to step without studying the finer points of engine construction.

Now that you are familiar with the general theory involved in engine design, it is desirable for you to study the parts of the basic engine in greater detail.

It is important for you to understand HOW the parts are built, of WHAT material, WHY they are built the way they are, and the PURPOSE each serves.

THE ENGINE BLOCK

The block serves as a rigid metal foundation for all parts of an engine. It contains the cylinders. In older engines, the valve seats, ports and guides are built into the block. It also supports

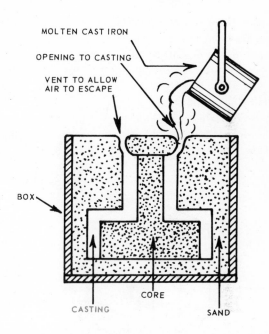

Fig. 2-2. Simple mold for casting.

the crankshaft and camshaft. Accessory units and clutch housing are bolted to it. Fig. 2-1.

Notice in Fig. 2-1, that the crankcase is formed with the block.

Blocks are made of either cast iron or aluminum. In some of the small one cylinder engines, the material is die cast metal.

Die cast metal is a relatively light, soft metal especially suited to the die casting process.

Blocks are commonly formed in two ways. One method is to pour molten cast iron, or aluminum, into a mold made of sand. A core is placed within the mold to form the hollows and passageways within the block. Fig. 2-2.

After the casting has cooled, it is removed from the mold, and the sand core is dissolved and washed out.

The second method is to use a mold of metal and force molten aluminum, or die-cast metal, into the mold under pressure. The pressure casting process has several advantages. It produces a block free of air bubbles (called voids), gives sharp

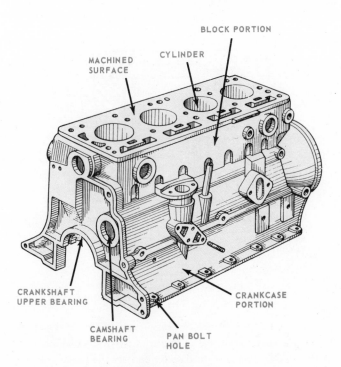

Fig. 2-1. Typical block construction for a four cylinder valve-in-head engine. (Datsun)

corners and a high degree of accuracy can be maintained. This reduces machining operations to a minimum. The same mold can be used over and over.

All parts of the aluminum or die cast block that are subjected to wear will have metal inserts either pressed in place or actually cast into the block.

The lighter the block, providing it has sufficient strength, the better. A more modern process, called Precision Thin Wall Casting, controls core size and placement much more accurately than the older casting process. This permits casting the block walls much thinner, thereby effectively reducing the weight. Since wall thicknesses are more uniform, block distortion during service is less severe.

CYLINDERS

The cylinder is a round hole formed in the block. It is first cast into the block then bored on a special machine and honed to a smooth finish. The cylinder dimensions must be kept extremely accurate. A good cylinder will not vary in diameter more than .0005 in. The paper on which this is printed is around .004 in. in thickness.

The cylinder forms a guide for the piston and acts as a container for intaking, compressing, firing and exhausting the fuel charge. Fig. 2-3.

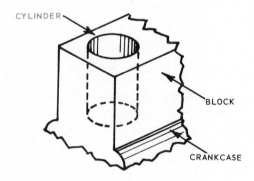

Fig. 2-3. Section of engine showing cylinder location.

Cylinders have been made of both steel and cast iron. Cast iron is by far the most popular. When steel cylinders are desired, they are in the form of cylinder sleeves. These may be either cast, or pressed into the block. A recent development die casts the block of aluminum that contains particles of silicon. After the cylinder is formed, an electro-chemical process is used to etch (eat) away the aluminum cylinder surface. This leaves the silicon exposed. The pistons are iron-plated aluminum and operate directly on the silicon surface of the aluminum cylinders.

CYLINDER SLEEVES

Some engines use removable cylinder sleeves. The sleeves are pressed into oversize cylinder holes. They are actually round, pipe-like liners.

When the cylinder has become worn, the sleeves may be pulled out and new ones pressed in. Fig. 2-4.

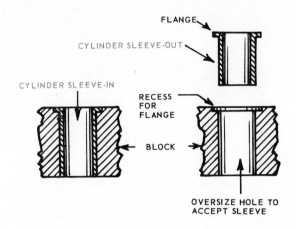

Fig. 2-4. Section of block showing typical cylinder sleeve.

Cylinder sleeves are widely used in heavy-duty truck and industrial engines. In the event the cylinder wall is badly damaged the old sleeve may be removed and a new one installed.

WET AND DRY SLEEVES

Cylinder sleeves are of two types. One type is called the DRY SLEEVE. This type is pressed into a hole in the block and is supported, and surrounded, over its full length. This type sleeve can be quite thin as it utilizes the block metal to give it full length support. Fig. 2-5.

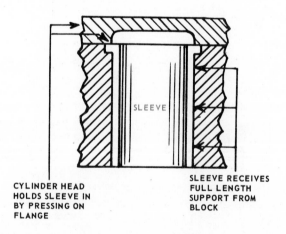

Fig. 2-5. "Dry" sleeve in place.

The WET SLEEVE is also pressed into a hole in the block. It is supported at the top and bottom only. The cooling water in the engine is allowed to directly contact the sleeve. The wet sleeve must be of heavier construction because it receives no central support from the block. Fig. 2-6.

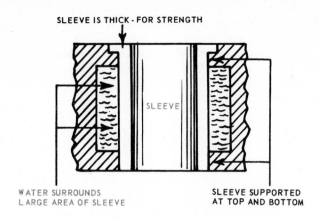

SLEEVE IS THICK - FOR STRENGTH

WATER SURROUNDS LARGE AREA OF SLEEVE

SLEEVE SUPPORTED AT TOP AND BOTTOM

Fig. 2-6. "Wet" sleeve in place.

SECURING THE SLEEVE

Sleeves can be secured in the block in several ways. Where a cast iron or steel sleeve is placed in an aluminum block to provide a wearing surface, it can be cast into place. Fig. 2-7.

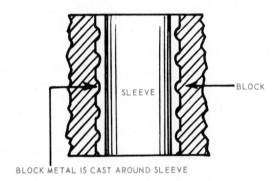

BLOCK METAL IS CAST AROUND SLEEVE

Fig. 2-7. Sleeve held by casting in block. Grooved sleeves prevent movement.

The removable sleeve may be held in place by friction alone. However, this requires a very tight fit and is not altogether dependable. Fig. 2-8.

A better method is to have a flange on the top edge of the cylinder sleeve that drops into a corresponding groove in the

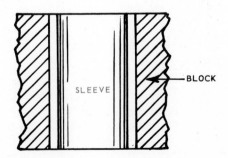

Fig. 2-8. Sleeve held in place by friction between sleeve and block.

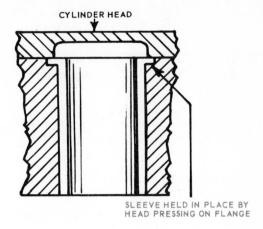

CYLINDER HEAD

SLEEVE HELD IN PLACE BY HEAD PRESSING ON FLANGE

Fig. 2-9. Flange on sleeve fits groove in block.

block. Fig. 2-9. When the cylinder head is bolted on, it presses on the flange and holds it in place. This type of sleeve can be fitted with a greater degree of freedom. Any sleeve must be a fairly snug fit so heat that is built up in the sleeve is conducted away by the surrounding block material.

PISTONS

The piston is literally a sliding plunger that rides up and down in the cylinder. It has several jobs to do in proper sequence.

The piston must move down through the cylinder to produce a vacuum to draw a fuel charge into the cylinder. It then travels up in the cylinder and compresses the mixture. When the mixture is fired, the pressure of the expanding gas is transmitted to the top of the piston. This drives the piston back down through the cylinder with great force. The piston must transmit the energy of this firing stroke to the crankshaft. The piston then travels up through the cylinder, and exhausts the burned fuel charge.

Study the piston in Fig. 2-10. Learn the names of all the parts of the piston.

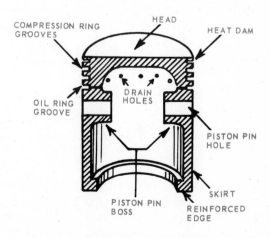

COMPRESSION RING GROOVES

HEAD

HEAT DAM

OIL RING GROOVE

DRAIN HOLES

PISTON PIN HOLE

PISTON PIN BOSS

SKIRT

REINFORCED EDGE

Fig. 2-10. Sectional view of typical piston.

The overall job the piston performs is a difficult one indeed. It is subjected to intense heat from the burning fuel. It must change directions at blurring speeds. It is hounded by friction against the cylinder walls. In addition to all this the piston receives the tremendous thrust of power on the firing stroke.

That a piston not only survives these forces, but will do so for many thousands of miles of driving, is a tribute to the engineering skill of engine manufacturers.

PISTON MATERIALS

Pistons are usually made of aluminum. Often aluminum pistons are tin-plated to allow a good breaking-in job when the engine is started. Aluminum pistons can be forged but are more commonly cast.

The aluminum piston is light and, for most purposes, this gives it an advantage over the cast iron type. A piston must change its direction of travel at the end of every stroke. At speeds sometimes in excess of four thousand revolutions per minute (rpm), it is obvious that the lighter the piston is, the more efficient it will be.

Cast iron is a good material for pistons used in a slow speed engine. It has excellent wear characteristics and will perform admirably in an engine suited to its needs. Pistons which are designed to operate in silicon aluminum cylinders are iron-plated aluminum.

EXPANSION PROBLEMS

Pistons must be carefully fitted into engine cylinders to prevent them from tipping from side to side. They must hold the burning fuel charges above the pistons and be tight enough to form a vacuum, compress and exhaust burned gases.

A piston will expand when it gets hot, so enough clearance must be left to allow for this. Aluminum pistons expand more than cast iron. Fig. 2-11.

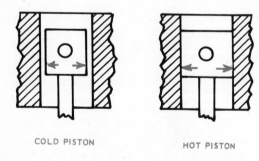

COLD PISTON   HOT PISTON

Fig. 2-11. Exaggerated view shows effect of heat on piston expansion.

The problem of fitting the aluminum piston close enough to prevent slapping, and still leave clearance enough for an oil film to separate the piston and the cylinder, has been solved in several ways. (Piston slap is caused by the piston tipping from side to side in the cylinder.)

SPLIT SKIRT

In a split skirt piston, skirt is either partially or, in some cases, completely split. When the piston warms and begins to expand, it cannot bind in the cylinder since the skirt merely closes the split. Fig. 2-12.

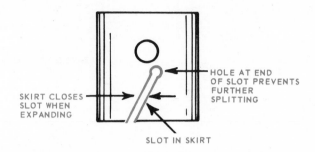

Fig. 2-12. Split skirt piston is designed to solve expansion problem.

T-SLOT

The T-slot piston is another variation of the split skirt. The top of the T tends to retard the transfer of heat from the head to the skirt of the piston. The vertical slot allows the skirt of the piston to close in when heated. Fig. 2-13.

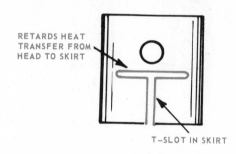

Fig. 2-13. T-slot piston is another means of controlling expansion.

STEEL STRUT

Steel braces and, in some cases, steel rings are cast into aluminum pistons. Steel expands less than aluminum and, as a result, the steel struts tend to control or minimize the expansion. Fig. 2-14.

CAM GROUND

The cam ground piston is a popular type. The piston, instead of being made round, is ground so it is elliptical or egg shaped. Notice in Fig. 2-15 that diameter A is larger than diameter B.

Diameter A is established so that the piston has a minimum amount of clearance in the cylinder. This clearance, around

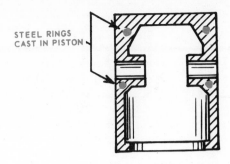

Fig. 2-14. Steel rings control or minimize expansion.

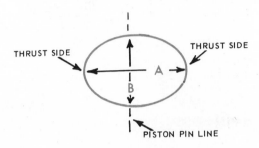

Fig. 2-15. Top view of cam ground piston — exaggerated.

.001 in., is necessary to allow oil to form a lubrication film between the piston and the cylinder wall. The larger diameter, A, is always at right angles to the block (in relation to the crankshaft line).

As the piston heats up, it will not expand much in diameter A, but will tend to expand diameter B. This will cause the piston to become round when fully heated. Fig. 2-16.

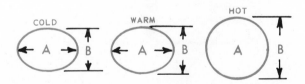

Fig. 2-16. Cam ground piston becomes round when hot. Diameter A remains constant.

Cam grinding, then, will give a minimum clearance on the thrust surfaces. The thrust surfaces are the two sides of the piston that contact the cylinder walls at right angles to the crankshaft. These are the surfaces that support the piston and prevent tipping. The thrust surfaces are also at right angles to the piston pin. The piston is wider across the thrust surfaces. Fig. 2-17.

PARTIAL SKIRT

In manufacturing partial skirt pistons, (also called SLIPPER SKIRT), cam grinding is used but a large area of the skirt is removed. This reduces piston weight and allows the piston to

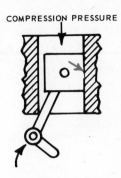

Fig. 2-17. Exaggerated view shows how a loose piston is pushed from one side of cylinder to other. Sides that rub cylinder wall are called THRUST surfaces.

approach the crankshaft more closely. As the nonthrust sides of the skirts are removed, the counterbalances on the crankshaft will not strike the piston. Since the non-thrust sides of a piston do not carry much of a load, so their removal is not detrimental. Fig. 2-18.

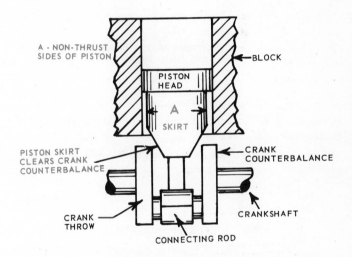

Fig. 2-18. Nonthrust sides of skirt are removed from partial skirt pistons. This allows piston to approach crankshaft more closely without striking.

PISTON TEMPERATURE

The piston head is subjected to the direct heat of the exploding fuel. This heat can raise the temperature of the piston crown (very top) somewhat above 600 deg. F. The temperature will lower as you go down the piston. The bottom of the skirt will be about 300 deg. F. Fig. 2-19.

The temperatures will vary according to engine design and work application. As the bottom of the skirt is the coolest, some pistons have the skirt slightly wider at the bottom. The top area of the skirt would be a trifle smaller in diameter.

HEAD CONSTRUCTION

It is obvious that the piston head is by far the hottest part

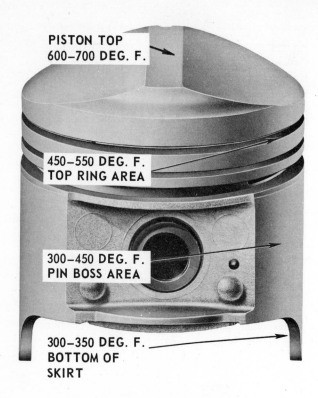

Fig. 2-19. Approximate temperature range is noted from piston head to skirt. (Sealed Power)

PISTON TOP
600–700 DEG. F.

450–550 DEG. F.
TOP RING AREA

300–450 DEG. F.
PIN BOSS AREA

300–350 DEG. F.
BOTTOM OF
SKIRT

of the piston. As a result, it expands more. In order to avoid having the head grow tight in the cylinder, the piston head is turned to a smaller diameter than the skirt of the piston (not cam ground). The head will generally be .030 to .040 in. smaller than the skirt. Fig. 2-20.

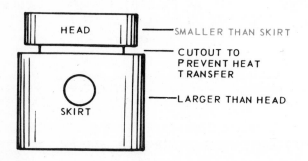

HEAD

SMALLER THAN SKIRT

CUTOUT TO PREVENT HEAT TRANSFER

SKIRT

LARGER THAN HEAD

Fig. 2-20. As piston head is hottest part of piston, it must be ground approximately .030 – .040 in. smaller than skirt.

HEAD SHAPE

Some pistons have flat-topped heads. Others are dome shaped. Still others have irregular shapes designed to help in exhausting burned gases, and also to assist in creating a rapid swirling to help break up gasoline particles on the compression stroke. One type forms the shape of the combustion chamber in the head of the pistons, thus allowing the use of a flat surfaced cylinder head. Fig. 2-21.

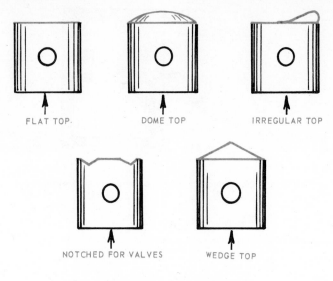

FLAT TOP.   DOME TOP   IRREGULAR TOP

NOTCHED FOR VALVES   WEDGE TOP

Fig. 2-21. Several types of piston heads.

PISTON PIN BOSS

The section of the piston that supports the piston pin must be thicker and stronger. This area is called the PIN BOSS, as shown in Fig. 2-22.

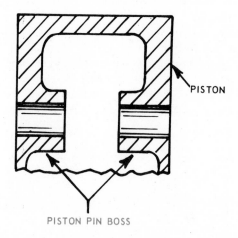

PISTON

PISTON PIN BOSS

Fig. 2-22. Pin boss must be strong to withstand directional changes each stroke of the piston.

THE PISTON CANNOT SEAL THE CYLINDER

The piston must have some clearance in the cylinder. If the skirt has .001 to .002 clearance and the head .030 to .040, it is obvious that the piston cannot seal the cylinder effectively. Fig. 2-23.

PISTON RINGS

The solution to the leakage problem is the use of PISTON RINGS. A properly constructed, and fitted ring will rub against the cylinder wall with good contact all around the

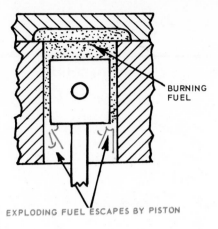

Fig. 2-23. Piston cannot seal by itself because clearance with cylinder wall must be maintained.

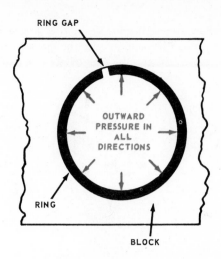

Fig. 2-25. Top view of cylinder shows sealing action of piston ring.

cylinder. The ring will ride in grooves that are cut into the piston head. The sides of the ring will fit the edges of the grooves quite closely. This side clearance can be around .002.

The rings will not contact the bottom of the ring grooves. Actually then, the ring will rub the cylinder wall at all times but will not be solidly fastened to the piston at any point. See Fig. 2-24.

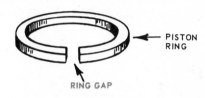

Fig. 2-26. Simplified piston ring showing ring gap.

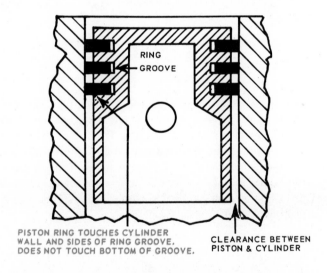

Fig. 2-24. Piston rings seal gap between piston and cylinder wall.

RING GAP

The ring is built so it must be squeezed together to place it in the cylinder. This will cause the ring to exert an outward pressure, thus keeping it tightly against the cylinder wall. See Fig. 2-25.

The ring is not solid all the way around but is cut through in one spot. This cut spot forms what is called the RING GAP. Fig. 2-26.

When the ring is in the cylinder, the cut ends must not touch. When the ring heats up, it will lengthen. Since it cannot

expand outwardly, it will close the gap. If there is not enough gap clearance, the ends will soon touch and as the ring continues to lengthen, it will break up into several pieces. This can ruin a good engine.

A general rule for ring gap clearance is to allow .003 to .004 per inch of cylinder diameter. For example, a four inch cylinder would require from .012 to .016 clearance in the gap. Fig. 2-27.

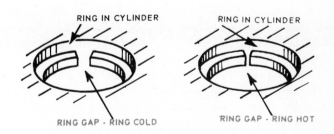

Fig. 2-27. Rings close gap when hot.

Many different types of joints have been used in an endeavor to stop leakage through the ring gap. THIS LEAKAGE IS COMMONLY REFERRED TO AS BLOW-BY. It has been found that the common butt joint is about as effective as any and is much simpler to adjust. Fig. 2-28 illustrates a few of the types of joints that have been used.

The ring is placed in the groove by expanding it out until it will slip over the piston head and slide down and into the ring groove.

Some engines use small pins in the ring grooves. The ring gap must be the normal amount plus the diameter of the pin. This pin is placed between the ends of the ring and thus will

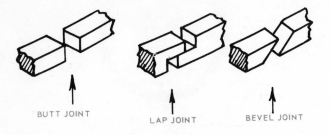

Fig. 2-28. Types of piston ring gap joints.

prevent the ring from moving around in its groove. This ring pinning technique is seldom used in auto engines. Fig. 2-29.

Fig. 2-29. In certain engines, piston ring is pinned in place.

TYPES OF RINGS

There are two distinct types of rings. One is called a COMPRESSION RING and the other an OIL CONTROL RING.

Most engines use three rings on each piston; two compression rings and one oil control ring. Others use two compression rings and two oil rings. Some diesel engines use five or more rings.

All rings may be above the piston pin; or a second oil control ring may be set into a groove near the bottom of the skirt. The compression rings are always used in the top grooves and the oil control rings in the lower grooves. Fig. 2-30.

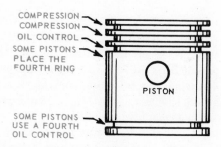

Fig. 2-30. Location of compression and oil control rings.

COMPRESSION RINGS

Compression rings are designed to prevent leakage between the piston and the cylinder. Fig. 2-31.

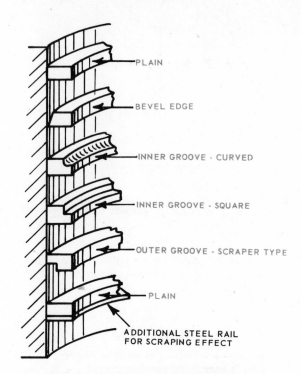

Fig. 2-31. Typical compression ring shapes, as they would look in a cylinder.

Various shapes are used to achieve this goal. Fig. 2-31. The idea behind the various grooves, bevels and chamfers is to create an internal stress within the ring. This stress will tend to force the ring to twist in such a fashion as to press the lower edge to the cylinder wall on the intake stroke. This will cause the ring to act as a mild scraper. The scraping effect will tend to assist in the removal of any surplus oil that may have escaped the oil control rings. Fig. 2-32.

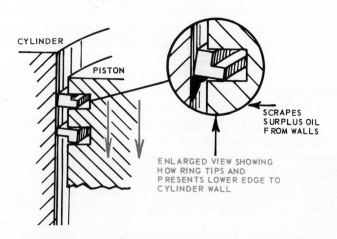

Fig. 2-32. Internal stress cause rings to tip and act as mild scrapers.

On compression and exhaust strokes, the rings will tend to slip lightly over the oil film. This will prolong the life of the ring. Fig. 2-33.

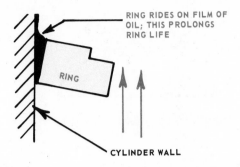

Fig. 2-33. On compression and exhaust strokes rings tip and slide easily on a film of oil.

On the firing stroke, pressure of the burning gases will force the top edges of the ring downward. This causes the ring to rub the wall with full face contact and to provide a good seal for the enormous pressure generated by the firing stroke. See Fig. 2-34.

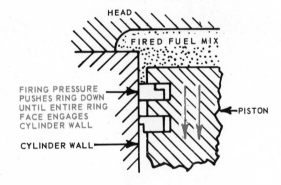

Fig. 2-34. Firing pressure forces ring face against wall to provide a good seal.

HEAT DAM

The compression rings, especially the top ring, are subjected to intense heat. In an endeavor to minimize the transfer of heat from the head of the piston to the top ring, a HEAT DAM is sometimes used. This is actually a thin groove cut into the head of the piston between the top ring groove and the top of the piston. The heat, instead of passing through the aluminum of the piston to the ring, encounters the heat dam. This helps to minimize heat transfer. Fig. 2-35.

TOP RING GROOVE INSERT

Some aluminum pistons have nickel-iron or comparable metal inserts cast into the piston heads. The top ring groove is

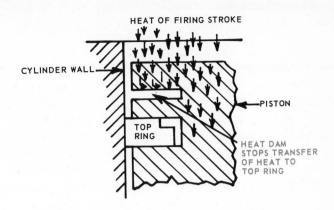

Fig. 2-35. Heat dam keeps top ring cooler.

cut in this metal. As top ring grooves in aluminum pistons pound out of shape, this insert groove will prolong the useful life of the piston and ring. Fig. 2-36.

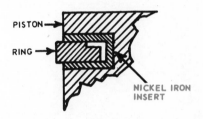

Fig. 2-36. Top ring insert groove, cast into piston.

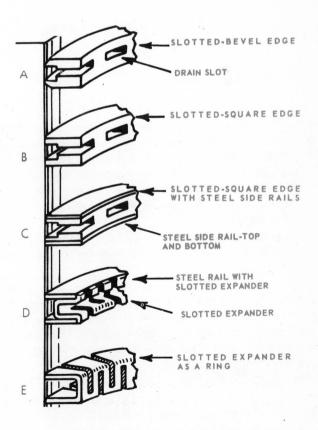

Fig. 2-37. Common types of oil control rings.

OIL CONTROL RINGS

The oil control ring is used to scrape the surplus oil from the cylinder walls. This is not an easy task, and much time and money has gone into the design and construction of oil rings. Fig. 2-37.

All oil rings are slotted and have scraping edges designed to scrape the surplus oil from the cylinder walls. The oil between scrapers passes through slots in the ring on through slots or drilled holes in the bottom of the ring groove. From there the oil drips down into the crankcase area. Fig. 2-38.

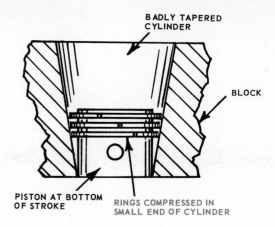

Fig. 2-39. Rings are compressed in bottom of cylinder.

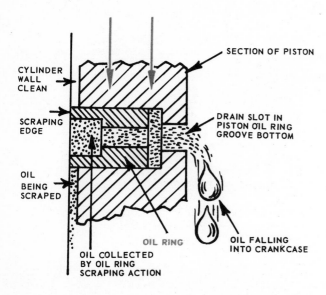

Fig. 2-38. Action of oil ring as it travels down cylinder wall.

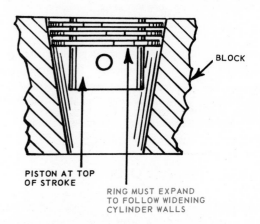

Fig. 2-40. Rings expand at top of cylinder.

EXPANDER DEVICES

Some ring sets, especially those designed for worn cylinders, utilize expanding springs between the bottom of the ring groove and the ring. This will force the ring outward against the cylinder wall.

If a cylinder is worn, the top is invariably wider than the bottom. When the head of the piston is on the bottom of its stroke, the rings will be squeezed in the smaller section of the cylinder. Fig. 2-39.

When the piston travels up the cylinder, the rings must expand outward to follow the ever-widening cylinder diameter. Fig. 2-40.

This makes it necessary for the rings to expand, then contract, for every stroke. If engine speed is high enough, the piston will travel up the cylinder and will be snapped back down before the rings have time to expand. This leaves the piston at the top of the stroke with the rings not touching the cylinder walls.

Expander devices used for rings designed for worn cylinders are stronger than those used for a new or rebored cylinder. Fig. 2-41 illustrates a common type of ring expander.

Expanders are generally used under the oil ring, or rings,

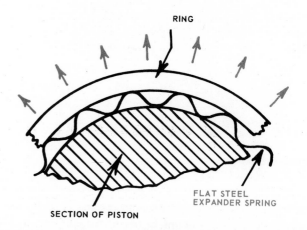

Fig. 2-41. Spring expander rests on bottom of ring groove and forces ring outward. This pressure will keep ring in constant contact with cylinder wall.

and the lower compression ring. The upper ring does not use one as the heat in this area would eventually destroy the temper in the expander spring.

Some expanders do not touch the bottom of the grooves. In this type, the ends butt together and when closing up the ring to get it into the cylinder, the expander is compressed within itself. This type is illustrated in D, Fig. 2-37. Notice in E, Fig. 2-37, that the expander has been developed into a complete ring.

Another type of expander is a round wire spring type. The ends butt together, and even though the spring does not touch the bottom of the groove, it still pushes out on the ring, as shown in Fig. 2-42.

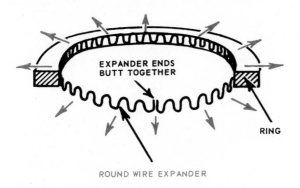

Fig. 2-42. Round wire spring expander does not touch bottom of ring groove. When ring is compressed to enter cylinder, spring is compressed within itself to keep constant tension on ring.

RINGS MUST FIT CYLINDER

In addition to following the cylinder walls, rings must make perfect contact all the way around. Fig. 2-43 illustrates a poorly fitted ring that would do a poor job of sealing.

Even with accurately bored cylinders, and new rings, it is impossible to secure a perfect fit when the rings are first installed. After the engine has operated for several hundred miles, the rings will wear into perfect contact with the cylinder walls.

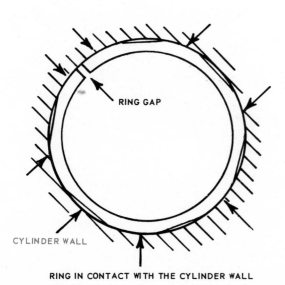

Fig. 2-43. Poor ring contact with cylinder wall is shown. Arrows indicate high spots that touch; area between does not touch.

WEAR-IN

To facilitate a fast job of wearing-in, the ring outer face is left rough. This surface feels smooth to the touch but a close inspection will show that fine grooves are left in the ring surface.

The cylinders may look smooth but final honing with fine stones leaves tiny surface scratches in the walls.

When the engine is started, the rings will be drawn up and down the cylinder. Since both the ring faces and the walls have fine scratches, this will cause some wear. Any high spots on the ring faces will soon be worn off and the ring will fit properly. Fig. 2-44.

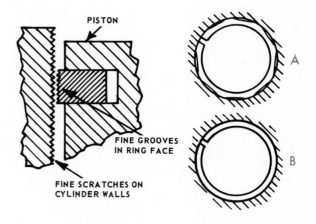

Fig. 2-44. Fine scratches in cylinder wall and grooves in ring face rub against each other to assist prompt wear-in. In detail A, ring touches cylinder only on high spots. Detail B, shows ring contact after wear-in.

The ring and wall surfaces must be designed with a degree of roughness that will cause the rings to wear-in. As soon as a near perfect fit is achieved, the initial roughness will be gone and excessive wear will cease.

SPECIAL COATINGS

To assist with a fast wearing-in period, ring faces are often coated with a soft, porous material. Materials such as graphite, phosphate and molybdenum are used for this purpose. See Fig. 2-45.

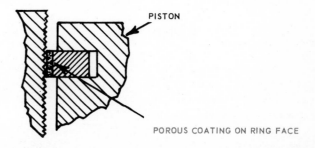

Fig. 2-45. Porous coating on ring face assures rapid and scuff-free break-in.

This soft, porous material also absorbs some oil and allows a gentle wear-in. Rings can also be tin coated.

All the expanding pressure of new rings is applied to the walls at the ring high spots. This can cause overheating and scuffing at these points. (Scuffing is a roughening of the cylinder wall. It is caused when there is no oil film separating the moving parts, and a hard metal-to-metal contact is made.) The porous coating wears quickly and, since it holds oil, the danger of scuffing is lessened.

Some rings have the outer edge chrome plated. This chrome surface will produce a ring that wears very well and stands up under severe operating conditions.

Chrome plated rings are generally finished somewhat smoother and to a higher degree of accuracy, giving less high points to retard wear-in. Fig. 2-46.

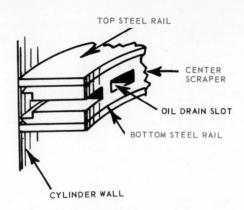

Fig. 2-47. Three-piece oil ring. Top and bottom rails are made of steel.

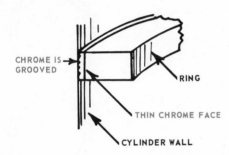

Fig. 2-46. Chrome face of chrome plated ring is grooved to aid break-in.

RING MATERIAL

Piston rings are made from high quality cast iron having excellent wearing properties. It also possesses a springy quality that will hold it out against the walls. The springiness of cast iron makes installation and handling a chore, that must be done carefully, to avoid breaking brittle rings.

Thin oil ring rails are made of steel. Some special expanding oil rings also utilize steel in their construction. Fig. 2-47. Some rings are made of stainless steel. Often, a molybdenum-filled cast iron ring is used in the top ring groove.

RING TYPES ARE NUMEROUS

Ring design is steadily improving. Many new designs are constantly appearing. All rings are designed to provide good sealing, long wear, quick break-in, excellent oil control and freedom from breakage.

PISTON PINS

Pistons are fastened to connecting rods by means of steel pins. These pins, called PISTON PINS, pass through one side of the piston, through the rod upper end, then on through the other side of the piston. Fig. 2-48.

The piston pin is usually hollow, to reduce weight. It is also casehardened to provide a long wearing surface. Casehardening is a process that hardens the surface of the steel but leaves the inner part fairly soft and tough to prevent brittleness. This

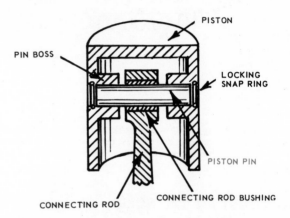

Fig. 2-48. Piston is fastened to the connecting rod with a piston pin.

hardness penetrates from .004 in. up to any depth desired. However, there would be little advantage in making the hard shell any deeper than a few thousandths of an inch. Fig. 2-49. Piston pins are ground to a very accurate size and are highly polished.

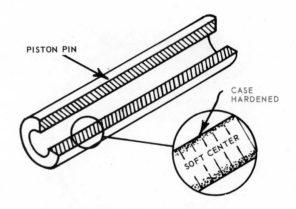

Fig. 2-49. Sectioned piston pin shows thin skin of casehardening.

PIN INSTALLATION

Piston pins are installed and secured to provide a bearing action in three separate ways.

One way is to have the piston pin fastened to the rod and use the piston bosses for bearings. At times, bronze bushings are pressed into the bosses to provide bearing surfaces. Current practice favors a press fit (friction holds the two pieces together) between the piston pin and the connecting rod, with the pin oscillating in the aluminum pin bosses. Fig. 2-50.

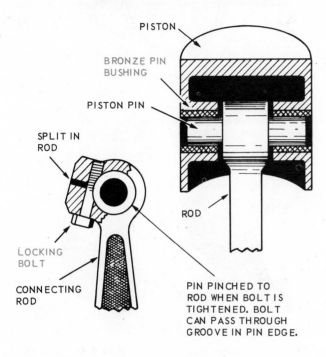

Fig. 2-50. Piston pin locked to rod turns in bronze bushings.

Another method is fasten the pin to one boss and let the rod oscillate on the pin. In this method, the upper connecting rod end must have a bronze bushing for a bearing surface. See Fig. 2-51.

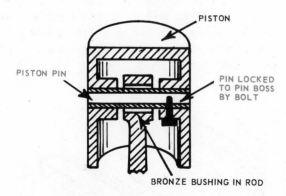

Fig. 2-51. Piston pin locked to one pin boss allows rod to oscillate on pin.

The third method is to have the pin held in place at each end by a SNAP RING that rides in a shallow groove cut in the end of each pin boss. The pin is free to turn in the bosses or in the rod. This is called a free floating pin. Fig. 2-52.

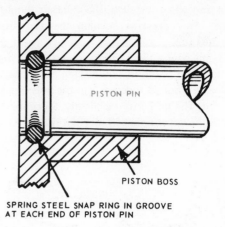

Fig. 2-52. Piston pin is free to turn in rod and in pin bosses. See Fig. 2-48 for full view.

CONNECTING RODS

As the name implies, connecting rods are used to connect pistons to the crankshaft. Fig. 2-53.

The upper end of a rod oscillates (swings back and forth), while the lower or big end bearing rotates, (turns).

As there is very little bearing movement in the upper end, the bearing area can be reasonably small. The lower (big) end rotates very fast, and the crankshaft journal turns inside the connecting rod. This rotational speed tends to produce heat and wear. To make the rod wear well, a larger bearing area is required.

The upper end of the rod has a hole through it for the piston pin. The lower end must be split so the rod can be installed on the crankshaft journal.

The upper and lower halves of the lower end of the rod are bolted together. The upper and lower halves should be

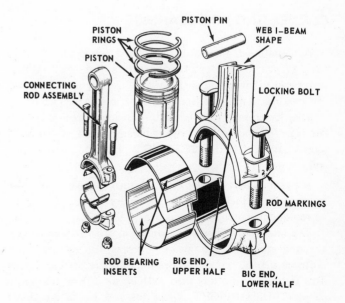

Fig. 2-53. Typical connecting rod; forged steel rod used in conjunction with aluminum piston.   (Jaguar)

numbered and when installed, the numbers should be on the same side. This prevents turning the cap around when installing the rod.

Turning the connecting rod cap around would make the rod bearing hole out-of-round. In making rods, the upper and lower halves are bolted together and the holes are bored to an accurate size. The hole may be slightly off-center. If the caps are crossed, the upper hole half may not line up with the lower hole. Fig. 2-54.

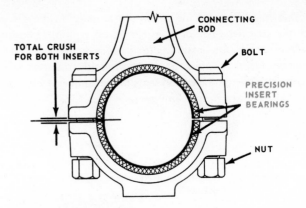

Fig. 2-55. Connecting rod precision insert bearing is properly fitted when bearing parting surfaces (ends) will touch before the rod halves meet. This provides CRUSH. (Sunnen)

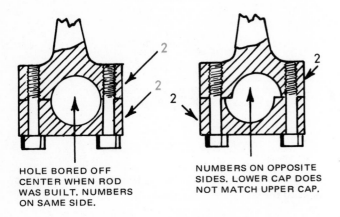

HOLE BORED OFF CENTER WHEN ROD WAS BUILT. NUMBERS ON SAME SIDE.

NUMBERS ON OPPOSITE SIDES. LOWER CAP DOES NOT MATCH UPPER CAP.

Fig. 2-54. Numbers on rod and cap must be kept together.

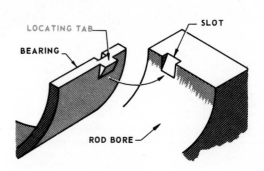

Fig. 2-56. Connecting rod precision insert is aligned and held in place by bearing locating tab engaging a slot in rod bore. (Clevite)

CONNECTING ROD CONSTRUCTION

Connecting rods are generally made of alloy steel. They are drop-forged to shape, then machine. The customary shape utilizes I-beam construction. Fig. 2-53.

Some rods are built of aluminum. Generally these are for small engines designed for light duty. Small engines often utilize the rod material for both upper and lower bearing surfaces. Special aluminum rods for high speed, high performance engines can be purchased from specialty machine shops.

CONNECTING ROD BEARINGS

As mentioned, the upper end of the connecting rod may use a bronze bushing for a bearing surface. If the rod is bolted, or pressed to the piston pin, it will not use any special bearing. The bearings in this case would be in the piston boss holes.

The lower end uses what is termed a precision insert bearing. The rod bearing hole is bored larger than the crank journal and an INSERT BEARING is placed between the rod and journal. Fig. 2-55.

A bearing insert usually does not turn in the rod. It is held in place by a locking lip on the insert that is placed in a corresponding notch in the rod. Figs. 2-53, 2-56.

The insert must fit the rod snugly in order to transfer heat to connecting rod. To insure a proper fit, the insert will protrude a small amount above the rod bore parting surface. This distance (from less than .001 in. to .001 or .002 in.) is called CRUSH HEIGHT. When the rod halves are drawn together, the inserts touch before the halves, forcing the insert tightly into place. Fig. 2-55.

INSERT CONSTRUCTION

An insert is started as a steel shell. This gives it shape and rigidity. The inner part of the shell that contacts the journal, is then coated to form a lining.

Some bearings have steel shells with a thin (.002 − .005 in.) babbit lining. Others use steel shells, then coat them with a copper-lead-tin matrix followed by a very thin (.001 in.) coating of pure tin. Others use aluminum coating. Fig. 2-57 illustrates a typical bearing insert.

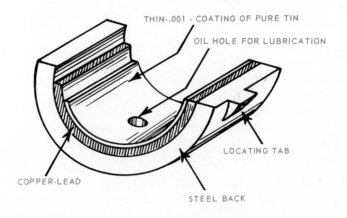

Fig. 2-57. Makeup of precision insert bearing is shown.

BEARING CHARACTERISTICS

The ideal bearing is not easy to construct. For every advantage, there seems to be a disadvantage that goes with it. A good bearing must have many characteristics.

LOAD STRENGTH

The bearing is subjected to tremendous stress from the firing stroke. There is pounding as the crank first pushes up, then pulls down. The bearing must not become fatigued and crack, nor must it spread out from the pounding force.

ANTISCUFFING

The material in the bearing should be such that in the event the oil surface skin is destroyed, the shaft journal will not be damaged by scuffing and scratching.

CORROSION

The bearing material must resist any tendency to corrode when exposed to vapors and acids in the crankcase.

CONFORMATION

No surface is perfectly true. Even with a highly accurate shaft, new rod and insert, we cannot expect the bearing to fit the journal without some minute imperfection. A good bearing is soft and ductible enough to shape to the journal after it has been used for some time.

IMBEDABILITY

If a small abrasive particle enters the bearing area, the bearing should allow the particle to imbed itself in the bearing material so it will not scratch the journal.

CONDUCTIVITY

All bearings produce heat. It is essential that the bearing material be of a type that will conduct heat to the rod. This is one reason why insert bearings must be a near-perfect fit.

TEMPERATURE CHANGE

The bearing strength must not be lessened when it reaches its operating temperature. It must have reasonable strength when both cold and hot. Fig. 2-58 illustrates various forces that work to destroy bearings.

Each insert must have a small hole to allow oil to enter for lubrication. Some inserts have shallow grooves in the surface to allow oil to spread out across the bearing. See Fig. 2-57.

CRANKSHAFT

The engine crankshaft provides a constant turning force to the wheels. It has throws to which connecting rods are

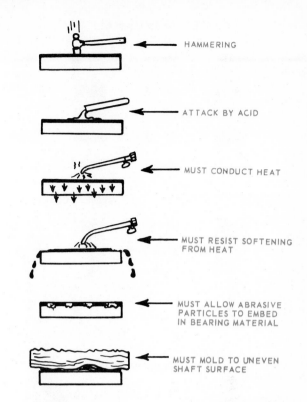

HAMMERING

ATTACK BY ACID

MUST CONDUCT HEAT

MUST RESIST SOFTENING FROM HEAT

MUST ALLOW ABRASIVE PARTICLES TO EMBED IN BEARING MATERIAL

MUST MOLD TO UNEVEN SHAFT SURFACE

Fig. 2-58. Bearing inserts must resist many destruction forces.

attached, and its function is to change reciprocating motion of the piston to a rotary motion to drive the wheels. Crankshafts are made of alloy steel or cast iron. Fig. 2-59.

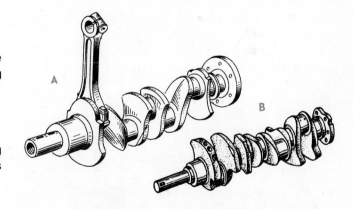

Fig. 2-59. A—Typical four cylinder engine crankshaft. B—Typical V-8 crankshaft. Two connecting rods are attached to each rod throw or journal of V-8 shaft.

MAIN BEARINGS

The crankshaft is held in position by a series of main bearings. The maximum number of main bearings for a crankshaft is one more than the number of cylinders. It may have less main bearings than the number of cylinders.

Most engines use precision insert bearings that are constructed like the connecting rod bearings, but somewhat larger. See Fig. 2-60.

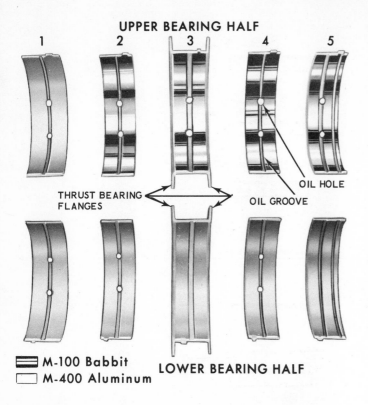

UPPER BEARING HALF

1 2 3 4 5

THRUST BEARING FLANGES

OIL HOLE

OIL GROOVE

▤ M-100 Babbit
▢ M-400 Aluminum

LOWER BEARING HALF

Fig. 2-60. Set of main bearings; some with an aluminum coating, others employ babbit. Note thrust flanges on thrust bearing. (Cadillac)

In addition to supporting the crankshaft, one of the main bearings must control the forward and backward movement (end play) of the shaft. This bearing has flanges on the edges that rub against a ground surface on the edge of the crankshaft main journal. This bearing is referred to as a THRUST BEARING. Fig. 2-60.

CRANKSHAFT THROWS

The crankshaft throws (part of the shaft the connecting rod fastens to) must be arranged in such a way as to bring the piston to top dead center at the proper time for the firing stroke. As the pistons in a multicylinder (more than one) engine do not fire one after the other, but fire in a staggered sequence, the throw position is very important. Fig. 2-61 illustrates an end view of four, six and eight cylinder crankshafts. Illustrations A, C, D and E are for in-line engines. An in-line engine is one in which all the cylinders are arranged one after the other in a straight row.

An eight cylinder in-line engine, Fig. 2-61D, requires eight connecting rod throws. A V-8 engine, B and F, has only four throws. Each throw services two connecting rods.

CRANK VIBRATION

To offset the unbalanced condition caused by off-center

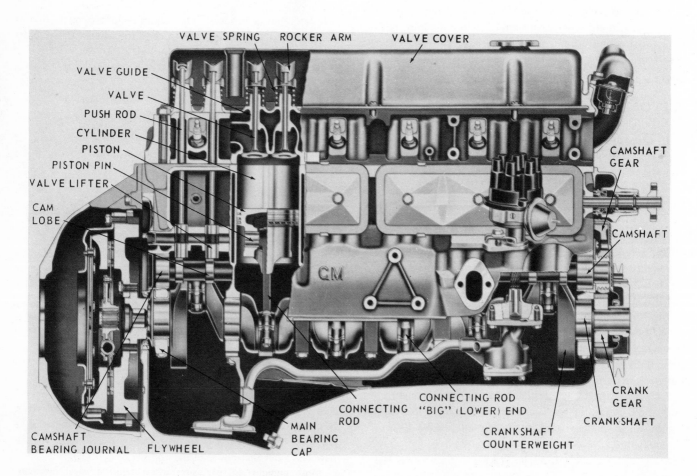

VALVE SPRING ROCKER ARM VALVE COVER

VALVE GUIDE

VALVE

PUSH ROD

CYLINDER

PISTON

PISTON PIN

VALVE LIFTER

CAM LOBE

CAMSHAFT BEARING JOURNAL

FLYWHEEL

MAIN BEARING CAP

CONNECTING ROD

CONNECTING ROD "BIG" (LOWER) END

CRANKSHAFT COUNTERWEIGHT

CRANK GEAR

CRANKSHAFT

CAMSHAFT GEAR

CAMSHAFT

Side view cutaway of a 6-cylinder, valve-in-head engine with 292 cu. in. displacement and a compression ratio of 8.01 to 1. Study the relationship of the various parts. (Chevrolet)

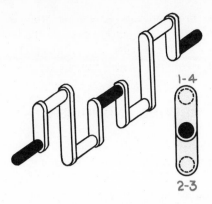

Fig. 2-61A. A four cylinder crankshaft will normally have throws spaced 180 deg. apart with cylinders 1 and 4 on the same side.

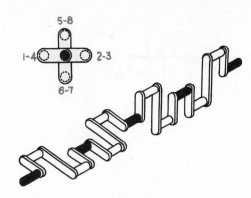

Fig. 2-61D. A straight eight crankshaft of 4-4 type resembles two four cylinder crankshafts connected end to end.

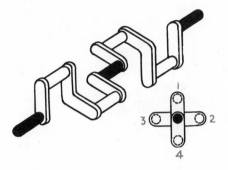

Fig. 2-61B. A V-8 engine may have the crankshaft throws arranged the same as a four cylinder engine, or method shown here which is widely used.

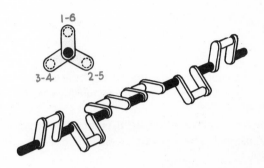

Fig. 2-61E. A left-hand crankshaft for a six cylinder engine will have No. 3 and 4 throws to the left of No. 1 and 6.

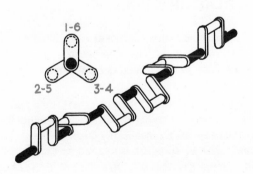

Fig. 2-61C. A righ-hand crankshaft for a six cylinder engine will have throws arranged like this.

Fig. 2-61F. A typical V-8 engine crankshaft has same arrangement of connecting rod throws as shown in Fig. 2-61B.   (Ford)

throws, many crankshafts use counterbalances to stop vibration. They may be forged as part of the crankshaft, or they may be bolted. Fig. 2-62.

VIBRATION DAMPER

When the front cylinders fire, power is transmitted through the rather long, crooked crankshaft. The pressure the piston applies can exceed 3,000 lbs. The front of the crankshaft receiving this power tends to move before the rear, causing a twisting motion. When the torque is removed from the front, the partially twisted shaft will unwind and snap back in the other direction. This unwinding, although minute, causes what is known as torsional vibration.

To stop the winding motion, a vibration damper, sometimes called a HARMONIC BALANCER, is attached to the front of the crankshaft. Basically the damper is built in two pieces. These pieces may be connected by rubber plugs, or spring-loaded friction discs, or a combination of the two. Fig. 2-63.

When a front cylinder fires and the shaft tries to speed up, it tries to spin the heavy section of the damper. In doing this,

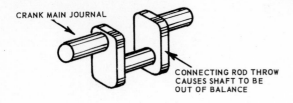

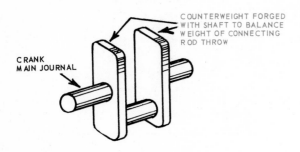

Fig. 2-62. Crankshaft needs counterweights to bring shaft into proper balance.

the rubber connecting the two parts of the damper is twisted. The shaft does not speed up as much with the damper attached.

The force necessary to twist the rubber and to speed up the heavy damper wheel, tends to smooth out the crankshaft operation.

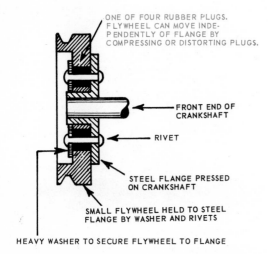

Fig. 2-63. Vibration damper utilizes small flywheel which moves a limited amount of bending rubber plugs. See A and B below.

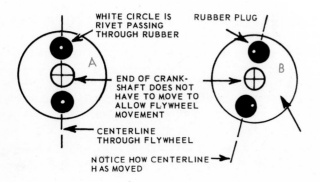

When the firing pressure is removed from the shaft, the shaft cannot spring back quickly because the twisted rubber attempts to keep the damper wheel turning. The unwinding force of the crankshaft cancels out the twist in the opposite direction.

DRILLED, GROUND AND POLISHED

The crankshaft is drilled so oil being fed to the main bearings will pass through the shaft to the connecting rod throws.

All bearing surfaces are precision ground and are highly polished. Fig. 2-64.

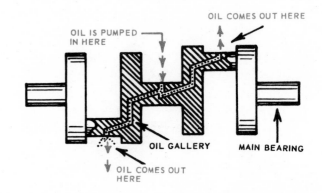

Fig. 2-64. Crankshaft partially cut away shows how oil is pumped into center main bearing. From there it enters an oil gallery drilled through the crankshaft to each rod journal.

TIMING GEAR OR SPROCKET

A timing gear or a chain sprocket is installed on the front end of the crankshaft. A chain sprocket is usually made of steel. In the case of the timing gear, the teeth are helical in shape.

A timing gear is used to turn a camshaft timing gear. Where a sprocket is used, the sprocket will drive a timing chain that operates the camshaft by means of a larger sprocket. Fig. 2-65.

A timing gear or sprocket is secured to the crankshaft by means of a metal key that rides in a slot cut in the crankshaft. The gear or sprocket also has a slot or groove in it. Fig. 2-66.

FLYWHEEL

A heavy flywheel is attached to the rear end of the crankshaft by means of bolts. The function of the flywheel is to smooth out engine speed and keep the crankshaft spinning between power strokes.

In some engines the flywheel also serves as a mounting surface for the clutch.

The outer rim of the flywheel has a large ring attached with gear teeth cut into it. The teeth of the starter motor engage this gear and spin the flywheel to crank the engine. When an automatic transmission is used, the torque converter assembly acts as a flywheel. Fig. 2-67.

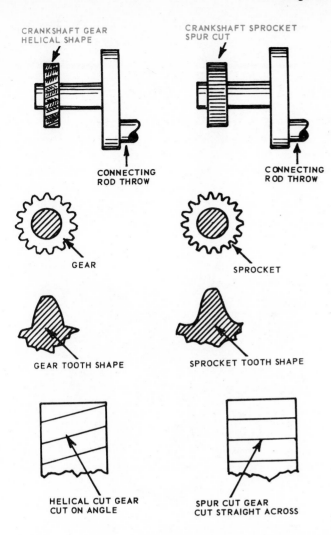

Fig. 2-65. Timing gear drive gear on crankshaft; also timing chain drive sprocket on crankshaft.

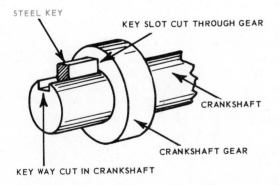

Fig. 2-66. A steel key, sliding in a slot in the crankshaft, passes through a similar slot in the gear to secure the gear to the shaft.

CAMSHAFT

An engine camshaft is used to open and close the valves. There is one cam on the camshaft for each valve in the engine. Generally only one camshaft is used in each engine. (On the elementary engine you designed in Chapter 1, you used two

camshafts and the valves were arranged in a T-fashion to better illustrate the action in relation to an end view of the crankshaft.) A camshaft has a series of support bearings along its length.

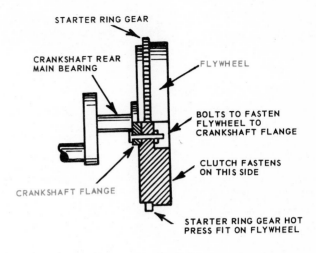

Fig. 2-67. Heavy flywheel is bolted to the crankshaft flange.

Camshafts turn at one-half crankshaft speed.

The cam lobes are usually not flat across the top as they might appear. The bottom of the valve lifter may be slightly crowned and the cam lobe tapered. This places the lobe-to--lifter contact to one side of center. Fig. 2-68.

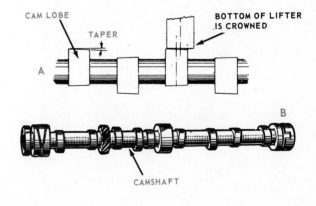

Fig. 2-68. A—Section of camshaft showing one method of tapering cam lobes. Lifter bottom is crowned. B—Typical camshaft design.

A gear cut into the camshaft is used to drive the distributor and oil pump.

The camshaft is kept in place by using a thrust washer behind the timing gear bolted to the block. Or, a spring loaded plunger is used to push on the end of the camshaft. An additional cam may be ground on the shaft to drive the fuel pump.

The typical camshaft is made of cast, or forged, steel. The cam surfaces are hardened for long life. Fig. 2-69 illustrates a typical camshaft with gear and thrust washer. The camshaft

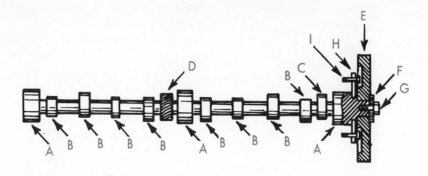

Fig. 2-69. Typical Camshaft. A—Main camshaft bearings. B—Cam lobes. C—Eccentric to drive fuel pump. D—Gear to drive distributor. E—Camshaft gear. F—Washer. G—Bolt to hold gear to shaft. H—Thrust washer bolts to block and keeps camshaft in place. I—Bolt to fasten thrust washer.

may be turned by a gear, timing chain, or a toothed timing belt. Figs. 2-70, 2-71, 2-72.

The camshaft gear can be made of cast steel, aluminum or a special pressed fiber. Chain sprockets are made of cast steel.

VALVES

Each engine cylinder ordinarily has two valves. However, some special racing engines use four valves per cylinder.

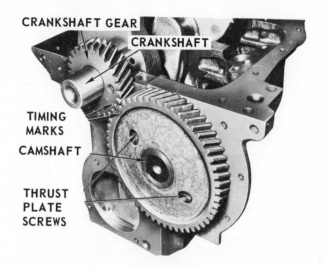

Fig. 2-71. Timing gear drive for camshaft also requires alignment of timing marks.
(Chevrolet)

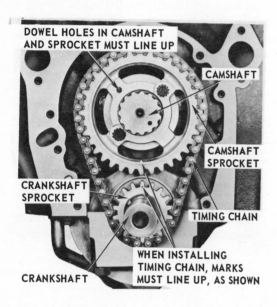

Fig. 2-70. Front view of typical timing chain drive. Shows alignment of valve timing marks.
(Cadillac)

Exhaust valves are made of heat resistant metal, because the head of a valve operates at temperatures up to 1300 deg. F. It is obvious that the steel used in valve construction must be of high quality.

In order to prevent burning, the valve must give off heat to the valve guide and to the valve seat. The valve must make good contact with the seat, and must run with minimum clearance in the guide.

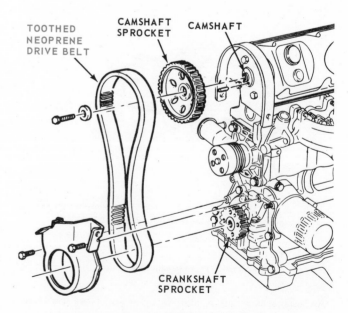

Fig. 2-72. This aluminum block, 4-cylinder engine has an overhead camshaft driven by a toothed, fiberglass reinforced neoprene belt.
(Chevrolet)

Some valves have special hardfacing on the face areas to increase their useful life. Others use hollow stems filled with metallic sodium. Fig. 2-73. At operating temperature, the sodium becomes a liquid and splashes up into the head. This draws the heat into the stem where it transfers to the valve guide.

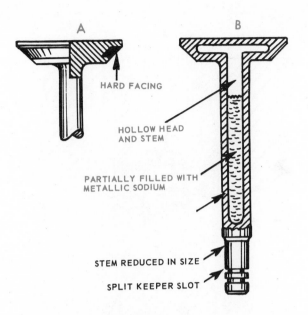

Fig. 2-73. A—Valve with special hardfacing to lengthen useful life. B—Special heat resistant valve, utilizing a hollow stem with metallic sodium.

VALVE SEATS

Two angles commonly used for valve seats are 30 and 45 deg. Some makers grind a one degree difference between the valve face and seat to permit fast seating. Fig. 2-74. The seat may be ground to 45 deg. and the valve, 44 deg., or vice versa (the other way around).

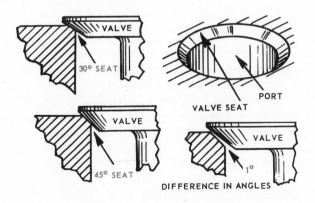

Fig. 2-74. Common valve seat angles.

Valve seats can be cut in the block metal (if cast iron), or special, hard steel inserts may be pressed into the block. Fig. 2-75. Seats can also be induction hardened.

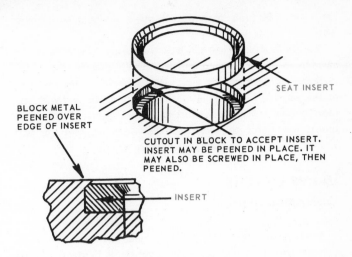

Fig. 2-75. Valve seat insert is often frozen to shrink diameter, then pressed into warm block to secure insert.

The valve face and seat must make perfect contact to insure efficient operation.

VALVE GUIDES

The valve guide can be an integral part of the cylinder head; or it may be made as a separate unit, pressed into a hole in the head or block, depending on which unit contains the valves.

The pressed-in type of valve guide is made of cast iron. The valve stem must fit this guide with about .002 — .003 in. clearance. Fig. 2-76.

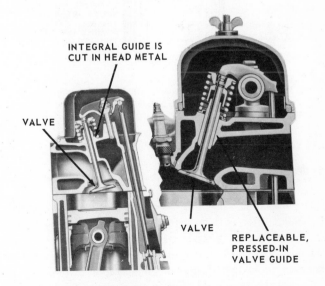

Fig. 2-76. Examples of integral (not removable) and pressed-in (removable) type valve guides are shown.

VALVE SPRINGS

Springs push the valves closed when the cams lower. Since the springs are compressed and expanded over 70,000 times per hour at 50 mph, they must be made of high quality spring wire.

You have already learned how the spring is fastened to the valve by means of a spring washer and split keepers.

Some springs have the coils closer together at one end than at the other. In an installation of this kind, the end with the more closely spaced coils must be placed against the head, or block, whichever the case may be. Some engines use two springs per valve, one spring inside of another.

VALVE LIFTERS
(Mechanical Type)

Mechanical lifters are usually made of cast iron. The bottom part that contacts the camshaft is hardened. Some lifters are hollow to reduce weight. A screw is placed in the top to adjust the clearance between the end of the valve stem and the lifter. Fig. 2-77.

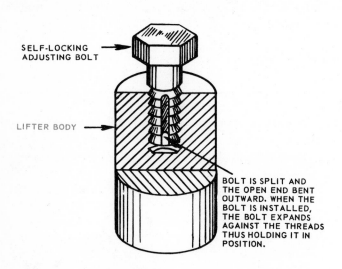

SELF-LOCKING ADJUSTING BOLT

LIFTER BODY

BOLT IS SPLIT AND THE OPEN END BENT OUTWARD. WHEN THE BOLT IS INSTALLED, THE BOLT EXPANDS AGAINST THE THREADS THUS HOLDING IT IN POSITION.

Fig. 2-77. Mechanical valve lifters usually are solid. Hollow lifters reduce weight.

VALVE LIFTERS
(Hydraulic Type)

The hydraulic valve lifter performs the same job as the mechanical lifter. The major difference is that the hydraulic lifter is self-adjusting, operates with no lifter-to-rocker arm clearance, Fig. 2-78, and uses engine oil under pressure to operate. Hydraulic lifters are quiet in operation.

To operate, engine oil under pressure enters the hydraulic lifter body. The oil passes through a small opening in the bottom of an inner piston, into a cavity beneath the piston. The oil raises the piston up until it contacts the push rod (the oil pressure is not high enough to open the valve).

When the cam raises the lifter, pressure is applied to the inner piston. The piston tries to squirt the oil back through the small opening but cannot do so as a small check ball seals the opening.

As the cam raises, the lifter becomes solid and lifts the valve. When the cam lowers, the lifter will be pushed down by the push rod. The lifter will then automatically adjust to remove clearances. Fig. 2-78 shows one type of hydraulic valve lifter. See Fig. 2-79 for complete lifter action.

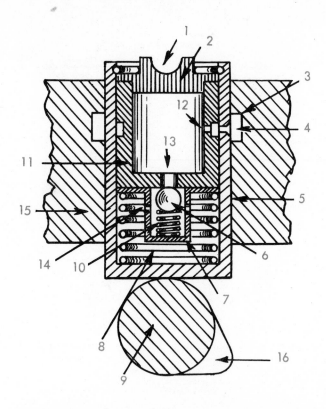

Fig. 2-78. Typical hydraulic valve lifter. Oil enters channel 3 in block, passes through hole 4 in lifter body 5, then enters hole 12 in inner piston. It fills inner piston cavity and passes through hole 13. This pressure will push check ball 6 off its seat. It now flows through hole 14 in the lifter cage 7. It fills the cavity under inner piston 11. The pressure raises inner piston up until it contacts valve push rod, or stem, whichever is used. Other parts are: 1—Push rod seat. 2—Inner piston cap. 15—Block. 8—Inner piston spring. 10—Ball check spring. 9—Camshaft. 16—Lobe.

TURNING HELPS

If the valve goes up and comes down in the same place time after time, carbon buildup between valve face and seat may cause the valve to remain partially open. This will cause valve burning. If the valve turns even a few degrees on each opening, a wiping action between face and seat will be developed. This keeps carbon from building up.

Turning also helps to prevent localized hot spots since the valve will keep moving away from the hottest areas.

RELEASE TYPE ROTATOR

There are several methods of causing valves to rotate as they open and close. One is through the use of a release type mechanism. This removes the spring tension from the valve, while open, and induces rotation from engine vibration. See Fig. 2-80.

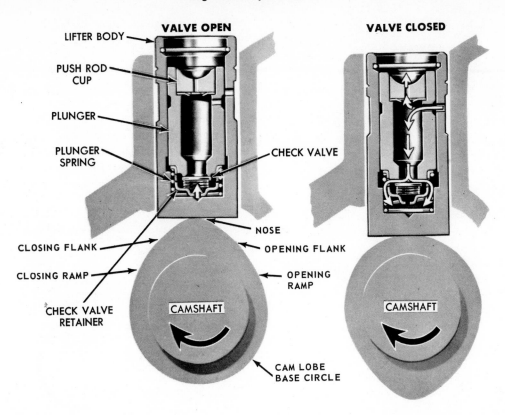

Fig. 2-79. Hydraulic valve lifter action during valve open and valve closed operation. Note use of a flat check valve or disc. (Ford)

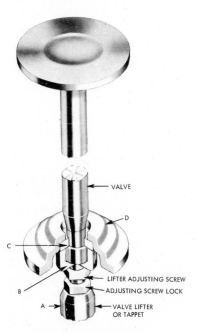

Fig. 2-80. Thompson Rotovalve—release type. (Thompson Products)

pressure on the seating collar causes the collar to press down on the flexible washer.

The washer then presses down on the balls, causing them to roll downward in their inclined races (small grooves guide ball). This rolling action causes the retainer and the valve to turn a few degrees.

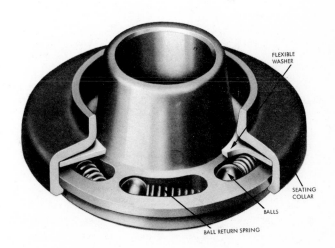

Fig. 2-81. Thompson Rotocap—positive type.

POSITIVE TYPE ROTATOR

An example of the positive type rotator is shown in Figs. 2-81 and 2-82. In this mechanism, when the lifter applies pressure to the valve stem or push rod, the valve spring

When the valve closes, lifter pressure is removed and the ball springs cause the balls to move back up the inclined races. The unit is then ready to function again.

45

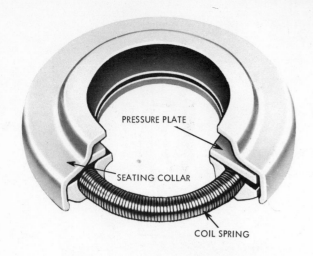

Fig. 2-82. Thompson Rotocoil—positive type.

ROLLER LIFTERS

Some heavy-duty engines, as well as those having a special camshaft, use valve lifters with rollers that contact the camshaft. The roller reduces wear on both the lifter and the cam. Fig. 2-83.

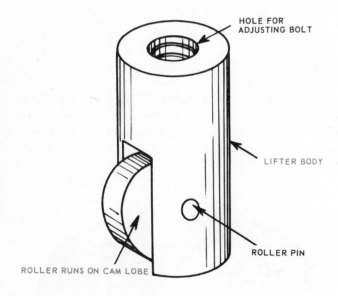

Fig. 2-83. Roller lifter is used in some heavy-duty applications.

VALVE TIMING

Both the intake and exhaust valves are open longer than it takes the piston to make a stroke. The exact number of degrees that a valve will open or close before top or bottom dead center varies widely, depending on engine design. The degrees shown in Fig. 2-84 are for one specific engine. You will note that the intake valve opens about 20 deg. before the piston starts down on the intake stroke. It closes about 67 deg. after the piston reaches the bottom of its stroke.

The exhaust valve opens about 69 deg. before the piston

reaches bottom dead center (BDC) on the power stroke. It does not close until about 27 deg. after top dead center (TDC) on the intake stroke.

The early opening and late closing of both valves greatly improves the intake of fresh fuel mixture and the thorough exhausting of burned gases.

The intake and exhaust valves, Fig. 2-84, are partially open at the same time. For example, the intake valve opens 20 deg. before TDC and the exhaust valve closes 27 deg. after TDC, on the same intake stroke. This situation is termed VALVE OVERLAP. It does not impair engine performance.

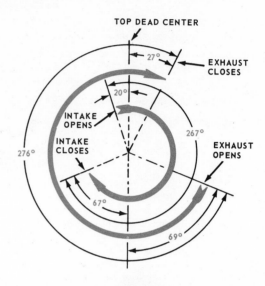

Fig. 2-84. Valve timing diagram. The angles will vary widely depending upon engine design. The length of time, in degrees, that a valve is held open is referred to as valve DURATION.

When a valve closes or opens, how fast it will rise, how long it will stay open, and how fast it will close depends on the shape of cam lobe and the position of the camshaft in relation to the crankshaft.

The two timing chain sprockets are generally marked to insure correct valve timing. When a line drawn through the center of both shafts bisects the timing marks, timing is correct.

The camshaft is mounted to one side of the crankshaft on most in-line engines (except engines using overhead camshafts). Fig. 2-85.

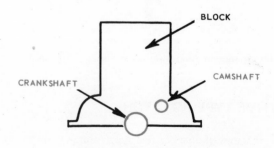

Fig. 2-85. Camshaft location in a typical in-line engine.

On V-type engines, the camshaft is generally mounted above the crankshaft in the center region of the block. See Fig. 2-86.

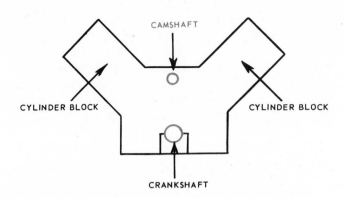

Fig. 2-86. Camshaft location in a valve-in-block or valve-in-head (OHV) V-type engine.

Engine valves are not always operated directly from a camshaft. The valve-in-head engine utilizes additional linkage to operate the valves. This will be discussed in the chapter on engine types.

CYLINDER HEAD

The cylinder head serves as a cover for the cylinders and forms the top of the combustion chamber. It may contain one or both valves. The head also holds the spark plugs.

If the cylinder head contains the valves, it is called a VALVE-IN-HEAD engine.

Cylinder heads are usually made of cast iron or aluminum. They must be strong and rigid. They are bolted to the block with HEAD BOLTS.

The surfaces of the head and block that make contact must be absolutely flat. Fig. 2-87.

OIL PAN

The oil pan acts as a reservoir for oil, and it also serves as a dust shield for the bottom of the engine. It is attached to the bottom of the block with cap screws.

The pan is generally made of thin steel stamped to shape. Fig. 2-88.

TIMING GEAR COVER

The timing gears must be covered to prevent the entrance of dirt and dust and to eliminate the loss of oil.

In addition to this function, the cover often contains an oil seal that allows the crankshaft to protrude through the cover and yet not leak oil.

Timing gear covers may be stamped from thin steel, or cast from aluminum or cast iron. Fig. 2-89.

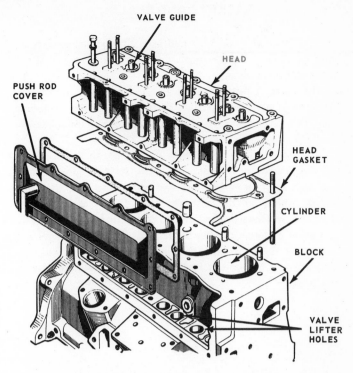

Fig. 2-87. Typical cylinder head for valve-in-head engine. Head and block surfaces must be smooth and true.   (Hillman)

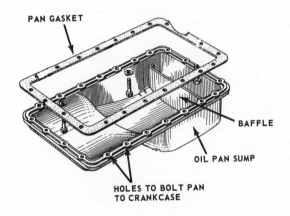

Fig. 2-88. Engine oil pan size and shape will vary widely. (Datsun)

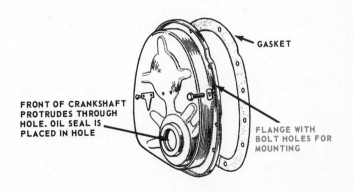

Fig. 2-89. One type of stamped steel timing gear cover. (Austin-Healey)

GASKETS AND SEALS

In an engine where machined parts fit together, gaskets are used to make the joints tight and to prevent leakage of oil, water and/or gasoline.

The cylinder heads must seal in the water of the cooling system and must also contain the pressure of the exploding fuel. Thin steel, copper and asbestos gaskets are used between the head and engine block.

It is very difficult to machine metal parts to the degree of accuracy necessary for leakproof joints. As the engine expands and contracts during warmup and cooling periods, there are minute shifts in the fastened parts. This, coupled with vibration, will loosen many parts to the point of leakage.

Gasket material is somewhat resilient (soft and springy) and will adapt itself to expansion and contraction. It will also conform to irregularities in the surfaces of the mating parts. Fig. 2-90.

Fig. 2-91 illustrates some of the common gaskets found in a car engine.

OTHER PARTS

Other engine parts that are required by the various systems — carburetion, ignition, lubrication and cooling will be discussed in other sections.

In this chapter you have studied about many engine parts. A thorough understanding of design, construction and application is an absolute essential for every good auto mechanic. Go through the questions on pages 48 and 49 carefully, and do not be satisfied until you can answer each and every one.

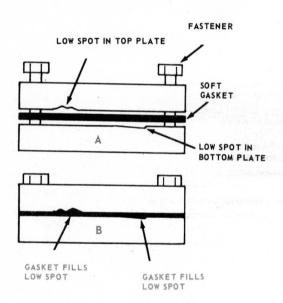

Fig. 2-90. Gaskets are designed to compress and seal irregular surfaces.

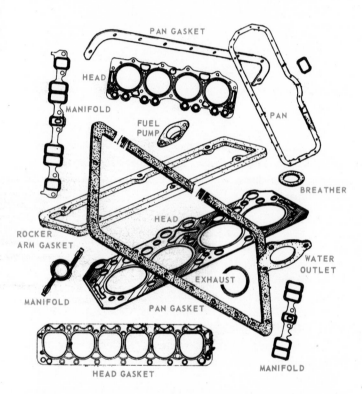

Fig. 2-91. A few of the typical gaskets used in an engine are shown and called out.

REVIEW QUESTIONS – CHAPTER 2

Think of these questions as a special sort of scale to weigh your progress. STEP ON AND SEE HOW HEAVY YOU ARE!

1. The unit that forms a basic foundation upon which the whole engine is built is called the _____.
2. What are cylinder sleeves?
3. The cylinders in a new engine should be glassy smooth. True or False?
4. A cylinder that is round to within .100 is accurate enough. True or False?
5. The main advantage of using aluminum in engine construction is that it is rustproof. True or False?

6. A piston must be _____ and yet _____.
7. What are three common ways of controlling piston expansion?
8. Cast iron pistons work best in a _____ speed engine.
9. Piston heads are of different shapes. What are some of the reasons for these shapes?
10. What is a partial or slipper skirt piston?
11. Temperature of piston heads often runs up to about _____ degrees.
12. What is placed on the piston to help it seal the cylinder?
13. What is a heat dam?
14. There are two types of piston rings. What are they?
15. Some rings are made of _____, while others are made of _____ _____.
16. What are ring expanders?
17. Some rings have special grooves cut into the inner

circumference. What are these grooves supposed to accomplish?

18. What are the approximate number of rings usually found on pistons used in auto engines?
19. What is a break-in period?
20. What is the advantage of chrome plated rings?
21. How can wear in top ring groove be lessened?
22. Are plain butt joints all right for piston rings?
23. One kind of ring is always used at the top of the piston. What type is it?
24. The bottom rings are always _____ rings.
25. Piston pins fasten the _____ to the _____ _____.
26. What is casehardening?
27. Connecting rods are used to connect the _____ to the _____.
28. The smaller connecting rod upper end bearing is _____, but the big end bearing is _____ to facilitate installation.
29. Connecting rods use precision _____ for the big end bearings.

30. Name five characteristics of a good bearing material.
31. Crankshafts are usually cast of aluminum. True or False?
32. Crankshafts are subject to vibration from being off balance and also from twisting. How are the vibrations controlled?
33. The flow of power is smoother in an eight cylinder engine than in a four. True or False?
34. Crankshafts have drilled passageways. What are these for?
35. Camshafts are driven in two ways. Name them.
36. Name three basic purposes of the flywheel.
37. Exhaust valves are made of cast iron. True or False?
38. A valve sheds its heat in two manners. What are they?
39. Two valve seat angles are _____ degrees and _____ degrees.
40. Explain how a hydraulic valve lifter works.
41. What are the advantages of the hydraulic lifter?
42. The oil pan holds _____ and provides _____ against dust, dirt, etc.
43. Gaskets are used to prevent the loss of oil, gasoline, water, etc. Why are they needed?

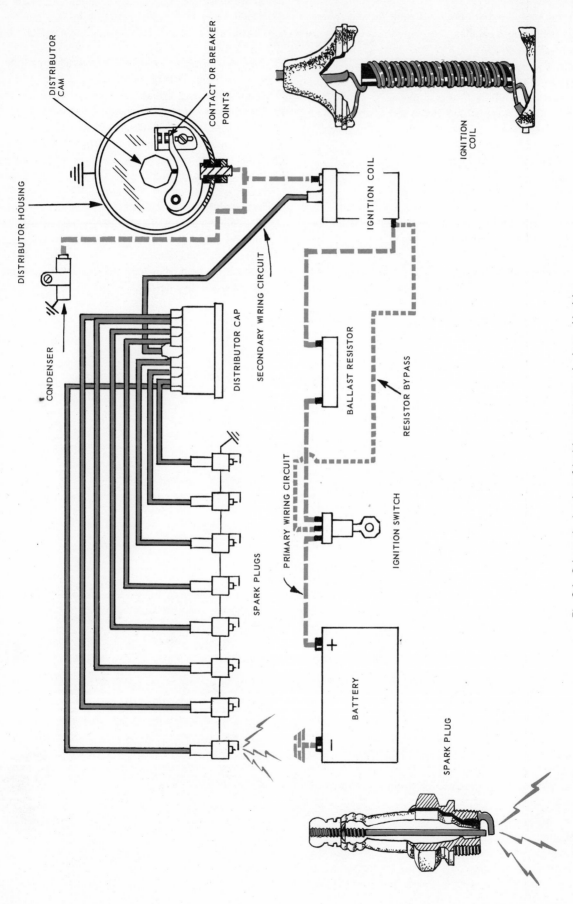

AUTOMOBILE IGNITION SYSTEM

DISTRIBUTOR CAM

CONTACT OR BREAKER POINTS

DISTRIBUTOR HOUSING

CONDENSER

DISTRIBUTOR CAP

SECONDARY WIRING CIRCUIT

SPARK PLUGS

IGNITION COIL

BALLAST RESISTOR

RESISTOR BYPASS

PRIMARY WIRING CIRCUIT

IGNITION SWITCH

BATTERY

SPARK PLUG

IGNITION COIL

Fig. 3-1. Schematic drawing of ignition system; spark plug and ignition coil details.

Chapter 3

IGNITION SYSTEMS

Chapters 1 and 2 made frequent reference to the use of a spark to ignite the fuel mixture. This chapter will provide you with a clear and concise description of the various units in the ignition system. The theory, design and construction of the parts will be discussed. You will also learn how they are combined to produce, control and distribute the spark.

READ FIRST PART OF CHAPTER ON ELECTRICAL SYSTEM

At this time it is recommended that you read the first eight pages of the chapter on Electrical Systems. This will provide you with a basic introduction to the field of electricity. Common electrical terms and theory will be discussed. SUCH KNOWLEDGE IS ESSENTIAL TO UNDERSTAND THIS SECTION ON IGNITION SYSTEMS.

HIGH VOLTAGE IS NECESSARY

For many years batteries were of 6 volt design. Increasing engine size and the addition of numerous electrical accessories, however, made it necessary to change to the 12V battery that is in common use today.

Voltage of 6 or 12V will not produce a spark that will jump across the electrodes of a spark plug in the combustion chamber.

In modern cars, the voltage available at the spark plug usually exceeds 20,000V. The ignition system must produce such voltage in a completely reliable manner. This means that the battery voltage of 12V must be upped many times.

PRIMARY AND SECONDARY CIRCUITS

The ignition system is divided into two separate circuits.

The PRIMARY circuit voltage is low, operating on the battery voltage of 12 volts. The wiring in this circuit is covered with a thin layer of rubber, or plastic, to insulate the wire and to prevent short circuits.

CAUTION: The spark delivered to the plugs may exceed 20,000V. If you touch a plug and are shocked, there is little real danger involved. The reason for this is that there is very little amperage. DO NOT CONFUSE THIS WITH VOLTAGE IN YOUR HOME ELECTRICAL SYSTEM. IN THE HOME UNDER CERTAIN CONDITIONS A CURRENT OF 120V

CAN, AND OFTEN DOES, ELECTROCUTE PEOPLE. ALWAYS TREAT ELECTRICITY WITH GREAT CAUTION AND RESPECT.

Study Fig. 3-1. Notice the two circuits. Learn the names of the various units and remember where they are inserted in the circuit.

SYSTEM COMPONENTS — BATTERY

To fully comprehend the functions of the parts or components of an ignition system, we will start at the battery and trace the flow of electricity through the system.

The battery is the source of the electrical energy needed to operate the ignition system. Fig. 3-2.

To function properly, the battery must be close to fully charged. (Fully charged means that the battery is in the proper condition to produce the highest output of electricity.)

The battery is generally located as close to the engine as feasible. This reduces the length of wiring necessary to connect the battery to the component parts. The close location will cut

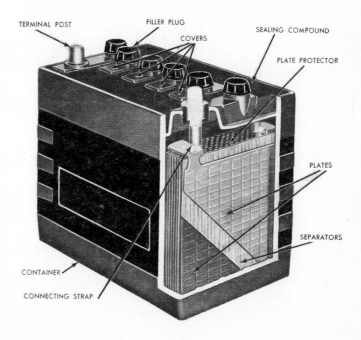

Fig. 3-2. Typical battery construction.   (Willard)

51

costs of excessive lengths of wire, and will reduce voltage drop. An under-the-hood location also makes the battery readily accessible for repair and maintenance.

A battery has two heavy lead terminals. One is positive and one is negative. Common practice in this country is to ground the negative terminal. The ground wire can be, and often is, uninsulated. The ground is fastened to the engine or to some other suitable metal location. Fig. 3-2.

IGNITION SWITCH

The primary circuit starts at the battery and flows through an insulated wire to the ignition switch. The usual ignition switch utilizes a key to operate it. It connects or disconnects the flow of electricity across the terminals. The ignition switch may have additional terminals that supply electricity to other units on the car when the key is turned on. Fig. 3-3.

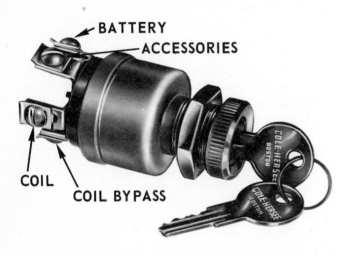

Fig. 3-3. Typical ignition switch. Coil bypass terminal furnishes current directly to coil (without passing through ballast resistor) during engine cranking. (Cole-Hersee)

RESISTOR

Electricity flows from the ignition switch to the resistor. From the resistor, the current travels to the coil.

The resistor controls the amount of current reaching the coil. The resistor may be either the simple or the ballast type.

There are resistors designed for specific jobs. One type, used on Ford V-8 engines for many years, was a simple unit used to reduce battery voltage reaching the coil. The primary circuit passed through the resistor, cutting down the voltage to a predetermined level. The type of coil utilized in these installations would overheat if the battery voltage reached the coil without being reduced.

This type of simple resistor was not temperature sensitive. It would deliver about the same voltage to the coil regardless of whether or not the coil had reached operating temperature. Fig. 3-4.

A modern adaptation of the resistor is used in high-compression, high horsepower engines. This type is known as the

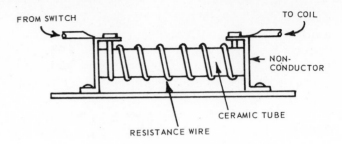

Fig. 3-4. Simple resistor in primary circuit of ignition system.

BALLAST resistor. It is a temperature sensitive, variable resistance unit.

A ballast resistor is designed to reduce the voltage available to the coil at low engine speeds, and to increase the voltage at the higher rpm range when it is needed. Without the resistor, the coil would require enough voltage to function efficiently at high speeds. Such voltage would cause excessive heating at low speeds. It would also cause oxidization of the contact points (a blue scale-like deposit).

The ballast type resistor tends to heat up at low engine speeds as the duration of flow in the primary circuit is longer, and with less interruption by the contact point opening. As it heats up, its resistance value goes up, causing lower voltage to pass the coil.

As engine speed increases, the points open and close more rapidly and the duration of current flow lessens. This causes a lowering of temperature. As the temperature drops, the resistor allows the voltage to the coil to increase. At high speeds, the coil received most all the battery voltage.

High engine speeds shorten the coil saturation period. (The saturation period is the length of time the points are closed and current is flowing through the primary windings of the coil.) It is obvious that this condition would require full voltage.

The ballast resistor is constructed of a special type wire, the properties of which tend to increase or decrease the voltage in direct proportion to the heat of the wire. Fig. 3-5.

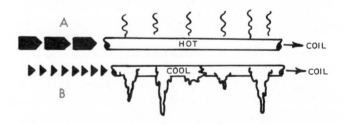

Fig. 3-5. Ballast resistor principle. A illustrates long pulsations of current passing through special ballast resistor wire at slow engine speeds. This heats special wire and lowers amount of current reaching coil. B shows short pulsations at high speed. This allows wire to cool, allowing a heavier current to flow to coil.

Current practice employs a simple calibrated resistance wire that lowers battery voltage to around 9 — 10½ volts during normal engine operation. A bypass wire runs from the ignition switch or starter solenoid to the coil. While the engine is cranking, the coil will receive full battery voltage. When the key is released, the circuit is through the resistance wire. Some systems, such as the transistor ignition system, use two ballast resistors to control coil voltage.

COIL CONSTRUCTION

The coil is constructed with a special laminated iron core.

Around this central core, many thousands of turns of very fine copper wire are wound. This fine wire is insulated by a thin coating of special insulation material. Fig. 3-6. One end of the fine wire is connected to the high tension terminal, and the other is connected to the primary circuit wire within the coil. All these turns of fine wire form what is called the SECONDARY circuit coil windings.

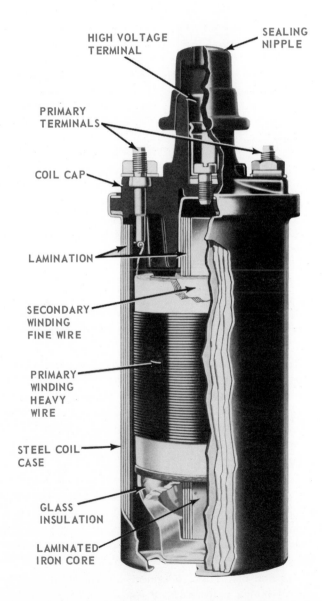

Fig. 3-6. Ignition coil. One end of secondary circuit wire is attached to high voltage terminal; other end is connected to primary circuit wire within coil.   (Chevrolet)

Several hundred turns of heavier copper wire are wrapped around the secondary coil windings. Each end is connected to a primary circuit terminal on the coil. These windings are also insulated. The turns of heavier wire form the PRIMARY circuit coil windings.

The core, with both the SECONDARY and PRIMARY windings attached, is placed inside a laminated iron shell. The job of the shell is to help concentrate the magnetic lines of force that will be developed by the windings.

This entire unit is then placed inside a steel, aluminum or Bakelite case. In some coil designs, the case is filled with a special oil. It is then hermetically sealed to prevent the entrance of dirt or moisture. Electrical outlets, such as the primary terminals, are generally contained in the coil cap. They are carefully sealed because the coil must withstand vibration, heat, moisture and the stresses of high induced voltages.

Coils vary in size and shape to meet the varying demands of different types of installations. Fig. 3-6.

COIL

The primary circuit leads from the resistor to the coil.

An ignition coil is actually a pulse-type transformer that raises the battery voltage up to, and exceeding, 20,000V.

HOW A COIL WORKS

When the ignition switch is turned on, the current flows through the primary windings of the coil, to the ground and back to the battery via the frame. Fig. 3-7.

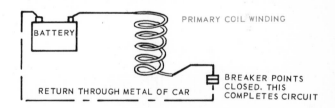

Fig. 3-7. Primary current flows through primary coil winding when points are closed.

A MAGNETIC FIELD IS DEVELOPED

As you know, when a current flows through a wire, a magnetic force field is built up around the conductor. As there are several hundred turns of wire in the primary windings, a rather strong magnetic field is produced. This magnetic field surrounds the secondary as well as the primary windings. See Fig. 3-8.

COLLAPSE OF THE FIELD

If there is a quick and clean interruption of current flow on its way to the ground after passing through the coil, the

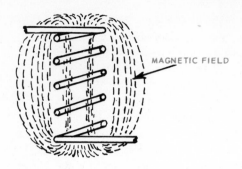

Fig. 3-8. Magnetic field is built up around coil primary when current is flowing.

magnetic field will collapse and pass quickly to the laminated iron core.

As you have learned, whenever a wire is passed through a magnetic force field, a current will be induced (set up) in the wire. By the same token, if the wire is held still and the magnetic field is passed around it, a current will still be induced in the wire. This is what happens when the magnetic field collapses and passes through the secondary windings on the way to the core. Fig. 3-9.

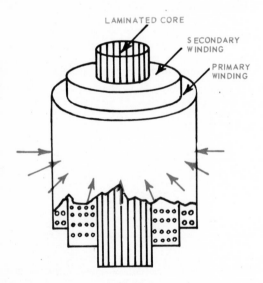

Fig. 3-9. Collapse of primary field. When primary circuit is broken, magnetic field will collapse through secondary winding to core. See arrows.

The magnetic field that passes through the secondary windings produces a tiny current in each turn. The windings possess thousands of turns, however, and as they are in series (connected together, positive to negative), the voltage of each turn of wire is multiplied by the number of turns. In the average coil, this will produce a voltage exceeding 20,000V.

As the field also passes through the primary turns, a voltage will be induced in the primary circuit also. As the turns are far less numerous, the voltage multiplication will be considerably less, somewhat over 200V.

The high voltage produced by the secondary windings is made available to the plugs at the high tension coil terminal.

POLARITY IS IMPORTANT

Most coils have primary terminals marked with (+) or (-). The plus sign indicates positive, and the minus indicates negative. The coil must be installed in the primary circuit according to the way the battery is grounded. If the battery has the negative terminal grounded, the negative terminal of the coil must be connected to the ground, or distributor side. If the battery positive terminal is grounded, the positive side of the coil must be connected to the distributor lead.

This is done to insure the correct polarity at the spark plug. The flow of electrons (current) is made easier if they flow from a hot surface to a cooler one. As the center electrode of the plug is always the hotter of the two, a negative polarity at the plug inner electrode will insure a flow of current from the center electrode. This will require less voltage and will lighten the load of the coil. Current practice grounds the NEGATIVE battery terminal.

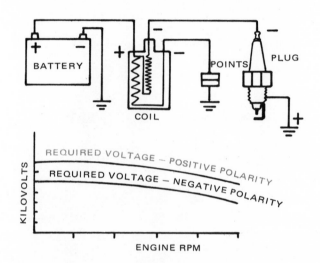

Fig. 3-10. Correct polarity is important. When NEGATIVE POLARITY is used, voltage necessary to fire plug diminishes. This makes task of coil easier.

The coil will build up enough voltage to cause a flow of current across the electrode gap. Even though a coil voltage can exceed 20,000V, it will build up voltage only high enough to produce an adequate spark. This will generally run between 2,000 and 10,000V. Fig. 3-10.

BREAKER POINTS

For efficient coil operation, the current flow through the primary windings must be interrupted (broken) instantly and cleanly with no flashover at the point of disconnection. (Flashover occurs when the current jumps or arcs across the space where the wire has been separated.) More on this later.

The unit that is used to connect and disconnect the flow of current in the primary circuit contains a set of BREAKER POINTS. The points open and close to make and break the circuit.

BREAKER CONSTRUCTION

Generally, breaker points are constructed as two separate pieces. Fig. 3-11. The stationary piece is fastened directly to the ground through the distributor breaker plate. This section does not move other than for an initial point adjustment.

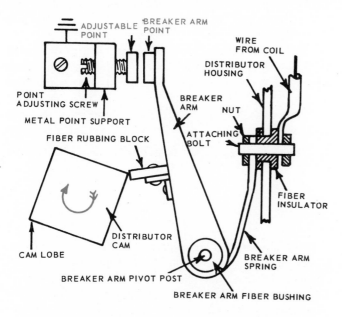

Fig. 3-11. Typical breaker point construction. Most, however, have adjustable point built into stationary support.

The second piece is the movable breaker point. It is pivoted on a steel post. A fiber bushing is used as a bearing on the pivot post. A thin steel spring (flat type) is used to press the movable breaker arm against the stationary unit, thereby causing the two contact points to bear firmly together.

The movable arm is pushed outward by a cam lobe. The cam lobe is turned at one-half engine rpm by the distributor shaft. The distributor shaft is turned by means of a gear that is meshed with an integral gear cut on the camshaft.

The breaker arm contacts the cam by means of a fiber rubbing block. This block is fastened to the breaker arm and rubs against the cam. A special high temperature lubricant is used on the block to prevent undue wear.

The movable breaker arm is insulated so that when the primary lead from the coil is attached to it, the primary circuit will not be grounded unless the contact points are touching. Fig. 3-11.

CONTACT POINTS MUST FIT

The breaker contact points must be carefully aligned so that the two points make perfect contact. They must be clean and free of oil, and open an exact amount. Each system, as developed by the manufacturer, must be held to specifications. If the point opening (gap) is specified as .020 in. it must be set to that amount. A variation in gap will upset the cam angle and ignition timing.

SPECIAL METAL

Contact points are made of tungsten steel. Tungsten is resistant to burning and will give good service.

The movable breaker arm must be rigid enough to hold the contact point in line. At the same time, it must be light enough to function at high speeds without undue spring pressure that would cause premature wear of the rubbing block. Too light a spring pressure, or too heavy a breaker arm, would cause the arm to float at high speeds. (Float is when the breaker arm is pushed out as the cam strikes the rubbing block; and before the arm can return after the cam has passed, the next cam strikes the block and reopens it. In other words, the breaker arm never has time to fully close before being opened again.)

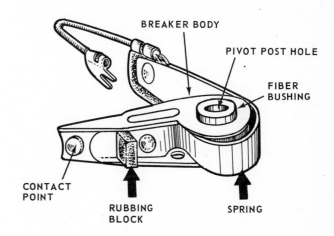

Fig. 3-12. One type of breaker arm design. Another design uses a flexible arm that does not pivot on a post. (Toyota)

Fig. 3-12 illustrates a typical breaker arm, or as it is sometimes called, a breaker lever.

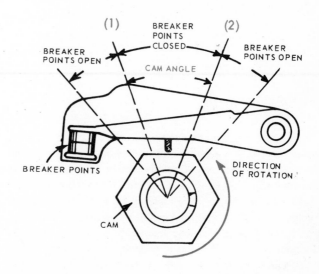

Fig. 3-13. Cam angle. Points close at 1 and remain closed as cam rotates to 2. The number of degrees formed by this angle determines cam angle.

CAM ANGLE

Cam angle, sometimes referred to as degrees of dwell, is the number of degrees the cam rotates from the time the points close until they open again. Fig. 3-13.

The cam angle is important. The longer the points are closed, the greater the magnetic buildup of the primary windings. If the cam angle is too small, the points will open and collapse the field before it has built up enough to produce a satisfactory spark.

When setting breaker point gaps, remember that when the gap is lessened, cam angle is increased. Then the gap is enlarged, cam angle is decreased.

DUAL POINTS

Some distributors, designed especially for high speed operation, utilize two sets of breaker points. The points are mounted opposite each other, and are designed so that when one contact set is snapped open, the other closes. Fig. 3-14.

In the single point setup, the points are held open as the rubbing block passes up on the toe of the cam, across the top and down the heel of the cam. This represents a considerable number of degrees of cam rotation. The longer the points are open, the less the cam angle can be.

The dual point set cuts down the open time to which the circuit is exposed. As the cam angle is increase, the engine can operate at much higher speeds and still produce a satisfactory spark.

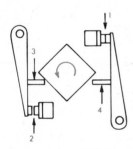

Fig. 3-14. Dual points. Rubbing block 3 is being contacted by cam lobe to open points at 2. Points 1 have just closed as rubbing block 4 comes off cam lobe.

POINTS WILL BURN

If you were to set up the ignition system as discussed so far, you would find that each time the points opened, there would be a heavy electrical arc across the points. Unless the points break the primary circuit quickly and cleanly, the magnetic collapse of the primary windings will be poor and a satisfactory voltage will not be produced in the secondary windings.

In addition to insufficient voltage, the points would soon burn beyond use. There is a tendency for a moving stream of electrons (current) to keep moving when the circuit is broken. When the contact points start to open, the magnetic field starts to collapse. This will induce extra voltage into the primary circuit.

Between the reluctance of the moving current to stop, and the added surge of induced voltage in the primary circuit, the circuit will not be broken cleanly, and a heavy spark will leap across the point opening. As mentioned, this will prevent the ignition system from functioning properly. Fig. 3-15.

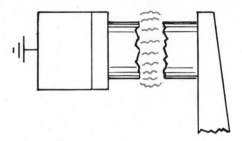

Fig. 3-15. Arcing at points. Notice how current leaps gap and burns contact face.

CONDENSER

When a condenser of proper size is inserted into the primary circuit, heavy arcing at the points will stop, and secondary voltage will reach satisfactory limits. This is accomplished by breaking the primary circuit cleanly. The condenser provides a place into which the primary current may flow when the points are opened.

CONDENSER CONSTRUCTION

Most condensers are constructed of two sheets of very thin foil separated by two or three layers of insulation. The foil may be lead or aluminum. The long narrow sheets of foil, separated by the insulation, are wound together into a cylindrical shape.

The cylinder is then placed in a small metal can or case. The cap, with condenser lead attached, is put on and the edges of

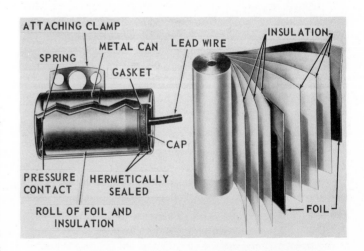

Fig. 3-16. Condenser construction. Unit is hermetically sealed in metal can. (GMC)

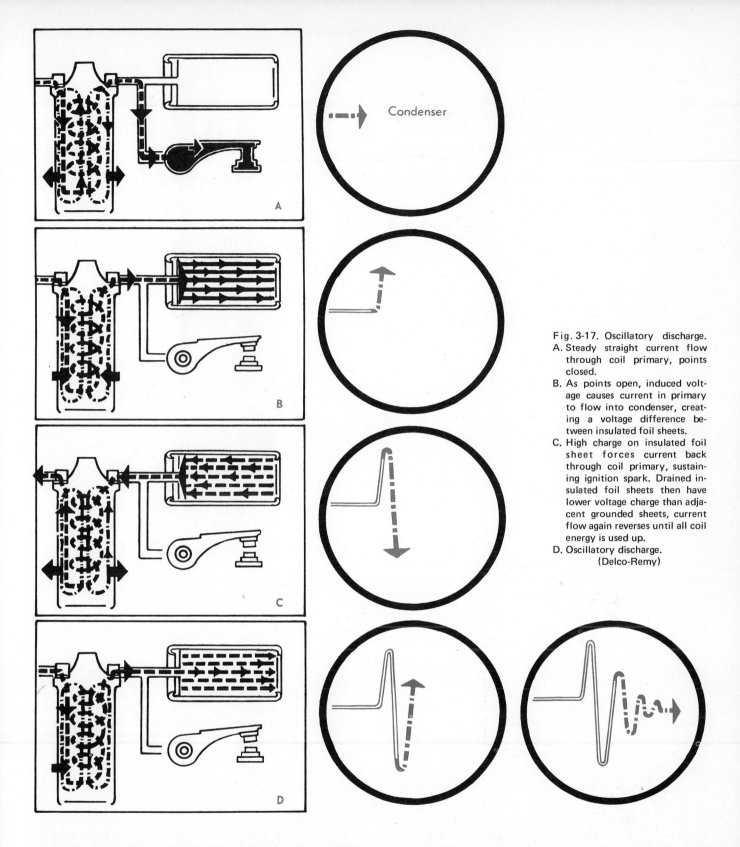

Fig. 3-17. Oscillatory discharge.
A. Steady straight current flow through coil primary, points closed.
B. As points open, induced voltage causes current in primary to flow into condenser, creating a voltage difference between insulated foil sheets.
C. High charge on insulated foil sheet forces current back through coil primary, sustaining ignition spark. Drained insulated foil sheets then have lower voltage charge than adjacent grounded sheets, current flow again reverses until all coil energy is used up.
D. Oscillatory discharge.
   (Delco-Remy)

the case pulled down, forcing the cap and seal into tight contact with the case. A spring is sometimes used at the bottom of the case to constantly push the condenser foil cylinder against the cap. This increases the seal pressure, eliminates vibration and series resistance caused by corrosion or vibration. The condenser is hermetically sealed to prevent the entrance of moisture.

Condensers will vary in size according to the rated capacities and if standard or heavy duty. An average capacity is around .20 microfarads (capacity measurement). Fig. 3-16 illustrates the construction of the typical condenser.

HOW A CONDENSER WORKS

A, Fig. 3-17, shows how a condenser is installed in the circuit. The primary circuit flows through the points to the ground and back to the battery. The primary field (magnetic force field) is strong.

In B, Fig. 3-17, the points have started to open. The primary field has started to collapse, but no current flows across the points. It has flowed into the condenser. The condenser cannot hold much of a charge but, by the time it has become fully charged, the points have opened too far for the current to jump. The primary field collapse will be strong and quick.

Latest theory is that at the instant the spark at the plug is formed, the condenser (because of its charge) will discharge back into the primary. It does this because its charge, due to the high induced voltage in the primary, is higher than the primary after field collapse is complete. See C, Fig. 3-17.

Each time the condenser discharges back into the primary windings, it produces another field in the opposite direction and collapses the one just built by its first discharge. The bouncing back and forth is termed oscillatory discharge. It will continue, weaker each time, until completely worn out, or stopped by the closing of the contact points. See D, Fig. 3-17.

PRIMARY CIRCUIT COMPLETE

You have now traced the flow of current through the primary system. After going to the ground, it returns to the battery through the metal parts of the car to which the battery is grounded.

Be sure you understand how each unit works and its relationship to the other parts.

SECONDARY CIRCUIT

You have seen how the coil produces a high voltage in the secondary circuit. The current flows from the coil to the spark plug. (How it is timed and distributed will be covered shortly.)

The current jumps from the center electrode of the plug to the side, or ground, electrode. When it jumps across, it produces a hot spark that ignites the air-fuel mixture.

SPARK PLUG

A spark plug is made up of three major parts: the electrodes, insulator and shell. Fig. 3-18.

Electrodes of a spark plug must be constructed of material that will be resistant to heat, oxidization and burning. A typical material is nickel alloy. Platinum, although expensive, is sometimes used. It makes top quality electrode material.

In an ordinary spark plug there are two electrodes, the center electrode and the side electrode. The space between the two is called the plug GAP. The gap varies from .025 to .040 in. for most cars.

The center electrode is usually negative to assist the flow of electrons. The electrons will then leave the hotter center electrode and pass to the side electrode. Air is more easily ionized (broken down from a nonconductor to a conductor) near the hotter center electrode.

The center electrode is insulated from the rest of the plug by a ceramic insulator. The top terminates in a snap-on terminal to which the plug wire may be attached. Fig. 3-18.

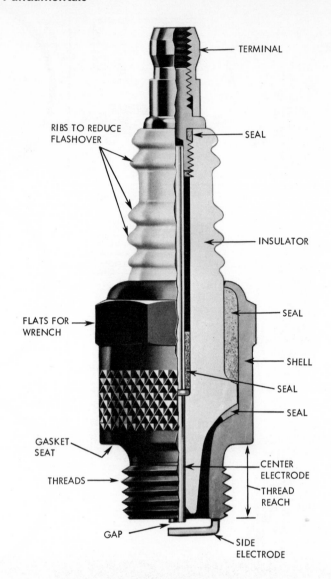

Fig. 3-18. Typical spark plug.
(Champion Spark Plug Co.)

SPARK PLUG INSULATORS

Insulators must have special properties. They must resist heat, cold and sudden temperature changes. Vibration and physical shock must leave them undamaged, and they must resist chemicl corrosion.

A common material used for making spark plug insulators is aluminum oxide, fired at high temperature. This produces a glassy smooth, dense and very hard insulator. The length, diameter and location of the insulator has a direct bearing on the heat range of the particular plug. This is covered under Plug Heat Range characteristics.

The top end of the insulator is often ribbed, or grooved, to prevent shorting. (Termed flashover.) Fig. 3-18.

SPARK PLUG SHELLS

The center electrode, surrounded by the insulator, is placed in a steel shell. The steel shell top is generally crimped over to bear against a seal. The crimping process grips the insulator tightly and also forms a seal pressure at both top and bottom insulator seals. This prevents combustion leaks.

The side electrode is welded to the steel shell.

The shell is threaded so it will screw into a threaded hole in the cylinder head. The shell threaded area will vary in length according to hole depths and ignition recommendations.

The shell forms a seal with the head by means of a copper gasket or a beveled edge that wedges against a similar bevel in the cylinder head. Fig. 3-19.

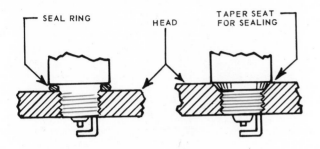

Fig. 3-19. Thread seals.

The thread seal is important since this is an area through which a great deal of the plug heat is transferred to the head metal. The insulator is subjected to tremendous temperatures. In order to prevent burning, it must get rid of surplus heat.

SPARK PLUG HEAT RANGE

Spark plugs are designed for various types of engines. It is important that the plug extend into the combustion chamber just the right amount.

The heat range is determined by the diameter and length of the insulator as measured from the sealing ring down to the tip. Fig. 3-20.

Notice, in Fig. 3-20, that the heat generated in the insulator must travel up the insulator until it can escape through the seal, to the shell and then to the head. The longer and thinner the insulator tip, the less efficiently it can transfer heat. As a result it will run hotter (hot plug). The short, heavy insulator carries heat well and will operate much cooler (cold plug).

All engine manufacturers specify heat ranges for the plugs to be used in their engines.

PROJECTED CORE NOSE

One spark plug design extends the insulator beyond the end of the shell, thus placing the electrode tip and insulator, further down into the combustion chamber. This lengthening, termed Projected Core Nose, places the insulator more directly in the path of the incoming fuel charge. As the intake stroke draws in fresh fuel, it passes over and around the plug insulator. This will cool it down. Its shape is not usually adaptable to L-head engines since there is insufficient room for piston clearance.

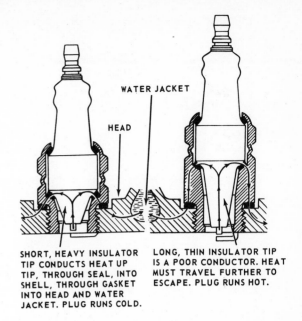

Fig. 3-20. Spark plug heat range.   (Ignition Mfgs. Institute)

The projected core nose, due to its long tip, tends to run hot at low engine speeds because the tip cannot transfer the heat fast enough. The incoming fuel charge is moving slowly and provides little cooling effect. This prevents slow speed fouling and assists in producing a spark without excessive voltage from the coil.

At high engine speeds, the insulator temperature is kept from exceeding acceptable limits by the rapid flow of the incoming fuel charges. The washing effect of the hot escaping exhaust tends to keep the plug clean. Fig. 3-21.

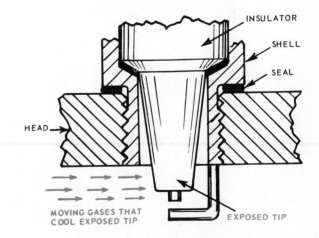

Fig. 3-21. Projected core nose plug.

PREIGNITION

If a plug runs too cold, it will foul. If it runs too hot (much above 1,700 deg.), it will cause PREIGNITION (lighting the fuel charge from red-hot deposits before the plug fires). The

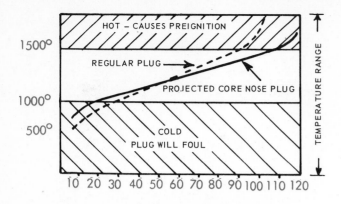

Fig. 3-22. Speed range performance graph.

projected core nose will perform satisfactorily within acceptable temperature limits over a wider car speed range. Fig. 3-22.

RESISTOR SPARK PLUG

The spark at the plug electrodes is delivered in two stages. The first part is called the CAPACITIVE portion; the second, and longest portion, the INDUCTIVE.

The voltage at the plug center electrode will rise rapidly until the voltage is sufficient to ionize the gap and cause the plug to fire. The first firing is the capacitive part. The voltage at the plug produces the capacitive portion.

The inductive portion is longer and follows the capacitive. It is produced by the remaining voltage in the system (coil residual and oscillatory action of the condenser).

The combustion process takes place during the capacitive

section. The inductive portion causes radio interference and is hard on electrodes. The inductive part, therefore is undesirable and attempts have been made to shorten it.

By placing a resistor (around 10,000 ohms) in the ceramic insulator, it is possible to shorten both the capacitive and inductive phases. It does not require any higher voltage and will lengthen electrode life as well as suppress radio interference. A resistor plug is shown in Fig. 3-23.

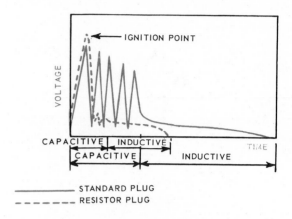

Fig. 3-24. Spark discharge graph. Notice how resistor plug reduces undesirable inductive portion of spark.

Fig. 3-24 illustrates the difference between the spark produced by a standard plug and a resistor type plug. Notice that no more voltage is required for the resistor plug. Late model cars use special resistance spark plug wires to cut down radio interference.

AUXILIARY GAP

Another feature of some spark plugs is the use of a small gap between the center electrode top and its terminal. This is designed to build up higher voltage and to reduce misfiring of dirty plugs. The terminal post is ventilated to allow ozone (produced by sparking at auxiliary gap) to escape. Fig. 3-25.

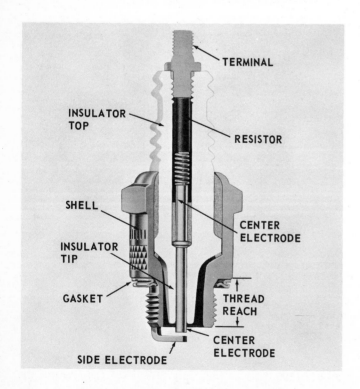

Fig. 3-23. A resistor spark plug. (Champion)

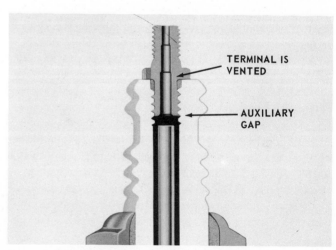

Fig. 3-25. Spark plug with an auxiliary gap.

You have seen how the necessary voltage is produced. How this voltage is delivered to each plug at the proper time will now be covered.

DISTRIBUTOR CAP

High voltage leaving the coil is carried via a heavily insulated wire to the center terminal of the distributor cap. Additional terminals, one per cylinder, will be arranged in a circle around the center terminal. Each one of these will have a heavily insulated wire connecting it with a spark plug.

The distributor cap is made of a plastic material, sometimes mica filled to reduce flashover tendency. Material used must provide excellent insulation. Fig. 3-26.

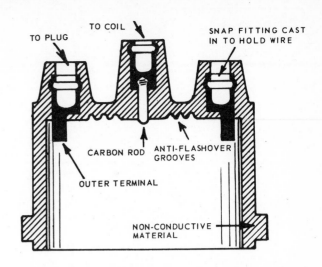

Fig. 3-27. Distributor cap — section view.

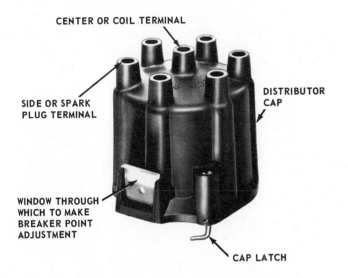

Fig. 3-26. A distributor cap for a six cylinder engine. (G.M.C.)

On the inside of the cup-like cap, the plug terminals have brass lugs extending down past the cap material. The center, or coil terminal, has a round ball, Fig. 3-28, or a carbon rod extending down. Note the fixed carbon rod used in Fig. 3-27.

ROTOR

To carry the secondary voltage from the center terminal to any one side terminal requires what is known as a ROTOR. This rotor is attached to the top of the distributor shaft. Its outer edge has a brass terminal that passes, as it turns, very close to the side terminals. This outer edge terminal is connected to a spring that rubs against the center coil terminal carbon post. When the current from the coil arrives at the center terminal, it travels down the carbon rod, through the spring, and out to the rotor outer edge terminal. From this point, it jumps the small gap between the rotor and side terminal and goes on to the plug. Fig. 3-28. Rotor is made of non-conductive material.

The cap, and whirling rotor, distributes the spark to each plug at the proper time. It is essential that the plug wires be arranged in the proper sequence. The firing order (order in which pistons reach TDC on the compression stroke) in a typical six cylinder engine is 1, 5, 3, 6, 2, 4.

The plug wires are arranged in distributor cap in that order. Direction of rotation of the rotor dictates whether a clockwise or counterclockwise order is required. Fig. 3-29.

Fig. 3-28. Distributor cap and rotor. Rotor transfers current from center terminal to outer terminal. (Oldsmobile)

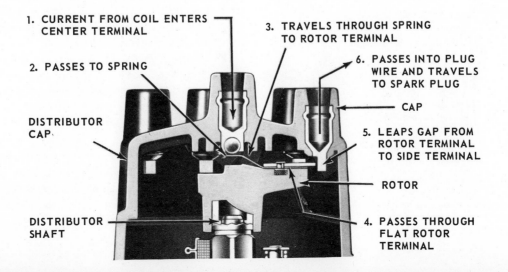

TIMING

For correct ignition timing, each cylinder should receive a spark at the plug electrodes as the piston nears the top (few degrees before TDC) of its compression stroke. This is made possible by the proper hookup of the distributor shaft to the camshaft, which turns at one-half crankshaft speed.

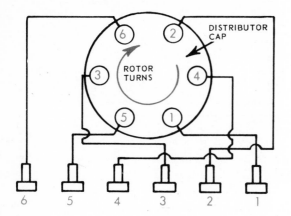

Fig. 3-29. Firing order — six cylinder engine: 1, 5, 3, 6, 2, 4. Notice how plug wires are arranged in distributor cap to produce this firing order.

The distributor shaft gear is meshed with the camshaft gear so that the breaker points are just opening when number one cylinder is ready to fire. The rotor will then point toward the number one cap plug terminal. Number one cylinder plug wire is attached to this terminal. Following the firing order, and with due regard to shaft rotation, the remaining wires are attached to the cap.

As the engine turns, the distributor shaft will revolve and cause the distributor cam lobes to open the points and collapse the primary field. The lobes are spaced equidistantly. Each time the distributor shaft has turned enough to cause the rotor to point to the next plug terminal, the cam lobe will once

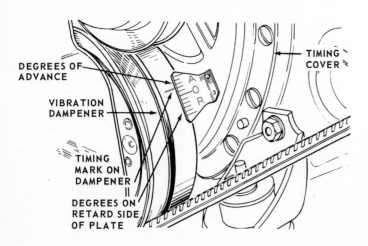

Fig. 3-30. Ignition timing marks. This setup has a timing mark cut into the vibration damper.

again break the points. This causes a secondary voltage surge from the coil to the cap center terminal, into the rotor and to the cap plug terminal. This cycle is repeated over and over.

The engine manufacturer specifies specific timing in regard to the number of degrees before TDC number one cylinder should fire. All other cylinders will fire at the same number of degrees before TDC.

SETTING BASIC TIMING

Most engines have the timing location, in the form of a line, dent, etc., marked on the rim of the vibration damper. A pointer is attached to the timing cover. When the mark is exactly under the pointer, the engine is ready to fire number one cylinder. The points should just break with the rotor pointing to the number one cap terminal. Fig. 3-30. Some engines have the timing mark on the flywheel.

The timing is generally set by using a strobe lamp (a light that is operated by high voltage surges from the spark plug wire). Fig. 3-31.

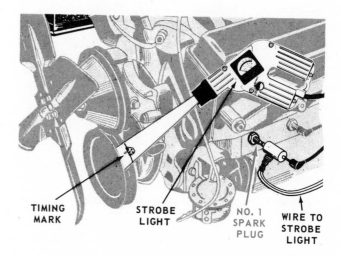

Fig. 3-31. Using a strobe light to time ignition. Every time No. 1 plug fires, strobe light will illuminate timing marks.
(Sun Electric)

After the contact point gap is accurately set, the strobe lamp is connected to the number one plug wire. The engine is started and idled quite slowly. Usually, the vacuum line to the distributor is removed.

The strobe light is set to shine on the pointer over the vibration damper. Every time number one plug fires, the strobe lamp lights. Each time it fires with the damper in the same position in relation to the pointer, the damper timing mark looks as though it were standing still.

To adjust the timing, the distributor hold-down clamp is loosened and the distributor is turned, by hand, one way or the other. As it is turned, the timing mark will move. When turned in the proper direction, the mark will line up with the pointer. When the two are aligned, the distributor hold-down is tightened and the engine is properly timed.

TIMING NEEDS VARY

As engine speed increases, it is necessary to fire the mixture somewhat sooner. If this is not done, the piston would reach TDC and start down before the air-fuel mixture can be properly ignited.

The pressure of the burning fuel charge should reach its peak with the piston somewhere between 15 and 25 deg. past TDC. With the engine at slow idle, very little advance is necessary. At the higher engine speeds, it is necessary to fire the mixture somewhat sooner. Notice in A, Fig. 3-32, that the

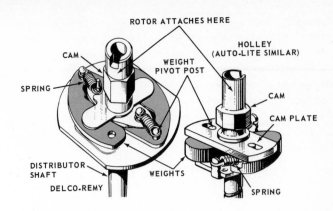

Fig. 3-33. Two different types of centrifugal advance mechanisms. (Automotive Electric Association)

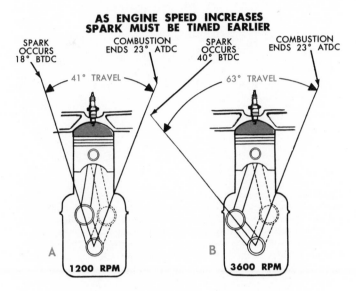

Fig. 3-32. When engine speed increases, spark must be timed sooner. Notice in "A" that only 41 deg. of crankshaft travel is required while in "B," at 3,600 rpm., 63 deg. are necessary. (Ford)

combustion cycle must start at 18 deg. before TDC in order to be complete by 23 deg. after TDC. In B, Fig. 3-32, engine speed has tripled. It is now necessary to ignite the charge 40 deg. before TDC in order to complete combustion by 23 deg. after TDC.

From this you can see that it is necessary to have some way of advancing (firing more degrees before TDC) the timing as the engine speed increases. It is also necessary to retard the timing at certain times in order to help control exhaust emissions.

CENTRIFUGAL ADVANCE

One method of advancing the timing is through the use of a centrifugal advance mechanism which is assembled in the distributor housing. Figs. 3-33 and 3-41. When the distributor shaft turns, it turns the centrifugal unit and the centrifugal unit turns the cam.

When the engine is idling, spring pressure keeps the two weights drawn together and the shaft and cam assume the position for low speed timing.

As the engine speeds up, the weights are drawn out by

centrifugal force. As the weights move apart, they force the cam to move, in an advance direction, in relation to the shaft. As the cam is advanced, the lobes strike the rubbing block sooner, causing the plugs to fire more degrees before TDC. The faster the engine turns, the farther apart the weights move, until they finally reach the limit of their travel.

As speed is decreased, the centrifugal pull on the weights is lessened and the springs pull the weights together, retarding the timing. By calculating the pull of the springs and the size of the weights, it is possible to properly advance the timing over a long rpm range. Fig. 3-34 illustrates how the weights control advance.

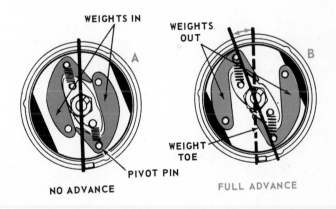

Fig. 3-34. Centrifugal advance unit action. In "A," the engine is idling. Springs draw weights in. Timing has no advance. In "B," engine is running at high speed. Centrifugal force has drawn weights outward. As they pivot, the weight toe ends force the cam plate to turn thus advancing the timing.

VACUUM ADVANCE

It has been found that at part throttle position, additional advance over and above that provided by a centrifugal mechanism, is desirable.

This is due to the fact that during part throttle (carburetor throttle valve part way open) operation, there is high vacuum in the intake manifold (pipes connecting the carburetor to the valve port openings in the head, or block). This high vacuum draws in a lesser amount of air-fuel mixture. The smaller amount will be compressed less and will burn slower.

To get the utmost economy from this part of the fuel charge, it is necessary to advance the timing beyond that provided by centrifugal weights. Any benefit from additional advance applies only to the part throttle position. During acceleration or full throttle operation, no vacuum advance is provided. Fig. 3-35.

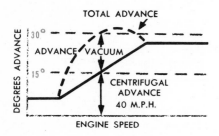

Fig. 3-35. Vacuum advance chart. Notice that part throttle vacuum advance is in addition to regular centrifugal advance.

HOW VACUUM ADVANCE MECHANISM WORKS

Vacuum advance may be accomplished in two ways. One method is to rotate the entire distributor housing against the direction of rotation of the distributor shaft. The breaker point rubbing block will then strike the oncoming cam sooner, advancing the spark timing.

Current practice mounts the breaker points on a movable breaker plate. This plate can revolve either on a center bushing, or on ball bearings on its outer edge. The hardened steel balls roll in a grove cut into the distributor housing.

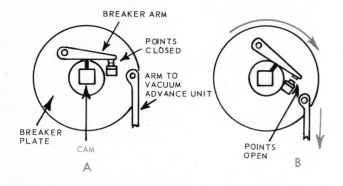

Fig. 3-36. Breaker plate action. In A, points are closed. When breaker plate is turned, as in B, the rubbing block will strike cam sooner and open the points, causing spark to be advanced. Notice that cam is in the same position in both A and B.

With either type of plate, an advance in timing may be obtained by turning the plate against distributor shaft rotation. Fig. 3-36.

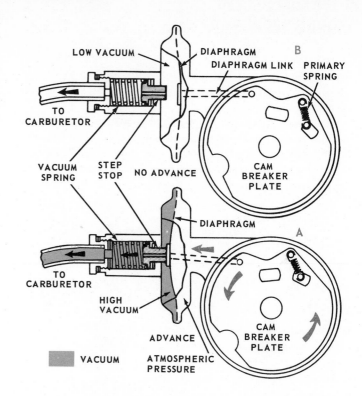

Fig. 3-37. Vacuum advance mechanism. In "A" the carburetor throttle valve is in part throttle position, thus drawing a heavy vacuum. With a vacuum on the left side, atmospheric pressure forces the diaphragm to the left. Diaphragm link will pull cam breaker plate around and advance timing. When throttle is opened and vacuum is lowered, (B), the breaker plate primary spring will pull the breaker plate back and retard the timing. The vacuum spring also controls the limits of advance.
(Ford)

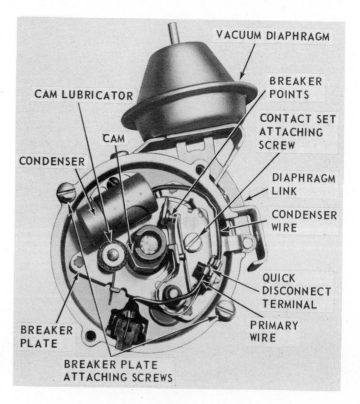

Fig. 3-38. Distributor with cap and rotor removed.
(GMC)

VACUUM ADVANCE DIAPHRAGM

The breaker plate is rotated by means of a vacuum advance diaphragm. This is a stamped steel container with a neoprene-coated cloth diaphragm stretched across the center. One end, airtight, is connected to the carburetor slightly above the closed position of the throttle valve. The other end is open to the atmosphere. The diaphragm has a lever rod attached to its center. The lever rod is connected to either the distributor housing, or to the breaker plate, depending on which type is used. Figs. 3-37, 3-38, 3-39 and 3-40.

When the throttle valve is in part throttle position, as shown in B, Fig. 3-37, there is high vacuum in the intake manifold. The vacuum pulls the diaphragm back toward the vacuum side. This in turn pulls the breaker plate around and advances the timing.

When the throttle is opened, the vacuum drops and the breaker spring draws the diaphragm back toward the distributor. This rotates the breaker plate in the retard direction. When the engine is idling, the throttle valve closes below the vacuum advance opening. This removes vacuum pull, and the spark will be retarded for idling. The vacuum advance mechanism is constantly working as the throttle valve position is changed.

SUMMARY

Current from the battery flows to the key switch. When the switch is turned on, current flows through a resistor to the primary windings in the coil. From the primary coil windings, the current travels to the breaker points, and on to ground where it returns to the battery via the metal parts of the car.

Current flow through the primary windings builds up a magnetic field that surrounds the secondary windings. When the points are opened by the distributor cam, current in the primary is stopped. A clean stoppage is made possible by use of a condenser. When the primary flow is broken, the magnetic field around the secondary windings collapses and induces a high voltage in the secondary circuit.

Trace the flow of current in both primary and secondary circuits in Fig. 3-1.

MAGNETO IGNITION

A magneto is an electrical device which generates and distributes electricity to ignite the combustible mixture in the combustion chamber. A magneto does not require a battery to operate.

The lawn mower, garden tractor and outboard boat engine are just a few of the many engines that use the magneto ignition system.

The magneto system is also practical for a variety of industrial and aircraft piston engines. It is compact and dependable.

PRINCIPLE OF MAGNETO OPERATION

Other than the lack of a battery to furnish current for the primary system, the magneto system is quite similar to the battery system. With no battery, a way must be found to impart voltage in the primary windings. Fig. 3-42.

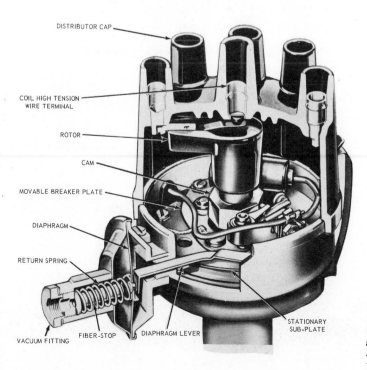

Fig. 3-39. Cutaway showing distributor head and vacuum advance construction.

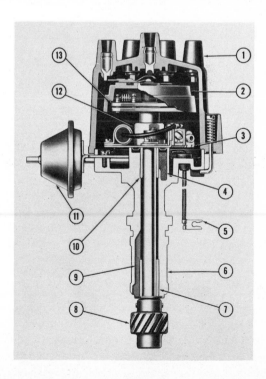

Fig. 3-40. Distributor with centrifugal advance unit above breaker plate assembly; rotor is round. 1—Cap. 2—Rotor. 3—Breaker plate assembly. 4—Permanent lubrication reservoir. 5—Primary terminal. 6—Housing. 7—Lower bushing. 8—Drive gear. 9—Passage for engine fumes. 10—Upper shaft bushing. 11—Vacuum advance. 12—Cam. 13—Centrifugal advance assembly. (Jeep)

The magnetic lines of force that travel from the north to the south pole of the permanent magnet provide the answer. If a coil is placed within these lines, and if these lines are suddenly collapsed, a current will be induced in the coil. See Fig. 3-42.

Notice in Figs. 3-43A, 3-43B, 3-43C, that as the rotating magnet is turned between the laminated coil core pole shoes,

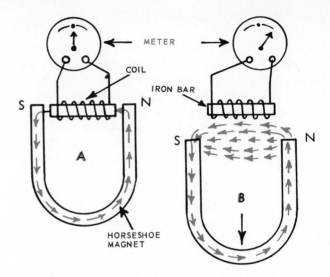

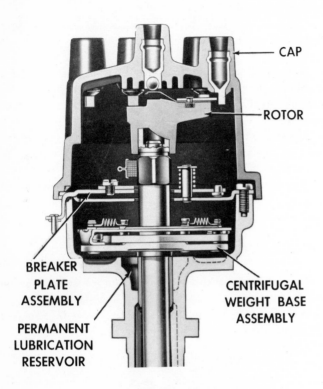

Fig. 3-41. Distributor with centrifugal advance unit under breaker plate assembly. (American Motors)

CAP

ROTOR

BREAKER PLATE ASSEMBLY

PERMANENT LUBRICATION RESERVOIR

CENTRIFUGAL WEIGHT BASE ASSEMBLY

Fig. 3-42. Using a magnet and coil to produce induced voltage. In A, magnetic lines of force are channeled through laminated iron bar. A coil of wire is wound around bar. Ends of coil are hooked to meter. No voltage is produced. In B, magnet has been jerked down away from coil. As lines of force cut across coil, voltage is induced in coil. Notice meter. You can also see lines of force when iron bar is removed from magnet. They are no longer channeled and spread out.

the magnetic field is built up, collapsed, and then built up again in the opposite direction. This gives the effect of moving a magnetic field through a coil of wire and induces voltage in the coil.

If we connect the primary wires through a set of breaker points, voltage will be built up when the field is established. When the magnet is turned and the field collapses, the field that has been built up by the primary itself, will be retained through electromagnetic action (tendency of a current to keep flowing).

As the magnet turns, an attempt is made to reestablish lines of force in the opposite direction. However, the magnetic field built up around the primary by the current flowing in it will resist field reversal. At this time, tremendous stress is set up in the field. If the points are opened, the field will collapse the lines of force in one direction, and instantly reestablish them in the opposite direction. This gives two fast flux, or force field, line changes and is sufficient to induce a high voltage in the secondary circuit. Figs. 3-43A, 3-43B, 3-43C.

By studying the drawings in Fig. 3-42, you can see how current is produced without a battery.

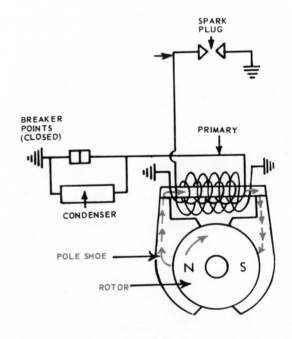

Fig. 3-43A. Action of rotating magnet. Force field traveling through coil core. Field is carried up pole shoe, through core, and down other pole shoe. Just as rotor (rotating magnet) arrived at this point, lines of force cut across coils. We now have an induced voltage.

FLYWHEEL MAGNETO

Examine the simple type of FLYWHEEL MAGNETO shown in Fig. 3-44. This is commonly used on small one and two cylinder engines.

Notice in Fig. 3-44, that the flywheel Q has a ring magnet A fastened to it. This will cause magnet A to spin around with the flywheel. As it spins, it passes close to the laminated core

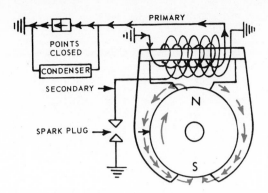

Fig. 3-43B. Action of rotating magnet. Magnet is in neutral position. Lines of force travel through pole shoes and not through core. Field will now collapse. Field built up by primary will continue to flow as points are still closed.

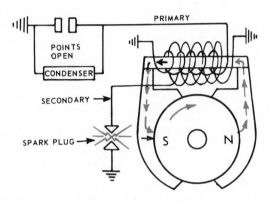

Fig. 3-43C. Action of rotating magnet. As magnet has now turned to this position, force field attempts to reverse itself. Old field in primary resists this and, as a result, tremendous stress is set up in field. When points are snapped open, old field will collapse and force lines will instantly reestablish themselves in opposite direction. This double flux field change will produce a high voltage in secondary and plug will fire.

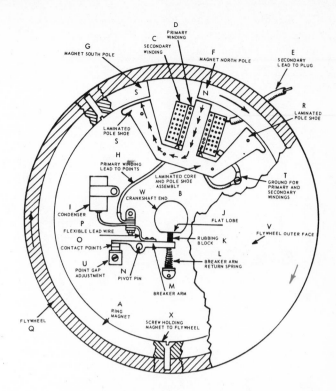

Fig. 3-44. Flywheel magneto.

and pole shoe assembly B. As you know, the magnetic lines of force flow from the north pole to the south pole of the magnet. Since they will bend to flow through an iron conductor, if it is nearby, see how they flow through laminated assembly B when the flywheel brings the poles in line. These lines surround both primary and secondary windings. As the primary breaker points are closed, the magnetic lines passing through the primary will build up a voltage in the primary circuit, (voltage is induced when lines first whip across coil-windings).

As the flywheel continues to spin, it brings the magnetic south pole in line with the coil core. The north pole moves in line with the laminated pole shoe R.

Once the points close, the original field set up in the primary will remain (through electromagnetic action). As the magnet poles move to the second position, and the field flow attempts to reverse itself, great stress is set up in the field.

At this time of maximum current flow in the primary (also time when cylinder is ready to fire), cam lobe J will strike rubbing block K and open the points. As in a battery system, when the current flow through the primary is broken, the

magnetic field around the primary and secondary windings will collapse and induce a high voltage in the secondary circuit sufficient to fire the plug.

This particular setup requires no distributor as it is timed to fire when the piston is ready. To provide for advanced timing for higher speeds, a centrifugal advance will move the rubbing block ahead against cam rotation to provide for increase degrees of timing.

Notice that the coil in this system has the primary windings on the inside of the secondary windings. Also notice that this system will produce one spark for every turn of the crankshaft. We thus have a spark on the firing stroke and another, called a "Maverick spark," at the end of the exhaust stroke. The Maverick spark can be eliminated by opening and closing the points with a gear driven cam to cut cam speed in half.

The flywheel, when in place, covers the entire assembly in most engines. Some have small openings cut into the flywheel to adjust the points. Some engines mount the breaker points and condenser outside to make servicing easier.

TRANSISTOR IGNITION SYSTEM

An important development in ignition systems makes use of a solid state electronic switching unit called a TRANSISTOR. This is a low voltage unit with a current carrying capacity superior to that of the conventional ignition breaker points.

The TRANSISTOR, Fig. 3-45, is actually of a fairly thin, wafer design. The thickness of the three parts in the illustration is exaggerated for clarity. The transistor shown is made up of two outer layers of POSITIVE materials with a center section of NEGATIVE material. The three are fused together.

The method of combining the two materials determines the polarity of the transistor. The one shown in 1, Fig. 3-45, would be a PNP type. Detail 3 shows the conventional wiring symbol, and detail 2 shows how the transistor connections are made.

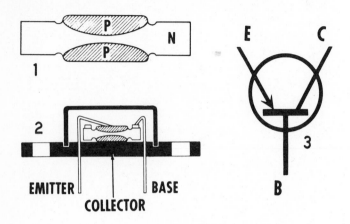

Fig. 3-45. 1—PNP transistor. 2—Transistor connections. 3—Wiring symbol for a transistor. B—Base. E—Emitter. C—Collector. (Delco-Remy)

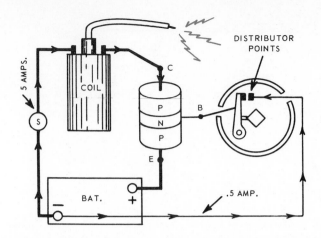

Fig. 3-46. Transistor as a switch.

Just before points opened, a .5 amp. current was passing from bat. neg. terminal through points to transistor B connection. From there it passed to E connection and on to bat. pos. terminal.

This caused a heavier 5 amp. current to flow from bat. neg. post through switch, through coil on to transistor C connection. From there through transistor to E connection to bat. pos. terminal, energizing the coil.

When points opened and broke B—E circuit, C—E flow also stopped, collapsing coil magnet field. High voltage was induced in coil secondary windings to fire the plug.

By connecting wires to the transistor, as shown in Fig. 3-46, we find that when a battery is placed in the circuit, current will flow through the transistor in two ways. A small BASE current (.5 amp.) will pass through BASE CIRCUIT E-B.

When the base current is flowing, the transistor will pass a much heavier COLLECTOR CURRENT (5 amps.) through the

COLLECTOR CIRCUIT C-E. When the base circuit E-B is broken, collector current instantly stops. This makes an excellent switch for the ignition system.

In the conventional ignition system, the contact points must carry current that has passed through the ignition coil. In order for the coil to function properly, it must carry a fairly

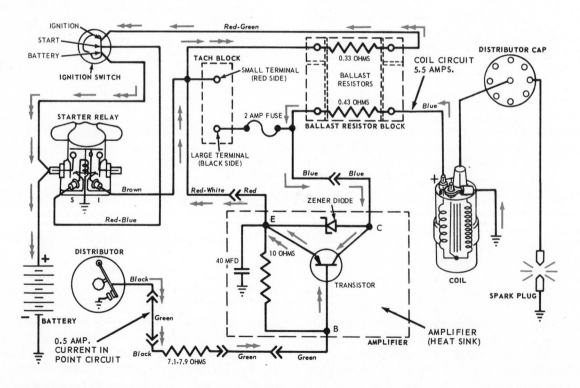

Fig. 3-47. Ford contact controlled transistor ignition system. Point circuit (double-head arrows) join coil circuit (single-head arrows) at E. (Ford)

heavy current (4 to 5 amps., based on rpm). This places a comparable load on the points. When they open and break the coil circuit, there is some arcing — despite the condenser. Eventually this causes the points to erode, pit and burn, which gives them a limited service life.

The transistor, on the other hand, offers a method of making and breaking the full coil current — with only a light load across the points. This lowers point current load and gives an extended life.

The system, shown in Fig. 3-46, is incomplete and will not function properly. Through the addition of HEAT SINK (AMPLIFIER HOUSING), RESISTORS, ZENER DIODE, CONDENSER, etc., a workable system is produced.

A wiring diagram of the Ford CONTACT CONTROLLED TRANSISTOR IGNITION SYSTEM is shown in Fig. 3-47. Note the addition of the parts just named. The term CONTACT CONTROLLED means that the conventional type breaker points are used to make and break the BASE circuit to trigger the transistor COLLECTOR circuit.

In the Ford system, Fig. 3-47, current (.5 amp.) flows from the battery through the ignition points (when closed), 7.1 — 7.9 ohm resistor, transistor base and emitter, common 0.33 ohm resistor and ignition switch, then back to the battery. This is termed the POINT CIRCUIT. Follow this circuit on the diagram (double head arrows ➤).

In that a small current is flowing through the transistor base-emitter (POINT) circuit, the transistor passes a heavier (5.5 amp. average) current from the battery through the coil, 0.43 ohm resistor, transistor collector and emitter, common 0.33 ohm resistor and ignition switch, back to the battery. This 5 amp. current energizes the coil magnetic field. Trace out this coil circuit on the diagram, (single-head arrows ➤).

When the breaker points open, the POINT CIRCUIT is broken, cancelling electron flow.

In that the coil current can flow through the transistor collector-emitter circuit only when the base-emitter point circuit is complete, the COIL CIRCUIT is also broken. This causes the primary winding field to collapse, producing high voltage to fire the plugs.

The ZENER DIODE (current passes one way only until a specific voltage is reached at which time it will pass current in the other direction) is located in the circuit to protect the transistor against excessive voltage. Before transistor voltage reaches a danger level, the Zener diode breaks down and shunts the current around the transistor. When voltage drops to acceptable limits, the diode once again resists current flow, causing the current to pass through the transistor. In this respect the Zener diode acts much as a fast action, no moving parts, voltage regulator. The breaking down does not harm the diode in any way.

The 7.1 — 7.9 ohm BASE RESISTOR (between the distributor and transistor base) provides necessary current limitation to protect the transistor.

The 5.6 ohm BASE-EMITTER resistor prevents the current from bypassing the base to emitter transistor circuit when the points are closed.

The COLLECTOR (.43 ohm) and EMITTER RESISTORS

(.33 ohm) both limit current and in conjunction with the other resistors already discussed, keep the point circuit at around .5 amp., while the coil circuit is allowed to reach about 5.5 amps. The point circuit current must pass through the BASE resistor, transistor base-emitter circuit, in addition to the emitter resistor.

The 40 mfd (microfarad) condenser is used to protect the transistor in the event the battery is disconnected while the engine is operating. This would cause generator or alternator voltage pulses that could destroy the transistor. Corroded or loose battery terminals would produce a high resistance in the system and the capacitors also protect against this hazard.

The transistor is extremely sensitive not only to excessive voltage but also to heat. The Zener diode is also heat sensitive. Both the transistor and diode, plus the base-emitter resistor and condenser, are housed in a finned aluminum HEAT SINK. The heat sink dissipates heat readily and protects the units from overheating. The heat sink is located in a cool area to further facilitate cooling.

OTHER TRANSISTOR SYSTEMS

In addition to the Contact-Controlled System, there are two other types: MAGNETICALLY CONTROLLED and CAPACITOR DISCHARGE CONTROLLED.

MAGNETICALLY CONTROLLED CIRCUIT

The magnetically controlled circuit has no contact points but uses a magnetic impulse generator (housed in the distributor head) that triggers the transistor through an amplifier unit. With no breaker points to burn and no rubbing block to wear, the gradual retarding of ignition timing and lowering of ignition system efficiency is eliminated. Fig. 3-49 illustrates the distributor assembly. Note rotating pole piece, stationary pole piece and magnetic pickup assembly.

CAPACITOR DISCHARGE CIRCUIT

This system utilizes breaker points to activate the transistor. (One version uses an impulse generator in distributor, eliminating contact points.) It also charges a capacitor that generally discharges to the primary coil circuit. This system has an extremely rapid rise time — 2 microseconds (millionths of a second) as compared to the conventional system with a rise time of around 100 to 150 microseconds. This gives the capacitor discharge system an unusual ability to fire fouled plugs.

The three systems, greatly simplified, are shown in Fig. 3-48. Study each one until you understand the basic circuits.

Transistor systems can be seriously damaged by the use of conventional tuning and checking procedures.

DO NOT UNDER ANY CIRCUMSTANCES ATTEMPT TO CHECK, ADJUST OR REPAIR A TRANSISTOR SYSTEM WITHOUT THE PROPER INSTRUMENTS — AND IN- STRUCTIONS!

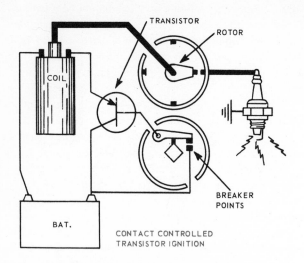

CONTACT CONTROLLED
TRANSISTOR IGNITION

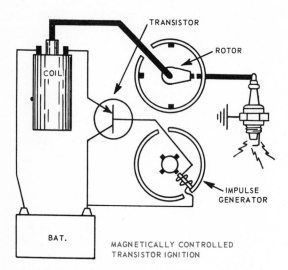

MAGNETICALLY CONTROLLED
TRANSISTOR IGNITION

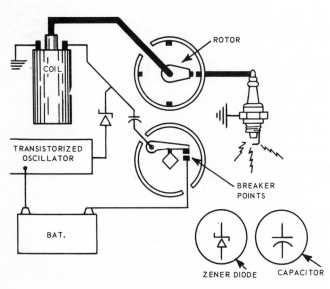

CAPACITOR DISCHARGE CONTROLLED
TRANSISTOR IGNITION

Fig. 3-48. Three different transistor ignition systems.
(Simplified)

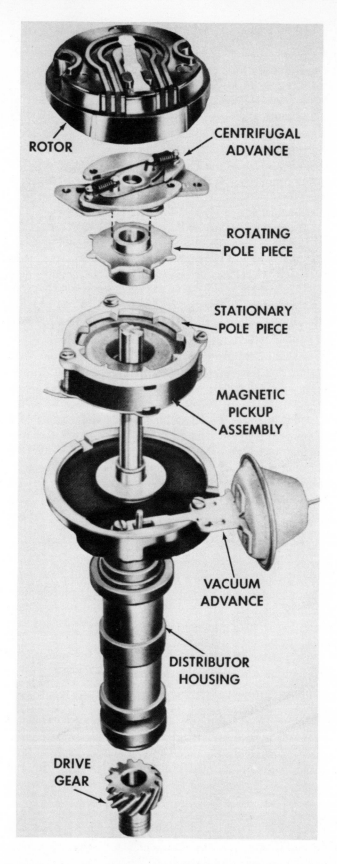

Fig. 3-49. One type of magnetically controlled transistor ignition system distributor. Note how rotating pole piece, stationary pole piece and magnetic pickup assembly have replaced condenser and points found in contact controlled systems. (Chevrolet)

REVIEW QUESTIONS — CHAPTER 3

1. An ignition system in good condition will produce over _____ volts.
2. Normal voltage requirements for successful plug firing are around _____ volts.
3. Name the two circuits in the ignition system.
4. What part does a ballast resistor play in the ignition system?
5. What is the procedure involved in determining how to hook up a coil?
6. Major duty of a coil is to reduce the primary voltage. True or False?
7. What happens when the primary field in the coil collapses?
8. Breaker points are placed in combustion chamber to fire fuel charge. True or False?
9. Breaker points commonly utilize _____ to provide a long service life.
10. What opens and closes the breaker points?
11. An average breaker point gap would run from _____ to _____.
12. What is the cam angle?
13. What is the condenser for?
14. Would an engine run without a condenser?
15. What are the three major parts of the spark plug?
16. What is meant by plug heat range?
17. What advantages are found in the projected nose core plug?
18. Spark plug gaps will vary from about _____ to _____.
19. What causes preignition?
20. The spark at the plug is divided into two stages. Name them.
21. What effect does the resistor plug have on the two stages?
22. Sometimes a plug will utilize an auxiliary gap to increase its efficiency. Just how does this work?
23. What function does the distributor cap serve?
24. When the high voltage reaches the distributor center terminal, how does it reach any given plug?
25. What is meant by timing an engine?
26. There are two common methods of advancing the timing while the engine is operating. Name them.
27. How do the mechanisms that advance and retard the timing function?
28. The major difference between a battery system and a magneto system is _____.
29. What happens when a magnetic field is passed over and around a coil of wire?
30. For what purpose is the coil core and shoe assembly used?
31. A magneto can be used only on one cylinder engines. True or False?
32. Take a blank piece of paper and sketch an entire battery ignition system on it.
33. How does the magnetically controlled transistor system differ from the contact controlled system?
34. The capacitor discharge system always uses contact points. True or False?
35. The transistor ignition system, regardless of type, is quite similar to the conventional system with the exception of the transistor and related parts. True or False?
36. The magnetically controlled transistor system uses an _____ to trigger or control the transistor.
37. Explain basically how the various transistor ignition systems operate.

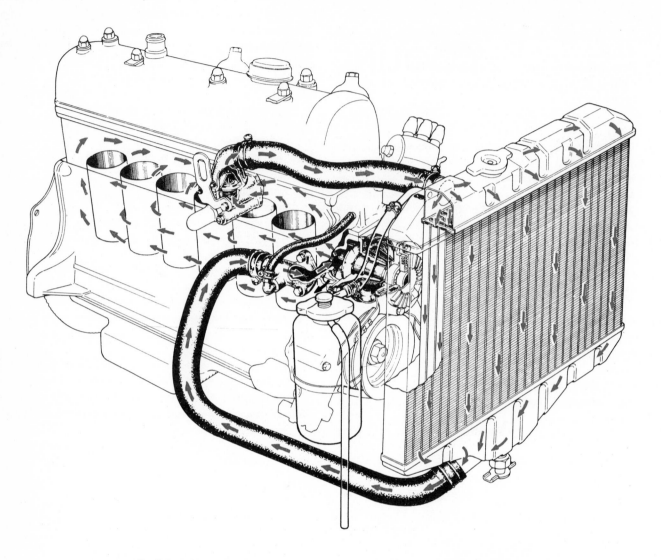

Typical coolant flow path in a system using a down-flow radiator. (Nissan Motors)

Chapter 4

COOLING SYSTEMS

When fuel is burned in an engine, a great deal of heat is produced. In fact, temperatures in excess of 4,000 deg. can be attained by the burning air-fuel mixture. Normal operating flame temperature is around 2,000 deg. F.

A considerable amount of this heat is carried off by the exhaust system, but the pistons, cylinder head, valves, cylinder walls, etc., will still absorb an excessive amount of heat.

As the temperature of the engine parts increases, a point is reached where the vital oil film will break down. When this happens, lubrication is no longer possible, and the engine will be ruined.

SOME HEAT IS ESSENTIAL

At the same time, an engine that is cold is very inefficient, runs poorly, contaminates the oil, forms deposits, increases wear, cuts down horsepower and will not achieve good fuel mileage. Study A and B, Fig. 4-1.

THREE JOBS FOR COOLING SYSTEM

The cooling system must remove surplus or unwanted heat. It must maintain an efficient temperature under all operating conditions. It must also bring an engine, when starting, up to operating temperature as soon as possible.

TWO METHODS

As we often do with something that is too hot, we either blow on it or pour water on it. These same two methods, highly refined, are utilized in cooling an internal combustion engine.

Most automobile engines use a water-cooling system. A few compact car engines, as well as a host of small one and two cylinder engines, use air-cooling systems. This chapter will deal with both.

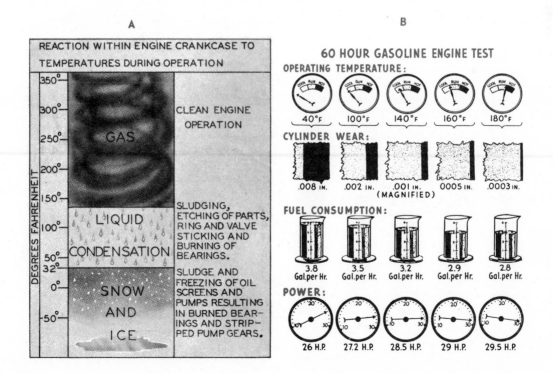

Fig. 4-1. The relationship between engine temperature and performance, life and economy.   (Continental Motors)

WATER-COOLED ENGINES

The water-cooled engine must have water passages through the cylinder block and head. Indirectly, the water must contact the cylinder walls, valve seats, valve guides, combustion chamber, etc. This is accomplished by literally taking the basic engine and building a container around it. In actual practice, the engine block is made in one piece, with water passages cast into the block and cylinder head. When water is placed in these passages, the engine is surrounded by a jacket of water. Hence the term, WATER JACKET. Fig. 4-2.

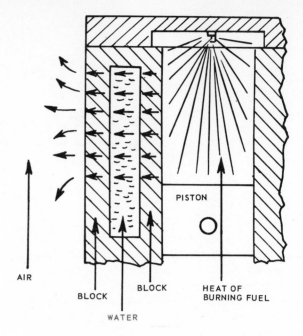

Fig. 4-3. Transfer of heat — water not circulating.

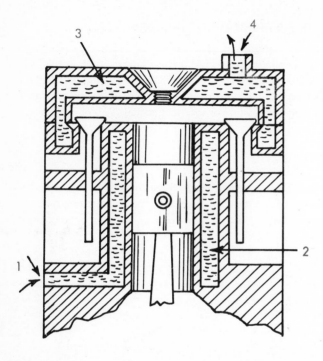

Fig. 4-2. Water jacket. 1, water enters block here, circulates around cylinder to 2, then flows up into the head 3, where it leaves at 4. Notice how water surrounds cylinder, valve seats, guides, head, plug and top of combustion chamber.

TRANSFER OF HEAT

Engine parts absorb the heat of the burning fuel. This heat is conducted through the part to the other side, which is surrounded with water.

The water then absorbs the heat from the part. In turn, the water transfers some of its heat to the outside of the engine. Fig. 4-3.

WATER MUST CIRCULATE

You will notice from Fig. 4-3, that the water will absorb just so much heat, then it will boil away. If the water stands still, it is of no value for cooling purposes. In order to function as a cooling agent, the water must circulate.

As the water is heated, it must move out of the engine and be replaced with cooler water. There must be no dead spots (areas in which there is little or no circulation). Fig. 4-4.

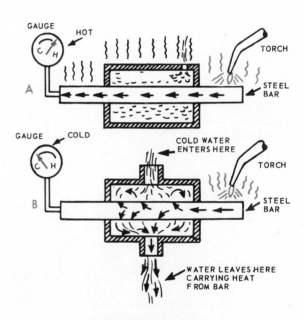

Fig. 4-4. Circulation is vital. Notice in A, that heat from torch has heated water and entire bar. The gage reads hot. In B, water is circulating through water jacket and carries heat away from bar. Gage now reads cold.

A PUMP IS NEEDED

To circulate water through the engine, a water pump is needed. A common type water pump is the centrifugal pump utilizing an impeller. Fig. 4-5.

The impeller consists of a round plate fastened to a rotating shaft to which a series of flat or curved vanes are attached.

When the impeller whirls, the water is thrown outward by centrifugal force.

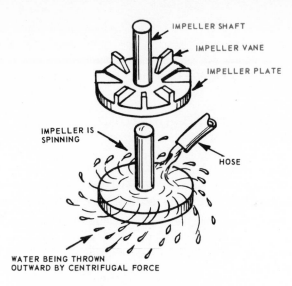

Fig. 4-5. Action of water pump impeller.

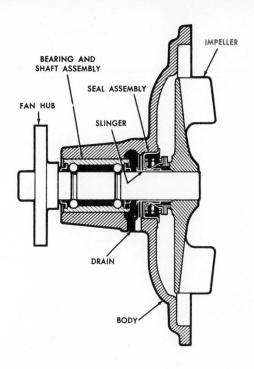

Fig. 4-7. Water pump in which impeller operates in a special shaped opening in the timing chain cover. Some applications place a pump similar to this in a properly designed hole in the front face of the engine block. The opening must be designed to produce the proper pumping action.   (Pontiac)

By placing the impeller in a closed housing and providing an inlet to allow water to contact the front face of the vanes, and an outlet to carry away water thrown outward, such a pump will move water efficiently. Fig. 4-6.

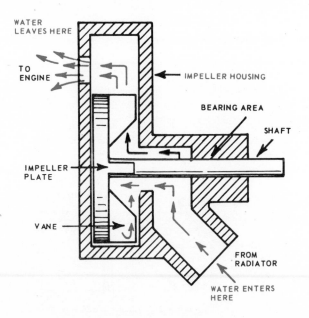

Fig. 4-6. Impeller housing. Water directed to center of impeller is thrown outward into housing. An outlet directs flow upon leaving.

PUMP CONSTRUCTION

The pump housing is cast either of iron or aluminum. The impeller shaft is supported on double row ball bearings, usually of the sealed type. A special seal is used to prevent water leakage. A hub is fastened to the front of the impeller shaft. A pulley and a fan are bolted to this hub. See Figs. 4-7 and 4-8 following.

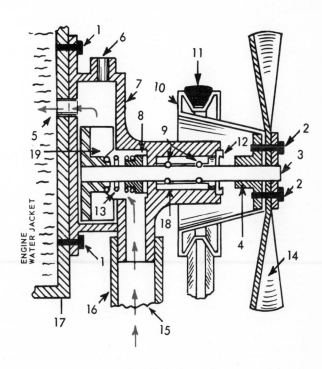

Fig. 4-8. Typical water pump.

1. Bolts-pump to block	8. Seal	15. Water inlet
2. Bolts-fan and pulley	9. Ball bearings	16. Hose
3. Pump shaft	10. Pulley	17. Block
4. Hub	11. V-belt	18. Bearing race
5. Water outlet	12. Dust seal	19. Impeller
6. Bypass	13. Seal spring	
7. Pump housing	14. Fan	

Some pumps place the impeller in the water jacket of the engine, Fig. 4-7, while others keep the impeller enclosed in the pump housing. Fig. 4-8 illustrates a typical water pump. The water pump is usually secured to the front of the engine by bolts.

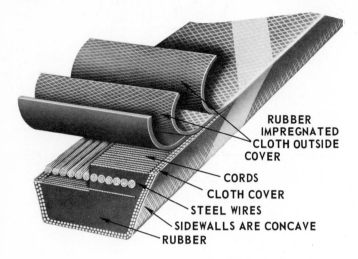

Fig. 4-9. V-belt construction. This particular belt uses steel wires and is designed to drive the air conditioning compressor. (Gates)

BELT DRIVE

Fig. 4-10 shows a V-belt being used to drive a pump. The belt is turned by a pulley on the front of the crankshaft. This same V-belt may also be used to drive the alternator (AC

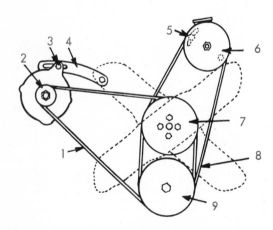

Fig. 4-10. V-belts installed. This shows one method of adjusting belt tension. Another method calls for depressing the belt between the two pulleys. The belt should press in a certain distance when shoved in with nominal finger pressure.

1. Fan and generator belt
2. Torque required to slip generator pulley when belt is tensioned properly 15-20 ft. lbs.
3. Bolt
4. Brace
5. Bolt (front) nut (rear)
6. Torque required to

slip power steering pump pulley when belt is properly tensioned is 40-45 lbs.
7. Fan and water pump
8. Power steering pump belt
9. Crankshaft pulley

generator). The belt must be in good condition and properly adjusted. Excessive tightness will cause premature bearing wear; looseness will allow the pump and fan to slow down and cause overheating. V-belt construction is illustrated in Fig. 4-9. V-belts installed, Fig. 4-10.

CONTROLLED CIRCULATION

It is not enough to pump water into the bottom of an engine and let it flow out the top. If this were done, some areas would receive rapid circulation, while others would contain "dead" water with resultant localized overheating.

The coolant may be channeled into an engine block by means of distribution tubes. This assures an even flow throughout the engine. Fig. 4-11 illustrates a typical distribution tube.

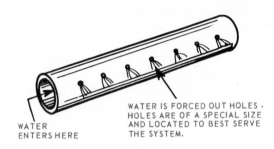

Fig. 4-11. Distribution tube. This may be in the block, head or both.

The tube is placed in the water jacket. The outlet holes line up with selected areas. This will give certain hard-to-cool areas the type of circulation they need.

If distribution tubes are not used, great care must be used in designing the water jacket and its passages.

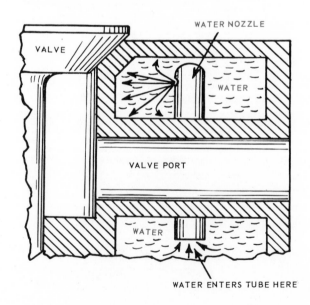

Fig. 4-12. Action of a water nozzle. Notice how water is pumped into nozzle tube. As it leaves small hole in top, it squirts out and causes a strong circulation around valve seat area.

WATER NOZZLE

In addition to distribution tubes, some engines employ water nozzles in the cooling system. Such nozzles speed up water circulation around the valve seats. Fig. 4-12.

COOLANT FLOW

The coolant must flow around the cylinders and pass up through the cylinder head on its way out. It must reach every part of the system and must keep moving with no "dead" spots. Figs. 4-13, 4-14 and 4-15 illustrate the path the water takes on its way through the engine.

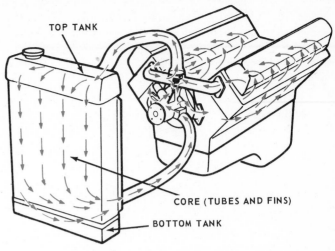

Fig. 4-15. Coolant flow path in a system using a down-flow radiator.

THE RADIATOR

When the coolant leaves the engine, it is quite hot — often well over 200 deg. If it were to be immediately pumped back into the engine, it would start to boil and would be ineffective as a coolant.

Therefore, before the water can be reused, it must be cooled by circulation of water and air. Fig. 4-16.

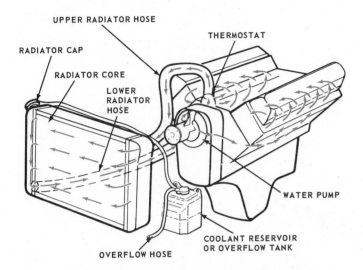

Fig. 4-13. Coolant flow path in a system employing a cross-flow radiator. This is a closed system. Note use of overflow tank. (Cadillac)

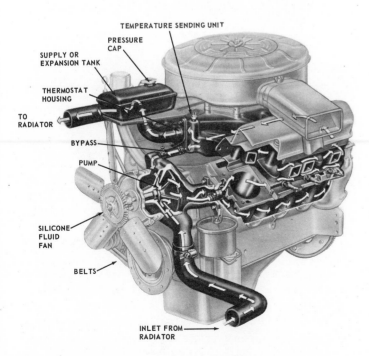

Fig. 4-14. Coolant flows through an engine. Note use of expansion tank. (Ford)

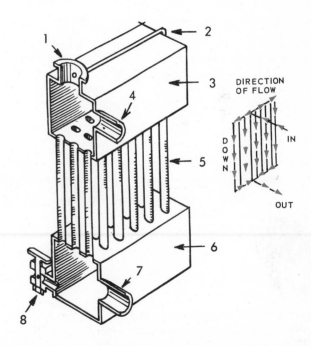

Fig. 4-16. Typical radiator. Water enters the top hose connection 4, then passes into top tank 3. From there it flows down through core tubes 5. When it reaches bottom tank 6, it has cooled. 1—Filler neck. 2—Overflow tube. 7—Lower hose connections. 8—Drain petcock. Not shown are thin fins soldered to core tubes.

Cooling is accomplished by passing the water, as it leaves the engine, into the top or side tank of a radiator.

From the top tank, it flows down through tiny copper tubes. The tubes have thin copper fins soldered over their entire length. As the water makes its way down through the tubes, it gives off heat to the tubes. Since copper is an excellent conductor, the tubes give off their heat, via the fins, to the air passing around the tubes.

By the time the water reaches the bottom tank, it is cool enough to reuse. Fig. 4-16 shows a typical radiator.

Many cars employ the cross-flow radiator, Fig. 4-13, in which the coolant enters the one side tank and then passes through the horizontal tubes to the opposite tank, then back into the engine.

Expansion tanks, Fig. 4-14, are sometimes used to provide additional coolant storage, as well as providing for coolant expansion.

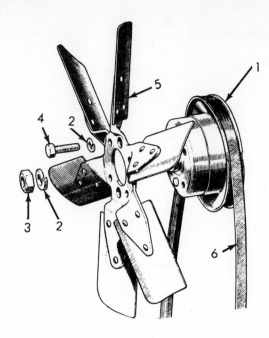

Fig. 4-18. Fan, six blade type. Some fans have two blades; others four, etc. 1—Pulley. 2—Lock washer. 3—Locknut. 4—Bolt. 5—Fan blade. 6—V-belt.   (M. G. Car Co.)

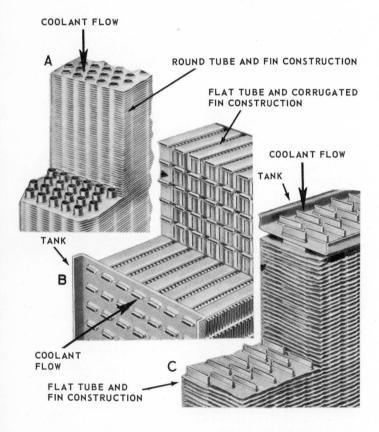

Fig. 4-17. Radiator construction. A—Round tube and flat fin. B—Flat tube and corrugated fin. C—Flat tube and flat fin.   (Harrison)

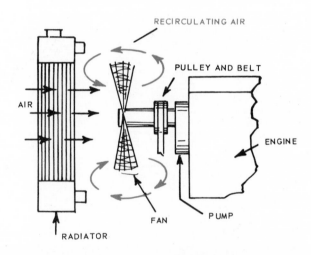

Fig. 4-19. Fan too far from radiator — allows recirculation. Some air is being pulled through radiator, but a great deal is moving through the fan, back around and through again.

FORCED AIR

The radiator may also contain a heat exchanger to provide cooling for the automatic transmission fluid.

A radiator is made of brass and copper to facilitate cooling, and to protect it from rust and corrosion. All joints are soldered. Aluminum is also being used in radiator construction.

The radiator is so placed that it can avail itself of the air that rushes through the grill of the car when it is in motion.

The radiator tubes, or core section, use various designs, all designed to cool the water quickly. Fig. 4-17 illustrates several different types.

To speed up the cooling action of a radiator, a fan is used to draw air through the radiator core. Fig. 4-18.

If a fan is set back some distance from the core, it will recirculate the same air. Fig. 4-19.

To forestall the loss of fan efficiency, some installations require the use of a radiator shroud. This prevents recirculation of air. Fig. 4-20.

The flow of the air over the outside of an engine will help, in a small way, to remove engine heat.

As the car gains speed on the road, a natural ram effect will force air through the radiator core. As a result, the need of a fan is not as great. Since, a fan requires engine power to spin, some fans have a thermostatic control that will allow the fan to slow up when not needed.

Another type uses a centrifugal unit that cuts fan speed when engine rpm is high. These units are designed to save horsepower and to cut down on objectionable fan noise.

Some fans use two blades; others use three, four, and occasionally more, depending on specific requirements. Blades are sometimes arranged in an uneven manner in an attempt to lessen noise.

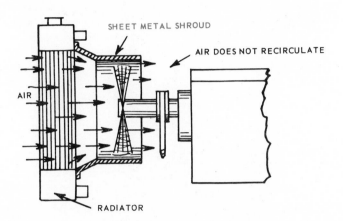

Fig. 4-20. Fan shroud stops recirculation. Thin sheet metal or plastic shroud is attached to radiator. Airflow is improved.

One type of fan employs flexible blades that reduce the pitch (angle at which blade is mounted) at higher speeds, permitting the fan to turn more easily.

The fan is often mounted as an intregal part of the water pump, or it can be mounted as an independent unit.

BEWARE!

A FAN CAN INFLICT SERIOUS INJURY. KEEP HANDS, TOOLS, CORDS, ETC., AWAY FROM THE FAN. DO NOT STAND IN LINE WITH A REVOLVING FAN — A BLADE CAN FLY OUT WITH LETHAL FORCE.

HOSES

The radiator is connected to the engine and pump by means of rubber hoses. These hoses are of different diameters, depending upon the particular engine.

There are three types of hoses in common use. Fig. 4-21.

HOSE CLAMPS

Hoses are secured to their fittings by clamps. There are two popular types. One is the adjustable, screw tightened type, and the other is the snap type.

The screw type has many adaptations. Two common screw types, as well as the snap type, are illustrated in Fig. 4-22.

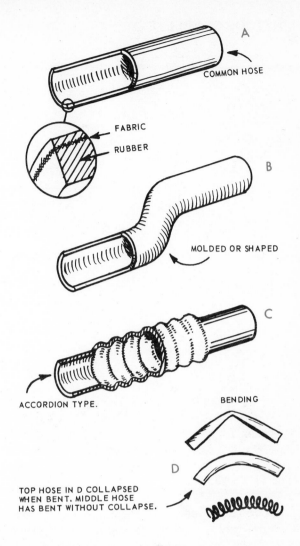

Fig. 4-21. Radiator hose types.
A. Common hose. Straight. Made of rubber with one or two layers of fabric for strenght. Will not stand much bending without collapsing.
B. Molded or shaped. Same construction as A. Since all necessary bends are molded in, this hose will not collapse.
C. Accordian type. Will stand severe bending without collapse. Reduces transfer of vibration from engine to radiator.
D. Top hose has collapsed when bent. Middle hose has bent with no sign of collapse. Middle hose has a spiral-spring-like wire inserted into it before bending. Wire is shown beneath middle hose.
All three types of hose may use the spiral wire in their construction. It can be molded in or placed in.

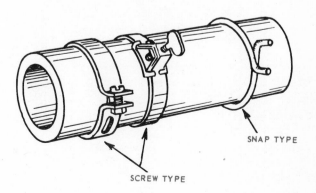

Fig. 4-22. Three types of hose clamps.

FIRST JOB COMPLETE

As mentioned earlier, the cooling system has three jobs. The first job is to remove surplus heat from the engine. With a water jacket, pump, radiator, hoses and fan, this job is complete. Fig. 4-23.

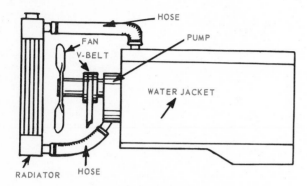

Fig. 4-23. Basic cooling system.

The second job is to maintain an even, efficient heat level. The third job is to bring a cold engine up to this termperature as soon as possible after starting. As yet, the system we have discussed will not perform the last two functions.

THE THERMOSTAT

The system, as studied so far, would cool at a constant rate. If the capacity is sufficient to cool the engine in hot areas under heavy loads, it would obviously overcool in cold weather.

Some type of temperature control is necessary.

To provide constant temperature control, a THERMOSTAT is installed in the cooling system. It is often placed directly beneath the top water outlet. The water outlet may have a bow-shaped bottom to house the thermostat.

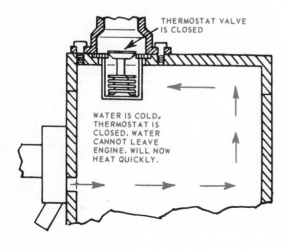

Fig. 4-24. Cold engine.

WHAT DOES A THERMOSTAT DO?

When the engine is cold, the thermostat shuts off the flow of water from the engine to the radiator. As the water is confined to the engine, it heats quickly. When it reaches a predetermined temperature, the thermostat opens and allows free circulation of coolant.

If it is quite cold, the thermostat may not open all the way. Water can circulate, but slowly, and the engine temperature can be kept to the proper level. The thermostat constantly changes the size of its opening depending on heat conditions.

This action will bring a cold engine up to proper heat in a minimum amount of time. It will also maintain an even, efficient temperature level under all conditions. See Figs. 4-24 and 4-25.

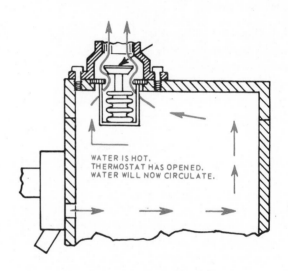

Fig. 4-25. Hot engine.

Note in Fig. 4-45, how the water is heated in three stages. A shows the thermostat closed, no flow to radiator and with the top area of engine (cylinder heads) heating. B illustrates stage 2, with the entire block heating up. C pictures the cooling system with the thermostat open and the coolant circulating throughout the system.

HOW IT WORKS — PELLET TYPE

The PELLET thermostat (named after the pellet of wax that operates it) is in common use today. It has superseded the older BELLOWS type whose operating action was upset by the modern pressurized cooling systems.

A small cylindrical case containing a pellet of copper impregnated (filled) wax operates the thermostat. The thermostat is placed in the water outlet (Fig. 4-26) so that the pellet case rests in the engine block coolant.

When the engine is cold, there is no pellet action and the spring holds the thermostat valve closed. See A, Fig. 4-26.

As the engine coolant warms, the wax pellet is heated. This causes the wax to expand and force the rubber tightly against the steel piston or pin. When the coolant temperature reaches a predetermined level, the wax will have expanded to the point that the pressure on the rubber is so great that it will force the case downward against spring pressure, thus pulling the valve open and allowing coolant to flow through the thermostat on into the radiator. B, Fig. 4-26.

There are numerous pellet thermostat designs. Many have a pull-push valve similar to that shown while others operate a flap type or a butterfly type valve.

The pellet thermostat is not pressure sensitive and works well in a fully pressurized system.

A small pin hole may be located in the thermostat valve. This acts as an aid when refilling the system as it permits the air trapped in the block to escape as the coolant flows upward in the block.

DIFFERENT HEAT RANGES

Thermostats are available in different heat ranges.

Common heat ranges are the 180 and 195 deg. thermostats. The 180 deg. thermostat opens near this temperature and is fully opened around 200 deg. The 195 deg. type starts opening near 195 and is fully open around 218 deg. F. Some thermostats range above 200 deg. F.

These higher range thermostats raise the engine operating temperature and, in so doing, aid in emission (exhaust pollution) control.

The exact thermostat temperature range used will depend on the engine, load requirements, weather, etc.

If alcohol (Methanol) antifreeze is used in the cooling system (ITS USE IS NOT RECOMMENDED!), a 160 deg. thermostat should be used. If a 180 deg. thermostat is used, boilover and coolant loss will result. The 180 deg. and higher thermostats work well with permanent (ethylene-glycol) base antifreeze.

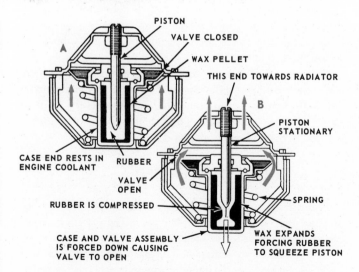

Fig. 4-26. Pellet type thermostat action. Pellet end rests in block coolant. A—Engine coolant cold, coolant cannot pass by closed thermostat valve. B—Coolant has reached operating temperature. Wax pellet expands, squeezing rubber against piston hard enough to force pellet case downward to open valve. (Toyota)

WHEN THERMOSTAT IS CLOSED

A thermostat will remain closed until the water in the engine reaches a specific temperature. Fig. 4-26.

Until this temperature is reached, the water in the engine still circulates, but only in the engine.

Two methods are used to allow the pump to circulate water through the engine and not through the thermostat.

One method uses a spring loaded bypass valve that opens and allows the water to reenter the pump. The bypass valve spring is weaker than that of the thermostat and will be forced open whenever the thermostat is closed. Fig. 4-27. Another version provides a similar setup but eliminates the bypass valve and provides a constant, metered flow of coolant.

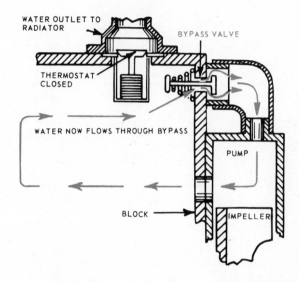

Fig. 4-27. One form of bypass construction. Water flows into block from pump. As thermostat is closed, water forces bypass valve open and water reenters the pump. When thermostat opens, bypass will close.

Coolant flow action through another bypass setup is pictured in A, B, Fig. 4-28. In A, thermostat is closed, bypass cover finger is raised and coolant moves from the pump, through the distribution tube, back out, down through the open bypass pipe on into the pump. Coolant continues to recirculate until the thermostat is fully open. In B, pellet thermostat case has moved down pulling valve open and at the same time, forcing finger to close off bypass pipe. Coolant now moves through thermostat to radiator. Fig. 4-28.

The other method uses a special bypass opening from the cylinder head, through the block, to the pump. This may contain a fixed type bypass that allows continuous recirculation of a small portion of the coolant.

PRESSURIZED COOLING SYSTEM

The modern, high speed, high horsepower engine produces terrific heat. It has been found that if the cooling system is pressurized, the boiling point of the water is raised. For this reason a pressurized cooling system provides a greater differ-

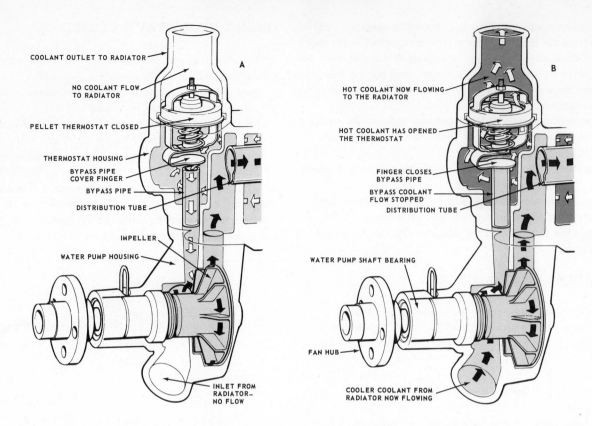

Fig. 4-28. Thermostat and bypass action. A—Coolant still cold, thermostat closed. Bypass pipe cover finger is UP exposing bypass pipe opening. Coolant circulates from pump, through distribution tube, down through bypass pipe to pump. B—Coolant has heated causing pellet thermostat to open. This closes bypass pipe and coolant flows through thermostat valve to radiator. (Volvo)

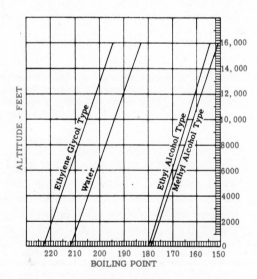

Fig. 4-29. Effect of altitude on boiling point of coolant. If you were driving a car with plain water in the cooling system, how high a mountain could you climb if your engine was running at 200 deg.? You can go up till it boils. How high? (Continental Motors)

ence between the boiling point of water and the desirable engine operating temperature. By permitting high coolant temperatures, the engine tends to produce less exhaust pollutants.

When operating a car at high altitudes, especially during the hot weather with heavy loads, the water in the cooling system will boil quickly. The reason is that as the car gains altitude, atmospheric pressure goes down.

The less the atmospheric pressure on the water, the sooner the water will boil. By adding pressurization to the system, it is possible to operate an engine at temperatures exceeding 250 deg. without boiling.

Pressurization also increases water pump efficiency by reducing cavitation (water in the low pressure area near pump blades turning to steam, allowing the pump impeller blades to spin in the air bubbles).

Each pound of pressure on the water in the cooling system will raise the boiling point around 3 deg. Study the chart in Fig. 4-29. Notice the difference between the boiling point of water, ethylene glycol, and methyl alcohol. Also notice that as you go up, the boiling point goes down.

HOW SYSTEM IS PRESSURIZED

Pressurizing a cooling system is done by placing a special pressure cap on the radiator filler neck. The system must, of course, be designed for this purpose.

The pressure cap presses firmly against a seat in the filler neck. As the water in the cooling system heats, it also expands. As it continues heating, and expanding, it squeezes the air on the top of the water.

The air cannot escape, as the pressure cap has sealed the opening. When the pressure is too great, the spring holding the pressure cap seal valve will no longer hold and allows the seal valve to rise. As soon as pressure (over what cap is built for) escapes, the spring closes the seal and keeps the predetermined pressure constant. The escaping air passes out through the radiator overflow pipe.

When the engine is shut off and cools down, another small valve in the cap opens. As the water cools, it contracts and tries to form a vacuum in the tank. This draws the vacuum valve open and allows air from the overflow pipe to enter the tank. Without a vacuum valve, the hose and the tank could collapse.

HOW CAP WORKS

The radiator filler neck has a flange near the bottom. When the pressure cap is put on, it has a spring loaded valve that seats against the flange.

The spring loaded valve has a valve in the center that pulls open when a vacuum is formed in the tank. This vacuum valve is spring-loaded also. The large spring is calibrated to open at a specific pressure (15 lbs. about average). Fig. 4-30 shows the construction of a pressure cap.

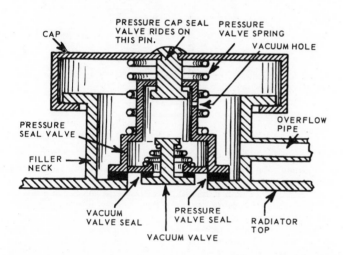

Fig. 4-30. Construction details of pressure cap, shown in place on radiator filler neck.

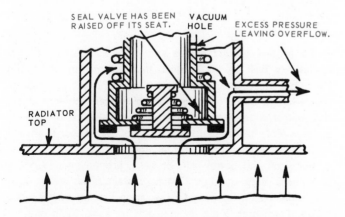

Fig. 4-31. Pressure cap in action. Water has expanded until pressure cap limit has been reached. Cap has opened and excess pressure is being released out the overflow.

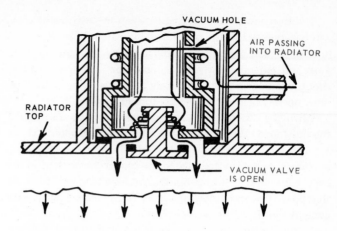

Fig. 4-32. Pressure cap in action. Water has cooled and contracted. This forms vacuum in radiator. Vacuum valve is drawn open and air passes into tank through overflow pipe.

Figs. 4-31 and 4-32 demonstrate the cap while working (simplified version).

CLOSED SYSTEM

The CLOSED cooling system is a standard cooling system with the addition of an overflow tank (also called a coolant reservoir) and a somewhat different pressure cap design. Figs. 4-13, 4-33.

The closed system pressure cap seals not only the bottom flange, as pictured in Figs. 4-31 and 4-32, but the top of the radiator filler neck as well. The bottom seal has a nonloaded (no spring pressure) center vent valve.

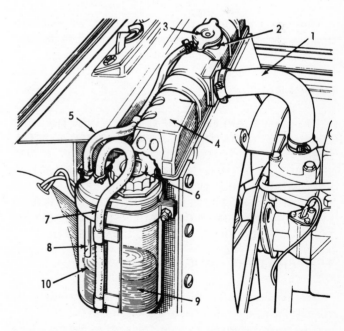

Fig. 4-33. One CLOSED cooling system setup. Excess coolant moves through overflow hose 5 into downpipe 8 and into overflow tank 9. When system cools, coolant will be drawn back into the system from the tank supply 10. Other components: 1—Top hose. 2—Radiator top tank filler neck. 3—Pressure cap. 4—Radiator top tank. 6—Tank cover cap. 7—Overflow tank vent hose.

The overflow pipe is located between the two sealed areas and is connected to the overflow tank via a rubber hose. The hose (or a rigid pipe) passes through the cover of the overflow tank and extends nearly to the bottom. The tank is filled about one-quarter full of coolant. The radiator, unlike the older open system, is filled to the top. Fig. 4-33.

When the coolant heats, it expands and flows up through the non-loaded center vent valve. This displaced coolant flows through the overflow pipe and on into the overflow tank. Any air that is in the system rises to the top of the coolant in the overflow tank and escapes out the overflow tank vent hose. At this point, the entire cooling system is completely filled with coolant and all air has been displaced.

As the coolant temperature nears the boiling point, the vent will be forced closed to create pressure on the system. At extremely high temperatures, the system may exceed the maximum cap pressure. When this happens, system pressure forces the bottom cap seal valve to raise off its seat and dissipate the pressure in the form of coolant flow into the overflow tank.

When the engine cools, the coolant contracts forming a mild vacuum. This draws the cap center vent valve open (as in Fig. 4-32) and coolant (not air) is drawn into the system from the overflow tank.

The closed system, in eliminating air, operates more efficiently, helps to reduce corrosion and allows a visual inspection of the system coolant level without removing the radiator cap. To replace any coolant lost to leakage, do not remove the pressure cap, merely bring the coolant in the overflow tank up to the indicated level. Study Fig. 4-33.

SYSTEM MUST BE IN GOOD REPAIR

Whenever a cooling system is pressurized, an additional strain is placed on all the system. Hoses must be in good condition, clamps tight and no coolant leaks anywhere.

DANGER!

BE VERY CAREFUL WHEN REMOVING THE CAP FROM THE RADIATOR OF A PRESSURIZED SYSTEM. OPEN THE CAP TO THE FIRST STOP — WAIT A FEW SECONDS AND THEN REMOVE. COVER CAP WITH A RAG DURING THIS OPERATION AND KEEP FACE A-WAY. WHEN THE PRESSURE IS SUDDENLY REMOVED FROM THE SYSTEM, IT CAN CAUSE SUPERHOT STEAM TO ESCAPE IN A VIOLENT FASHION. BE CAREFUL!

TEMPERATURE GAUGE

The temperature of the coolant can be determined by observing the temperature gauge or indicator on the dash.

Two types of gauges are used. One uses an electric sender and receiver; the other uses a gas-filled capillary tube, bulb and Bourdon tube instrument. Both instruments are quite effective and will give good service. The electric gauge has one advantage. It can span greater distances from engine to instrument panel by merely adding enough wire to reach. Figs. 4-34 and 4-35.

CARE OF COOLING SYSTEM

It is essential that the cooling system receive proper care. Good care will insure long and trouble free operation, and which will add greatly to the efficiency and life of the engine.

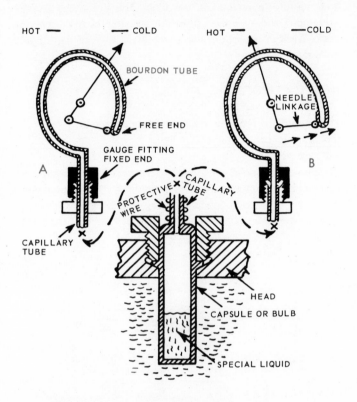

Fig. 4-34. Bourdon Tube Temperature Gauge.
Bourdon tube is made of thin brass. One end is fixed to gauge fitting and free end is fastened to indicator needle.
When liquid in bulb is heated, it will vaporize and create pressure. This pressure is transmitted to the hollow Bourdon tube through capillary tube.
When Bourdon tube receives pressure, it will spring apart — the free end swings away from fixed end; see B. This movement causes needle to swing across face of gauge.
The bulb is placed in water jacket of engine.
Capillary tubing is strengthened by wire wrapped around it for its entire length.
When water cools, pressure is removed from tube and needle moves back to cold.

To protect against engine and radiator freeze-up in the winter, as well as to protect the system from rust and corrosion, manufacturers install a permanent type of anti-freeze (about 50 percent water and 50 percent ethylene glycol plus corrosion and rust inhibitor chemicals) in the cooling system. Most recommend one or two year drain, flush and refill cycles. Always refill with recommended coolant.

Keep fan and pump belt, or belts, properly adjusted and also check their condition. Make certain hoses are in good condition. Watch for irregular operation that might indicate a faulty thermostat. Keep all connections tight. Keep radiator core clean and free of bugs, dirt, etc. Check the level of the coolant often. Check when cold. Check pressure cap for proper operation.

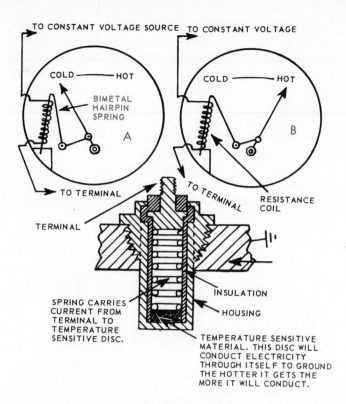

Fig. 4-35. Electric Temperature Gauge.
Current from battery flows through wire coil that surrounds bimetal (two different metals fastened together) hairpin. It then passes on to the terminal of the sender unit, traveling from the terminal through the spring to the temperature sensitive disc.

As water heats disc, the disc will carry more electricity to ground. Resistance coil on the hairpin will start to heat up. This will transfer heat to hairpin.

Expansion of two dissimilar metals in hairpin causes it to bend as it heats. This bending actuates needle and causes it to move to hot. See B.

When water cools, disc will slow down flow of electricity and hairpin will cool, bending back to its original shape. See A.

AIR-COOLED ENGINES

The air cooling of engines has found a wide application in the field of small engines. It is also applicable to large engines. Several small auto engines use air cooling. Familiar examples are the German Volkswagen, and the discontinued Chevrolet Corvair.

Air cooling for engines is efficient and dependable. Generally an air-cooled engine warms up more quickly and operates at a somewhat higher temperature than the water-cooled engine.

Basically an air cooling system consists of fins on the engine head and block, a fan to force air, a shroud to direct the air over the fins, and in some cases a thermostatically adjusted air control.

Air cooling is most efficient when applied to engines that are constructed of aluminum. Aluminum is a better conductor of heat than cast iron. In fact, aluminum will conduct heat about three times better than cast iron. However, air cooling is successfully used on cast iron engines.

It is essential that the engine block and head have a number of rather deep cooling fins. The fins are cast right into the

engine parts. The theory behind the use of fins is simple. Heat generated inside a metal container must pass through the walls of the container and be carried away by flow of the air.

Let us assume that we have a square cylinder block of aluminum 6 in. wide and 6 in. high. Inside the block is our engine combustion chamber.

As heat is generated by the burning fuel mixture, it will be conducted through the walls of our block and will be carried away by the air. If we produce enough heat in the combustion chamber, it is obvious that the block will not be able to dissipate the heat into the air fast enough and will, in time, melt. Fig. 4-36.

The rate at which the aluminum block will be able to cool itself will, to a considerable degree, be determined by the amount of block surface that is exposed to the air.

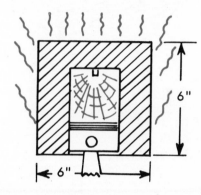

Fig. 4-36. Heat is conducted through walls of block and is carried away by surrounding air.

Computing the number of lineal inches on the top and two sides of the block, we find three sides (6 in. per side) equals 18 lineal in. In other words, the air moving around the block can draw heat only from these 18 in. of exposed surface.

By adding fins to the sides and top, see how much more surface the air can strike. Fig. 4-37.

Fig. 4-37 shows that by adding 2 in. fins to the top and sides, the air can now circulate over 162 lineal in. of block surface. The block now has nine times the cooling surface.

So far we have considered lineal inches as viewed on three

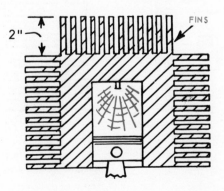

Fig. 4-37. Fins added to block increase contact surface for air.

sides. If we consider our cylinder block to be square, this would give us five sides upon which to calculate the number of SQUARE INCHES of contact area. We will ignore the bottom side, as this would be fastened to the crankcase and would not be finned. Fig. 4-38.

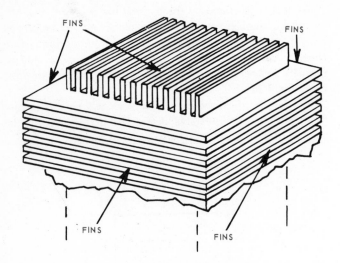

Fig. 4-38. Block with fins on five sides.

FINS ARE NECESSARY

By adding fins as shown in Fig. 4-38, we would now have a total of about 1,800 sq. in. of contact area. This would be compared to five sides of the unfinned block at 180 sq. in. This represents a sizeable increase in cooling efficiency.

THE AIR MUST BE MOVED

In addition to adding fins, it is essential that we provide a fan or blower to produce a moving stream of air. The air, if not moved, will soon heat up and be unsuitable for further cooling of the finned surfaces.

Moving of the air may be done in several ways. One way is to have a separate fan or blower. Another method is to cast

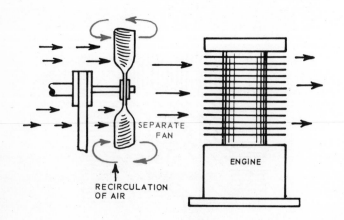

Fig. 4-39. Separate blower for cooling.

fan fins on the flywheel. Some engines, such as aircraft, depend on the forward speed of the engine through the air to provide sufficient airflow for proper cooling. Automobile engines use a separate fan, while most small engines use the flywheel fan, Figs. 4-39 and 4-40.

From the drawing, you can see that the flywheel fan throws air in a fashion similar to the way a water pump impeller throws water. Fig. 4-40 demonstrates that even though the fan is throwing air, it is not directed onto the fins.

The separate fan in Fig. 4-39 does a better job, but it is not always possible to line up the fan with the fins. The fan is also losing efficiency by recirculating some of the air near the outer edges of the blades.

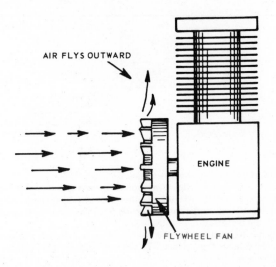

Fig. 4-40. Flywheel fan. Such a fan is very inefficient without a shroud.

These problems are overcome by the use of a sheet metal air shroud. The shroud prevents recirculation and also guides the air into the fins. Figs. 4-41 and 4-42 illustrate the use of

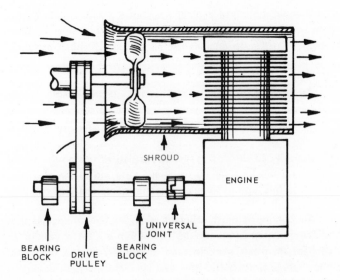

Fig. 4-41. Separate fan with sheet metal shroud. Notice that shroud prevents recirculation and also guides air to the fins.

shrouds. Some shrouds are designed to form a plenum chamber (an enclosed area in which the air pressure is higher than that of the surrounding atmospheric pressure).

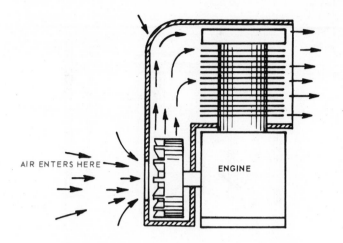

Fig. 4-42. Flywheel fan with sheet metal shroud. Shroud collects air thrown off by the fan and directs it to fins. Fan is acting as a centrifugal blower.

A THERMOSTAT IS SOMETIMES USED

It is possible to have an air-cooled engine run too cool. To prevent this, a thermostatically controlled shutter, or valve, to limit airflow, is sometimes used. The thermostat is located in a spot that will allow it to accurately measure engine temperature. The bellows thermostat is the type commonly used.

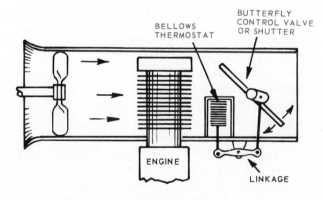

Fig. 4-43. Schematic of temperature control for an air-cooled engine. When engine is cold, thermostat contracts and will close shutter. As engine heats up, air passing engine will warm thermostat, causing it to expand and open shutter.

The bellows is filled with butyl alcohol, under a vacuum. If the bellows leaks, or in any other way becomes inoperative, the shutter will then return to the full open position. This eliminates the possibility of the shutter remaining in the closed position with resultant engine failure. Fig. 4-43 demonstrates the use of such a bellows.

SOME ENGINES NEED NO FAN

Engines designed to move rapidly through the air, such as motorcycles and aircraft, often use no fan, depending entirely upon forward momentum to provide a cooling stream of air. Such equipment should not be operated for prolonged periods while standing still.

ALL UNITS MUST FUNCTION

In the air-cooled engine, it is essential that the fan turn at the proper speed. Also, metal shrouds must be in place, cooling fins clean, drive belt properly adjusted and, if a thermostat is used it must operate properly.

SUMMARY

All internal combustion gasoline engines must be cooled, either with air or a liquid; the liquid commonly used being a 50/50 mix of water and ethylene-glycol.

In the case of the liquid-cooled engine, the coolant must surround all the areas that are subjected to heat from the burning fuel charge. The coolant is contained in an area referred to as the water jacket.

When the engine is first started, the thermostat is closed and the coolant in the jacket cannot circulate. Pressure from the water pump then forces water through the bypass. This allows coolant to circulate from the pump into the block, then back into the pump.

As soon as the coolant in the engine reaches a high enough temperature, the thermostat will open and allow the coolant to be forced from the engine into the top, or the side, radiator tank, depending on radiator type used. From this tank, the coolant flows through the radiator core to the other radiator tank. The coolant is then pumped back into the engine. The coolant repeats this circulation pattern over and over.

COOLING SYSTEM UNITS

The main units in the system are:
1. Radiator for cooling.
2. Water pump for circulation.
3. Thermostat for temperature control.
4. Pressure cap for pressurization of the entire system.
5. Bypass valve to allow recirculation while heating up.
6. Fan to provide a moving stream of air through the radiator.

Study Fig. 4-44. Learn the names of all the parts. If in doubt as to the function of any part, refer to that specific item in this chapter.

SUMMARY — AIR-COOLED

When the engine is started, the block heats quickly. The heat is transferred by conduction to the fins. Air, moved by a fan, passes through a shroud surrounding the fins. Moving air removes the heat from the cooling fins. The amount of air can be controlled by a thermostatic shutter.

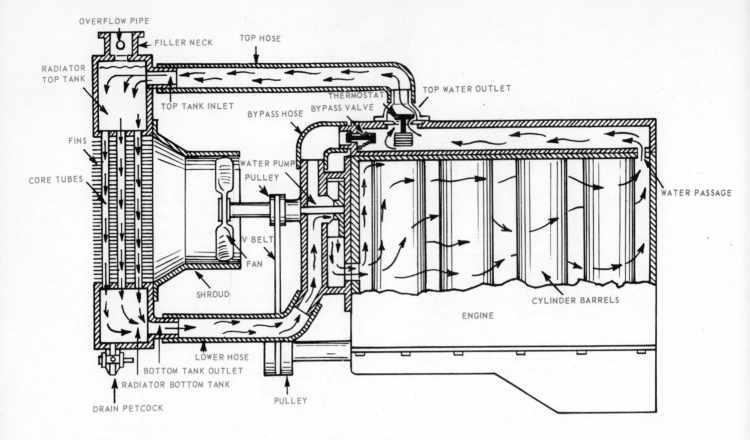

Fig. 4-44. Cutaway view of complete cooling system. Notice how water moves to rear of block, passes through water passage to head, then moves forward. Since thermostat is open, it does not enter bypass, but passes into top tank. Radiator pressure cap is not shown.

REVIEW QUESTIONS – CHAPTER 4

If you sweat trying to answer these questions, YOU need cooling — Cooling System review.

1. Give six reasons why an engine must not run too cool.
2. What are the three jobs that the cooling system must perform?
3. There are two popular methods of cooling the engine. Name them.
4. Water pumps are generally of the gear type. True or False?
5. Water pumps are usually chain driven. True or False?
6. To prevent dead spots and to produce a better than normal circulation in some vital areas, cooling systems use _____.
7. A radiator is generally made of copper and brass to aid in cooling. True or False?
8. Name two common core types.
9. How many blades are there on the ordinary auto engine fan?
10. A fan will inflict a serious injury if you come in contact with it. True or False?
11. List three types of water hose.
12. What is the purpose of the thermostat?
13. Name two types of thermostats.
14. How do thermostats work?
15. What is meant by the heat range of a thermostat?
16. What is the function of a bypass valve?
17. Why are cooling systems pressurized?
18. How does a pressure cap work?
19. List the general points to be considered in the proper care of the cooling system.
20. Of what value are fins on an air-cooled engine?
21. What two types of units are used to produce a stream of moving air?
22. What method is used to control the temperature of the air-cooled engine?
23. Name principal units in an engine cooling system.

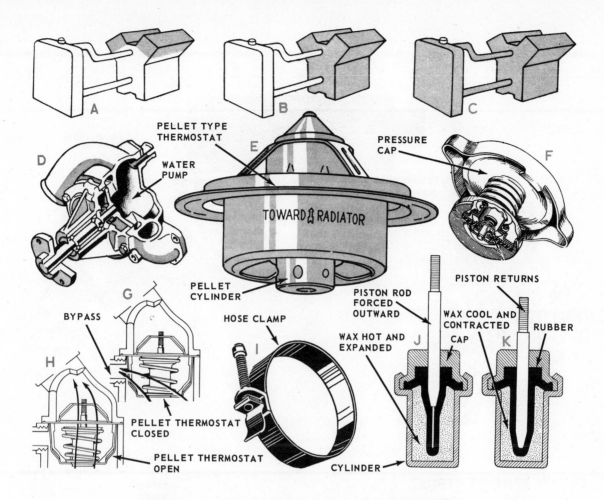

Fig. 4-45. A, B, and C illustrate how cooling system comes up to operating temperature in three stages. D—Cutaway of a typical water pump. E—One form of pellet thermostat. F—Cutaway of a radiator pressure cap. G—Pellet type thermostat in closed position. H—When pellet heats, thermostat opens. I—One form of hose clamp. J—Pellet is hot. Note how special wax has expanded, squeezing rubber together and forcing piston away from cylinder. K—When pellet cylinder cools, wax contracts and rubber returns to its normal position, drawing piston with it.

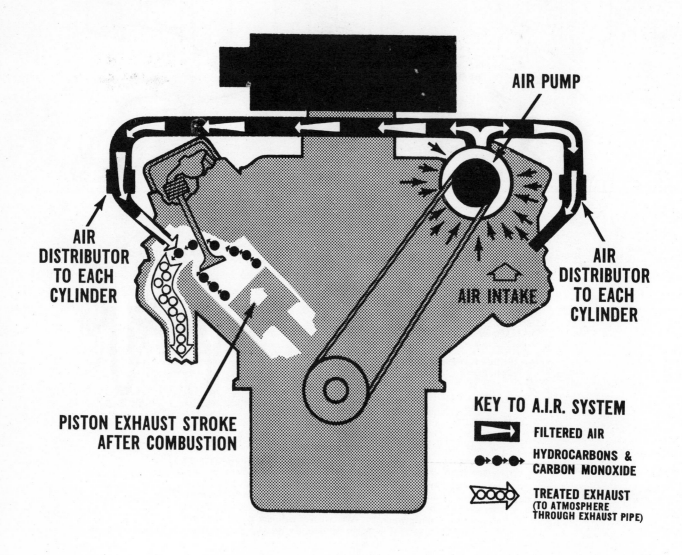

AIR PUMP

AIR
DISTRIBUTOR
TO EACH
CYLINDER

AIR
DISTRIBUTOR
TO EACH
CYLINDER

AIR INTAKE

PISTON EXHAUST STROKE
AFTER COMBUSTION

KEY TO A.I.R. SYSTEM

FILTERED AIR

HYDROCARBONS &
CARBON MONOXIDE

TREATED EXHAUST
(TO ATMOSPHERE
THROUGH EXHAUST PIPE)

Air Injection System feeds fresh air into the still burning fuel charge as it leaves the exhaust valve port. This intensifies and maintains the burning to assist in lowering emission levels. (Chevrolet)

Chapter 5
LUBRICATION SYSTEMS
EMISSION CONTROL

If you were to start and run an engine without oil, it would last only a short time. The bearing surfaces in the engine would heat up and scuff, and finally the engine would freeze. (Freezing occurs when lack of lubrication causes two parts in moving contact to overheat, expand, tear bits of metal from each other, then stick and refuse to move.) Fig. 5-1.

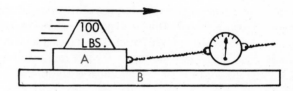

Fig. 5-2. Friction. It takes force to pull block A along the surface of block B.

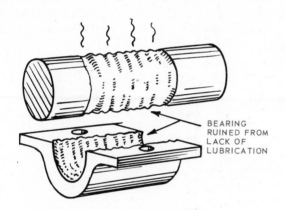

Fig. 5-1. Ruined journal and bearing. Oil would have prevented this. Size exaggerated.

BEARING RUINED FROM LACK OF LUBRICATION

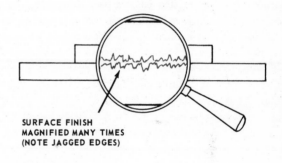

SURFACE FINISH MAGNIFIED MANY TIMES (NOTE JAGGED EDGES)

Fig. 5-3. Microscopic view (exaggerated) of block contact surfaces.

The force that causes wear and heat, and robs the engine of a portion of its potential power, is called FRICTION.

FRICTION DEFINED

Friction may be defined as the resistance to movement between any two objects when placed in contact with each other. The amount of friction depends on the type of material, surface finish, amount of pressure holding the two objects together, and relative amount of movement between the objects.

You can see in Fig. 5-2, that it takes a force to pull block A over block B.

If you were to examine the contact surfaces of the two blocks under a microscope, regardless of how smooth they were polished, you would find a series of sharp points, grooves, etc. When the two objects contact each other, these jagged surfaces touch each other. Fig. 5-3.

FRICTION CAUSES WEAR

As soon as block A starts to move, the edges bump into each other and impede the movement. If there is considerable pressure holding the two blocks together (and in the engine, bearing pressures can soar to over 1,000 lbs.) and the movement is continued, two things happen.

Each block will literally tear pieces from both itself and the other block. These tiny particles roll and bang along under block A until they work out the back edge. Removal of these pieces is continual and will result in wear.

FRICTION GENERATES HEAT

A fire may be started by rubbing sticks together. Dragging brakes heat up and smoke. These are examples of heat produced by friction.

In an engine, the rubbing speeds and pressures are tremendous. This will build up enough heat to melt bearings.

FRICTION CANNOT BE ELIMINATED

It is impossible to completely eliminate friction between moving parts. If this were possible, perpetual motion machines could be easily constructed.

All that can be done is to reduce friction to an absolute minimum.

HOW FRICTION IS REDUCED

Friction is of two kinds, DRY friction and WET friction. So far we have dealt with DRY friction.

Dry friction is when both rubbing parts are clean and free of other materials.

Wet friction is when rubbing parts have other material or materials placed between them.

Have you ever tried playing a fast game of basketball on a well-waxed dance floor? This is a good example of DRY friction being changed into WET friction by placing something between your shoes and the floor — in this case, wax. It is obvious that the engine must operate under conditions of WET friction.

BEARING CONSTRUCTION

Bearing construction also plays a part in determining the amount of friction involved. There are two principal types of bearings used in engines. One is called a FRICTION bearing. The friction bearing is literally two smooth surfaces sliding against each other. The only motion involved is a sliding effect. The babbitt insert and the crankshaft journal are examples of this type.

The other type is called ANTIFRICTION bearing. This bearing utilizes balls or rollers between the moving parts.

Both types of bearings must be used wet.

The FRICTION type, as it is less expensive and cheaper to service, in addition to being satisfactory, is generally used in all major auto engine bearings. Ball and roller bearings are used in many small, high speed engines, especially the two-cycle types. Fig. 5-4 illustrates both bearing types.

OIL REDUCES FRICTION

It has been found that high quality lubricating oil successfully reduces friction to an acceptable level.

Engine oils are derived from crude oil, or petroleum, that is removed from the earth by means of wells.

PETROLEUM MAKES MANY THINGS

Crude oil is heated and is then pumped into a tall distillation tower. As the crude is heated, the lighter or more volatile (easily evaporated) components turn to a vapor and "flash" or travel to the top of the tank. As heat is increased, the heavier or less volatile parts will vaporize.

As all these component parts have a different specific gravity, and evaporate at different heats, it is possible to connect a series of pipes entering the distillation tower at different heights. If the pipes are arranged correctly, and other operating conditions are correct, a different vapor will travel out of each pipe.

The pipes enter a cooling tank and the vapors within condense and run down into storage tanks.

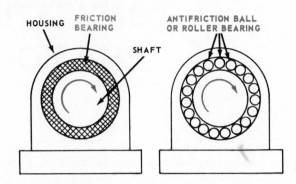

Fig. 5-4. Friction and antifriction bearings.

Liquefied petrolium gas (LP-Gas) is the most volatile part and leaves first. In fact, it is so volatile it will remain in the form of a gas until pressurized, at which time it will become a liquid.

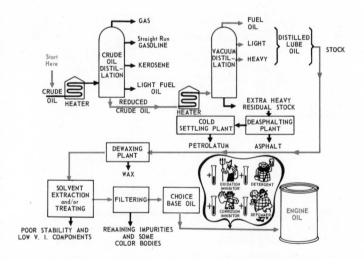

Fig. 5-5. Refining of lubricating oil is a somewhat complicated task. Study the various steps involved.   (John Deere)

DISTILLATION PROCESS

The production of lubricating oil is quite complicated. After distillation, it must have the impurities removed. This process takes several steps. Various additives such as oxidation inhibitors, detergents, defoamers, corrosion inhibitor, pour point depressants, viscosity index (VI) improvers, etc., may be added to produce oil for specific requirements. Study the schematic diagram, Fig. 5-5, showing how lubricating oil is refined.

ENGINE OIL SERVICE CLASSIFICATION

Currently, there are five basic oils adapted for use in gasoline engines. The API. (American Petroleum Institute) has classified these oils as SA, SB, SC, SD and SE. These replace the OLD designations of ML, MM and MS.

The letter "S" indicates the oil is basically for Service Station (gasoline engine) use. Commercial and fleet (diesel engines) oils use the letter "C" and are classified as CA, CB, CC and CD.

SA (Utility Gasoline and Diesel Engine Service). This oil is for use in utility engine operation under highly favorable conditions, i.e., light loads, moderate speeds and clean conditions. It generally contains no additives.

SB (Minimum Duty Gasoline Engine Service). This oil is for gasoline engines operating under mild conditions where little oil compounding (addition of other materials to the oil) is required. They will provide resistance to oil oxidation, bearing corrosion and scuffing (parts scratching each other).

SC (1964 Gasoline Engine Warranty Maintenance Service). Oil similar to that used for service in auto engines manufactured from 1964 through 1967. These oils have additives and offer control of wear, rust, corrosion and high and low temperature deposits.

SD (1968 Gasoline Engine Warranty Maintenance Service). An oil for service typical of that required for 1968 through 1970 auto and light truck engines. The oil is also applicable to some 1971 and later auto engines when so stated by the auto maker. This oil provides additional protection (as compared with SC) against wear, rust, corrosion and high and low temperature deposits.

SE (1972 Gasoline Engine Warranty Maintenance Service). This is currently the top ranked oil and is generally specified for 1972 and later car and light truck engines. It provides maximum protection against rust, corrosion, wear, oil oxidation and high temperature deposits that can cause oil thickening. Obviously, this oil can be used in all engines in which SD or SC is normally recommended.

Study the chart in Fig. 5-6. Note the relationship between the new oil classification numbers and that previously used.

In addition to classifying oil according to the severity of service, (SA, SB, SC, etc.), oil for SE service may also be required to pass certain actual use tests to determine how well the oil prevents engine rusting, oxidation, corrosion, scuffing, wear, thickening and the formation of sludge and varnish. These laboratory engine tests are called ENGINE SEQUENCE TESTS. They are also sometimes referred to as "ASTM Sequence Tests" or "Car Manufacturers Sequence Tests."

Oils may also be classified according to MILITARY SPECIFICATIONS (specifications set up for oils to be used by the Military). MANUFACTURERS SPECIFICATIONS (developed by some engine makers) can also be used.

OIL HAS SEVERAL JOBS

In addition to lubricating the engine, the oil must also assist in COOLING. The steady flow of oil through the bearings carries away a great deal of the heat that is generated. Oil strikes the piston head and carries heat with it. As the oil is returned to the sump before being recirculated, it is much cooler by the time it reaches the bearings again.

Some high speed and heavy-duty engines, plus many air-cooled engines, run the oil through a cooler (similar to the

REFERENCE CHART		
NEW API ENGINE SERVICE CLASSIFICATIONS	PREVIOUS API ENGINE SERVICE CLASSIFICATIONS	RELATED DESIGNATIONS MILITARY AND INDUSTRY
SERVICE STATION ENGINE SERVICES		
SA	ML	Straight Mineral Oil
SB	MM	Inhibited Oil
SC	MS (1964)	1964 MS Warranty Approved, M2C101-A
SD	MS (1968)	1968 MS Warranty Approved, M2C101-B, 6041 (Prior to July, 1970)
SE	None	1972 Warranty Approved, M2C101-C, 6041-M (July 1970)
COMMERCIAL AND FLEET ENGINE SERVICES		
CA	DG	MIL-L-2104A
CB	DM	Supp. 1
CC	DM	MIL-L-2104B
CD	DS	MIL-L-45199B, Series 3

Fig. 5-6. Chart shows the relationship between the old and the new API Service Classifications.

regular radiator only smaller). Others use fins on the oil pan to aid in the cooling.

Oil must SEAL. Piston rings would not be too effective without oil as a final sealing agent. It also assists in sealing around valve stems.

A good oil will CLEAN. The detergents in the oil assist in removing sludge, gum, some carbon, etc.

OIL CLEANS, COOLS, SEALS AND LUBRICATES

SE oils have many extra additives (compounds designed to improve the oil) added to them. These additives fight bearing corrosion; help to keep small foreign particles in suspension; reduce oxidization; minimize carbon, lacquer and gum formation; and reduce engine wear.

SE oils have a high detergent quality that tend to keep the engine clean by preventing the formation of sludge and gum. Sludge is produced by engine oil combining with water and dust. When these three ingredients get together and are thoroughly mixed, they produce the thick, black, greasy substance known as sludge.

OIL VISCOSITY

Oil viscosity refers to the thickness of fluidity of the oil. In other words, it is a measure of an oil's ability to resist flowing.

The Society of Automotive Engineers (SAE) have set low (0 deg.) and high (210 deg. F.) temperature requirements for oil.

Oils that meet special SAE low temperature requirements have the letter W following the viscosity rating, i.e., (SAE 10W). Oils that meet the high temperature requirements have no such letter, but are merely designated as (SAE 30), etc.

Viscosity is determined by taking any measured sample of oil and heating it to a certain point. The heated oil is then allowed to flow out of a very precise hole. The length of time

it takes to flow determines its viscosity rating. The faster it flows, the lower the rating. This testing device is called a VISCOSIMETER. Fig. 5-7.

VISCOSITY INDEX

Oils thin out when heated and thicken when cooled. As a good engine oil must be reasonably thin for cold starts, as well as thick enough at high temperatures, the Viscosity Index (VI) is important.

When oils are heated, they tend to lose their body (resistance to oil film failure). If a given oil is thin at cold temperatures and when heated, does not thin out much more, it is said to have a high VI. If the heating produces a great change in viscosity, it will have a low VI. The VI, then, is a measure of the ability of an oil to resist changes in viscosity when heated.

MULTI-GRADE OILS (MULTI-VISCOSITY)

These are oils that meet SAE requirements for both the low temperature requirements of a light oil and the high temperature requirements of a heavy oil. They meet the viscosity and performance requirements of two or more SAE grades and as such, are marked SAE 5W–20, SAE 5W–30, SAE 10W–20W–30, SAE 10W–30, SAE 10W–40, SAE 10W–20W–40, etc. They are often referred to as multi-viscosity, all-season, and all-weather.

CHOOSING OIL VISCOSITY

Always follow the manufacturer's recommendation for viscosity rating. The most commonly used oils are SAE 10W, SAE 20W, SAE 20, SAE 30, SAE 10W–30 and SAE 10W–20W–40. These recommendations are based on engine design and the lowest anticipated temperatures.

WHEN OIL SHOULD BE CHANGED

The answer to this popular problem depends on so many variables that a general recommendation is very difficult to make.

Many manufacturers give recommendations of from 1,000 to 6,000 miles between oil changes.

Any quality of oil will retain proper lubricating properties longer if:
1. The engine is operated for reasonably long periods each time it is started.
2. The engine is in good mechanical condition.
3. The engine operates at a temperature between 180 and 200 deg. F. or higher.
4. The engine has a good oil filter.
5. The crankcase ventilation system is open and the PCV valve is functioning properly.
6. The engine is operated at moderate to fairly high speeds.
7. The engine is provided with an efficient air cleaner.
8. The carburetion system is properly adjusted.
9. The engine is operated under clean conditions.

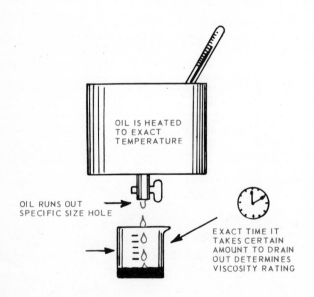

OIL IS HEATED TO EXACT TEMPERATURE

OIL RUNS OUT SPECIFIC SIZE HOLE

EXACT TIME IT TAKES CERTAIN AMOUNT TO DRAIN OUT DETERMINES VISCOSITY RATING

Fig. 5-7. Principle of the viscosimeter.

There are other considerations but these are the most important.

The above conditions are ideal. If your engine operates under them, using the oil for the maximum recommended time is possible.

Engines that are started and stopped frequently, that are driven short distances, that never warm up, that are operated under dusty conditions, that are in poor mechanical condition, that have no filter system, should have the oil changed at least every 500 miles.

Very few engines are operated at either extreme. Knowing the conditions surrounding the operation of any particular engine will allow you to determine, in an intelligent manner, when to change the oil.

When you do change it, drain only when the engine is hot. Do not use kerosene as a flushing agent.

Do not be confused by the color of the oil. Oils containing detergents will keep small particles in suspension and will LOOK dirty but in reality will still be usable. Base your oil change on mileage — after considering all the factors involved.

HOW OIL WORKS

When oil is injected into a bearing, it is drawn between the two bearing surfaces. The oil then separates the parts, and they revolve without actually touching each other. When the shaft stops revolving, and oil is no longer supplied, the shaft will settle down and squeeze the oil film apart and will rest on the bearing. Some residual oil (oil that clings to the parts) will be retained. When the engine is started, the shaft will have only residual oil for the first few seconds of operation. This is why engines should run slowly, in neutral, for a minute when first started. Fig. 5-8 shows the action of oil in a bearing.

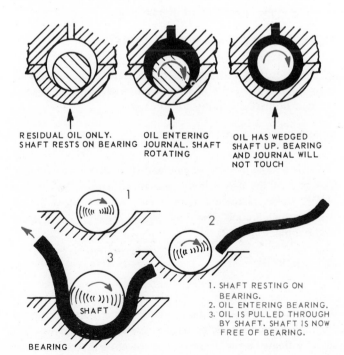

RESIDUAL OIL ONLY. SHAFT RESTS ON BEARING

OIL ENTERING JOURNAL. SHAFT ROTATING

OIL HAS WEDGED SHAFT UP. BEARING AND JOURNAL WILL NOT TOUCH

1. SHAFT RESTING ON BEARING.
2. OIL ENTERING BEARING.
3. OIL IS PULLED THROUGH BY SHAFT. SHAFT IS NOW FREE OF BEARING.

Fig. 5-8. How oil enters a bearing.

OIL SLIPS

The top of the film that contacts the shaft journal, and the bottom of the film that contacts the bearing, resist movement to some extent. The sliding effect of the oil is achieved nearer the center of the film. Fig. 5-9.

This slipping action allows the journal to spin easily and still not contact the bearing surface.

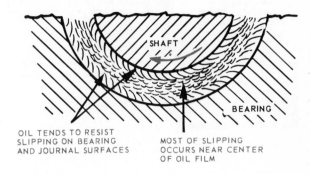

OIL TENDS TO RESIST SLIPPING ON BEARING AND JOURNAL SURFACES

MOST OF SLIPPING OCCURS NEAR CENTER OF OIL FILM

Fig. 5-9. Action of oil in bearing.

BEARING CLEARANCE

It is important that all bearings have the proper clearances. The ideal being loose enough for oil to enter, but tight enough to resist pounding. Oil clearances will average around .002 in.

HOW OIL GETS TO THE BEARINGS

Engines are lubricated in three ways:
1. Full pressure oiling.
2. Splash oiling.
3. Combination splash and pressure.

The full pressure system will be discussed first.

FULL PRESSURE SYSTEM

The pressure system draws oil from the pan sump by means of an oil pump. The pump then forces oil, through special passages, to the crankshaft and camshaft journals. The crankshaft is drilled to permit oil to flow to the connecting rod journals. In some engines the rods are drilled their full length to allow oil to pass up to the wrist pin bushings. Bearing throw-off may be helped, by spurt holes in the rod, to lubricate the cylinder walls. Timing gears, lifters, and rocker arms, where used, are also oiled. In the pressure system, all bearings are oiled by either pumping oil into the bearings, or squirting or dripping it on. Fig. 5-10.

SPLASH SYSTEM

The splash system supplies oil to moving parts by attaching dippers to the bottom of the connecting rods. These dippers can either dip into the sump itself, or they can dip into shallow trays. The whirling dippers splash oil over the inside of

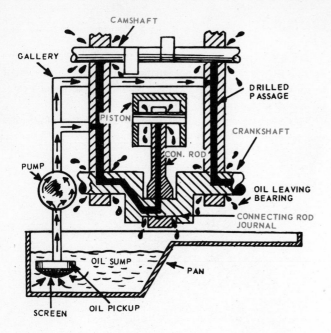

Fig. 5-10. Schematic of pressure system. Notice how oil travels to all major bearings. When it passes through bearing, it is thrown around creating an oil spray and mist. Oil then returns to sump in pan.

the engine. The dipper is usually drilled so that oil is forced into the connecting rod bearings.

An oil pump may be used to keep oil trays full at all times. The splash system is used primarily on small one cylinder engines. Fig. 5-11.

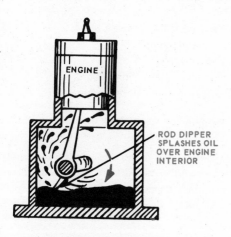

Fig. 5-11. Simple splash oiling system.

COMBINATION SPLASH AND PRESSURE

This system uses dippers and trays. An oil pump supplies oil to the camshaft and crankshaft bearings. On overhead engines, oil is pumped to the rocker arms and valves.

Small pipes are sometimes used to squirt oil over the trays. The whirling dipper is drilled to permit oil to enter the connecting rod bearing. When the dipper strikes the stream of

oil, it throws oil over the interior of the engine. At the same time, oil is driven up the dipper hole into the rod bearing. When idling, the oil stream may not quite reach the dippers but will keep the trays full. Fig. 5-12.

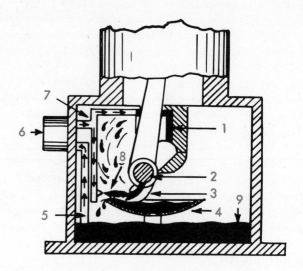

Fig. 5-12. Combination splash and pressure oiling system. Oil pump 6 draws oil up pipe 5 from sump 9. It then pumps oil to main bearings 1 through discharge pipe 7. Oil is also pumped down discharge to nozzle 8 where oil sprays out and strikes dipper 3. This rams oil into rod bearing 2. Tray 4 also holds oil for dipper to strike.

FOURTH METHOD OF OILING

Most two-cycle engines (discussed later) draw gas and air into the crankcase. In the case of these engines, lubricating oil is mixed with the gasoline and upon entry into the crankcase, provides ample lubrication for the bearings.

PRESSURE SYSTEM WIDELY USED

The pressure system best meets the needs of the modern car engine and is therefore in almost universal use. This chapter will deal primarily with the pressure system.

SUMP

Oil is poured into the engine and flows down into the pan. One end of the pan is lower and forms a reservoir called a SUMP. Baffle plates are sometimes used to prevent the oil from sloshing around. The drain plug is placed in the bottom of the sump. Fig. 5-13.

OIL PICKUP

Some engines have a rigid pipe leading down into the oil in the sump. It does not come too close to the bottom to avoid picking up sediment. Some engines use a floating pickup. This actually floats on top of the oil and therefore draws only from the top of the sump. As the oil level drops, the pickup will also drop, to maintain a constant supply of oil.

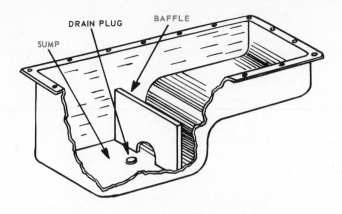

Fig. 5-13. Oil pan with sump and baffle. There are many designs for sumps and baffles.

Both types of oil pickup use screens to exclude the larger particles that may be drawn into the system. In the event the screen becomes clogged, there is a small valve that will be drawn open, thus allowing the pump to bypass the screen and obtain oil. Fig. 5-14.

OIL PUMPS

There are four principal types of oil pumps that have been used on engines. They are the GEAR, ROTARY, VANE, and

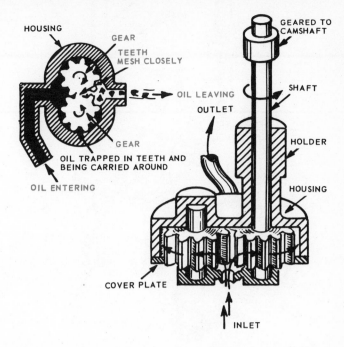

Fig. 5-15. Gear oil pump. Notice how oil is caught in teeth and carried around. As pump whirls rapidly, it will produce a steady stream of oil.

PLUNGER pumps. The two most widely used are the GEAR and the ROTARY types.

GEAR PUMP

The gear pump uses a driving and a driven spur gear as the moving force to pump oil. These gears are placed in a compact housing. The driven gear revolves on a short shaft, while the driving gear is fastened to a longer shaft that is turned by a spiral gear meshed with a similar gear on the camshaft.

The teeth of the two gears must mesh with a minimum amount of clearance. The housing just clears the top, bottom, and sides of both gears.

An inlet pipe allows oil to be drawn in by the whirling gear teeth. Each tooth catches what it will hold and carries it around. When the oil reaches the opposite side, it cannot come back through the center of the gears as they are closely meshed.

As each tooth carries a load of oil around and deposits it on the outlet sides, the oil has no where to go but through the outlet to the oil channels. The gear pump is efficient and produces oil pressure needed. Figs. 5-15 and 5-16.

ROTARY PUMP

A rotary pump uses a housing, an inner rotor, and an outer rotor. The outer rotor is cut away in the form of a star with rounded points.

The inner rotor is shaped in the form of a cross with rounded points that fit in the outer rotor star. The inner rotor is much smaller. It is driven by a shaft turned by the camshaft.

The inner rotor is mounted off center so that when it turns, the rounded points "walk" around in the star-shaped outer

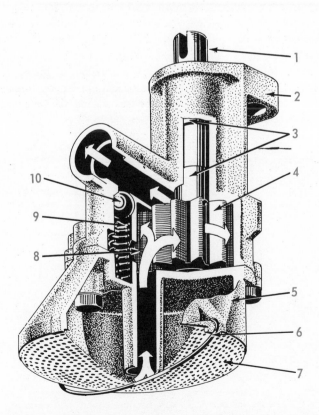

Fig. 5-14. Rigid type oil pickup. Note screen 7 over pump inlet. Study oil flow: 1—Pump gear shaft. 2—Pump housing. 3—Bushings. 4—Driving gear. 5—Cover. 6—Retainer. 7—Screen. 8—Driven gear. 9—Relief valve spring. 10—Relief valve ball.   (Volvo)

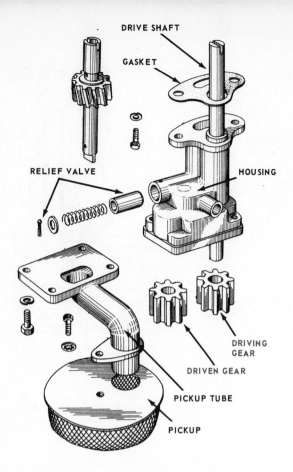

Fig. 5-16. Exploded view of a gear type oil pump.
(Datsun)

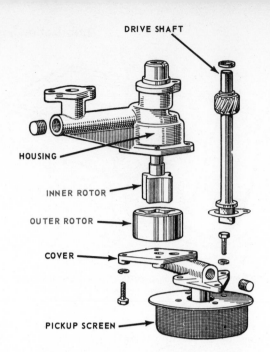

Fig. 5-18. Exploded view of a rotor type oil pump. Note gear on drive shaft. This gear will mesh with another gear on the camshaft.
(Austin-Healey)

rotor. This causes the outer rotor to revolve also. As the inner rotor "walks" the outer rotor around, the outer rotor openings pick up oil at the inlet pipe and pull it around until it lines up with outlet. It is then forced out as the inner and outer rotor points close together. The oil cannot make a circuit around, as the inner rotor fits snugly against the outer rotor at one spot. Figs. 5-17 and 5-18 illustrate the rotor pump.

VANE PUMP

This pump uses a round housing with a round rotor placed off center. This rotor contains two or more VANES, or rotor blades, that fit into slots in the rotor. The vanes are held out against the round housing by springs between the vanes.

As the shaft turns, the vanes are forced in and out to keep contact with the housing. As they swing out, they pick up a charge of oil and carry it around. As the vane is forced in, this places a squeeze on the oil and it is pushed out of the pump. As a steady stream of charges are picked up, carried around, and squeezed out, the pump produces a good stream of oil. Figs. 5-19 and 5-20.

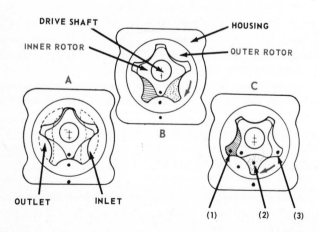

Fig. 5-17. Rotor oil pump action. A—Pump inlet and outlet areas. B—Drive shaft turns inner rotor causing it to walk around in outer rotor. Outer rotor will also turn. C—Rotor pulls around a charge of oil and forces it through the outlet (1). While this is happening, another charge is moving around (2) and another is being picked up at the inlet (3). (Jaguar)

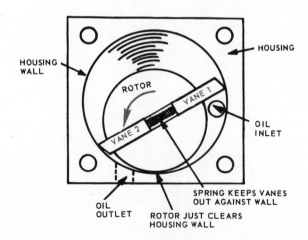

Fig. 5-19. Vane oil pump. Vane 1 has just passed inlet hole. As it continues by, it will form a vacuum that will continue to draw oil in until vane 2 passes inlet. As vane 2 passes inlet it will push oil around and at the same time, draw a vacuum behind to pull oil in for vane 1.

As each vane, pushing a chrage of oil, arrives near outlet, spaces grow small and the oil is squeezed out of pump.

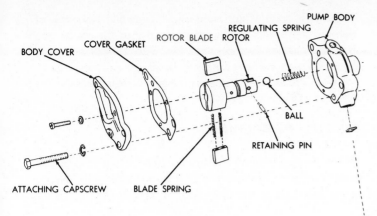

Fig. 5-20. Exploded view of vane type oil pump. (Allis Chalmers)

PLUNGER PUMP

The plunger type pump is used in many small one cylinder engines to maintain the oil level in the dipper trays.

It utilizes a cylinder, plunger, and check valves. It is usually driven by a cam on the camshaft.

As the cam releases the plunger, it is pushed upward by a strong spiral spring. The inlet check valve opens and oil is

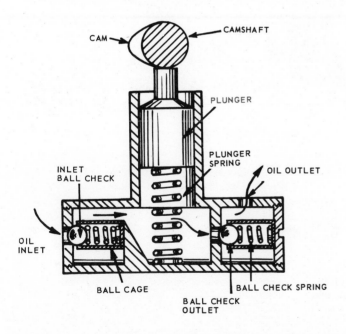

Fig. 5-21. Plunger oil pump. When plunger travels up it forms vacuum within pump. This opens inlet check and closes outlet. Cam pushes plunger down to create pressure. This shuts intake check and forces oil through outlet check valve.

drawn in. As the cam continues to rotate, it will press the plunger down. This closes the inlet valve and opens the outlet. The process is repeated over and over, producing a steady stream of oil. Fig. 5-21 illustrates one type of plunger pump.

PRESSURE RELIEF VALVE

An oil pump will produce pressures far beyond those necessary. The pumps are designed to carry a large flow of oil but when the usable pressure limits are reached, a pressure relief valve opens and allows oil to return to the sump.

The relief valve will allow the pressure reaching the bearings to remain at any predetermined level. By varying spring tension, the pressure can be raised or lowered.

The relief valve can be located right at the pump, Fig. 5-16, or at any spot between the pump and the bearings. Figs. 5-22 and 5-23 illustrate a relief valve located in the side of the block.

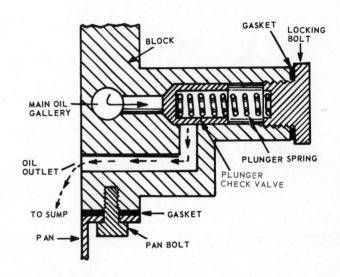

Fig. 5-22. Oil pressure relief valve. Oil in main oil gallery is under pressure from oil pump. When pressure exceeds setting for relief valve, it will push plunger check valve back against plunger spring pressure. When valve is forced back enough to expose oil outlet channel, oil will pass from main oil gallery past valve, through oil outlet and back into pan sump.

Fig. 5-23. Exploded view of an oil pressure relief valve. Valve is located in side of block. 1—Valve. 2—Spring. 3—Gasket. 4—Locking bolt. (Austin)

OIL GALLERIES AND LINES

Oil that is allowed to pass the pressure relief valve is carried to the bearing through tubes and oil galleries (passageways cast in the block or made up of large tubing). The main oil gallery runs lengthwise in the block and most other lines connect with it. Most engines have a plug at each end of the main gallery. When overhauling the engine, the plugs may be removed and the gallery cleaned.

Lines from the main gallery are drilled to the main bearings. The crankshaft is drilled to carry oil from the main journals to the rod journals. Figs. 5-24 and 5-25.

The valve train is lubricated by oil spray and in the case of overhead valve engines, by oil pumped into the rocker arm shaft, through rockers to valves, Fig. 5-25. Oil can also reach the rocker arms via hollow push rods.

Camshafts receive oil under pressure to their journals. The lobes are lubed by oil spray.

OIL ENTERS BEARING

The bearing receives oil from a hole drilled through the bearing support. The insert has a hole that aligns with the one in the bearing support. The oil passes through and engages the

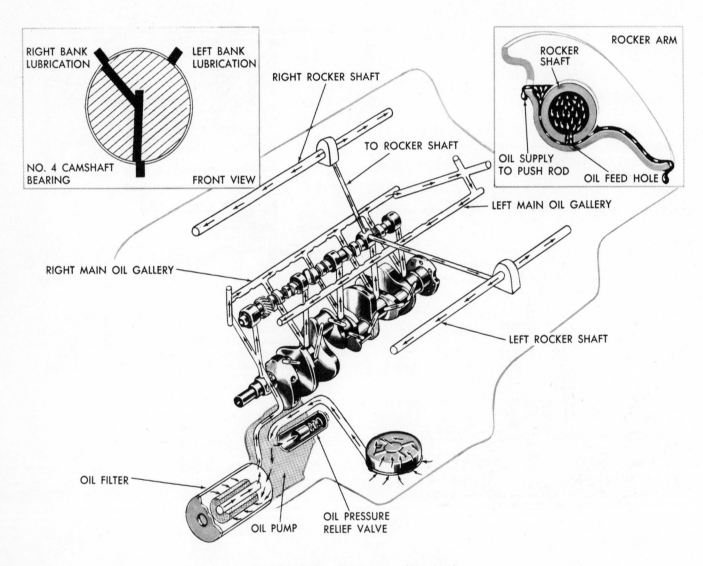

Fig. 5-24. Engine oiling system.   (Plymouth)

Some rods are drilled full length to carry oil to the wrist pins. Others have tiny holes that line up with the crank journal once every revolution. When the two holes align, oil is squirted through the "spit" hole and onto the cylinder wall. Spit holes are also used to lubricate the camshaft lobes in some engines.

The timing chain is oiled either by spray or by oil discharged from the No. 1 camshaft bearing.

turning journal and is pulled around between the insert and journal.

Some inserts have shallow grooves cut in them to assist in spreading the oil.

Oil is constantly forced through the bearings. Oil leaving a bearing is thrown violently outward. This helps to produce the fine oil mist inside the engine that is useful in lubricating

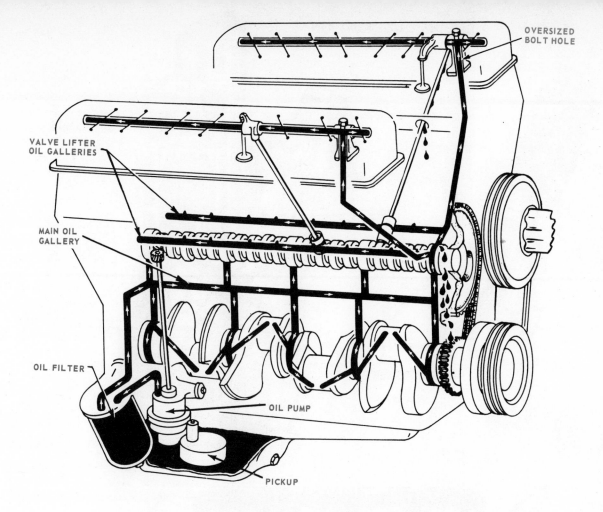

Fig. 5-25. Schematic diagram showing oil flow through engine lubrication system.   (Buick)

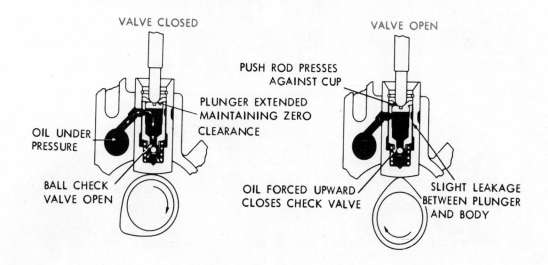

VALVE CLOSED

VALVE OPEN

PUSH ROD PRESSES
AGAINST CUP

PLUNGER EXTENDED
MAINTAINING ZERO
CLEARANCE

OIL UNDER
PRESSURE

BALL CHECK
VALVE OPEN

OIL FORCED UPWARD
CLOSES CHECK VALVE

SLIGHT LEAKAGE
BETWEEN PLUNGER
AND BODY

Fig. 5-26. Hydraulic valve lifter, showing tappet action — cutaway views.

hard-to-reach spots. All oil eventually flows back into the sump. Figs. 5-27 and 5-28.

HYDRAULIC VALVE LIFTERS

Hydraulic type valve lifters are used on many engines. One type of hydraulic lifter is shown in Fig. 5-26. In this lifter, oil under pressure is forced into the tappet when the valve is closed. This extends the plunger in the tappet so valve clearance is eliminated.

When the camshaft lobe moves around and starts to raise the tappet, oil is forced upward in the lower chamber of the tappet. This closes the ball check valve and oil cannot escape. Since oil does not compress, the valve is actually lifted on the

ENGINE LUBRICATION

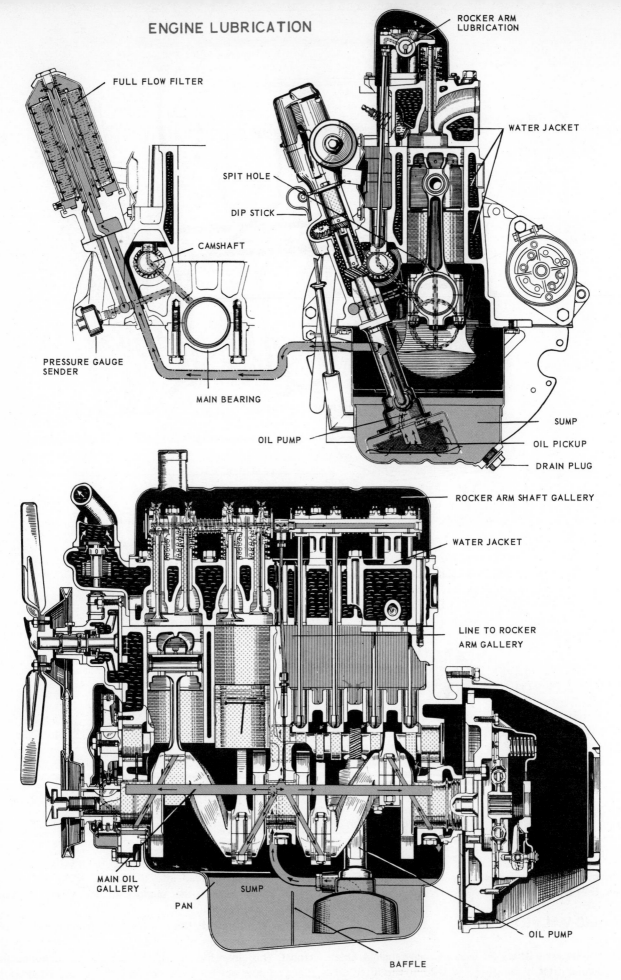

FULL FLOW FILTER

ROCKER ARM LUBRICATION

WATER JACKET

SPIT HOLE

DIP STICK

CAMSHAFT

PRESSURE GAUGE SENDER

MAIN BEARING

OIL PUMP

SUMP

OIL PICKUP

DRAIN PLUG

ROCKER ARM SHAFT GALLERY

WATER JACKET

LINE TO ROCKER ARM GALLERY

MAIN OIL GALLERY

PAN

SUMP

BAFFLE

OIL PUMP

Fig. 5-27. Above. Front section. Solid blue indicates area receiving oil under pressure. Fig. 5-28. Below. Longitudinal section. Note blue dots indicate area lubricated by splash, spray or mist. (Sunbeam-Talbot)

column of oil and is supported by the oil while the valve is open.

When the cam lobe moves from under the tappet and the spring causes the valve to close, pressure in the lower chamber of the tappet is relieved. Oil loss from the lower chamber is replaced by oil pressure from the lubricating system of the engine.

STUDY THE COLORED CROSS SECTION VIEWS

Study Figs. 5-27 and 5-28. They show a typical four cylinder, valve-in-head engine. The solid blue lines indicate lubrication by direct pressure. The dotted blue areas are lubricated by splash, dripping, spray or oil mist.

OIL NEEDS CLEANING

Engine oil soon becomes contaminated with dust, bits of carbon, metal dust, and sludge. If these impurities are allowed to remain in the oil, it will hasten engine wear. Most modern engines are equipped with an oil filtering system to remove these harmful impurities.

FILTERS

Many materials have been tried in an effort to find a satisfactory filtering medium. Cotton waste, special fibers, metal, clay and paper have all been used.

For auto engines, cotton waste and paper filters are in general use. Heavy-duty engines sometimes employ a metal filter that is very good where contaminants build up quickly.

All filters eventually clog up and must be replaced. Replacement varies but will average 4,000 to 6,000 miles per filter. A, B and C, Fig. 5-29, illustrates the filtering principle used in metal, cotton and paper filters.

FILTER OPERATION

Some oil filters use a permanent metal container that holds the filtering element. Others use a disposable type in which the entire filter, container and all, are thrown away.

Oil is pumped into the container and surrounds the filter. The oil passes through the filtering element to the center

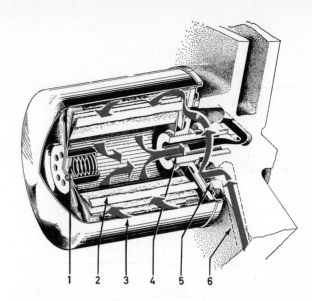

Fig. 5-30. Full-flow type of oil filter. When a change is required, the entire filter is replaced. 1—Relief valve. 2—Filter element (folded paper). 3—Body. 4—Connecting pipe. 5—Gasket. 6—Engine block. (Volvo)

outlet pipe. It is then returned to the sump, or to the bearings, depending on the type of system.

Most bypass filters (only part of oil supplied by pump goes through the filter) use either flexible hose or tubing to connect them to the engine. The full-flow type (entire output of pump goes through filter) generally fastens directly to the engine. Figs. 5-30 and 5-31.

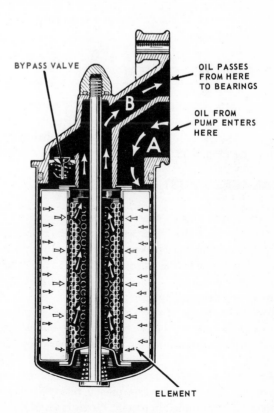

Fig. 5-31. Cutaway of full-flow oil filter. Note bypass valve which will open and supply oil to the bearings in the event the filter clogs. (Rootes)

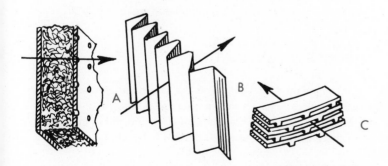

Fig. 5-29. Oil filter materials. A—Cotton waste between thin perforated metal sheets. B—Paper, treated with resins. C—Cuno type. Thin layers of metal sheet held apart by bumps or bars.

FULL-FLOW FILTER SYSTEM

In the widely used full-flow filter system, the oil is pumped directly to the filter from the pump. After passing through the filter, the oil then goes to the oil galleries. In this way, all oil is filtered before reaching the bearings.

In the event the filter becomes clogged, a small relief valve opens and allows the oil to flow directly to the bearings. This avoids the possibility of starving the bearings when the filter becomes unusable. Fig. 5-32.

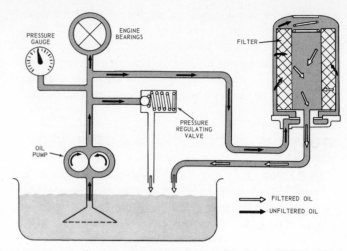

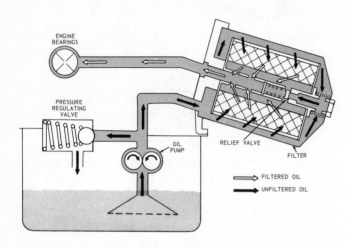

Fig. 5-32. Full-flow oil filter system. Note built-in relief valve to allow oil to flow even if filter becomes clogged. This system filters all oil before it reaches the bearings. (John Deere)

Fig. 5-33. Bypass filter system. This setup filters some of the oil all of the time, but does not filter oil before it reaches the bearings. (John Deere)

BYPASS FILTER SYSTEM

The bypass system differs from the full-flow in that it does not filter the oil before it reaches the bearings. It is constantly filtering some of the oil from the system and then returning it to the sump. Fig. 5-33.

CRANKCASE MUST BE VENTILATED

Even in new engines there is a certain amount of hot gas spurting by the piston rings and on into the crankcase. This causes pressure in the crankcase and unless it is vented, will force oil to escape by the oil seals.

It is also important to eliminate these gases, as they form moisture and deposit sulphuric acid in the oil.

A certain amount of raw gasoline will also pass the pistons and enter the oil. As the oil becomes hot, the moisture and gasoline will evaporate and the ventilation system will remove them.

An entrance, called a breather, is provided on the engine at the top or side. A cover cap, usually filled with oil soaked metal wool, is furnished to prevent the entrance of dirt.

The whirling engine parts create on air disturbance within the engine. An outlet pipe at the opposite end runs down below the engine and when the car is in motion, a partial

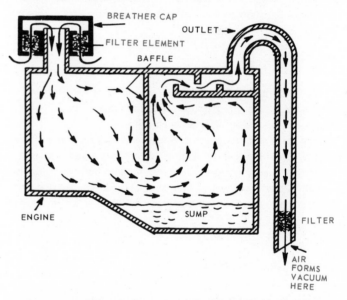

Fig. 5-34. Airflow in a road draft crankcase ventilation system. Notice filter material in both inlet and outlet. Good ventilation is important.

vacuum is formed at the end of the pipe. This draws air through the engine. A baffle is provided to prevent oil vapor from being drawn through the road draft tube. Fig. 5-34.

POSITIVE CRANKCASE (FORCED) VENTILATION

The method of ventilating the crankcase used on todays cars, utilizes engine vacuum to draw a stream of fresh air through the engine. This will be covered in detail in the section of this chapter devoted to EMISSION CONTROL.

OIL SEALS

Oil is retained in the engine by gaskets and oil seals. The seals are placed at the front and rear of the crankshaft. Figs.

5-35, 5-36 and 5-37 illustrate a gasket, seals and an oil slinger. The slinger catches what might escape past the seal and as it is whirling, throws the oil outward into a catch trough that returns the oil to the sump. Many engines do not use oil slingers.

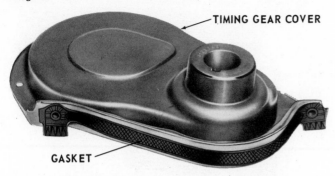

Fig. 5-35. A neoprene gasket is used to prevent oil leakage between the timing gear cover and engine pan. (Chevrolet)

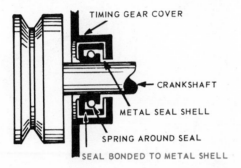

Fig. 5-36. Front crankshaft oil seal. Seal is pressed into timing chain cover. Open end of seal faces inward so oil pressure tends to squeeze seal against shaft. Spring pulls seal against shaft. Seal will be neoprene or leather.

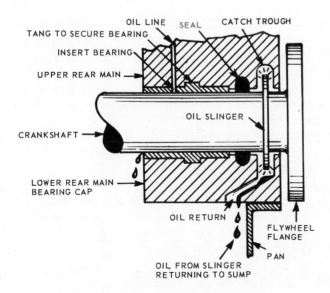

Fig. 5-37. Rear crankshaft oil seals. Oil that escapes past bearing is stopped by seal. Any that passes seal is thrown out into the catch trough and returned to sump. Some engines use a seal only. One seal is made of asbestos rope, oil impregnated and covered with graphite. Others are made of synthetic rubber.

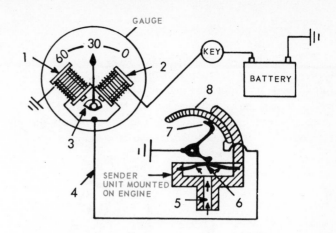

Fig. 5-38. Schematic diagram of electric oil gage. Electricity travels from battery, through key switch to coil 2, then across wire 4 to coil 1 and on to ground. Both coils are connected by wire 4 to resistance type unit 8, mounted on engine.

Oil entering 5 will raise diaphragm 6 causing sliding contact 7 to slide further out on resistance unit. This will increase resistance.

When resistance is increased, coil 1 will strengthen magnetic pull on armature 3. This will pull armature and indicating needle over to read more pressure.

When pressure drops, sliding contact will come back and decrease resistance. This strengthens coil 2 and weakens coil 1. Coil 2 will now pull armature and needle down to read less pressure.

OIL PRESSURE GAGES

One type of gage utilizes the Bourdon tube gage connected to the oil gallery via a tube. This type gage was covered in the Cooling section.

An electric pressure gage uses a different sender unit than that used by the water temperature gage. Fig. 5-38 shows a typical electric oil pressure gage setup.

Some cars use a light to indicate when the pressure drops below a certain point. The light warns of system failure but does not indicate actual pressure. Fig. 5-39.

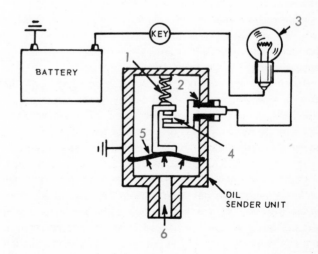

Fig. 5-39. Schematic diagram of oil pressure indicator light. Oil enters 6 and raises diaphragm 5. This separates points 4, breaking circuit through indicator light 3. When pressure drops, spring 1 will close points and light will burn. Insulation 2 prevents circuit from grounding out. Indicator will light when engine stops or when pressure drops to around 7 lbs.

DIPSTICK

A long metal rod, commonly referred to as a "dipstick," is used to check the amount of oil in the crankcase sump. The rod is held in a tube, the end of which protrudes into the sump. It may be withdrawn and the level of the oil in the sump checked by examining the rod to see where the oil reached. Fig. 5-40 shows a dipstick and its markings.

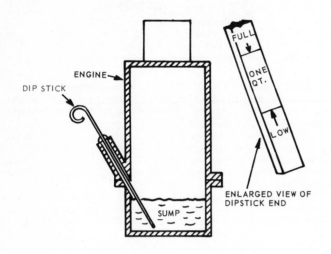

Fig. 5-40. Dipstick. Keep oil level above LOW or ADD oil level. DO NOT FILL ABOVE FULL MARK.

EMISSION CONTROL DEVICES

Approximately 99.9 percent of the exhaust from the average car engine (gasoline powered) consists basically of water, nitrogen, hydrogen, carbon dioxide and carbon monoxide.

About 0.1 percent (one tenth of one percent) of the exhaust consists of hydrocarbons (unburned gasoline).

Hydrocarbon emission into the atmosphere, where it reacts with the sunlight, tends to produce what may be called photochemical (sunlight causing chemical reactions or changes) smog.

Automobile hydrocarbon emission can occur through the exhaust, from the crankcase ventilation system, from carburetor leakage and from gas tank fumes. The exhaust and crankcase emissions make up the major source. In order to reduce the hydrocarbon (HC), oxides of nitrogen (NOx) and carbon monoxide (CO) emission to an acceptable level (measured in ppm — parts per million parts of air), positive crankcase ventilation has been brought into use. In addition, an air injection system and certain carburetor and timing modifications have been developed to reduce the hydrocarbon ppm in the exhaust itself. The future will undoubtably see the development of more sophisticated and efficient devices.

POSITIVE CRANKCASE VENTILATION — PCV

By connecting a hose between the engine interior and the

intake manifold, engine vacuum will draw the crankcase fumes out of the engine and into the cylinders where the gases will be burned along with the regular fuel charge. This is termed POSITIVE CRANKCASE VENTILATION. Fig. 5-41.

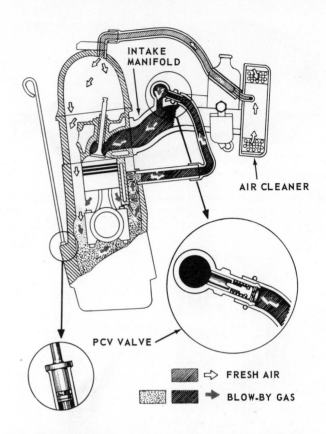

Fig. 5-41. Positive Crankcase Ventilation (PCV) system. Vacuum in intake manifold draws air (from air cleaner) through engine interior. (Nissan)

Positive crankcase ventilation (PCV) systems are of two types — OPEN and CLOSED. In the open type, C, Fig. 5-42, air enters the crankcase through a ventilated (open) breather cap. During periods of low manifold vacuum (heavy acceleration, high road speed), the crankcase fumes may back up and pass out the breather cap. This permits some air pollution and is therefore undesirable.

The CLOSED PCV system prevents this by drawing air into the crankcase via a hose attached to the carburetor air cleaner, A, Fig. 5-42.

During periods of low vacuum, the fumes will pass back into the cleaner where they will join incoming air to the carburetor and thus find their way into the cylinders. As soon as vacuum is restored, circulation will be reestablished. The CLOSED system is now used on all cars.

PCV CONTROL VALVE NEEDED

In order to avoid upsetting the basic carburetor air-fuel ratios for various operating conditions, it is essential that a PCV control valve, B, Fig. 5-42, be incorporated into the system to meter the flow of crankcase gases.

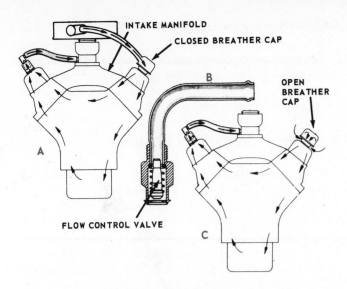

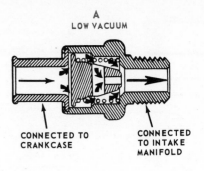

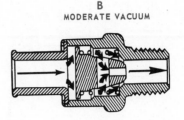

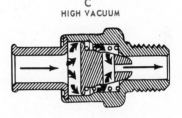

Fig. 5-42. Two types of positive crankcase ventilation systems. A—Closed type that draws air in through air cleaner. B—Airflow control valve. C—Open type that draws air in through breather cap.
(Plymouth)

A typical PCV flow control valve is shown in Fig. 5-43.

PCV FLOW CONTROL VALVE ACTION

When engine speed is quite high, the intake manifold vacuum will drop, allowing the control valve spring to force the valve away from the seat thus permitting a maximum amount of valve opening. A, Fig. 5-44.

In B, Fig. 5-44, engine speed has been reduced, vacuum

Fig. 5-44. PCV control valve action.
(Ford)

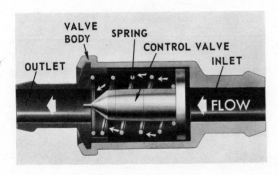

Fig. 5-43. Typical PCV control valve.
(Chevrolet)

increased, thus drawing the valve closer to the seat and restricting the flow of crankcase fumes to the manifold.

During periods of low engine speed, C, Fig. 5-44, the vacuum reaches maximum and pulls this particular valve into contact with the seat, thus allowing fumes to pass through the center opening only, with a valve such as pictured in Fig. 5-43, the tapered valve moves to its deepest position during high vacuum.

In the event of a backfire, (flame traveling back through the

intake manifold), the reverse pressure will cause the valve to move away from the seat until the rear shoulder strikes the valve body. This effectively seals the crankcase from the backfire.

PCV systems require periodic servicing to make certain the flow control valve is functioning and that the inlet air filter is clean. Some PCV valves may be taken apart for cleaning while others must be replaced at recommended intervals. A clogged PCV valve will cause heavy sludging and serious engine damage.

AIR INJECTION MAKES COMBUSTION MORE COMPLETE

Hydrocarbon and carbon monixde emission via the exhaust can be reduced by a more complete burning of the fuel charge.

Following the power stroke, the exhaust valve opens to permit exhausting of the burned fuel charge. As the charge passes the exhaust valve, it is still burning. By injecting a stream of fresh air into the burning gases, just as they pass the exhaust valve, the burning will be both intensified and prolonged. This increased burning action will tend to burn up a great deal of the normally unburned hydrocarbons, as well as changing a significant portion of the carbon monoxide (poisonous) into harmless carbon dioxide.

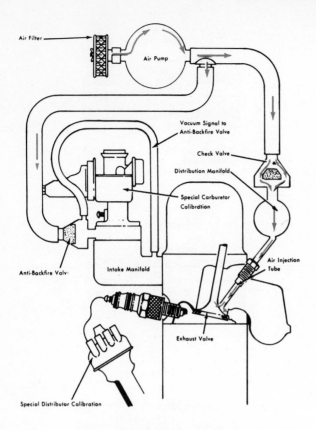

Fig. 5-45. Typical air injection exhaust emission control system. (American Motors)

CARBON MONOXIDE IN EXHAUST

Car engines, despite the use of emission controls, produce poisonous carbon monoxide. Use the same precautions regarding ventilation and a sound exhaust system that apply to a car not equipped with controls.

AIR INJECTION SYSTEM ACTION

A typical air injection system (called Air Injection Reactor, Thermactor, Air Guard, etc.) is illustrated in Fig. 5-45.

Note that an AIR PUMP, Fig. 5-45, draws fresh air in through a FILTER. This air is then mildly compressed. The compressed air is then directed, via a hose, through a CHECK VALVE and into a DISTRIBUTION MANIFOLD. From the distribution manifold, the air passes through the AIR IN-JECTION TUBES into the exhaust valve port area.

The air pump or compressor, is of the positive displacement, vane type. Fig. 5-46. Relief valve limits the amount of pressure the pump can develop.

Spring-loaded carbon shoes permit sliding contact between the vanes and rotor. The pump is belt driven by the engine. Figs. 5-46 and 5-47.

As each rotating pump vane passes the intake chamber, Fig. 5-47, a charge of air is drawn in. The next vane propels the air charge around into the compression chamber. The compression chamber reduces the volume of the charge thus

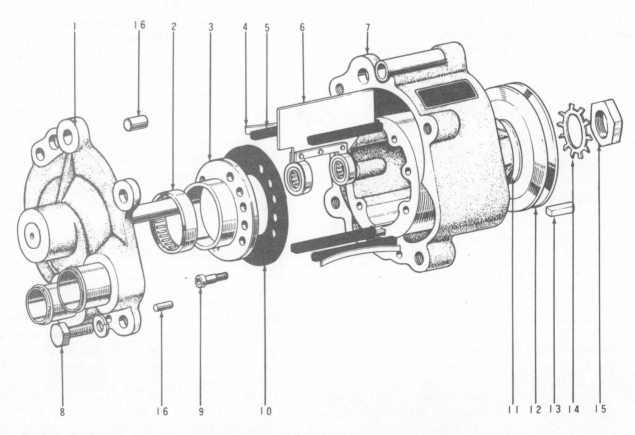

Fig. 5-46. Exploded view of an air injection pump. 1—Cover. 2—Bearing. 3—Rotor ring. 4—Shoe spring. 5—Carbon shoe. 6—Vane. 7—Rotor housing. 8—Attaching bolt. 9—Rotor ring cap screw. 10—Rear seal. 11—Pulley hub. 12—Pulley. 13—Key. 14—Lock washer. 15—Nut. 16—Knock pin. (Toyota)

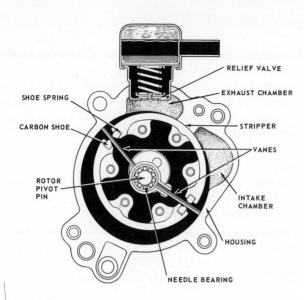

Fig. 5-47. End view of an air injection pump.
(Toyota)

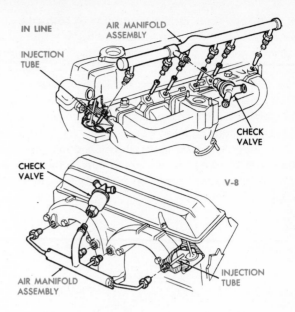

Fig. 5-48. Air distribution manifold and air injection tube setup for both in-line and V-type engine use.
(Chevrolet)

placing it under compression. Pressurized air is then discharged through the exhaust chamber. Note that the intake and exhaust chambers are separated by the section of housing termed the "stripper," Fig. 5-47.

CHECK VALVE

The check valve, Fig. 5-45, is forced open by the air passing through to the distribution manifold. In the event of pump failure or at any time that exhaust pressure may equal or surpass pump pressure, the check valve will seat and prevent backfeeding of hot exhaust into the hoses or pump.

DISTRIBUTION MANIFOLD AND AIR INJECTION TUBES

An air distribution manifold and air injection tube setup for both an in-line and V-type engine is shown in Fig. 5-48. Other engines incorporate the distribution tube and injection tubes as an integral part of the cylinder head.

ANTIBACKFIRE VALVE — DIVERTER VALVE

Quick closure of the throttle valve permits a momentary continuation of fuel flow. This fuel flow produces a mixture so rich that it will leave a large amount of unburned gas following the power (combustion) stroke.

The exhaust, rich in unburned fuel, upon striking the injected stream of fresh air, will flare up into a violent burning action, causing a backfire.

The antibackfire valve, Figs. 5-45 and 5-49, will admit a swift rush of air from the air pump, into the intake manifold for a short time (about 1 - 2 seconds) whenever the throttle is suddenly closed. This air leans out the fuel mixture and prevents backfiring. This setup also prevents the engine from

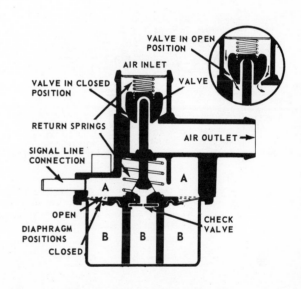

Fig. 5-49. Cross section of one type of air injection system antibackfire valve. (Chevrolet)

stalling at intersections, eliminating the need of a throttle-closing dashpot on cars with automatic transmissions.

The diverter valve prevents backfiring in a somewhat different way, Fig. 5-50.

OPERATION OF THE ANTIBACKFIRE VALVE

The air pump is connected to the INLET side of the antibackfire valve. The valve outlet is connected to the intake manifold. Normally, the valve is held closed by spring action. Fig. 5-49.

By connecting the valve (via the signal line connection) to the intake manifold, a vacuum is formed in chamber A and through the small orifice in the check valve, in chamber B as well. In that an equal vacuum now exists on both sides of the valve diaphragm, the valve is kept closed by the return springs.

Upon a sudden closure of the throttle, a very powerful vacuum is instantly formed in chamber A. In that the increase in vacuum has not had a chance to effect chamber B as quickly, chamber B will, at this instant, have a lower vacuum than chamber A. This forces the diaphragm to open the check valve against spring pressure.

If the throttle remains in the released position, the stronger vacuum will be applied to chamber B through the check valve orifice. With an equal vacuum on each side, the return springs will force the valve shut.

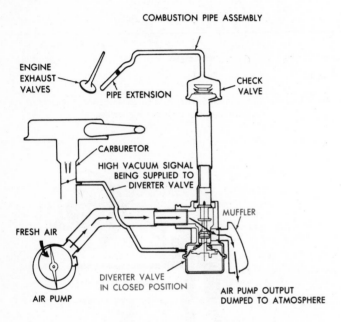

Fig. 5-50. The diverter valve in this air injection reaction system prevents backfiring by momentarily diverting the air stream into the atmosphere during periods of sudden increase in manifold vacuum. Note use of muffler on diverter valve.
(Chevrolet)

When the car is suddenly accelerated, immediately following sudden deceleration that opened the antibackfire valve, it is important that the flow of air to the intake manifold stop instantly to prevent upsetting the mixture required for proper acceleration.

The acceleration causes a vacuum drop in chamber A, leaving chamber B with a higher vacuum. This draws open the diaphragm check valve, allows spring action to snap the backfire valve closed and cuts off the flow of air and permits smooth acceleration. Some systems use a diverter or bypass valve in place of the described antibackfire valve. Instead of admitting a "gulp" of air to the intake manifold during deceleration, it momentarily diverts or bypasses the air from the air pump into the atmosphere. The rich mixture formed during deceleration then, is not subjected to the normal stream of air from the air injection tubes. This prevents backfiring. Fig. 5-50 illustrates a typical system employing a diverter valve. Note muffler on valve to minimize sound of diverted air.

ENGINE MODIFICATIONS

A further reduction in engine emissions has been effected by various engine modifications.

Changes in basic engine design have improved combustion chamber shape for more complete burning. Reducing compression ratios from around 10.5 to 1 down to around 8.5 to 1, have made it possible to use a regular (not premium) no-lead or low-lead fuel to reduce discharge of lead into the atmosphere.

Increased camshaft valve overlap has reduced the amount of NOx (oxides of nitrogen) emission.

The intake manifold design has been altered to provide less condensation of the fuel mixture as it passes from the carburetor to the cylinders.

CARBURETION MODIFICATIONS

In general, carburetors are more carefully calibrated to produce a greater degree of control over the air-fuel mixture at all speeds. The mixture is somewhat leaner (more air, less fuel). Chokes have been redesigned so that they move more quickly to the off position, avoiding an overly rich mixture for a prolonged period of time.

Carburetor idle screws have been equipped with "limiter" caps to restrict the amount they may be opened. In some applications, the orifice (hole) at the base of the idle mixture screw, is of a fixed size so that regardless of how far the idle needle is opened, only a set amount of mixture is delivered.

To maintain proper air-fuel mixture control during deceleration, one type of carburetor utilizes a vacuum diaphragm operated auxiliary valve to admit fresh air along with a metered amount of fuel, Fig. 5-51.

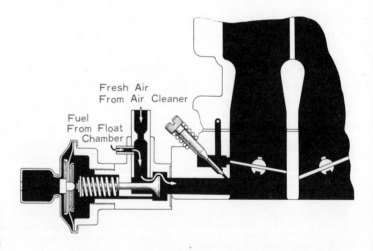

Fig. 5-51. Vacuum operated auxiliary valve admits fresh air along with a metered amount of fuel. This prevents a sudden rich fuel mixture during deceleration. (Toyota)

Some cars are using electronic fuel injection systems to assist in highly accurate mixture control for all speeds and loads.

HEATED INTAKE AIR

In order to assist in proper engine operation when using a lean fuel mixture, the air passing to the air cleaner is often heated. This heated air also reduces carburetor icing and speeds engine warmup.

A thermostatically controlled, vacuum diaphragm operated damper unit (flapper valve) in the air cleaner intake tube can admit air from either the exhaust manifold heat stove or from the underhood area. Fig. 5-52 shows one such setup.

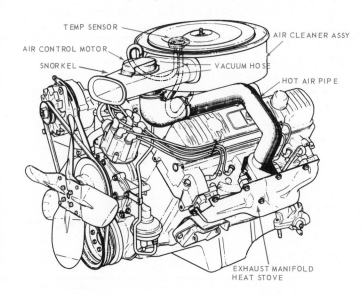

Fig. 5-52. Thermostatically controlled air cleaner is designed to provide heated intake air. (Buick)

THERMOSTATICALLY CONTROLLED AIR CLEANER

In A, Fig. 5-53, the engine is shut off. With no vacuum in the vacuum chamber, the diaphragm spring pushes the control damper down, closing off the hot air pipe from the manifold heat stove.

In B, Fig. 5-53, the engine is operating but the underhood air temperature is below 85 deg. F. The temperature sensing (thermostatic) spring has closed the air bleed valve, admitting full engine vacuum to the vacuum chamber. This draws the diaphragm upward and lifts the damper so that outside air is cut off and hot air is admitted.

When the underhood temperature rises above 128 deg. F., the sensor bimetal spring opens the air bleed valve. This reduces vacuum in the vacuum chamber to the point the diaphragm spring tilts the damper downward, completely closing off the hot air entrance. All air is then taken in through the "snorkel," C, Fig. 5-53.

When underhood temperature is between 85 deg. F. and 128 deg. F., the sensing spring holds the air bleed partially open. This permits both outside and hot air to be drawn into the carburetor, D, Fig. 5-53.

The actual operating temperature can vary somewhat depending on system design. The idea is to attempt to keep the inlet air temperature somewhere around 100 deg. F. to 115 deg. F.

If, at any time, the engine is heavily accelerated, vacuum to the chamber will instantly drop off permitting the spring to snap the damper downward cutting off the hot air and permitting maximum air flow through the snorkel. Fig. 5-53.

Fig. 5-53. Operation of the thermostatically controlled air cleaner. Study each view and note the action of the various controls in determining the position of the control damper. (Chevrolet)

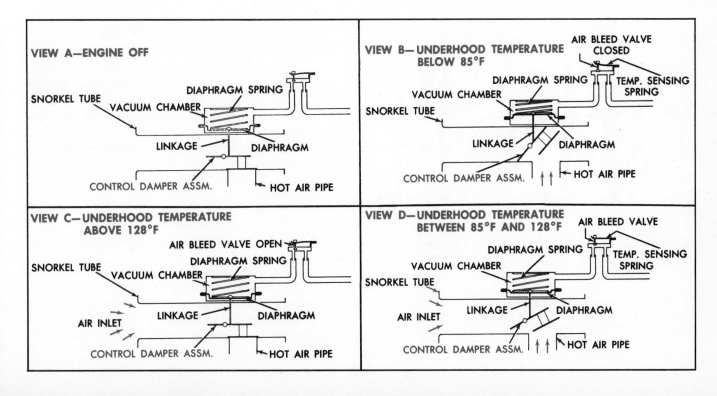

COOLING SYSTEM MODIFICATIONS

By installing high temperature thermostats (around 195 deg. F.), the operating temperature of the engine has been increased. This improves both fuel vaporization and burning. The walls of the combustion chamber and cylinder operate somewhat hotter.

However, engines run cooler in the "quench" area where the incoming fuel mixture makes contact, which cools and tends to cause a small amount of the fuel vapor to return to the liquid state. When the mixture is ignited and the flame reaches this area, it is quenched (put out) and fails to burn this small amount of fuel. The unburned fuel (hydrocarbons) is then passed to the atmosphere on the exhaust stroke.

Now, with the increase in engine operating temperature, much less fuel is left unburned.

IGNITION SYSTEM MODIFICATIONS

In the conventional ignition system, contact points wear and burn and rubbing block wear takes place. This causes a reduction in spark intensity, alters ignition timing and changes the dwell angle. This increases engine emissions.

Some makers are using TRANSISTOR ignition systems. By the nature of their design, they will operate longer without the undesirable changes mentioned above. This effectively reduces emissions.

RETARDED INITIAL IGNITION TIMING

Another emission control technique is to retard the ignition timing during idle and low speed operation. On engines having this type of control, the initial ignition timing (point in crankshaft rotation at which the spark occurs during idle with no centrifugal or vacuum advance) is affected.

When the timing is retarded (fires less degrees before TDC), the carburetor throttle valve must be opened wider to maintain the same idle speed. This action admits more air to the engine, thus reducing hydrocarbon emission levels during idle and deceleration periods. Fig. 5-54 illustrates how, by changing initial timing from 8 deg. BTDC to 4 deg. BTDC, the throttle valve or plate opening must be increased to maintain the same idle rpm.

PORTED IGNITION VACUUM ADVANCE CONTROL

To keep the timing retarded at idle, the distributor vacuum advance connecting hole (called a "port") in the carburetor air horn, can be located just above the throttle valve (in the idle position). With the valve in the closed position, there is no vacuum above the valve and the distributor vacuum advance remains in the retard position. As soon as the throttle valve is opened above idle, the vacuum advance port is exposed to vacuum and ignition timing is advanced. Fig. 5-55 shows this setup.

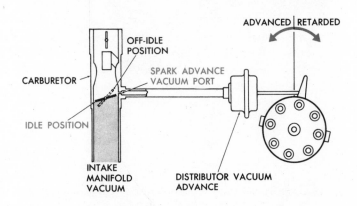

Fig. 5-55. Ported vacuum advance unit remains in the retarded (no vacuum) position when the throttle valve is in the idle position. (Chevrolet)

SPEED CONTROLLED IGNITION VACUUM ADVANCE

The speed controlled vacuum advance setup provides retarded ignition timing at idle and in the lower speed ranges under certain conditions. Fig. 5-56 shows the component parts of one such system.

When the ambient air temperature is below 49 deg. F., the circuit is deenergized and the distributor modulator valve admits carburetor vacuum to the distributor vacuum advance, providing full advance.

When the ambient (surrounding air) temperature exceeds 65 deg. F., the circuit is energized and carburetor vacuum is cut off from the distributor vacuum advance. The ignition timing is now retarded.

As the car gains road speed, a point will be reached, depending on engine and application (23 to 35 mph), where the speed sensor will be spinning fast enough to generate a signal of ample strength to trigger the control amplifier. This deenergizes the distributor modulator valve and full vacuum is

IDENTICAL ENGINES IDLING AT 700 RPM

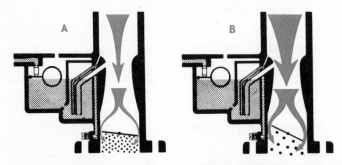

INITIAL TIMING 8 DEG. BTDC
RELATIVELY LITTLE AIR FLOW PAST CARBURETOR THROTTLE PLATE WHICH IS IN NEARLY CLOSED POSITION.

INITIAL TIMING 4 DEG. BTDC
LARGER THROTTLE PLATE OPENING IS NEEDED TO ALLOW MORE AIR INTO THE FUEL TO ACHIEVE 700 RPM.

Fig. 5-54. Initial ignition timing has a direct relationship to the amount of exhaust emissions. Slightly retarded timing in B, causes more air to be admitted to maintain same idle speed. Mixture is leaned out and burns cleaner. (Chevrolet)

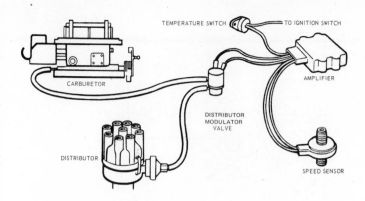

Fig. 5-56. Speed controlled ignition vacuum advance.   (Ford)

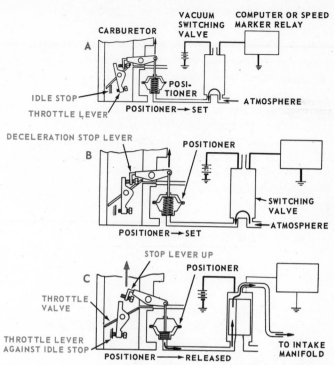

Fig. 5-57. One form of "dual" throttle stop (one position for idle — another for deceleration) control.   (Toyota)

restored to the vacuum advance unit. From this speed upward, the engine operates on full vacuum advance.

As vehicle speed drops to around 18 mph, the speed sensor causes the amplifier to once again deenergize the modulator valve and the timing is retarded. Fig. 5-56.

Instead of a speed sensor, some retard-advance systems use either a mechanical or pressure switch operated by the transmission. This generally provides for retarded timing in the first two gears. When the car is shifted into high, the switch restores full vacuum advance.

DUAL THROTTLE STOP CONTROL

Another carburetor modification consists of providing two separate closed throttle positions — one for idle with the car at rest and another (throttle held open wider than for idle) for deceleration.

During deceleration, the mixture reaching the cylinders tends to be overly rich, and hydrocarbon and carbon monoxide emission levels rise sharply. By providing a special throttle stop for these periods, more air is admitted, the mixture burns more completely and emissions drop. One such system is shown in Fig. 5-57.

In A, Fig. 5-57, the car is cruising. The throttle valve is partly open and the throttle lever is free of both deceleration and idle stops. The speed sensor triggers the switching valve so that it will admit atmospheric pressure to the positioner diaphragm. The diaphragm spring then pushes up on the right end of the deceleration stop lever, causing the left end of the lever to drop down.

When the driver lets up on the accelerator pedal, B, Fig. 5-57, the throttle valve can only close until it strikes the left end of the deceleration stop lever. This holds the valve partially open during deceleration to provide the proper fuel-air mixture.

When the car is stopped or traveling at very low speeds, as in C, the speed sensor triggers the switching valve. This again causes it to admit intake manifold vacuum to the positioner. This draws the positioner diaphragm down pulling the right end of the deceleration stop lever with it. This draws the left end UP and out of the way, allowing the throttle lever to engage the idle stop and return to normal slow idle. Fig. 5-57.

SPEED SENSOR

One type of speed sensor is pictured in Fig. 5-58. It is driven by the speedometer gear. The speedometer cable is affixed to the other end of the sensor so that, as the speedometer gear whirls the sensor, it also turns the speedometer cable. The sensor produces voltage proportional to car speed. The varying voltage actuates the amplifier which, in turn, triggers a switching valve to perform such tasks as advancing or retarding the timing, moving a decelerator stop lever. Figs. 5-57 and 5-58.

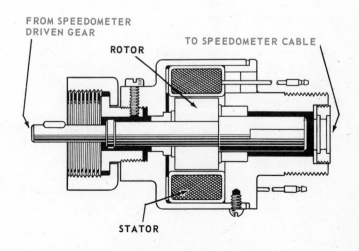

Fig. 5-58. Typical speed sensor. Sensor voltage is proportional to speed of the car.
(Toyota)

HOT ENGINE OVERRIDE BYPASS VALVE

Retarding the ignition timing tends to cause the engine to operate at a somewhat higher temperature. In the event of extremely hot weather, heavy loads, constant air conditioner use or prolonged idling in traffic, it is possible for the engine temperature to rise to excessive levels. To guard against this, many systems use a temperature operated bypass valve.

When engine temperature exceeds a predetermined level (225 deg. F. at idle in the case of the system shown in Fig. 5-59), the bypass valve routes intake manifold vacuum directly to the distributor vacuum advance. This overrides all other controls and advances the ignition timing, speeding up idle rpm. This speedup, along with the advanced timing itself, returns the engine temperature to safe levels. Fig. 5-59.

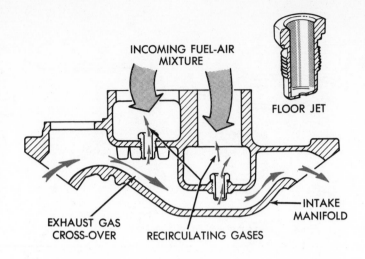

Fig. 5-60. One method of recirculating a portion of the exhaust gases. This system lowers NOx emissions. (Chrysler)

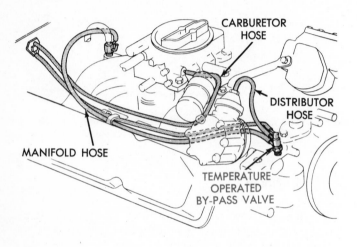

Fig. 5-59. Temperature operated bypass valve applies full vacuum advance when engine temperature exceeds a certain level. (Plymouth)

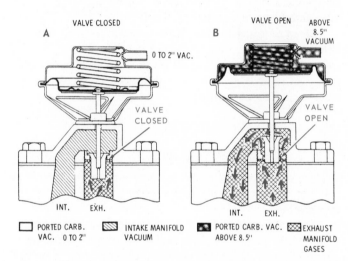

Fig. 5-61. Action of the vacuum diaphragm operated EGR valve. Note how the engine vacuum is used to open the valve in B. (Oldsmobile)

EXHAUST GAS RECIRCULATION (EGR)

When peak combustion chamber flame temperature exceeds 2,500 deg. F., the nitrogen in the air mixes with the oxygen to produce oxides of nitrogen (NOx).

By introducing a portion of the hot exhaust gases into the intake manifold, the incoming fuel charge will be diluted and flame temperatures lowered. This, in turn, reduces the level of NOx emission. Fig. 5-60 shows one method of admitting exhaust gas into the intake manifold (note how gas flows through floor jets).

The amount of exhaust gas that is recirculated can be controlled by an EGR valve such as that pictured in Fig. 5-61. This valve has a vacuum diaphragm operated valve. The vacuum tube is connected to a port in the carburetor air horn that is located just above the throttle valve in the closed or idle position. When the throttle is opened, the vacuum port is exposed and vacuum lifts the valve admitting hot exhaust gas into the intake manifold. Fig. 5-61.

A temperature sensitive control is placed between the EGR valve and the carburetor. Below certain temperatures, the EGR valve is rendered inoperative.

Generally, the EGR valve is closed (no recirculation) during engine warmup, during idle and during full throttle operation.

EVAPORATION CONTROL SYSTEM (ECS)

The Evaporation Control System is designed to prevent the release of either liquid gasoline or gasoline vapor into the atmosphere.

The system is composed basically of a closed (sealed) fuel tank, a tank fill limiter and/or a vapor separator, a charcoal canister, a purge valve and the necessary tubing.

TYPICAL EVAPORATION CONTROL SYSTEM

A typical ECS is illustrated in Fig. 5-62. Note that this system vents both fuel tank and carburetor float bowl into the charcoal canister. Study the parts and learn their names.

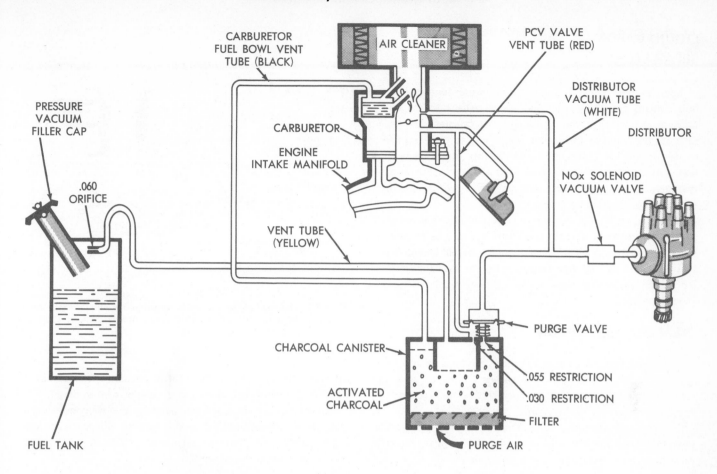

Fig. 5-62. One type of Evaporation Control System (ECS). Learn the names of the various parts.   (Chrysler)

ECS OPERATION

The fuel tank is sealed with a pressure-vacuum cap. This is a special cap, using both a pressure and a vacuum relief valve that prevents excessive pressure or vacuum in the tank in the event of system failure.

The tank "breathes" through the vent tube as fuel is consumed or when it expands and contracts from heating and cooling. Air can also enter the tank through the cap vacuum relief valve.

The tank shown in Fig. 5-62 is designed so that when it is filled to capacity, there is a sufficient air dome above the fuel to permit thermal expansion without forcing liquid gasoline into the vent lines. When filling the tank, the .060 in. orifice in one tank vent slows down the venting of the tank while filling so that an air dome is present, thus limiting fill capacity.

Any vapors in the tank travel through the vent tube to the charcoal canister. Any vapors present in the carburetor fuel bowl may also travel to the canister.

CANISTER ACTION

The canister is filled with activated charcoal granules. When the fuel vapors contact the charcoal, they are absorbed and stored.

A purge (cleanout) valve is placed at the top tank outlet.

(Some canisters do not use purge valves, but vent directly to the carburetor.) On this specific setup, two openings, one with a .030 in. restriction and the other with a .050 in. restriction, connect the purge valve with the canister interior.

With the engine off, any vapors from either the fuel tank or carburetor bowl move through the vent tubes to the bed of charcoal where they are absorbed.

When the engine is started, and kept at idle speed, engine vacuum will draw purge air (fresh air) through the canister filter. It will pass through the charcoal, through the .030 in. restriction and into the PCV (Positive Crankcase Ventilation) tube where the purge air is drawn into the cylinders for burning.

As the purge air travels through the charcoal pellets, the vapors leave the charcoal, join the airstream and is burned upon entering the cylinders.

At higher speeds, the distributor vacuum tube is exposed to carburetor air horn vacuum, which causes the purge valve to open. This exposes the larger .055 in. restriction and permits both restrictions to pass purge air to the cylinders. This action soon removes all vapors from the canister.

When the engine is stopped, any vapors that form once again move through the vent tubes into the canister where they are trapped until the engine is once again started.

Study Fig. 5-62 carefully and trace the flow of vapors from their source to the carburetor.

LIQUID-VAPOR SEPARATOR

Some evaporation systems use a liquid-vapor separator. This unit collects any liquid fuel and drains it back to the tank, preventing it from entering the charcoal canister. The separator is connected to three vent lines — one to each top front corner of the fuel tank and one to the top, center back. The rear vent standpipe in the separator has a hole near the bottom that permits any liquid fuel that may have been forced into the separator to drain back into the tank.

The liquid-vapor separator is mounted ABOVE the gas tank so that regardless of how steep a hill the car may be on or how much it may be tipped to the side, one vent will be above the fuel level and will be free to pass vapor to the canister. Fig. 5-63 pictures an evaporation control system using a liquid-vapor separator.

DIESELING OR RUN-ON

Some engines — due basically to lean mixtures, high operating temperatures and fast idle speeds — tend to continue running after the ignition key is turned off. This is commonly known as "dieseling" or "run-on."

To prevent dieseling, some engines employ an electric solenoid. When the key is on, the solenoid keeps the throttle valve open far enough for normal idle. When the key is turned

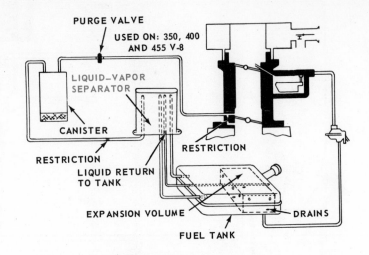

Fig. 5-63. This evaporation control system employs a liquid-vapor separator to prevent liquid fuel from being forced into the charcoal canister. (Pontiac)

off, the solenoid allows the throttle valve to COMPLETELY CLOSE. This shuts off all air and the engine cannot diesel.

Another setup automatically engages the air conditioning compressor drive clutch for several seconds whenever the key is turned off. This places the engine under a small load causing it to slow down and stop without dieseling.

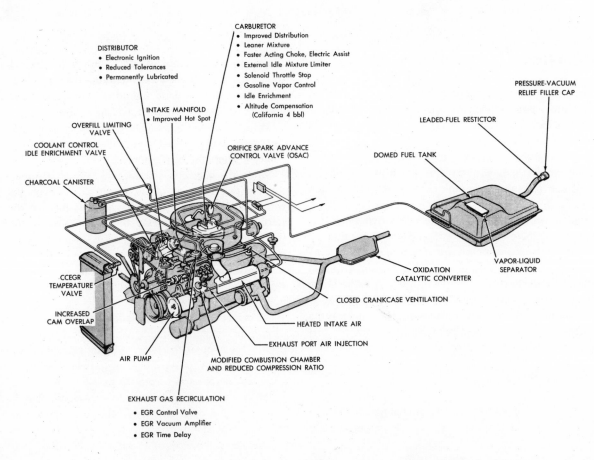

Various modifications employed to reduce exhaust emissions. Note the catalytic converter. (Plymouth)

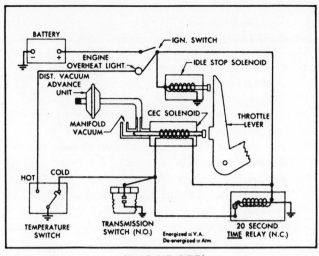

(ENGINE OFF)

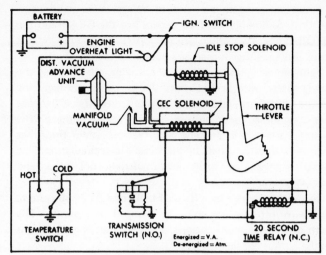

(COLD OVERRIDE AND TIME RELAY ENERGIZED)

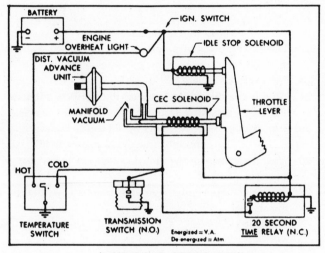

(LOW GEAR OPERATION)

(HIGH GEAR OPERATION)

Another type of emission control. It provides for retarded timing when needed, dual throttle stop control, EGR, antidiesel (idle stop solenoid) control, etc.   (Pontiac)

COMBINATION SYSTEM

Most cars incorporate a great number of the modifications mentioned to reduce emission levels to the lowest point. With a combination control system, such as that shown in Fig. 5-64, a very significant reduction in hydrocarbon (HC), oxides of nitrogen (NOx) and carbon monoxide (CO) is achieved.

Study Fig. 5-64 and see how many emission control modifications you can find.

OTHER DEVELOPMENTS

Intensive research is being conducted in all areas (engine design, engine type, combustion chamber shape, materials used, reactors, fuels, etc.) in an endeavor to either improve existing modifications or to develop additional ones.

Thermal reactors, designed to raise the temperature of the exhaust gases and to add oxygen to sustain complete burning, are being tested.

Catalytic converters (reactors) are being installed on some vehicles. Figs. 5-64 and 5-65. They contain a catalyst of pellets or a honeycomb block of ceramic material chemically treated with a metal such as platinum. Exhaust gases passing through the catalyst are converted into nitrogen, water and carbon dioxide. Some sulphuric acid mist is also produced, so the usefulness of catalytic converters is being carefully evaluated.

SUMMARY

Friction, the enemy of all engines, is controlled through the use of lubricating oil.

Lubricating oil must cool, clean, seal and lubricate. The VISCOSITY (SAE 20W, 30, 40, etc.) and the SERVICE RATING (SA, SB, SC, SD, SE) are important factors in choosing the proper oil for a specific engine.

Engine oil is drawn from a crankcase sump, through a strainer, by the oil pump. It is then pumped to all the major bearings through a network of oil galleries and tubes.

Maximum oil pressure is controlled by a pressure relief valve. The most commonly used oil pumps are the GEAR and the ROTOR types. Two others are the VANE and the PLUNGER pumps. All are either turned or actuated by the camshaft.

Upon entering the bearings, the engine oil separates the moving parts with a fine film. As it passes through the bearings, it is thrown outward in every direction. Oil drops down on some parts and is sprayed against others. The whirling parts create an oily mist that lubricates the other areas. It is then returned to the sump for recirculation. The system is ventilated to remove moisture, hot gases and excessive pressure.

Oil can be filtered by a FULL-FLOW or a BYPASS system. Even with filtering, oil must be changed periodically. The length of time between changes depends on driving conditions and the number of miles driven.

The oil pressure is determined by means of an electric or a Bourdon tube gage. The amount of oil in the sump can be checked by using the dipstick.

Crankcase hydrocarbon emission is controlled by utilizing a PCV system.

Exhaust emissions (hydrocarbons, carbon monoxide and oxides of nitrogen) are controlled by a number of modifications or systems such as higher engine temperatures, leaner fuel mixtures, faster operating chokes, recalibrated carburetors, retarded ignition timing, camshaft alterations, heated intake air, engine design, exhaust gas recirculation, air injection, fuel system evaporation control, etc. Additional research is being conducted on emission controls such as thermal reactors and catalytic reactors to further reduce emission levels.

REVIEW QUESTIONS – CHAPTER 5

1. What is meant by the term FRICTION?
2. For purposes of illustration, friction was defined or divided into two kinds. Name them.
3. What is a FROZEN bearing?
4. Name two kinds of bearings that can be used in an engine.
5. Explain how oil reduces friction.
6. Oil for an engine is produced from crude oil by a process known as _____.
7. The American Petroleum Institute has classified oil for auto engines into five types. Name them.
8. What is a detergent oil?
9. Sludge is made up of _____, _____ and _____ .
10. A good engine oil has four tasks. What are they?
11. Define the word VISCOSITY.
12. If an oil has this marking on the can, SAE 20W, what would it mean?
13. What would SAE 10W–30, mean?
14. If an oil is very thick when cold, and very thin when hot, this oil would have a _____ VI.
15. List some of the factors to be considered in determining when an oil should be changed.
16. What would an average bearing oil clearance be?
17. Name three ways of lubricating an engine.
18. How are two-cycle engines lubricated?
19. The most popular type of lubrication system is the _____ _____ system.
20. Name four types of oil pumps and explain how each one works.
21. The two most widely used types of oil pumps are the _____ and the _____ .
22. The oil pressure in a system cannot exceed a predetermined pressure. How is this pressure controlled?
23. What is an oil gallery?
24. What is the difference between the full-flow filter system and the bypass filter system?
25. Two popular filter elements would be the _____ _____ and the _____ elements.
26. What happens, in the full-flow system, when the filter becomes clogged?
27. Why is it important to provide adequate ventilation for the engine?
28. How does the engine crankcase ventilation system (PCV) work?
29. In addition to gaskets, oil _____ are placed on the front and rear of the crankshaft.
30. The oil pressure gage tells you what?
31. How do you determine how much oil there is in the oil sump?
32. Name the major bearings in the engine, and tell how each is lubricated, i.e., spray, drip, mist, or direct pressure.
33. Automobile exhaust emissions tend to add to air pollution by passing _____ , _____ _____ _____ and _____ _____ into the atmosphere.
34. List five or more modifications or systems used to help reduce emission levels.

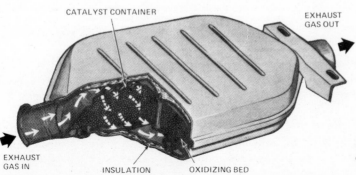

Fig. 5-65. One type of catalytic converter. The oxidizing bed consists of ceramic pellets coated with platinum and palladium. (Oldsmobile)

CATALYST CONTAINER

EXHAUST GAS OUT

EXHAUST GAS IN

INSULATION

OXIDIZING BED

Chapter 6

FUEL SYSTEMS

As you have learned in a preceding chapter, an engine uses a mixture of gasoline and air for its fuel. This mixture is drawn into the engine by the vacuum formed during the intake stroke. It is then compressed, fired and exhausted. This cycle is repeated over and over in each cylinder or combustion chamber.

You will recall that gasoline is derived from crude oil or petroleum, and that it is made up of hydrocarbons (hydrogen and carbon). Gasoline also has various additives designed to prevent rust, resist freezing, control vapor lock and resist detonation.

MIXTURE PROPORTIONS

The proportion of gasoline and air necessary to secure proper combustion will vary according to engine speed, load, temperature and design of the engine. An average mixture for a moderate cruising speed could be 15 to 17 parts of air to one part of gasoline. These proportions are based on weight. For a mixture that produces more power, 12 to 13 parts of air to one of gasoline would be average.

If you were to consider VOLUME instead of WEIGHT, you will find that one gallon of gasoline would require around 9,000 gallons of air (for part throttle operation). Fig. 6-1.

CHEMICAL CHANGES

Oxygen in the air is what is used for combustion. Nitrogen is not used. Air is about 21 percent oxygen and 79 percent nitrogen. Other elements are present but in small amounts.

The engine draws in air (nitrogen and oxygen) and gasoline (hydrogen and carbon). When this mixture is ignited and burned, it produces several things. For each gallon of gasoline that is burned, about one gallon of water (steam) is produced. Other products include carbon dioxide, water, hydrogen, nitrogen, carbon monoxide (deadly gas), hydrocarbons (unburned gasoline), oxides of nitrogen, hydrochloric acid, lead oxides and sulfurous acid. Fig. 6-2.

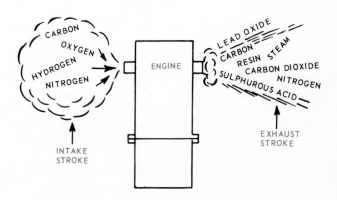

Fig. 6-2. Combustion brings about chemical changes.

The exhaust stroke rids the engine of most of the combustion products. As you learned, some slip by the rings and enter the crankcase. Unless disposed of, these can raise havoc with the engine.

A DIFFICULT TASK

It is the job of the fuel system to constantly deliver the

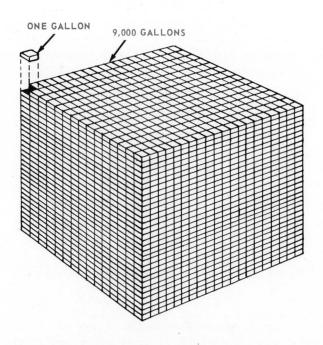

Fig. 6-1. Ratio of gasoline to air, by volume. If this were 9,000 one gallon cans, 8,999 would be filled with air and 1 with gasoline. This is an average fuel mixture ratio.

proper air-fuel mixture to the engine. The mixture and amount must be variable to meet the ever changing needs of the engine.

The basic key to an adequate fuel system is the carburetor itself. Here, the actual mixing of the fuel and air takes place. There are other parts, too, and all play an important role. For a complete understanding of the system and the role of each part, it is perhaps best to start with the source of supply — the gasoline tank.

GASOLINE TANK

Gasoline tanks are located in just about every conceivable spot — in the rear, under the seat, in the front, etc. An ideal location is a spot where the tank is protected from flying stones, where it will not bottom on the road and where, in event of an accident, it will not burst. The overall design features of a car often dictate the location of the tank.

The tank is usually constructed of thin sheet steel, coated with a lead-tin alloy to protect against rusting. Capacity generally is around 18-26 U. S. gallons. Internal baffles are arranged to prevent sloshing. A filler neck, reaching to a convenient spot, is attached either directly or by a neoprene hose. The filler cap normally seals the tank but does have a vacuum and pressure relief valve to allow air to either enter or leave the tank in the event the emission control vent lines were to become inoperative. The tank is vented to a charcoal canister (see section on emission controls — Chapter 5) that in turn is vented to the carburetor. The tank must be vented to allow fuel to be drawn out. The tank is securely strapped into place. It is often protected on the outside by an application of undercoating material. A drain plug is usually provided at the bottom. Fig. 6-3.

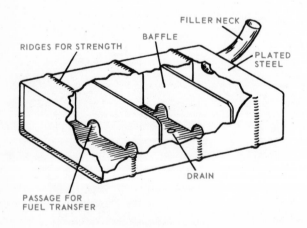

Fig. 6-3. Construction of a typical gasoline tank.

CAUTION:

A GASOLINE TANK, EVEN WHEN APPARENTLY EMPTY, IS AS DANGEROUS AS A BOMB! NEVER WELD, CUT, SOLDER OR BRAZE A GASOLINE TANK UNTIL YOU HAVE BEEN THOROUGHLY TRAINED IN THE SAFETY PRECAUTIONS NECESSARY. IN FACT, KEEP ALL SPARKS, HEAT AND OPEN FLAMES AWAY FROM THE TANK. REMEMBER — IT CAN KILL YOU!

GASOLINE PICKUP PIPE

A tube enters the tank to draw off fuel for the engine. This tube can enter from the top, side or bottom. The tube end is generally located about one-half inch from the tank bottom. This allows considerable water and sediment to form before being drawn into the pickup tube.

Several types of filters are used on the ends of gasoline pickup tubes, including plastic, screen and sintered bronze (small particles of bronze pressed together, leaving the mass porous). An effort is made to design screens that will not clog and yet filter out sediment and most water, Fig. 6-4.

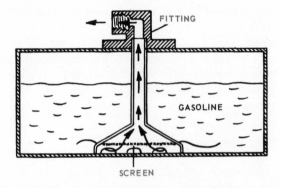

Fig. 6-4. Gasoline pickup pipe. Notice that screen is above bottom of tank. Pickup tube is fastened to bottom of tank.

WATER FORMS IN TANK

When a gasoline tank is only partially full, water tends to form within the tank. The moisture in the air within the tank condenses on the tank walls, then runs under the gasoline.

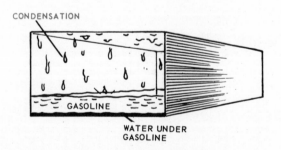

Fig. 6-5. Gasoline tank showing condensation when tank fuel level is low.

Since water is heavier than gasoline, it will remain at the bottom where it will fail to evaporate as it is not exposed to the air. This contamination of the gasoline through condensation is a slow process but can cause trouble. Fig. 6-5.

TANK GAGE UNIT

The gasoline tank contains a float that is attached to a resistance device quite similar to that used in an electric oil pressure sender. In this case, the travel of the float, up and down, causes the resistance to vary, depending on the depth to which the tank is filled.

The gage on the instrument panel may be a double coil and armature type (described in the Lubrication chapter), or it can use a bimetal receiver.

Most gages are designed to read empty when one or two gallons of gasoline are left in the tank. This is a concession to human nature as people are inclined to drive until the tank does read empty. Figs. 6-6 and 6-7 illustrate a typical tank sender and bimetal receiver. When the float drops, current flow to ground through the bimetal wire coil will be less,

because it must travel through more resistance wire. This cools the bimetal hairpin and pulls it together. When the tank is filled, the contact slides up, cutting out resistance and current flow increases. This heats the hairpin and the ends separate, causing the needle to move toward the full mark.

HOW GASOLINE LEAVES THE TANK

Fuel tanks usually are lower than the carburetor, so the gasoline will not flow under its own power. Some method of drawing it from the tank to the carburetor must be used. Two types of pumps are commonly used. One is mechanically operated, the other electrically.

MECHANICAL FUEL PUMP

A mechanical pump is mounted on the engine and is operated by the camshaft. Basically it consists of an air chamber divided in the center by a flexible diaphragm. The diaphragm is made of several layers of cloth, impregnated (filled) with a special compound that makes it airtight and resistant to petroleum products. The diaphragm is pinched between the halves by a series of machine screws that hold the pump halves together. Fig. 6-8.

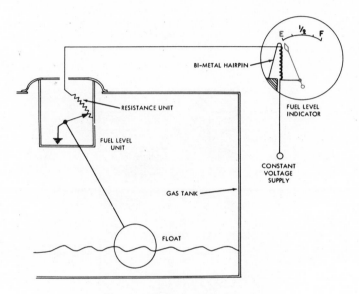

Fig. 6-6. Fuel gage operation with tank empty. (Plymouth)

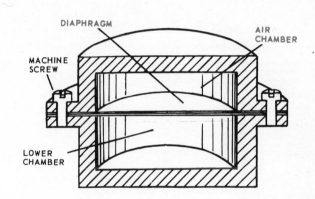

Fig. 6-8. Air chamber and diaphragm.

CHECK VALVES

Two check valves, one inlet and one outlet, are now added to the top chamber. The inlet is connected to the fuel tank and the outlet to the carburetor via copper, steel or plastic tubing. These checks use flat washer type valves held with a fine spring. The valve and the spring are contained in a perforated (full of holes) housing. Fig. 6-9.

PULL ROD AND RETURN SPRING

A pull rod is now fastened to the diaphragm. A large, cupped edge washer is placed on the top and bottom of the diaphragm, and the pull rod is riveted to the center. A fairly stiff return spring pushes upward, thus flexing the diaphragm in an upward direction. Fig. 6-10.

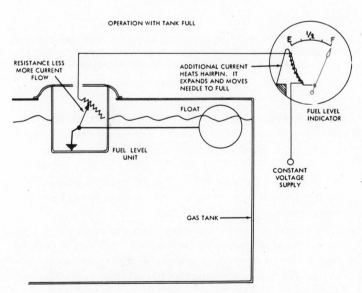

Fig. 6-7. Fuel gage operation with tank full.

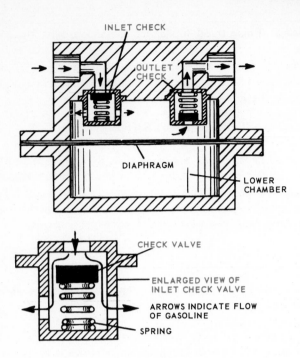

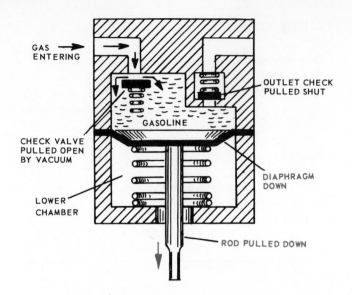

Fig. 6-11. Intake stroke. When pull rod is drawn down, diaphragm is flexed downward. This forms vacuum in air chamber. Intake check will open and gasoline will be drawn into pump.

Fig. 6-9. Check valves added to top chamber. Small spring holds valve shut until vacuum or pressure overcomes spring tension.

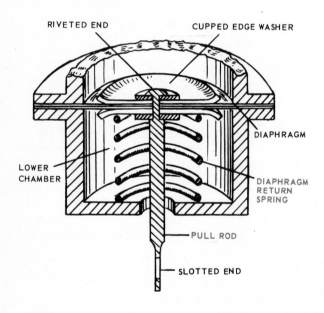

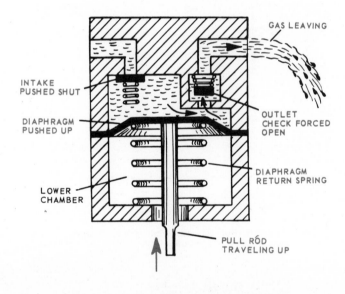

Fig. 6-12. Outlet stroke. When pull rod is released, diaphragm is pushed up by return spring. This creates pressure, forcing inlet valve shut and opening outlet. Gasoline will be forced from outlet.

Fig. 6-10. Pull rod and return spring added to diaphragm.

HOW DOES IT WORK?

If you were to connect the inlet of a fuel pump to a container of gasoline, then run the outlet to another container, the pump could be worked by pulling down on the pull rod with your fingers. Figs. 6-11 and 6-12.

ROCKER ARM AND LINK

In order to pull the diaphragm down, it is necessary to install a link and rocker arm. The link hooks in the slotted end

of the pull rod and the rocker arm extends out where it rides on a cam on the camshaft. The link and rocker arm are pivoted together so when the cam lobe pushes up on the rocker arm, it pulls the link and pull rod down. When the cam turns, the rocker follows the lobe down. The link and pull rod are raised by the spring — not the rocker arm. A small spring keeps the rocker firmly against the cam lobe.

It is necessary to extend metal from the lower chamber to form a housing for the rocker and link, and also to prevent engine oil from running out where the rocker enters the engine. The fuel pump mounting flange is part of this housing. Fig. 6-13.

You will notice in Fig. 6-13 that a seal has been placed

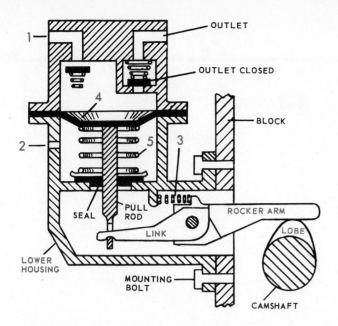

Fig. 6-13. Lower housing, rocker and link added. Notice that cam lobe has raised rocker arm, pulling diaphragm 4 down. This opens inlet pulling in fuel. When cam lobe turns, spring 3 will force rocker to follow. Diaphragm spring will push diaphragm up, close inlet, and expel gasoline through outlet. Vent 2 is provided so diaphragm can flex without compressing air in lower chamber. Seal keeps engine oil from entering lower chamber. All link and rocker rubbing surfaces are hardened to retard wear.

around the pull rod. This keeps hot engine oil and fumes from entering the lower chamber. A vent is provided in the lower chamber so the diaphragm can be pulled down without compressing the air in the chamber. Current practice is toward

the use of a solid rocker arm as shown in Fig. 6-14, instead of an articulated link. Note that when the diaphragm is held down by fuel pressure in the carburetor, the rocker arm can slide up and down on the diaphragm pull rod.

COMPLETE BASIC FUEL PUMP

Fig. 6-13 shows the entire basic fuel pump. Study the parts, learn their names and functions. As illustrated, this fuel pump will move gasoline efficiently but will not filter nor dampen the pulsating movement of gasoline leaving the outlet.

FUEL PUMP EFFICIENCY MEASUREMENTS

The amount of pressure the fuel pump exerts on the fuel line to the carburetor depends on the strength of the diaphragm spring. Pump pressures can be designed in by the manufacturer according to need. An average would be around 4 to 6 lbs. pressure. The normal pump will discharge over a quart per minute.

FILTERS ARE ESSENTIAL

The tank pickup screen will not filter out the very fine particles. It is necessary to have an additional filter either between the tank and the pump or between the pump and the carburetor.

Some pumps have a filter built as an integral part of the assembly. The fuel pump illustrated in Fig. 6-14 uses a screw-in, nonserviceable pleated paper fuel filter. Fig. 6-15 shows an exploded view of the same pump. Note the solid rocker arm. Note also the vapor return valve that allows a

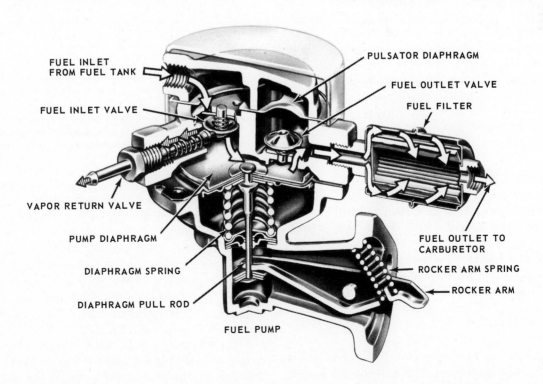

Fig. 6-14. Cutaway of a typical fuel pump. Note valve action, paper filter and vapor return or discharge valve. (Ford)

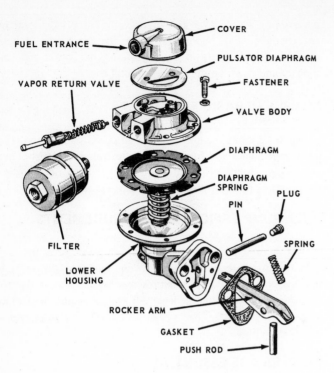

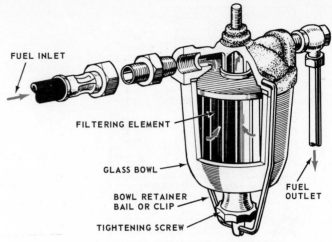

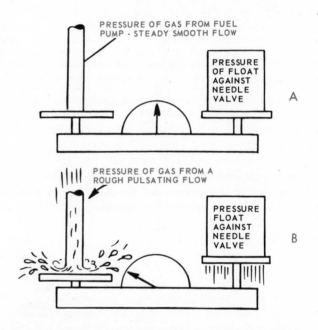

Fig. 6-15. Exploded view of fuel pump pictured in Fig. 6-14. Study relationship of various parts. (Ford)

Fig. 6-16. Cutaway of a fuel filter. This filter uses a glass sediment or settling bowl and a paper filtering element. (Toyota)

Fig. 6-17. Pulsation effect. Notice how float pressure controls flow of gasoline in A. In B, fuel is being delivered in a jerky manner, called PULSATION. See how it slams against the scale and upsets needle pressure.

certain amount of vapor and gasoline to return to the tank through a separate line. This prevents vapor lock during periods of high underhood temperature.

TYPES OF FUEL FILTERS

Many materials have been used in an attempt to find an ideal fuel filter. A fine screen, metal plate (cuno), ceramic, treated paper and sintered bronze elements have all been used. All filters need periodic cleaning or replacement. Fig. 6-16 illustrates an inline (placed in the fuel line) filter. This particular type uses a glass settling bowl (to hold dirt, water, etc.) along with a replaceable cleaning element. All fuel moving to the carburetor must pass through the cleaning element. A special pleated paper is widely used for cleaning elements. Fig. 6-16.

FUEL PULSATION

A fuel pump delivers fuel to the carburetor in little spurts. Even though it forms a solid stream, the pressure pulsates (rises and falls) as the pump operates. This moving column of gasoline can create a slamming effect against the float needle valve, causing it to come off its seat and flood the carburetor. Fig. 6-17 illustrates this effect.

PULSATION DAMPER

To smooth out the flow of gasoline, many systems are equipped with pulsation dampers. Such a damper can be built into the pump or mounted in the line as a separate unit.

A pulsation damper consists of a flexible diaphragm stretched across an air compartment. The trapped air acts as a spring against which the diaphragm can flex. Some pulsation devices use a spring.

When the pump ejects fuel into the line, instead of ramming the fuel column, the ejected gasoline bluges the diaphragm outward, absorbing the blow. While the pump is getting ready to eject another stream, the diaphragm will return to its normal position to continue the pressure. In this manner, the fuel flow is smoothed out. Fig. 6-18.

Fig. 6-19 shows a fuel pump with an integral pulsation damper. A filter is also attached. Notice the use of a magnet in the filter to trap any metallic particles. Figs. 6-14, 6-15 and 6-22 also show a pulsation damper setup.

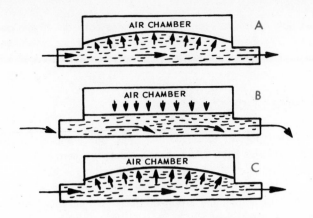

Fig. 6-18. Pulsation dampener principle. A—Pump ejects fuel. Diaphragm pushes up against trapped air in air chamber. This absorbs slamming effect. B—No fuel from pump. Trapped air pushes diaphragm down, continuing some pressure during lull. C—Pump ejects another spurt. Again it is absorbed by diaphragm.

ELECTRIC FUEL PUMP

An electric fuel pump is quite efficient. It has a couple of advantages not offered by the mechanical type pump.

The electric pump will fill the carburetor merely by turning on the key. Another feature is the electric pump's adaptability to most any location. This allows the pump to be mounted away from the heat of the engine to reduce the chance of vapor lock. Some electric pumps are run submerged inside the fuel tank.

The basic difference between the electric and the mechanical pump is that the electric pump uses an electromagnet (a magnet produced by electricity flowing through a coil) to operate a metal bellows that alternately forms a vacuum and then pressure. Some electric pumps use an electromagnet to work a regular diaphragm. Other models drive either a vane or impeller type pump with a small electric motor.

When the key is turned on, the electromagnet is energized (filled with electricity by passing an electric current through it) and attracts a metal armature that is attached to the bellows, or to a diaphragm. This stretches the bellows and forms a vacuum. Gasoline is drawn in through the inlet check valve.

As the armature continues downward, it strikes a pair of contact points and opens them. This breaks the circuit to the electromagnet and it releases its pull on the bellows. A return spring shoves the armature and bellows up. As the bellows close, gasoline is forced out the outlet check. When the armature has traveled up far enough to expel the gasoline, the points close and it is drawn down again. This process is repeated over and over.

As in the case of the mechanical fuel pump, when the carburetor float bowl is full, the return spring cannot collapse the bellows and the pump will be inoperative until some of the gasoline has been drawn from the bowl.

An electric fuel pump, using a bellows to produce the pumping action, is shown in Fig. 6-20. The diaphragm type (diaphragm instead of bellows) and the centrifugal type electric pumps are also used.

SERVICEABLE AND NONSERVICEABLE FUEL PUMPS

Many makers are using a nonserviceable fuel pump. Note in Fig. 6-21, that the pump body and cover are crimped together. If the pump is defective, it is discarded.

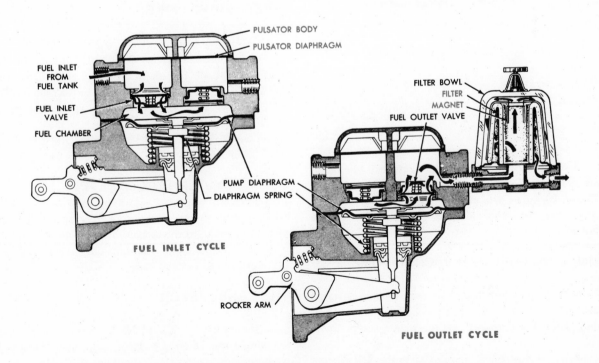

Fig. 6-19. Pump with pulsation damper. Notice magnet, in addition to ceramic filter, to trap any metal particles. (Lincoln)

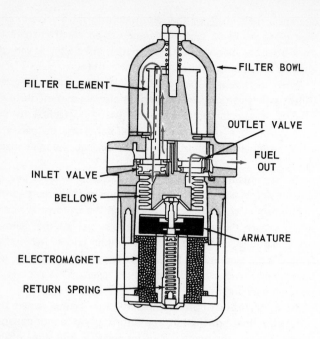

Fig. 6-20. Electric fuel pump utilizing a bellows to produce the pumping action. (John Deere)

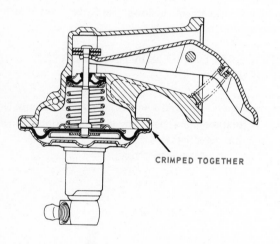

Fig. 6-21. Cutaway of a nonserviceable fuel pump. Note how lower housing is crimped over upper section. (Plymouth)

A serviceable (may be taken apart and repaired) fuel pump is shown in Fig. 6-15. Note the various parts making up the pump. Fig. 6-19 shows a serviceable pump in cutaway form. Study fuel flow. Fig. 6-14 illustrates a similiar pump. Note pulsator diaphragm.

COMBINATION FUEL AND VACUUM PUMP

For a number of years, many cars used vacuum-operated windshield wipers. When the engine was pulling hard, such as in passing, the vacuum dropped and the wipers would almost stop.

To overcome this problem, a combination fuel and vacuum pump was used. It is similiar to the regular mechanical fuel pump and the two are placed back to back in one main

assembly. A single rocker arm operates both diaphragm pull rods. One diaphragm pumps fuel, the other air, forming a vacuum to keep the wipers running when engine vacuum is low.

VAPOR LOCK

It is important to shield fuel lines, pump and carburetor from excessive heat. When gasoline reaches a certain temperature, it will turn to a vapor. Depending on the amount and location of the vaporization, the fuel flow can either be reduced or completely stopped. Upon cooling, fuel flow will resume at the normal rate.

Baffles or shields in the form of metal plates, asbestos gaskets and covering for lines made of asbestos tubing are used to prevent vapor lock. Gasoline itself is controlled to eliminate this tendency as much as possible. A fuel pump vapor discharge (return) valve, Fig. 6-14, can be incorporated to prevent vapor lock when underhood temperatures are high. The vapors return, via a separate line, to the gasoline tank where they condense and rejoin the liquid fuel. Cars equipped with air conditioning often utilize vapor return devices to offset the danger of vapor lock due to increased underhood temperatures.

FUEL LINES

Gasoline is carried from the tank to the carburetor in plated steel, copper or, in some cases, plastic lines. When plastic is used, it is often from the pump to the carburetor. All joints must be tight. All lines should be secured to prevent destructive vibration and rubbing.

FIRE! FIRE! FIRE!

BE THOROUGH IN YOUR WORK. A LOOSE CONNECTION OR UNSECURED GASOLINE LINE CAN COME LOOSE, BREAK OR LEAK. IF THIS HAPPENS, A GREAT DEAL OF GASOLINE CAN BE SPRAYED OVER THE ENGINE. IF IGNITED, THIS GAS WILL CAUSE A DANGEROUS FIRE.

ALWAYS HAVE A FIRE EXTINGUISHER HANDY IN THE SHOP AND CARRY ONE IN YOUR VEHICLE!

SUPPLY SYSTEM IS COMPLETE

At this point you are able to supply a steady, smooth flow of clean gasoline to the carburetor. How this gasoline is mixed with air, in the correct proportions, and delivered to the cylinders, will be covered next.

THE CARBURETOR

In the early days of automobiles, the carburetor was called a mixing valve. This was a rather appropriate name in that its main function was plain mixing of gasoline and air.

The modern carburetor, although it still mixes gasoline and air, has become a highly developed and complex unit. There

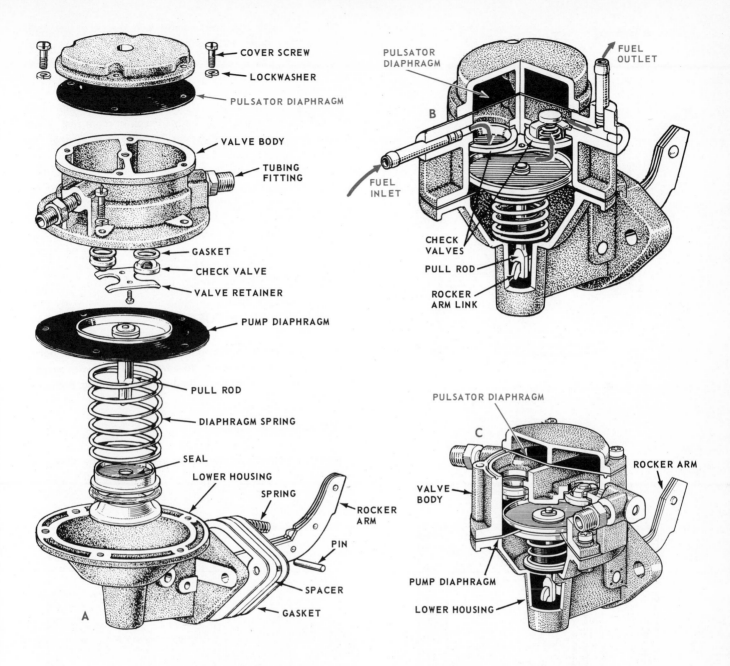

Fig. 6-22. Typical serviceable fuel pump construction. Learn part names. Study fuel flow in B. (Nissan)

are many types and sizes; however, they all share the same basic concepts.

BUILD A CARBURETOR

As you have done with an engine, you will now "build" a basic carburetor on paper. You have already learned that the gasoline must be mixed with air. The proportion should be anywhere from 9 to 1 up to 16 to 1 by weight, depending on engine temperature, speed, load and design.

A rapidly moving stream of air enters each cylinder on the intake stroke. You will use this airstream to make your carburetor work.

Each intake port feeds into the intake manifold. The intake manifold runs from the ports to the carburetor. More on the manifold later.

AIR HORN

You will start with a plain round metal tube. This tube is called the air horn. The tube will have a flange that bolts to the intake manifold. Fig. 6-23.

FUEL BOWL

A fuel bowl is attached to the air horn to hold a supply of gasoline. Inside of the bowl, a float will be hinged. This float will control a tapered pin (needle valve) that will open and

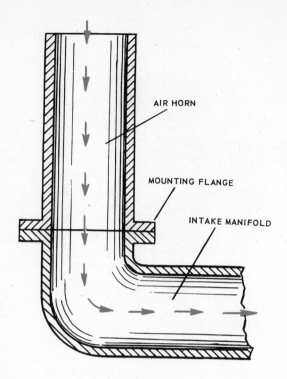

Fig. 6-23. Air horn. Arrows indicate flow of air through horn, into manifold and on into cylinders.

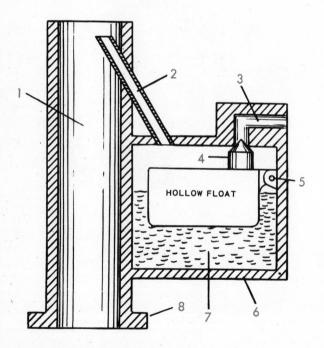

Fig. 6-24. Float bowl, float, needle valve and bowl vent added to carburetor air horn. Notice how float has pushed needle valve into its seat to shut off flow of gasoline 1—Air horn. 2—Bowl vent or balance tube. 3—Fuel inlet. 4—Needle valve. 5—Hinge. 6—Float bowl. 7—Gasoline. 8—Flange.

enters. The float and needle valve will keep the fuel level constant. The bowl must have a hole (vent) to allow air to enter or leave as the fuel level changes. The bowl may be vented to the air horn, Fig. 6-24, to the atmosphere, or to both.

If the air cleaner becomes clogged, a stronger vacuum will result in the air horn. This tends to suck in more fuel unless the bowl is vented into the air horn. If it is vented, the pressure on the fuel in the bowl will decrease in the same proportion and fuel intake will tend to remain constant. This type is called a balanced carburetor. A common setup employs the internal bowl vent plus an external vent that opens during periods of engine idle. This type of external vent helps prevent hot (engine overheated) starting problems by venting fuel vapors, to the atmosphere, that would otherwise be forced into the carburetor air horn. The external vents are now channeled to the emission control charcoal canister.

MAIN DISCHARGE TUBE

In order to allow the gasoline to be drawn into the air horn, it is necessary to connect the fuel bowl and horn with a small tube. This tube is called the main discharge tube. Fig. 6-25.

A VACUUM IS NEEDED

Notice in Fig. 6-25, that the gasoline level in the discharge tube is slightly below the nozzle. In order to get the gasoline

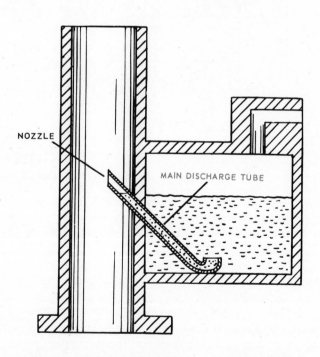

Fig. 6-25. Main discharge tube. Note fuel level in discharge tube.

close to admit gasoline. When the bowl fills, the float will rise and shove the needle valve against its seat to stop the flow of gasoline. When the fuel level drops, the float releases its pressure, the needle valve lifts from the seat and more gasoline

to flow out the nozzle and mix with the airstream, a vacuum is necessary. Some vacuum created by the moving air, but it is not strong enough.

VENTURI

By building a restriction (venturi) in the air horn, the air will be forced to speed up where it strikes the venturi. As the air suddenly speeds up, it tends to be "stretched." In other words, a vacuum is created. The more the air speeds up, the stronger the vacuum. This vacuum will be strongest in the throat of the venturi but will still exist for a short distance below it.

Fig. 6-26 illustrates an air horn with a venturi. Three tubes containing mercury enter through the horn at different points. Notice the variations in vacuum at different spots in the horn. The higher the air velocity, the more vacuum formed. This is evident by examining the distance the atmospheric pressure has pushed the mercury up the tube. Fig. 6-26.

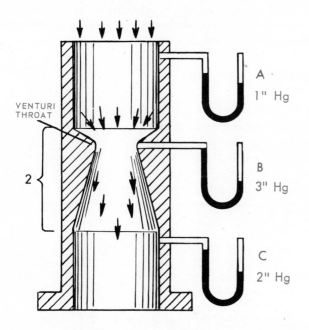

Fig. 6-26. Venturi increases vacuum. A—Air rushing into air horn near top produces some vacuum. B—As it travels down into venturi throat it speeds up. This causes an increase in vacuum. C—After leaving venturi throat, air expands and slows down somewhat. This reduces vacuum.

SECONDARY VENTURI

The vacuum can be increased still more by constructing a smaller venturi in the center of the air horn and positioned so that its outlet end is even with the throat of the main (primary) venturi. The primary venturi forms a vacuum at the outlet of the secondary venturi, causing the air to really whip through the secondary venturi. This produces a sizable increase in vacuum. The secondary venturi is sometimes called a "booster" venturi. Some carburetors use a third booster venturi just above the secondary.

The main discharge tube nozzle enters the secondary venturi. When the engine is started, the rushing air draws off drops of gasoline where they will join the passing air. See Fig. 6-27.

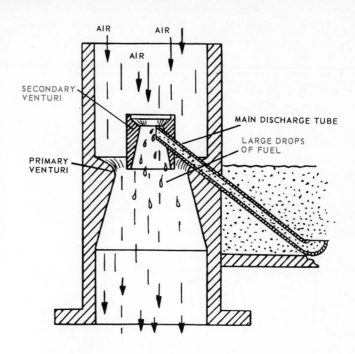

Fig. 6-27. Secondary venturi. This increases vacuum. Note large drops of fuel being drawn from nozzle. An air bleed is needed to break up drops.

AIR BLEED

The setup so far would administer gasoline to the air, but the drops would be rather large and would tend to stick to the nozzle. By adding a tiny stream of air to the fuel as it travels through the nozzle, the drops will leave readily and will be quite small. This added air is termed "air bleeding." The tube through which the air travels is called an air bleed passage. See Fig. 6-28.

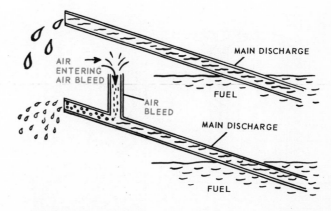

Fig. 6-28. Air bleed principle. Notice large drops drawn out of top tube. By adding an air bleed as shown below, droplets are much smaller.

JETS

There are numerous jets used in a modern carburetor. Their job is to control the passage of gasoline and air through a very carefully calculated size hole. The jets are constructed of brass and screw into the ends of the various passageways in the carburetor. Some of the permanent types are pressed into place and are not removed for routine maintenance. A jet is placed at the lower end of the main discharge tube to control the amount of gasoline that can pass through the tube. See Fig. 6-29.

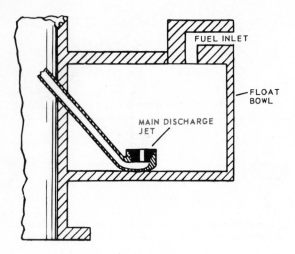

Fig. 6-29. Main discharge jet. This jet controls amount of fuel entering discharge tube.

ENGINE WILL NOT START

If you were to try to start an engine when cold, the carburetor as built so far would not work. A cold engine needs a mixture rich with gasoline (8 or 9 to 1). Our main discharge jet will furnish a mixture suitable for running. For starting only, the mixture must be much richer.

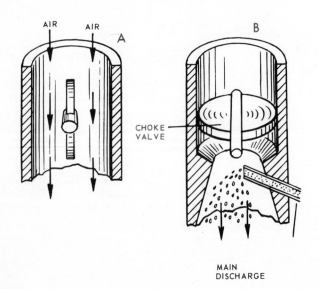

Fig. 6-30. Choke valve. A—Choke valve open. Air passing through air horn. B—Choke valve closed. Vacuum from intake drawing on discharge nozzle. Note heavy amount of fuel being delivered.

CHOKE IS NEEDED

If a choke valve (sometimes called a "butterfly" valve) is placed near the top of the air horn, it will be possible to close it and shut off the air supply. As the intake stroke forms a vacuum that cannot be relieved through the air horn, the vacuum will pull on the main discharge nozzle and draw out a large quantity of gasoline. This provides the proper starting mixture. The choke valve can be opened or closed either manually or automatically. More on this later. Fig. 6-30.

ENGINE STARTS BUT RUNS WILD

The engine will now start by closing the choke. As soon as it starts, the choke will have to be partially opened. When the engine is hot, the choke should be fully opened.

There is one trouble yet, and that is the engine will run "wild" or wide open. We have not provided a way to control the speed. A throttle valve is needed.

THROTTLE VALVE

A throttle valve is similar to a choke valve but is placed between the bottom of the primary venturi and the mounting flange. This will give us control of the amount of air-fuel mixture reaching the cylinders. It is controlled by linkage connected to the accelerator pedal. Fig. 6-31 illustrates a throttle valve, and Fig. 6-32 shows the necessary linkage.

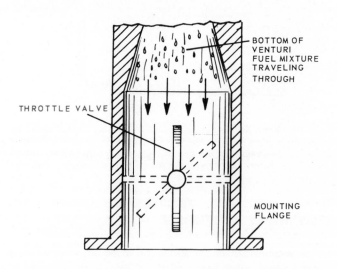

Fig. 6-31. Throttle valve. Throttle valve controls flow of fuel mixture. Shown in wide open, half open, and closed positions.

BASIC CARBURETOR IS COMPLETE

Study the basic carburetor in Fig. 6-33. Notice the position of the choke and throttle valves. Study the venturi design and notice the relationship of the secondary to the primary venturi.

A basic carburetor as shown could be used to operate an

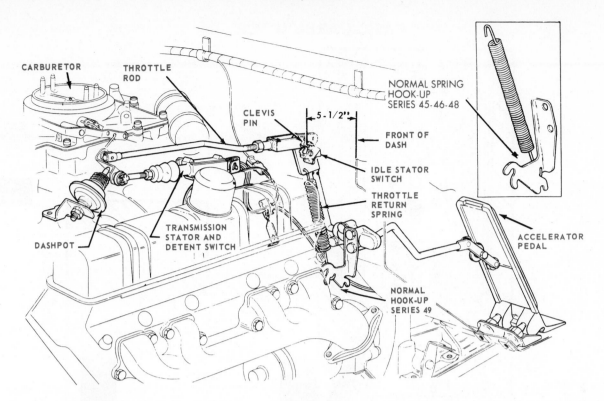

Fig. 6-32. One type of carburetor throttle control linkage. Linkage arrangements vary widely. (Buick)

engine. The choke is open (engine hot), the throttle at about one-half speed, and the main discharge nozzle is metering fuel into the airstream. If the main discharge jet was of the correct size, the engine would run efficiently as long as this certain engine speed was maintained.

MANY FAULTS

The basic carburetor, however, has many faults. The engine will not idle without stalling. The engine will have little power at speeds just above an idle. Whenever the throttle valve is quickly opened to full throttle, the engine will have a flat spot (engine seems to die, power fades and for a brief second there is no response). The engine will then catch and pull smoothly. The engine will not get sufficient gasoline to operate at top speed with full power. The choke must be operated by the driver instead of automatically. The carburetor would be subject to vapor lock. There is no air bleed to the main discharge tube.

This basic carburetor would look rather hopeless. As it stands, it is; but it is the foundation upon which all carburetors are built. With further developments, it will operate quite satisfactorily.

WHAT ARE THE DEVELOPMENTS?

To function at all speeds under varying conditions of heat, load, etc., the carburetor must have a number of basic CIRCUITS, or SYSTEMS. The circuits actually are passageways in the carburetor designed to perform certain functions.

Some mechanical parts in addition to the passageways are used. When all the necessary circuits are incorporated, the carburetor becomes efficient.

CARBURETOR CIRCUITS — WHAT THEY DO

The job of the various carburetor circuits is to produce the proper air-fuel ratio to meet all possible needs of the engine. For example: when starting, ratios can be 8—1; when idling 12—1; part throttle (30 to 65 mph) 16—1; full throttle 13—1; and upon full acceleration 12—1. These ratios are merely averages and will vary from engine to engine.

The circuits can be listed as:
1. Idle circuit.
2. Low speed circuit.
3. High speed circuit.
4. High speed full power circuit.
5. Acceleration circuit.
6. Choke circuit.
7. Float circuit.

There are other devices that are used in the construction of carburetors, but they are not classified as circuits. These will be discussed in succeeding pages.

HOW CIRCUITS WORK

Thinking back to the faults of the basic carburetor, let us see how each circuit overcomes a specific problem or fault. The problem will be given first, followed by the manner in which a particular circuit functions.

131

BASIC CARBURETOR

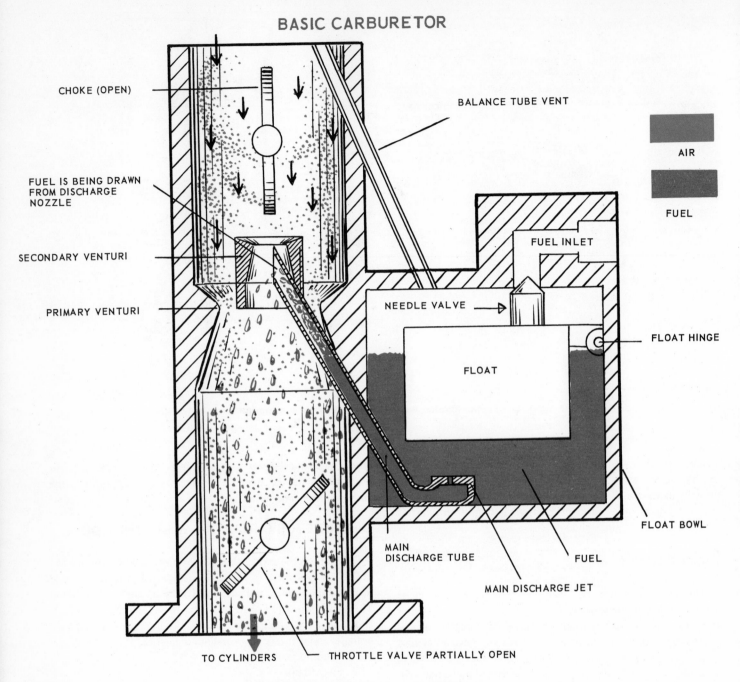

CHOKE (OPEN)

BALANCE TUBE VENT

AIR

FUEL

FUEL IS BEING DRAWN
FROM DISCHARGE
NOZZLE

FUEL INLET

SECONDARY VENTURI

NEEDLE VALVE

FLOAT HINGE

PRIMARY VENTURI

FLOAT

FLOAT BOWL

MAIN
DISCHARGE TUBE

FUEL

MAIN DISCHARGE JET

TO CYLINDERS

THROTTLE VALVE PARTIALLY OPEN

AUTOMOBILE FUEL SYSTEM

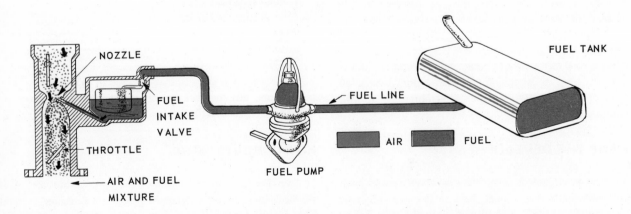

NOZZLE

FUEL TANK

FUEL
INTAKE
VALVE

FUEL LINE

THROTTLE

AIR

FUEL

AIR AND FUEL
MIXTURE

FUEL PUMP

Fig. 6-33. Above. Basic carburetor. Below. Automobile fuel supply system.

PROBLEM — ENGINE WILL NOT IDLE

When the throttle valve is closed to slow the engine to idle speed, the main discharge nozzle stops supplying fuels, and the engine will stall. The nozzle fails to feed gasoline because, when the throttle is closed, the rush of air through the air horn is cut off.

SOLUTION — IDLE CIRCUIT

Notice in Fig. 6-34, that an additional passage has been added. It branches off the main discharge tube, runs up through the low speed jet, across, down through the economizer jet and on to the idle port. From the idle port it runs down to the idle screw port, where it enters the air horn slightly beneath the throttle valve.

Since there is no vacuum above the throttle but a strong vacuum below, the vacuum will pull at the idle screw port.

This vacuum will draw fuel from the bowl up through the low speed jet, which calibrates or controls the amount that passes. As the fuel is being pulled toward the economizer, the vacuum is also drawing air in at the bypass and air bleed openings. The fuel is thoroughly mixed with this air and travels down through the economizer, which further meters the flow and tends to mix it better. The fuel and air proportions are controlled by the size of the air bleed, bypass and low speed jet openings.

Moving down through the passage and the idle port, the mixture arrives at the idle screw port. The amount that can be drawn into the engine is determined by the position of the idle adjustment screw. When screwed in, it reduces the amount and when backed out, it admits larger quantities. To adjust, the idle screw is turned in until the engine misses and then turned out until the engine "rolls." The screw is then set halfway between these two positions. The engine will now idle smoothly. This is the IDLE CIRCUIT.

PROBLEM — ENGINE WILL NOT OPERATE AT PART THROTTLE

As soon as the throttle valve is opened past the idle position, more fuel mixture is necessary. The idle screw port will not furnish enough and airflow through the venturi is not strong enough to cause the fuel to be drawn from the main discharge nozzle. The engine will not operate properly in this range.

SOLUTION — LOW SPEED CIRCUIT

Going back to Fig. 6-34, you will see a slot-shaped opening just above the throttle. This is called the idle port. It opens into the idle fuel passage.

When the throttle valve starts to open, this port is uncovered and it will begin to supply fuel mixture to the engine. This is in addition to what is already supplied through the idle screw port.

This additional mixture will furnish enough extra fuel to allow the engine speed to increase to the point where the main

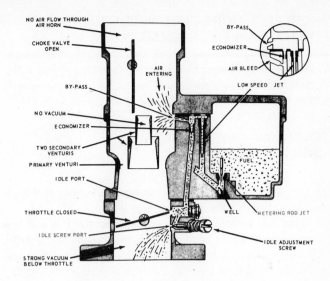

Fig. 6-34. Carburetor idle circuit. Follow flow of fuel from metering rod jet to idle screw port. Study position of choke and throttle and note that there is no vacuum above throttle. (Carter Carburetor)

discharge nozzle will function. Instead of the slot-shaped opening, some carburetors use a series of small round holes. As the throttle opening grows wider, it keeps uncovering extra holes. This type is illustrated in Fig. 6-35. The individual holes

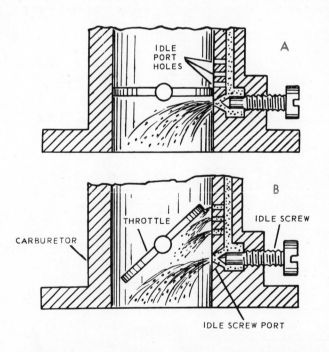

Fig. 6-35. Part throttle circuit — low speed. A—Throttle is closed. Idle screw port feeding fuel. Idle port holes inoperative. B—Throttle part way open. Idle port holes uncovered and feeding fuel mixture. Idle screw port also feeding mixture.

perform the same function as the slot-shaped idle port. This is the LOW SPEED CIRCUIT. Fig. 6-35 illustrates the low speed circuit in action.

PROBLEM — MORE FUEL NEEDED FOR HIGH SPEED CRUISING

As the throttle continues to open, the idle and low speed circuits will not furnish enough fuel. Unless another source begins feeding additional fuel, the engine will not gain speed.

SOLUTION — HIGH SPEED CRUISING CIRCUIT

This problem is solved with your basic carburetor as it feeds the engine through the main discharge jet.

When the throttle valve is opened past the idle and low speed ports, it allows air to start flowing through the air horn. This will build a vacuum in the venturi, and the main discharge nozzle will begin to feed fuel. At this time, the idle and low speed ports are still feeding fuel.

As the throttle valve opens further, venturi vacuum becomes quite strong, and the main nozzle supplies ample fuel. With the throttle in this position, vacuum is lowered at the idle screw port, and both it and the idle port will stop supplying fuel until the throttle is closed again. Fig. 6-36.

Notice that the throttle is well open, and there is a strong vacuum in the venturi. See how the fuel is being drawn from the bowl through the main discharge tube metering rod jet (which controls fuel flow). Fuel is then pulled up the discharge tube and out the nozzle where it is thoroughly mixed with the passing air. This is the HIGH SPEED CRUISING CIRCUIT.

ANTIPERCOLATOR

The antipercolator plug is shown just above and to the right of the main discharge nozzle. Fig. 6-36. When gasoline becomes heated, it will form bubbles of vapor which expand and shove on the surrounding gasoline. If vapor were to form in the discharge tube, or in the well, the vapor would push gasoline out the discharge nozzle, causing an overly rich mixture.

The antipercolator plug hole will allow these vapors to escape before they form sufficient pressure to force the fuel out. The circled cut to the left of the air horn in Fig. 6-36, shows an enlargement of the antipercolator plug. Notice the vapors leaving the fuel.

The antipercolator plug is especially useful in hot weather when the engine temperature climbs and heats the carburetor. Some carburetors use a very thick asbestos gasket between the carburetor flange and the intake manifold.

PROBLEM — NOT ENOUGH FUEL FOR HIGH SPEED — FULL POWER

When the throttle reaches the wide open position, and venturi vacuum peaks off (reaches its highest point). Under this condition, engine top speed will not be sufficient since not enough fuel can pass the metering rod jet.

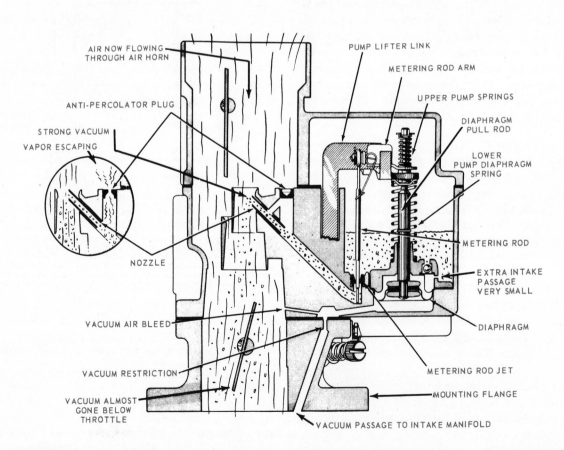

Fig. 6-36. High speed cruising circuit. Choke open and throttle is almost wide open. Main discharge nozzle is feeding fuel mixture into venturi and idle circuit is inoperative. (Carter Carburetor)

SOLUTION — HIGH SPEED —
FULL POWER CIRCUIT

You will see in Fig. 6-36, that a stepped rod, called the metering rod, is placed in the center of the metering rod jet. This metering rod partially plugs the opening, thereby limiting the fuel flow. Fig. 6-37.

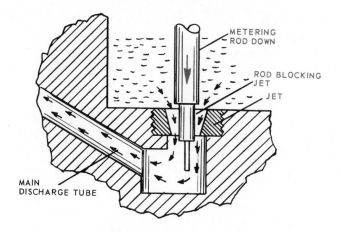

Fig. 6-37. Metering rod controls size of jet opening. Large step of metering rod now in jet. When rod is raised, small step will block jet.

To obtain more fuel for our high speed — full power range, it will be necessary to lift the metering rod up far enough so the larger step is removed from the jet. This leaves the small step in the jet, and it will allow more gas to flow. Fig. 6-38.

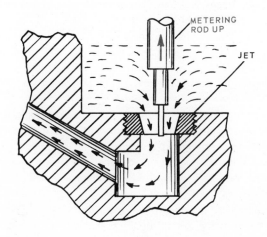

Fig. 6-38. Metering rod in raised position. Small step allows more fuel to pass.

HOW METERING ROD IS CONTROLLED

The metering rod may be lifted in two ways. One method hooks the top of the rod to linkage (actuating rods hooked together to work something) connecting the rod to the throttle lever. When the throttle valve is moved to the wide

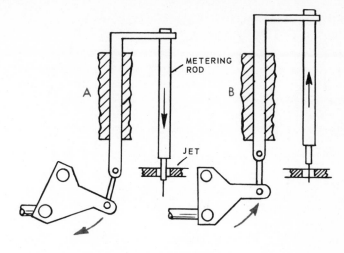

Fig. 6-39. Mechanical control of metering rod. A—Throttle closed. B—Throttle fully opened.

open position, it will lift the metering rod up. This mechanical control must be designed to raise metering rod at exactly the right spot. Fig. 6-39.

A mechanically controlled metering rod is fine for wide open throttle operation. If an engine is laboring under a hard pull, it needs a richer mixture. The throttle valve may not be open far enough to raise the metering rod, however, and the fuel mixture will not be rich enough. To overcome this fault, the metering rod is often lifted by a vacuum controlled piston.

VACUUM CONTROLLED METERING ROD

The metering rod is hooked to a vacuum piston or, in some cases, to a diaphragm. The piston is spring loaded (pushed by a spring) to keep it in a certain position; in this case UP. When vacuum is applied to the piston, it overcomes the spring pressure and pulls the metering rod down, partially closing the jet.

This vacuum passage reaches from the piston, or diaphragm, down through the carburetor, through the mounting flange and into the intake manifold.

As the car moves along with the throttle in high speed cruising range, there is a fairly strong vacuum in the intake manifold. This vacuum keeps the metering rod piston in the down position. When the engine begins to labor and the throttle valve is opened a little more, vacuum in the manifold will drop. This will reduce the pull on the piston, so the spring will shove it up, thereby lifting the metering rod. Fig. 6-40.

DUAL METERING ROD CONTROL

Some carburetors control the metering rod mechanically; some by vacuum only; others use both. The carburetor in Fig. 6-36 uses both means. Notice that the metering rod is hooked to the diaphragm pull rod. This pull rod is being held down by vacuum from the manifold acting on the diaphragm. The pump lifter also passes under the metering rod arm. Either loss of vacuum or full opening of the throttle valve will cause the metering rod to be lifted.

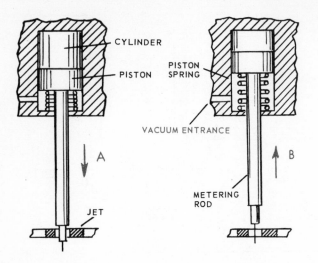

Fig. 6-40. Vacuum control of metering rod. A—No load on engine, intake vacuum high. Piston is held down against spring. B—Vacuum gone, spring raises piston and raises rod.

HIGH SPEED — FULL POWER BYPASS JET

Some carburetors do not use a metering rod. The main discharge jet remains a fixed size. Another jet, called a power jet, will open and allow additional fuel to pass through the main discharge jet. This additional fuel enters the discharge tube via a bypass opening. The power jet can be controlled mechanically, by vacuum, or both. Figs. 6-41 and 6-42 illustrate the power jet in closed and open positions. The power jet or a metering rod makes up the HIGH SPEED — FULL POWER CIRCUIT or SYSTEM.

PROBLEM — A FLAT SPOT UPON ACCELERATION

The carburetor, as discussed so far, would still allow a "flat spot" when sudden acceleration is attempted. This is especially true at speeds below 30 mph. When the throttle valve is

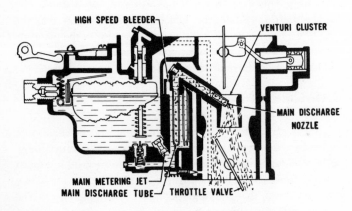

Fig. 6-41. Power jet operation. Vacuum is strong (part throttle) and holds actuating rod up. This leaves power bypass jet closed.

thrown open, the engine will hesitate, seem to die, may even pop, then catch and accelerate. The momentary flat spot is the problem.

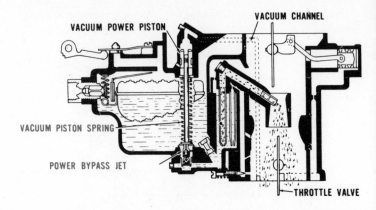

Fig. 6-42. Power jet operation. Vacuum is low, actuating rod is pushed down by vacuum piston spring and power bypass jet is opened. (Dodge)

SOLUTION — ACCELERATION PUMP CIRCUIT

An accelerator pump using either a piston or a diaphragm is incorporated in the carburetor. In the case of the diaphragm type, Fig. 6-43, the diaphragm is connected to a pull rod which, in turn, is fastened to a lifter link.

A spring is placed on the pull rod, both below the link and above. The lower spring is the strongest and is designed to keep the diaphragm up against the pull of the vacuum chamber below the diaphragm. Normal vacuum, however, will overcome the spring tension and hold down the diaphragm.

As in the case of the metering rod, the moment intake manifold vacuum drops, the pull rod spring will draw the diaphragm up. This up action reduces the size of the pump gasoline reservoir, the intake check will be forced shut, the outlet pushed open and fuel will be sprayed into the air horn.

When vacuum increases, the diaphragm will be drawn down. This shuts the discharge check and draws gasoline through the intake check, filling the pump reservoir.

The very same action takes place when the throttle valve is opened. This raises the pump link, mechanically, and forces the diaphragm up.

The pump can discharge fuel into the air horn either through the main fuel nozzle or through one or more special pump discharge jets. Some carburetors control the pump diaphragm or piston with linkage only, others use linkage and vacuum. The pump illustrated in Fig. 6-43 is controlled by both.

Notice in Fig. 6-43 that in addition to an intake check, there is also a tiny intake passage. This passage will let a measured quantity of fuel return to the bowl during the pump stroke. Some carburetors do not have this feature.

The springs on the pull rod play an important role. If the pump lifter link is suddenly raised by slapping the accelerator to the floor, this will not smash the diaphragm against the fuel

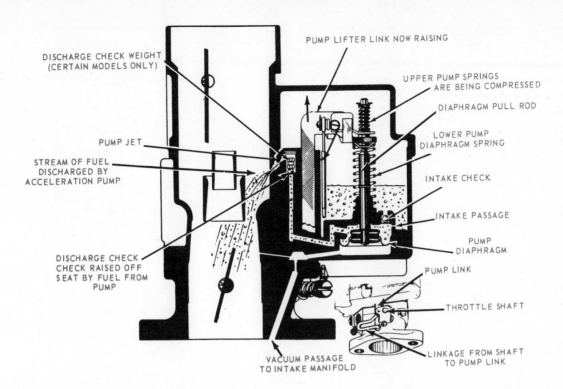

Fig. 6-43. Accelerator pump circuit. Pump lifter link compressing pump upper spring will raise diaphragm pull rod, compressing fuel above diaphragm. This fuel will be forced out of pump jet. Study names of various parts. (Carter)

in the reservoir. On the contrary, it allows the link to raise and compress the top spring. The tension of this spring will then quickly and smoothly, draw the diaphragm up. Instead of a short heavy burst of fuel, a long smooth stream will be discharged.

When the link is suddenly dropped by releasing the

accelerator, the diaphragm will not be squashed down, since the link will slide down the pull rod and the diaphragm will be pulled down by vacuum. The top spring acts as a bumper during this down movement.

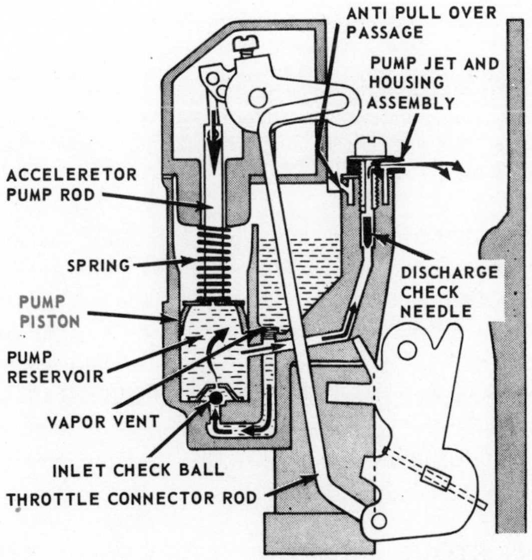

MECHANICALLY OPERATED PISTON PUMP

Fig. 6-44 shows a mechanically operated piston type accelerator pump. It uses one spring that is placed above the piston and attempts to shove it down. It cannot go down as the pump lever is held in the up position by the pump rod which is connected to a throttle lever.

The reservoir below the piston is filled with fuel through the inlet check. It is ready to spray gasoline into the air horn as soon as the throttle valve is opened.

As the throttle valve opens, the throttle connector rod releases the upward pressure on the accelerator pump rod, allowing the pump piston spring to force the piston down. This closes the inlet check ball and forces gasoline past the discharge check, through the pump jet and into the air horn. When the throttle is closed, the connector rod will raise the piston and refill the reservoir through the inlet check.

PROBLEM — TO SECURE A RICH MIXTURE FOR COLD STARTING

Fig. 6-44. Mechanically operated accelerator pump piston. When piston is forced down, it will compress gasoline in pump reservoir, closing inlet check and forcing gasoline by outlet check. Pullover passage vents discharge area to prevent vacuum from drawing gasoline from accelerator discharge jet during normal operation. (Ford)

As mentioned, when an engine is cold it must have a rich mixture to start. As it warms up, the mixture must lean out and, when hot, the mixture must be even leaner.

SOLUTION — AUTOMATIC CHOKE CIRCUIT

The basic carburetor had a choke valve, but it was necessary to operate it by hand. Although some cars use the manually (by hand) operated choke, most modern carburetors are equipped with an automatic choke control.

CHOKE THERMOSTATIC COIL

The choke valve shaft extends through the carburetor into a round housing. Inside the housing, there is a thermostatic coil spring (a bimetallic spring formed of two dissimilar metals that cause spring to wind, or unwind, depending on temperature change). One end of the spring is attached to the housing cover and the other end is attached to a lever on the choke shaft. When it is cold, the coil will hold the choke valve in the closed position. When the coil is hot, it will tend to lose its tension and allow the choke valve to open. Fig. 6-45.

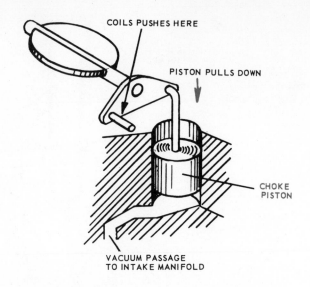

Fig. 6-46. Choke vacuum piston. Vacuum in manifold draws on bottom of piston, causing it to pull down against force of coil and partially open choke valve.

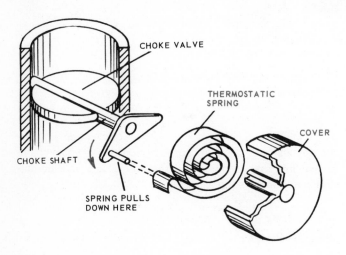

Fig. 6-45. Thermostatic coil and cover. When cold, spring pulls down in direction of arrow. Cover can be turned to increase or decrease spring tension.

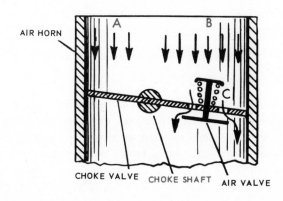

Fig. 6-47. Choke shaft is offset on valve. Less air pushes on side A and, as a result, side B is pushed down and partially opened against pull of thermostatic coil. Sometimes a small auxiliary air valve is set in choke valve C.

CHOKE VACUUM PISTON

A small vacuum-operated piston is connected, by means of a link, to the choke shaft. As soon as the engine starts, the vacuum will pull on this piston and cause it to attempt to open the choke valve against the pull of the thermostatic coil. Fig. 6-46. Fig. 6-52 shows a vacuum break diaphragm that is used to open the choke.

OFFSET CHOKE VALVE

The choke shaft does not grasp the choke valve in the center, but is positioned so that it is slightly off center. When the engine starts, the velocity of air trying to enter the carburetor will push more on the long side and partially open the valve. Fig. 6-47. A small inlet valve is sometimes placed in the choke valve to admit extra air.

A BALANCED CONDITION

The cold thermostatic coil closes the offset choke valve. When the engine starts, air entering the carburetor will open the valve slightly. The vacuum piston will also pull to open the choke valve a little more. Between the two, they will be able to offset the pull of the coil and open the valve enough to permit the engine to run. See choke action in Fig. 6-48.

If the engine is accelerated, the vacuum pull on the piston will diminish and allow the choke valve to close more tightly and enrich the mixture.

CHOKE STOVE

When the engine starts, tiny slots or grooves in the walls of the choke piston cylinder are uncovered. The slots open up inside the choke housing. A tube, or heating pipe, runs from the housing, down to the exhaust manifold where it enters a

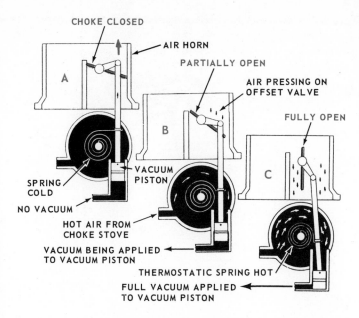

Fig. 6-48. Thermostatically controlled choke action. A—Choke closed, engine stopped. B—Engine started, vacuum piston and offset choke valve force choke open a small amount. C—Engine at operating temperature, thermostatic spring allows choke to fully open.   (Toyota)

heating compartment (choke stove). Fig. 6-49.

With the engine running, the vacuum will draw air through the tube from the stove up and around the thermostatic coil. As the engine heats up, the air passing the tube becomes hotter and heats the coil. The coil will release tension on the choke valve, allowing it to start to open. When the engine is fully heated up, the coil tension is so weak that the piston and passing air keep the valve fully open.

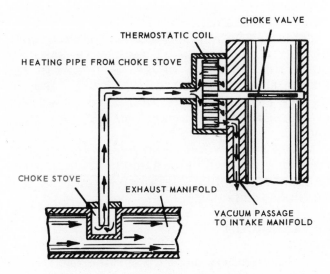

Fig. 6-49. Choke stove. Notice how exhaust gases warm choke stove. Clean, heated air is then drawn up through tube, across thermostatic coil and down through vacuum passage. This soon heats coil causing it to unwind.

When the engine is stopped, the coil cools off and closes the choke for another start. Fig. 6-49.

FLOODING

In case the carburetor floods (puts too rich a mixture in cylinders) the engine, use is made of a choke "unloader," an additional lever attached to the choke shaft. In the event of flooding, the throttle valve is opened all the way, causing linkage from the throttle lever to strike the unloader trip lever and open the choke to allow air to enter the cylinders.

FAST IDLE

When the engine is warming up, it is necessary to idle the engine somewhat faster to prevent stalling. This is accomplished by connecting the choke shaft to a fast idle cam.

When the choke is on, the fast idle cam swings out in front of the throttle idle speed adjustment screw. This holds the throttle partially open to speed up the engine. As the choke opens, the fast idle cam swings down and the throttle returns to its regular hot idle position. The fast idle cam is sometimes curved or notched. Fig. 6-50.

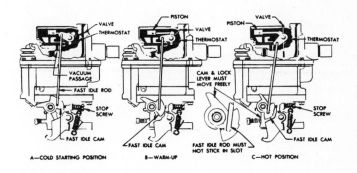

Fig. 6-50. Action of fast idle cam during engine warmup. As choke opens, fast idle rod will lower fast idle cam. As cam lowers, it will allow idle stop screw to close throttle more and more. This will reduce engine speed. When choke is fully opened, engine will idle at normal rpm.
(Buick)

A complete carburetor, with the automatic choke attached, is shown in Fig. 6-51. Study the complete choke circuit to determine exactly how it works. Also note the normal three sections of a carburetor:

1. Air horn assembly.
2. Main body.
3. Throttle body.

OTHER TYPES OF CHOKES

There are other automatic choke arrangements. One type uses an electrically heated coil to speed up the opening of the choke thermostatic spring. Another employs an electromagnet that is energized when the engine is cranked. The electromagnet helps hold the choke CLOSED. When the starter is released, the thermostatic spring takes over. Another passes

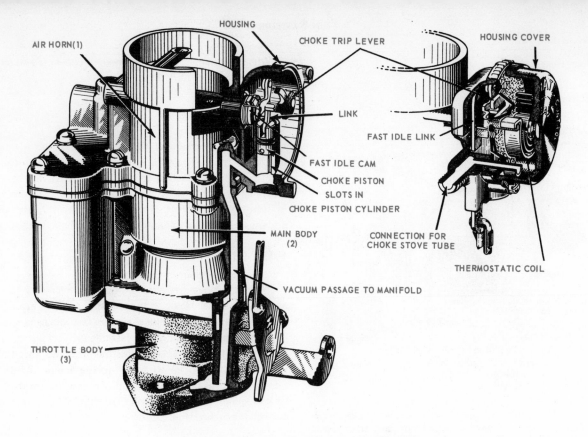

Fig. 6-51. Complete carburetor with choke attached.

engine coolant around the choke housing.

Another type places the thermostatic coil in a well in either the exhaust manifold or in the exhaust crossover (V-type engine) section of the intake manifold. Linkage connects from the well control to the choke lever. Note that the choke circuit

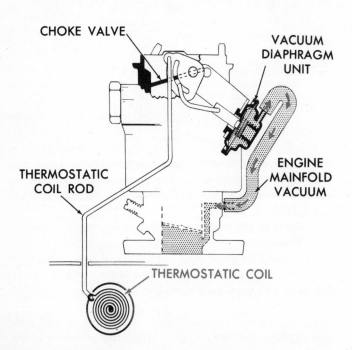

Fig. 6-52. Automatic choke, well type. Thermostatic coil may be placed in exhaust crossover (hot) area of intake manifold. This setup uses a diaphragm unit instead of a piston to partially open choke when engine starts. (Chevrolet)

shown in Fig. 6-52, utilizes a vacuum break diaphragm unit to provide initial opening of the choke.

MANUAL CONTROL

The choke can be controlled by hand instead of automatically. This is done by using a flexible cable control. One end of the control is attached to the choke valve shaft lever and the other ending in a knob on the dash.

WATCH OUT!

CARBURETORS ARE OCCASIONALLY PRIMED, WHEN DRY, BY POURING A SMALL AMOUNT OF GASOLINE DOWN THE AIR HORN. NEVER DO THIS WITH THE ENGINE BEING CRANKED. DO NOT SPILL GASOLINE ON THE ENGINE. BEFORE TRYING TO START, PUT AWAY THE PRIMING GASOLINE CAN AND STAND WELL BACK.

NEVER PLACE YOUR HAND OVER THE AIR HORN TO CHOKE THE ENGINE. A BACKFIRE THROUGH THE CARBURETOR COULD TURN YOUR HAND INTO A MOST UNSAVORY HAMBURGER — WELL DONE!

FLOAT CIRCUIT

The basic carburetor used a typical float circuit. Fig. 6-53 shows the float, float bowl, bowl vent, needle and needle seat. Correct float settings are very important. If the fuel level in the bowl is too high it will cause excessive amounts of gasoline to be delivered to the engine (rich mixture). If the setting is

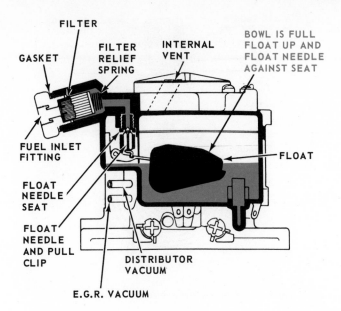

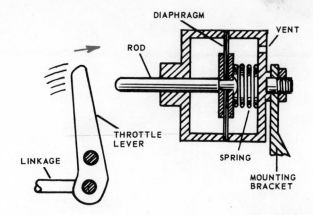

Fig. 6-53. Float circuit. Study all parts and learn their names. Trace flow of fuel from inlet into bowl. Note use of internal vent. (Buick)

Fig. 6-55. Throttle return dashpot. When throttle lever strikes dashpot rod, rod must push diaphragm back. Before it can move back, air must be forced out vent. This causes diaphragm to move slowly. When throttle lever is moved away, spring will force diaphragm out. Air will be drawn in vent and dashpot is ready to work again.

too low, not enough gasoline will be administered (lean mixture). The float may be constructed of thin brass or of closed-cell plastic.

OTHER CARBURETION FEATURES — IDLE COMPENSATOR

To prevent stalling when idling in very hot weather, some carburetors utilize a thermostatically controlled air valve. Fig. 6-54. When the carburetor becomes hot, the bimetallic strip will curl upward and open the hot idle air valve. This feeds extra air to the manifold below the throttle valve. The extra

air compensates for an overly rich idle mixture, due to heat-produced vapors entering the carburetor.

THROTTLE RETURN DASHPOT

On cars equipped with automatic transmissions, the drag placed on the engine when the throttle is suddenly returned to idle will result in stalling the engine. The dashpot is arranged so the throttle lever will strike the dashpot just before reaching idling range. The dashpot diaphragm rod will slowly collapse and allow the throttle to slowly return to the idling position. This stalling occurs when the car is stopped. Fig. 6-55. Some dashpots are built as an integral part of the carburetor. Some engines, using air injection systems for smog control, do not need throttle return dashpots.

IDLE SPEED—UP CONTROL FOR AIR CONDITIONING

To compensate for the increased load on the engine when the air conditioning compressor clutch is engaged, an idle speed control is often used. The speed-up during idle also provides increased airflow for better cooling.

When the compressor clutch is energized, electricity is fed to a small solenoid on the side of the carburetor. The solenoid then opens an air valve, thus permitting increased airflow below the throttle plates. Fig. 6-56 illustrates the idle speed-up control as well as the hot idle compensator valve.

Another type setup uses a vacuum diaphragm to control idle speed-up when the air conditioner is on. This unit forces the throttle open enough to raise the idle rpm to the desired level.

WARMING THE FUEL CHARGE

When the air-fuel mixture enters the manifold during the warm-up period, gases from the exhaust manifold are directed to the base of the intake manifold where the carburetor is

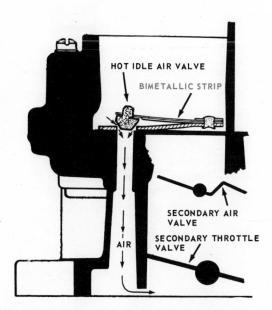

Fig. 6-54. Hot idle compensator. Note how heat has caused bimetallic strip to open air valve. (Pontiac)

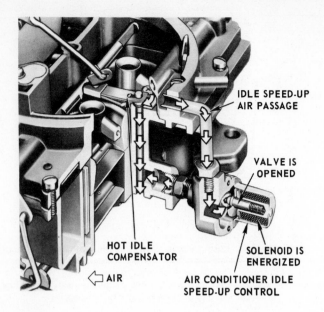

Fig. 6-56. Air conditioner idle speed-up control. When control solenoid is energized, valve is opened to admit air below throttle plates. (Ford)

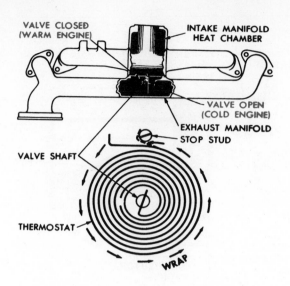

Fig. 6-58. Manifold heat control valve. (Plymouth)

attached. This warms the fuel charge. When the engine is hot, the gases are directed away from the intake manifold. This action is controlled by the thermostatic MANIFOLD HEAT CONTROL VALVE. Figs. 6-57, 6-58 and 6-59. Carburetor intake air is also heated by means of a thermostatically controlled air cleaner. Fig. 6-81. See Emission Controls, Chapter 5, for additional detail.

HEAT CROSSOVER — V-8 ENGINE

On a V-8 engine, the manifold heat control valve is located in the exhaust manifold on one bank. When cold, the heat control closes and directs some of the hot gases up through the intake manifold. After passing through, the gases flow out the exhaust manifold on the other side. When hot, the valve opens and both exhaust manifolds expel gas directly into the exhaust system. Figs. 6-60 and 6-61.

WATER HEATED MANIFOLD

Some engines pass cooling system water (coolant) through

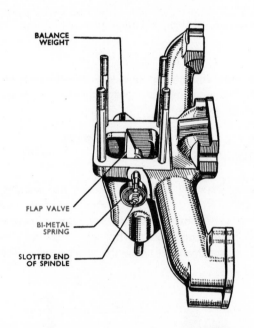

Fig. 6-59. Exhaust manifold with heat control valve. Note bimetal spring that closes valve when cold. (Hillman Motor Car Co.)

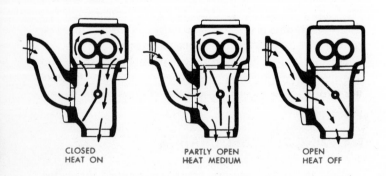

Fig. 6-57. Operation of manifold heat control valve. Notice how hot gases are directed up and around intake manifold passages. (Buick)

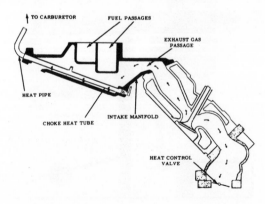

Fig. 6-60. V-8 engine heat crossover. Heat control is closed and hot gases are being directed up through passages in intake manifold. Notice different type of choke stovepipe used. (Oldsmobile)

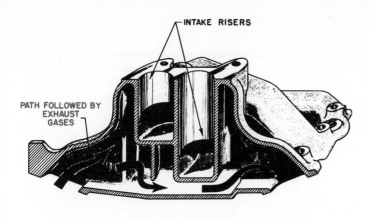

Fig. 6-61. Sectioned intake manifold (V-8), showing passage for exhaust gases. Carburetor bolts on top of intake risers. (Dodge)

passages in the intake manifold. This provides a controlled temperature for the fuel mixture. Fig. 6-62 illustrates a water-heated intake manifold for a V-8 engine. The top view

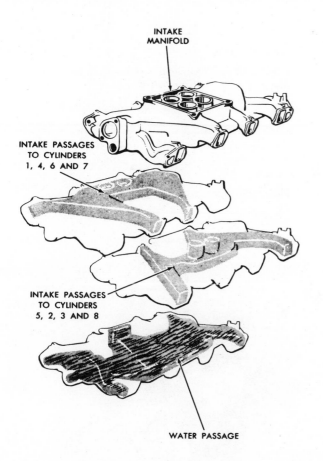

Fig. 6-62. Water heated intake manifold. Water replaces hot gases in this installation. (Lincoln)

shows the complete manifold (for a four-barrel carburetor). The second sketch shows the intake passages (shaded portion). The bottom sketch shows the flow of water through the manifold.

CARBURETOR ICING

It is possible, especially during the warm-up period, for ice to form on the throttle valves. As the nozzles feed fuel into the air horn, it turns into a vapor. This vaporizing action chills the air-fuel mixture and the throttle bores in the throttle valve area.

When the moisture-laden mixture contacts the chilled throttle valve, it condenses, then freezes. This pileup of ice can reach the point where, when the accelerator is released, the engine will stall because no air can pass around the throttle valve.

Some heat crossover systems channel a small portion of the hot gases directly through a passage in the carburetor flange. This heats the throttle valve and idle port areas and prevents the formation of ice.

Fig. 6-63 shows the bottom of a four-barrel carburetor. On this model, the vacuum draws air for the automatic choke from a tube running through the heat crossover passage similar

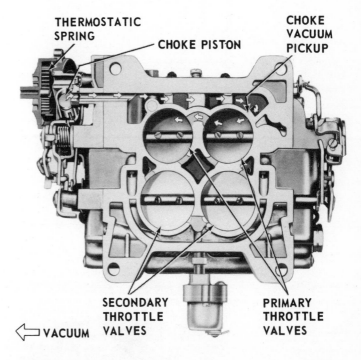

Fig. 6-63. One type of carburetor icing control. Hot air from choke housing (passes by choke piston through slots in cylinder wall) is drawn through carburetor mounting flange and discharged around base of throttle valves.

to that in Fig. 6-60. The heated air passes through the choke housing, into the carburetor flange, around the primary bores where it is finally drawn into the manifold. This prevents icing. Heated intake air also helps.

OTHER CARBURETOR FUNCTIONS

The carburetor furnishes an opening for the distributor vacuum advance line, linkage to operate kickdown switches for some automatic transmissions and overdrives, vacuum switches to operate accelerator pedal controlled starters, linkage (TV rod) to actuate transmission controls, a device to open the throttle a bit when starter is engaged, feeds a portion of the exhaust gases (EGR) into the air-fuel mixture, plus other special functions on specific installations.

GOVERNORS

Some carburetors are equipped with various types of governors to limit engine speed. These are usually found on heavy equipment such as trucks and buses. Fig. 6-64 shows a schematic view of a centrifugal type governor to control engine speed.

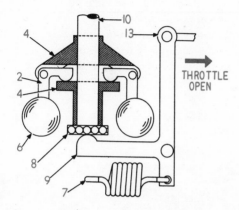

Fig. 6-64. Centrifugal governor. Shaft 10 revolves, causing assembly 4 to rotate entire unit. As rate of spin increases, weights 6 swing outward, drawing rounded ends of 2 downward. This shoves 4 down and causes it to press on 9. Friction is lessened by ball bearing 8. As weights continue outward, 9 is depressed and throttle lever is pulled back slowing engine. Return spring 7 attempts to keep throttle shaft 13 in open position. (Continental)

DUAL CARBURETOR

Many engines employ the dual carburetor (referred to as the TWO-BARREL). Basically this carburetor is nothing more than single-barrel carburetors mounted side by side. Each barrel has its own discharge nozzle, idle port, accelerator discharge jet and throttle valve. A fuel bowl, air horn, choke and throttle shaft common to both is used.

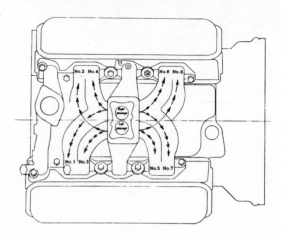

Fig. 6-65. V-8 intake manifold for a two-barrel carburetor. (Oldsmobile)

Each barrel feeds into a separate intake passage. Fig. 6-65 illustrates a typical dual carburetor intake manifold for a V-8 engine. Notice that one barrel feeds cylinders 1, 4, 6, 7; the other 2, 3, 5, 8.

A typical two-barrel carburetor is shown in Fig. 6-66.

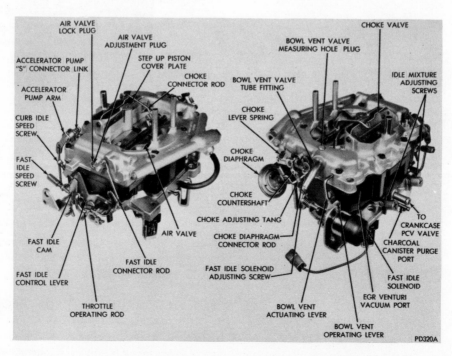

Two views of a modern four-barrel carburetor. (Plymouth)

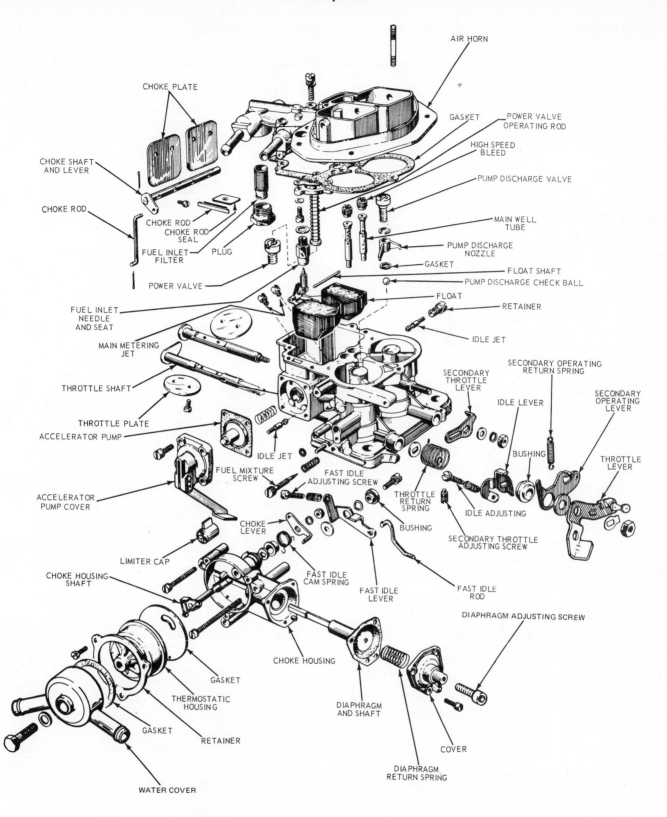

Fig. 6-66. Typical two-barrel carburetor. Notice the three major parts. 1—Throttle body. 2—Main body. 3—Air horn. (Dodge)

FOUR—BARREL CARBURETOR

Basically, a four-barrel carburetor is a cluster of four single-barrel carburetors. They are arranged in two groups. One group of two barrels is called the PRIMARY side, and the other two are termed the SECONDARY side.

There is wide variation in four-barrel carburetor design. The one illustrated in Figs. 6-67, 6-68 and 6-69 is typical.

Fig. 6-67 illustrates the primary and secondary float circuits. Fuel enters the strainer in the secondary bowl

145

opening. It flows through a passage to the primary side. Notice that the needle valve in the secondary side is spring loaded. The secondary float bowl supplies less gasoline. As it has less drain, it is subject to flooding when road bumps shake the float up and down. The spring pressure allows the float to bounce without the needle valve moving. Some four-barrel carburetors have a single float circuit that serves both sides. Fig. 6-67 illustrates a double float circuit.

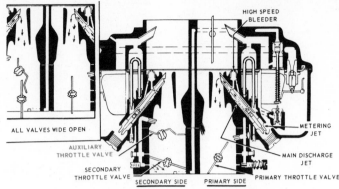

Fig. 6-69. Intermediate and full throttle circuits. Note position of auxiliary throttle valve during full power operation, (small figure on left side).

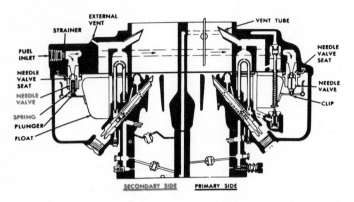

Fig. 6-67. Four-barrel primary and secondary float circuits. Notice the spring-loaded needle valve used in secondary circuit. (Buick)

A four-barrel idle circuit is shown in Fig. 6-68. Both secondary and primary sections feed fuel during idle. Only the primary idle discharge side is adjustable. Some models feed idle fuel through the primary side only.

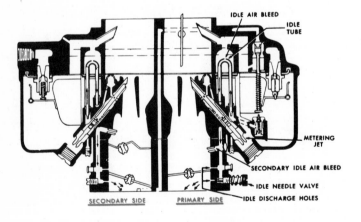

Fig. 6-68. Four-barrel idle circuits. Both primary and secondary sides feed fuel for idling. Primary side is adjustable; secondary is fixed. (Buick)

Both intermediate and full power throttle positions are illustrated in Fig. 6-69. Notice that in the right-hand view, the primary side is supplying fuel for the intermediate speed ranges. Both the throttle auxiliary valve and the secondary throttle valves are still closed. As long as they remain in this position, fuel will flow from the idle discharge, but none from the secondary main discharge nozzle.

As speed is increased, or the throttle is fully opened, the secondary throttle valves will open. The valves can be opened by a vacuum or by mechanical linkage. They normally open when the primary throttle valves have been opened about 40 deg.

When the secondary valves are open and engine speed is around 1400 rpm, the airflow through the secondary barrels reaches an intensity that will force the auxiliary valves open. The auxiliary valves are mounted off-center. When they open, fuel will flow from the secondary discharge tubes. Secondary idle flow will cease (see small left cross section, Fig. 6-69).

Notice that the primary barrels have a choke valve. None is needed in the secondary since they are closed when starting. Fig. 6-69.

A four-barrel carburetor gives the advantage of a small venturi area at medium speeds, but when full power is needed, the advantages of the large venturi area are available.

Fig. 6-70 is an exploded view of another four-barrel carburetor.

INTAKE MANIFOLDS

Intake manifolds are designed to carry fuel from the carburetor to the valve ports. They can be made of cast iron or aluminum.

The V-8 engine mounts the manifold between the cylinder heads. The L-head engines bolt the manifold to the side of the block. The I-head bolts the manifold to the cylinder head.

Gas exhaust and water heating of the manifold has been discussed under WARMING THE FUEL CHARGE.

MANIFOLD DESIGN

It is not enough just to carry fuel from the carburetor to the cylinders. Each cylinder should get the same type and amount of fuel mixture. To design an intake manifold to deliver equal amounts of fuel to all cylinders is not easy.

It is essential that the passages in the manifold be as nearly equal in length as possible. In addition, they should be smooth with gentle curves. Fig. 6-65, shows one specific V-8 intake

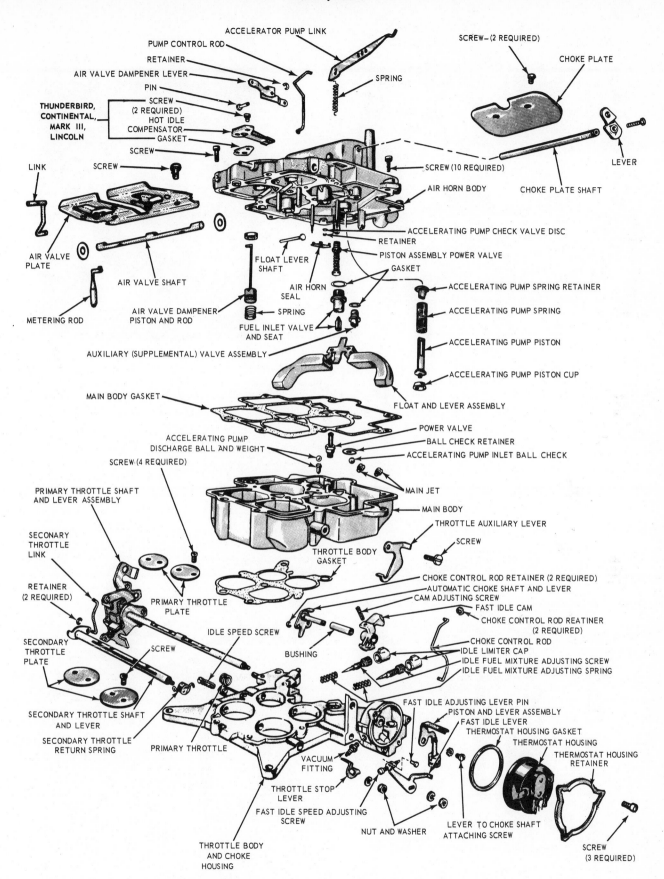

Fig. 6-70. Exploded view of one type of four-barrel carburetor. (Ford)

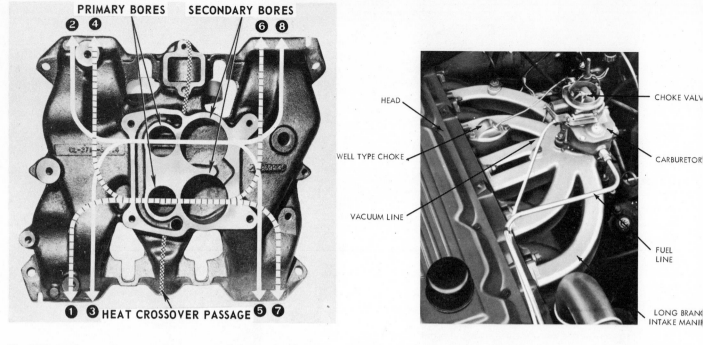

Fig. 6-71. Intake manifold for a V-8 engine. Manifold is designed for a single four-barrel carburetor. Note very large secondary bores. (Cadillac)

Fig. 6-72. This six cylinder intake manifold design provides a separate passage for each cylinder. In addition, equal fuel distribution is aided by placing carburetor some distance from engine. (Plymouth)

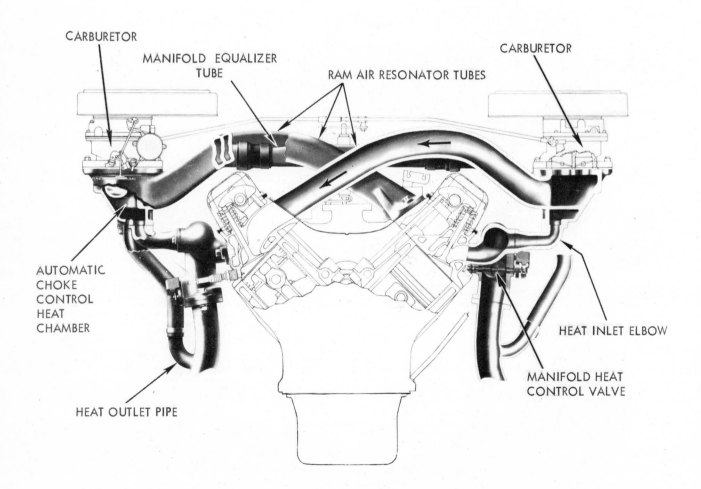

Fig. 6-73. Ram induction intake manifold as used on V-8 engine. Carburetors mounted off to one side allow long intake passages across to opposite bank of cylinders. (Plymouth)

manifold design for a dual carburetor. Another V-8 intake manifold for a four-barrel carburetor, is pictured in Fig. 6-71. Note how large the secondary bores are in relation to the primary.

A six cylinder manifold passage design is shown in Fig. 6-72.

RAM INDUCTION

By offsetting the carburetors to one side of the engine (V-8 engine), and feeding the fuel mixture through extra long intake manifold passageways, a RAM effect is produced. Due to the inertia of the long moving stream of fuel mixture, it will flow into the cylinder up to the last second the intake valve is open.

A compression wave will travel back and forth through the passage at the speed of sound. If the manifold length is correct, the compression wave will be timed to the opening of the intake valve and the air will "ram" itself into the cylinder. This gives the effect of mild supercharging.

The ram induction manifold will produce this effect only at certain engine speeds, depending upon the length of the manifold. The manifold is designed to produce this beneficial effect at the normal operating speeds. Fig. 6-73 illustrates the ram type manifold for a V-8 engine.

EXHAUST SYSTEM

Burned gases leave through the exhaust valves and enter the exhaust manifold. From there they enter the exhaust pipe, travel to the muffler and leave the car via the tail pipe.

Exhaust manifolds should be smooth inside; all bends should be gentle. This improves the exhaust flow.

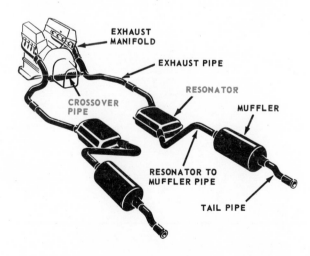

Fig. 6-74. A dual exhaust system employing a crossover pipe and two resonators in addition to two mufflers. (Mercedes)

Some engines use a dual exhaust system with two or four mufflers. The two small mufflers are called resonators. Other engines use a single exhaust system with one or two mufflers. Fig. 6-74 shows a dual system.

MUFFLER DESIGN

There are two basic muffler designs. One uses a series of baffled chambers to dampen the sound. Fig. 6-75.

The second type, termed a Glass Pack, Steel Pack, etc., is of a straight-through design. This passes the exhaust through a perforated pipe surrounded with metal shavings or glass wool. The straight-through muffler is designed to eliminate as much back pressure (impeded gas flow) as possible. Fig. 6-76 illustrates one type of straight-through, fiber glass muffler.

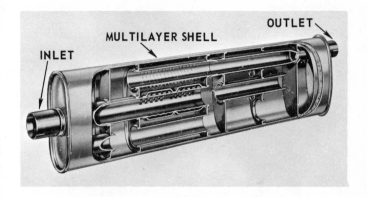

Fig. 6-75. A muffler using baffle principle. (Walker)

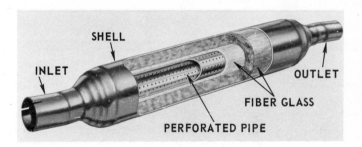

Fig. 6-76. A "straight-through" muffler. Heavy fiber glass packing is used around the perforated pipe. (Walker)

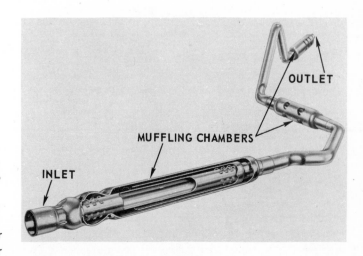

Fig. 6-77. A "chambered pipe" exhaust system in which a series of muffling chambers are incorporated into system. (Walker)

A modification of the straight-through principle is pictured in Fig. 6-77. Instead of using fiber glass, the perforated sections of the pipe are surrounded with sound deadening chambers. This design, termed "chambered pipe," requires a minimum of space beneath the car.

AIR CLEANERS

It is essential that air entering the carburetor is clean air. A good air cleaner, properly serviced, will add greatly to the useful life of the engine.

There are three types of air cleaners in general use.

OIL–WETTED POLYURETHANE AIR CLEANER

An older type (not very efficient) oil-wetted filter is shown in Fig. 6-78. This design draws the air through oil-dampened copper mesh.

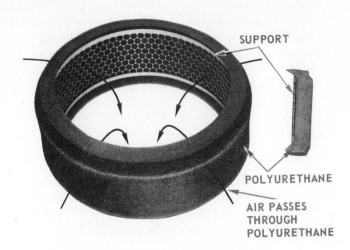

Fig. 6-79. Polyurethane air cleaner element. This filer can be rinsed in solvent, squeezed dry, dipped in oil, squeezed to remove excess oil and placed back in service. (Chevrolet)

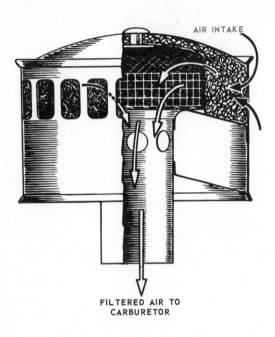

Fig. 6-78. Oil-wetted air cleaner. Hollow section below cleaning mesh acts as a silencer. (Standard)

The modern oil-wetted cleaner utilizes a POLYURETHANE (soft, porous, sponge-like material) element. The polyurethane, when dampened with oil, effectively filters the air. When dirty, it may be cleaned. Fig. 6-79.

OIL BATH AIR CLEANER

An oil bath air cleaner pulls the air down a passageway, forces it to make a sharp bend over a puddle of oil, back up through a mesh, and over and down into the carburetor. The air gains considerable speed as it travels down into the cleaner. When it makes the sharp bend, just above the oil, the heavier dust particles cannot make the corner and are slammed into the oil where they settle to the bottom. The wool-like mesh stays wet with oil and cleans the air even more.

Note how the air is forced to change directions just above the oil bath in Fig. 6-80. This filter may be cleaned by draining the oil, flushing the element, cleaning the base and refilling to the correct level with the recommended grade of engine oil.

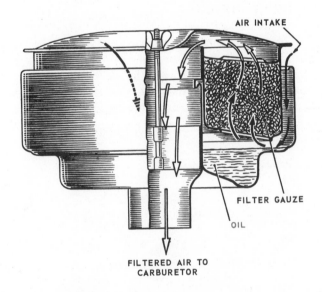

Fig. 6-80. Oil bath air cleaner. This type cleaner often incorporates a hollow section below the oil to assist in muffling noise of incoming air. (Standard)

DRY TYPE AIR CLEANER

A dry type air cleaner draws air through a specially treated paper. The paper is pleated accordian style. Fig. 6-81 illustrates a modern filter of this type having a series of ducts (passages) and a thermostat designed to furnish heated air to the cleaner during engine warm-up. The air is drawn from around the exhaust manifold. When the engine is heated the thermostat will close off the source of heated air.

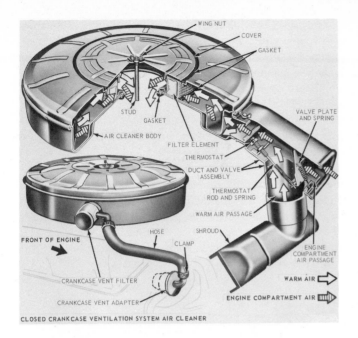

Fig. 6-81. Modern air cleaner with paper type air filter elements. Filter must be cleaned or replaced quite often when operated under dusty conditions. If paper element becomes clogged, it restricts incoming air and produces a choking effect on engine. (Lincoln)

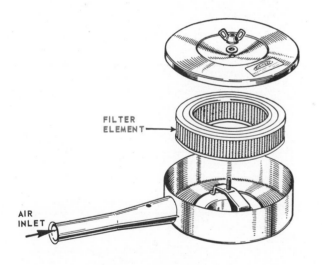

Fig. 6-82. Exploded view of a paper element air cleaner. (Hillman Motor Car Co.)

Fig. 6-82 shows an exploded view of a typical paper element air cleaner that does not have the "heated air" feature.

SUPERCHARGERS

Superchargers are used to increase engine horsepower by forcing more air-fuel mixture into the cylinders. The normal engine depends upon the vacuum created by the pistons to draw the mixture into the cylinders. The supercharger, or as it is sometimes called, "blower," assists the vacuum by raising the pressure of the incoming air. It does this with one of three types of superchargers.

Superchargers can be mounted between the carburetor and manifold or ahead of the carburetor. Due to the increase of air-fuel mixture delivered to the engine (with a resultant increase in power), it is essential that the engine compression and design lend itself to supercharging.

ROOTES TYPE SUPERCHARGER

In a Rootes supercharger, a pair of rotors whirling within an enclosure seize air and carry it around and out the opposite side. This type runs about one and a half times faster than the engine. It is a positive displacement type. Fig. 6-83.

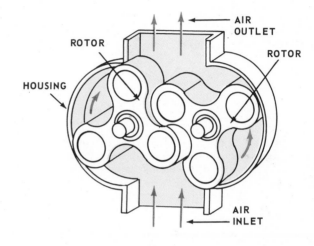

Fig. 6-83. Rootes type super charger. Rotors have minimum clearance between themselves and housing. (John Deere)

VANE TYPE SUPERCHARGER

The vane type supercharger is also a positive displacement type and works like a vane oil pump. Its operational speed is comparable to that of the Rootes type. Fig. 6-84.

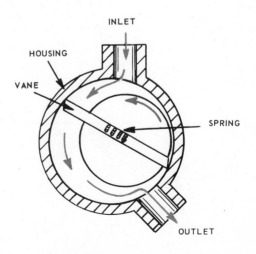

Fig. 6-84. Vane type supercharger. Air is drawn in behind one vane and forced around and out by the other.

CENTRIFUGAL TYPE SUPERCHARGER

The centrifugal supercharger is a nonpositive type. It has an impeller with a series of curved fins or vanes. As the impeller is whirled (at speeds in excess of 25,000 rpm), air is drawn in at the center and thrown off the rim. The air is trapped in a circular housing, which is small at one end and gets progressively larger. Fig. 6-85.

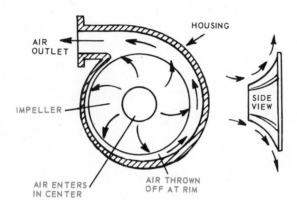

Fig. 6-85. Centrifugal type supercharger. Spinning impeller draws air in at center and throws it off at rim. Impeller uses series of vanes or fins to throw air.

OTHER TYPES OF FUEL AND FUEL SYSTEMS

FUEL INJECTION

A fuel injection system sprays fuel either directly in the cylinder or in the intake manifold.

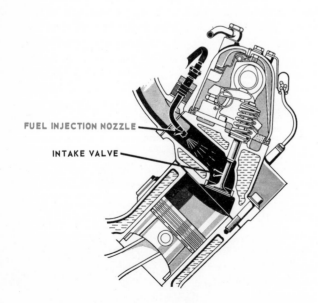

Fig. 6-86. Cutaway showing location and arrangement of fuel (gasoline) injection nozzle in the intake port. (Mercedes)

Two types of injection systems are in use. The older system, in use for a number of years, employs an engine driven mechanical injection pump that injects spurts of gasoline into the intake manifold. Fig. 6-86.

A more recent development, termed ELECTRONIC FUEL INJECTION, uses an electric fuel pump to provide pressure, electromagnetically operated injection valves (one per cylinder) to admit fuel into intake manifold near intake valves, a master electronic control that determines amounts of fuel to be injected and various sensors, such as engine temperature, speed, intake manifold vacuum, air temperature, etc., that feed information to the master control unit. This type of fuel injection provides extremely accurate control of air-fuel mixture under all operating conditions. Fig. 6-87 shows a schematic of a typical electronic fuel injection system.

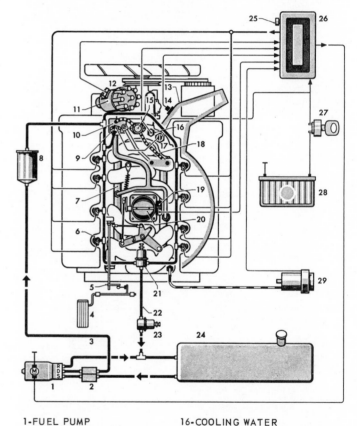

1-FUEL PUMP
2-DAMPER CONTAINER
3-FUEL PRESSURE LINE
4-ACCELERATOR PEDAL
5-REGULATING LINKAGE
6-INJECTION VALVES
7-IDLING SPEED AIR DUCT
8-FUEL FILTER
9-STARTING VALVE
10-IDLING SPEED AIR DISTRIBUTOR
11-IMPULSE TRIGGER
12-IGNITION DISTRIBUTOR
13-SUCTION NOISE DAMPER
14-AIR TEMPERATURE FEELER
15-SUPPLEMENTARY AIR VALVE
16-COOLING WATER TEMPERATURE FEELER
17-THERMOSTATIC TIME SWITCH IN COOLING WATER CIRCUIT
18-SUPPLEMENTARY AIR LINE
19-THROTTLE VALVE SWITCH
20-THROTTLE VALVE
21-FUEL PRESSURE REGULATOR
22-FUEL RETURN LINE
23-DIAPHRAGM DAMPER
24-FUEL TANK
25-IDLING SPEED CONTROL, SET SCREW
26-ELECTRONIC CONTROL UNIT
27-IGNITION STARTING SWITCH
28-BATTERY
29-PRESSURE SENSOR

Fig. 6-87. Schematic drawing showing layout of one type of electronic fuel injection system. (Mercedes-Benz)

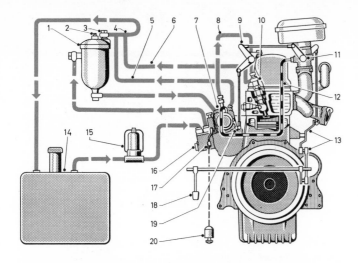

Fig. 6-88. Mercedes-Benz automotive diesel engine fuel injection system. 1—Main fuel filter. 2—Vent screw. 3—Hollow screw with throttle screw. 4—Fuel return line. 5—Overflow line. 6—Injection nozzle leakage line. 7—Injection pump. 8—Pressure line from injection pump to injection nozzle. 9—Angular lever for auxiliary mechanical control. 10—Injection nozzle. 11—Venturi control. 12—Vacuum line with throttle screw. 13—Linkage and lever for accelerator pedal control. 14—Fuel tank. 15—Fuel pre-filter. 16—Fuel pump with hand lever. 17—Adjusting lever. 18—Accelerator pedal. 19—Lever for auxiliary mechanical control. 20—Heater plug (glow plug) starting switch with starting and stopping cable. (Mercedes-Benz)

DIESEL

The use of diesel fuel is limited primarily to commercial equipment. There is a very limited use of diesel engines in automobiles at the present time. Fig. 6-88.

The true diesel engine does not use a spark plug to ignite the fuel. Air is drawn into the cylinder and highly compressed. The compression is so high that the compressed air will reach temperatures of 1,000 deg. F. At the precise time the piston has completed the compression stroke, diesel fuel is sprayed into the combustion chamber. The intense heat of the compressed air ignites the fuel, and the power stroke follows.

The diesel fuel is sprayed by means of special injector nozzles. The pressure at the nozzle must be very high — 3,000 to 20,000 lbs. — depending on the system used.

The amount of fuel being delivered by the injector is controlled by means of a foot throttle connected to an actuating mechanism that in turn controls the pump output. Fig. 6-88.

LIQUEFIED PETROLEUM GAS

LP-Gas (mixture of butane and propane) is only in the liquid state when highly compressed. When pressure is reduced, it turns into a vapor.

Storage and transportation of LP-Gas is expensive, due to the heavy container necessary to contain and transfer it. It is generally used in a fleet (large groups of vehicles) operation.

Its use is favored by some and certain advantages are claimed, i. e., clean burning, low cost, high octane rating, no oil sump contamination and longer engine wear, etc.

LP-Gas is stored in a heavy tank under pressure. It passes

from the tank through a filter to a high pressure reducing valve. Gas leaves this valve at pressures from 4 — 15 lbs. The semiliquid (half vapor, half liquid) fuel then moves into a vaporizer that warms it up and changes it into all vapor. The vapor travels to a second pressure reducing valve where it is reduced to atmospheric pressure. From here it travels to the carburetor (actually a mixing valve) and on into the cylinders when the engine is started. Fig. 6-89 shows the operational theory of the commonly used "liquid withdrawal" type of LP-Gas system.

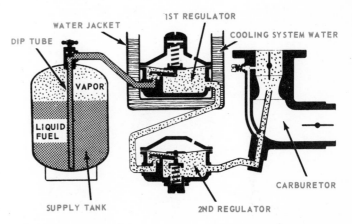

Fig. 6-89. Schematic LP-Gas fuel system using liquid withdrawal technique (liquid drawn from tank — not vapor). Liquid fuel leaves tank and after passing through two regulators, plus being heated, it is turned into an atmospheric pressure vapor. Vapor enters mixing valve (carburetor) and is drawn into cylinders. (Bendix—Zenith)

SUMMARY

FUEL FLOW

A fuel pump, by means of a diaphragm and check valves, draws gasoline from the tank. The gasoline is drawn out through a pickup tube with a screen on the end. Gasoline then passes through lines to the pump, where it is expelled through a pulsation dampener chamber through a filter to the carburetor fuel bowl.

When the fuel level in the bowl raises, a float will begin to close the float needle valve. At a predetermined level, the float will completely shut the float needle valve, and fuel will no longer flow to the carburetor. When fuel is consumed and the level starts to drop, the fuel pump will again fill the bowl.

The fuel pump is operated electrically or mechanically and may have a filter either on, before or after it.

MIXTURE

For efficient operation an engine such as used in a passenger car requires an air-fuel ratio that is constantly changing. This varied mixture is produced and delivered by the carburetor or a fuel injection system.

The carburetor contains a number of circuits designed to produce a proper air-fuel mixture for all engine needs. The

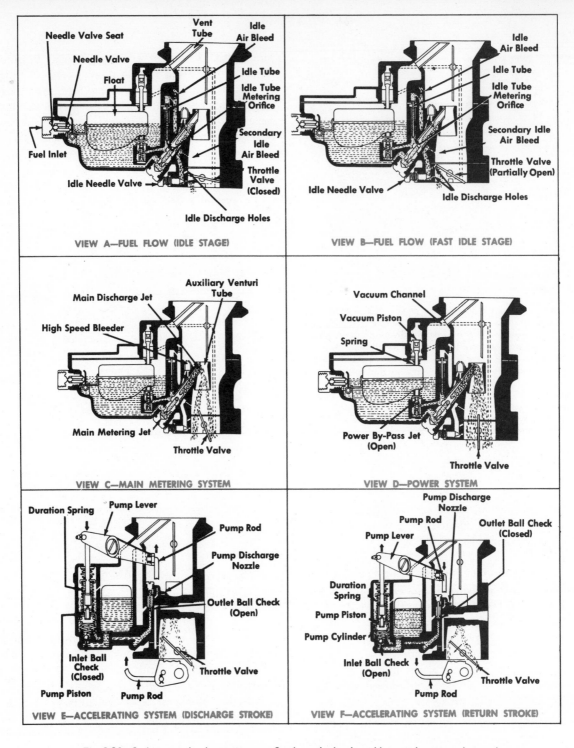

Fig. 6-90. Carburetor circuits or systems. Study each circuit and be certain you understand exactly how it operates and what particular parts are concerned.   (GMC)

general carburetor function is as follows:

When the engine first starts, the choke valve is closed. Vacuum is strong in the area beneath the choke valve, resulting in a very rich mixture by reducing the air supply and increasing the withdrawal of fuel from the bowl. As soon as the engine starts, the choke partially opens. As the engine is idled to warm-up, fuel is fed into the passing airstream by the idle screw port. The choke continues to open. When the engine is hot, it will be fully opened.

When the accelerator opens the throttle valve, the idle screw port fuel delivery is assisted by additional fuel coming from the idle port or part throttle holes just above the idle screw port.

As the throttle continues to open, the air speed through the air horn and venturi increases until fuel begins to feed from the main discharge nozzle. At very high speeds where the throttle is wide open or when the car is pulling very hard, additional fuel is administered by a power valve or by a metering rod. During the cruising and full power range, no fuel is fed by the idle port or idle screw port.

To assist in smooth acceleration, an accelerator pump is provided to feed additional gasoline during acceleration only.

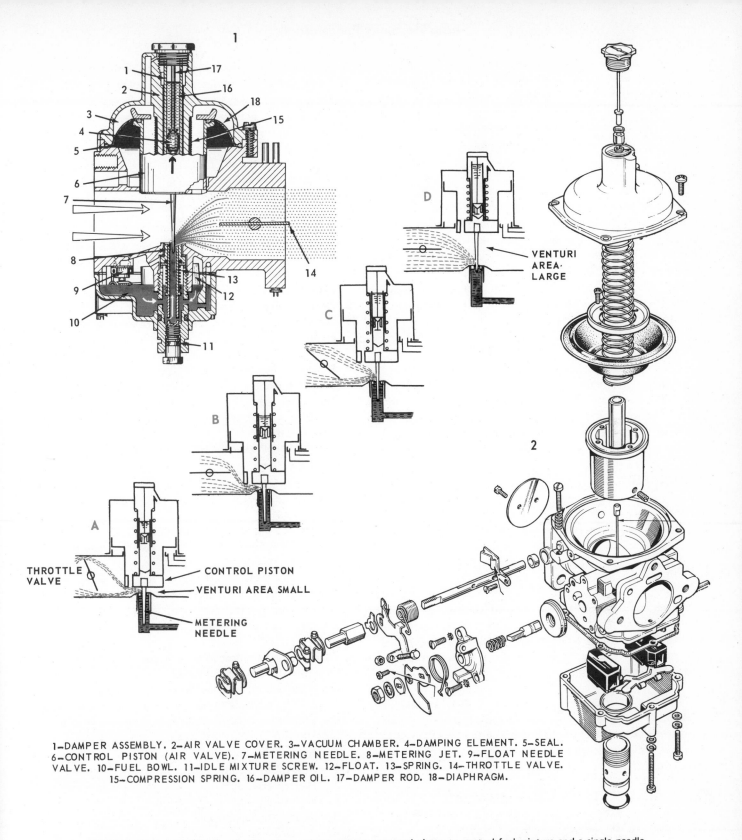

1—DAMPER ASSEMBLY. 2—AIR VALVE COVER. 3—VACUUM CHAMBER. 4—DAMPING ELEMENT. 5—SEAL. 6—CONTROL PISTON (AIR VALVE). 7—METERING NEEDLE. 8—METERING JET. 9—FLOAT NEEDLE VALVE. 10—FUEL BOWL. 11—IDLE MIXTURE SCREW. 12—FLOAT. 13—SPRING. 14—THROTTLE VALVE. 15—COMPRESSION SPRING. 16—DAMPER OIL. 17—DAMPER ROD. 18—DIAPHRAGM.

Fig. 6-91. Side draft (SU Type) carburetor, using a vacuum-operated piston to control fuel mixture and a single needle valve that provides proper mixture for all operating conditions. Venturi size is variable, depending on control piston position. Study control piston and metering needle action in A, B, C and D. Note that as control piston and metering needle move upward, venturi area is increased, allowing a greater flow of air along with increased fuel flow made possible when tapered metering needle was raised out of its seat or jet. A—Idling. B—Throttle fully opened — low speed. C—Intermediate speed. D—High speed operation.   (Jaguar, Toyota, Mercedes-Benz)

Engine speed is controlled by the position of the throttle valve in the carburetor.

Study Fig. 6-90. Follow through each circuit and study the parts and operational principle involved.

The air and fuel mixture is delivered to the combustion chambers by means of an intake manifold. The temperature of this mixture is somewhat controlled by heat crossover passages, heated air intake and, in some cases, by running the coolant through passages in the intake manifold. Even distribution of fuel to all cylinders is very important.

EXHAUST

The burned gases are expelled through an exhaust manifold that bolts onto the block, or on the cylinder head, depending on the engine design. The manifold should have gentle bends with no back pressure causing kinks. From the manifold the gases pass through the exhaust pipe to the muffler and out through the tail pipe. The exhaust system should muffle engine noise and at the same time provide an easy and smooth exit for the burned gases.

CARBURETOR TYPES AND KINDS

There are downdraft, updraft and sidedraft carburetors. Fig. 6-91. Some are single-barrel, some double-barrel, and others have four barrels. All carburetors — despite shape, size and type — attempt to provide as nearly perfect an air-fuel mixture as possible for all engine needs.

OTHER FUELS

For passenger cars, there is a limited use of diesel fuel, as well as LP-Gas. At the present time, the standard carburetion system utilizing gasoline for fuel dominates the industry.

AIR INTAKE

Air entering the carburetor must be cleaned. Failure to clean the air will result in rapid deterioration of the engine. Air is generally cleaned by either an oil bath cleaner or a dry type that utilizes special paper as the filtering element. Cleaners must be serviced regularly.

There are many DOORS TO SUCCESS. They are seldom marked "WELCOME — ENTER." You are more apt to find the doors securely locked. Knowledge is the most versatile key. STUDY and LEARN — don't stay "locked out" for the rest of your life.

REVIEW QUESTIONS — CHAPTER 6

Try your "key" on the following questions:

1. Air is largely made up of _____ and _____.
2. An average fuel-air ratio for normal operation is _____ parts of air to _____ part of fuel.
3. Name the parts that make up a fuel delivery system.
4. A fuel pump uses a diaphragm and one check valve. True or False?
5. The most commonly used fuel pump is the electric type. True or False?

6. Name three elements commonly used to filter fuel.
7. Explain how fuel pulsations are controlled.
8. What is vapor lock?
9. What is a carburetor venturi and how does it work?
10. How is the fuel level in the carburetor controlled?
11. What is a secondary or booster venturi? How does it help?
12. Of what use is the air bleed in a carburetor?
13. What are carburetor jets?
14. Engine speed is controlled by the _____
15. What is carburetor linkage?
16. Explain how the idle circuit functions.
17. What is provided to assist the idle circuit when the throttle is opened just beyond an idle?
18. What circuit provides fuel for normal cruising speeds? How does it work?
19. Of what value is an antipercolator?
20. Describe the operation of the high speed — full power circuit.
21. What is a "flat spot," and what may be done to cure it?
22. How does the cure for the flat spot operate?
23. Explain the operation of the automatic choke.
24. During the warm-up period, engines must idle fairly fast. A device is provided for this purpose. What is it called and how does it work?
25. If a cold engine becomes "flooded," what mechanism will force the choke valve open?
26. Cars with automatic transmissions usually have a _____ to prevent stalling when the throttle is closed.
27. Of what value is the idle compensator?
28. What are some features that are found in a well designed manifold (intake)?
29. How can ice form in a carburetor? How is it prevented?
30. What is a dual or double-barrel carburetor?
31. What is a quad or four-barrel carburetor?
32. List two important things that a good exhaust system will do.
33. What are the two most common types of air cleaners?
34. Of what value is a supercharger?
35. Name the two types of mufflers.
36. The following should be matched. Above are the names of the carburetor circuits. Below are problems. Match the problem with the circuit that will cure it.
    1. Idle circuit.
    2. Low speed circuit.
    3. High speed circuit.
    4. High speed — full power circuit.
    5. Acceleration circuit.
    6. Choke circuit.
    7. Float circuit.
    Engine cold, will not start.
    Bad "flat spot."
    Fuel level too high.
    No power at top speed.
    Accelerator up — engine won't run.
    Engine very rough at speed just above idle.
    Engine rough at cruising speeds.
37. List three safety hints dealing with the fuel system.

Chapter 7

ENGINE CLASSIFICATION, PARTS IDENTIFICATION

You have constructed a basic engine, studied parts and their uses and added the ignition, lubrication, cooling and fuel systems. This completes the basic engine.

This chapter is devoted to a study of the various ways in which the basic engine theory and parts are utilized to produce multicylinder engines of several types. Various methods of engine classification as well as different kinds of engines will be discussed.

CYCLE CLASSIFICATION

FOUR—STROKE CYCLE ENGINE

Your basic engine was a four-stroke cycle. For a brief review, study Fig. 7-1. This demonstrates that it takes two complete revolutions of the crankshaft to complete the cycle. As a result, the crankshaft receives one power stroke and must then coast until the next power stroke arrives. The vast majority of automobile engines use the four-stroke cycle engine.

TWO—STROKE CYCLE ENGINE

The two-stroke cycle engine performs the intake, compression, firing and exhaust sequence in ONE REVOLUTION of the crankshaft.

This is accomplished by eliminating the valves as used in the four-stroke cycle engine. In place of the valves, two ports enter the cylinder wall. One is used for the intake of fuel and the other for exhaust.

FOUR-STROKE CYCLE ENGINE

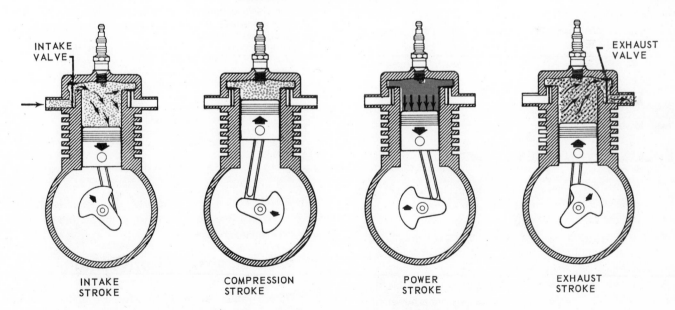

INTAKE VALVE

EXHAUST VALVE

INTAKE STROKE

COMPRESSION STROKE

POWER STROKE

EXHAUST STROKE

Fig. 7-1. Four-stroke cycle principle: INTAKE—As piston moves down or toward crankcase, intake valve opens and partial vacuum is created in cylinder. Vaporized air-fuel mixture is forced into cylinder by atmospheric pressure.

COMPRESSION—Intake valve closes. As crankshaft rotates, piston moves up and compresses air-fuel mixture.

POWER—Ignition system fires spark plug to ignite mixture just before piston reaches top of its travel. Expanding gases, which result from burning of fuel, force piston down to turn crankshaft.

EXHAUST—After fuel charge is burned, exhaust valve opens. Burned gases are forced out of cylinder by upward movement of piston.

This series of events, called a four-stroke cycle, is then repeated. A four-stroke cycle engine fires on every fourth stroke. A cycle requires two turns of the crankshaft.

(McCulloch Motors)

FUEL MIXTURE CONTAINED IN THE CRANKCASE

The fuel-air mixture is drawn into the crankcase through a REED or a ROTARY valve. Oil is usually mixed with the gasoline to provide lubrication. A passage connects the intake port to the crankcase interior.

SEQUENCE OF EVENTS IN ONE REVOLUTION

Imagine that a charge of fuel is in the cylinder and that the piston is at the bottom of its stroke. As the piston travels up, it closes off both intake and exhaust ports. As it continues upward, it compresses the fuel charge and, at the same time, forms a vacuum in the airtight crankcase.

The vacuum pulls open a small flap-like reed valve and fuel mixture enters the crankcase.

When the piston reaches the top of its stroke, it has compressed the charge and filled the crankcase with fuel mixture. This has all happened in one-half revolution of the crankshaft.

The plug now fires the charge and drives the piston down. Several things happen as the piston plunges downward. The mixture in the crankcase is compressed and the reed valve shuts. The piston is also performing the power stroke.

When the piston has traveled down the cylinder far enough to expose the exhaust port, the exhaust will flow out of the port. As the piston travels down a trifle more, the intake port is uncovered. As the crankcase fuel mixture is compressed, it forces combustible mixture to flow through the intake port into the cylinder.

The piston top is generally shaped to deflect the incoming mixture upward. This helps to fill the cylinder with fresh fuel as well as assist in scavenging (cleaning out) the remaining exhaust gases.

This has all taken place in one revolution of the crankshaft. Every time the piston reaches the top of its travel, it will be driven downward by a power stroke. Study Fig. 7-2.

CERTAIN ADVANTAGES

The two-stroke cycle engine has several advantages that have made it popular for small engine applications — outboard motors, chain saws, lawn mowers, go-carts, etc.

By eliminating the valves, it also eliminates valve springs, cam followers, camshaft and gears. This enables the engine to be quite compact and light. Machining costs are reduced.

More power per pound of engine weight is available as it fires every stroke. It does not produce twice the power of a four-stroke cycle of comparable size, as some power is lost through compressing crankcase mixture as well as poorer volumetric efficiency (ability to intake fuel).

It can also operate at very high speeds, as there are no valves to "float."

In order to completely scavenge the cylinder, some of the intake fuel also leaves the cylinder. This makes for less fuel economy. Despite this fact, the two-stroke cycle is popular.

TWO-STROKE CYCLE ENGINE

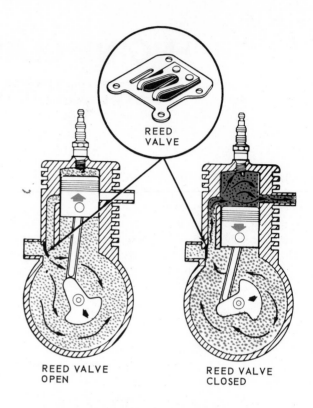

Fig. 7-2. Two-stroke cycle principle:

In left-hand view, piston is traveling upward, compressing air-fuel charge and drawing mixture into crankcase through reed valve. Fuel vapor contains suspended droplets of oil that lubricate surfaces of moving parts.

In right-hand view, air-fuel mixture has fired and piston is traveling down. It has uncovered intake and exhaust ports, allowing burned gases to escape and a fresh charge to enter. Air-fuel mixture flows through intake port because piston, while traveling down, compressed mixture in crankcase.

In two strokes, requiring one revolution, engine has performed all necessary functions to enable it to receive a power stroke for every crankshaft revolution.

A two-cycle engine that utilizes an exhaust valve and no mixture in the crankcase will be shown in the section on diesel engines later in this chapter.

WANKEL ROTARY PISTON ENGINE

The Wankel rotary piston engine, although under development for a number of years, has arrived on the automotive scene in significant numbers.

Named after its German inventor, Dr. Felix Wankel, the Wankel rotary piston engine shows considerable promise.

This engine is also often referred to as a Wankel Engine, Rotary Piston Engine, Rotary Combustion Engine or just Rotary Engine. In the interest of brevity, we shall refer to the engine as a "rotary engine."

CERTAIN ADVANTAGES

The rotary engine has several advantages as compared to the reciprocating piston engines now in popular use. It is considerably smaller and lighter. It is an extremely simple engine, using many less parts. It has no reciprocating (back and forth) motion, produces less vibration and offers significantly more horsepower than a reciprocating engine of comparable size.

Just how widely used the rotary engine will become depends on a great many factors, not the least of which are possible improvements to the standard piston engine, adaptability to future emission control requirements, other possible engine types and public acceptance of the engine.

ROTARY ENGINE CLASSIFICATION

Rotary engines are not classified as four cylinder, six cylinder, etc. In fact, they have no cylinders in the usual sense.

The number of ROTORS used in the engine is the basis for one classification — i.e., single-rotor, double-rotor, triple-rotor, etc. They are also referred to as 1-rotor, 2-rotor, 3-rotor, etc.

Although a number of 3-rotor and 4-rotor experimental engines have been built, common usage at this time employs either a 1-rotor or a 2-rotor engine. The 1-rotor engines are used mostly for small gasoline engine applications (snowmobiles, garden tractors, outboards, etc.) and the 2-rotor models for automobiles.

ROTARY CONSTRUCTION

Our coverage will relate to the 2-rotor engine. Remember that the 1-rotor operates exactly the same way; the only difference is that it has just one rotor.

Basically, there are just three moving parts in the 2-rotor engine — the eccentric shaft (also called mainshaft) and the two rotor assemblies. The remainder of the engine consists of two rotor housing, two end housing, two fixed gears and a center housing. All the housings are held rigidly together with a series of bolts.

ROTOR HOUSING

The rotor housing, usually made of aluminum or nodular iron, has a curved, oblong inner shape. Fig. 7-3. This shape is known as a 2-node EPITROCHOID curve. The space between the inside epitrochoid curve and the exterior is hollowed out to allow passage of coolant. The series of small round holes are for the draw bolts that will secure the center or intermediate housing and the end housings. Fig. 7-3.

The epitrochoid curve area of the rotor housing is commonly plated with a thin layer of hard chrome, nickel alloy, etc., to produce a smooth, long wearing contact surface for the rotor seals.

This particular housing contains holes for TWO spark plugs and has a peripheral exhaust gas port. Some rotary engines employ ONE spark plug per housing.

ECCENTRIC SHAFT

The ECCENTRIC SHAFT (similiar in action to a crankshaft) has TWO eccentric rotor journals. Note that they face exactly opposite (180 deg.) each other. Fig. 7-4. The shaft has two main bearing journals and is drilled for the passage of lubricating oil.

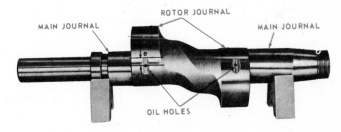

Fig. 7-4. Two-rotor engine eccentric shaft. Note oil holes.

The eccentric shaft passes through the center of both rotor housings and is supported by the main journal bearings located in the end housing. Fig. 7-4.

INTERMEDIATE HOUSING

The rotor housings are separated by an INTERMEDIATE housing. This housing has a center hole through which the eccentric shaft passes. Both sides are ground to provide a smooth, flat surface upon which the rotor side seals operate. The edges are hollow for passage of coolant. On the

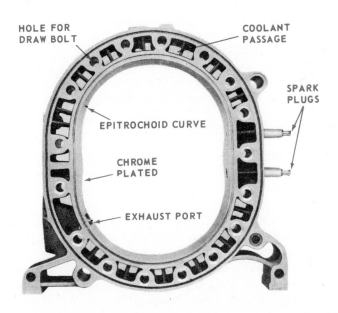

Fig. 7-3. Typical rotor housing. This particular housing is made of aluminum. The inner curve area is covered with a hard chrome plating a few thousandths of an inch thick. (Mazda)

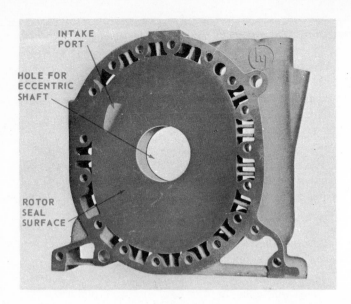

Fig. 7-5. Intermediate (center) housing that separates the two rotors. (Mazda)

intermediate housing shown in Fig. 7-5, a fuel mixture intake port is provided on each side. Fig. 7-5.

END HOUSINGS

The open outer end of each rotor housing is covered by a END HOUSING (also called side plates). These housings are hollow to permit the passage of coolant. The side facing the rotor is ground to provide a smooth, flat surface for the rotor outer side seals. It also contains a fuel intake port. Note the fixed (non-moving) gear attached to the center of the side housing shown in Fig. 7-6.

A closeup view of the end housing fixed gear is pictured in Fig. 7-7. This gear bolts to the end housing (both end housings

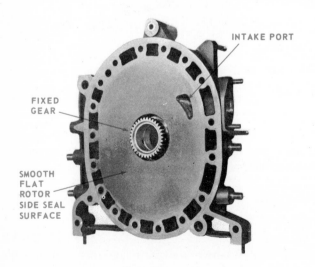

Fig. 7-6. One form of end housing. Note fixed gear, flat seal contact surface and the inlet port.

have a fixed gear). It also contains the main bearing and as such, supports the eccentric shaft. Fig. 7-7.

ROTOR

A triangular shaped ROTOR made of special cast iron is used in each rotor housing. The rotor has an inner bearing surface that rides on the eccentric shaft rotor journal. The rotor also has an internal gear (fixed to the rotor) that is in constant mesh with the end housing fixed gear.

The three rotor faces each contain a depression (cutout) that forms the combustion chamber. Both rotor sides as well

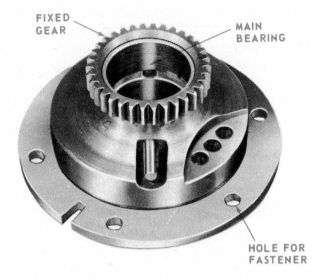

Fig. 7-7. One fixed gear, such as this, is bolted to each side housing. The eccentric shaft journals ride in main bearings in the center of each fixed gear. (Mazda)

as each apex (end or tip) are grooved for seals. Fig. 7-8 illustrates a typical rotor along with the seals.

The rotor, mounted on the eccentric shaft, is shown in position in the rotor housing in Fig. 7-9. All three rotor tips just clear the rotor housing epitrochoid surface. Fig. 7-9.

ASSEMBLED ROTARY ENGINE

A cutaway of a typical 2-rotor rotary engine is pictured in Fig. 7-10. Systems, such as carburetion, cooling, lubrication, etc., are much like that used on the reciprocating piston engine.

Study this illustration very carefully and fix the relative locations of the rotors, eccentric shaft, fixed gear and housing, firmly in mind. Fig. 7-10.

Another 2-rotor rotary engine is featured in Fig. 7-11. This engine uses two spark plugs per rotor. See how the rotor internal gear engages the end housing fixed gear. Fig. 7-11.

ROTARY ENGINE OPERATION

The rotary engine is a four-cycle engine and performs the

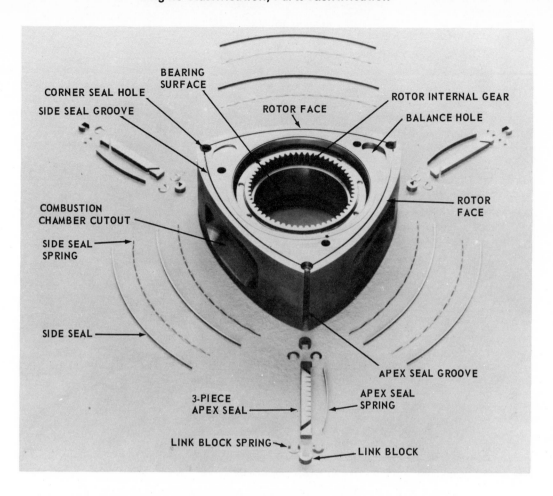

Fig. 7-8. Rotor and rotor seals. (NSU)

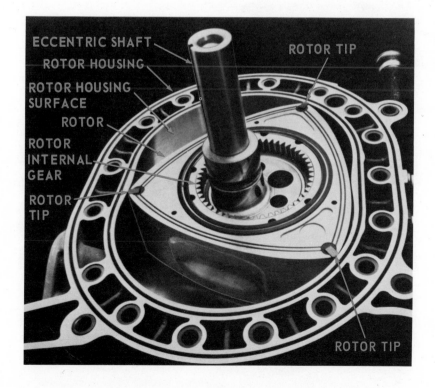

Fig. 7-9. Rotor mounted on eccentric shaft and in place in the rotor housing. You can see that each rotor tip almost touches the housing. (Mazda)

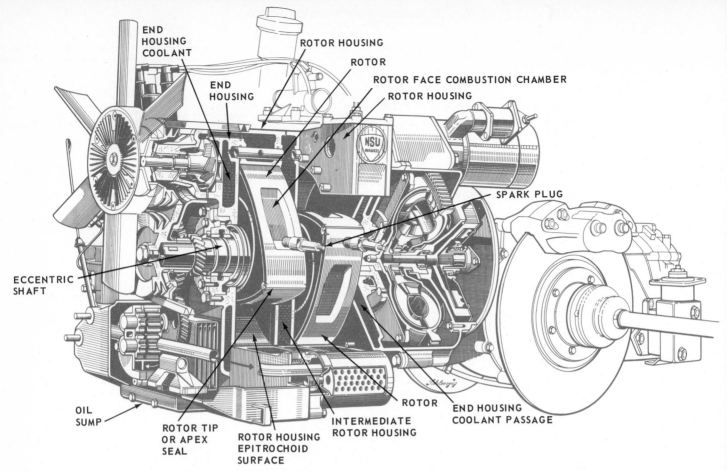

Fig. 7-10. Two-rotor Wankel rotary piston engine. Study part names and location. Note combustion chamber shape formed in each rotor face. (NSU)

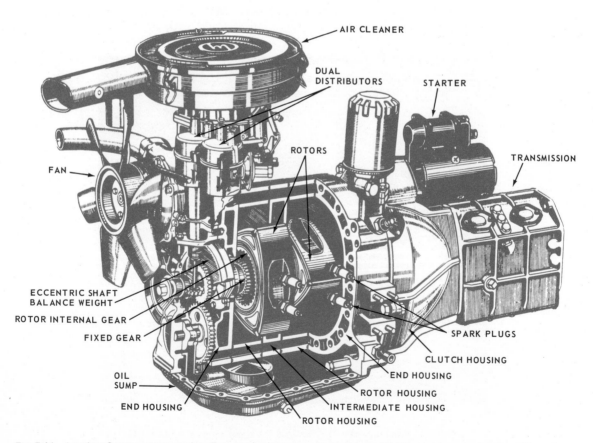

Fig. 7-11. Another 2-rotor rotary engine. Two spark plugs per rotor and a dual ignition system are used on this engine. One plug (the leading plug) fires a fraction of a second before the trailing plug. (Mazda)

four distinct strokes involved, i.e., intake, compression, power and exhaust.

Each face of the triangular rotor acts much as a conventional piston in that it draws in the mixture, compresses it, applies the power of the burning charge to the eccentric shaft and exhausts the burned mixture. Instead of a reciprocating action, however, the rotor continually revolves in the same direction as the eccentric shaft.

FIXED GEAR FORCES ROTOR TO REVOLVE AT ONE—THIRD ECCENTRIC SHAFT SPEED

If the rotor were mounted freely on the eccentric shaft rotor journal, the rotor would merely whirl as a fixed part of the shaft and the tips could not follow the epitrochoid rotor housing surface.

By meshing the rotor internal gear with a fixed gear attached to the end housing, when the eccentric shaft turns, it forces the rotor to "walk" around the fixed gear. This "walking" action causes the rotor to revolve around the eccentric shaft rotor journal in the same direction as the eccentric shaft and, at the same time, keeping all three rotor tips constantly close to epitrochoidal wall of rotor housing.

This planetary motion of rotor constantly changes chamber volume between each rotor face and rotor housing.

Study the rotor action in A, B, C, D and E, Fig. 7-12. In A note that the red line on the eccentric shaft and the red dot on the lower rotor lobe are aligned. In B, the eccentric shaft nas moved to the right (clockwise) — see red line. The rotor has also moved clockwise but at a slower speed. Follow the alignment of the red line and red rotor dot from A through E. Actually, the eccentric shaft must make THREE complete revolutions in order to turn the rotor ONE full turn. The rotor

then, turns at ONE-THIRD eccentric shaft speed. Also note how the chamber volume for each rotor face is changed as the rotor orbits about the fixed gear. Fig. 7-12.

PERFORMING THE FOUR—STROKE CYCLE

To better understand just how these different strokes or phases (we shall refer to them as phases so as to differentiate between them and the "strokes" of the reciprocating engine) are accomplished, study the larger bottom sketches, 1 through 5, in Fig. 7-12. We will follow ONE rotor face through the complete cycle. Note that this face lies between rotor tips A and B.

In 1, Fig. 7-12, rotor face A-B has rotated past the intake port. As tip B passed the intake port, the chamber volume rapidly increased. This action formed a vacuum causing fuel mixture to flow into the enlarged chamber area. Note that the chamber is packed with fuel mix at this point. This is the INTAKE phase.

In 2, Fig. 7-12, the rotor has turned further to the right and in so doing, chamber size has started to reduce. In that the fuel mixture is now trapped in this rapidly diminishing chamber, COMPRESSION takes place and will continue until rotor face A-B has turned to the position shown in 3. At this point, maximum compression is applied. Note how the chamber size has been greatly reduced, 2 and 3, Fig. 7-12.

When compression is at the maximum, the spark plug ignites the fuel charge, 3, Fig. 7-12. When the charge ignites, the rapidly expanding gases apply pressure to the A-B face of the rotor forcing it to continue walking around the fixed gear. As it rotates, it transmits the power of the burning gas to the eccentric shaft rotor journal causing the eccentric shaft to spin, 4, Fig. 7-12. This is the POWER phase.

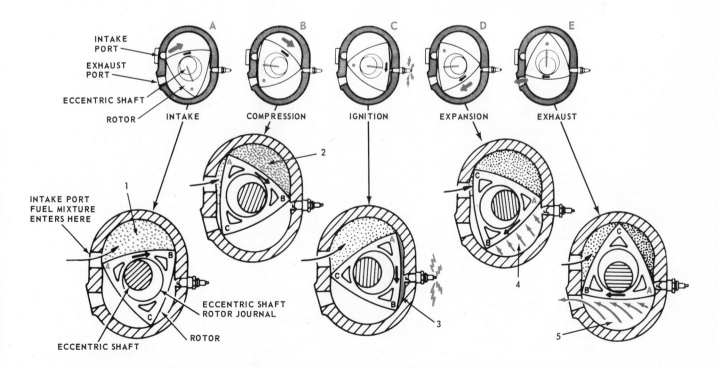

Fig. 7-12. Rotor action within the rotor housing. Rotor internal gear and housing fixed gear are not shown.   (NSU)

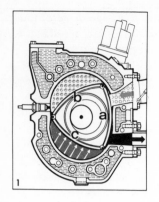
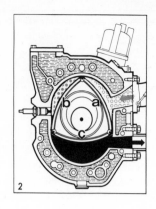
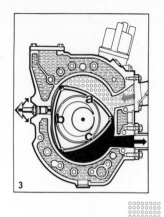
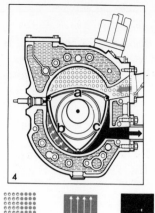

INTAKE   COMPRESSION   POWER   EXHAUST

Fig. 7-13. All three rotor faces (A—B—C) are constantly working in one phase (intake, compression, power, exhaust) or the other.

The burning gas will continue to revolve the rotor until tip B uncovers the exhaust port. When this happens, the still burning gas, under considerable pressure, will rush out the exhaust port as in 5, Fig. 7-12. This is the EXHAUST phase.

Follow rotor face A-B through 1 to 5 and note how, as the ports are uncovered and the chamber volume changes, each phase of a four-cycle operation is performed.

ALL THREE ROTOR FACES "WORK"

As mentioned, each of the rotor faces (also called flanks) perform a true four-cycle as shown in Fig. 7-12, and as such, become "working" faces. Study the four figrues in Fig. 7-13. Note that the engine rotation, as viewed from this end, is COUNTERCLOCKWISE. Fig. 7-12 showed CLOCKWISE rotation.

In 1, Fig. 7-13, face A has just finished the exhaust phase and as the intake port is opend, is set for the INTAKE. Face B is in the COMPRESSION phase and face C is the POWER phase.

In 2, Fig. 7-13, face A is now in the INTAKE phase, face B well through COMPRESSION and face C has started the EXHAUST.

As the rotor turns, 3, face A is well into the INTAKE, face B has reached maximum COMPRESSION and face C is in the middle of the EXHAUST phase.

Further rotation, as in 4, Fig. 7-13 has caused face A to reach full INTAKE. Face B is now under POWER and face C is completing the EXHAUST phase.

You can easily see from Fig. 7-13 that each rotor face, in turn, will perform the intake, compression, power and exhaust phases and that as each face passes the spark plug, it will have taken in a charge of fuel mix, compressed it and will be ready to apply POWER to the eccentric shaft.

In that the rotor turns one-third eccentric shaft speed, the eccentric shaft will receive ONE power impulse for every ONE revolution of the shaft.

A 2-rotor rotary engine, such as pictured in Figs. 7-10 and 7-11, will produce TWO power impulses for each revolution of

the eccentric shaft. This, coupled with the fact that there is no reciprocating motion, produces a very smooth running engine. However, it still must be balanced.

ENGINE MUST BE BALANCED

In order to eliminate vibration, a balance weight on the front of the engine as well as one incorporated in the flywheel, is used to counterbalance the rotor assemblies. Fig. 7-14 illustrates the use of such weights.

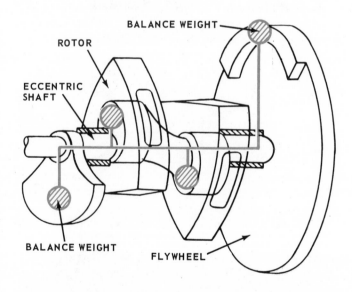

Fig. 7-14. Counterbalance weights offset that of the rotor assemblies. The relatively heavy flywheel also smooths out torsional (twisting action) vibration caused by intermittent power impulses.

COMPRESSION RATIO

The compression ratio of the rotary engine is computed much the same as that for the reciprocating piston engine. The minimum chamber capacity (full compression) is divided into the maximum chamber (full intake) capacity. Fig. 7-15 shows the minimum and maximum chamber capacities for both a rotary and a reciprocating piston engine. If the minimum chamber capacity was 4 cubic inches and the maximum 32 cubic inches, the compression ratio would be 8 to 1 (minimum into maximum). Fig. 7-15.

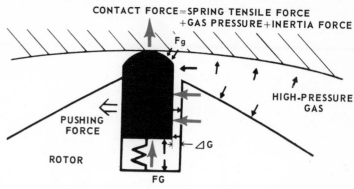

CONTACT FORCE=SPRING TENSILE FORCE +GAS PRESSURE+INERTIA FORCE

Fig. 7-16. The apex seal is subjected to spring, gas and inertia pressure. (Mazda)

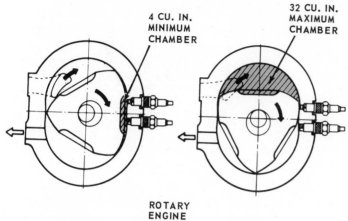

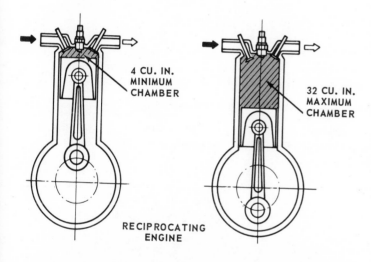

Fig. 7-15. Rotary and reciprocating engine compression ratios are computed the same way. The MINIMUM chamber capacity is divided into the MAXIMUM chamber capacity. These particular engines would have a ratio of 8 to 1. (Mazda)

ROTOR APEX SEALS

Tip or apex seals remain in contact with the curved rotor housing walls at all times. They slide across the surface at very high speeds.

In addition to the normal apex seal spring pressure, the seals are forced against the housing wall by gas pressure working in the seal groove and also by the strong inertia force generated by the spinning rotor that literally jams the seals outward at high rpm.

A cross section view of a typical apex seal is pictured in Fig. 7-16. Note the spring, gas and inertia forces working on the seal.

Despite the problems faced by the apex seals, careful design and the use of special materials such as graphite, carbon, silicon nitride, etc., produce seals with a very long service life.

An apex seal and spring are shown in Fig. 7-17.

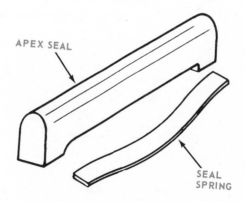

Fig. 7-17. One type of apex seal and spring.

ROTOR SIDE SEALS

Both sides of the rotor have spring loaded seals to close the very small running clearance between rotor sides and the housing surface. The seals provide a barrier against compression loss and also control oil consumption. The seals can incorporate a rubber O-ring at the bottom of the seal groove to prevent oil transfer between seal and seal groove.

The side seals operate under conditions much less severe than the tip or apex seals and as such can be made of regular cast iron similiar to the material used for conventional piston rings. Typical side seals are shown in Fig. 7-8.

A section of a rotor, illustrating an apex seal, corner seal and side seals, is illustrated in Fig. 7-18.

ROTARY ENGINE CARBURETION AND EXHAUST

The rotary engine uses a conventional fuel system (carbur-

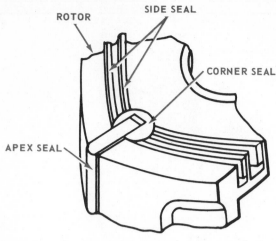

Fig. 7-18. Section of rotor tip showing juncture of apex, side and corner seals. (Mazda)

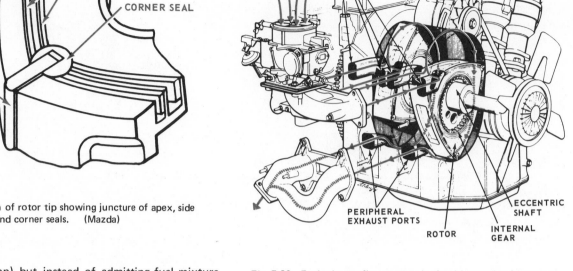

Fig. 7-20. Fuel mixture flow pattern in the side port intake system. (NSU)

etor or fuel injection) but instead of admitting fuel mixture and exhausting burned gases through valves, ports (holes) are used. The rotor tips, as they revolve, open and close the ports.

SIDE PORT INTAKE

One type of rotary engine uses side ports for the intake system. Each rotor side housing, as well as the intermediate housing, has an intake port. The rotor side seals slide across the port opening. A schematic of the side port design is portrayed in Fig. 7-19.

housing while the exhaust ports are located in the rotor housings. Fig. 7-20.

PERIPHERAL PORT INTAKE

A cutaway view of a rotary engine using PERIPHERAL intake ports (intake port, as well as exhaust, is loacted in the periphery of the rotor housings), is illustrated in Fig. 7-21.

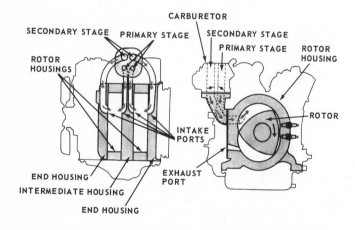

Fig. 7-19. Schematic illustrating SIDE PORT intake. At low speeds, the primary stage ports feed fuel mix. At high speeds, the carburetor secondary system opens and fuel mix enters through both primary and secondary ports. Exhaust port is in the rotor housing.

A cutaway view of a SIDE port intake system is shown in Fig. 7-20. Follow the fuel mixture flow from carburetor, into rotor chamber and on out the exhaust ports. Note the intake ports are located in the side housing and in the intermediate

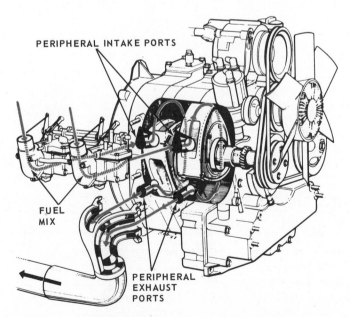

Fig. 7-21. Rotary engine using PERIPHERAL INTAKE ports. The left hand rotor apex seal is just passing over the intake port. The upper rotor face is just completing its intake phase. The back rotor face is under compression and the bottom face is finishing the exhaust phase. (NSU)

Study the fuel flow. In this design, the apex seals slide across both the intake and exhaust ports.

COOLING THE ROTARY ENGINE

Many small rotary engines for snowmobiles, etc., utilize air cooling. In these engines, a blast of cooling air is forced across a series of cooling fins cast into the engine.

The incoming fuel mixture is sometimes employed to assist in air cooling. The fuel mixture is passed from the carburetor through one end housing, through openings in the rotor and eccentric shaft, through the other housing and back around and into the rotor housing through a peripheral port. The fuel mixture, as it flows through the openings, tends to cool both rotor and eccentric shaft. This technique is shown in Fig. 7-22.

Automotive applications of the rotary engine utilize water cooling. The system incorporates a radiator, water pump, thermostate, and cooling passages through end and rotor housings. The system functions the same as that on a reciprocating piston engine.

A schematic view of a typical rotary engine lubrication system is illustrated in Fig. 7-23. In addition to lubricating all

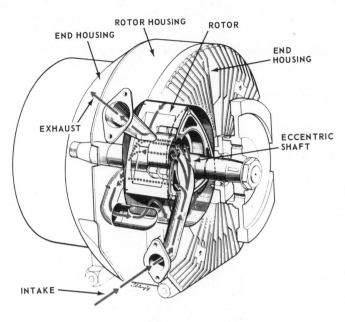

Fig. 7-22. Using the incoming fuel mixture to cool the rotor and eccentric shaft. Note how mixture passes through opening in rotor and shaft.

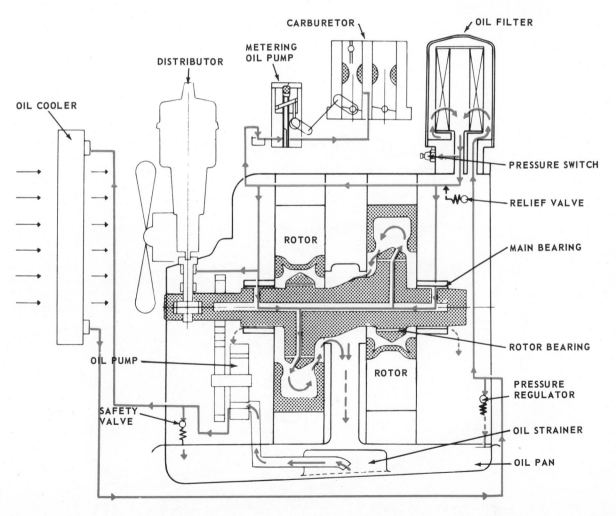

Fig. 7-23. Schematic of a rotary engine lubrication system. An oil cooler is employed. A metering oil pump injects a carefully measured amount of oil into the fuel so as to provide a positive and controlled amount of lubrication for the apex seals. (Mazda)

moving parts, the oil helps greatly to keep the rotor within the proper temperature range. By tracing the path of the oil in Fig. 7-23, you will see that oil is pumped through hollow cavities in the rotor and will thus carry off the excess heat.

A cross sectional view of a rotary engine is shown in Fig. 7-24. Note how the rotor is hollowed out for the passage of oil for cooling.

ROTARY ENGINE EMISSION CONTROL

As with the reciprocating engine, the rotary engine utilizes various emission controls. A typical emission control system for a rotary engine is portrayed in Fig. 7-25.

VALVE LOCATION CLASSIFICATIONS

Engines are also classified according to the valve location. Various arrangements have been used, but modern practice has settled on the I-head, F-head and L-head. The I-head is the most popular. The names are derived from the resemblance to letters of the alphabet. This will be demonstrated in forthcoming sketches.

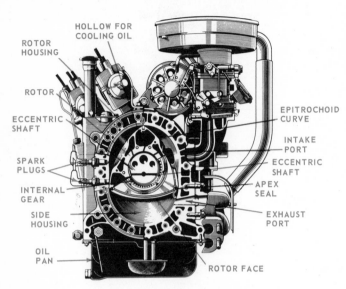

Fig. 7-24. Transverse cross-sectional view of a rotary engine. Note hollows in rotor for cooling oil.

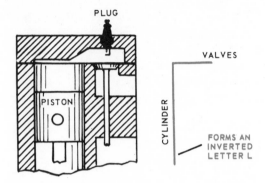

Fig. 7-26. L-head engine. Valves are in the block; both on same side.

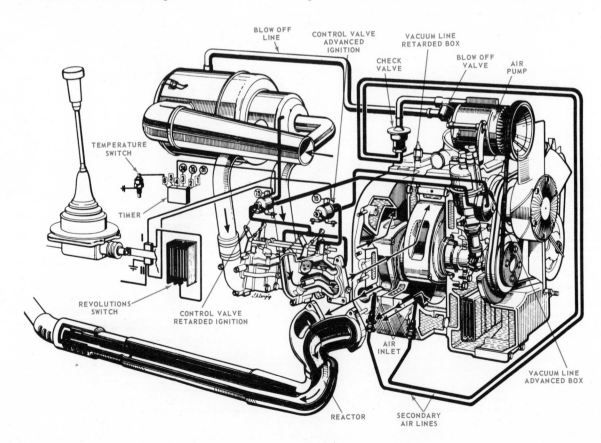

Fig. 7-25. Typical rotary engine emission controls.   (NSU)

L—HEAD

The L-head engine has both valves in the block, and on the same side of the cylinder. A line drawn up through the cylinder and across to the valves will produce the letter L in an inverted (upside down) position. Fig. 7-26 illustrates the L-head design. Fig. 7-27 shows a cross section of an actual L-head engine.

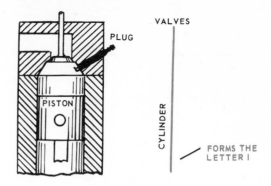

Fig. 7-29. I-head engine. Both valves are in the head.

I—HEAD

In the I-head engine both valves are located in the cylinder head. They may be arranged in a straight line or staggered. Figs. 7-29 and 7-30. This is commonly called a "valve-in-head" engine.

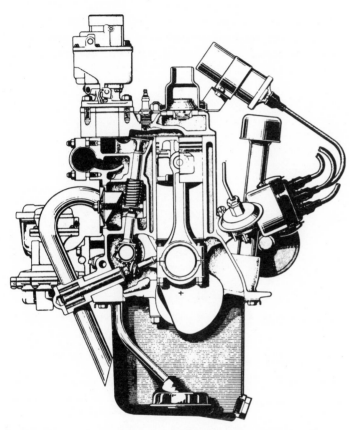

Fig. 7-27. Transverse cross section of an L-head engine. Both valves are in the block and on same side. (American Motors)

F—HEAD

The F-head places one valve in the head and one in the block. Fig. 7-28.

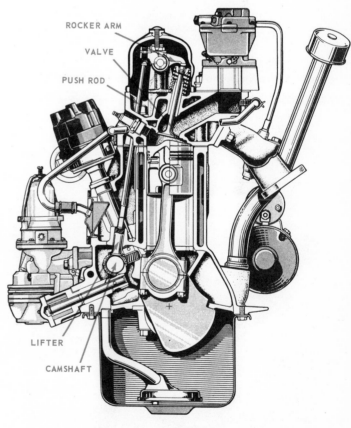

Fig. 7-30. Transverse cross section of an I-head (overhead valve) engine. Both intake and exhaust valves are in the head. (American Motors)

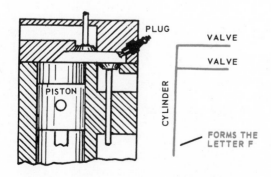

Fig. 7-28. F-head engine. One valve is in the block; one in the head.

T—HEAD

Your basic engine was of T-head design — both valves in the block and on opposite sides of the cylinder. This design was

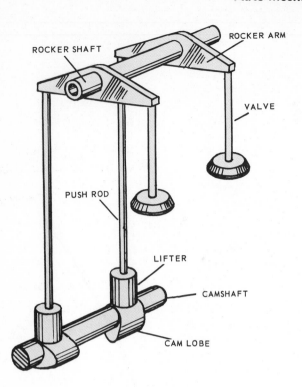

Fig. 7-31. Schematic shows simplified valve train for one cylinder of overhead valve engine.

used to make the illustrations clear. It is not currently being used in automobile engines.

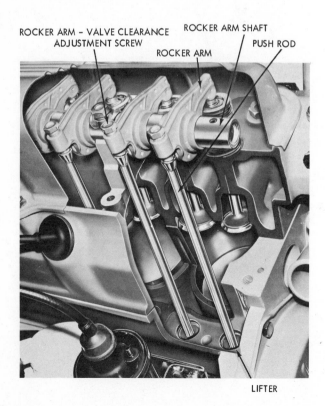

Fig. 7-32. Cutaway view shows a typical valve train used to operate overhead valves. (Plymouth)

DIFFERENT VALVE TRAIN

Your basic engine was of a type that requires the same valve train as the L-head engine in Fig. 7-27.

When valves are placed in the head, as in the I-head, they are inverted and a somewhat different train is required to operate them. Fig. 7-31 illustrates a typical valve-in-head engine valve train. Fig. 7-32 shows a cutaway view of the valve train in a modern engine.

Fig. 7-33 depicts the overhead valve train utilizing a hydraulic lifter. Notice the different type rocker arm. Instead of oscillating on a rocker arm shaft, this type uses a ball arrangement on which the rocker arm pivots. A detailed view of the ball type rocker is shown in Fig. 7-34.

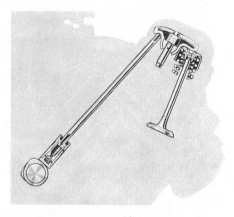

Fig. 7-33. Overhead valve train utilizing hydraulic lifter. This view shows one side of a V-8 engine. Notice absence of rocker arm shaft. This particular engine uses ball type rocker arm. (Pontiac)

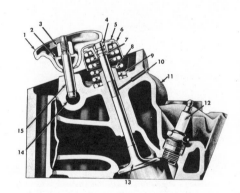

1 Rocker Arm
2 Rocker Arm Ball
3 Ball Retaining Nut
4 Valve Key
5 Spring Cap
6 Shield
7 Intake Valve Seal
8 Inner Spring
9 Outer Spring
10 Valve Guide Vent
11 Cylinder Head
12 Spark Plug
13 Intake Valve
14 Oil Gallery in Head
15 Rocker Arm Stud

Fig. 7-34. Detailed cross section showing ball type rocker arm. Rocker arm pivots on rocker arm ball. Entire unit is attached to rocker arm stud. This stud is not threaded into head, but is PRESSED into place. (GMC)

OVERHEAD VALVE ADJUSTMENT

Some installations provide an adjusting screw in the push rod end of the rocker arm. This screw is moved in or out until the recommended clearance exists between the valve stem and the rocker arm. Fig. 7-35.

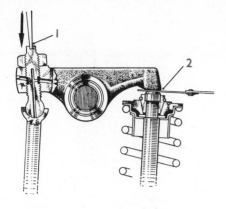

Fig. 7-35. Valve clearance adjustment. Screwdriver (1) is being used to adjust rocker arm screw to produce proper clearance at (2). A feeler gage is inserted between rocker arm and end of valve stem to determine when clearance is correct. (Austin)

Where hydraulic lifters are used, there is no clearance. The only adjustment necessary is the initial setting to place the lifter plunger at the halfway point in its travel. Some engines make the initial setting with adjustable rocker arms, while others use push rods that, when installed, provide the proper setting. Push rods of slightly different lengths are available to make minor changes. Fig. 7-36 shows such a setup.

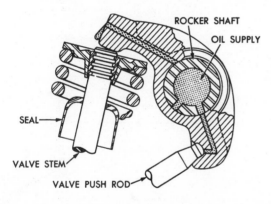

Fig. 7-36. Nonadjustable rocker arm. This type depends upon precise parts so that, upon assembly, hydraulic lifter plunger will be at mid point of travel. (Dodge)

HOW DOES THE COMBUSTION CHAMBER AFFECT VALVE LOCATION?

The shape and size of the combustion chamber is very important. A well designed chamber will allow high compression ratios without the occurrence of damaging detonation (violent explosion of the fuel charge). Compression ratios of around 8.5 or 10.5 to 1 were used for many years. Current practice has lowered the ratio to around 8 or 8.5 to 1. This aids emission control.

High compression necessitates squeezing the air-fuel charge into a small area. In order to have this small area, and still allow room for the valves to open, it is necessary to locate the valves in the head. In this position, they may open down into the cylinder. Some pistons have small indentations cut into their tops to allow the valves to clear. Fig. 7-37.

With the L-head engine, sufficient space must be left above the valves to allow full opening. This space naturally becomes part of the combustion chamber. When the piston is on TDC on the compression stroke, this space lowers the compression ratio.

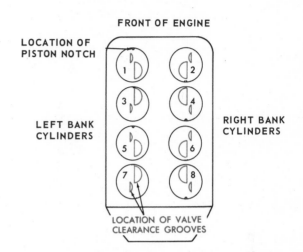

Fig. 7-37. Piston indentations used for valve clearance. Each piston has two small indentations to allow valve heads to clear. (Chevrolet)

WHAT CAUSES DETONATION?

When the highly compressed air-fuel charge is fired, the flame travels from the plug in an outward direction. As the charge continues to burn, pressure greatly increases.

If the pressure in any part of the chamber raises high enough, before the flame reaches it, this area will be hot enough to fire itself. This will produce two flame fronts and thereby increase the burning rate. Instead of firing smoothly, there is a violent explosion that literally slams the piston down. This is very damaging to an engine.

Detonation can be caused by an excessively high compression ratio or by a low grade of gasoline. An overheated valve or bit of carbon, etc., can ignite the "end gas" and cause the formation of the second flame front. Fig. 7-38.

PREIGNITION

Preignition is caused when a glowing bit of carbon, an overheated spark plug, etc., ignites the fuel charge before the spark plug fires. When preignition occurs, the premature explosion attempts to drive the piston down against the direction of rotation.

COMBUSTION CHAMBER DESIGN

Many years ago a man named Ricardo developed a combustion chamber that is still used in the L-head engine.

Ricardo found that by causing a violent turbulence in the

171

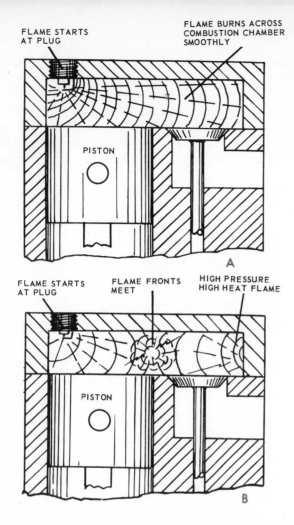

Fig. 7-38. Detonation. A—Proper flame travel. B—Two flame fronts increase burning rate and cause detonation.

compressed fuel charge, and by locating the spark plug near the center, flame travel through the charge was smooth, and compression could be raised considerably without detonation.

He located the combustion chamber over the valves and allowed the piston to come very close to the head. As the piston travels up on compression, it forces the fuel into the

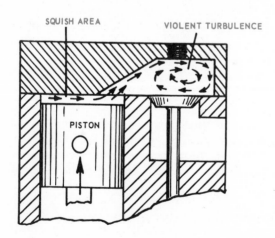

Fig. 7-39. Ricardo principle. When top of piston nears head, trapped air-fuel mixture in "squish" area shoots into combustion chamber, causing turbulence.

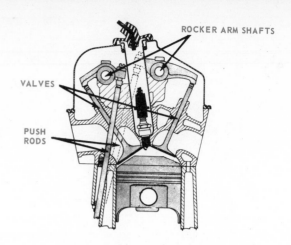

Fig. 7-40. Hemispherical combustion chamber. Valves set in two planes. This particular setup requires two rocker shafts. (Plymouth)

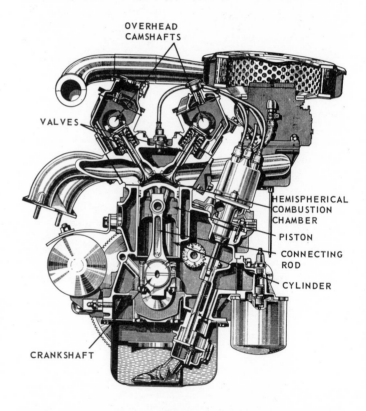

Fig. 7-41. Cross section view showing engine using a hemispherical combustion chamber. Lobes on the double overhead camshafts operate directly against the valve stem caps. (Fiat)

area above the valves. As the piston comes close to the head, the remaining mixture is rapidly squeezed, causing it to shoot into the combustion chamber, setting up a rapid turbulence. The area between the piston top and the head is referred to as the "squish" or "squash" area. Fig. 7-39.

HEMISPHERICAL COMBUSTION CHAMBER

The hemispherical type of combustion chamber is compact and allows high compression with very little detonation. By placing the valves in two planes, it is possible to use larger valves, thereby improving breathing.

A HEMISPHERICAL combustion chamber is highly efficient and despite its somewhat expensive nature (shaping chamber, more complex valve mechanism, etc.), it is used on a number of engines. Figs. 7-40, 7-41 and 7-43.

WEDGE SHAPE COMBUSTION CHAMBER

Another widely used combustion chamber shape is the WEDGE. When properly designed, it is quite efficient.

When the wedge shape is incorporated in the head, a flat top piston will produce a desirable "squish" effect. When the top of the engine block is machined at an angle of around 10 deg., it is possible to build the cylinder head flat. The combustion chamber wedge is formed by the angled cylinder and a piston with an inverted V top. This setup will also produce a "squish" effect.

The spark plug is generally located at one corner of the wedge. This provides a smooth flame travel and loads (presses down) the piston gradually. You can see by studying Fig. 7-42, that when the charge is first ignited, it presses on only a portion of the piston. This partial pressure starts the piston down smoothly. As soon as the piston has moved a little, the flame front spreads across the entire piston dome and full

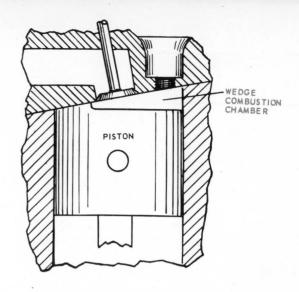

Fig. 7-42. Wedge combustion chamber. Notice that head is flat and cylinder block is machined at an angle — approximately 10 deg. Valves are set in a plane.

thrust is applied. This provides a smooth loading feature.

The engines in Figs. 7-30, 7-45 and 7-47 use a wedge chamber formed in the head.

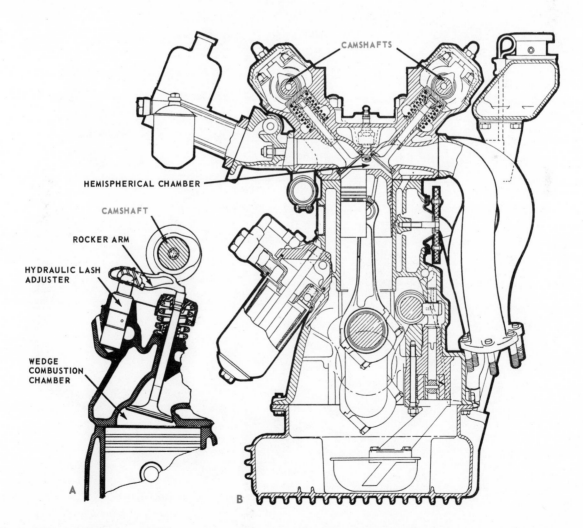

Fig. 7-43. Engines using overhead camshafts. A—uses a wedge combustion chamber with all valves in one plane. Hydraulic lash adjuster maintains zero clearance at all times. B—uses a hemispherical combustion chamber. Two camshafts are employed. Note how cams actuate valves.

NOTICE COMBUSTION CHAMBER

As you study the various engines illustrated in this chapter, notice the types of combustion chambers used. Study the valve arrangement and piston shape.

OVERHEAD CAMSHAFT

Some engines mount the camshaft over the head. Either one or two shafts can be used. This eliminates the added weight of the rocker arms and push rods that may tend to make the valves "float" at high speeds. Figs. 7-41 and 7-43.

CYLINDER CLASSIFICATION

Cylinder placement gives rise to another method of engine classification. Many arrangements have been used, but only those in popular use today will be covered in this text.

IN—LINE ENGINE

In in-line engines, the cylinders are arranged one after the other in a straight line. They are in a vertical position, or near vertical. Fig. 7-44. The number of cylinders most commonly used is either four or six. Until the popular adoption of the V-8, an eight cylinder in-line was used.

The eight cylinder in-line is no longer in common use due primarily to its great length and the fact that under high compression operation, it had a tendency to actually flex or bend. Torsional vibration would set up unless the crankshaft was unusually heavy and the block itself would tend to twist.

V—TYPE ENGINES

The V-type engine places two banks or rows of cylinders at an angle to each other — generally 90 deg.

The V-type engine has certain advantages: short length;

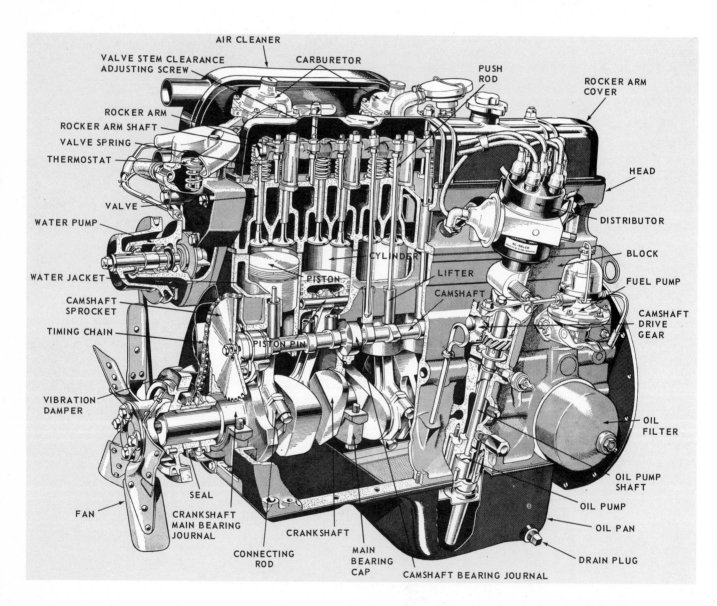

Fig. 7-44. Cutaway view of a typical six cylinder, in-line, overhead valve engine. Valves are actuated by pushrods operating rocker arms. (Triumph)

Fig. 7-45. Cross section of a typical V-type engine. Angle between cylinder banks is usually 90 deg. (Ford)

extreme block rigidity; short, heavy crankshaft; low profile that is conducive to low hood lines.

Extremely high compression ratios can be used in the V-type engine without block distortion under load. The short, heavy crankshaft is resistant to torsional vibration. The shorter block permits a reduction in car length with no sacrifice in passenger room.

For many years V-8, V-12 and V-16 engines were of the flat-head (L-head) design, but current practice on V-6, V-8 and V-12 (V-16 engines are not currently in production) engines is to use the valve-in-head (I-head) setup. Figs. 7-45 and 7-64.

HORIZONTAL—OPPOSED ENGINES

This type of engine is like taking a V-type engine and flattening the V until both banks lie in a horizontal plane.

The advantage here is an extremely low overall height, which makes this engine ideally suited to installations where vertical space is limited. Fig. 7-46 shows a horizontal-opposed engine. It is an air-cooled, six cylinder version.

SLANT OR INCLINED TYPE ENGINE

In reality an inclined type engine is an in-line engine with the cylinder block slanted to one side. The engine in Fig. 7-47 illustrates this type. Notice that by slanting the engine, space is provided to utilize a long branch, ram type intake manifold.

OTHER CYLINDER ARRANGEMENTS

There have been many methods used in locating cylinders. Fig. 7-48 shows one other type. This is termed a RADIAL engine, which has been very popular for aircraft use.

FIRING ORDER

Firing order refers to the sequence in which the cylinders fire. Firing orders are given from front to rear. The order in which the engine must fire is determined by the number of cylinders and the arrangement of the crankshaft throws. Firing

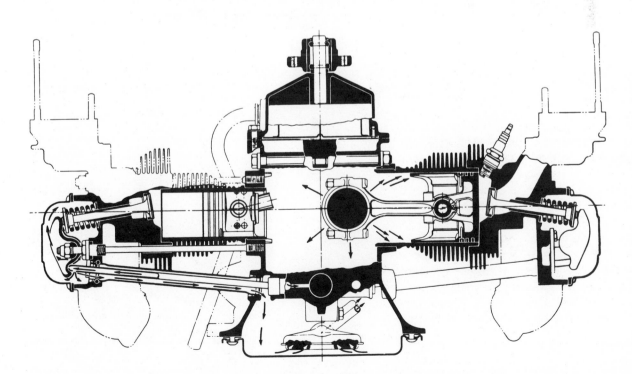

Fig. 7-46. Cross section of horizontal-opposed engine. Air-cooled engine has a very low vertical profile, making it adaptable to limited space.

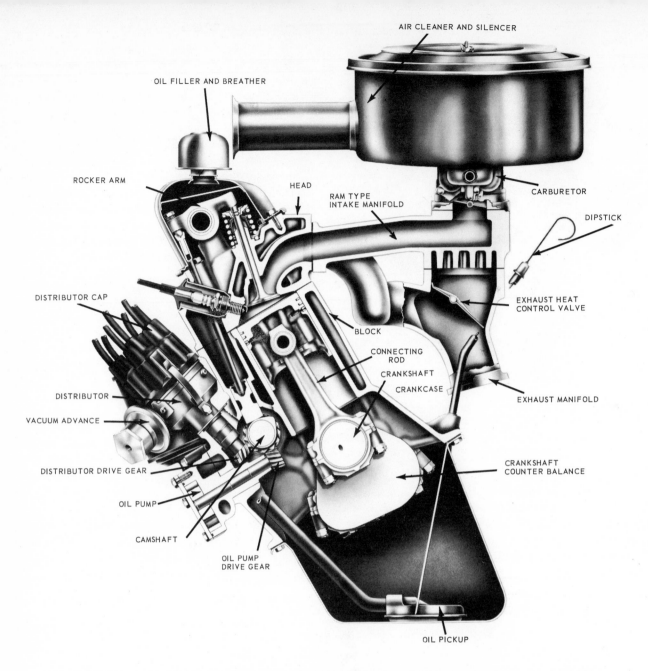

AIR CLEANER AND SILENCER

OIL FILLER AND BREATHER

ROCKER ARM

HEAD

RAM TYPE
INTAKE MANIFOLD

CARBURETOR

DIPSTICK

DISTRIBUTOR CAP

EXHAUST HEAT
CONTROL VALVE

BLOCK

CONNECTING
ROD

CRANKSHAFT

DISTRIBUTOR

CRANKCASE

EXHAUST MANIFOLD

VACUUM ADVANCE

DISTRIBUTOR DRIVE GEAR

CRANKSHAFT
COUNTER BALANCE

OIL PUMP

CAMSHAFT

OIL PUMP
DRIVE GEAR

OIL PICKUP

Fig. 7-47. Cross section of slant or inclined engine. By slanting engine, overall height is reduced and room is available to use a long branch, ram type intake manifold. (Plymouth)

order for Chevrolet cars for example is: four cylinder, 1-3-4-2; six, 1-5-3-6-2-4; V-8, 1-8-4-3-6-5-7-2. The Jeep V-6 firing order is 1-6-5-4-3-2.

COOLING CLASSIFICATION

As you have learned, engines are either liquid-cooled or air-cooled. Both types of cooling are used in American and foreign made engines. Countless numbers of air-cooled engines are used to operate saws, mowers, scooters, etc.

Both types have advantages and disadvantages but, when properly designed, either will function efficiently.

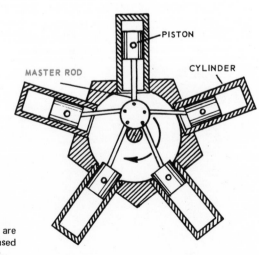

PISTON

MASTER ROD

CYLINDER

Fig. 7-48. Radial engine. All connecting rods are fastened to master rod. This design has been used for many years in aircraft — very successfully.

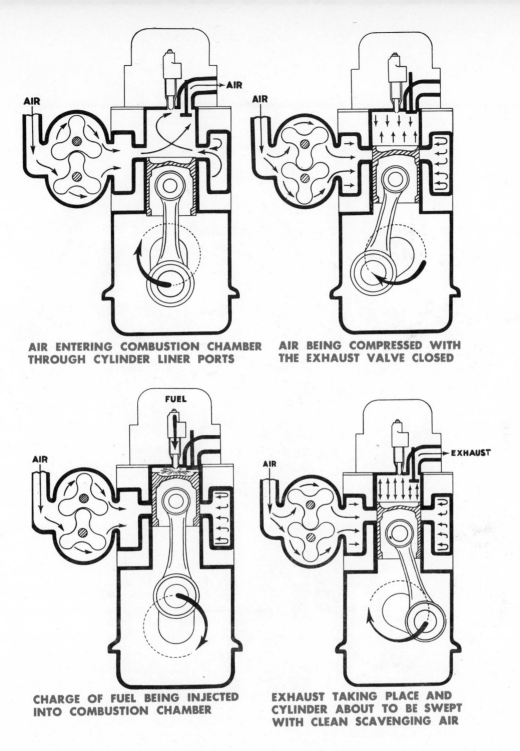

AIR ENTERING COMBUSTION CHAMBER
THROUGH CYLINDER LINER PORTS

AIR BEING COMPRESSED WITH
THE EXHAUST VALVE CLOSED

CHARGE OF FUEL BEING INJECTED
INTO COMBUSTION CHAMBER

EXHAUST TAKING PLACE AND
CYLINDER ABOUT TO BE SWEPT
WITH CLEAN SCAVENGING AIR

Fig. 7-49. Two-stroke cycle diesel engine operation.
(Detroit Diesel Engine Div. — General Motors)

FUEL CLASSIFICATION

You have also learned that automobile engines can use gasoline, LP-Gas or diesel fuel. Gasoline powers the vast majority of cars. The diesel engine has found wide application in the trucking field.

Both gasoline and LP-Gas utilize the same basic type of engine. One of the principal differences is in carburetion. The true diesel does not use a carburetor and needs no ignition system. Due to the pressures and stresses involved, diesel engines are very sturdily built. The theory of diesel engines was covered in the FUEL SYSTEM section. Diesel engines are either two or four-stroke cycle types.

TWO—STROKE CYCLE DIESEL

A two-stroke cycle diesel engine is somewhat different than the gasoline engine of this type discussed earlier in this chapter. The diesel two-stroke cycle engine utilizes an exhaust valve. No fuel mixture enters the crankcase, and it uses a Rootes type supercharger to force air into the cylinder intake ports. Fig. 7-49 shows this type of two-stroke cycle diesel

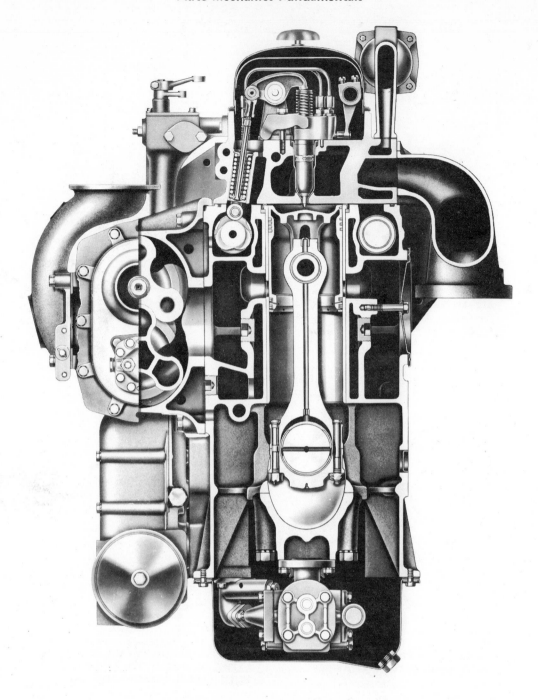

Fig. 7-50. End cross section of series 71 diesel engine.
(Detroit Diesel Engine Div. — General Motors)

engine operation. Notice that air rushing in through cylinder liner ports helps scavenge the cylinder by flowing out the exhaust passage for a portion of the piston travel. As air only enters, there is no loss of fuel as in the gasoline two-stroke cycle engine.

Fig. 7-50 shows an end cross section view of a typical GM series 71 diesel engine. Notice the heavy crankcase webbing to add rigidity and strength.

GAS TURBINE ENGINE

A great deal of research is being carried on in an endeavor

to adapt the gas turbine engine to meet modern automotive needs. The gas turbine has proved to be effective for both aircraft and heavy trucking use.

The big advantage of the gas turbine is its freedom from excessive vibration and its simplicity of design. With proper development and adequate adaptation, the gas turbine has great potential. Fig. 7-51 is a schematic of a gas turbine.

Notice that the turbine is divided into two basic sections — the GASIFIER section and the POWER section.

The compressor, in the manner of a centrifugal supercharger, forces air into the burner section. As the fuel is burned, it leaves the burner area at high speed. It passes

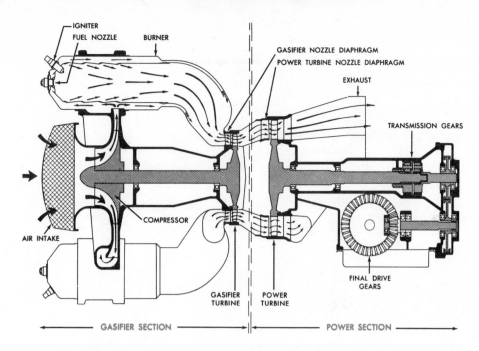

Fig. 7-51. Schematic showing operation of a gas turbine engine.

through the gasifier nozzle diaphragm and is directed against the gasifier turbine blades.

This drives the gasifier turbine at high speed. Since it is on the same shaft as the compressor, the gasifier turbine drives the compressor at the same high rpm.

After leaving the gasifier turbine, the hot gases are directed through a second set of stationary blades. These blades direct the gases against the power turbine, causing it to spin at high speed.

The power turbine drives transmission reduction gears to lower the rpm to a usable level.

If desired, the gases can be carried through other power turbines until all residual (what is left) power is utilized.

EXPERIMENTAL ENGINES

The search for better engines is never ending. Many ideas have been tried and adopted, others have failed. There are several types of engines such as the Stirling, gas turbine, steam, stratified charge, different adaptations of the rotary, etc., under development at the present time. Much experimentation and testing is being done on electrically driven cars. Whether or not these types of engines will find successful application as automobile power plants, only time will tell.

ENGINE MOUNTING

It is common practice to mount the engine on resilient rubber pads. This isolates engine vibrations and prevents their transfer to the car body. It also eases torque (twisting strain) loading of the drive line.

The "three point" suspension is popular. This method uses one mount on either side of the engine near the front and one at the back, usually under the transmission. An alternate plan

is two; one on either side near the back, with one at the front. The front mount would be near the center of the engine.

It is advisable to support the engine somewhat up on the sides to minimize torque deflections. Fig. 7-52 illustrates a typical mounting pattern.

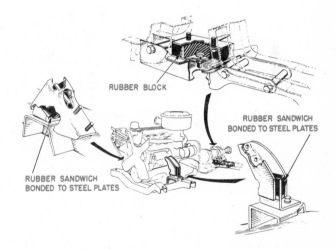

Fig. 7-52. Engine mounts and mounting pattern. This is a typical three point suspension pattern. (Plymouth)

DANGER!

ENGINE REMOVAL AND INSTALLATION CAN BE QUITE DANGEROUS. MAKE CERTAIN THAT ALL LIFT EQUIPMENT IS IN GOOD CONDITION. SECURE THE LIFT STRAP OR CHAIN WITH STRONG BOLTS, WITH THE BOLT THREADS TURNED IN DEEPLY. IF A STUD IS

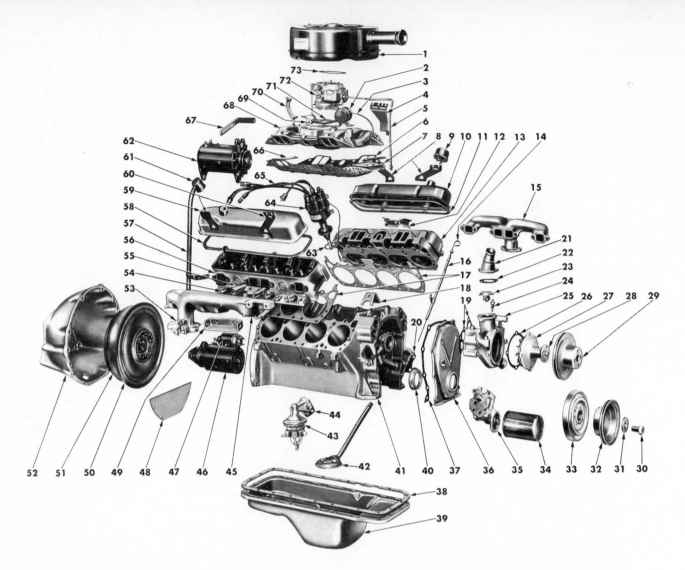

Fig. 7-53. Exploded view of a V-8 engine external parts: 1—Air cleaner. 2—Coil. 3—Vacuum line. 4—Plug wire holder. 5—Fuel line. 6—Gasket. 7—Gasket. 8—Plug wire holder. 9—Breather cap. 10—Rocker arm cover. 11—Gasket rocker cover. 12—Bracket. 13—Cylinder head. 14—Dipstick. 15—Exhaust manifold. 16—Dipstick tube. 17—Cylinder head gaskets. 18—Distributor hold down. 19—Water pump gaskets. 20—Gasket. 21—Water outlet fitting. 22—Gasket. 23—Thermostat. 24—Engine sender — water temp. 25—Water pump. 26—Gasket. 27—Water pump impeller assembly. 28—Fan and pulley flange. 29—V-belt pulley. 30—Securing bolt. 31—Washer. 32—Crankshaft pulley. 33—Vibration damper. 34—Oil cleaner. 35—Oil pump and cleaner base. 36—Timing chain cover. 37—Gasket. 38—Pan gasket. 39—Pan. 40—Front crankshaft seal. 41—Block. 42—Oil pickup. 43—Fuel pump. 44—Gasket. 45—Exhaust manifold. 46—Starter motor. 47—Starter motor solenoid. 48—Flywheel housing dust cover. 49—Bracket. 50—Flywheel ring gear. 51—Torque converter. 52—Flywheel housing. 53—Exhaust heat control. 54—Spark plug. 55—Bracket. 56—Cylinder head. 57—Breather system outlet pipe. 58—Gasket. 59—Rocker arm cover. 60—Plug wire holders. 61—Breather pipe attachment cap. 62—Generator. 63—Gasket. 64—Distributor. 65—Plug wires. 66—Gasket. 67—Generator adjustment arm. 68—Intake manifold. 69—Well type automatic choke. 70—Bracket. 71—Gasket. 72—Carburetor. 73—Air cleaner gasket. (Plymouth)

USED, MAKE SURE THE NUTS ARE SCREWED ON FAR ENOUGH TO HOLD.

MOUNT THE PULLER STRAP TO A SPOT THAT WILL WITHSTAND THE PULLING STRAIN AND ALLOW THE ENGINE TO BALANCE.

WHEN PULLING, OR INSTALLING, KEEP CLEAR OF THE ENGINE AT ALL TIMES. WATCH HANDS AND FINGERS WHEN GUIDING IN OR OUT.

PLACE ENGINE ON THE FLOOR OR ON AN ENGINE STAND IMMEDIATELY. NEVER WORK ON AN ENGINE THAT IS SUPPORTED IN MIDAIR BY THE LIFT.

REMEMBER ENGINES ARE HEAVY AND MUST BE HANDLED WITH UTMOST CARE!

IDENTIFICATION AND CLASSIFICATION

The remainder of this chapter is devoted to engine classification and parts identification. Study ALL the engine figures in this chapter very carefully. Make certain you know the names and uses of all parts. It is most important that you build a complete and accurate mental image of the various types of engines, and the relationship of one part to the other.

As you study each figure, ask yourself these questions:

1. Is this engine air or water-cooled?
2. Is this a V-type, in-line or opposed engine?
3. Does it burn gasoline or diesel fuel?
4. If gasoline, does it have an updraft, downdraft or side

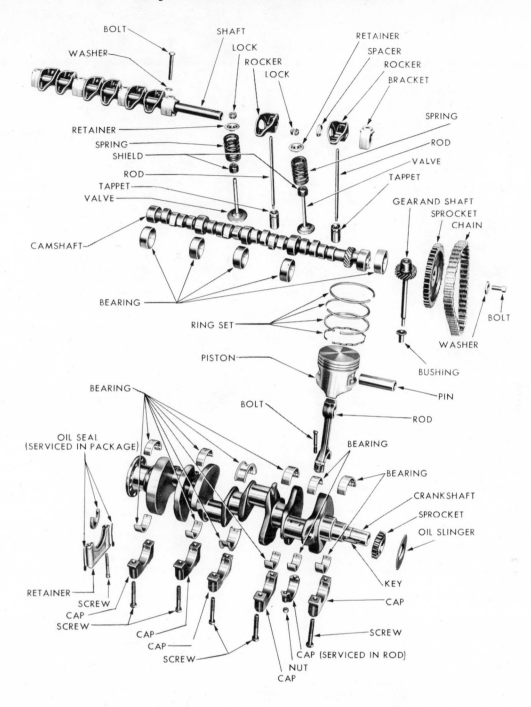

Fig. 7-54. Exploded view of V-8 engine internal parts.   (Plymouth)

draft carburetor?

5. Is it an L-head, F-head or I-head?
6. What type of combustion chamber design is used?
7. Is it a four-stroke cycle or a two-stroke cycle?
8. How many cylinders does it have?
9. Do I know the names and uses of all the parts?
10. Do I know the working relationship of one part to another?

Be certain to test your knowledge of the various systems by indentifying the cooling, lubrication, ignition and fuel systems.

SUMMARY

Engines can be classified in many ways. The most commonly used classifications are by CYCLE, VALVE LOCATION, CYLINDER ARRANGEMENT, FUEL USED, COOLING and NUMBER OF CYLINDERS.

The vast majority of auto engines are of the four-stroke cycle type. The two-stroke cycle is generally used in small utility engines. Both have advantages and disadvantages.

The I-head (overhead valve or valve-in-head) engine is in

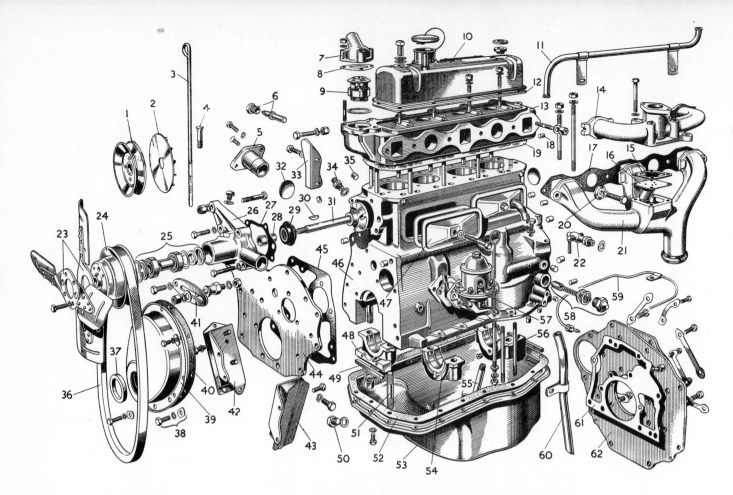

Fig. 7-55. Exploded view of a four cylinder in-line engine, external parts: 1—Generator pulley. 2—Generator cooling fan. 3—Dipstick. 4—Dipstick guide. 5—Distributor shaft bushing. 6—Water temp gage element. 7—Thermostat housing. 8—Washer. 9—Thermostat. 10—Rocker cover. 11—Heater pipe. 12—Rocker cover gasket. 13—Cylinder head. 14—Intake manifold. 15—Gasket. 16—Intake manifold heating plate. 17—Intake and exhaust manifold gasket. 18—Core plug. 19—Cylinder head gasket. 20—Manifold washer. 21—Exhaust manifold. 22—Cylinder block drain. 23—Fan blades. 24—Fan and water pump pulley. 25—Water pump bearings and seals. 26—Water pump. 27—Gasket. 28—Spacer. 29—Seal. 30—Water pump shaft key. 31—Shaft. 32—Welch plug. 33—Generator bracket. 34—Oil pressure gage connection. 35—Core plug. 36—Fan belt. 37—Oil seal. 38—Timing cover bolt. 39—Timing chain cover. 40—Gasket. 41—Generator swinging link. 42—Engine right hand front mount. 43—Engine left hand front mount. 44—Engine front plate. 45—Gasket. 46—Crankcase and cylinder block. 47—Tappet cover. 48—Front main bearing cap. 49—Cork seal. 50—Pan drain plug. 51—Front main bearing stud. 52—Pan gasket. 53—Sump. 54—Center main bearing cap. 55—Oil return pipe. 56—Rear main bearing. 57—Fuel pump. 58—Oil pressure relief valve. 59—Vaccum advance line. 60—Breather pipe. 61—Gasket. 62—Engine rear plate. (Austin)

Fig. 7-56. Transverse and longitudinal view of a single overhead camshaft V-8 engine. This engine uses fuel injection. (Mercedes-Benz)

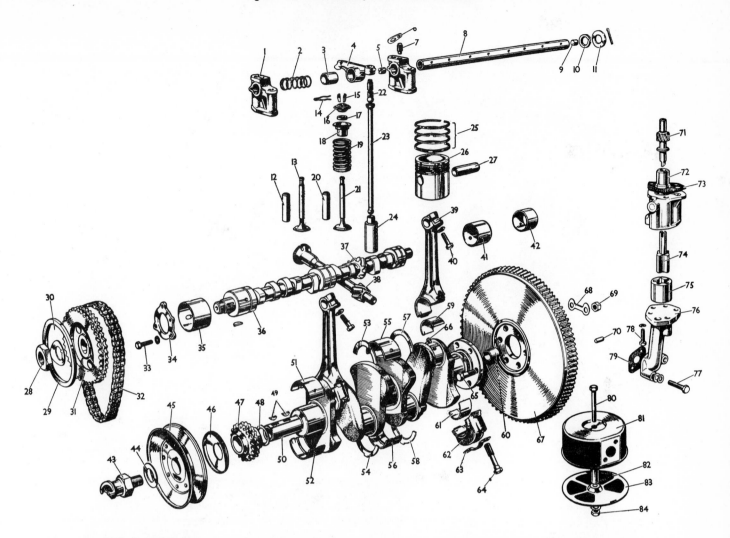

Fig. 7-57. Exploded view of a four cylinder in-line engine, internal parts: 1—Rocker arm support. 2—Spring. 3—Rocker arm bushing. 4—Rocker arm. 5—Rocker shaft end plug. 6—Locking plate. 7—Locking plate locating screw. 8—Rocker shaft. 9—Plug. 10—Spring washer. 11—Flat washer. 12—Exhaust valve guide. 13—Exhaust valve. 14—Valve circlip. 15—Keepers. 16—Keeper spring washer. 17—Oil seal. 18—Oil seal. 19—Oil seal retainer. 20—Intake valve guide. 21—Intake valve. 22—Rocker arm adjusting screw. 23—Push rod. 24—Lifter. 25—Piston rings. 26—Piston. 27—Piston pin. 28—Camshaft gear nut. 29—Chain tensioner. 30—Lock washer. 31—Camshaft gear. 32—Timing chain. 33—Bolt. 34—Camshaft locating plate. 35—Camshaft front bearing. 36—Camshaft. 37—Distributor gear. 38—Distributor shaft. 39—Connecting rod. 40—Piston pin locking bolt. 41—Camshaft center bearing. 42—Camshaft rear bearing. 43—Retaining bolt. 44—Lock washer. 45—Crankshaft pulley. 46—Oil slinger. 47—Crankshaft gear. 48—Washer. 49—Keys. 50—Crankshaft. 51—Top half front main bearing shell. 52—Lower half front main bearing shell. 53—Front upper thrust washer. 54—Front lower thrust washer. 55—Top half center main bearing shell. 56—Lower half center main bearing shell. 57—Rear upper thrust washer. 58—Rear lower thrust washer. 59—Big end upper half bearing shell. 60—Clutch shaft pilot bearing. 61—Lower half big end bearing shell. 62—Big end bearing cap. 63—Lock washer. 64—Big end bearing cap bolt. 65—Flywheel bolt. 66—Oil restriction nipple. 67—Flywheel. 68—Lock washer. 69—Flywheel bolt nut. 70—Dowel. 71—Oil pump gear. 72—Oil pump body. 73—Gasket. 74—Inner rotor. 75—Outer rotor. 76—Oil pump bottom cover. 77—Bolt. 78—Bolt. 79—Gasket. 80—Bolt. 81—Strainer. 82—Spacer. 83—Strainer bottom cover. 84—Strainer center bolt nut.   (Austin)

almost universal use today.

The most commonly used combustion chamber shapes are the WEDGE and the HEMISPHERICAL.

Detonation during high compression can be stopped by proper fuel and combustion chamber design. Ricardo, in early experiments, set forth some basic principles of design that are still incorporated in modern engines. He found that it was essential to provide turbulence in the compressed charge.

The V-8, V-6, in-line four and six cylinder, and four or six cylinder horizontal-opposed engines, dominate the field of engine design and use.

The diesel two-stroke cycle engine differs from the small gasoline engine of the same type. The diesel incorporates an exhaust valve, a supercharger and a conventional crankcase.

Experiments with new and different types of engines are always in progress. Further development of existing engines is also taking place. Considerable research is being done on steam, gas turbine and electric drive autos.

The Wankel rotary piston engine uses a triangular shaped rotor that revolves inside an epitrochroid shaped housing. It operates on the four-cycle principle. It has no reciprocating action and as a result, is very smooth running. The rotary is a simple engine with very few (rotor and eccentric shaft) moving parts in the basic engine.

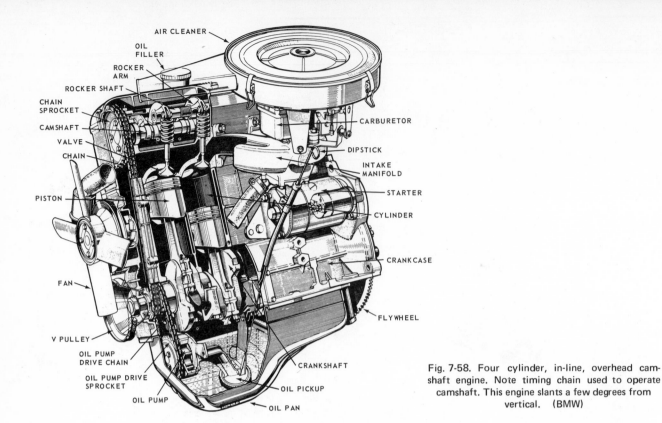

AIR CLEANER

OIL FILLER

ROCKER ARM

ROCKER SHAFT

CHAIN SPROCKET

CAMSHAFT

VALVE

CHAIN

PISTON

FAN

V PULLEY

OIL PUMP DRIVE CHAIN

OIL PUMP DRIVE SPROCKET

OIL PUMP

OIL PAN

OIL PICKUP

CRANKSHAFT

FLYWHEEL

CRANKCASE

CYLINDER

STARTER

INTAKE MANIFOLD

DIPSTICK

CARBURETOR

Fig. 7-58. Four cylinder, in-line, overhead camshaft engine. Note timing chain used to operate camshaft. This engine slants a few degrees from vertical. (BMW)

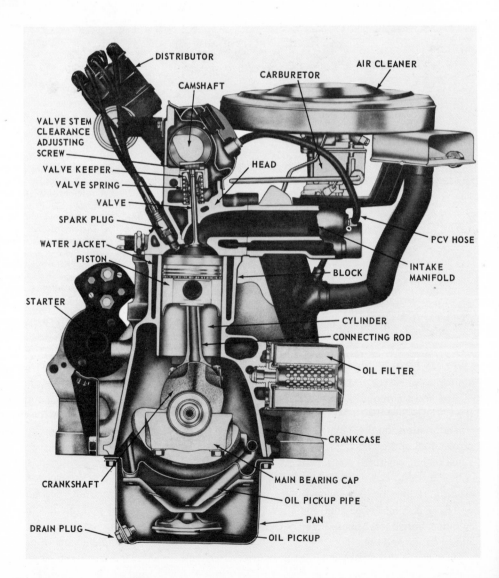

Fig. 7-59. Cross section view of an aluminum block, four cylinder in-line engine. The iron plated aluminum pistons operate directly on the silicon-aluminum cylinder bores. The cylinders are honed and electro-chemically treated to etch away the aluminum surface thus exposing the tiny particles of silicon. The overhead camshaft is belt driven. (Chevrolet)

DISTRIBUTOR

CAMSHAFT

CARBURETOR

AIR CLEANER

VALVE STEM CLEARANCE ADJUSTING SCREW

VALVE KEEPER

VALVE SPRING

VALVE

SPARK PLUG

WATER JACKET

PISTON

STARTER

HEAD

PCV HOSE

INTAKE MANIFOLD

BLOCK

CYLINDER

CONNECTING ROD

OIL FILTER

CRANKCASE

CRANKSHAFT

MAIN BEARING CAP

OIL PICKUP PIPE

PAN

DRAIN PLUG

OIL PICKUP

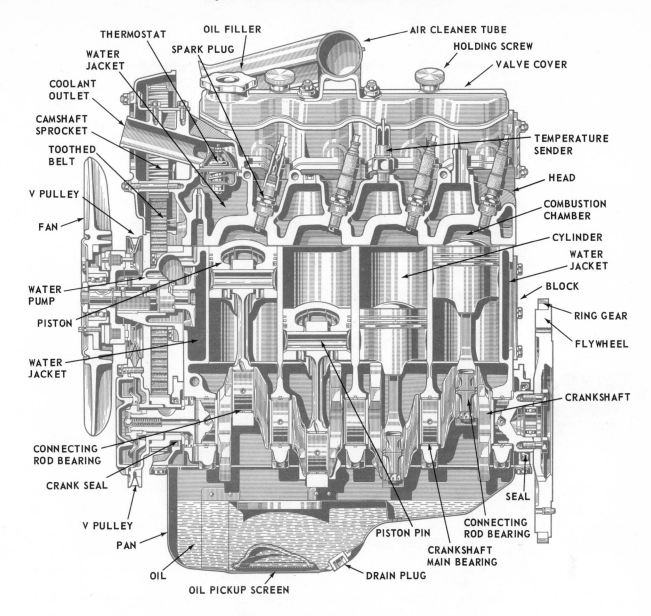

THERMOSTAT
OIL FILLER
AIR CLEANER TUBE
WATER JACKET
SPARK PLUG
HOLDING SCREW
COOLANT OUTLET
VALVE COVER
CAMSHAFT SPROCKET
TEMPERATURE SENDER
TOOTHED BELT
HEAD
V PULLEY
COMBUSTION CHAMBER
FAN
CYLINDER
WATER JACKET
WATER PUMP
BLOCK
PISTON
RING GEAR
FLYWHEEL
WATER JACKET
CRANKSHAFT
CONNECTING ROD BEARING
CRANK SEAL
V PULLEY
SEAL
PAN
CONNECTING ROD BEARING
PISTON PIN
CRANKSHAFT MAIN BEARING
OIL
DRAIN PLUG
OIL PICKUP SCREEN

Fig. 7-60. Four cylinder, in-line engine.   (Fiat)

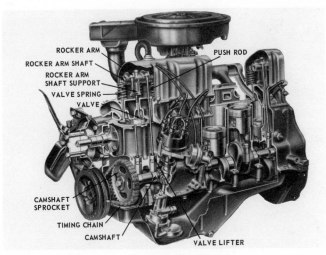

ROCKER ARM
ROCKER ARM SHAFT
ROCKER ARM SHAFT SUPPORT
VALVE SPRING
VALVE
PUSH ROD
CAMSHAFT SPROCKET
TIMING CHAIN
CAMSHAFT
VALVE LIFTER

Fig. 7-61. Cutaway of a Ford six cylinder engine. Study the valve train.

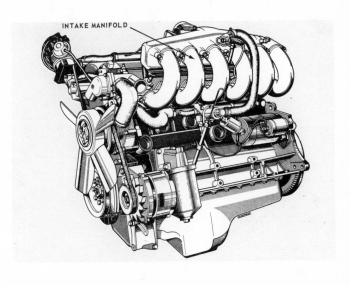

INTAKE MANIFOLD

Fig. 7-62. Fuel injected, in-line six cylinder engine.   (BMW)

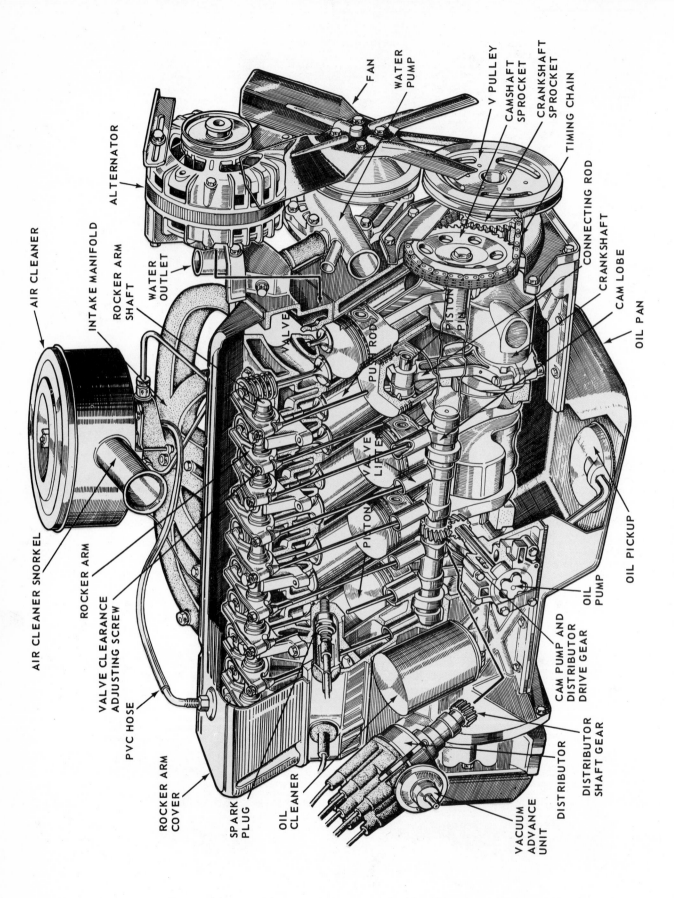

Fig. 7-63. This is the Plymouth Slant Six engine. Notice how it is inclined from the vertical.

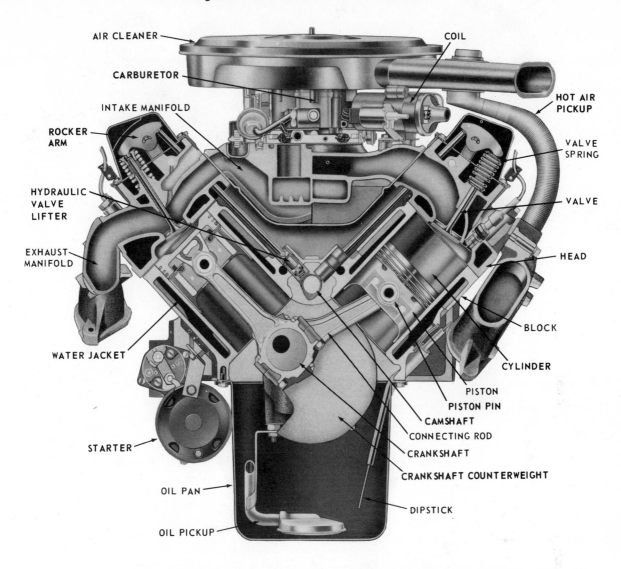

AIR CLEANER

COIL

CARBURETOR

HOT AIR PICKUP

INTAKE MANIFOLD

ROCKER ARM

VALVE SPRING

HYDRAULIC VALVE LIFTER

VALVE

EXHAUST MANIFOLD

HEAD

BLOCK

CYLINDER

WATER JACKET

PISTON

PISTON PIN

CAMSHAFT

STARTER

CONNECTING ROD

CRANKSHAFT

CRANKSHAFT COUNTERWEIGHT

OIL PAN

DIPSTICK

OIL PICKUP

Fig. 7-64. Above. Cross section of a Cadillac 472 cu. in. V-8 engine. Fig. 7-65. Below. Ford 427 V-8 high performance engine.

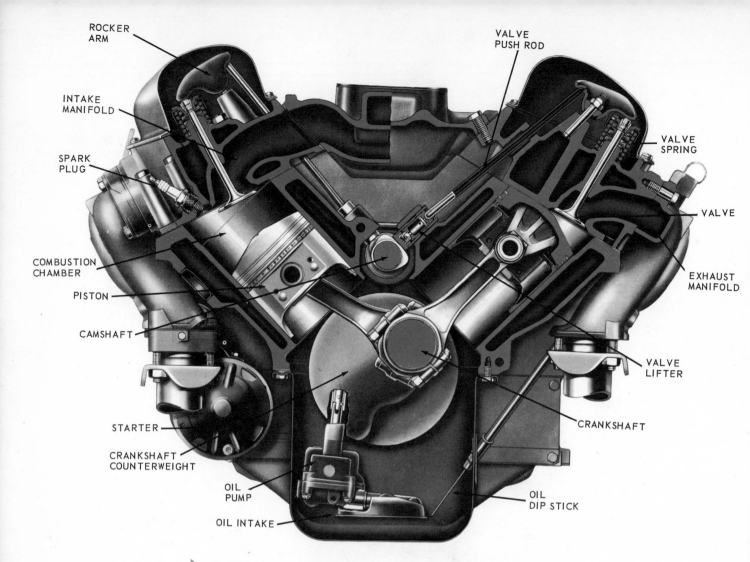

Fig. 7-66. Chevrolet V-8, front section view, with principal parts identified.

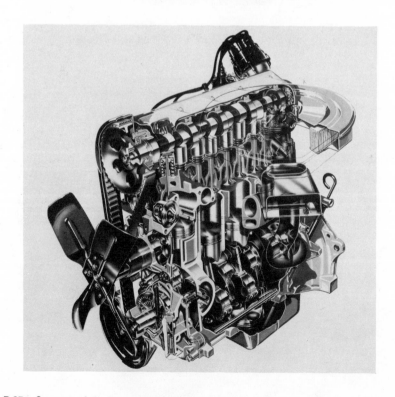

Fig. 7-67. Cutaway of the Chevrolet aluminum block, four cylinder overhead camshaft engine.

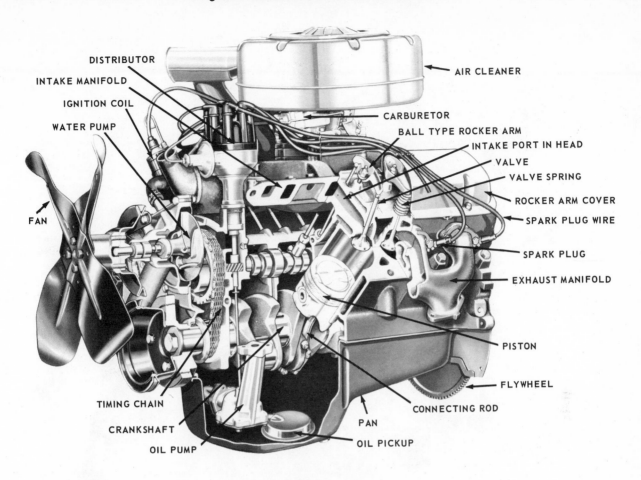

DISTRIBUTOR
INTAKE MANIFOLD
IGNITION COIL
WATER PUMP

FAN

AIR CLEANER

CARBURETOR
BALL TYPE ROCKER ARM
INTAKE PORT IN HEAD
VALVE
VALVE SPRING
ROCKER ARM COVER
SPARK PLUG WIRE

SPARK PLUG

EXHAUST MANIFOLD

PISTON

FLYWHEEL

TIMING CHAIN
CRANKSHAFT
OIL PUMP

PAN
OIL PICKUP

CONNECTING ROD

Fig. 7-68. Above. Cutaway view of a typical Ford V-8 valve-in-head engine. Fig. 7-69. Below. Cutaway of a Chevrolet 400 cu. in. V-8 engine.

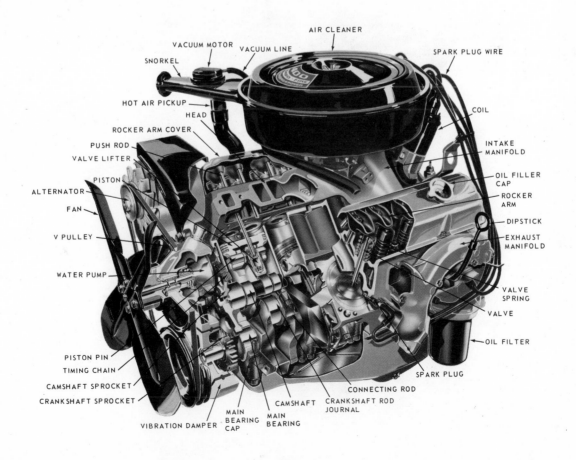

VACUUM MOTOR
SNORKEL
VACUUM LINE
AIR CLEANER
SPARK PLUG WIRE

HOT AIR PICKUP
HEAD
ROCKER ARM COVER
PUSH ROD
VALVE LIFTER
PISTON
ALTERNATOR
FAN
V PULLEY
WATER PUMP

COIL
INTAKE MANIFOLD
OIL FILLER CAP
ROCKER ARM
DIPSTICK
EXHAUST MANIFOLD
VALVE SPRING
VALVE
OIL FILTER
SPARK PLUG

PISTON PIN
TIMING CHAIN
CAMSHAFT SPROCKET
CRANKSHAFT SPROCKET
VIBRATION DAMPER
MAIN BEARING CAP
MAIN BEARING
CAMSHAFT
CONNECTING ROD
CRANKSHAFT ROD JOURNAL

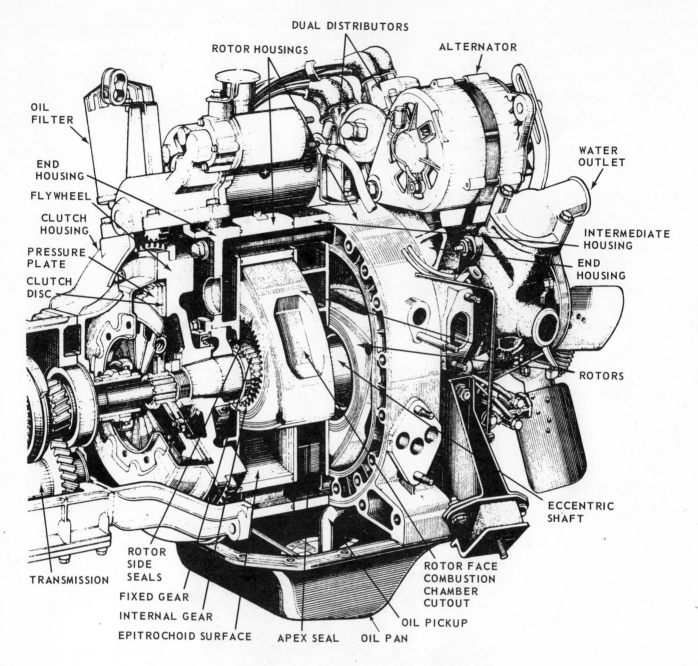

DUAL DISTRIBUTORS

ROTOR HOUSINGS

ALTERNATOR

OIL
FILTER

END
HOUSING

FLYWHEEL

CLUTCH
HOUSING

PRESSURE
PLATE

CLUTCH
DISC

WATER
OUTLET

INTERMEDIATE
HOUSING

END
HOUSING

ROTORS

ECCENTRIC
SHAFT

TRANSMISSION

ROTOR
SIDE
SEALS

FIXED GEAR

INTERNAL GEAR

EPITROCHOID SURFACE

APEX SEAL

OIL PAN

OIL PICKUP

ROTOR FACE
COMBUSTION
CHAMBER
CUTOUT

Fig. 7-70. Mazda 2-rotor Wankel rotary piston engine.   (Mazda)

Fig. 7-71. Assembled view, complete engine. Driver's side.   (Cadillac)

Engines are mounted on rubber, generally using the three point suspension system.

Engine removal and installation can be dangerous — be careful!

Mechanics, like engines, are classified in many ways. A good engine is developed by hard work, study, experimentation and experience. Top-notch mechanics are developed the same way. If you desire to be the BEST, apply these rules. If not, you may end up in the low horsepower, "blockhead" classification.

REVIEW QUESTIONS — CHAPTER 7

Check YOUR horsepower rating by answering the following questions:

1. List the strokes in a four-stroke cycle engine — in their proper order.
2. List the strokes needed in the two-stroke cycle engine. Explain the sequence of events during these strokes.
3. What are the basic differences between four-stroke cycle and two-stroke cycle engines?
4. Explain where the valves are located in I-head, F-head, and L-head engines.
5. For what development in engine design is Ricardo well known?
6. Using cylinder arrangement, list the three most popular engines in use today.
7. What is detonation and how is it caused?
8. What is preignition and how is it caused?
9. List some factors that must be used when developing a good combustion chamber.
10. What combustion chamber is in widespread use today?
11. Why do some engines use an overhead camshaft?
12. What determines an engine's firing order?
13. Engines are mounted on_____.
14. Most engines are mounted using the _____ _____ suspension principle.
15. In fuel classification, what three principal fuels are concerned?
16. How does the two-stroke cycle diesel differ from its gasoline counterpart?
17. List five important points to consider in the SAFE removal and installation of an engine.
18. Did you study all of the illustrations in this chapter? Can you identify the various systems? Do you know the names and function of all the parts?
19. In the rotary engine, the rotor turns _____ to every _____ turns of the eccentric shaft.
20. Explain the basic difference between the rotary and the reciprocating piston engine.

DID HE GET THE JOB?

Imagine that you are a garage owner and that you have just interviewed a man seeking a position as mechanic in your shop. If you had asked him the above questions, and if he gave the answers that you did, WOULD YOU HAVE HIRED HIM?

Let us hope he got the job. If, for some reason he failed, what would you tell him?

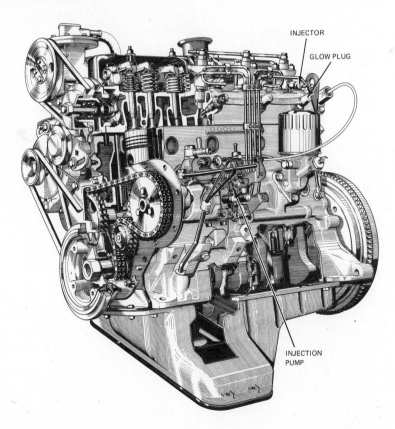

Fig. 7-72. Cutaway of a four cylinder automotive diesel engine.   (Peugoet)

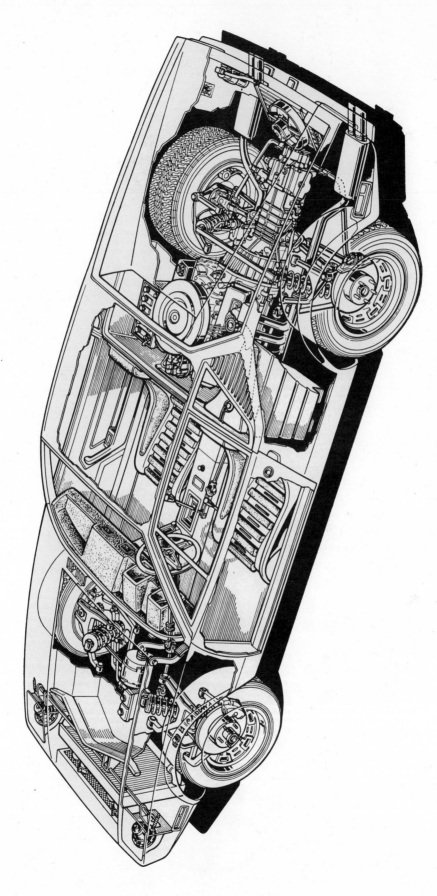

A mid-engined Ford Pantera.   (Ford)

Chapter 8

ENGINE TESTS, MEASUREMENTS

Everyone has heard the word horsepower, but very few people, including many mechanics, understand exactly what it means and how it is computed.

WHAT IS HORSEPOWER?

Many years ago men realized the necessity of devising some sort of yardstick with which to measure the work-producing ability of the gasoline engine. Since man had used horses to do his work for so long, it was natural to compare the power developed by the early engine to that produced by a horse.

The work ability of an average draught horse had long been the standard unit of work measured in foot-pounds. WORK can be defined as a force applied to a body that causes the body to move. The amount of work done can be computed by multiplying the distance the body moved by the force (weight of body). This answer would be in FOOT-POUNDS.

It was found, for example, that the average draught horse could lift one hundred pounds, three hundred and thirty feet in one minute. If one pound is lifted one foot in one minute,

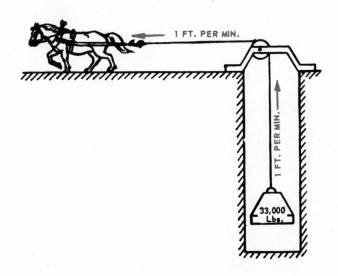

Fig. 8-1. Horsepower. Ability to perform work at rate of 33,000 ft. lbs. per min. = 1 horsepower.

one FOOT-POUND of work would be done. The horse lifted 100 pounds 330 feet in one minute so using the work formula (distance moved x force weight) we find the horse performed work at the rate of 33,000 FOOT-POUNDS per minute.

The RATE at which work is performed is termed POWER. This then became the standard measure for one HORSE-POWER. Fig. 8-1.

HORSEPOWER FORMULA

To find engine horsepower, the total rate of work in foot-pounds accomplished is divided by 33,000. If a machine lifts 100 lbs. 660 ft. per minute, its total rate of work would be 66,000 ft. lbs. Dividing this by 33,000 ft. lbs. (1 horsepower) you will find that the machine is rated as 2 horsepower (hp). Your formula would be:

$$\frac{\text{Rate of work in ft. lbs.}}{33,000 \ (1 \ hp)} = \frac{\text{Distance moved x force}}{33,000} = \frac{D \times W}{33,000} = hp$$

This formula may also be used to determine the horsepower required to perform a specific task. Let us assume that you wanted to lift buckets of earth from an excavation. The loaded bucket weight is 5,000 lbs., and you want to raise it 15 ft. per minute. Using your formula:

$$hp = \frac{D \times W}{33,000} = \frac{15 \times 5,000}{33,000} = \frac{75,000}{33,000} = 2.27 \ hp$$

The minimum amount of horsepower (disregarding friction) required would be 2.27.

POTENTIAL HORSEPOWER

If a certain engine burns a specific fuel at a certain rate, and if all the heat of the burning fuel could be converted into useful power, you would have the ideal engine. Unfortunately, engines have not reached this level. In fact, even the best engines are relatively inefficient in that the usable power falls far short of the potential.

Engine efficiency can be broken down into THERMAL EFFICIENCY and MECHANICAL EFFICIENCY. But any comparison between different engines on an efficiency basis must be conducted under identical situations. Weather, altitude, temperature, speed, etc., all affect the test results.

THERMAL EFFICIENCY

Engine thermal efficiency (heat efficiency) is based on how much of the energy (ability to do work) of the burning fuel is converted into useful horsepower.

The heat generated by the burning fuel drives the piston down, producing the power stroke. Much of this heat is lost to the cooling system, some to the lubrication system and a great deal to the exhaut system. The thermal efficiency of the average engine is around 25 percent. This means that, discounting frictional losses, the engine is losing about 70 percent of the heat energy.

FORMULA FOR THERMAL EFFICIENCY

The generally accepted formula for computing BRAKE THERMAL EFFICIENCY is:

$$\text{Brake Thermal Efficiency} = \frac{\text{Brake Horsepower (bhp)} \times 33{,}000}{778 \times \text{Fuel Heat Value} \times \text{Weight of Fuel Burned Per Minute}}$$

The 778 is Joule's equivalent, the fuel heat (calorific) value is based on the Btu (British thermal unit) per pound.

MECHANICAL EFFICIENCY

Mechanical efficiency is based on the relationship of power developed within the engine, and actual brake horsepower delivered at the crankshaft. Fortunately, mechanical efficiency is better — around 90 percent. This means that 10 percent is lost to friction within the engine. Engine friction is not constant but increases with speed.

FORMULA FOR MECHANICAL EFFICIENCY

The formula is:

$$\text{Mechanical Efficiency} = \frac{\text{Brake horsepower}}{\text{Indicated Horsepower}} = \frac{\text{bhp}}{\text{ihp}}$$

PRACTICAL EFFICIENCY

When dealing with auto engines, you are primarily interested in horsepower delivered at the drive wheels.

Every pound of gasoline that enters the engine has the ability to do a certain amount of work. There are many factors that rob energy before it gets to the drive wheels. If you figure the potential energy in a pound of fuel as 100 percent, you

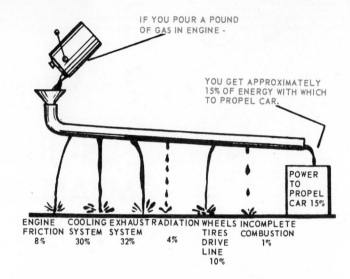

Fig. 8-2. Practical efficiency. Energy leaks out between engine and drive wheels, demonstrating some factors involved. Percentages are approximate and change from engine to engine, and also from conditions under which engine is operated.

may get as little as 15 percent delivered to the drive wheels. Fig. 8-2 illustrates some of these energy losses.

INDICATED HORSEPOWER

The indicated horsepower (ihp) is a measure of the power developed by the burning fuel within the cylinders.

In order to compute ihp, it is necessary to find the pressures within the cylinder for the intake, compression, power and exhaust strokes. A measuring device gives a constant reading of the pressures. This is placed on an indicator graph.

The next step is to establish a mean effective pressure (mep). This is done by taking the average pressure during the power stroke and subtracting the average pressure during the other three strokes. Once the mep is determined, the following formula is used to determine the indicated horsepower:

$$\text{Indicated Horsepower (ihp)} = \frac{\text{PLANK}}{33{,}000}$$

P — mep in lbs. per sq. in.
L — length of the stroke in feet
A — cylinder area in sq. in.

N — power strokes per min. $\frac{\text{rpm}}{2}$

K — number of cylinders

BRAKE HORSEPOWER

Brake horsepower (bhp) is a measurement of actual usable horsepower delivered at the engine crankshaft. Brake horsepower is not constant but increases with speed. An early method of measuring crankshaft horsepower made use of a device called a PRONY BRAKE. Fig. 8-4.

Engine Tests, Measurements

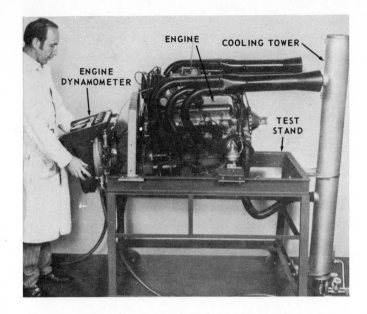

Fig. 8-3. Attaching a portable engine dynamometer. Engine is operated with wide open throttle. Dynamometer water brake is adjusted until engine load pulls rpm down to desired level, at which point gages will indicate bhp being developed. (Go-Power)

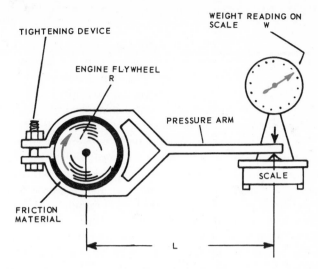

Fig. 8-4. Prony brake. Prony brake is used to measure bhp. A clamping device grasps spinning flywheel. Clamp is tightened until a specific rpm, with throttle wide open, is reached. At this point, weight reading on the scale is taken. Knowing scale reading (W), distance from center of crank to arm support (L), and rpm of flywheel (R), it is possible to compute bhp.

The formula for determing brake horsepower on the Prony Brake is:

$$bhp = \frac{2\pi \times R \times L \times W}{33,000} \quad \text{or reducing it further}$$

$$bhp = \frac{R \times L \times W}{5,252}$$

R — Engine rpm
L — Length from center of drive shaft to point where beam presses on scale.
W — Weight as registered on the scale.

The modern method of determining brake horsepower is by using an engine DYNOMOMETER or a CHASSIS DYNAMOM-ETER.

One type of engine dynamometer uses a large generator (dynamo) to which the engine is attached. The engine drives the generator and by placing an electrical load in the circuit, bhp at various rpm can be determined. Fig. 8-5.

Another popular type of engine dynamometer utilizes a water brake power absorbtion unit that attaches directly to the engine. Fig. 8-3.

The chassis dynamometer is used to measure bhp at the drive wheels. The car drive wheels are placed on two rollers and the engine is engaged. As the wheels turn the rollers, the rollers are loaded (made increasingly difficult to turn). By computing the amount of loading the wheels can handle, it is possible to determine bhp at the drive wheels. This is practical efficiency that is meaningful to the average motorist.

Chassis dynamometers are becoming quite popular. Many of the larger garages are installing them for use in engine tuning and testing. Fig. 8-6.

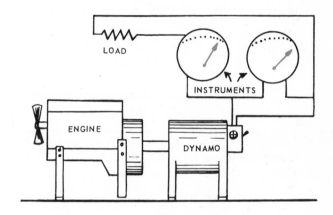

Fig. 8-5. Engine dynamometer. This schematic demonstrates general idea of dynamometer. Generator dynamo can be loaded until engine rpm is at a certain point with throttle wide open. By using instrument readings, horsepower can be calculated.

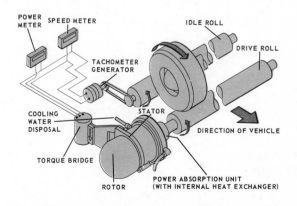

Fig. 8-6. Schematic showing operation of a typical chassis dynamometer. This unit measures horsepower delivered to driving wheels. (Clayton)

195

GROSS AND NET HORSEPOWER RATINGS

Under S.A.E. (Society of Automotive Engineers) specifications, a GROSS brake horsepower rating is the maximum horsepower developed by an engine equipped with only basic accessories needed for its operation. This would include the built-in emission controls, oil, coolant pump and fuel pump. The fan, air cleaner, muffler, generator, etc., can be left off.

The NET horsepower rating is the maximum horsepower developed by the engine when equipped with ALL the accessories such as fan, generator, muffler, air pump, air cleaner, fuel pump, coolant pump, etc., that it would have when installed in the car. The NET rating is a more informative figure for the car buyer in that it shows the horsepower that will actually be produced in normal usage.

The specified test condtions (29.38 in. Hg. barometric pressure, 85 deg. F ambient air temperature, 0.38 in. Hg. water vapor pressure or humidity), are the same for both GROSS and NET ratings.

Some other nations use a DIN horsepower rating. This is a NET rating done under slightly different conditions and specifications, so it will vary from ratings used in the United States.

FRICTIONAL HORSEPOWER

Frictional horsepower (fhp) is a measurement of the amount of hp lost to engine friction. It is computed by this formula:

$$fhp = ihp - bhp$$

As mentioned, frictional horsepower is not constant but increases with engine speed. Fig. 8-7.

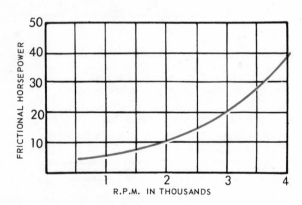

Fig. 8-7. Frictional horsepower graph. Notice how an increase in engine rpm causes more friction.

TORQUE

Engine torque (twisting motion) is the ability of the crankshaft to impart a twisting or turning force. It is measured in pounds-feet. If a pipe wrench is placed on a pipe and a

pressure of one pound is exerted one foot from the center of the pipe, the torque (Torque — Force x Distance) applied to the pipe would be one POUND-FOOT. Actually, when measuring TORQUE, the reading is given in lbs. ft. When measuring WORK, the reading is given in ft. lbs. Fig. 8-7.

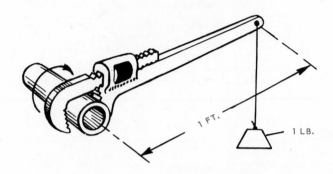

Fig. 8-8. Torque. TORQUE or twist applied to this pipe would be one pound-foot. This is torque in excess of wrench weight.

TORQUE AND BHP ARE NOT THE SAME

Engine bhp increases with engine speed. Engine torque increases with engine speed up to the point where, all factors considered, the engine is drawing in the maximum amount of fuel mixture. Torque is greatest here and any additional increase in rpm will cause torque to diminish. Fig. 8-9 compares torque and bhp.

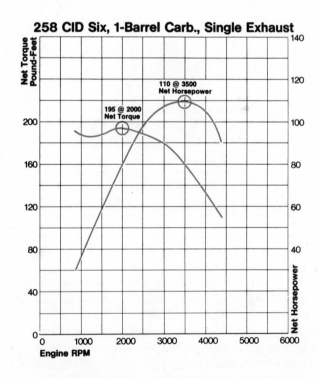

Fig. 8-9. Bhp and torque graph for ONE SPECIFIC ENGINE. Notice how horsepower increases almost to maximum speed while torque drops off at a much lower rpm. Relationship between torque and bhp will vary depending on engine design. (American Motors)

VOLUMETRIC EFFICIENCY

Volumetric efficiency (sometimes referred to as breathing ability) is the measure of an engine's ability to draw fuel mixture into the cylinders. It is determined by the ratio between what is ACTUALLY drawn in and what COULD be drawn in if all cylinders were completely filled.

As engine speed increases beyond a certain point, the piston speed becomes so fast that the intake stroke is of such short duration that less and less fuel mix is drawn in. As torque is greatest when the cylinders receive the largest amount of fuel mix, torque will drop off with a decrease in volumetric efficiency.

Many factors influence volumetric efficiency. Engine speed, temperature, throttle position, intake system design, atmospheric pressure, etc., all help in its determination.

It can be improved by adding a supercharger; straighter, smoother and ram length intake manifold; larger intake valves; better exhaust flow, etc.

A formula used to determine volumetric efficiency is:

$$\text{Volumetric Efficiency} = \frac{\text{Total volume of the charge}}{\text{Total cylinder volume (displacement)}}$$

All tests between engines must be under identical conditions. See Figs. 8-10 and 8-11.

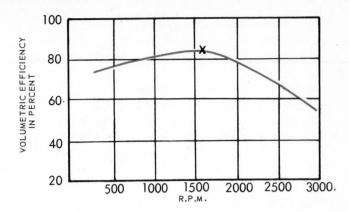

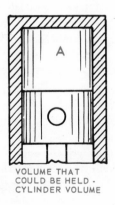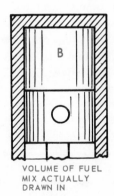

Fig. 8-10. Volumetric efficiency.

ENGINE SIZE

Engine size (not physical dimensions) is related to piston displacement. Piston displacement refers to the total number of cubic inches of space made available in the cylinder when the piston moves from the top of its stroke (TDC) to the bottom (BDC). Fig. 8-12.

All related factors being equal, the bigger the engine displacement, the more power produced. To calculate piston displacement, find the area of the cylinder (0.7854 x Diameter^2) and multiply that answer by the total piston travel

Fig. 8-11. Volumetric efficiency at various speeds. This graph is only approximate; volumetric efficiency will vary from engine to engine. X marks point of highest efficiency and highest torque.

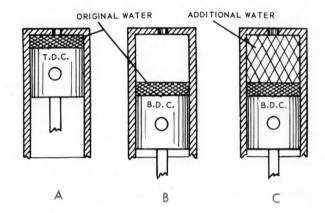

Fig. 8-12. Piston displacement. A—If you were to fill cylinder with water when piston is on TDC; B—Next, place piston on BDC; C—Then measure additional water required to fill cylinder, you would have cylinder displacement in cu. in. Space from top of piston, when piston is on TDC is called clearance volume. This is space occupied by water in A.

(stroke) from TDC to BDC. This answer is then multiplied by the number of cylinders and you have the total piston displacement in cubic inches. The formula:

$$\text{Piston Displacement} = 0.7854 \times D^2 \times \text{Travel} \times \text{Number of Cylinders}$$

For example you want to find the displacement of a V-8 engine have a cylinder diameter (bore) of 3.875 and a piston travel (stroke) of 3 inches.

0.7854 x D^2 (15015) x Travel (3 inches) x Number of Cylinders (8) = 283.0248 cu. in.

CYLINDER BORE AND PISTON STROKE

As you noticed in the foregoing section, the displacement hinged on the cylinder BORE (diameter) and the piston STROKE. The bore is the diameter of the cylinder. The stroke is the length the piston moves from TDC to BDC.

Auto Mechanics Fundamentals

When bore and stroke are equal, the engine is referred to as SQUARE. Modern practice uses a stroke that is shorter than bore diameter. This is referred to as an OVER SQUARE engine.

By reducing the stroke, piston speed is decreased. This prolongs the life of cylinders, pistons and rings.

Let us say that you want to compute the distance a ring slides up and down during one mile of driving. Your hypothetical (make believe) car will have a tire diameter of 30 in., the engine drives 1 to 1 through the transmission (standard transmission — no overdrive), and the differential (rear end) reduces drive line speed 4 to 1 (drive line turns four times for each revolution of the wheels.)

The circumference of the tire (6.283 x Radius) would be 94.245 inches. Reducing circumference to feet, you would have 7.85. If the tire rolls 7.85 feet per revolution, it will make 672.61 revolutions per mile.

The engine rpm would be four times that of the wheel rpm, so this would give 2690.44 engine rpm per mile. As the rings slide up and down for each stroke (3 in.), this would be a total of 6 in. per engine revolution. With 2690.44 rpm per mile, the rings would slide 1345.22 ft. for every mile of travel.

The same engine with a 6 in. stroke would slide the rings twice as far. This shows that the short stroke decreases wear. Not only is the sliding distance less, but the speed at which the rings slide is greatly reduced.

COMPRESSION RATIO

Compression ratio is the relationship between the cylinder volume (clearance volume) when the piston is on TDC, compared to the cylinder volume when the piston is on BDC. Fig. 8-13.

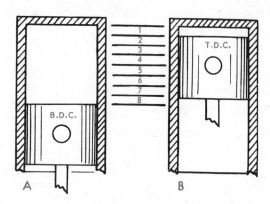

Fig. 8-13. Compression ratio. Cylinder volume in A totals 8. When piston is on TDC, cylinder volume totals 1. Volume in A is 8, volume in B is 1; compression ratio is 8 to 1.

Compression ratios range from 5 or 6 to 1 to slightly over 11 to 1 for gasoline engines, 16 to 1 for diesels.

To a certain degree, the higher the compression ratio in an engine, the more power it will develop. Such being the case, the question is, "Why not raise the compression a great deal on all engines?"

One reason this is impractical is that engine fuels of today will not let us. When a fuel mixture is compressed, heat energy is developed. The temperature at which the air-fuel mixture will ignite itself sets a limit on the amount it can be satisfactorily compressed. Another trouble is that when the compression ratio of an engine is too high, the result is detonation, or knocking of the engine.

Effective emission control, at least at the present time, has required that compression ratios be lowered from 10 or 10.5 to 1 down to about 8 or 8.5 to 1.

SUMMARY

Horsepower is a measure of the rate at which work is performed. One horsepower is the ability to perfrom work at the rate of 33,000 foot pounds per minute.

Engines are relatively inefficient. Efficiency is based on THERMAL and MECHANICAL efficiency. Thermal refers to heat efficiency which is not good. Most engines are about 25 percent efficient under the thermal basis, and about 90 percent efficient on a mechanical basis.

On a practical efficiency basis, only about 15 percent of the total potential energy of the fuel is used to drive the wheels.

Energy is lost to the cooling, lubrication and exhaust systems as well as to engine, drive line and wheel friction.

Indicated horsepower is a measure of the power developed by the burning fuel within the cylinders. This ihp is not what is delivered by the crankshaft. Crankshaft horsepower is termed brake horsepower and is measured by a dynamometer. Frictional horsepower is a measure of the horsepower lost to engine friction. Net bhp ratings are taken with all normal accessories on the engine. Gross bhp figures are with some accessories (fan, cleaner, generator, etc.) removed.

Engine brake horsepower and torque are not the same. Brake horsepower increases with rpm until maximum is reached. Torque increases to the point where volumetric efficiency is greatest and any additional speed will cause it to diminish. Torque is a measure of the engine's twisting or turning force at the crankshaft.

Volumetric efficiency refers to how well the engine is able to draw in fuel mixture. It is different from engine to engine, speed to speed, and is influenced by a number of other factors.

Engine size is usually referred to in terms of cubic inches. The number of cubic inches of cylinder space is computed on the basis of piston displacement. The cylinder bore and stroke are the determining factors in cubic inch computations.

Compression ratios are important in that they, (when high) improve power and economy. Current ratios have been lowered somewhat (to improve emission control) to around 8 or 8.5 to 1.

The science of engine testing and measurement is very complicated and in its true sense lies in the realm of engineering. To expect even the best mechanics to be completely familiar with it would be most foolhardy.

It is, however, important that the student of auto mechanics be familiar with basic items in the field. The chapter you have just read will provide you with a knowledge of the more common factors. Make certain you understand them and, for your own benefit, make up some problems that will utilize the formulas given.

REMEMBER

The training and background of the superior mechanic represents hundreds of hours of study in all phases of auto mechanics. He not only knows HOW, but WHEN and WHY as well. He can be distinguished by his never-ending search for information and knowledge. The auto mechanic is a man to be respected and admired.

REVIEW QUESTIONS — CHAPTER 8

1. What is horsepower?
2. How is horsepower computed? Give the formula.
3. Engine efficiency can be classified as _____ and _____ .
4. Give the formula for both efficiency classifications.
5. What is indicated horsepower?
6. What is brake horsepower?
7. The _____ is now used to determine brake horsepower.
8. Explain the difference between GROSS and NET brake horsepower.
9. Does the S. A. E. rating indicate true brake horsepower?
10. The difference between indicated horsepower and brake horsepower is classified as _____ horsepower.
11. What is torque?
12. Are brake horsepower and torque one and the same?
13. To what does the term volumetric efficiency refer?
14. Give the formula for volumetric efficiency.
15. Piston displacement means _____ .
16. What is meant by bore and stroke?
17. How is compression ratio figured?
18. What is WORK?
19. The rate at which work is done is referred to as _____ .
20. If someone asked, "What is the cubic inch rating of this engine?" What would he mean?
21. What is ENERGY?

If you answered all the above questions correctly — even if you had to refer back to do it — you have the stuff it takes. CONGRATULATIONS!

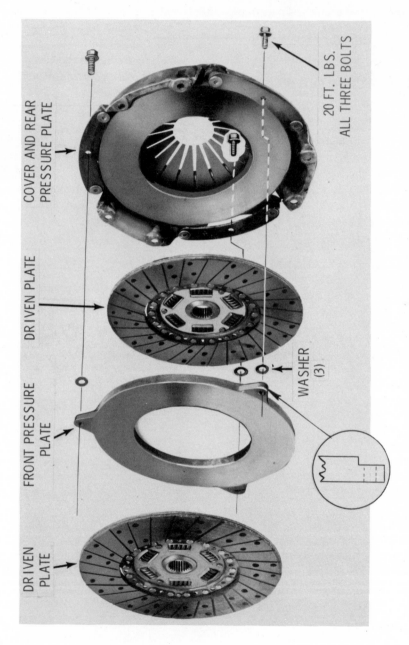

DRIVEN PLATE

FRONT PRESSURE PLATE

DRIVEN PLATE

COVER AND REAR PRESSURE PLATE

WASHER (3)

20 FT. LBS. ALL THREE BOLTS

This heavy duty, double disc clutch employs TWO pressure plates and TWO clutch discs (driven plates). (Oldsmobile)

Chapter 9

ENGINE CLUTCHES

WHAT IS A CLUTCH?

A clutch is a mechanism designed to connect or disconnect the transmission of power from one working part to another.

There are many types of clutches. All accomplish the same general purpose. This chapter concerns the clutch used on modern automobiles with standard transmissions. Its function is to connect or disconnect the flow of power from the engine to the transmission and drive line.

OF WHAT USE IS A CLUTCH?

A clutch is used to transmit engine power to the transmission in a manner that will allow the car to start out in a smooth fashion. The clutch is used to facilitate the shifting of

transmission gears. It also can be used to allow the engine to operate, when the car is stopped, without placing the transmission in neutral (out of gear).

HOW IS A CLUTCH MADE?

The clutch in popular use is the single plate, dry disc type. It consists basically of six major parts — flywheel, clutch disc, pressure plate, springs, pressure plate housing or cover and the linkage necessary to operate the clutch. To best illustrate basic clutch construction and operation, each part will be discussed in order.

FLYWHEEL

In addition to its use to smooth out engine operation and provide a base for the starter ring gear, the flywheel forms the foundation on which the entire clutch is attached.

The clutch side is machined smooth to provide a friction surface. Holes are drilled into the flywheel to provide a means of mounting the clutch assembly. A hole in the center of the flywheel allows a bearing to be installed (this hole in many cases is actually drilled in the crankshaft).

The bearing in the center of the flywheel will act as a support for the outboard end of the transmission input shaft (sometimes called "clutch shaft"). It is referred to as a "pilot bearing." This may be either a ball bearing or a bronze bushing. Both must be provided with lubrication. Fig. 9-1.

CLUTCH DISC

The clutch disc is round and constructed of thin, high quality steel, with a splined (grooved) hub placed in the center. The hub splines engage splines on the transmission input shaft. The clutch disc can move back and forth on the shaft, but when the disc is turned, the shaft must turn also.

Both sides of its outer edge are covered with friction material. It is often made of asbestos, cotton and copper wires either woven or molded together. It is riveted to the disc.

To assist in smooth engagement, the disc outer edges are often split and each piece cupped. The friction material is riveted to these segments. When the disc is compressed, these cupped segments act as a spring-like cushion.

The inner hub and thin outer disc are fastened together in such a manner as to allow a certain amount of radial (circle

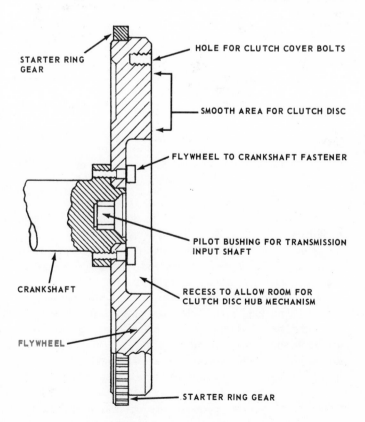

STARTER RING GEAR

HOLE FOR CLUTCH COVER BOLTS

SMOOTH AREA FOR CLUTCH DISC

FLYWHEEL TO CRANKSHAFT FASTENER

PILOT BUSHING FOR TRANSMISSION INPUT SHAFT

CRANKSHAFT

RECESS TO ALLOW ROOM FOR CLUTCH DISC HUB MECHANISM

FLYWHEEL

STARTER RING GEAR

Fig. 9-1. Cross section view shows typical flywheel design. Flywheel provides foundation on which clutch assembly is built.

around the center) movement between them. This movement is controlled by stop pins. Coil springs act as a drive unit between hub flange and outer disc. These springs act as a dampening device, soften the torque thrust when the outer disc is pressed against the flywheel, and transmit this thrust to the hub. Torsional vibration is controlled by a molded friction washer between clutch hub flange and outer disc.

Fig. 9-2 illustrates the edges of the cupped segments. Notice how the facing (friction material) is riveted to the segments. A typical clutch disc is shown in Fig. 9-3.

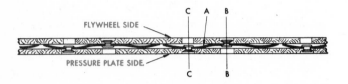

Fig. 9-2. Edge view of clutch disc. Notice how segments are cupped to produce cushion effect that assists in smooth clutch engagement. A—Cupped segment. B—Rivet. C—Rivet.

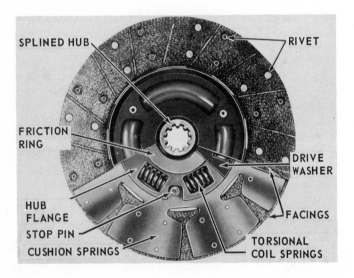

Fig. 9-3. One type of clutch disc. Note splined hub and arrangement of torsional coil springs that cushions shock of engagement. (GMC)

PRESSURE PLATE

The clutch pressure plate is generally made of cast iron. It is round and about the same diameter as the clutch disc. One side of the pressure plate is machined smooth. This side will press the clutch disc facing area against the flywheel. The other side has various shapes to facilitate attachment of springs and release mechanisms. Fig. 9-4.

PRESSURE PLATE ASSEMBLY

In addition to the pressure plate itself, there are a number of coil springs, (or diaphragm), a cover and release levers. The

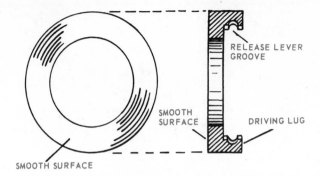

Fig. 9-4. Pressure plate. Pressure plate is spring loaded and presses clutch disc against flywheel.

springs are used to provide pressure against the pressure plate. Various numbers of springs are used, depending on the type of service for which the assembly is designed. The springs push against the clutch cover and pressure plate. Fig. 9-5.

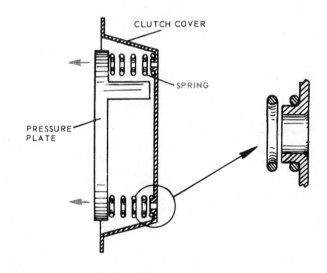

Fig. 9-5. Clutch springs — coil type. Size and number of springs vary with clutch design.

The release levers are designed to draw the pressure plate away from the clutch disc. One end of the release lever engages the pressure plate. The other end is free and is designed to be pressed inward. Between the two ends, the lever is hinged to the clutch cover by means of an eyebolt.

There are generally three release levers. Some levers are of the semi-centrifugal type. This type has a weight added to its outer end. As the clutch assembly spins, centrifugal force will act upon these weights and cause them to exert additional pressure against the pressure plate. This action assists the springs. Release lever location and action is shown in Fig. 9-6.

The clutch cover bolts to the flywheel and acts as a base for the springs. The pressure plate release lugs pass through slots in the cover, allowing the cover to drive the pressure plate.

Most flywheel and pressure plate assemblies are marked, so that upon reassembly they may be bolted together without

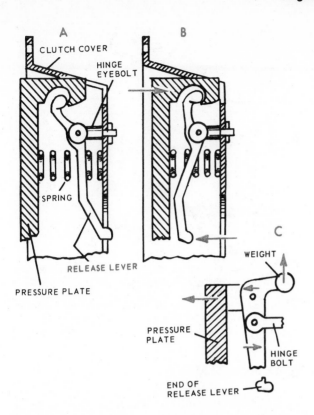

Fig. 9-6. Release lever action. A—Release lever free, pressure plate engaged. B—Release lever pushed inward, drawing pressure plate away from flywheel. C—Weight added to centrifugal type release lever. When clutch revolves, weight is thrown outward, adding to spring pressure.

disturbing the flywheel balance. At the factory, the flywheel and clutch pressure plate assembly are bolted together and balanced. After balancing, they are marked so the balance may be retained. The flywheel, clutch disc, pressure plate, release levers, springs and cover are shown in their respective positions in Fig. 9-7.

Fig. 9-7. Clutch units in their relative positions. Arrows indicate balance marks that must be aligned when assembling. (American Motors)

CLUTCH RELEASE LEVER OPERATING MECHANISM

Another essential part of the clutch is the release lever operating mechanism. This consists of a ball bearing referred to as a "throw-out" bearing. This bearing is mounted on a sleeve or collar that slides back and forth on a hub that is an integral part of the transmission front bearing retainer.

The throw-out bearing is filled with grease at the factory and ordinarily need not be serviced during its useful life. Another type throw-out bearing that is used on many foreign

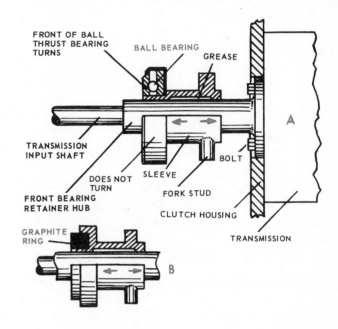

Fig. 9-8. Clutch throw-out bearings. A—Ball bearing type. B—Graphite type.

cars is the graphite type. This type employs a ring of graphite to press against a smooth plate fastened to the clutch levers. Fig. 9-8.

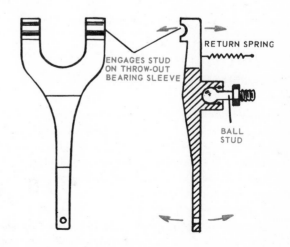

Fig. 9-9. Clutch throw-out fork. Fork engages throw-out bearing, pivots on ball stud.

The throw-out bearing sleeve is moved in and out by a throw-out fork. This fork is usually pivoted on a ball head stud. A return spring pulls the fork back toward the transmission. Fig. 9-9.

When the free end of the throw-out fork is pulled rearward, the inner end will push the throw-out bearing against the clutch release levers. If sufficient pressure is applied to the fork, the release levers will draw the pressure plate back, releasing the clutch disc. As the throw-out bearing contact the whirling release fingers, the bearing will rotate. The sleeve is held still by the clutch throw-out fork. The action of the entire throw-out assembly is shown in Fig. 9-10.

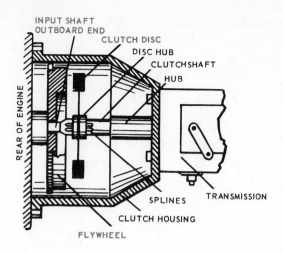

Fig. 9-11. Clutch housing and input shaft, showing flywheel, clutch disc and hub. Remainder of clutch mechanism is now shown. This housing forms full enclosure. Notice how outboard end of input shaft is supported by pilot bearing pressed into flywheel.

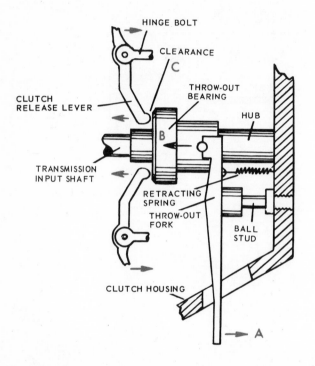

Fig. 9-10. Clutch throw-out assembly. When clutch throw-out fork is moved in direction A, throw-out bearing is moved in direction B. When throw-out contacts release levers, pressure plate will retract. When clutch throw-out fork is fully released, there must be clearance at C to prevent bearing from being turned constantly by whirling levers.

CLUTCH HOUSING AND INPUT SHAFT (CLUTCH SHAFT)

The clutch housing, made of cast iron or aluminum, bolts onto the engine. The housing surrounds the flywheel and clutch mechanism. The bottom section of some housings is open, while other housings form a complete enclosure. Housings usually have openings in them to allow circulation of air to cool the clutch. The transmission can either bolt to the housing, or be an integral part of it.

The transmission input shaft passes through the clutch mechanism, and its outboard (outer) end is supported by the clutch pilot bearing. The shaft is splined near the end, and the clutch disc rides on it at this point. Fig. 9-11.

COMPLETE ASSEMBLY

Fig. 9-12 shows the entire clutch mechanism in an exploded view. Notice the use of the graphite throw-out bearing and the full enclosure clutch housing. The transmission input shaft is not shown.

Fig. 9-13 pictures a cross section of a clutch fully assembled. The transmission clutch shaft is shown in place. Both pilot and throw-out use the antifriction ball bearing setup.

HOW DOES A CLUTCH WORK?

Remember that the flywheel, clutch cover, release levers and pressure plate all revolve as a single unit. The transmission input shaft and the clutch disc are splined together, forming another unit.

The only time the clutch disc will turn (car standing still) is when it is pinched between the flywheel and pressure plate. When the release levers draw the pressure plate away from the flywheel, the clutch disc will stand still while everything else continues to revolve. When the pressure plate travels back toward the flywheel, the disc is seized and forced to turn the transmission input shaft. Fig. 9-14.

DIFFERENT PRESSURE PLATE SPRING

Instead of a number of coil springs, another type of clutch uses a diaphragm type spring. General clutch construction is the same. The basic difference is in the pressure plate spring.

The diaphragm spring is round and quite thin. It is made of high quality, heat-treated steel. It is constructed with a dished profile to produce the necessary spring effect. A number of fingers radiate from the center where they terminate in a solid edge. One application cuts six of the fingers somewhat shorter to assist in cooling. The remaining twelve are left full length. Fig. 9-15. Another version bends the six fingers upward.

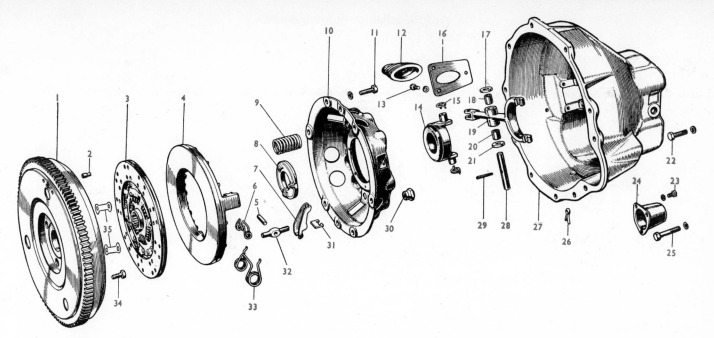

Fig. 9-12. Exploded view — clutch assembly. Only one release lever is shown — there are three. Number of springs used will vary from five to seven in average clutch. 1—Flywheel. 2—Locating pin. 3—Clutch plate with lining. 4—Pressure plate. 5—Release lever pin. 6—Release lever retainer. 7—Release lever. 8—Release lever plate. 9—Pressure plate spring. 10—Clutch cover. 11—Cover set pin. 12—Fork and lever seal. 13—Retaining plate screw. 14—Release bearing. 15—Release bearing retainer spring. 16—Seal retaining plate. 17—Fork and lever thrust washer. 18—Fork and lever shaft bush. 19—Clutch fork and lever. 20—Fork and lever shaft bush. 21—Fork and lever thrust washer. 22—Clutch to gearbox set pin. 23—Starter cover screw. 24—Cover. 25—Clutch to gearbox set pin. 26—Split pin for drain hole. 27—Clutch housing. 28—Fork and lever shaft. 29—Taper pin. 30—Eyebolt nut. 31—Release lever strut. 32—Eyebolt. 33—Antirattle spring. 34—Flywheel to crankshaft bolt. 35—Lockwashers.   (Austin)

Weights are attached and they produce a centrifugal effect to assist spring pressure at high rpm.

The outer edge of the diaphragm engages the pressure plate. The fingers point inward and dish out slightly. Two pivot rings are placed a short distance from the outer edge. The pivot rings are secured by means of a stud to the clutch cover. One

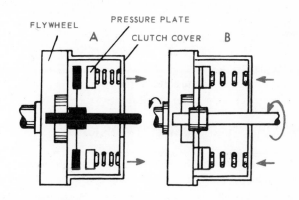

Fig. 9-14. Clutch operation — simplified. A—Flywheel, pressure plate, springs, and clutch cover (in white) are revolving. Clutch disc and clutch shaft (in black) are stopped as pressure plate is disengaged. B—Pressure plate engaged — seizes clutch disc and entire unit revolves.

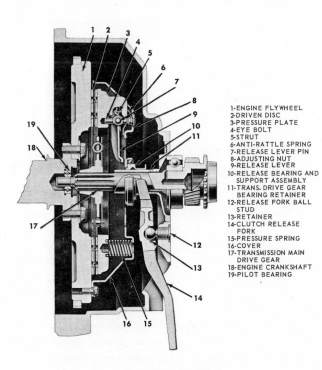

1-ENGINE FLYWHEEL
2-DRIVEN DISC
3-PRESSURE PLATE
4-EYE BOLT
5-STRUT
6-ANTI-RATTLE SPRING
7-RELEASE LEVER PIN
8-ADJUSTING NUT
9-RELEASE LEVER
10-RELEASE BEARING AND
   SUPPORT ASSEMBLY
11-TRANS. DRIVE GEAR
   BEARING RETAINER
12-RELEASE FORK BALL
   STUD
13-RETAINER
14-CLUTCH RELEASE
   FORK
15-PRESSURE SPRING
16-COVER
17-TRANSMISSION MAIN
   DRIVE GEAR
18-ENGINE CRANKSHAFT
19-PILOT BEARING

Fig. 9-13. Cross section — assembled clutch. Study this and learn names of clutch parts.   (GMC)

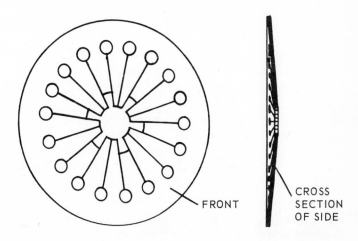

Fig. 9-15. Diaphragm spring. Notice short fingers for cooling.

205

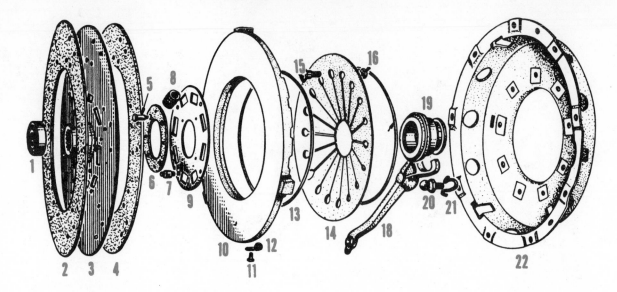

Fig. 9-16. Exploded view — diaphragm spring clutch. This type of clutch uses fingers of diaphragm to act as release levers. 1—Pilot bearing. 2—Facing. 3—Disc. 4—Facing. 5—Dampener spring plate rivet. 6—Dampener spring outer plate. 7—Driven member stop pin. 8—Dampener spring. 9—Dampener spring plate. 10—Pressure plate. 11—Retracting spring bolt. 12—Pressure plate retracting spring. 13—Spring pivot ring. 14—Clutch spring. 15—Spring retainer bolt. 16—Spring retainer bolt nut. 17—Spring pivot ring. 18—Throw-out fork. 19—Release bearing and support assembly. 20—Fork ball head stud. 21—Ball retainer. 22—Cover.

pivot ring is on the outside of the diaphragm, one on the inside.

The pressure plate is driven by three double spring steel straps. The straps are riveted to the cover and bolted to the pressure plate.

The throw-out bearing contacts the ends of the fingers. The fingers then act in the place of release levers. Fig. 9-16.

HOW DIAPHRAGM CLUTCH WORKS

When the throw-out bearing applies pressure to the diaphragm fingers, the entire diaphragm is bent inward. The action can best be described as the same sort of effect as squeezing the bottom of an oil can. The inner pivot ring prevents the outer edge of the diaphragm from moving inward, so when the center of the diaphragm is pressed inward, the outer edges pivot outward. When the fingers are released, the diaphragm resumes its original postion. Study Fig. 9-17. which demonstrates diaphragm action.

Fig. 9-18 shows the position of the diaphragm when the clutch is engaged. Notice how the fingers point outward, causing the diaphragm to pivot on the outer pivot ring. This action presses the diaphragm outer edge tightly against the pressure plate. The clutch disc is held firmly.

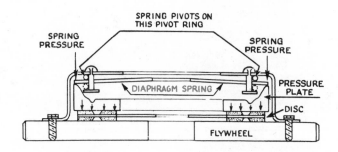

Fig. 9-18. Diaphragm spring clutch in engaged position. Disc is pinched between pressure plate and flywheel. Note spring pressure produced by diaphragm at outer edges.   (Chevrolet)

Fig. 9-17. Diaphragm spring action. Dotted line shows diaphragm in disengaged position. Notice how outer edge moves back and forth, depending on position of fingers.   (Chevrolet)

Now study Fig. 9-19. The throw-out bearing has pressed the fingers in. The diaphragm now pivots on inner pivot ring, and the outer edges lift the pressure plate away from the clutch disc. The disc is free and the clutch is disengaged.

Pressure plate, cover and diaphragm assembly are pictured

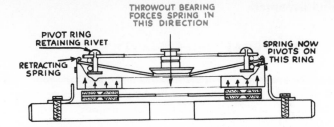

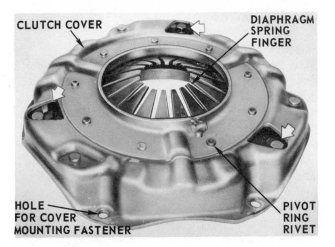

Fig. 9-19. Diaphragm spring clutch in disengaged position. Throw-out bearing has bent clutch diaphragm fingers inward. This pulls outer edge back, retracting pressure plate and leaving clutch disc free.   (Chevrolet)

Fig. 9-20. Diaphragm clutch pressure plate assembly. Arrows indicate retracting springs and fasteners.   (Chevrolet)

in Fig. 9-20. A complete diaphragm clutch assembly is shown in Fig. 9-21.

OTHER TYPES OF CLUTCHES

There have been a great number of clutches designed and used over the years. The two shown in this chapter completely dominate the automotive field today. Some heavy-duty type car and truck clutches utilize extra springs and two clutch discs with a second pressure plate, smooth on both sides, sandwiched between them.

CLUTCH PEDAL LINKAGE

There are two principal methods used to actuate the clutch throw-out fork. One system uses a series of links, levers and rods connected between the clutch pedal and throw-out fork. When the pedal is depressed, the force is transmitted to the fork. Fig. 9-22 illustrates a fairly simple hookup.

The other system utilizes hydraulic control of the fork. On many cars, due to design, the mechanical linkage required to operate the fork would be extremely complicated. The hydraulic control simplifies the job.

When the clutch is depressed, it actuates a small master cylinder. Fig. 9-23. The pressure created in the master cylinder is transmitted to a "slave" cylinder bolted near the fork. The slave cylinder is connected to the fork with a short adjustable rod. When pressure is applied to the slave cylinder, it operates the fork.

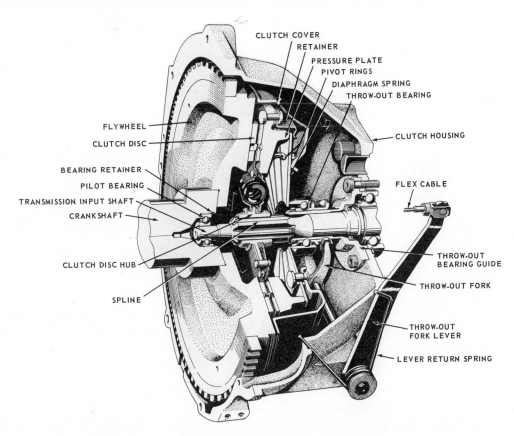

Fig. 9-21. Cutaway of a complete clutch assembly. This unit uses a diaphragm type spring.   (Volvo)

207

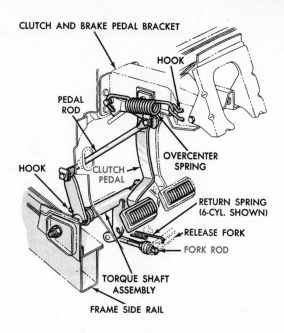

Fig. 9-22. One mechanical clutch linkage arrangement. When clutch pedal is depressed, clutch throw-out fork is actuated. (Plymouth)

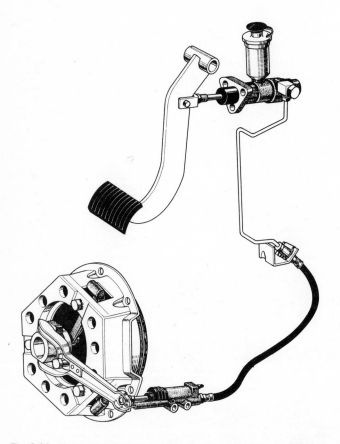

Fig. 9-23. Hydraulic clutch control. Clutch pedal movement builds pressure in the master cylinder that in turn actuates the slave cylinder. (Toyota)

Both master and slave units are of simple design. Regardless of position or obstructions, they are easily connected with hydraulic tubing. The principles of hydraulics will be discussed in the chapter on BRAKES.

CLUTCH PEDAL "FREE TRAVEL"

Free travel of the clutch pedal is most important. When the clutch pedal is released, the throw-out bearing must not touch the release levers. It must clear for two reasons. As long as it touches, it will continue to rotate. This will shorten its useful life. Also, if the throw-out bearing is not fully released, it may bear against the release levers hard enough to partially disengage the clutch. Any removal of pressure plate force will cause the clutch to slip. When slippage occurs, the clutch facing will overheat and burn up.

All manufacturers specify the amount of free travel that must be allowed. Free travel, or pedal travel, means how far the clutch pedal can be depressed before the throw-out bearing strikes the release levers.

Free travel is easy to check. Depress the clutch with one finger. You will find that it moves very easily for about an inch or inch and a half. It will then seem to strike something, and you will have to push harder to move it past this point. The hard point was where the release levers and throw-out bearing made contact.

The distance you moved the pedal from the full out position until it became hard to push is the free travel. Free travel is set by adjusting the clutch linkage. Do not adjust the clutch assembly. Fig. 9-24.

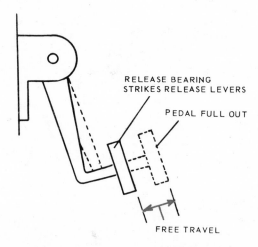

RELEASE BEARING STRIKES RELEASE LEVERS

PEDAL FULL OUT

FREE TRAVEL

Fig. 9-24. Clutch free travel. Follow manufacturer's recommended setting.

CLUTCH COOLING

By their very nature, clutches generate considerable heat. You will notice that all clutch covers have openings and that the clutch housings are also provided with openings to facilitate air circulation.

RIDING THE CLUTCH

Some drivers develop the habit "riding the clutch" (resting left foot on clutch pedal). The natural weight of the leg will cause the throw-out bearing to strike the release levers. Any additional pressure may even cause the clutch to slip. Clutch pedals are to be used only when needed. Keep foot off!

WILL A CLUTCH EXPLODE?

Clutches in racing and other competition engines can, and sometimes do, explode. The high rpm sets up tremendous centrifugal forces. If the clutch fails, parts are thrown outward in a violent fashion. The clutch housing will be shattered and the driver endangered.

WARNING!

IN ANY VEHICLE USED FOR COMPETITION, IT IS IMPERATIVE THAT A SCATTER SHIELD BE USED TO PROTECT THE DRIVER IN THE EVENT OF CLUTCH FAILURE!

A scatter shield is a piece of heavy steel plate that covers the clutch housing. It should extend down the sides to prevent parts from striking anyone standing nearby. It must be bolted securely into place.

BE CAREFUL!

NEVER WORK ON A CLUTCH ASSEMBLY, WHEN IN THE CAR, WITHOUT FIRST DISCONNECTING THE BATTERY. OTHERWISE, IF THE STARTER IS ACCIDENTALLY OPERATED, SEVERE DAMAGE COULD BE INFLICTED ON YOUR HANDS.

SUMMARY

A clutch is necessary to assist in smooth starting, shifting gears and allowing the car to be stopped without putting the transmission in neutral.

The clutch is designed to connect or disconnect the transmission of power from one working part to another — in this case, from the engine to the transmission.

Automobile clutches in use today are of the single plate, dry disc type. One type uses a coil spring as a means of loading the pressure plate. Another type uses a diaphragm spring.

The flywheel, clutch disc, pressure plate, springs, clutch cover and operating mechanism linkage make up the primary parts of the typical clutch.

In operating position (engaged), the clutch disc is held firmly between the flywheel and pressure plate. When the flywheel turns, the disc turns the transmission input shaft to which it is splined.

To disengage the clutch, the clutch pedal is depressed. Either through mechanical or hydraulic linkage, the throw-out fork pushes the throw-out bearing into contact with the pressure plate release levers. The levers draw the pressure plate away from the clutch disc to release it. When the disc is freed, the flywheel and pressure plate assembly continue to rotate, even through the disc and input shaft are not moving.

There are two principal types of throw-out bearings, the ball bearing type and the graphite type. Most cars use the ball bearing type.

The clutch disc is designed to assist in smooth engagement by torsion springs in the hub assembly as well as cupped segments to which the friction facing is riveted.

Clutch free pedal should be adjusted according to the manufacturer's specifications. Never ride the clutch.

REVIEW QUESTIONS — CHAPTER 9

Every chapter in this book is an important link in your chain of learning. Remember: A CHAIN IS NO STRONGER THAN ITS WEAKEST LINK!

KNOW the answers to the following questions:

1. Define the word clutch.
2. What are the three main uses of the clutch?
3. Name the major parts that make up the clutch.
4. Of what use is the pressure plate?
5. What is the clutch disc?
6. What part does the flywheel play in the clutch assembly?
7. Release levers are pivoted on an _____ bolt.
8. Two types of springs are used to load the pressure plate. Name them.
9. The clutch cover is bolted to the _____ .
10. The pressure plate is driven by the_____.
11. Name two types of throw-out bearings.
12. What is a semicentrifugal release lever?
13. The_____ operates the throw-out bearing.
14. What are the two ways of actuating the throw-out fork?
15. Explain the operation of the diaphragm spring.
16. What kind of release levers are used in the diaphragm clutch?
17. What is clutch pedal free travel?
18. How is free travel adjusted?
19. Even though most competition engines use heavy-duty "beefed up" clutches, what additional unit must be installed to safeguard driver and spectators?
20. Working on the clutch, while in the car, can be dangerous. It is wise to always_____ before starting the job.
21. Most clutches have aligning punch marks. What are they for?
22. What is meant by riding the clutch?
23. What is the transmission input shaft?
24. How is the outboard end of the transmission input shaft supported?
25. What are splines?
26. What two parts of the clutch and transmission are splined together?
27. Of what use is the clutch housing?
28. Explain how the clutch disc is designed to assist in smooth clutch engagement.
29. From what type material is the facing used on the clutch disc made?
30. Explain HOW a clutch works during engagement and disengagement — step by step, naming the parts involved and their function.

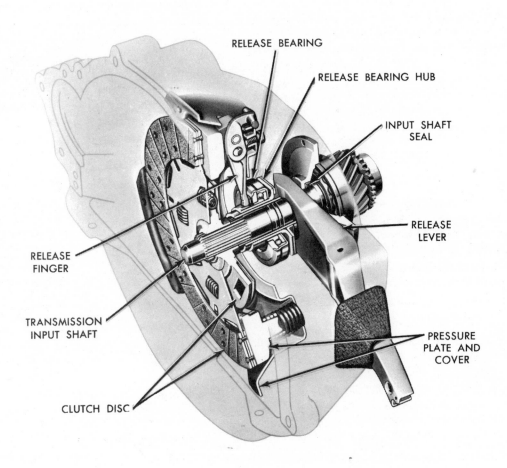

RELEASE BEARING

RELEASE BEARING HUB

INPUT SHAFT
SEAL

RELEASE
LEVER

RELEASE
FINGER

TRANSMISSION
INPUT SHAFT

PRESSURE
PLATE AND
COVER

CLUTCH DISC

Ford clutch assembly.

Chapter 10

STANDARD TRANSMISSIONS

The average engine would have power enough to operate a car without a transmission provided the car operated on reasonably level roads and maintained sufficient speed. When the car must be started from a standstill, or when attempting to negotiate steep grades, the engine would not provide sufficient power and the car would stall.

Much less torque is required to move a car rolling on level ground than is necessary to drive the same car up a steep hill. Fig. 10-1.

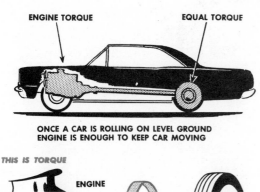

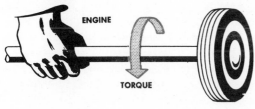

Fig. 10-1. Torque required — car on level ground. Torque is the twisting force on axle — as shown by hand.   (Chevrolet)

THE ENGINE NEEDS HELP

To enable the engine to increase torque to the drive line, it is obvious that a torque multiplier (transmission) is needed. Fig. 10-2.

By using a torque multiplier, it is possible to adapt the available power of the engine to meet changing road and load conditions.

HOW IS TORQUE MULTIPLIED?

If a man (engine) exerted all his strength (power) and still failed to lift a certain object (drive the car), he would multiply

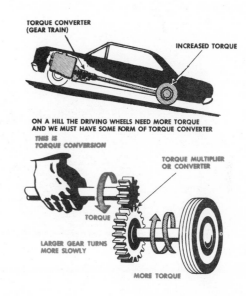

Fig. 10-2. Torque required — car on steep grade. Original amount of torque is now insufficient, so it is necessary to increase torque by means of a torque converter or multiplier. The twisting force of hand is multiplied through use of transmission gears.   (Chevrolet)

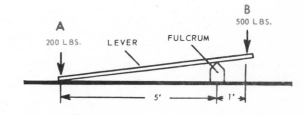

Fig. 10-3. Lever and fulcrum. Man A is able to raise weight B, because he has multiplied his strength by using a lever and fulcrum.

his strength by using a lever and fulcrum. Fig. 10-3.

If the man weighed 200 lbs. and stood on the lever 5 ft. from the fulcrum, he would exert a force of 1,000 ft. lbs. If the weight of 500 lbs. rested on the lever 1 ft. from the fulcrum, it would exert a force of 500 ft. lbs. The man, by using a simple torque converter or multiplier, can now raise 500 lbs. with ease. How much could the man in Fig. 10-4 raise?

In order for the man in Fig. 10-4 to lift 4,000 lbs., it was necessary for him to stand 20 ft. from the fulcrum. In order to

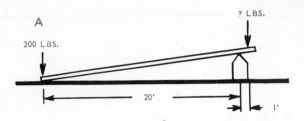

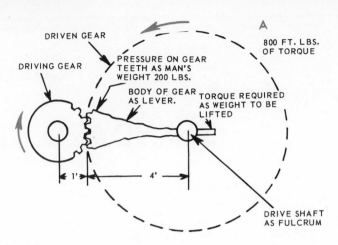

Fig. 10-4. Problem: How much weight will man A be able to lift in this problem?

raise the weight a short distance, it is mandatory that he depress the end of the lever a distance 20 times as much as distance the weight has been raised. Fig. 10-5.

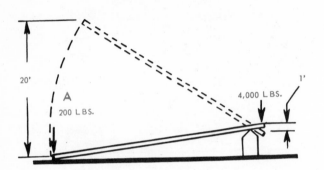

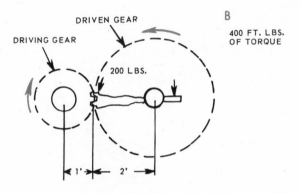

Fig. 10-5. Lever travel. If speed with which the man depresses lever remains constant, the more he multiplies his strength and the slower the weight will raise. To raise 4,000 lbs., it was necessary to depress lever 20 ft. to raise weight 1 ft.

Fig. 10-6. Principle of gear torque multiplication. The larger the driven gear in relation to the driving gear, the greater the torque multiplication. As torque is increased, driving gear must turn more times to turn driven gear once.

You will note that as the man's (engine's) strength (torque) is increased, the speed with which the weight (car) is moved becomes slower. This is true, as it is necessary for the man to depress the lever a greater distance as he moves further out from the fulcrum.

From a study of this simple lever and fulcrum, it is possible to make these deductions:

1. If the speed of an engine remains constant and its torque is multiplied, it will lift more weight but will take longer to do it.

2. If the speed of an engine remains constant and its torque is reduced, it will lift less weight but will lift it faster.

Therefore, as engine torque multiplication increases, the car road speed will decrease and as torque multiplication decreases, the car road speed will increase — (engine rpm remaining constant).

TORQUE MULTIPLICATION THROUGH GEARS

The same principle used in the lever and fulcrum example is used in automotive transmission gears. Consider the fulcrum as the car drive shaft; the amount of weight to be lifted as the torque required to turn the shaft; the engine torque as the man's weight; the body of the gear as the lever. Then it is easy to see how the gear multiplies the torque. Fig. 10-6.

You can see that as the size of the gear is increased, the same amount of pressure on a tooth will produce an increase in torque. In A, Fig. 10-6, the driving gear tooth is 1 ft. from the center of the shaft. If it were pushing down with a 200 lb. force, the driving tooth would be exerting a 200 ft. lb. push on the tooth of the driven gear. As the driven tooth is 4 ft. from the center of the drive shaft, it would exert an 800 ft. lb. force on the drive shaft. In B, the force of the driving gear would be the same, but as the driven tooth is now 2 ft. from the center, it would exert a force of only 400 ft. lbs. on the drive shaft.

TRANSMISSION GEAR RATIOS

The average car transmission provides the driver with a choice of three forward speeds: DIRECT DRIVE (no torque multiplication), SECOND GEAR (approximate ratio of 2 to 1) and a LOW GEAR (approximate ratio of 3 to 1). By selecting one of the ratios, it is possible to operate the car under all normal conditions.

Ratios vary from car to car, depending on engine horsepower and car weight. A reverse gear is also incorporated. The reverse ratio is about the same as the low gear. (Engine turns three times to transmission output shaft once.)

Torque is multiplied even more through the differential gears. This will be discussed in a later chapter.

Many cars are equipped with a four-speed transmission to provide even a wider range of gear ratios. Typical ratios could be: 1st gear, 2.52 to 1; 2nd gear, 1.88 to 1; 3rd gear, 1.46 to 1; 4th gear, 1.00 to 1; reverse, 2.59 to 1.

TRANSMISSION GEARS

Transmission gears are made of high quality steel, carefully heat-treated to produce smooth, hard surface gear teeth with a softer but very tough interior. They are drop-forged (machine hammered into shape) while at a red heat. The teeth and other critical areas are cut on precision machinery.

Gear teeth of transmission gears are of two principal types, spur and helical. The helical gear is superior in that it runs more quietly and is stronger because more tooth area is in contact. Helical gears must be mounted firmly, since there is a tendency for them to slide apart due to the spiral shape. See Fig. 10-7.

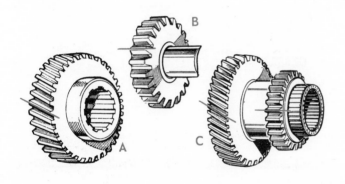

Fig. 10-7. Gear tooth shapes. A—Helical. This tooth is cut at an angle to the centerline of the gear. B—Spur. Cut straight across (parallel to gear centerline). C—Left side, helical; right side, spur. (Land-Rover)

There must be some clearance between the gear teeth to allow for lubrication, expansion and possible size irregularity. This clearance is very small (a few thousandths of an inch). Fig. 10-8 illustrates backlash (clearance between face and flank

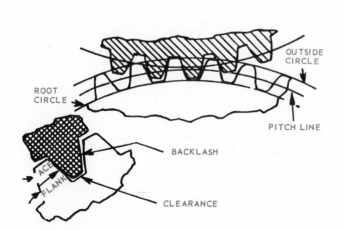

Fig. 10-8. Elementary gear tooth nomenclature. Pitch line or pitch diameter is true diameter used in calculating exact ratios. Outside circle is full diameter of gear. Root circle is minimum diameter of gear.

areas) and also gives gear tooth nomenclature.

There are many other types of gears. Those that are used on the automobile will be discussed where they apply to a unit under study.

TRANSMISSION CONSTRUCTION AND OPERATION

A typical transmission consists of a cast iron or an aluminum housing, four shafts, bearings, gears, synchronizing device and a shifting mechanism. Fig. 10-9 shows the four

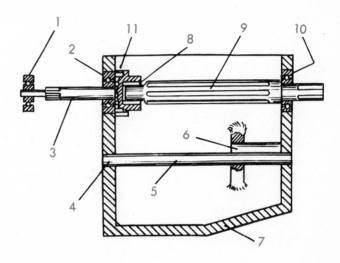

Fig. 10-9. Housing shafts and bearings. 1—Clutch pilot bearing. 2—Input shaft bearing. 3—Input shaft. 4—Countershaft supported in housing. 5—Countershaft. 6—Reverse idler shaft supported in housing. 7—Housing. 8—Output shaft roller bearing. 9—Output shaft. 10—Output shaft ball bearing. 11—Input shaft gear teeth. Input and output shafts turn. Countershaft and reverse idler shaft remain fixed.

shafts and bearings in their relative positions in the housing.

To show how the transmission works, the four shafts with gears are shown in Fig. 10-10. By studying the illustration, you

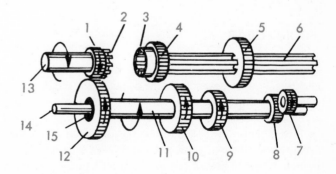

Fig. 10-10. Transmission in neutral. Input shaft 13, is revolving. Input gear 1 is driving entire cluster gear. Cluster gear 8 is driving reverse idler gear 7. Output shaft 6 is not moving as both sliding low and reverse gear 5 and sliding high and second gear 4 are in neutral positions. 2—Splined stub. 3—Splined recess on sliding gear 4. 6—Output shaft is splined to allow gears to be shifted back and forth. 8—Cluster reverse idler gear. 9—Cluster low gear. 10—Cluster second gear. 11—All cluster gears connected to this cluster body. 12—Large constant mesh cluster gear. 14—Countershaft. 15—Cluster bearing.

will see that the transmission is in neutral. Notice that the input or clutch shaft is turning, but the output or main shaft is standing still. Also note that the input shaft gear is always driving the counter or cluster gear. The cluster gear revolves on roller bearings.

The cluster gear assembly has four gears all cut on the same steel blank. They are: large gear — drive gear, next in size — second speed gear, next — low speed gear, smallest — reverse gear. When the input shaft is turning, the cluster gear also turns, but in an opposite or counter direction.

The reverse idler gear turns when the cluster gear is turning. The reverse idler rides on a bushing between it and the reverse idler shaft. There are two gears that are always in contact: input shaft gear and cluster drive gear; cluster reverse gear and reverse idler gear. These are called constant mesh gears.

The output shaft is splined and two gears are placed on it. Both gears are free to move back and forth on the splines. The smaller of the two is the second speed gear. Notice that it has a splined recess in its face. This splined recess is designed to slip over a splined stub protruding from the input shaft. Notice the action of this splined recess when the transmission is placed in high gear. Fig. 10-13.

The larger sliding gear is the low and reverse gear. It can be shifted into mesh with either the reverse idler gear or the low cluster gear. Study Fig. 10-10, and trace the flow of power (indicated by arrows).

FIRST OR LOW GEAR

The clutch is depressed, thus releasing the clutch disc. The input shaft and cluster gear will stop turning. The output sliding low gear is shifted forward, and meshed with the cluster low gear. The output second gear is left out of mesh.

When the clutch is engaged, the input shaft will turn the cluster gear. The cluster gear will revolve at a slower speed. The cluster low gear, being smaller than the output shaft low sliding gear, will turn the output shaft at even a slower speed. The output shaft speed will be roughly one-third that of the input shaft, giving a gear reduction of three to one. This reduction multiplies the engine torque but gives less forward speed. Fig. 10-11.

SECOND GEAR

When ready for second gear, the clutch is depressed to break the flow of power from the engine. Low sliding gear is moved to a neutral position, and second sliding gear is meshed with the cluster second gear. The clutch is then released and the flow of power, with less torque multiplication, is resumed.

Notice that the cluster second gear is larger in proportion to the sliding second gear. The gear reduction is now less (around 2 to 1) and torque falls off but speed picks up. Fig. 10-12.

HIGH GEAR

When the car attains sufficient speed, the clutch is again depressed. Sliding second gear is moved forward until the splined recess has slipped over the splined stub on the input shaft. Low sliding gear remains unmeshed. When the clutch is

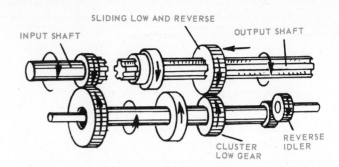

Fig. 10-11. Transmission in low gear. Sliding low and reverse gear has been shifted forward and is meshed with cluster low gear. Output shaft is now turning. All gears that are in mesh with another gear are shown with simulated teeth.

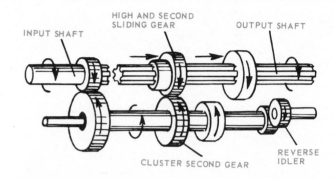

Fig. 10-12. Transmission in second gear. Sliding low gear has returned to neutral position. Sliding second gear moved into mesh with cluster second gear. Output shaft now turns faster.

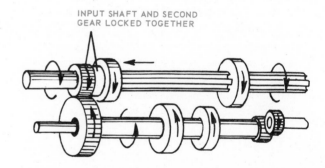

Fig. 10-13. Transmission in high gear. Sliding second gear has moved toward input until splined recess in gear engages splined stub on input shaft. Sliding low gear is in neutral position. Input and output shafts are locked together and drive is direct.

released, the transmission is in high gear with no torque multiplication. The input and output shafts are locked together through the splined stub and recessed second gear. This gives direct drive through the transmission. Fig. 10-13.

REVERSE

The sliding second gear is placed in the neutral position. The low reverse sliding gear is moved back until it engages the reverse idler gear. As the reverse idler is revolving in the same direction as the input shaft, it will impart a reverse rotation to the output shaft. The gear reduction or torque multiplication is comparable to that of the low gear, around three to one. Fig. 10-14.

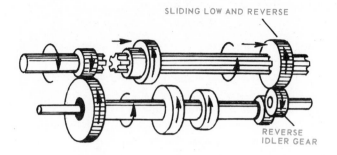

Fig. 10-14. Transmission in reverse gear. Sliding second gear has returned to neutral. Low and reverse sliding gear are meshed with reverse idler gear. Output shaft now rotates in a reverse direction. Ratio is about same as low gear.

SYNCHRONIZING MECHANISM

Once the car is in motion, the drive line will turn the output shaft continuously. As a result, the sliding gears will be whirling. When an attempt is made to mesh them with any of the cluster gears (which tend to stop when the clutch is depressed), the gear teeth will be subjected to damaging impact forces. The sound of "grinding" gears, when shifting, results from the sliding gear teeth literally smashing against the cluster gear.

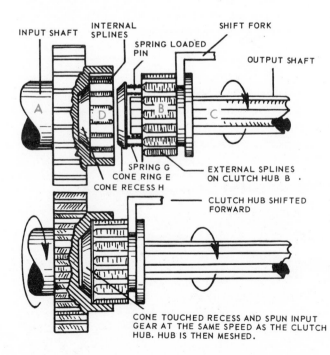

Fig. 10-15. Simplified synchronizer. As clutch hub B is shifted toward splined recess D, cone ring E will strike cone recess and spin input shaft. Hub B is then meshed into splined recess D without grinding. Cone ring pins and springs are forced into clutch hub.

It is obvious that for one gear to mesh with another quietly and without damage, both gears must be rotating at nearly the same speed. Most transmissions are equipped with a device called a synchronizer. Its job is to move ahead of the unit that is to be meshed, seize the other unit and bring the rotational speed of both units together. Once both units are rotating at the same speed, they may be meshed.

Fig. 10-15 is a simplified version, illustrating the principle of the synchronizing device. A is the input shaft and it is stopped with the clutch disengaged. In order to mesh whirling splined hub B on output shaft C with the splined recess D in input shaft, it will be necessary to start input shaft turning. Notice cone ring E that is an extension of hub B. The ring is supported on pins that fit into holes in the hub. The ring is held out by springs G.

As sliding hub B is moved toward splined recess D, ring E will touch cone-shaped bottom H of recess. As the hub is moved closer to recess, the springs G will force cone ring E tightly against the bottom of cone recess. As the cone ring is whirling, it will impart torque to the input shaft and start it turning. As the hub moves closer, spring pressure increases and cone ring will be jammed into cone recess with enough force to spin input shaft at nearly the same speed of whirling hub B. The hub may then be meshed into recess with no grinding or shock. Fig. 10-15.

The simplified synchronizer as shown in Fig. 10-15, would be for direct (high) drive. By putting a cone ring (called blocking ring or synchronizing cone) on both sides of the hub, Fig. 10-16, it could be shifted the other way and engage splined recess in the output shaft second gear. For this arrangement, the second gear would no longer be a sliding gear and would not be splined to the output shaft.

The second gear would become a constant mesh gear and would be turned whenever the cluster gear is revolving. It would rotate on the output shaft either on a bushing or roller bearings. Fig. 10-16.

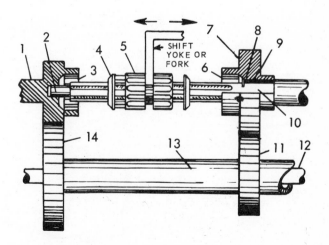

Fig. 10-16. Constant mesh second gear with synchromesh in high and second gears. Output shaft second gear 7, is free to rotate on output shaft 10. A roller bearing or bushing is provided at 9. A snap ring 8 keeps second gear from moving back and forth. When synchromesh hub 5 is moved into second gear splines 6, second gear is locked to output. 1—Input shaft. 2—Supported inner end of output shaft. 3—Input splines. 4—Cone synchronizer ring. 11—Cluster second gear. 12—Countershaft. 13—Cluster assembly. 14—Cluster drive gear.

When high gear is desired, the sliding clutch hub would be shifted into the input shaft recess as shown in Fig. 10-15. For second gear, the clutch hub would be shifted back until meshed with the second gear. When the hub engages the second gear, the output shaft would be locked, through the clutch hub, to the second gear; and the output shaft would be driven by the cluster second speed gear driving the locked output shaft second gear. Both shifts would be quiet, as the synchronizer would equalize or synchronize their speeds before meshing. Fig. 10-16.

Study the synchronizer illustrated in Fig. 10-17. This popular type, splines the hub to the shaft but does not permit

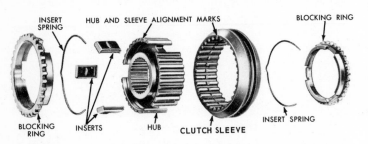

Fig. 10-17. One popular type of synchronizer unit. Hub is splined to output shaft. Clutch sleeve moves back or forth on hub as needed to connect hub with proper gear. (Ford)

end movement of the hub. The clutch sleeve is shifted forward or backward on the hub. This action forces (by means of inserts) the blocking rings to equalize gear speeds before the clutch sleeve engages the other gear to lock gear to hub through clutch sleeve.

Low gear is generally used for starting. Since gears are stopped when the car is not in motion and the clutch is depressed, synchronizing the low gear is not absolutely

necessary. Current practice, however, is toward full synchronization of all forward gears. This permits low or 1st gear to be engaged while the vehicle is still in motion.

VARIOUS TYPES OF SYNCHRONIZER DEVICES

There are several types of synchronizer units used today. All use a cone (blocking ring) that precedes the movement of either the clutch hub or, in many cases, the clutch sleeve. The cone engages a tapered surface on the part to be engaged to synchronize the speed before meshing.

As you study the transmissions in this section, determine what type of synchronizer is used and how it works.

THREE—SPEED SYNCHROMESH TRANSMISSION

NEUTRAL

Fig. 10-18 shows a typical synchromesh transmission in neutral. Notice that the input shaft drives the cluster gear, which in turn drives the constant mesh second speed gear. As the second gear turns on the output shaft, no power is transmitted. This transmission uses the strut-type synchronizer. Notice that the clutch sleeve is in a neutral position. The sliding low and reverse gear is also in a neutral position.

LOW GEAR

Fig. 10-19 illustrates the transmission in low or first gear. The clutch sleeve is still in the neutral position, but the sliding low and reverse gear has been shifted into mesh with the cluster low gear. Power is transmitted (follow arrows) from the input shaft to cluster, to sliding low gear, to output shaft.

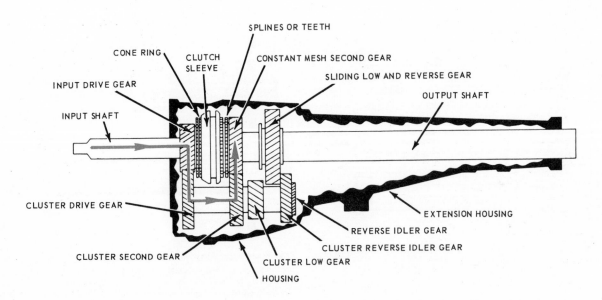

Fig. 10-18. Synchromesh transmission in neutral. Power flow is from input gear to cluster drive gear to output constant mesh second gear. As synchromesh clutch sleeve is in neutral, constant mesh second gear will revolve on output shaft and no power will be transmitted beyond this point. (Pontiac)

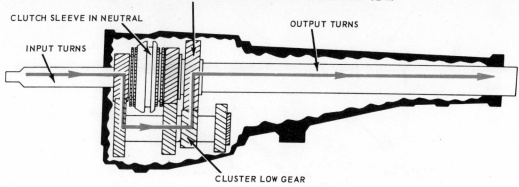

Fig. 10-19. Synchromesh transmission in low gear. Power flow is from input gear to cluster drive gear, from cluster low gear to output sliding low and reverse gear. As sliding gear is meshed, power is transmitted to output shaft.   (Pontiac)

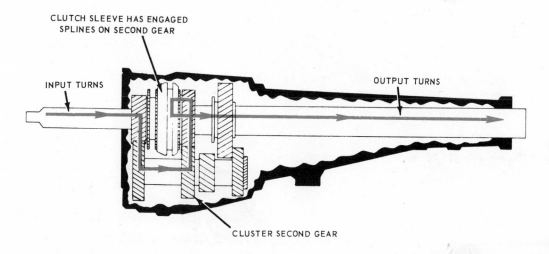

Fig. 10-20. Synchromesh transmission in second gear. Power flow is from input gear to cluster drive gear. From cluster drive to cluster second gear to constant mesh output second gear. As clutch hub has engaged teeth or splines on second gear, second gear is locked to output and will transmit power.   (Pontiac)

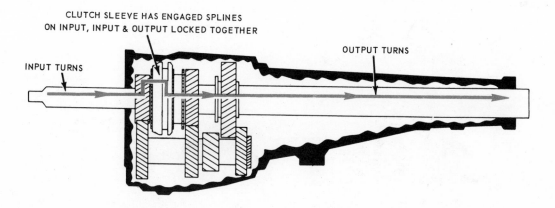

Fig. 10-21. Synchromesh transmission in high gear. Clutch hub has now engaged splines on the end of input shaft. This will lock input and output shafts together and power is transmitted straight through. Cluster is driven but will not transmit power.   (Pontiac)

SECOND GEAR

Fig. 10-20 shows the transmission in second gear. The sliding gear has been moved to a neutral position. Notice that the clutch sleeve has been shifted back, until it engages the splines or small teeth on the second gear. As the clutch sleeve is splined to the clutch hub which, in turn, is splined to the output shaft, the second gear is locked to the output shaft. The input shaft drives the cluster gear, the cluster drives the second speed gear, and the second speed gear drives the output shaft through the clutch sleeve and hub.

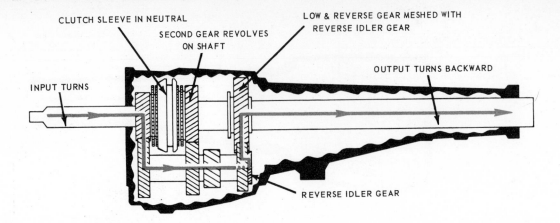

Fig. 10-22. Synchromesh transmission in reverse gear. Clutch sleeve has now been returned to a neutral position. Sliding low and reverse gear has been meshed with reverse idler gear. Input gear drives cluster, cluster drives reverse idler and reverse idler drives output shaft in reverse direction. (Pontiac)

HIGH GEAR

Fig. 10-21 shows the transmission in high gear. The sliding gear is still in a neutral position. The clutch sleeve has been shifted away from the second gear and has engaged the small teeth on the input shaft. As this locks the input and output shafts together, the power flow is direct. The cluster is driven and still turns the second gear which is free to turn on the output shaft.

REVERSE GEAR

Fig. 10-22 illustrates the transmission in reverse gear. The clutch sleeve is in the neutral position. The sliding gear has been moved into mesh with the reverse idler gear. The input shaft turns the cluster; the cluster drives the reverse idler; the idler drives the sliding gear in a direction opposite to engine rotation. This will cause the output to drive the car backward.

SHIFT MECHANISM

The transmission is shifted by means of shifter forks that ride in a groove cut into the clutch sleeve and sliding gear. The forks are attached to a cam and shaft assembly. Spring loaded steel balls pop into notches cut in the cam assembly to hold the shift mechanism into whatever gear is selected. The shafts pass through the housing or housing cover and are fastened to shift levers. Fig. 10-23.

The shift levers can be operated via linkage attached to a steering wheel shift control or to a floor shift stick. Some transmissions operate the shift forks by means of a floor shift stick that enters the transmission housing and engages sliding

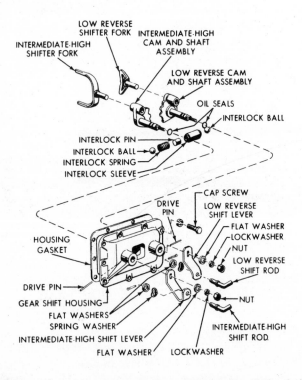

Fig. 10-23. Gearshift housing assembly — exploded view. Shifter forks engage low and reverse sliding gear and clutch sleeve. Forks are actuated by moving shift levers. Each fork must be returned to neutral before other can be moved. This is accomplished by interlock pin. Interlock balls hold cam into whatever position it has been placed. (Mercury)

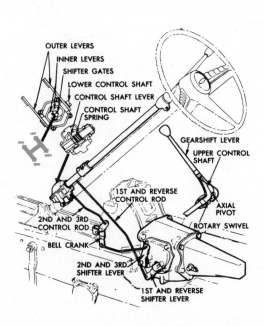

Fig. 10-24. Steering column gearshift. Study all parts and trace movement from gearshift lever to transmission shift lever. (Cheverolet)

shift fork rails. The shift stick is pivoted in a ball-like assembly. Figs. 10-24, 10-25 and 10-26 illustrate these types.

Transmissions have also been shifted by means of electric solenoids, oil pressure and by vacuum. The three shown are in universal use.

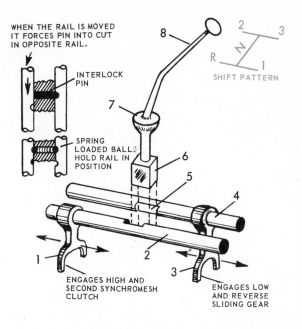

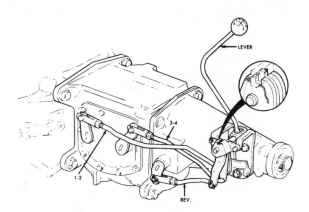

Fig. 10-25. Four-speed transmission floor shift. Notice use of additional lever to actuate reverse gear. (Chevrolet)

Fig. 10-26. Floor shift. This type of shift mechanism is incorporated into transmission top cover. 1—Shift fork. 2—Shift rail. 3—Shift fork. 4—Shift rail. 5—Slot for end of shift stick. 6—Shift stick end. 7—Pivot ball. 8—Stick. When end of stick 6 enters slot 5, it must be pulled to one side in order to move shift rail. Interlock pin arrangement prevents both rails from moving at once. Each rail, after being shifted, must be returned to neutral before other rail will move. Another setup, consisting of spring-loaded detent balls, holds each rail into whatever gear is selected.

COMPLETE TRANSMISSION — CROSS SECTION

Fig. 10-27 illustrates a typical three-speed transmission. Both high and second gear have synchromesh. Notice the bearing locations and the method used to secure all shafts. Second is a constant mesh gear turning on a bushing on the output or main shaft. Fig. 10-28 illustrates the gear drive positions in a fully synchronized, three-speed transmission.

ASSEMBLED TRANSMISSION — CUTAWAY

The transmission shown in Fig. 10-29 is of a heavy duty, four-speed, fully synchronized design. All gears are in constant mesh, and shifting is accomplished by actuating the sliding clutch sleeves through means of shifter forks.

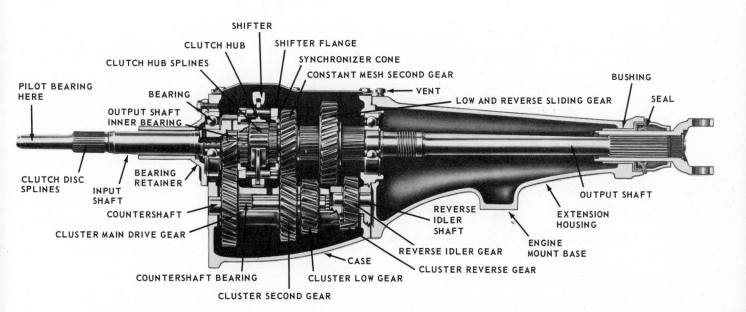

Fig. 10-27. Typical three-speed synchromesh transmission. Note synchronizer construction. Pressure is applied to cones by shifter flange. When pressure reaches a certain point, clutch hub will move into engagement with either high or second gear splines or teeth. (Dodge)

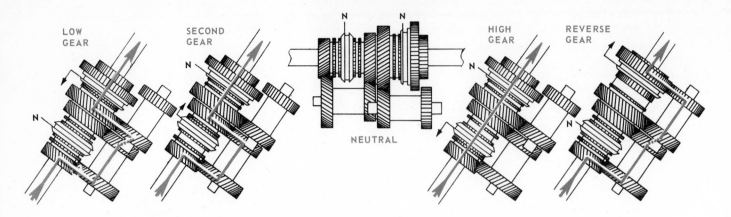

Fig. 10-28. Gear drive positions in a fully synchronized (all forward gears) transmission. N—Synchronizer clutch sleeve in neutral position. (Pontiac)

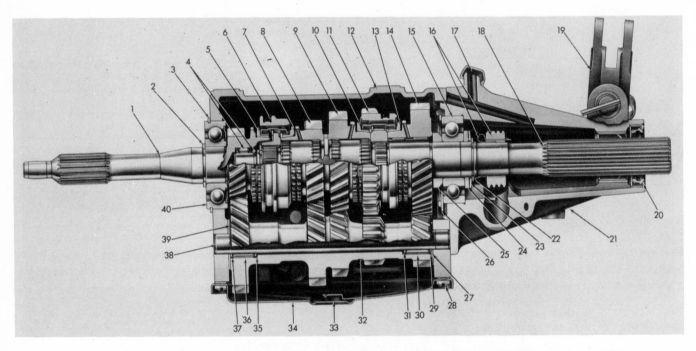

Fig. 10-29. Cutaway of a fully synchronized four-speed transmission. 1—Input shaft. 2—Retaining ring (brg. to shaft). 3—Retaining ring (brg. to case). 4—Pilot bearing. 5—Fourth speed blocker ring. 6—3-4 synchronizer. 7—Third speed synchronizer blocker ring. 8—Third speed gear. 9—Second speed gear. 10—Second speed blocker ring. 11—1-2 synchronizer. 12—Transmission case. 13—First speed synchronizer blocker ring. 14—First speed gear. 15—Rear bearing. 16—Speedometer gear. 17—Vent cap. 18—Output shaft. 19—Intermediate shifter lever. 20—Rear oil seal. 21—Rear extension. 22—Retaining ring (brg. to shaft). 23—Belleville washer. 24—Spacer. 25—Spacer. 26—Retaining ring (brg. to extension). 27—Thrust washer. 28—Cover screw. 29—Spacer. 30—Roller bearing. 31—Spacer. 32—Countergear or cluster gear. 33—Magnet. 34—Cover. 35—Spacer. 36—Roller bearing. 37—Spacer. 38—Cluster gear shaft. 39—Thrust washer. 40—Input shaft main drive gear bearing. (Chevrolet)

THREE—SPEED TRANSMISSION — EXPLODED VIEW

An exploded view of a three-speed, fully synchronized (all forward speeds) transmission is shown in Fig. 10-30. Study all the parts and their construction. Learn their names.

FOUR—SPEED TRANSMISSION

Fig. 10-31 pictures an exploded view of a four-speed transmission. All forward gears are synchronized. Gear teeth

are helical cut, with the exception of the reverse sliding gear and rear reverse idler gear. Learn the location, names and use of all the parts. Study the gear drive positions for the various speeds as shown in Fig. 10-32.

SUMMARY

The engine must have a transmission (torque multiplier) to adapt its available torque to meet changing road and load conditions.

Torque multiplication is accomplished through the use of

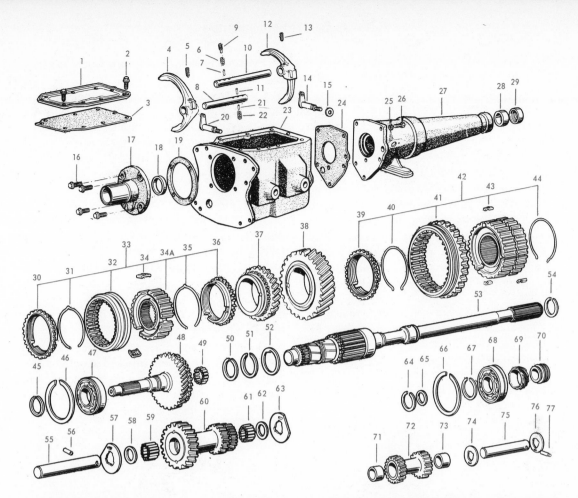

Fig. 10-30. Three-speed transmission — exploded view. 1—Cover. 2—Cover fastener. 3—Gasket. 4—2nd and 3rd gear shifter fork. 5—Setscrew. 6—Interlock spring. 7—Interlock pin. 8—Shift rail. 9—Setscrew. 10—Shift rail. 11—Interlock pin. 12—1st and reverse gear shifter fork. 13—Setscrew. 14—Shift lever. 15—Seal. 16—Retainer fasteners. 17—Bearing retainer. 18—Seal. 19—Gasket. 20—Shift lever. 21—Pin. 22—Spring. 23—Case. 24—Gasket. 25—Lock washer. 26—Extension to case fastener. 27—Extension housing. 28—Bushing. 29—Seal. 30—Blocking ring. 31—Synchronizer spring. 32—Clutch sleeve. 33—Second and high synchronizer assembly. 34—Insert. 34A—Synchronizer hub. 35—Spring. 36—Blocking ring. 37—Second gear. 38—First gear. 39—Blocking ring. 40—Spring. 41—Reverse sliding gear and first speed clutch sleeve. 42—First and reverse gear synchronizer assembly. 43—Insert. 44—Spring. 45—Bearing retaining snap ring. 46—Snap ring. 47—Input shaft bearing. 48—Input shaft and main drive gear. 49—Output shaft inner support bearing. 50—Snap ring. 51—Snap ring. 52—Thrust washer. 53—Output shaft. 54—Snap ring. 55—Countershaft. 56—Lock pin. 57—Thrust washer. 58—Thrust washer. 59—Cluster gear bearing. 60—Cluster or counter gear. 61—Cluster gear bearing. 62—Thrust washer. 63—Thrust washer. 64—Snap ring. 65—Snap ring. 66—Snap ring. 67—Snap ring. 68—Output shaft bearing. 69—Speedometer drive gear. 70—Seal. 71—Reverse idler gear bushing. 72—Reverse idler. 73—Bushing. 74—Thrust washer. 75—Reverse idler shaft. 76—Thrust washer. 77—Lock pin. (Pontiac)

Fig. 10-31. Four-speed transmission — exploded view. This transmission provides synchromesh for 1st, 2nd, 3rd and 4th gear. In this synchronizer design, synchronizer hub does not move. When clutch hub sleeve is shifted, it forces cone-shaped blocking ring with it. Blocking ring engages other part on a similar taper and synchronizes the two. Additional pressure on clutch sleeve causes it to slip on clutch hub inserts (these are what move blocking rings) and it engages splines on other part. When disengaged, it snaps back into notch in middle of hub inserts, causing them to release pressure on blocking ring. Learn names of all parts. Study them and be certain you know their functions. Note vent to allow air to escape, or to enter, as transmission heats up and cools down. (Ford)

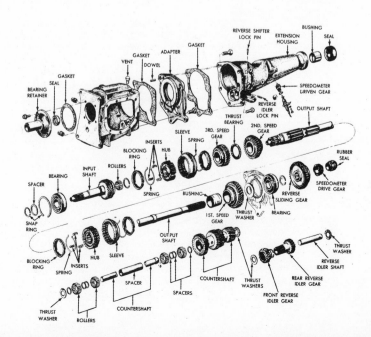

transmission gears. By arranging a movable set of various sized gears, it is possible to secure any reasonable degree of torque multiplication.

Most transmissions give three forward gear ratios and one reverse. Some sports-type or high performance cars, as well as some cars with small engines, utilize four-speed transmissions.

Transmissions are generally equipped with helical gears to assure longer life and quieter operation. One gear (low) may be of the sliding type, or all forward gears may be of constant mesh design. A synchromesh unit is designed to provide smooth gear shifting without clashing or grinding. The transmission consists of the input and output shaft assemblies, the cluster gear and the reverse idler assembly. All are housed in a strong steel or aluminum housing. The gears are selected either by a floor shift or steering wheel column shift control.

The shafts revolve on roller and ball bearings. The second speed gear and reverse idler gear are sometimes mounted on bronze bushings. The input and output shafts are secured by means of snap rings and bearing retainers. End play in the cluster gear is controlled by the use of bronze and steel thrust washers.

The transmission housing can be bolted to the clutch housing or cast as an integral part.

The transmission housing often acts as a base for the rear engine mount.

The transmission is lubricated by filling the housing partially full of heavy transmission gear oil. As the gears whirl, they throw the oil to lubricate all parts. Leakage is controlled by oil seals at both front and rear ends.

NOTE:

When shifting the transmission, it is wise to hesitate in the neutral range for a brief moment. The synchromesh unit depends upon a limited amount of friction to synchronize the speeds. By hesitating a moment, and then shoving the stick smoothly into the desired gear, the synchromesh unit will have time to function.

So-called "speed shifting" is very injurious to the synchromesh gears and, for that matter, the entire drive train. Speed shifting may be justified when a vehicle is entered in competition. At any other time, it is considered the mark of a poor driver, generally one who knows nothing of how his car is built. Good drivers and good mechanics (good mechanics are almost always excellent drivers) have a sincere respect for the work and skill involved in good machinery.

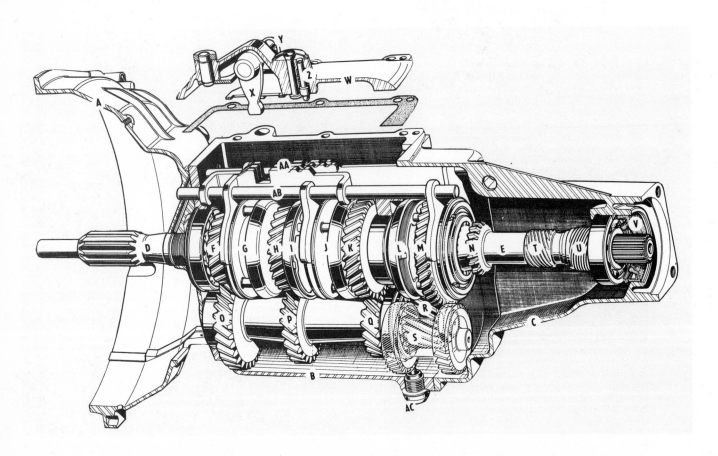

Cutaway view of a fully synchronized four-speed transmission showing shifter forks in place. A—Clutch housing. B—Transmission case. C—Extension housing. D—Input shaft. E—Output shaft. F—Input shaft main drive gear. G—Fourth gear (high) syncronizer. H—Third speed gear. I—Third gear synchronizer. J—Second speed synchronizer. K—Second speed gear. L—First gear synchronizer. M—First and reverse sliding gear. N—Reverse gear dog clutch. O—Cluster main drive gear. P—Cluster third speed gear. Q—Cluster second speed gear. R—First and reverse idler gear. S—Reverse gear. T—Auxiliary drive gear. U—Speedometer drive gear. V—Rear oil sea. W—Gearshift cover. X—Forks shift lever. Y—Gearshift lever. Z—Lock pin. AA—Shift fork. AB—Shift fork shafts. CC—Magnetic drain plug.

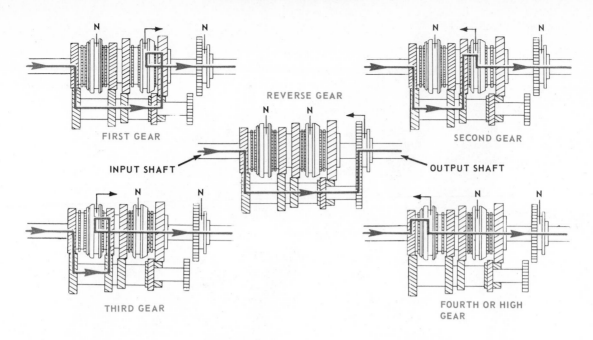

Fig. 10-32. Gear drive positions in a typical four-speed transmission.
Forward gears are all synchronized.   (Buick)

REVIEW QUESTIONS – CHAPTER 10

Do not leave this chapter until you can answer the following questions. If you must go back and reread certain sections, this may slow you down; but remember — as torque goes up, speed goes down.

1. Of what value is a transmission?
2. How does a transmission multiply torque?
3. Why is gear action very similar to a lever and fulcrum?
4. Name the various gears that are available in a standard transmission.
5. What are the approximate ratios of these gears?
6. Explain how the transmission works in each gear.
7. Name the various parts of a gear tooth.
8. Two types of gear teeth are generally used. They are the _____ type and the _____ type.
9. Name the four shafts used in the transmission.
10. How are these shafts turned or driven?
11. What is a synchromesh unit?
12. How does synchromesh work?
13. What lubricates a transmission?
14. What is a constant mesh gear?
15. Explain how gears are shifted.
16. What secures the various shafts to prevent them from working in or out?
17. How is the end play in the reverse idler gear and cluster gear controlled?
18. What supports the inner end of the output shaft?
19. What holds the shifter fork cam in place after a shift is made?
20. Why are four-speed transmissions sometimes used?
21. How is the outer end of the input shaft supported?
22. The inner bearing on the input shaft is a bronze bushing. True or False?
23. Reverse idler gears are always supported on roller bearings. True or False?
24. When torque is multiplied, forward speed is _____.
25. When torque is decreased, forward speed is _____.
26. Low gear is not always provided with a synchromesh. Why is it less important to have low gear synchronized?
27. Transmission housings are made of _____ or _____.
28. The transmission housing often provides a base for the _____.
29. Why is speed shifting considered foolish by expert drivers?
30. If the tooth of the driving gear were pushing on the tooth of the driven gear with a force of 100 ft. lbs., how many ft. lbs. of torque would be applied to the driven gear shaft if the driven gear pitch circle diameter is 1 ft.?

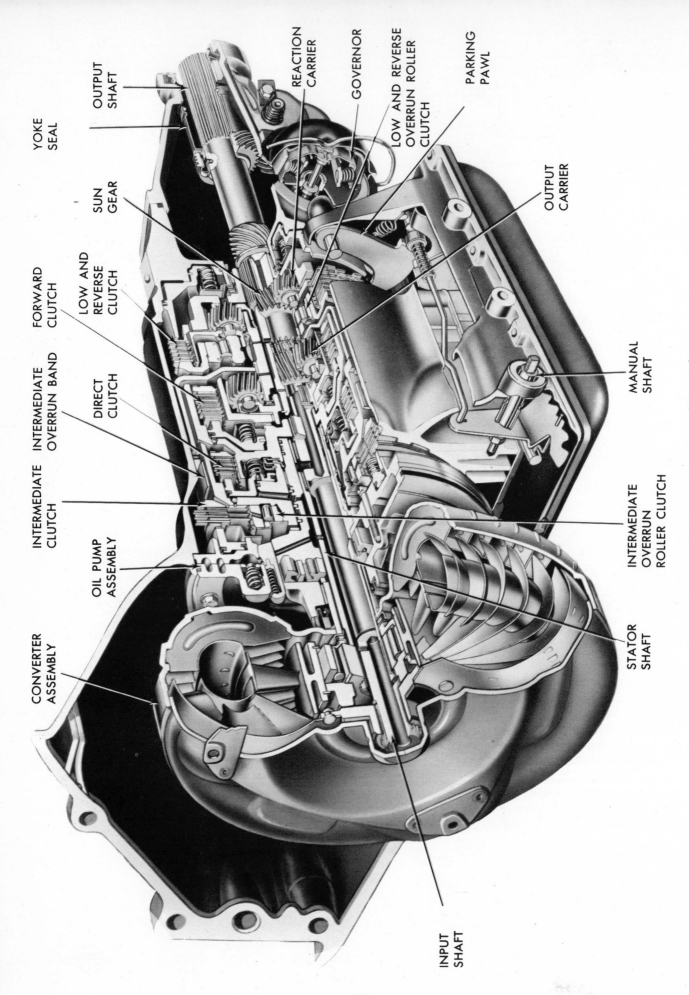

YOKE SEAL

OUTPUT SHAFT

REACTION CARRIER

GOVERNOR

LOW AND REVERSE OVERRUN ROLLER CLUTCH

PARKING PAWL

SUN GEAR

OUTPUT CARRIER

FORWARD CLUTCH

LOW AND REVERSE CLUTCH

INTERMEDIATE OVERRUN BAND

DIRECT CLUTCH

MANUAL SHAFT

INTERMEDIATE CLUTCH

OIL PUMP ASSEMBLY

INTERMEDIATE OVERRUN ROLLER CLUTCH

CONVERTER ASSEMBLY

STATOR SHAFT

INPUT SHAFT

Turbo-Hydramatic 350 automatic transmission.   (Chevrolet)

224

Chapter 11

TRANSMISSION OVERDRIVES

When the standard transmission has been shifted into high gear, the ratio between engine and drive shaft is one to one. When cruising at 60 mph (depending on the size and differential gear ratio), the engine would be turning about 2,600 rpm. Given sufficient horsepower, a road that is fairly level and a reasonable road speed, the engine could propel the car with a lower overall gear ratio. To take advantage of this fact, a unit called an OVERDRIVE is offered by several car makers.

WHAT IS AN OVERDRIVE?

An overdrive is a mechanical unit that bolts to the rear of the standard transmission. The transmission output shaft drives the overdrive; the overdrive output shaft turns the drive shaft. Between these two shafts, a series of units, including a planetary gearset, function to provide a lower overall gear ratio of approximately .7 to 1. Fig. 11-1 shows the ratio in low, second, high and overdrive gear.

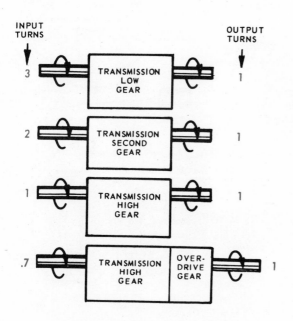

Fig. 11-1. Various gear ratios. Notice how addition of over-drive lowers ratio.

VALUE OF AN OVERDRIVE

The first advantage of the overdrive is that it reduces engine speed around 30 percent while still maintaining the same road speed. The reduction in engine speed gives increased fuel milage, reduces engine noise and prolongs engine life. An overdrive works satisfactorily provided:

1. That the engine has sufficient power to propel the car in overdrive gear without "lugging" (straining to pull).
2. That reasonable road speed is maintained.
3. That the road is fairly level.

An overdrive will not guarantee higher top speeds (illegal on the highway). As car speed increases, wind resistance raises quickly. At very high speeds, the wind will produce a resistance to forward motion comparable to that of climbing a long grade (hill). This can easily upset the advantage of a higher gear, and performance could be improved by shifting into direct (high) gear.

OVERDRIVE CONSTRUCTION AND OPERATION

PLANETARY PINION CARRIER

The standard transmission output shaft is made somewhat longer so it will extend into the overdrive housing. It has a splined end that drives a planetary pinion carrier with three planet pinion gears. The term planetary comes from the fact that the pinions revolve, in orbital fashion, around a central or sun gear, like planets revolve around the sun. Fig. 11-2 illustrates the carrier and pinions. For study purposes, only two of the three pinions are shown and the unit is somewhat simplified.

SUN GEAR

A sun gear (termed sun because planet pinions revolve around it) assembly is installed on the output shaft. One end has helical teeth that are meshed with the pinion gears. This is the sun gear proper. The sun gear assembly is NOT splined to the transmission output shaft but is free to turn. The other

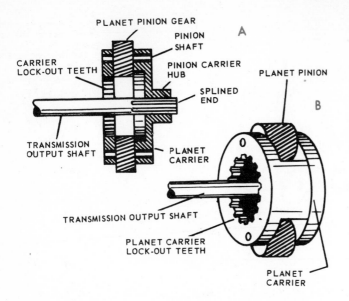

Fig. 11-2. Planet pinion carrier. A—Cross section view of carrier. Note that carrier hub is splined to transmission output shaft. B—View of entire carrier. Two of three pinions are shown.

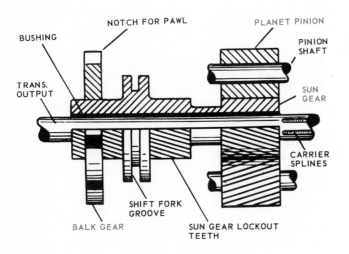

Fig. 11-3. Sun gear assembly. Sun gear meshes with planet pinions. Balk gear is splined on. Sun gear assembly can turn on transmission output shaft when desired. Planet pinions shown are part of planet carrier assembly.

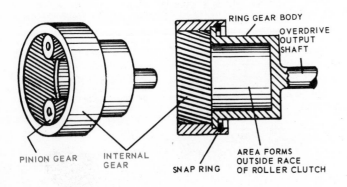

Fig. 11-4. Internal gear. Internal gear surrounds and meshes with planet pinion gears. Only two are shown. Ring gear is held to ring gear body by snap ring. Both units spline together.

end has a continuation of the helical teeth upon which a balk (stop) gear is placed. When required, it is possible to stop the sun gear by holding the balk gear with a steel plunger termed a pawl. (More on this later.)

An additional set of teeth, similar to the sun gear proper, are cut on the sun gear assembly. A flange is also cut on the assembly, to provide a spot for a shift fork to move the sun gear unit in toward the planet carrier far enough so the additional teeth will mesh with internal teeth cut in the transmission side of the planet carrier. When the sun gear assembly is shifted toward the carrier, these two units will mesh, thus locking carrier and sun gear units together. The sun gear proper is always in mesh with the planet pinion, even when the sun gear lock-up teeth are removed from mesh with the carrier. Fig. 11-3.

INTERNAL GEAR

A large ring gear with internal teeth surrounds the planet carrier and meshes with the planet pinions. The internal gear is splined to, and becomes a solid part of, the overdrive output shaft. Fig. 11-4.

PLANETARY GEARSET

You now have the essential parts for what is termed a planetary gearset. It consists of a sun gear, planet pinion gears and carrier, and an internal gear. Fig. 11-5.

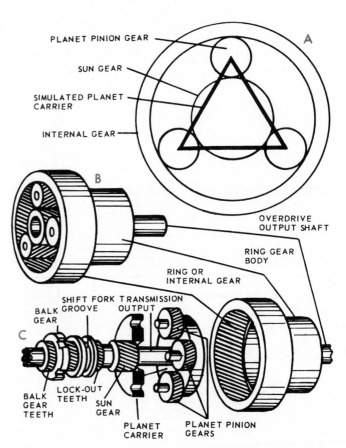

Fig. 11-5. Planetary gearset. A—End view of gearset. B—Showing ring gear body and output shaft. C—Exploded view.

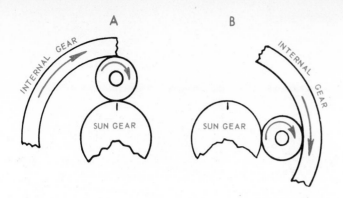

Fig. 11-6. Overdrive effect. A—Sun gear stopped. Pinion gear walking around sun gear, forcing internal gear to move in same direction. B—Pinion has walked 90 deg. Notice how internal gear has gained. It is traveling faster than pinion carrier.

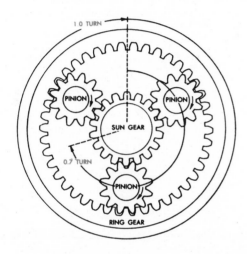

Fig. 11-7. Relationship between rpm of pinion carrier and internal gear. Every time transmission output shaft turns pinion carrier .7 of a turn, internal gear is forced to rotate a full turn. This causes overdrive output shaft to revolve faster than transmission output shaft. (Chevrolet)

USING THE PLANETARY GEARSET TO PRODUCE AN OVERDRIVE RATIO

When the pawl (steel plunger) stops the balk gear, the sun gear is held stationary. The planet pinion carrier will be turned by the transmission output shaft. As the planet carrier revolves, the pinions are forced to walk (roll around) around the sun gear. As they are also meshed with the internal gear (which is part of overdrive output shaft), the internal gear is forced to rotate faster than the carrier. By the time the carrier has made .7 of a turn, the internal gear will have made a complete turn. As the carrier is on transmission output shaft, and the internal gear is on overdrive output shaft, the planetary gearset has allowed engine speed to drop 30 percent and still maintain the same road speed. Figs. 11-6 and 11-7.

OTHER FUNCTIONS OF THE PLANETARY GEARSET

The planetary gearset can be used to multiply torque, decrease torque, drive direct with no multiplication or produce

a reverse rotation. How this is done will be covered in the next chapter since the automatic transmission makes full use of the planetary gearset.

OVERDRIVE LOCKED OUT

When the use of the overdrive is not wanted, it may be placed in the LOCKED-OUT stage. When any two of the three planetary members are locked together, the planetary action is stopped, and the entire assembly revolves as a solid unit.

The overdrive is locked out by shifting the sun gear assembly until its additional lock-out teeth mesh with internal teeth on engine side of the planet carrier. Fig. 11-8.

SOMETHING MISSING?

If the sun gear assembly is shifted out until the lock-out teeth are free of the pinion carrier, and the pawl is not holding the sun gear stationary, what type of drive would we have?

Trace the power flow. The transmission output shaft will

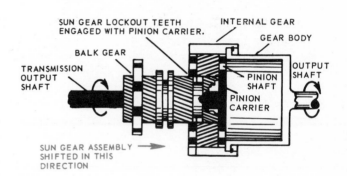

Fig. 11-8. Overdrive locked out. Sun gear assembly has been shifted until sun gear lock-out teeth have engaged carrier lock-out teeth. This locks two members of planetary gearset together and gearset will drive as a solid unit with no planetary action.

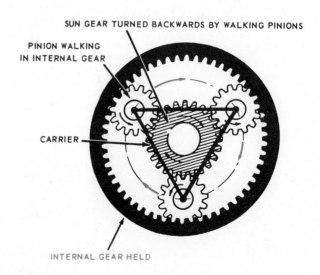

Fig. 11-9. Unit will not drive. Internal gear is being held by output shaft. Engine is spinning pinion carrier. Pinions are walking around in internal gear, causing sun gear to revolve backwards. No torque is transmitted to internal gear and car will not move. (Chevrolet)

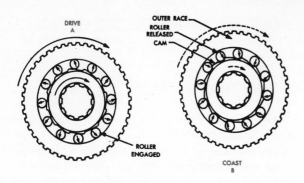

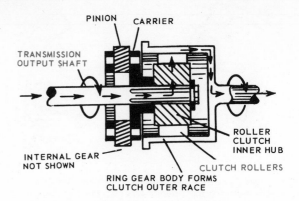

Fig. 11-10. Roller clutch action. In A, hub is being driven by transmission output shaft. Rollers have traveled up cams and have seized outer race thus driving it. In B, accelerator has been released and inner hub has slowed down. This causes rollers to travel back down cams and release outer race. Outer race will now free-wheel or over-run hub.
(Chevrolet)

Fig. 11-12. Overdrive roller clutch. Drive is through transmission output shaft, to clutch hub, to rollers, to ring gear body, to overdrive output shaft. When accelerator is released, transmission output shaft will slow down, rollers will retract, and overdrive output will free-wheel.

turn the pinion carrier. It will take torque to move the internal gear, as it is part of the output shaft. As a result, the whirling pinions will walk around in the internal gear, and the sun gear will be spun backward. Type of drive — NONE! Fig. 11-9.

MISSING PART — ROLLER CLUTCH

Fig. 11-10 illustrates the principle of the roller clutch. Notice that it consists basically of three parts, the splined inner hub with tapered cams on the outside surface, a series of hardened steel rollers (one per cam surface) and an outer ring or race. As used in an overdrive, the rollers are contained or held in position with a steel cage. A spring is incorporated to assist in the functioning of the unit as a clutch.

The inner hub is the driving part; the outer race is the driven part. When the cam hub receives driving torque, the rollers are forced up the cam surfaces until they are wedged tightly against the outer race. The outer race will then be driven. When the driving torque is removed from the hub, the rollers disengage and return to their low positions, freeing the outer race. The race may continue to turn while the hub turns slowly or even stops. This action is referred to as FREE-WHEELING. The outer race will be driven only when the speed of hub exceeds that of race. Figs. 11-10 and 11-11.

OVERDRIVE ROLLER CLUTCH

A roller clutch is splined to the transmission output shaft in back of the splined pinion carrier. The internal gear body forms the outer race. When the transmission output shaft is

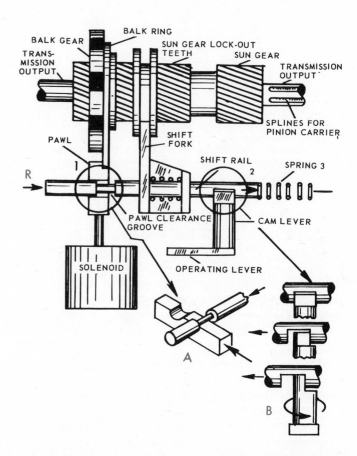

Fig. 11-13. Basic overdrive controls. Top sketch — Sun gear assembly has been shifted into lock-out position. Cam lever has been turned until it has drawn shift rail and fork to right, 2. Pawl is blocked by rail at 1. When sun gear assembly moves, balk gear and ring remain fixed. They slide on balk gear teeth. A—Showing how rail clears pawl when moved to left. B—Showing action of cam lever in shifting rail. Spring 3, furnishes force to keep rail moving to left. R—When transmission is placed in reverse, pressure on rod at R, forces sun gear into lock-out position.

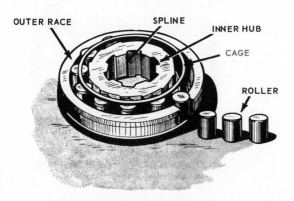

Fig. 11-11. Roller clutch (often called "overrunning clutch"). Notice how rollers are contained in thin steel cage. This forces them to act as a unit. Outer race has been cut away to show hardened steel rollers.
(Sunbeam—Talbot)

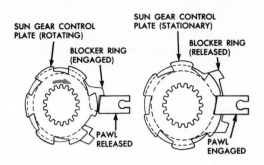

Fig. 11-14. Balk gear and balk ring action. Sketch at left shows pawl being held out by balk ring (also called blocker ring). Balk gear (also called sun gear control plate) is still being turned. Right hand sketch shows how balk ring has been pulled up when balk gear began reverse rotation. Opening in balk ring is lined up to allow pawl to snap in and lock balk gear. This stops sun gear assembly and produces overdrive effect. (Chevrolet)

turning faster than the internal gear, and the sun gear is free, the clutch rollers will drive the internal gear body. When the sun gear is stopped, the planetary action will drive the internal gear and body faster than the carrier and clutch hub, and the clutch will remain disengaged. Fig. 11-12.

OVERDRIVE CONTROLS

Sun gear unit is shifted by fork, shift rail and cam lever arrangement. Overdrive may be locked out (sun gear splines or lock-out teeth shifted into mesh with carrier) by driver, by moving control lever via mechanical linkage arrangement.

The shift rail extends into the transmission. It is so arranged that when the car is shifted into reverse, the overdrive shift rail is moved, locking out the overdrive. This is necessary to back the car when the overdrive is in the overdrive position. A slot in the overdrive shift rail will also prevent the pawl from engaging the blocking gear when the overdrive is locked out.

The pawl is operated by an electric solenoid. The solenoid is energized by an electric current actuated by a governor controlling a relay. When car speed is around 30 mph, the governor closes a set of contact points, energizes the relay which in turn energizes the solenoid, causing it to attempt to push the pawl into the balk gear. Fig. 11-13.

BALK RING

A balk ring, Fig. 11-14, is attached to the balk gear hub in such a fashion as to allow the gear to rotate while the ring is

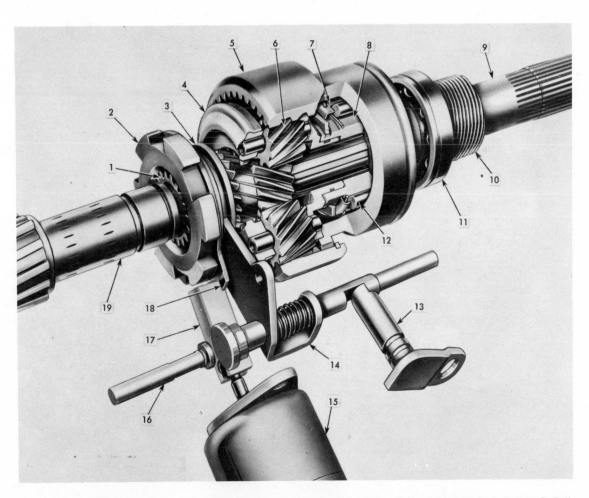

Fig. 11-15. Complete overdrive unit. Learn names of parts and proper locations: 1—Sun gear teeth for balk gear. 2—Balk gear. 3—Sun gear shift groove or collar. 4—Pinion carrier assembly. 5—Internal gear. 6—Planet pinion gear. 7—Roller clutch with roller cage. 8—Roller clutch inner hub. 9—Overdrive output shaft. 10—Speedometer drive gear. 11—Output shaft bearing. 12—Roller clutch roller. 13—Cam lever rod. 14—Shift fork. 15—Solenoid. 16—Shift rail. 17—Pawl. 18—Balk ring. 19—Transmission output shaft. (Chevrolet)

held stationary by the tip of the solenoid pawl. There is a controlled amount of friction between the gear and the ring.

When in overdrive, and before the pawl has stopped the sun gear, the drive is through the roller clutch. As the balk gear is turning, the block or balk ring will rotate until stopped by the pawl. When the governor energizes the solenoid, the pawl will shove in. However, it will be unable to move since the balk ring is interferring with its movement. When the driver releases the accelerator, the roller clutch will disengage. As the internal gear walks the pinions around the sun gear, the sun gear will slow down. When the transmission input shaft rpm is 70 percent that of the overdrive output shaft, the sun gear will be stopped. As engine speed decreases, the sun gear will begin to

rotate in a reverse direction. Since there is friction between the balk gear and balk ring, the ring will be pulled backward to expose an opening through which the pawl will snap. The balk gear will stop and the transmission is in overdrive. All this happens in a second or so.

COMPLETE OVERDRIVE UNIT

Fig. 11-15 shows a cutaway of a complete overdrive unit. Study all parts and learn their names. Remember that when desired, the sun gear rotates on the transmission output shaft. Number 1 in the illustration shows the teeth upon which the balk gear 2 rides. These teeth are cut full length of the sun gear

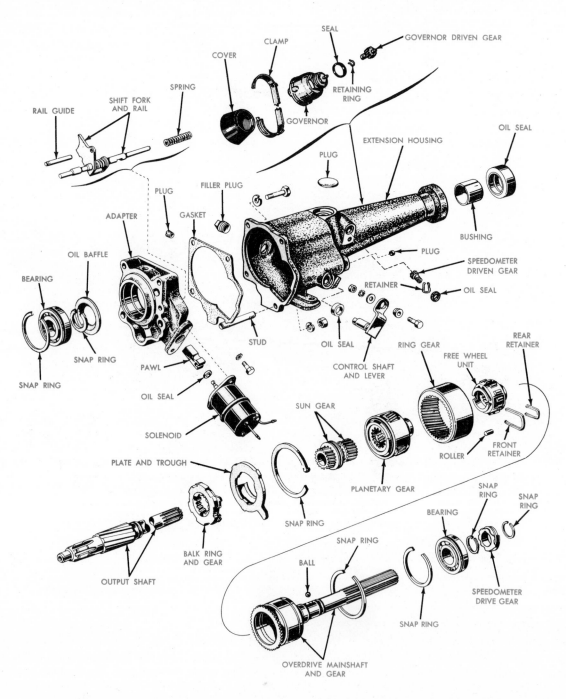

Fig. 11-16. Exploded view of a typical overdrive. Study each part, its name location and construction.   (Ford)

unit with a portion removed. In this way, the teeth serve as sun gear teeth, lock-out teeth, and balk gear teeth.

Note the action of the shift rail. The lever 13 has been moved to allow the rail to move into overdrive position with sun gear lock-out teeth clear of carrier, and solenoid pawl free of shift rail.

EXPLODED VIEW — COMPLETE OVERDRIVE

Fig. 11-16 illustrates the overdrive in an exploded view. Notice the solenoid and governor assemblies. The retractor spring forces the shift rail and fork to the left, when the shift lever is moved to overdrive. The internal gear and internal gear body are splined together and are held with the large snap ring.

OVERDRIVE DRIVE STAGES

LOCKED—OUT

When overdrive is not desired, it is placed in the locked-out position. Fig. 11-17. Locking out is accomplished by the driver, who moves a dash control to operate the shift rail lever.

The lever causes the overdrive shift rail and fork to move the sun gear lock-out teeth into engagement with the pinion carrier. A switch in the governor circuit is held open to prevent solenoid energization at speeds above 30 mph. Drive is direct, as though the overdrive were a solid unit.

DIRECT, FREE—WHEELING

Assume that the driver has placed the control lever in the overdrive position. The lever moves the sun gear lock-out teeth out of the pinion carrier. The solenoid pawl is disengaged and the governor switch circuit is closed. Now the sun gear can move, no two planetary members are locked together, and the drive is through the roller clutch. When the accelerator pedal is released at speeds below 30 mph, the clutch rollers retract, and the car coasts or freewheels. Fig. 11-18.

NEVER COAST

Never descend a hill with the transmission in neutral or the overdrive in free-wheeling because all stopping must be done with the brakes. If the brakes fail, it may be difficult to get the car back into gear. Never coast — it is DANGEROUS!

OVERDRIVE

With all controls in the same position as that of direct, free-wheeling, the overdrive is ready to function in overdrive gear. When the car reaches a speed of around 30 mph, the governor points close and the overdrive electrical relay is energized. The relay completes the circuit to the solenoid, then attempts to shove the pawl into the balk gear. Since it cannot do this, the pawl is held out by the balk ring. The car will remain in direct, free-wheeling as long as the engine power flow is uninterrupted.

When the car has reached a speed of about 30 mph, and the

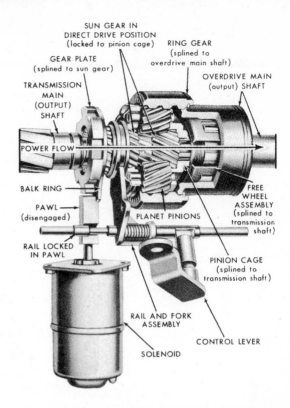

Fig. 11-17. Power flow with overdrive in locked-out stage. Pawl disengaged, sun gear assembly shifted so that sun gear lock-out teeth are meshed with carrier lock-out teeth thus locking sun gear and carrier together. Power flow is direct as overdrive functions as a solid unit. (Mercury)

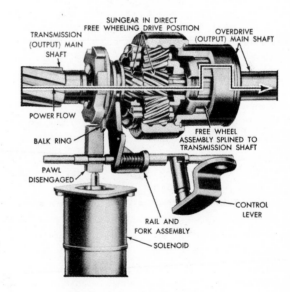

Fig. 11-18. Power flow with overdrive in direct, free-wheeling stage. Drive is direct, through roller clutch. When accelerator is released, car will coast or free-wheel. (Mercury)

driver desires the use of overdrive gear, he releases the accelerator pedal. The roller clutch will then disengage; the sun gear slows down, then reverses. As it reverses, it pulls the balk ring with it. This action exposes an opening for the pawl, which pawl pops in and stops the balk gear. The sun gear can

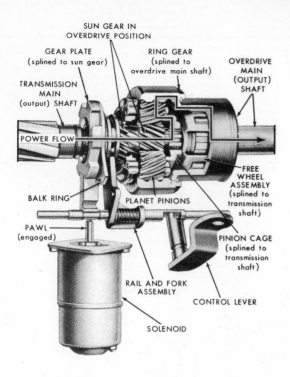

Fig. 11-19. Power flow with overdrive in overdrive stage. Sun gear stopped by pawl. Planet pinion carrier revolving. Pinions walking around sun gear, driving internal gear faster than carrier. (Mercury)

no longer turn, and the drive is through the carrier pinions walking around the sun gear. This drives the internal gear faster than the carrier to produce the overdrive effect. See Fig. 11-19.

REMOVING FROM OVERDRIVE

The car may be taken out of overdrive in three ways:

1. The driver may pull the manual control to locking out the overdrive.

2. When the car road speed drops 4 or 5 miles under 30 mph, the governor switch points open, causing the relay to break the solenoid circuit. The pawl is withdrawn by spring action. This leaves the overdrive in direct, free-wheeling until the road speed is again increased to 30 mph.

3. If the driver desires extra power for passing, he may move the accelerator pedal to the floor, engaging the kickdown switch. This will break the governor circuit which, in turn, causes the relay to break the solenoid circuit. The solenoid spring could not withdraw the pawl, since it would be held in the balk gear by the heavy torque from the accelerating engine.

At the same time the kickdown switch breaks the governor circuit, it also shorts out the distributor side of the coil primary by shunting the circuit to ground in the solenoid housing. This causes the engine to quit, and as soon as the torque thrust is removed, the pawl pops out and frees the sun gear. As the solenoid pawl assembly pulls into the solenoid body, the ground circuit from the coil is broken, and the engine begins to run again. The drive is now direct through the roller clutch. This whole operation, after the kickdown switch

is depressed, takes but a fraction of a second.

After passing, the accelerator is released. The kickdown switch reestablishes the governor circuit. The governor energizes the relay, the relay energizes the solenoid, and the pawl once again stops the sun gear. The unit is then in overdrive.

Fig. 11-20 shows a schematic drawing of overdrive electrical parts and circuits. Study it carefully. Notice the ground points A are open when the solenoid is inactive.

At B, the wire to the solenoid divides into two coil circuits, the PULL-IN coil and the HOLD-IN coil. When the solenoid is energized, the pull-in coil draws the pawl plunger down. Once down, the pull-in winding circuit is broken and the plunger is held down by the hold-in coil.

SUMMARY

The overdrive is a unit that is attached to the rear of the transmission. It reduces engine speed about 30 percent while maintaining the same road speed. It accomplishes this advantage by using a planetary gearset.

The overdrive proper consists of the transmission output shaft, overdrive output shaft, roller clutch, sun gear unit, planet pinion carrier, planet pinions, and the internal gear that is splined to the output shaft.

A series of mechanical and electrical controls are provided to enable the driver to actuate the overdrive.

There are three drive stages:

1. LOCKED-OUT stage in which the sun gear lock-out teeth are meshed with the planet carrier, causing the unit to function as a solid piece.

2. DIRECT, FREE-WHEELING where the sun gear lock-out teeth are free of the carrier, the pawl is disengaged and drive is transmitted through the roller clutch. When power is removed, the car coasts or free-wheels.

3. OVERDRIVE stage in which the pawl engages the balk gear, stopping the sun gear. Sun gear lock-out teeth are clear of the pinion carrier. Drive is through the planet pinion carrier to pinions, to internal gear, to output shaft.

The overdrive may be shifted into direct at any time by depressing the accelerator pedal to the floor. At speeds below 30 mph, the overdrive will drop into direct, free-wheeling.

REVIEW QUESTIONS — CHAPTER 11

Knowledge is life's planetary gearset. It lets you slow down a little, while still maintaining the same "road speed." As you would with the overdrive — NEVER COAST.

Intelligence is the ability to learn. Knowledge is what is learned. Intelligence without knowledge is like a car without an engine — WORTHLESS! Learn all you can!

By knowing the answers to the following questions, you will have made your understanding of the next chapter much easier. It deals with automatic transmissions which make full use of the planetary gearset.

1. What is an overdrive?
2. What does it do?
3. Is an overdrive suited to ALL cars?
4. If so or if not — why?

5. The overdrive will give a speed reduction (for the engine) of about _____ percent.
6. For what is the sun gear assembly used?
7. How does the internal gear transmit power?
8. Why is the gearset called a planetary gearset?
9. The unit connecting the sun gear and the internal gear is called a _____.
10. The balk gear is splined to the _____.
11. What is the use of the balk gear?
12. Why is it necessary to have a roller clutch?
13. How does the roller clutch function — and when?
14. The pinion carrier is splined to the _____ _____.
15. The roller clutch is splined to the _____ _____.
16. The solenoid is used to actuate the _____.
17. When the sun gear lock-out teeth engage the pinion carrier, the overdrive is in direct, free-wheeling. True or False?
18. When the pawl is disengaged and the sun gear lock-out teeth are free of the pinion carrier, the overdrive is locked out. True or False?
19. At about what speed will the overdrive kick in or become operative?
20. Name the three ways in which the unit may be taken out of overdrive.
21. What is the kickdown switch for?
22. The overdrive governor operates the overdrive shift fork and rail. True or False?
23. Draw a diagram of the overdrive electrical circuits. Indicate and name the units involved.
24. To back the car, it is necessary to _____ the overdrive.
25. Where is the overdrive fastened?
26. Explain the complete operation involved in the three overdrive stages.

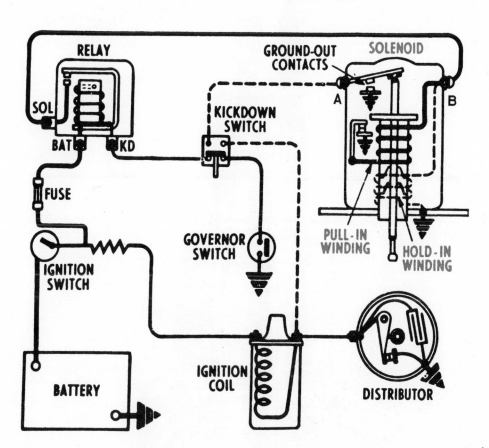

Fig. 11-20. Schematic of typical overdrive electric control circuit. Examine each circuit in diagram. Note two windings in solenoid. (Chevrolet)

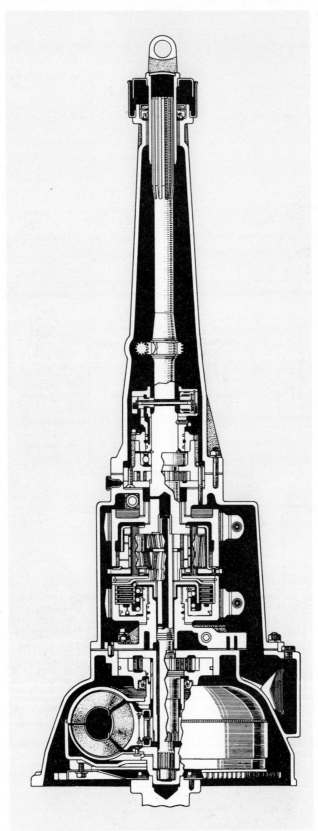

Cross section view of a two-speed Toyoglide automatic transmission.   (Toyota)

Chapter 12

AUTOMATIC TRANSMISSIONS

Like the standard transmission, the automatic transmission is designed to adapt the power of the engine to meet varying road and load conditions. In this case, however, the transmission does this automatically. Instead of three set forward gear ratios, it can produce an infinite number of ratios between engine and wheels.

The automatic transmission eliminates the clutch pedal and, instead of a solid type conventional clutch, it utilizes a fluid coupling between the engine and transmission.

After the driver has selected the necessary range (speed or ratio selection) by shifting a lever or pushing a button, the transmission shifts itself up or down depending on road speed, throttle position and engine loading.

BASIC CONSTRUCTION

Basically, in addition to the metal housing, the automatic transmission assembly consists of:

1. A fluid coupling or torque converter to transmit power from the engine to the transmission proper.

2. One or more planetary gearsets and shafts to secure the necessary forward and reverse speeds.

3. A series of brake bands and multiple disc clutches designed to control the planetary gearsets.

4. Hydraulic servos and pistons to actuate the bands and clutches.

5. One or more oil pumps to provide the necessary hydraulic pressure.

6. Numerous valves (control, check, pressure regulation, balanced, transition, downshift, etc.), all designed to control, direct and administer hydraulic pressures throughout the transmission.

7. Some means of cooling the oil.

8. Manual control system used by the operator to select certain speed ranges: high, low, etc.

FLUID COUPLING

The fluid coupling utilizes a special lightweight oil to transmit engine power from one member to another.

For purposes of illustration, imagine cutting a hollow steel doughnut down the center, Fig. 12-1. You now have two halves, each shaped like the unit shown in Fig. 12-2.

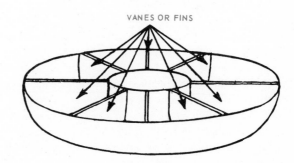

Fig. 12-1. Cutting a hollow steel doughnut.

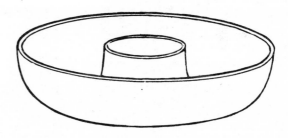

Fig. 12-2. One half of the steel doughnut.

VANES OR FINS

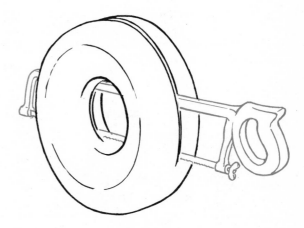

Fig. 12-3. Vanes are placed in each torus. Notice that vanes are straight and equally spaced. In an actual torus, a greater number of vanes are used.

TORUS

Each half, or member, is called a TORUS. Vanes or fins are placed in each torus. The vanes are straight and are spaced at equal intervals. Fig. 12-3.

The two torus members are fastened to shafts. One member to an input, the other to an output shaft. They are placed face to face with a slight clearance between them. Fig. 12-4.

As indicated in Fig. 12-4, the flywheel holds the driving torus. A thin housing from the flywheel also extends around the driven member. An oil seal is placed between the transmission shaft and this housing. The housing is filled with oil.

You will also notice that the driven member is splined to the transmission or output shaft. The shaft is supported by a pilot bushing in the crankshaft. When the crankshaft turns the flywheel, the driving member and housing will spin. The driven member is entirely free since there is no mechanical connection between it and the driving member.

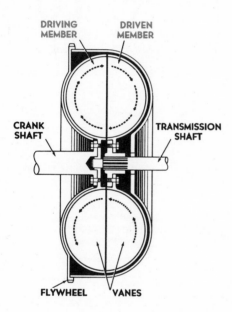

Fig. 12-4. Torus members in place. One torus member is fastened to each shaft. Notice how transmission shaft is supported by bearing in end of crankshaft.

HOW FLUID COUPLING FUNCTIONS

Imagine two fans facing each other, at reasonably close range. If one fan were plugged in, the air blast from the driving fan would cause the other fan to spin. Air is the medium of power transfer. Since the two fans are not in an enclosed area, the power transfer would not be very efficient. Fig. 12-5.

The fluid coupling works on somewhat the same principle as the fan. However, oil is used instead of air, and since the driving and driven torus are mounted very close together in an enclosed space, power transfer is excellent.

If oil were placed in a horizontal torus, and if the torus were spun, the vanes would carry the oil around at high speed. This motion of oil would be referred to as a ROTARY motion.

Fig. 12-5. Coupling drive using fans and air. Air from driving fan will cause other fan to spin.   (Chevrolet)

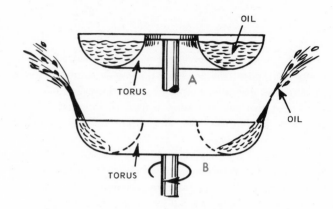

Fig. 12-6. Oil action in torus. A—Cutaway of torus at rest. Oil is level. B—Torus is spinning. Oil is thrown outward and is directed upward by curved torus edge.

The centrifugal force set up by the whirling would cause the oil to fly outward and upward. Fig. 12-6.

If the driven torus were placed above the driving torus, the oil that is thrown out would strike the vanes in the driven torus. After striking the vanes it would travel up, over and back down into the driving torus. As the driving torus is whirling, it would be thrown outward and upward back against the vanes of the driven torus. This around and around motion of the oil is termed VORTEX flow. The same action takes place when the torus members are in the vertical position. Figs. 12-7, 12-8 and 12-9.

As the speed of the driving torus is increased, the rotary and vortex flow of the oil becomes more violent. This will cause the driven torus to turn faster and faster.

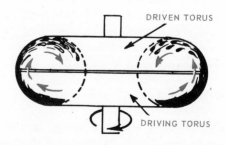

Fig. 12-7. Start of vortex flow. When driving torus starts to spin, oil is thrown out, up into driven torus, around and back into driving torus. This action, around and around, is termed vortex action.

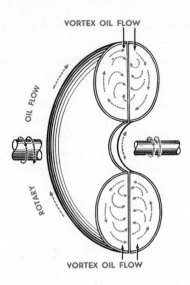

Fig. 12-8. Rotary and vortex oil flow.

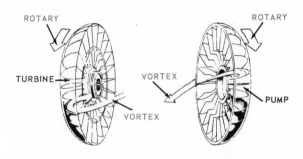

Fig. 12-9. Rotary and vortex flow. (Pontiac)

speed is low, there is no power transfer. This is the same as having a conventional clutch in the disengaged position. As engine speed is increased, power transfer becomes more effective. All jerking and roughness is eliminated by the use of the fluid coupling. This provides smooth takeoff and reduces wear and strain on the drive train.

The fluid coupling cannot increase torque above that produced by the crankshaft.

When the engine is accelerating, the driving torus transmits engine torque to the driven torus. When coasting, the driven torus becomes the driver and allows the engine to act as a brake.

TORQUE CONVERTER

A torque converter is similar to the fluid coupling, but there is one very important difference. The fluid coupling can transmit all available engine torque, but it cannot multiply this

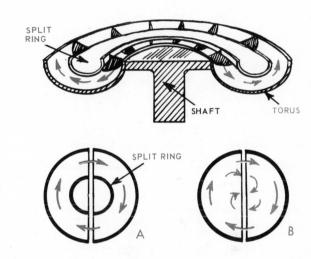

Fig. 12-10. Use of a split ring to smooth vortex flow. Top view illustrates split ring as installed in a torus. Vanes are cut away to receive ring. A shows how oil vortex flow is guided and smoothed by split ring. B shows turbulence present in center when split ring is not used.

Finally the speed of the driver and driven members will become equal. At this point the rotary flow is constant, but there is little or no vortex flow. You will remember that vortex flow is caused by centrifugal force throwing the oil outward. If both members are whirling at the same speed, each member will attempt to throw oil into the other, cancelling the vortex flow.

SPLIT RING

To assist the oil in maintaining a smooth vortex flow, a hollow ring is sometimes placed in each member. The oil is guided around, and there is less tendency for the oil to work against itself in the center area. Fig. 12-10.

FLUID COUPLING EFFICIENCY

The fluid coupling effects a smooth transfer of power from the engine to the transmission.

At high engine speeds, the coupling is very efficient. It gives a one to one ratio between driven and driving members. At medium speeds the coupling is not quite as effective. At low engine speeds there is little power transfer.

This allows the coupling to act as a clutch. When engine

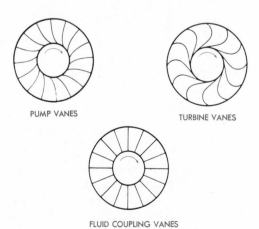

Fig. 12-11. Fluid coupling and torque converter vanes. Torque converter vanes are curved. Fluid coupling vanes can be straight. (Chevrolet)

torque. The torque converter can and does multiply engine torque. The amount of multiplication depends on the type and design of converter used, as well as road and engine speeds.

The ability to multiply torque makes it possible to reduce the number of gears in the automatic transmission.

The torque converter uses a driving and driven torus but, in this case, they are generally referred to as the pump (driver) and the turbine (driven). In some cases, more than one pump and turbine are used. The pump is sometimes referred to as the impeller.

VANES

The converter blades or vanes are curved to facilitate proper vartex flow. The pump vanes are curved one way, and the turbine blades the other. Fig. 12-11.

STATOR

The secret of torque multiplication lies in the use of one or more STATORS. Fig. 12-12.

A stator is a small, wheel-like arrangement that is inserted between the pump and turbine. The job of the stator is to intercept the oil thrown off by the turbine and redirect the path of this oil so it will enter the pump smoothly. The stator is mounted between the pump and the turbine. As the pump begins to spin, oil is thrown outward into the curved vanes of the turbine. The oil then circulates around through the turbine vanes. Instead of being discharged back into the pump vanes (as in fluid couplings), the oil is first passed through the stator.

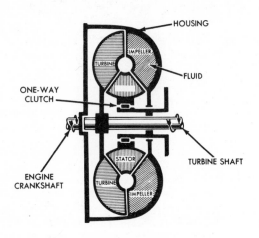

Fig. 12-12. One type of stator. (Ford)

In the fluid coupling, which uses no stator, torque transmission and multiplication efficiency is reduced by having turbine oil enter the pump at various angles, depending on speed and load conditions. This poor entry angle can work against the pump and absorb valuable torque.

The stator is mounted on a stationary shaft, and is allowed to spin in the direction of the pump. Any attempt to rotate the stator against pump rotation, causes it to lock to the shaft.

SPRAG AND ROLLER ONE—WAY CLUTCHES

Stator locking action is produced by using either a roller or a sprag one-way (overrunning) clutch. The roller clutch is described in Chapter 11, Fig. 11-10.

The sprag clutch is similar to the overrunning type used in the overdrive. Instead of using round rollers that walk up a tapered step to lock up the clutch, the sprag clutch uses a flat unit with curved edges. When the outer race is turned one way, the sprags tip and present their narrow diameter. This allows the outer race (fastened to stator) to turn.

When the stator attempts to rotate in the opposite direction, the sprags tip the other way and present the long diameter. This causes them to jam between the two races, locking them together and preventing rotation in this direction. Fig. 12-13. Note location of stator one-way clutch. Fig. 12-12. See roller clutch in Fig. 12-20.

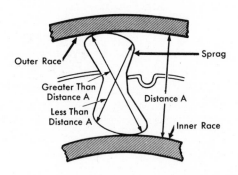

Fig. 12-13. Cross section showing sprag clutch construction. (Cadillac)

TORQUE MULTIPLICATION

As the pump discharges oil into the vanes of the turbine, the force of the oil tends to turn the turbine. In, A, Fig. 12-14, you see a wheel (turbine) with cups (vanes) attached. Oil is being sprayed into the cup by a pressure gun (pump).

The oil strikes the cup (vanes) and exerts some force. The force, however, is not enough to cause the heavy wheel to turn. You can see that the oil is striking one side of the cup, then is directed around and out (vortex action). Obviously a great deal of the potential force of the oil is being lost.

In B, Fig. 12-14, a stationary curved piece (stator) has been placed so the oil is redirected into the cup, (vane). The oil can now strike the cup and, instead of leaving, will continue to circulate back into the cup. The torque produced by the oil in A, has been increased in B, by adding the extra torque that was lost when the oil left the cup. By redirecting the stream of oil, the torque has been multiplied. This is the job of the stator.

As the wheel (turbine) turns faster and faster, it will finally approach the speed of the oil stream (pump). C, Fig. 12-14. When this happens, the cup will be moving at about the same speed as the oil stream. In this case the oil will not fly out of the cup and cannot be redirected. Torque multiplication is no

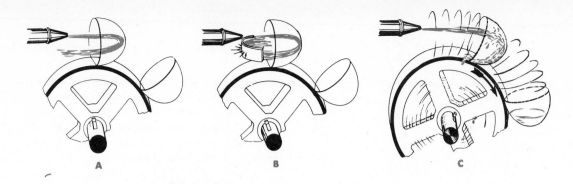

Fig. 12-14. Torque multiplication by using a stator.

longer possible. TORQUE MULTIPLICATION IS ONLY POSSIBLE WHEN THERE IS A DIFFERENCE IN THE SPEED OF THE PUMP AND TURBINE! THE GREATER THE DIFFERENCE — THE GREATER THE MULTIPLICATION, A, B, and C, Fig. 12-14.

NEWTON'S LAW

Newton's law says: "For every action there is an equal and opposite reaction."

Applying Newton's law to the torque converter, we find three areas of action and reaction caused by the oil flow. A—Pump, B—Stator, C—Turbine. Fig. 12-15.

As the stator blades are held in a fixed position during the torque multiplication period, the reaction of these blades to the oil flow is to direct the force in the same direction as that of pump A.

Using Newton's law, the reaction of the turbine blades on the oil stream will be equal to the force exerted by both pump A and stator B. If A is the action force of the pump, B the reaction force of the stator, and C the reaction force of the turbine, then: C = A + B.

The original force A has been increased by B, giving a reaction force of C that is greater than A. In short, original torque A has been multiplied to C. Fig. 12-15.

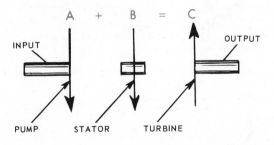

Fig. 12-15. Newton's law as applied to converter forces. Force of pump A, plus force of stator B, combine to produce force C, which is greater than original force A.

WHY CONVERTER BLADES ARE CURVED

The pump vanes are curved in such a manner as to direct the vortex flow leaving the vanes in a direction that will exert as much force as possible against the turbine vanes. The leading edge (part of the vane where oil first enters), or inner portion of the pump vane, is curved to assist smooth reentry of oil being discharged by the stator.

The turbine blades are curved in an opposite direction. This curve produces two favorable functions. When pump oil is thrown against the turbine vanes, the curved leading edge (that portion where oil first enters) assists in a smooth entry. This reduces breakup of the smooth oil stream and prevents a buildup or damming effect that reduces entry oil velocity.

The last portion of the turbine vane is curved to increase the turbine energy absorption by deflecting the oil stream as much as possible. The more it is deflected, the more energy it will give up to the deflecting object.

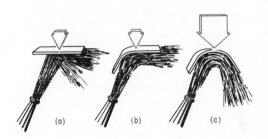

Fig. 12-16. Oil stream deflection. Bending direction of oil stream so a maximum amount of oil stream energy will be absorbed by deflecting object. (Dodge)

Fig. 12-16 illustrates this principle. Note in A, Fig. 12-16, that a stream of oil is striking a flat vane. Notice how the stream is broken up. This imparts only a small force to the vane. In B, Fig. 12-16, the leading edge of the vane has been curved. The oil stream is now entering in a smooth fashion and will retain its velocity. However, no greater force is imparted to the vane. In C, Fig. 12-16, the last portion of the vane has been curved to provide a much greater deflection of the oil stream.

As the oil stream velocity was maintained by the curved leading edge, oil stream deflection by bending the last portion will now provide a much greater absorption of energy by the vane.

STATOR IS NECESSARY

Since the curved turbine blades discharge oil in a direction opposite to pump rotation, the work of the pump would be

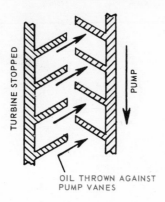

Fig. 12-17. Oil transfer without stator. As oil leaves turbine blades and attempts to reenter pump, oil works against pump rotation. Oil is entering at wrong angle.

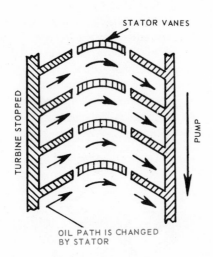

Fig. 12-18. Oil transfer with a stator. As oil leaves turbine vanes, it strikes stator vanes. As stator is locked, vanes bend oil stream so it enters pump at proper angle. It will now assist pump as it will enter with some initial velocity.

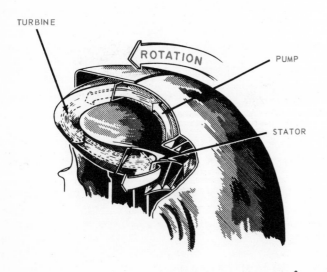

Fig. 12-19. Stator action. Stator bends oil stream leaving turbine blades in direction that will allow it to enter pump with an initial velocity. (Dodge)

hampered since oil could not reenter the pump blades in a smooth manner. As a result, vortex flow would be impaired and torque transmission would fall off. Fig. 12-17.

The stator blades are curved at an angle that will intercept the oil leaving the turbine blades, then redirect it in a direction that allows the oil stream to enter the pump blades smoothly. At the same time, this adds to the force of the pump and increases vortex velocity. Fig. 12-18.

Fig. 12-19, shows how the oil stream from the curved turbine blades is redirected by the stator so it enters the pump vanes with an initial velocity.

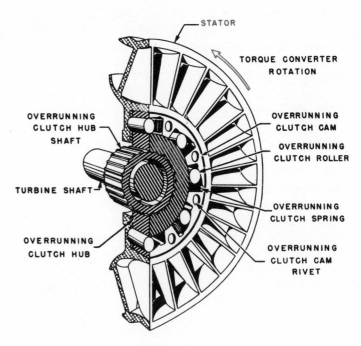

Fig. 12-20. Stator. One form of stator. This unit uses a one-way roller clutch. Notice that stator can turn in one direction only.

Stator vanes, as well as a roller overrunning or one-way clutch, are shown in Fig. 12-20. The one-way clutch hub is stationary and the stator can turn only in the direction shown by the arrow.

VARIABLE PITCH STATOR

To further increase the usefulness of the stator, and to adapt it to different load requirements, one type of stator is built with adjustable vanes. It is referred to as a VARIABLE PITCH stator. Fig. 12-21.

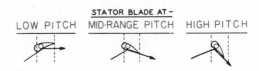

Fig. 12-21. Variable pitch stator. Pitch or angle of stator blades can be changed to meet different requirements. (Buick)

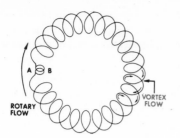

Fig. 12-22. Torque converter — vortex flow. Vortex flow is very rapid in torque converter. This is especially true at stall. (Chevrolet)

VORTEX FLOW

Vortex flow in the torque converter is much faster than that of the fluid coupling. This is accomplished by: the initial velocity of pump oil reentry provided by the stator; the curved vanes; smooth turbine oil entry. The higher the vortex flow velocity, the greater the torque transfer.

As road speed builds up, the turbine speed approaches that of the pump. Vortex flow is slowed down and the torque converter acts much like the fluid coupling. Figs. 12-22 and 12-23 illustrate the vortex flow during torque multiplication, and during the fluid coupling condition.

OTHER VARIATIONS

Some converters use a pump, turbine and one stator. Others use a pump, turbine and two or more stators. Another type uses a pump, three turbines, a small inner stator and a larger outer stator. Some use two pumps. All are designed to facilitate torque transfer and multiplication. The pump, turbine and one stator design is in common use today.

TORQUE CONVERTER IN ACTION

When the car is standing still with the transmission in DRIVE range and engine idling, there is very little transfer of torque from pump to turbine.

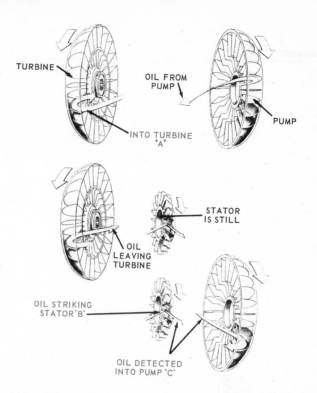

Fig. 12-24. Oil travel path. Oil is thrown from pump into turbine at A. Oil leaving turbine strikes stator at B, and is deflected into pump at proper angle, and with initial velocity at C. (Pontiac)

AT STALL

As the engine is accelerated, pump speed rapidly increases. Oil is thrown into the turbine with increasing force, A, Fig. 12-24. Leaving the turbine, the oil strikes the stator, B, Fig. 12-24. As the stator is forced backward by the oil, the one-way clutch will lock it up. The oil is then diverted, and enters the whirling pump, C, Fig. 12-24. As this vortex flow increases in speed, more and more torque is applied to the turbine. Fig. 12-24.

The maximum torque multiplication is delivered when the

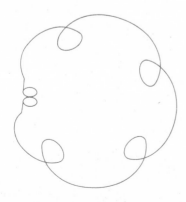

Fig. 12-23. Vortex flow at coupling point. Vortex flow has almost stopped when coupling point is reached. Coupling point is when pump and turbine are traveling at same speed and torque multiplication is no longer possible. (Chevrolet)

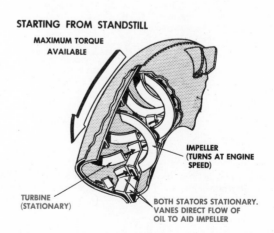

Fig. 12-25. Torque converter — at stall. Stall condition, as illustrated, will provide maximum amount of torque multiplication. Turbine and both stators are stopped and pump is spinning very fast. (Plymouth)

pump has reached its highest velocity, and the turbine is at stall (standing still). This condition is shown in Fig. 12-25. Notice that this converter uses two stators. Both are in operation.

INCREASING SPEED

As the turbine begins to turn, torque multiplication tapers off. When turbine speed increases, the oil leaving the trailing edges of the turbine will change its angle so it begins to strike the rear face of the primary stator.

This change of angle is caused by the fact that even though the oil is being thrown toward the stator, the turbine is moving by the stator faster than the oil is moving toward the stator.

If you parked your car (turbine) alongside a target (stator), you could throw a rock (oil) backward and strike the target (stator blade). Say that you threw the rock at 50 mph., it would strike the target (stator blade) on the front face at 50 mph.

Then, if you drove the car (turbine) past the target (stator blade) at 100 mph and threw a rock, instead of striking the target, the rock would actually travel away from the target at about 50 mph. If there was a row of targets (stator blades), instead of striking the front face of the one you aimed at, the rock would actually strike the rear face of one of the other targets.

A similar action takes place with the turbine oil. As the turbine rotary speed surpasses the vortex velocity, the oil strikes the back of the stator blades causing the stator to start free-wheeling. This causes the primary stator sprag clutch to uncouple, and the primary stator will free-wheel with the turbine. As the secondary stator angles still deflect the oil, it remains stationary. Fig. 12-26.

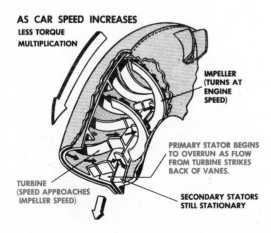

Fig. 12-26. Torque converter — increasing car speed. Turbine is starting to approach pump or impeller speed. One stator is free-wheeling and other is still stopped and is deflecting oil into pump. (Plymouth)

CRUISING SPEED

As car speed reaches the cruising range, the turbine speed approaches that of the pump.

The angle of oil discharge from the turbine causes the oil stream to strike the back of both primary and secondary stators. Both stators are now free-wheeling and there is no torque multiplication.

Vortex flow is slow and the converter is performing much like a fluid coupling. This condition is known as the COUPLING POINT. Fig. 12-27.

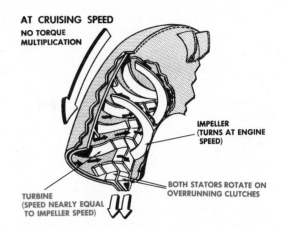

Fig. 12-27. Torque conveter — cruising speed. Pump and turbine are turning at about same speed and both stators are free-wheeling. Vortex flow is almost nil. This is referred to as coupling point. (Plymouth)

TORQUE MULTIPLICATION CURVE

The ratio between torque multiplication with the turbine at STALL and the COUPLING POINT is illustrated in Fig. 12-28. Notice how torque multiplication drops off as turbine speed increases.

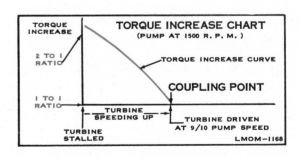

Fig. 12-28. Torque curve chart. (Mercury)

INFINITE GEAR RATIOS

The torque converter provides an infinite number of ratios as opposed to the standard transmission's three or four ratios. This provides a smooth flow of power that automatically adjusts to varying load conditions, (within limit of unit). See Fig. 12-29.

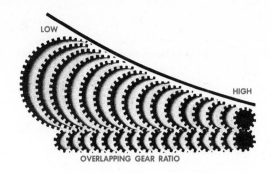

Fig. 12-29. Torque converter provides an extensive number of possible gear ratios. (Chevrolet)

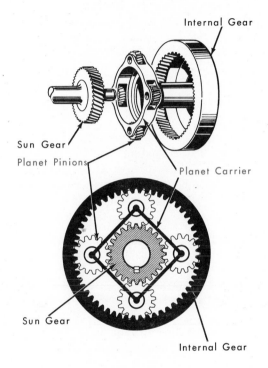

Fig. 12-30. Typical planetary gearset. (Cadillac)

PLANETARY GEARSETS

As you learned in the overdrive section, a planetary gearset can be used to step up the speed of the output shaft to reduce torque. The planetary gearset can also be used to reduce output speed, thereby increasing torque. The gearset can transmit motion at the same torque, and can be used to reverse torque. The automatic transmission makes full use of these features of the planetary gearset.

The planetary gearset is compact. Its gears are in constant mesh. It is quiet and strong. Fig. 12-30 illustrates a typical planetary gearset. Learn the names of all parts.

TORQUE INCREASE — LARGE

If the planetary gearset internal gear is held, and power is applied to the sun gear, the planet pinions are forced to turn and walk around in the internal gear. This action causes the pinion carrier to revolve at a lower speed than the sun gear. Torque increase is large. Fig. 12-31.

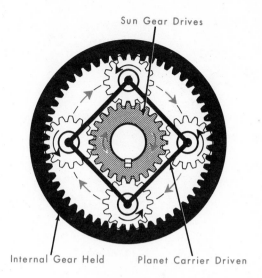

Fig. 12-31. Planetary gearset in large torque increase. (Cadillac)

TORQUE INCREASE — SMALL

If the sun gear is held, and the internal gear is driven, the planet pinions must walk around the sun gear. This causes the pinion carrier to move slower than the internal gear, and torque will increase. Fig. 12-32.

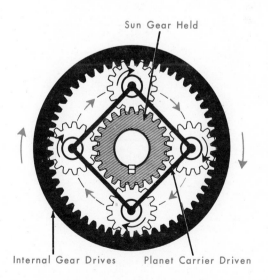

Fig. 12-32. Planetary gearset in small torque increase. (Cadillac)

TORQUE TRANSMISSION — CONSTANT

When any two members of the gearset are locked together, planetary action is stopped. Under these condtions, the gearset

Two Members Locked Together

Fig. 12-33. Planetary gearset in direct drive.
(Cadillac)

will revolve as a solid unit, providing a means of torque transmission with no increase or decrease. Fig. 12-33.

TORQUE TRANSMISSION — REVERSE

By holding the planet carrier and driving the sun gear, the planet pinions are forced to rotate about their pins. This causes them to drive the internal gear in a reverse direction at a reduced speed. Fig. 12-34.

Sun Gear Drives

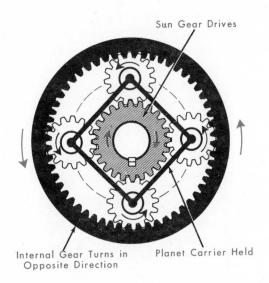

Internal Gear Turns in   Planet Carrier Held
Opposite Direction

Fig. 12-34. Planetary gearset in reverse.
(Cadillac)

USE OF THE PLANETARY GEARSET IN A TYPICAL AUTOMATIC TRANSMISSION

Detail 1, Fig. 12-35, shows the start of a planetary gearset as used in a two-speed transmission. Notice that the primary sun gear and shaft are one piece, and they are splined to the converter turbine. The arrow indicates the constant direction of rotation for the sun gear and shaft.

In detail 2, Fig. 12-35, the output shaft, planet carrier (one piece), one long and one short pinion have been added. The long and short pinions always rotate in the direction of the arrows. The long pinions are in constant mesh with both the sun gear and the short pinions. At this point, drive is not yet possible.

Detail 3, Fig. 12-35, illustrates the addition of a forward sun gear and flange. The integral flange and forward sun gear are free to turn independently of the primary sun gear and shaft. The forward sun gear is in constant mesh with the short pinions.

A brake drum has been placed on the forward sun gear flange, and a brake band surrounds the drum. If the band were applied (squeezed against the drum), the drum would be stopped. This would stop the forward sun gear. As the shaft turns the primary sun gear, it drives the long pinions counterclockwise. The long pinions turn the short pinions clockwise. The short pinions then walk around the stationary forward sun gear, turning the pinion carrier and output shaft at a ratio of 1:75 to 1 in this particular transmission.

BRAKE BANDS

A brake band is a steel band that encircles the drum. It is usually faced with a bonded friction lining. One end of the band is secured. The other end is fastened to a servo actuating rod. When the band is tightened, it will stop the drum.

SERVO OPERATES BAND

A hydraulic piston called a SERVO is used to apply the band. It consists of a cylinder in which is placed a piston. An opening is provided at one end to admit oil. The piston is held in the released position by a spring pressing against the piston.

When oil, under pressure, is admitted to the cylinder, the piston is forced forward, either actuating the band through direct contact or by actuating linkage. When the oil pressure is reduced, the spring will return the piston and the band will release.

Some servos have openings on both sides of the piston. In this case, oil can be used to assist in servo release. Other servos have two pistons.

Fig. 12-36 is a schematic drawing of a typical servo. This unit acts directly on the band through a piston rod or stem.

The servo in Fig. 12-37 is of the direct acting type. Note the band adjusting screw and how the piston may be applied and released by oil pressure.

ACCUMULATOR PISTON

Often an accumulator piston is used to assist the servo in engaging the band quickly and smoothly. Fig. 12-38 illustrates one type. Note use of lever and strut design.

Oil entering the cylinder presses on the large apply piston. As the apply piston is held back by a spring, oil enters a small check valve and pushes on the accumulator piston.

Being small in diameter, the accumulator piston is not held back by a spring. It moves outward against the actuating lever

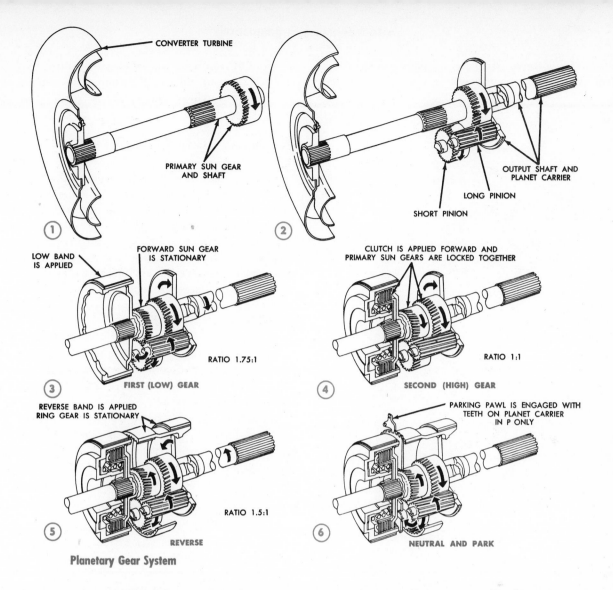

CONVERTER TURBINE

PRIMARY SUN GEAR AND SHAFT

① ②

OUTPUT SHAFT AND PLANET CARRIER

LONG PINION

SHORT PINION

LOW BAND IS APPLIED

FORWARD SUN GEAR IS STATIONARY

CLUTCH IS APPLIED FORWARD AND PRIMARY SUN GEARS ARE LOCKED TOGETHER

RATIO 1.75:1

RATIO 1:1

③ FIRST (LOW) GEAR

④ SECOND (HIGH) GEAR

REVERSE BAND IS APPLIED RING GEAR IS STATIONARY

PARKING PAWL IS ENGAGED WITH TEETH ON PLANET CARRIER IN P ONLY

RATIO 1.5:1

⑤ REVERSE

⑥ NEUTRAL AND PARK

Planetary Gear System

Fig. 12-35. Step by step construction of two-speed automatic transmission gearset. (Ford)

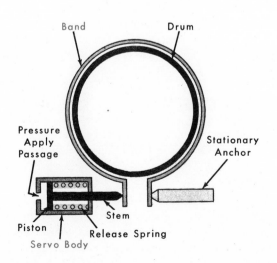

Band

Drum

Pressure Apply Passage

Stationary Anchor

Piston

Stem

Release Spring

Servo Body

Fig. 12-36. Schematic of one type of band servo. (Cadillac)

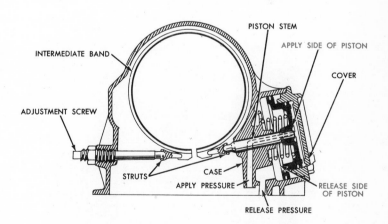

INTERMEDIATE BAND

PISTON STEM

APPLY SIDE OF PISTON

COVER

ADJUSTMENT SCREW

STRUTS

CASE

APPLY PRESSURE

RELEASE SIDE OF PISTON

RELEASE PRESSURE

Fig. 12-37. Direct acting servo. Note that oil pressure may be applied to either side of piston, causing it to move in direction desired. (Ford)

rapidly but with light pressure. After the actuating lever has started to engage the band, it takes additional pressure to provide further band application. The large apply piston will

move forward, add its pressure to that of the accumulator and the band will be applied firmly.

The accumulator piston gives a fast, soft application that is immediately followed by additional heavy pressure for firm engagement. Not all servos use the accumulator piston.

Oil pressure and its control will be covered later in this chapter.

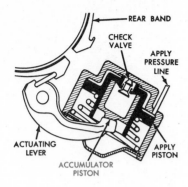

Fig. 12-38. One type of accumulator piston.

CLUTCH

Detail 4, Fig. 12-35, shows the addition of a multiple disc clutch. The clutch hub is splined to the primary sun gear shaft. A stack of clutch discs (the number varies) are arranged in alternate fashion. The driving discs are splined to the clutch hub; the driven discs are splined to the clutch drum.

A piston in the clutch drum squeezes the plates together when oil pressure from the transmission is applied to the piston. A heavy spring releases the discs when oil pressure drops.

When the clutch is applied, the discs are pressed together to lock the clutch hub and drum together. The clutch hub is splined to the primary sun gear shaft, and the drum is part of the forward sun gear. With the clutch applied, both primary

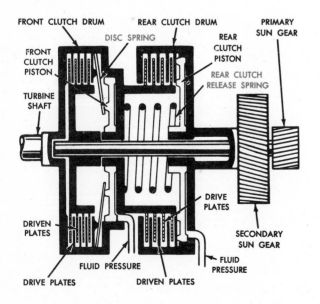

Fig. 12-39. Pair of multiple disc clutches. One is released by a coil spring, other is released by disc type spring.   (Ford)

and forward sun gears will be locked together. This stops all gear action and the parts revolve as a solid unit. The transmission is in direct or high gear. A typical clutch is shown in Fig. 12-39.

INTERNAL GEAR FOR REVERSE

Detail 5, Fig. 12-35, shows the addition of a ring or internal gear that is in constant mesh with the short pinions. An additional band and actuating servo are provided to either stop or release the internal gear. This is the reverse band and servo.

With low band and high gear clutch release, and reverse band applied, the internal gear will be held. This causes the short pinions to walk around, pulling the pinion carrier with them and imparting reverse motion to the output shaft.

NEUTRAL AND PARK

Detail 6, Fig. 12-35, illustrates the transmission in Neutral and Park. Notice the parking pawl that engages teeth on the planet carrier. This locks the output shaft, and prevents the car from rolling when parked.

The use of a torque converter makes it possible to have only two forward ratios, direct and low. Common practice, however, favors the use of three forward ratios or ranges for most cars. Study the illustration in Fig. 12-35 until you have learned the action in each gear.

PRINCIPLES ARE THE SAME

There have been numerous types of automatic transmissions. Some utilize a fluid coupling, others have torque converters. Design encompassed two, three and four speed ranges. Even though construction and design have varied, and still does, the general operating principles are much the same. Each automatic transmission uses a torque converter, or fluid coupling, to transmit torque to one or more planetary gearsets. These gearsets are controlled by servos and clutches that are actuated by oil pressure. The pressure is controlled by various devices.

CONTROL SYSTEM

For purposes of illustration, you will now "build" a simple control system for a two-speed automatic transmission.

OIL PUMP

There have been several types of pumps used in automatic transmissions. One is the variable output, VANE type. Output is altered by moving the outer vane cylinder. Fig. 12-40.

Another, in current use, is the ROTOR pump. Study the exploded view in Fig. 12-41.

Still another widely used pump is the "internal-external" GEAR pump. This design is pictured in Figs. 12-71 and 12-72.

The start of your oil actuating and control system will be the gear oil pump drawing oil from the transmission oil sump or pan. The pressure output of the pump is controlled by a

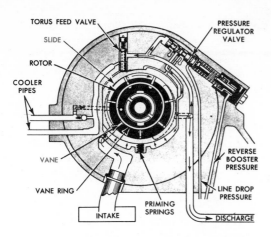

Fig. 12-40. Variable output vane oil pump. Pump is shown in position that will produce maximum output. Note that slide is up. (Pontiac)

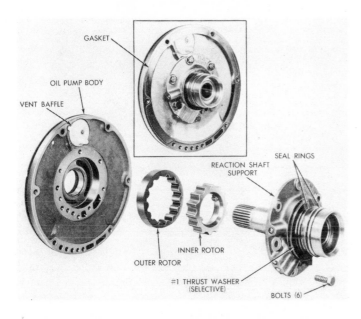

Fig. 12-41. A rotor oil pump. (Plymouth)

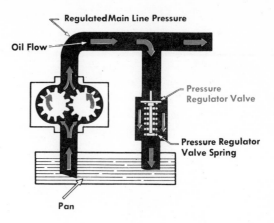

Fig. 12-42. Gear oil pump and pressure regulator. (GMC)

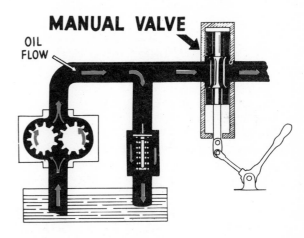

Fig. 12-43. Manual control valve. Valve is shown in open position.

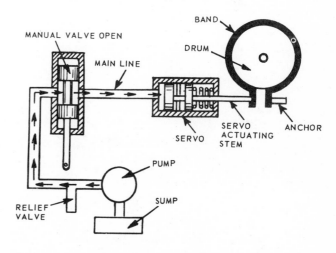

Fig. 12-44. Low range. When manual valve is opened, oil under pressure from pump travels via main line to servo. Oil will press on servo piston, causing it to tighten band. Car is now in low range.

pressure regulator valve. When pressure exceeds the amount required, the valve opens and bypasses oil back to the sump. This produces constantly regulated main line pressure. See Fig. 12-42.

MANUAL CONTROL VALVE

The next addition is a manually operated control valve that will allow the driver to cut off, or open, the main line pressure. Fig. 12-43.

LOW RANGE

From the manual control valve, the main line will go to the low range band servo. To place the car in the low range, the driver merely opens the manual valve via linkage. As the engine turns the input shaft, the oil pump will build up pressure. The oil pressure will travel to the servo and tighten the band. When the car is accelerated, torque will be imparted and the car will move forward in the LOW range. Fig. 12-44.

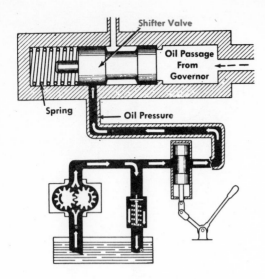

Fig. 12-45. Schematic of shifter valve. Valve is shown in closed position. (GMC)

HIGH RANGE

Another line must be added. This will run from the main line to a shifter valve. It will leave the shifter valve and branch out. One line will go to the high range clutch unit, another to the low range servo. In Fig. 12-45, you can see that the main line pressure has traveled to the shifter valve. The low range circuit is not shown.

HOW SHIFTER VALVE IS OPENED

GOVERNOR

You will note an opening on the right end of the shifter valve marked "oil passage from governor."

As the car moves forward in low range, the output shaft spins a governor unit. Oil pressure from the main line is piped

to the governor unit, then to the shifter valve. When the car is not moving, the governor weight is in and oil pressure is stopped at the governor. Fig. 12-46.

As the car gathers road speed, the governor unit will spin faster and faster. At a predetermined speed, the governor valve will open and oil pressure will travel to the shifter valve. See A, Fig. 12-47. Note governor in B, Fig. 12-47.

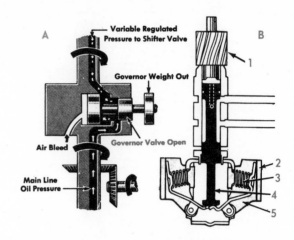

Fig. 12-47. A—Schematic of governor valve in open position. Note main line oil pressure is now passing through valve. B—Cross section of a different type of governor. 1—Driven gear. 2—Primary weight. 3—Spring. 4—Valve. 5—Secondary weight. (GMC)

When governor pressure reaches a certain level, the tension of the shifter spring will be overcome and the shifter valve will open. Fig. 12-48 shows the governor pressure building up against the shifter valve.

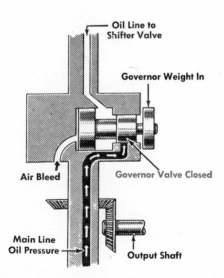

Fig. 12-46. Schematic of governor valve in closed position. Note that main line oil pressure cannot pass governor valve. (GMC)

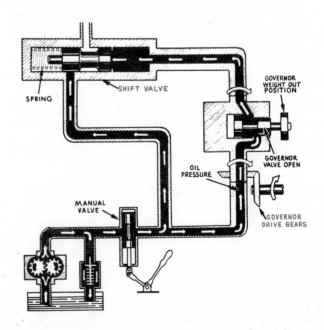

Fig. 12-48. Governor pressure acting on shifter valve. Governor weight has been thrown out by centrifugal force. Governor valve is open and main line pressure is passing through governor to shifter valve. (Lincoln)

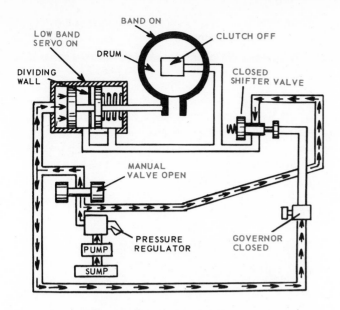

Fig. 12-49. Low range — schematic. Manual valve open. Pressure to low band servo. Pressure to governor. Pressure to shifter valve. Governor closed; no pressure to open shifter valve. Shifter valve closed; no pressure to off sides of low ban servo or clutch.

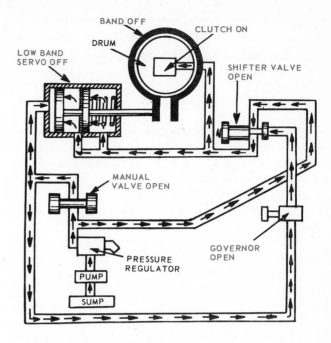

Fig. 12-50. High range — schematic. Road speed has opened governor. Governor pressure opens shifter valve. Main line pressure now flows through shifter valve to OFF side of both servo pistons and to clutch. Servo releases and clutch applies. Transmission is now in high range.

SHIFT RANGE CHANGES

When the manual valve is opened, oil pressure actuates the low band servo. The car will then move forward in low range. Oil pressure also travels to the shifter valve and the governor. Fig. 12-49.

When the car reaches a predetermined road speed, the governor will have opened enough to allow oil pressure to flow to the shifter valve and force it open. When the shifter valve opens, oil pressure flows to the clutch and band servo units.

As the pressure builds up in the clutch, a corresponding pressure builds up in the servo. As the pressure from the shifter valve is working on two pistons of the servo, as opposed to the pressure from the manual valve working on one servo piston, the servo will be forced to move to the left and free the low band.

When the servo pistons stop moving to the left, the pressure in the clutch unit will build up rapidly, thus locking the clutch unit. The transmission is now in high range and will remain there until car speed falls below the shift point. When this happens, governor pressure is lost, clutch disengages and manual valve pressure reapplies low band servo. All this is done automatically. Fig. 12-50 shows the circuit in HIGH range.

SHIFT POINT SHOULD VARY

The point at which the transmission upshifts into high and downshits into low range is controlled by the governor. These points are always constant.

It is very desirable to have the shift points vary according to load requirements. For rapid acceleration, the transmission should stay in low range longer. When attempting to pass at low speeds just above the downshift point, it is desirable to have the downshift occur when needed.

DELAYING THE UPSHIFT

You will notice a spring in the shifter valve in Fig. 12-48. As mentioned, the spring tension is overcome by governor pressure at a certain road speed. By placing a regulator plug to the left of the spring, and then connecting linkage to the throttle, it is possible to vary the spring tension.

When the throttle is depressed, the regulator plug is forced

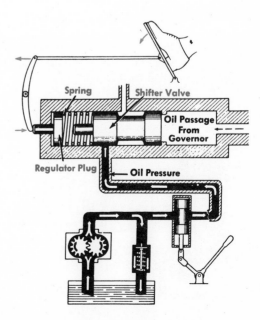

Fig. 12-51. Shifter valve regulator plug. Throttle pressure will change spring tension and cause shifter valve to open under various pressures. (GMC)

against the spring, increasing its tension. The car must gain more road speed to boost governor pressure, to overcome shifter spring tension. This will delay the upshift from low to high range. Fig. 12-51.

THROTTLE VALVE

Instead of moving the regulator plug by mechanical means, a throttle valve is used. The throttle valve is connected to the foot throttle through linkage. This rod is often referred to as the TV (throttle valve) rod. When the throttle is depressed, the throttle valve is moved and transmits oil pressure to the regulator plug. The amount of pressure depends on the distance the throttle is moved.

The throttle valve is an application of the spool balanced valve. Pressure is admitted to the valve at A, Fig. 12-52. It travels around the small spool section and exits through B. It is bypassed to chamber C at the spool end D. The pressure pushes on the shifter valve regulator plug and on the spool at D, Fig. 12-52.

Spring pressure E is controlled by the TV rod. When the throttle opening is small, spring pressure E is light. Oil pressure at D forces the spool to the right against the light spring pressure.

As the spool moves to the right, it starts to block opening A. As opening A size is reduced, the oil pressure entering the valve is also reduced.

If the spring is pushing to the left with a pressure of 15 lbs., the spool will be forced to the right until the entrance A is small enough to admit oil at 15 lbs. pressure. The valve will then be balanced. If spring pressure is increased, it will force the spool to the left, increasing opening A and oil pressure will build up in C. This will force the spool to the right until opening A is again closed far enough to balance oil pressure with spring pressure.

FORCED DOWNSHIFT

Another valve (it can be part of the throttle valve) is provided to admit high main line pressure to the shifter valve. This valve is sometimes called the T valve.

When the driver depresses the accelerator almost to the floor, the throttle valve strikes a detent plug and spring. When the accelerator is pressed beyond this point, the TV rod forces the T valve open. This directs main line pressure to the offside of the shifter valve. The shifter valve will close, cutting off pressure to the clutch and offside of the band servo. Now the servo will be actuated by manual valve pressure, and the low range band will tighten. This places the car in low range before road speed is slow enough to cause sufficient drop in governor pressure. This is called a forced downshift. Fig. 12-53.

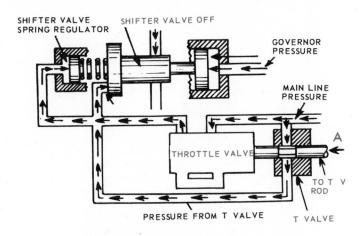

Fig. 12-53. Forced downshift. When TV rod forces throttle valve rod in at A, main line pressure passes through T valve to off side of shifter valve. This additional pressure overcomes governor pressure and shuts shifter valve. Transmission will now downshift.

VACUUM MODULATOR

Makers now use a vacuum modulator in preference to the TV rod. The modulator utilizes engine vacuum, a very reliable indicator of engine loading. Therefore, the modulator provides very accurate means of controlling the throttle valve (also called modulator valve).

The vacuum modulator consists of a container separated into two areas by a flexible diaphragm. A spring forces the diaphragm rod to apply pressure to the throttle valve to

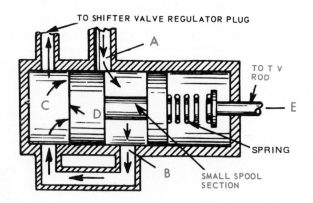

Fig. 12-52. Throttle valve. Oil pressure balances spring pressure by moving spool until opening A admits proper pressure. TV rod adjust spring pressure.

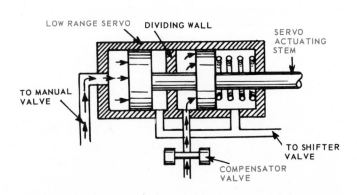

Fig. 12-54. Compensator valve. Heavy throttle pressure opens compensator valve, admitting additional pressure to operate servo. Compensator pressure acts on ON side of second servo piston.

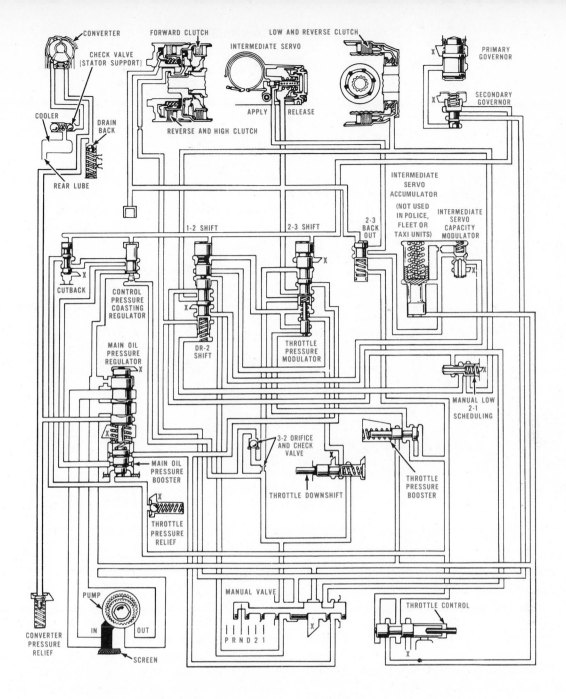

Fig. 12-55. Complete hydraulic control for Ford C6 transmission, illustrated in Fig. 12-62. (Ford)

increase TV pressure. In operation, engine vacuum causes the diaphragm to move away (against spring pressure) from the throttle valve, lowering the pressure. As engine loading varies, vacuum fluctuates. So the modulator is constantly altering throttle valve pressure in accordance with engine vacuum. See A, Fig. 12-68.

An altitude sensitive modulator is illustrated in B, Fig. 12-68. It incorporates an evacuated (air removed) bellows. The collapsing bellows tends to force the diaphragm toward the throttle valve. At sea level, the collapsing force is highest, adding to diaphragm pressure and raising TV pressure. As altitude increases, the collapsing force lessens and TV pressure is decreased.

The altitude compensating bellows provides a more uniform shift feel regardless of elevation. See B, Fig. 12-68.

COMPENSATOR VALVE

To add to the pressure on the band during heavy acceleration, a compensator valve is often used. This valve is controlled by foot throttle pressure on the throttle valve. During rapid acceleration (when torque is heavy), the compensator valve admits additional pressure to the band servo. Fig. 12-54.

COMPLETE CONTROL CIRCUIT

Fig. 12-55 shows the complete hydraulic system as used in the Ford C6 automatic, dual-range transmission. This transmission is illustrated in Fig. 12-62. Trace each hydraulic circuit and study its construction and application.

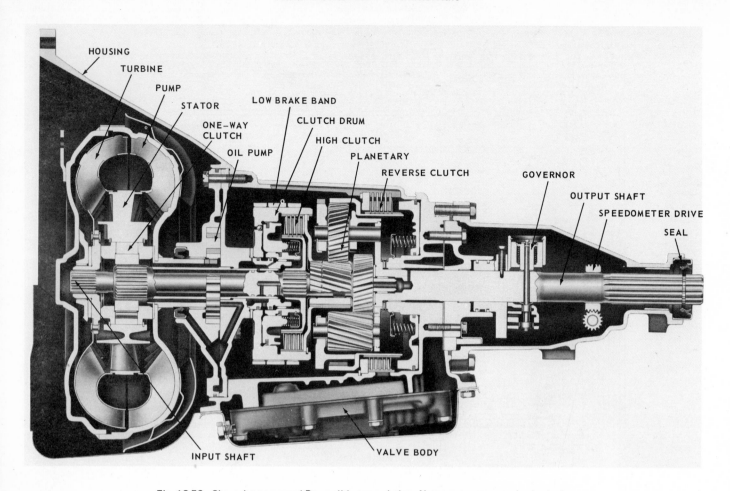

Fig. 12-56. Chevrolet two-speed Powerglide transmission. Note compact control valve body.

HYDRAULIC CONTROL CIRCUITS

The various oil circuits are extremely compact, and every possible means is used to avoid open tubing. Many of the shafts and other units are drilled to carry oil. Fig. 12-57. Most of the control valves are housed in one compact control valve body. Fig. 12-56. An exploded view of the many components of a control valve body is shown in Fig. 12-58.

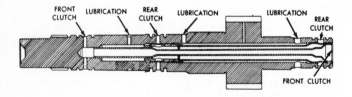

Fig. 12-57. Sun gear shaft showing drilled passageways.
(Ford)

OIL COOLING

Transmission oil is generally cooled by circulating it, via tubes, to a unit housed in the lower radiator. In cross-flow radiators, the tubes are connected to the right or left hand side tank. Fig. 12-59.

LUBRICATION

The transmission is lubricated by special oil which circulates through the unit. Never use any oil but AUTOMATIC TRANSMISSION FLUID, TYPE "A", SUFFIX "A." The container will be marked AQ - ATF - (several numbers) - A. Some makers specify an exact fluid (such as Dexron, Ford Type F, etc.,) for their own transmission. ALWAYS FOLLOW THESE RECOMMENDATIONS!

SHIFTING

The transmission may be placed in any desired range by moving a gearshift lever, usually located on the steering column. This moves a selector lever on the transmission. The range indicator with a sliding arrow shows the operator which range is "in." Fig. 12-60.

A console floor shift mechanism is shown in Fig. 12-61. A push-button shift control also has been used.

DIFFERENT TYPES OF AUTOMATIC TRANSMISSIONS

Automatic transmission design has grown quite sophisticated, resulting in a more compact, smoother acting and highly reliable unit.

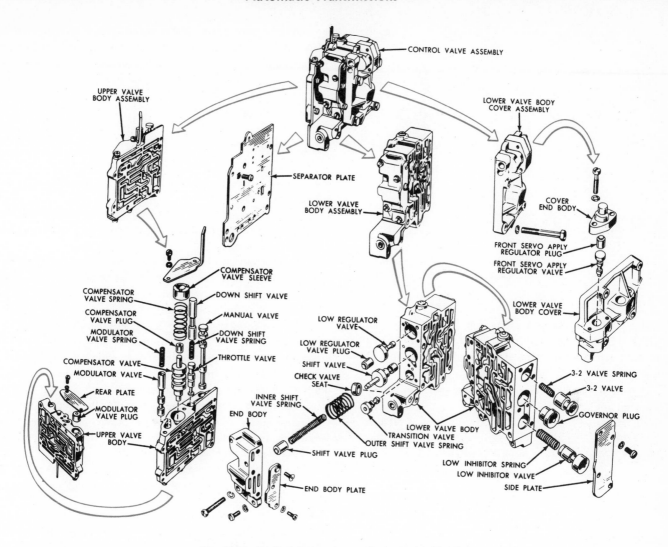

Fig. 12-58. Exploded view of control valve body. (Ford)

The use of the straight fluid coupling, once popular, has been replaced with the more efficient torque converter.

Torque converters have used both multiple turbines and stators. Current practice uses the basic converter design consisting of one pump, one turbine and a single stator. The stator is generally of FIXED PITCH (angle cannot be changed) construction. The transmissions in Figs. 12-62, 12-63, 12-69 and 12-75 employ fixed stators. Figure 12-73 shows a VARIABLE PITCH (angle can be changed) stator unit.

The transmission and torque converter, as used in two front wheel drive cars, are separated. The torque converter drives the transmission via a link belt assembly. This unit utilizes a

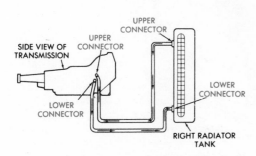

Fig. 12-59. Method of cooling transmission oil. This side view shows oil cooler located in right-side radiator tank. (Cadillac)

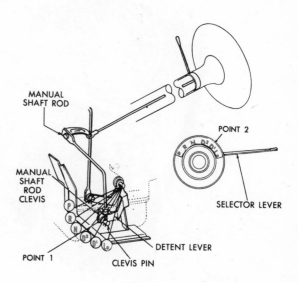

Fig. 12-60. Column shift linkage. (Mercury)

253

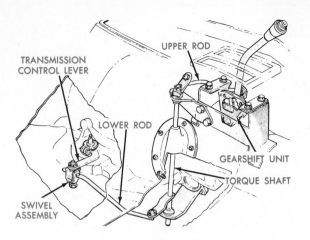

Fig. 12-61. A console (floor) gearshift linkage arrangement. (Plymouth)

variable pitch (two position) stator. Fig. 12-70.

Figs. 12-64, 12-65, 12-66 and 12-67, illustrate the clutch and band action in one type of two-speed transmission. Study the power flow in each range.

A cross-sectional view of a three-speed transmission is illustrated in Fig. 12-69.

A cutaway view of a two-speed transmission is pictured in Fig. 12-73.

Fig. 12-74 shows a cutaway of a three-speed transmission.

Another three-speed transmission design is illustrated in Fig. 12-75. Fig. 12-77 shows the same transmission in exploded form.

The clutch and band action for the transmission in Fig. 12-62 is depicted in Fig. 12-76. Study the action and power flow in each speed range.

SUMMARY

The automatic transmission shifts in accordance with speed and load conditions.

It uses one or more planetary gearsets to produce the desired gear ratios. Multiple disc clutches and bands are used to control gearsets.

The clutches and bands are actuated by hydraulic pressure produced by one or two oil pumps. The oil pressure is directed and controlled by a series of valves. Distribution is via oil passageways in the various units.

Definite speed ranges such as Drive, Low, Reverse, etc., can be selected by the driver through mechanical linkage connected to a control valve in the transmission.

Power transmission from engine to transmission is through a fluid coupling or a torque converter.

The torque converter can multiply the engine torque, thereby giving a wide range of power. The amount of multiplication is dependent on converter design and position between "stall" and "coupling point."

Fig. 12-62. Ford C6, three-speed transmission. (Ford)

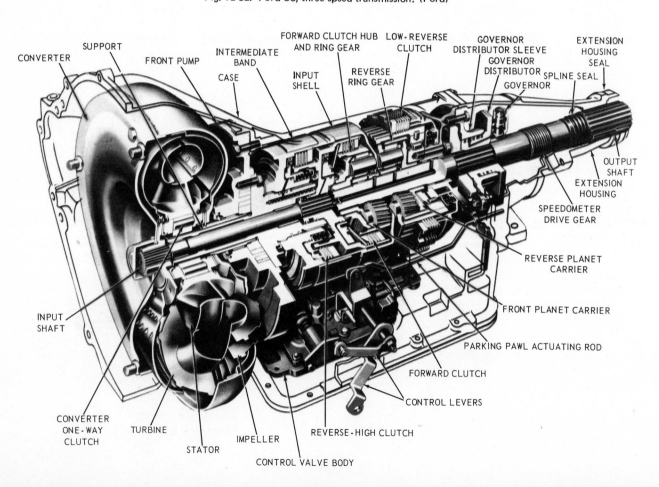

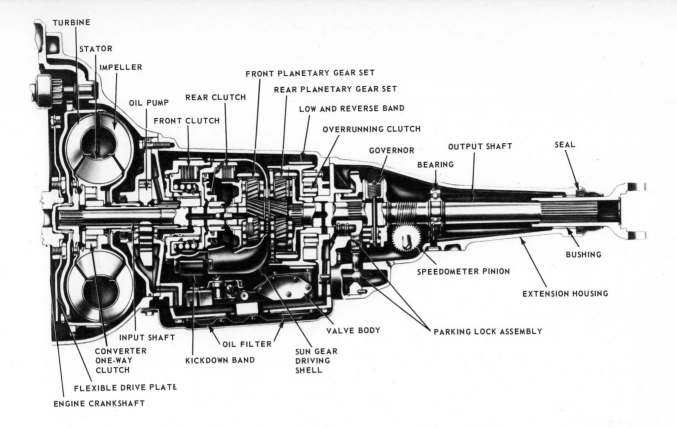

Fig. 12-63. Plymouth Torque Flite, three-speed transmission. (Plymouth)

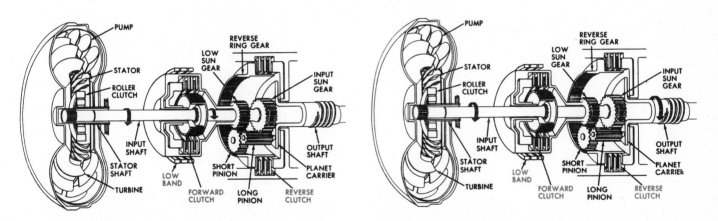

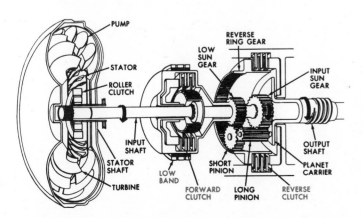

Fig. 12-64. Schematic illustrates Pontiac two-speed transmission action in NEUTRAL. Clutches and band are released.

Fig. 12-66. Transmission in DIRECT DRIVE range. Forward clutch applied. Low band and reverse clutch released.

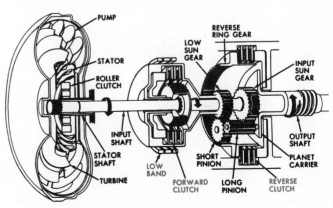

Fig. 12-65. Transmission in LOW range. Note power flow. Low band is applied while clutches are released.

Fig. 12-67. Transmission in REVERSE. Reverse clutch applied. Low band and forward clutch released.

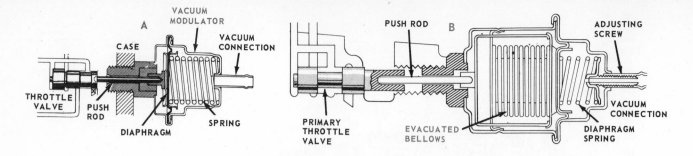

Fig. 12-68. A—Vacuum modulator controlling throttle valve by varying pressure on the push rod. B—Altitude sensitive modulator. Note evacuated bellows. (Ford)

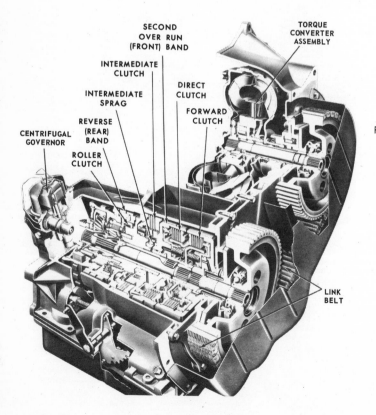

Fig. 12-69. Cutaway of an automatic transmission as used by B.M.W. How many parts can you identify? (B.M.W.)

Lubrication is provided by special transmission oil as it circulates through the unit.

Oil is cooled by piping it to a cooler unit in one of the radiator tanks.

A complete study and mastery of the automatic transmission involves a great deal of material. This chapter has endeavored to give you a working knowledge of the basic principles involved. For further study, manufacturers' shop manuals are an excellent source of information.

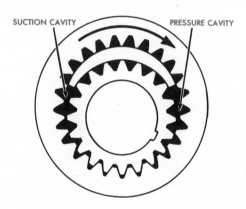

Fig. 12-71. Transmission internal-external gear oil pump. See exploded view in Fig. 12-72. (GMC)

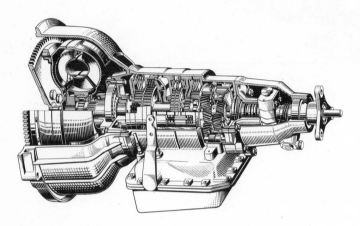

Fig. 12-70. Transmission and torque converter are separated on the front wheel drive Cadillac Eldorado. Note link belt connecting the two units. (Cadillac)

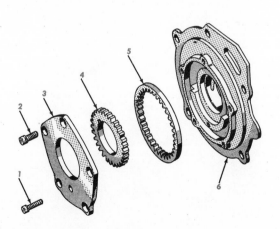

Fig. 12-72. Exploded view of an internal-external gear pump. 1—Fastener. 2—Fastener. 3—Cover. 4—Drive gear. 5—Driven gear. 6—Pump. (Jeep)

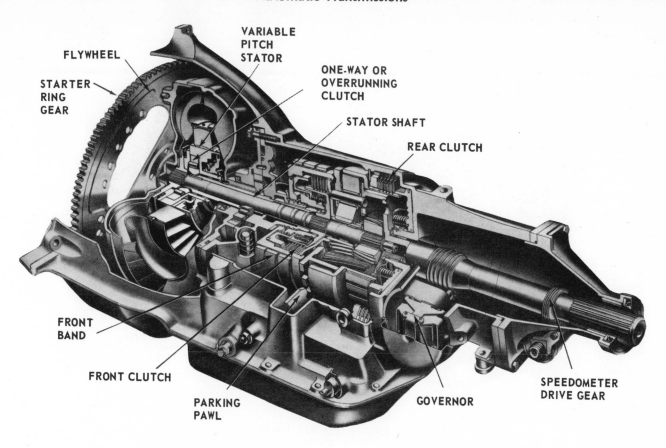

FLYWHEEL

STARTER RING GEAR

VARIABLE PITCH STATOR

ONE-WAY OR OVERRUNNING CLUTCH

STATOR SHAFT

REAR CLUTCH

FRONT BAND

FRONT CLUTCH

PARKING PAWL

GOVERNOR

SPEEDOMETER DRIVE GEAR

Fig. 12-73. Cutaway view of a Oldsmobile Jetaway, two-speed transmission.

Fig. 12-74. Cutaway view of a Chevrolet Turbo Hydra-Matic, three-speed transmission.

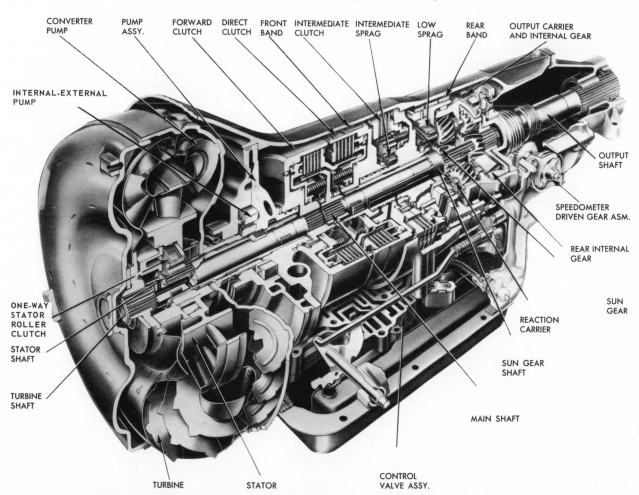

CONVERTER PUMP

PUMP ASSY.

FORWARD CLUTCH

DIRECT CLUTCH

FRONT BAND

INTERMEDIATE CLUTCH

INTERMEDIATE SPRAG

LOW SPRAG

REAR BAND

OUTPUT CARRIER AND INTERNAL GEAR

INTERNAL-EXTERNAL PUMP

OUTPUT SHAFT

SPEEDOMETER DRIVEN GEAR ASM.

REAR INTERNAL GEAR

SUN GEAR

REACTION CARRIER

SUN GEAR SHAFT

ONE-WAY STATOR ROLLER CLUTCH

STATOR SHAFT

TURBINE SHAFT

TURBINE

STATOR

MAIN SHAFT

CONTROL VALVE ASSY.

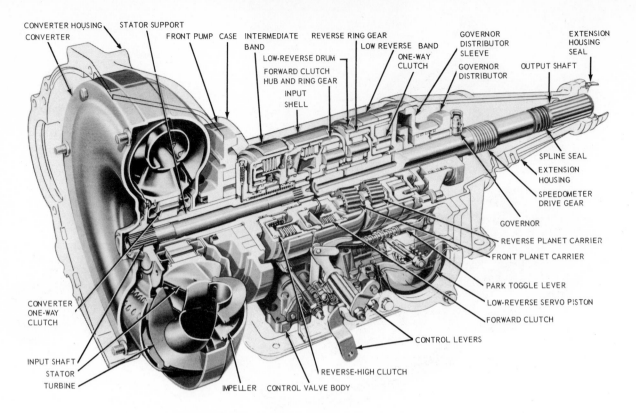

CONVERTER HOUSING
CONVERTER
STATOR SUPPORT
FRONT PUMP CASE
INTERMEDIATE BAND
LOW-REVERSE DRUM
FORWARD CLUTCH HUB AND RING GEAR
INPUT SHELL
REVERSE RING GEAR
LOW REVERSE BAND
ONE-WAY CLUTCH
GOVERNOR DISTRIBUTOR SLEEVE
GOVERNOR DISTRIBUTOR
EXTENSION HOUSING SEAL
OUTPUT SHAFT
SPLINE SEAL
EXTENSION HOUSING
SPEEDOMETER DRIVE GEAR
GOVERNOR
REVERSE PLANET CARRIER
FRONT PLANET CARRIER
PARK TOGGLE LEVER
LOW-REVERSE SERVO PISTON
FORWARD CLUTCH
CONTROL LEVERS
REVERSE-HIGH CLUTCH
CONTROL VALVE BODY
IMPELLER
TURBINE
STATOR
INPUT SHAFT
CONVERTER ONE-WAY CLUTCH

Fig. 12-75. Ford C4, three-speed transmission.

Fig. 12-76. Transmission action during various speed ranges in Ford C6, three-speed transmission.

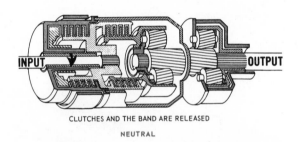

CLUTCHES AND THE BAND ARE RELEASED

NEUTRAL

THE INTERMEDIATE BAND IS APPLIED. THE REVERSE AND HIGH CLUTCH DRUM, THE INPUT SHELL AND THE SUN GEAR ARE HELD STATIONARY.

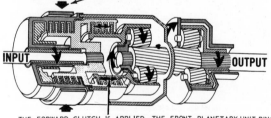

THE FORWARD CLUTCH IS APPLIED. THE FRONT PLANETARY UNIT RING GEAR IS LOCKED TO THE INPUT SHAFT.

SECOND GEAR

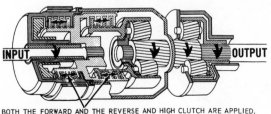

BOTH THE FORWARD AND THE REVERSE AND HIGH CLUTCH ARE APPLIED. ALL PLANETARY GEAR MEMBERS ARE LOCKED TO EACH OTHER AND ARE LOCKED TO THE OUTPUT SHAFT.

HIGH GEAR

THE FORWARD CLUTCH IS APPLIED. THE FRONT PLANETARY UNIT RING GEAR IS LOCKED TO THE INPUT SHAFT.

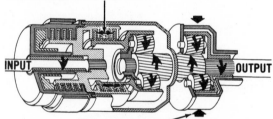

THE LOW AND REVERSE CLUTCH (LOW RANGE) OR THE ONE-WAY CLUTCH (D1 RANGE) IS HOLDING THE REVERSE UNIT PLANET CARRIER STATIONARY.

FIRST GEAR

GEAR RATIOS			
FIRST 2.46:1			
	SECOND 1.46:1		
		HIGH 1.00:1	
			REVERSE 2.17:1

THE REVERSE AND HIGH CLUTCH IS APPLIED. THE INPUT SHAFT IS LOCKED TO THE REVERSE AND HIGH CLUTCH DRUM, THE INPUT SHELL AND THE SUN GEAR.

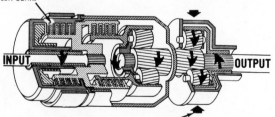

THE LOW AND REVERSE CLUTCH IS APPLIED. THE REVERSE UNIT PLANET CARRIER IS HELD STATIONARY.

REVERSE

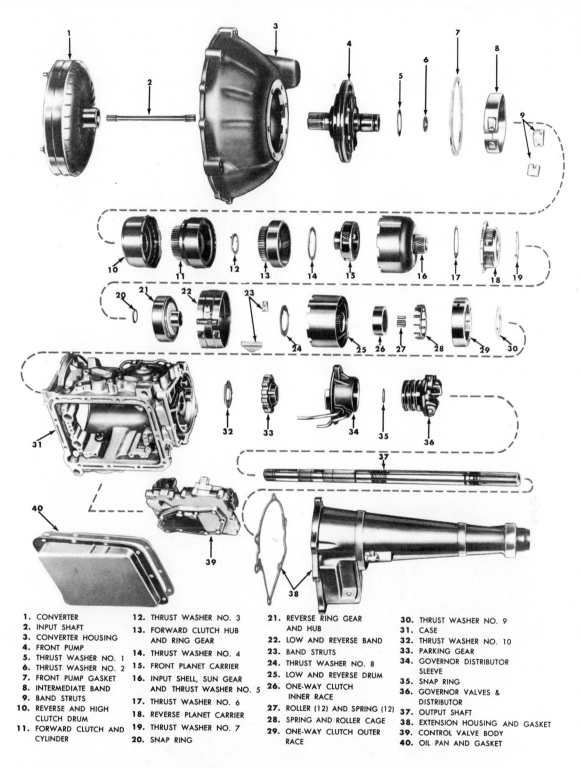

Fig. 12-77. Exploded view showing major units in the Ford C4 transmission.

1. CONVERTER
2. INPUT SHAFT
3. CONVERTER HOUSING
4. FRONT PUMP
5. THRUST WASHER NO. 1
6. THRUST WASHER NO. 2
7. FRONT PUMP GASKET
8. INTERMEDIATE BAND
9. BAND STRUTS
10. REVERSE AND HIGH CLUTCH DRUM
11. FORWARD CLUTCH AND CYLINDER

12. THRUST WASHER NO. 3
13. FORWARD CLUTCH HUB AND RING GEAR
14. THRUST WASHER NO. 4
15. FRONT PLANET CARRIER
16. INPUT SHELL, SUN GEAR AND THRUST WASHER NO. 5
17. THRUST WASHER NO. 6
18. REVERSE PLANET CARRIER
19. THRUST WASHER NO. 7
20. SNAP RING

21. REVERSE RING GEAR AND HUB
22. LOW AND REVERSE BAND
23. BAND STRUTS
24. THRUST WASHER NO. 8
25. LOW AND REVERSE DRUM
26. ONE-WAY CLUTCH INNER RACE
27. ROLLER (12) AND SPRING (12)
28. SPRING AND ROLLER CAGE
29. ONE-WAY CLUTCH OUTER RACE

30. THRUST WASHER NO. 9
31. CASE
32. THRUST WASHER NO. 10
33. PARKING GEAR
34. GOVERNOR DISTRIBUTOR SLEEVE
35. SNAP RING
36. GOVERNOR VALVES & DISTRIBUTOR
37. OUTPUT SHAFT
38. EXTENSION HOUSING AND GASKET
39. CONTROL VALVE BODY
40. OIL PAN AND GASKET

REVIEW QUESTIONS — CHAPTER 12

1. What unit eliminates the conventional clutch as used on standard transmission installations?
2. What is a planetary gearset?
3. Name the parts in a planetary gearset.
4. A planetary gearset can be used to _____ or _____ torque and it can also be used to _____ the direction of rotation.
5. What unit is used to stop the planetary internal gear?
6. The planetary gearset can be used to transmit torque as a solid unit by _____ any two _____ of the gearset together.
7. The action in question 6 is accomplished by applying oil pressure to what unit?
8. What is a fluid coupling?
9. How does a fluid coupling work?
10. What is the principal difference between a fluid coupling and a torque converter?
11. The fluid coupling can multiply torque. True or False?
12. The vanes in a torque converter are straight. True or False?
13. Fluid motion in both the torque converter and the fluid coupling is in two directions. One motion is called the _____ flow; the other the _____ flow.
14. Of what value is the split ring used in the fluid coupling and torque converter?
15. What is a stator?
16. How does a stator work?
17. The fluid coupling uses a stator. True or False?
18. The stator can revolve in one direction only. What stops its rotation in one direction?
19. Why must the stator revolve in one direction only?
20. What is the stall point?
21. Explain what we mean by coupling point.
22. Why are torque converter vanes curved?
23. What is a servo and what does it do?
24. A multiple disc clutch uses a servo to actuate it. True or False?
25. What value is there in the use of an accumulator piston?
26. There are two types of oil pumps in general usage in the automatic transmission. Name them.
27. What is a shifter valve?
28. What opens the shifter valve?
29. What controls the source that opens the shifter valve?
30. Explain the action when the simple two-speed automatic transmission is placed in Low Range; in High Range.
31. What unit is used to delay the upshift?
32. How is downshift accomplished?
33. Draw a sketch showing how a spool balance valve works.
34. To apply extra pressure on the brake band during heavy acceleration, a _____ valve is often used.
35. What lubricates an automatic transmission?
36. SAE 20W MS engine oil will work well in an automatic transmission. True or False?
37. Explain step by step how a typical two-speed automatic transmission functions. Start with the shift lever in neutral.

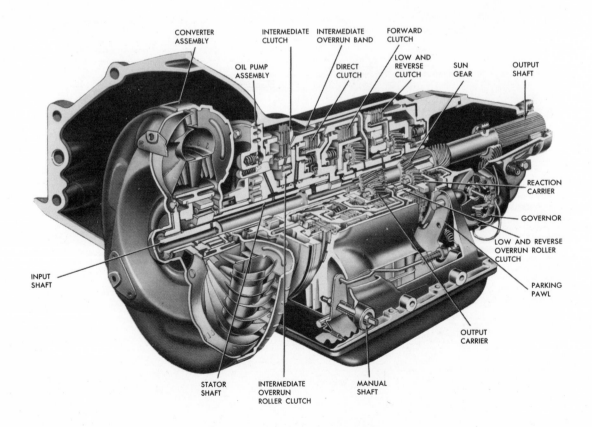

Turbo-Hydra-Matic 350 automatic transmission. This is a three-speed unit employing four multiple disc clutches, two planetary gearsets, two roller clutches and a intermediate overrun band. (Chevrolet)

Chapter 13

DRIVE LINES

In order for the transmission output shaft to drive the differential pinion gear shaft, it is necessary to connect these two units by means of a propeller shaft. Fig. 13-1.

PROPELLER SHAFT MUST BE FLEXIBLE

When the rear wheels strike a bump or a hole, the rear axle housing moves up or down in relation to the frame. As the

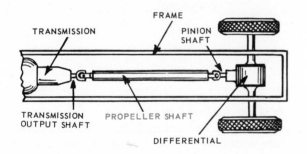

Fig. 13-1. Connecting transmission and differential. Top view showing propeller shaft hookup.

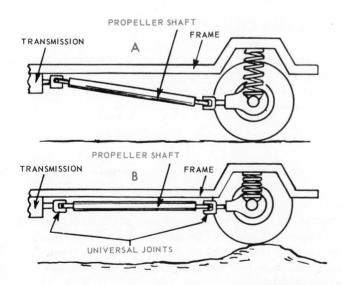

Fig. 13-2. Propeller shaft must flex. A represents normal operating angle of propeller shaft. B shows how bump forces differential up. Notice how propeller shaft angle changes. Since transmission is fixed to frame, propeller shaft must flex.

transmission is more or less solidly fixed to the frame, it becomes obvious that to allow the rear axle to move up and down, the drive line or propeller shaft must be able to flex. Fig. 13-2.

UNIVERSAL JOINT IS NEEDED

It is clear that a solid drive line would be bent and finally broken as the angle of drive changes. To allow the drive line, or propeller shaft, to move without breaking, a flexible joint is used. This is called a UNIVERSAL JOINT. Fig. 13-3.

By using a universal joint, torque can be transmitted constantly even though the rear axle is moving up and down. Fig. 13-4.

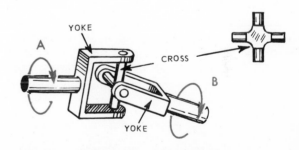

Fig. 13-3. Simple universal joint. Shaft A is transmitting torque in horizontal plane. Shaft B is transmitting same torque, but at different angle. Yokes swivel around ends of cross, imparting a degree of flexibility to universal joint.

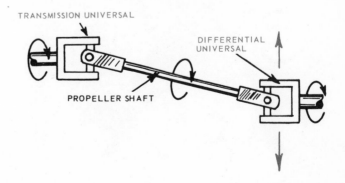

Fig. 13-4. Universal joints in action. As rear axle moves up and down, universals allow drive angles to change without binding propeller shaft.

TYPES OF UNIVERSAL JOINTS

CROSS AND ROLLER

The cross and roller universal joint is in common use. It consists of a center cross (sometimes referred to as a SPIDER) and two yokes (sometimes called KNUCKLES). The yokes surround a needle bearing roller. The roller can be retained in the yoke by using snap rings, a U-bolt or bolted plate. The roller needles surround the cross ends (called TRUNNIONS).

This setup allows the yokes to swivel on the cross with a minimum amount of friction. Fig. 13-5 shows an exploded view of a typical cross and roller universal joint.

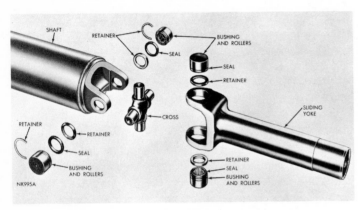

Fig. 13-5. Typical cross and roller universal joint. (Plymouth)

An enlarged view of the cross or spider is shown in Fig. 13-6. Notice that the trunnions are smoothly finished. A hollow in each trunnion leads to the center of the cross. A grease fitting is placed in the center. When lubricated, grease will flow out to each needle roller.

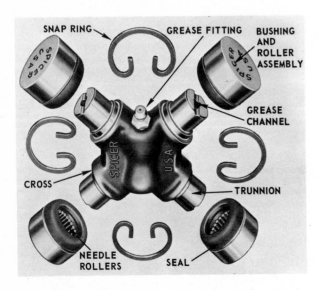

Fig. 13-6. Universal joint cross with bushing and roller assemblies. Note grease channel in each trunnion. (Spicer)

A phantom view of a cross and roller joint is shown in Fig. 13-7. Typical installations are pictured in Fig. 13-16.

BALL AND TRUNNION UNIVERSAL JOINT

Another type of universal, called the ball and trunnion, is sometimes used. This universal, in addition to allowing the drive line to bend, also allows fore and aft (forward and backward) movement of the propeller shaft.

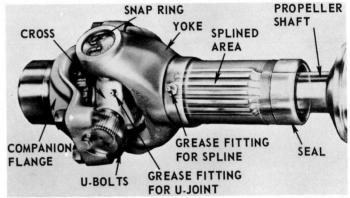

Fig. 13-7. Phantom view of a cross and roller universal joint.

A ball and trunnion joint is illustrated in Fig. 13-8. The cross pin is inserted in the hole in the ball on the end of the propeller shaft. The ball and roller assembly mounts on the ends of the cross pins. The round ball slides back and forth in a round hole in the center of the joint body. The rollers also slide and tip in grooves in the outer body.

The cutaway, Fig. 13-9, shows the ball and trunnion in the assembled position.

DRIVE LINE CHANGES LENGTH

As the rear axle moves up and down, it swings on an arc that is different from that of the drive line. As a result, the distance between transmission and rear axle will change to some extent.

SLIP JOINT

To allow the drive line to adjust to these variations in length, a slip joint is often used. The forward end of the propeller stub shaft has a series of splines cut into it. The universal yoke sleeve has corresponding splines cut into it. The yoke sleeve is slipped over the splined propeller shaft stub, and the universal is bolted to the transmission output shaft. The shaft can move fore and aft by sliding in the yoke sleeve splines. Fig. 13-10.

As mentioned, the ball and trunnion universal allows fore and aft movement, so a slip joint is not necessary when this type of joint is used.

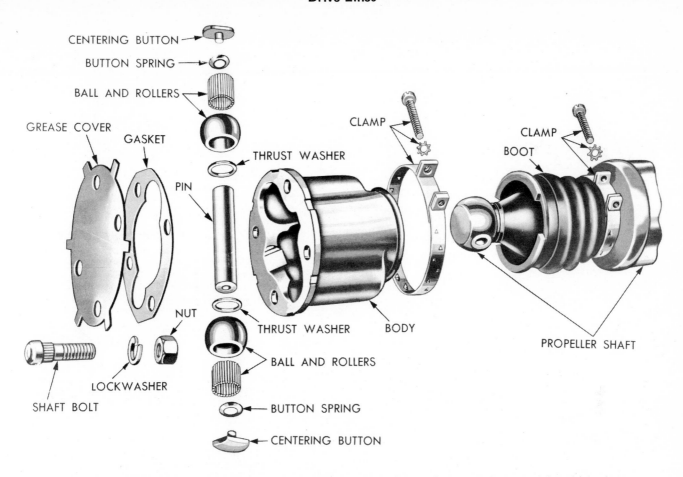

CENTERING BUTTON
BUTTON SPRING
BALL AND ROLLERS
GREASE COVER
GASKET
PIN
THRUST WASHER
CLAMP
CLAMP
BOOT
NUT
THRUST WASHER
BODY
LOCKWASHER
BALL AND ROLLERS
SHAFT BOLT
BUTTON SPRING
CENTERING BUTTON
PROPELLER SHAFT

Fig. 13-8. Exploded view of ball and trunnion universal joint. Boot, clamp and grease cover protect universal joint from water and dust. Centering buttons prevent end movement of pin, ball and needle rollers. (Plymouth)

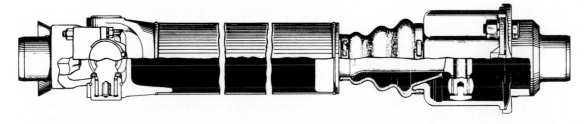

Fig. 13-9. Assembled ball and trunnion universal joint. Ball and trunnion joint is shown on right end. Left joint is conventional cross and roller.

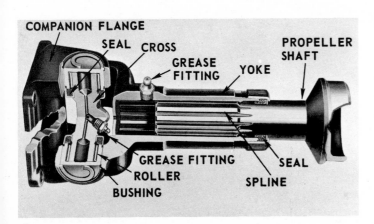

COMPANION FLANGE
SEAL CROSS
GREASE FITTING
YOKE
PROPELLER SHAFT
GREASE FITTING
ROLLER
BUSHING
SEAL
SPLINE

Fig. 13-10. Universal joint incorporating a slip joint in yoke.

PROPELLER SHAFTS

The propeller shaft is generally hollow to promote light weight and of a diameter sufficient to impart great strength. Quality steel is used in its construction. A rubber mounted torsional damper may be used,

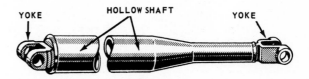

YOKE
HOLLOW SHAFT
YOKE

Fig. 13-11. Typical propeller shaft. Yokes are welded to hollow shaft. Some shafts have a splined stub on one end, yoke on other. (Mercury)

263

The universal yoke and splined stub (where used) are welded to the ends of a hollow shaft. The shaft must run true, and it must be carefully balanced to avoid vibrations. The propeller shaft is often turning at engine speeds. It can cause great damage if bent, unbalanced or if there is wear in the universal joints. Fig. 13-11.

HOW DO THE WHEELS DRIVE THE CAR?

When the propeller shaft turns the differential, the axles and wheels are driven forward. The driving force developed between the tires and the road is first transferred to the rear axle housing. From the axle housing, it is transmitted to the frame in one of three ways:

1. Through leaf springs that are bolted to the housing and shackled to the frame.

2. Through control or torque arms shackled (bolted, but free to swivel) to both frame and axle housing.

3. Through a torque tube that surrounds the propeller shaft, which is bolted to the axle housing and pivoted to the transmission by means of a large ball socket (sometimes referred to as a torque ball). Fig. 13-12.

HOTCHKISS DRIVE

The Hotchkiss type of drive is used when the rear axle drive thrust is transmitted to the frame through the springs or via control arms.

The Hotchkiss drive consists of an open propeller shaft secured to the transmission output shaft and differential pinion gear shaft. Front and rear universal joints are used. If

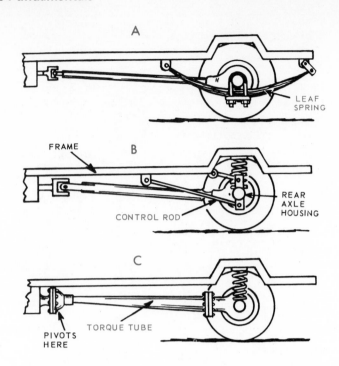

Fig. 13-12. Methods used to transfer driving force from rear axle housing to car frame. A—Force transmitted through leaf springs. B—Transmitted through control rods. C—Transmitted through torque tube.

both joints are of the cross and roller type, a slip joint must be provided.

The main task performed by the Hotchkiss drive is that of transferring output shaft torque to the differential. Wheel drive thrust is NOT transmitted to the drive line. A typical Hotchkiss drive pictured in Figs. 13-1, B in 13-12 and 13-13. The Hotchkiss drive is in widespread use today.

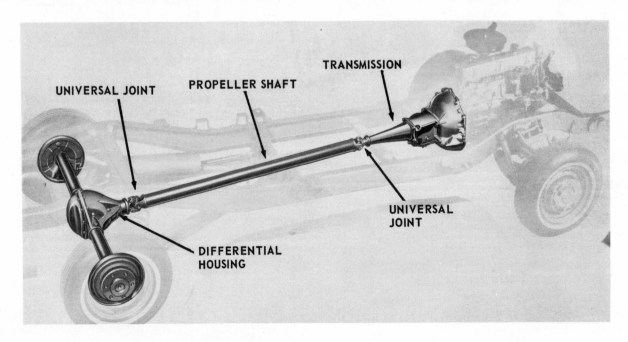

Fig. 13-13. Typical Hotchkiss drive setup. In this particular car, drive thrust from rear axle housing is transmitted to frame via control arms. (Chevrolet)

TWO—PIECE HOTCHKISS

To reduce excessive propeller shaft length (with resultant problems of vibration and installation), a two-piece propeller shaft is often used. In this type of installation, it is necessary to have a center support bearing for the shafts. This support consists of a ball bearing held in a dustproof housing that is, in turn, supported by the frame. Rubber is generally incorporated in the mount to reduce noise and vibration transfer to the frame. The rubber will also allow some movement of the shaft without binding.

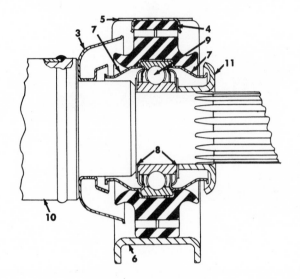

Fig. 13-14. One type of propeller shaft center support bearing. A dust protected ball bearing, mounted in rubber, is used to support two shafts where they meet. (GMC)

Fig. 13-14 illustrates a typical center support bearing. Note that the front propeller shaft rests in the rubber cushioned bearing. The splined stub allows slip joint action at this point.

Both front and rear shafts in a two-piece Hotchkiss drive are pictured in Fig. 13-15. The center support is also on rubber mounts. Note the use of three universal joints.

TORQUE TUBE DRIVE

When the rear axle housing thrust is not carried to the frame through the springs or by control rods, a torque tube drive is used.

The torque tube is a hollow steel tube that reaches from the transmission to the rear axle housing. One end of the torque tube is secured to the axle housing by a ring of bolts. The other, or forward end, is fastened to the transmission through a torque ball arrangement.

The propeller shaft runs through the torque tube. The transmission end has a universal joint, located inside the torque ball. The rear end of the propeller shaft is splined to the pinion gear shaft.

All housing drive thrust is carried through the torque tube to the torque ball, to the transmission, to the engine, and then on to the frame through the engine mounts. The car is literally being shoved forward by the torque tube pressing on the engine.

The torque tube is often braced by metal strut rods fastened to the axle housing, and extending out some distance to the torque tube. On some installations, a center support bearing is located midway in the torque tube to steady the propeller shaft.

The universal joint located inside the torque ball receives lubrication from the transmission oil. Oil is kept from running into the torque tube by an oil seal. See C, Fig. 13-12. Also Fig. 13-19.

One make of car has used a two-piece torque tube that allows a bend in the tube to facilitate a beneficial drive angle. This type uses a front universal in the torque ball and a second joint where the two torque tubes are bolted together. The torque tube is little used today. Fig. 13-17.

UNIVERSAL JOINT CAN CAUSE FLUCTUATING SHAFT SPEED

If a conventional universal joint is driving at an angle and the speed of the driving shaft is constant, the speed of the driven shaft will rise and fall. Even though the driven shaft is turning the same number of turns per minute, its speed will

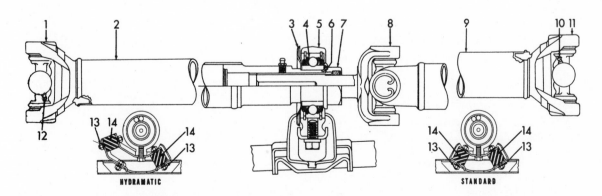

Fig. 13-15. Two-piece Hotchkiss dirve — assembled. This setup uses three universal joints; two on one shaft, one on other. Rubber cushioned center support is used where shafts connect.

(American Motors)

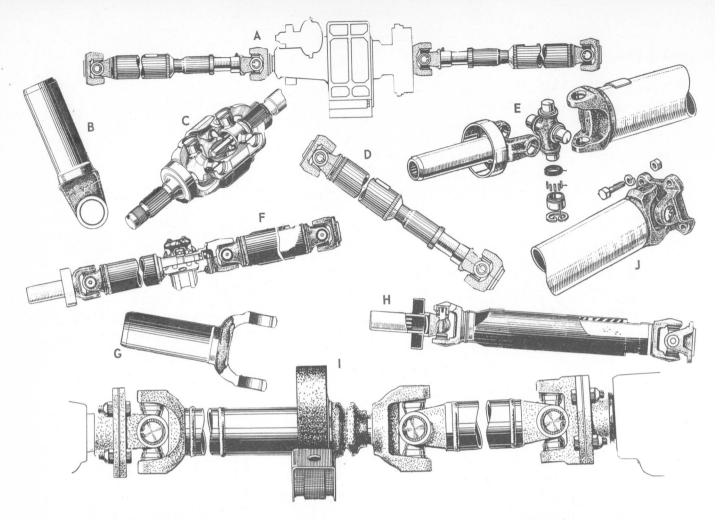

Fig. 13-16. Various propeller shaft and universal joint assemblies. A—Four wheel drive. B—Yoke, top view. C—Front wheel drive axle shaft constant velocity joint. D—One-piece shaft. E—Exploded view of cross and roller joint. F—Two-piece propeller shaft. G—Yoke, side view. H—Shaft made up of two separate tubes, bonded together with rubber to help eliminate torsional vibration. I—Two-piece shaft. J—Universal joint assembled.   (Volvo, Toyota, Land Rover, Renault)

fluctuate. Actually, the driven shaft speed will rise and fall two times for every revolution. This is due to the fact that the angle of drive is not equally divided between the two shafts. Fig. 13-20.

UNIVERSAL JOINT YOKES MUST BE IN SAME PLANE

When using a universal joint on each end of a propeller shaft, the disturbing shaft speed fluctuation can be greatly reduced if both universal joint yokes are in the same plane (lined up). Fig. 13-21 illustrates speed fluctuations by means

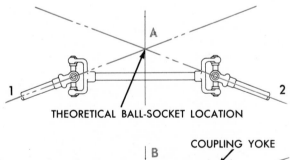

THEORETICAL BALL-SOCKET LOCATION

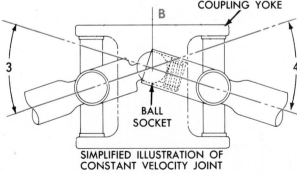

SIMPLIFIED ILLUSTRATION OF CONSTANT VELOCITY JOINT

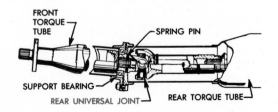

Fig. 13-17. Two-piece torque tube. A second universal, located where two torque tubes meet, is used. Front universal is not shown.   (Buick)

Fig. 13-18. Schematic showing principle of constant velocity universal joint. "A" illustrates necessary location for ball socket that will keep angle of drive evenly divided between shafts "1" and "2". "B" shows coupling yoke and ball socket. Note how both shafts form same angle — "3" and "4." (Chevrolet)

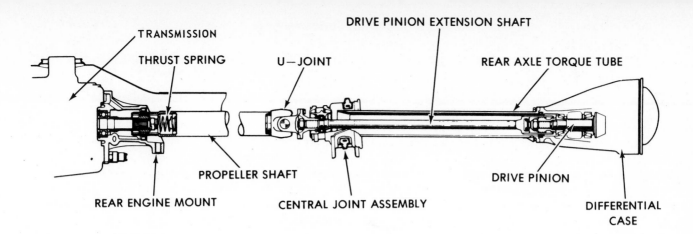

Fig. 13-19. One form of torque tube drive. This setup uses a short torque tube from differential housing to the central joint assembly (where tire drive thrust is transmitted to the frame). A conventional Hotchkiss drive is used from the central joint to the transmission. (Opel)

of a speed fluctuation chart.

With the joint yokes in the same plane, the torque of the driving shaft will be transmitted to the pinion shaft in a smoother manner. The propeller shaft speed will rise and fall from the front joint action, but the fluctuation will be cancelled out by the action of the rear joint. Fig. 13-22.

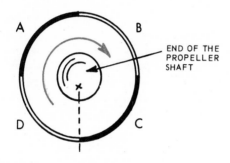

Fig. 13-20. Shaft speed fluctuates as drive angles change. Were you to chart speed of X as it moves around one revolution, you would find that it accelerates through D, decelerates through A, accelerates through B and once again decelerates through arc or distance C. These speed fluctuations caused by drive angle changes do not divide angle equally between both shafts connected by universal joint.

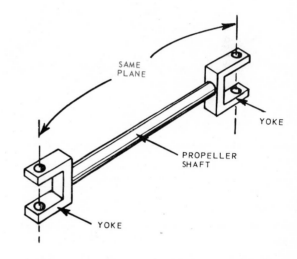

Fig. 13-22. Universal joint yokes must be in same plane.

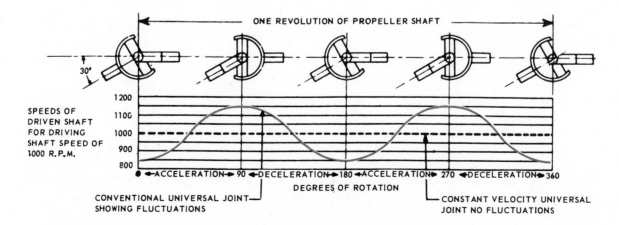

Fig. 13-21. Speed fluctuation chart. As universal joint turns, angle formed by cross changes. It never divides angle equally between two shafts, (except for one brief moment as cross angle shifts from one extreme to other). This causes acceleration-deceleration fluctuation that transmits torque in a jerky fashion. Dotted line illustrates even speed when constant velocity (later) universal joint is used. Undulating line (waving) represents speed fluctuation achieved with conventional type universal joint.

CONSTANT VELOCITY UNIVERSAL JOINT

There are several different types of universal joints designed to produce even transfer of torque without speed fluctuations. They are commonly referred to as constant speed or constant velocity universal joints. These joints provide an extremely smooth power transfer.

Constant speed transfer is made possible by special design of the joint that automatically divides angle of drive equally between two shafts. Fig. 13-23 shows one widely used design. Fig. 13-24 pictures a one-piece propeller shaft utilizing two constant speed universal joints. Also see Fig. 13-18.

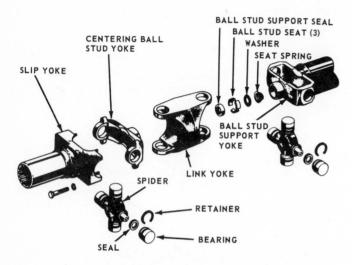

Fig. 13-23. Exploded view of a constant velocity universal joint. (American Motors)

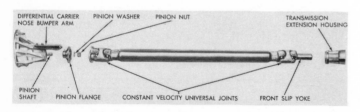

Fig. 13-24. One-piece propeller shaft employing two constant velocity joints for very smooth torque transfer. (Cadillac)

SUMMARY

Drive lines can be of two general types, Hotchkiss and torque tube. The torque tube currently enjoys little use.

When the rear axle housing drive thrust is carried by the springs or control arms, an open shaft Hotchkiss drive may be used. It will use two or more universal joints. The propeller shaft may be of two-piece design.

If the drive line must carry the drive thrust, the torque tube system is employed. A propeller shaft, using one or more

universal joints, is enclosed in a steel tube. The tube carries the thrust.

Universal joints are of two basic types, the cross and roller and ball and trunnion.

As drive line angles change, it is necessary to utilize a slip joint to allow fore and aft movement of the propeller shaft.

Drive lines must operate at high speeds. Therefore, they must be accurately constructed and carefully balanced.

Some universal joints are lubricated upon construction and are sealed for life. Other types utilize a grease fitting for periodic lubrication.

To prevent speed fluctuation transfer to the pinion shaft, it is essential that both front and rear propeller shaft yokes be in the same plane.

Constant velocity universal joints are used to provide smooth torque transfer without speed fluctuations.

REVIEW QUESTIONS – CHAPTER 13

1. What is the name of the unit that connects the transmission output shaft to the differential pinion gear shaft?
2. As the rear axle housing moves up and down in relation to the transmission, what unit allows the necessary flexing of the drive line?
3. What are two common types of drive?
4. When the drive thrust is through the springs, a _____ type of drive may be used.
5. When the drive thrust is not carried by the springs or control arms, the _____ drive must be used.
6. Name two types of universal joints.
7. The _____ and _____ universal joint will allow fore and aft movement of the propeller shaft.
8. When the _____ and _____ universal joint is used, it is necesssary to incorporate a _____ _____ in the drive system.
9. Most propeller shafts are solid. True or False?
10. Wheel drive thrust is carried to the frame in three ways. Name them.
11. What advantage is there in making the propeller shaft in two pieces?
12. What is a torque ball and what is its purpose?
13. Some propeller shafts use a constant velocity universal joint on both ends. True or False?
14. What would happen if the two propeller shaft universal yokes were not in the same plane?
15. Why must the drive line be carefully balanced?
16. In a two-piece drive line, a _____ bearing must be used.
17. What is a constant velocity universal joint? Where is it commonly used?
18. How does the constant velocity universal joint stop shaft speed fluctuations?
19. Universal joint friction is reduced by using _____ _____ in the yoke bushings.
20. How are universal joints lubricated?

Chapter 14

REAR AXLE ASSEMBLIES

The rear axle assembly has several important functions. It must hold the two rear wheels on, keep them upright and drive them forward or backward. It must drive both wheels in such a manner that one can turn faster than the other, yet both must receive torque.

The rear axle assembly must absorb the driving force of the wheels, and transmit it to the frame through springs, control rods or a torque tube. It provides an anchorage for springs, supports the weight of the car and forms the foundation upon which the rear wheel brakes are constructed. Obviously, the rear axle assembly must be well constructed with quality materials.

CONSTRUCTION

The rear axle basic assembly may be broken into several major sections: housing, axles, differential. Combined, these form the basic unit.

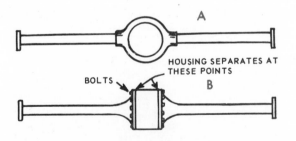

Fig. 14-1. Axle housings. A—One-piece banjo type (so named because of resemblance to musical instrument). B—Three-piece split type housing. Sections are bolted together.

HOUSING

The axle housing is usually made of stamped steel parts welded together. Or, the center section of the housing may be made of cast steel. Two basic types have been used: the BANJO type housing (in wide use) and the SPLIT housing (little used) consisting of two or more pieces. Both types are illustrated in Fig. 14-1.

AXLES

Two steel axles are placed inside the housing. Their inner ends almost touch; and, in some cases, they do touch. The outer ends protrude out of the housing and form a base upon which the wheel hubs are attached. The inner ends are splined and are supported by the differential assembly. The outer ends are supported in roller or ball bearings. Fig. 14-2.

ATTACHING WHEEL HUBS

Two methods are employed to secure the wheel hubs to the axle. One method employs a taper on the axle end; the other forms the axle end into a flange upon which the wheel is bolted. Fig. 14-3.

TYPES OF AXLES

Live axles (axles that turn with the wheels) are of three basic types: full-floating, three-quarter-floating and semi-floating. Most cars utilize the semifloating axle. Most trucks have full-floating axles. If the axle breaks, the wheel will not come off. The three types are illustrated in Fig. 14-4.

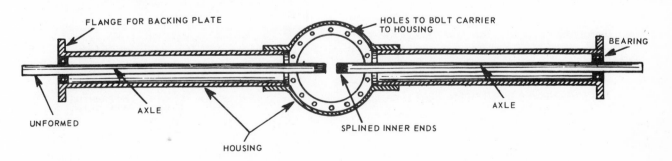

Fig. 14-2. Rear axles. Axle inner end supports are not shown. Housing is banjo type. Axle outer ends have not yet been formed.

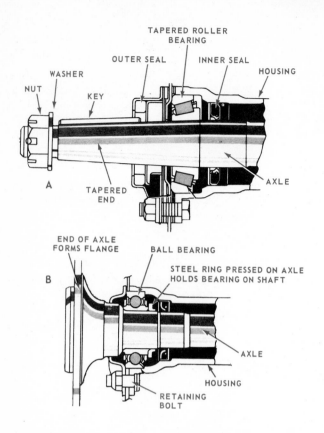

Fig. 14-3. Attaching wheels and hubs. A—Tapered end of axle will wedge into a tapered hole in wheel hub. Key will prevent axle rotation in hub. B—Flanged axle end. Wheel is bolted directly to axle flange. This method is widely used.   (John Deere)

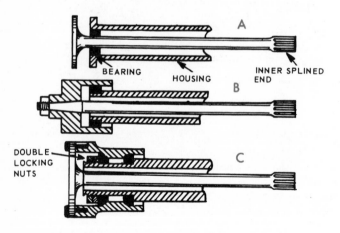

Fig. 14-4. Axle types. A—Semifloating. Axle drives, retains and supports wheel. Outer end is supported by bearing. B—Three-quarter-floating. Axle drives and retains wheel. Housing supports load via a bearing in hub that contacts housing. C—Full-floating. Axle drives wheel. Housing-hub bearings support load and retain wheel. Note double locking nuts that draw hub and tapered roller bearings together. All inner ends are supported in differential.

DRIVING THE AXLES

The rear wheels of a vehicle must turn at different speeds when rounding the slightest corner (outside wheel must roll

farther). Therefore, it is necessary to employ a unit called a DIFFERENTIAL to drive the axles so both axles receive power, yet they are free to turn at different speeds.

THE DIFFERENTIAL

A splined axle side gear is placed on the inner splined end of each axle. The axle side gear is supported by the differential case. The side gear is free to turn in the case. Fig. 14-5.

As shown in Fig. 14-5, the differential case may be turned. It will revolve about the axle side gears. The differential pinion shaft will turn with the case, but the axle side gears will not be driven.

By bolting a large ring gear to the differential case, and connecting it to a ring gear pinion gear and shaft, it will be possible to turn the case. The propeller shaft will be attached to the ring gear pinion shaft.

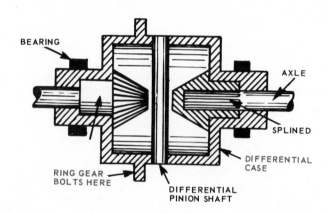

Fig. 14-5. Axle side gears. Axle side gear is free to turn with axle inside differential case. Pinion shaft turns with case. Axle side gear is splined to axle end. See right-hand axle gear.

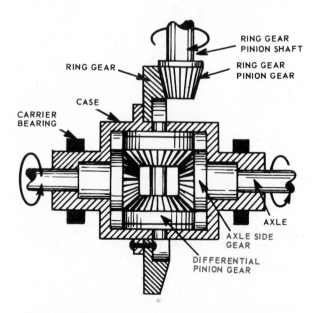

Fig. 14-6. Differential unit. Two differential pinions are pulled around by differential pinion shaft. When case is revolved, pinions will turn axle side gears. Since these are splined to axles, axles will now turn with case.

When the propeller shaft turns the ring gear pinion, the pinion will turn the ring gear. The ring gear, in turn will revolve the differential case and pinion shaft. The axle side gears will still NOT TURN. By adding two differential pinion gears (the differential pinion shaft will pass through these gears) that mesh with the side gears, the revolving case will turn the axle side gears with it. Fig. 14-6.

DIFFERENTIAL ACTION

The propeller shaft turns the ring gear pinion shaft. The ring gear pinion turns the ring gear which, in turn, revolves the differential case. When the case turns, the differential pinion

shaft turns with it. As the differential pinions are mounted on this shaft, they are forced to move with the case. Being meshed with the axle side gears, the pinions will pull the axle side gears along with them.

When the car is moving in a straight line, the ring gear is spinning the case. The differential pinions and axle side gears are moving around with the case, with no movement between the teeth of the pinions and axle side gears. The entire movement is like a solid unit.

When rounding a turn, the case continues whirling and pulling the pinions around on the shaft. As the outer wheel must turn faster, the outer axle side gear is now moving faster than the inner axle side gear. The whirling pinions not only

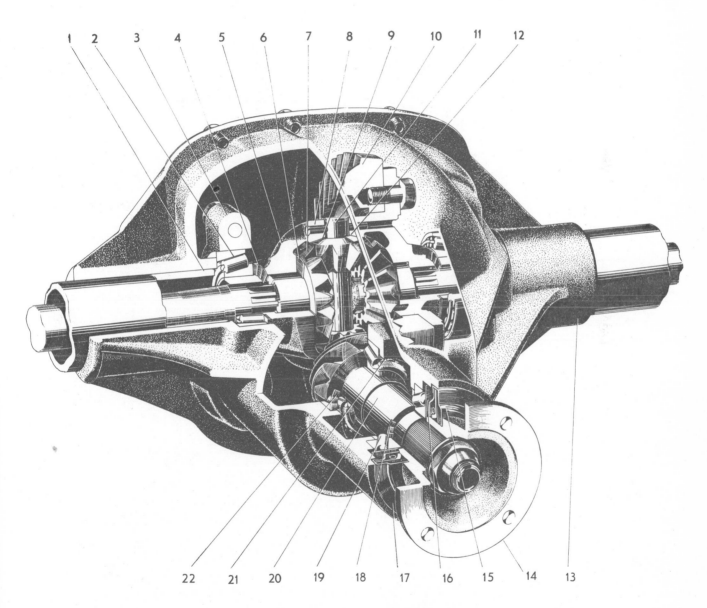

Cutaway view of a rear axle differential unit.

1. Tubular shaft	7. Differential side gear	13. Rear axle casing	19. Front pinion bearing
2. Differential carrier bearing	8. Lock pin	14. Flange	20. Pinion
3. Bearing cap	9. Differential pinion	15. Dust cover plate	21. Rear pinion bearing
4. Shims	10. Crown wheel	16. Oil slinger	22. Shims
5. Differential carrier	11. Shaft	17. Oil seal	
6. Thrust washer	12. Thrust washer	18. Shims	

pull on both axle side gears, but now begin to rotate on their shaft, meanwhile walking in the axle side gears. This allows them to pull on both axle side gears, while at the same time, compensating for the difference in speed by rotating around their shaft. This action is illustrated in Fig. 14-7.

You can see in Fig. 14-7, detail A, the car is moving in a straight line. The pinion is pulling both gears, but it is not turning. In B, the right side axle gear is moving faster than the left axle gear. The pinion gear is still moving at the same speed. It is still pulling on both gears, but has now started to turn on the pinion shaft. This turning action, added to the forward rotational speed of the shaft, has caused the right-hand side gear to speed up and actually begin to pass the pinion shaft.

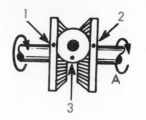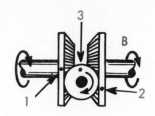

Fig. 14-7. Differential action. A—Axle side gears, pinion gear and shaft revolve as a unit — straight line driving. B—Axle side gears move forward but at different speeds. Pinion gear 3 is turning on shaft. Notice that dot 2 has moved ahead of pinion shaft. Dot 1, even though it is moving forward, is doing so more slowly. It is moving slower than pinion shaft. This would be the axle on inside of turn.

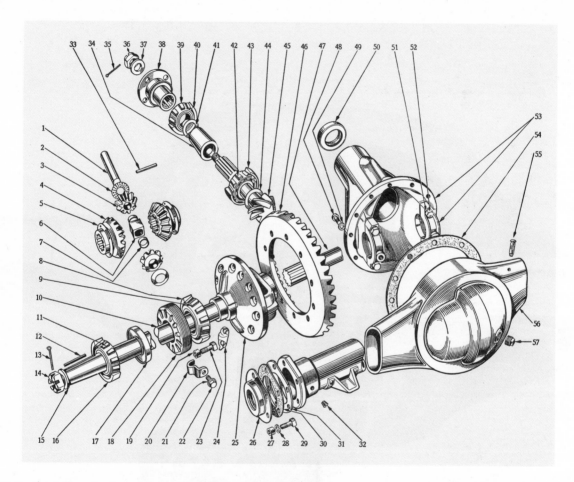

Fig. 14-8. Exploded view of typical rear axle assembly. Study all parts and learn names. Notice relationship of one part to other.   (Dodge)

1. Differential pinion shaft
2. Differential pinion thrust washer
3. Differential pinion gear
4. Axle side gear
5. Side gear thrust washer
6. Axle thrust block
7. Thrust block spacer
8. Carrier bearing cup
9. Carrier bearing cone and rollers
10. Axle
11. Axle outer bearing
12. Wheel hub key
13. Cotter pin
14. Axle nut
15. Washer
16. Axle outer bearing cup
17. Grease seal
18. Carrier bearing adjuster
19. Ring gear to case bolt nut

20. Carrier bearing adjuster lock
21. Lock bolt split washer
22. Lock bolt
23. Ring gear to case bolt
24. Ring gear to case bolt lock
25. Differential case
26. Grease seal
27. Nut
28. Lock washer
29. Bolt
30. Gasket
31. Shim
32. Grease fitting plug
33. Differential pinion shaft lock pin
34. Bearing spacer
35. Cotter pin
36. Lock nut
37. Washer
38. Pinion shaft flange

39. Pinion shaft bearing
40. Pinion shaft bearing cup
41. Adjusting shim
42. Pinion shaft bearing
43. Pinion shaft bearing cup
44. Shim
45. Ring gear pinion gear
46. Ring gear
47. Axle
48. Carrier to housing bolt (cap screw)
49. Housing bolt lock washer
50. Pinion shaft grease seal
51. Carrier bearing cap lock washer
52. Carrier bearing cap screw
53. Differential carrier and carrier bearing cap
54. Carrier to housing gasket
55. Housing vent
56. Housing
57. Fill level plug

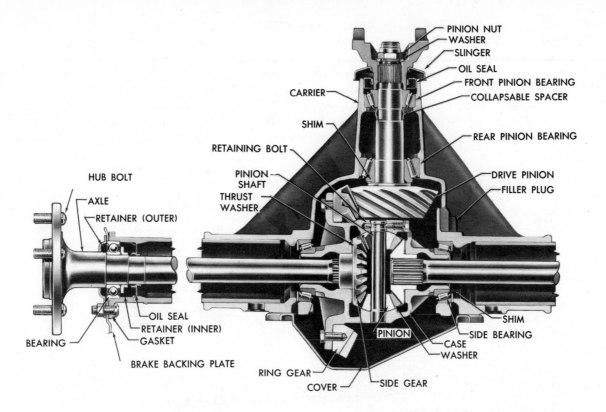

Fig. 14-9. Cross section of a rear axle assembly. Note thrust washers behind axle side gears and differential pinions. Differential carrier is an integral part of axle housing. Axle end is flanged for wheel attachment.
(Oldsmobile)

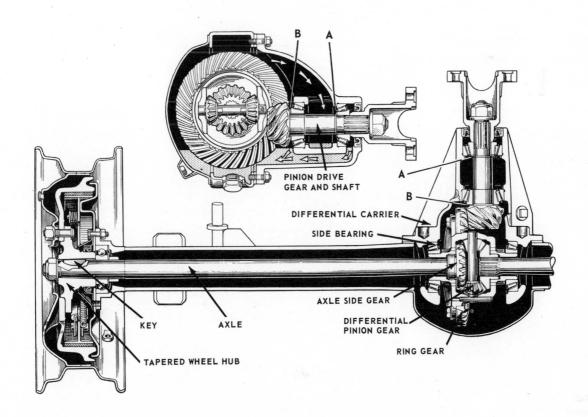

Fig. 14-10. Hypoid drive rear axle assemble — banjo housing, semi-floating axle, tapered and keyed wheel hub, demountable differential carrier. (Hillman)

The reverse walking effect on the left-hand side gear, has caused it to slow down. The differential action adjusts itself to any axle speed variation.

If one wheel begins to slip, the axle on firm ground will stand still. The case continues spinning the pinions, but they will merely walk around the stopped axle gear and impart the torque to the spinning axle. A special traction differential is often used to overcome this tendency. It will be covered later in this chapter.

DIFFERENTIAL CARRIER AND BEARINGS

A heavy and rigid section is bolted to the housing. It contains the pinion gear, shaft and bearings. This is termed the DIFFERENTIAL CARRIER. Two large bearing holders are provided to support the spinning differential case. These are termed CARRIER BEARINGS. In some applications, the carrier is made as a solid part of the axle housing.

All the component parts of the axle housing, axles and differential, are shown in Fig. 14-8.

An assembled view of a differential is shown in Figs. 14-9 and 14-10.

SPECIAL TRACTION DIFFERENTIAL

To avoid the loss of driving force that occurs when one wheel begins to slip, special differentials are designed to automatically transfer the torque to the wheel that is not slipping. This enables the car to continue its forward motion. Although there are several variations, all employ the principle of a friction device (clutch plates or a cone clutch) to provide some resistance to normal differential action.

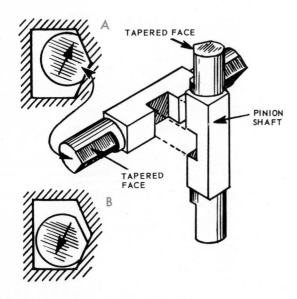

Fig. 14-11. Pinion shafts and case ramps. Notice that shafts fit together but are free to move outward. A—End view of a pinion shaft at rest in bottom of case ramp. B—When case pulls on pinion shaft, pinion end slides up ramp. This forces pinion shaft in an outward direction and moves pinion gear outward thus locking clutch. Both shafts react but in opposite directions.

CHRYSLER SURE—GRIP

The Chrysler Sure-Grip differential is basically a standard model, but with several important additions. The axle side gears are driven not by two differential pinions but by four. This requires two separate pinion shafts.

The two shafts cross, but are free to move independently of each other. The shaft outer ends are not round, but have two flat surfaces that form a shallow V. These ramp-like surfaces engage similar ramps cut in the differential case. Fig. 14-11.

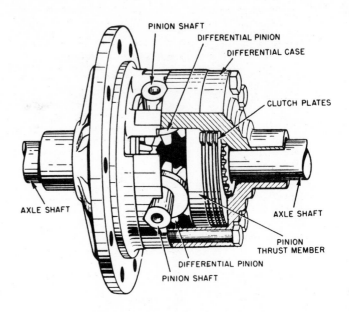

Fig. 14-12. Cutaway of Sure-Grip differential. See how pinion shafts fit in case ramps. Note clutch unit (one on each side) splined to case and to axle.   (Chrysler)

A series of four clutch discs are used in back of each axle side gear thrust member. Two of these discs are splined to the differential case, and two are splined to the thrust member. The thrust member is splined to the axle. When the thrust members push outward, the clutch discs are forced together, locking the axle to the case. Figs. 14-12 and 14-13.

SURE—GRIP DIFFERENTIAL OPERATION — BOTH AXLES TURNING AT THE SAME SPEED

When the propeller shaft drives the pinion gear, the torque thrust is transmitted to the ring gear. As the ring gear drives the differential case, the pinion shafts are forced to rotate with the case.

The differential pinions encounter resistance when they attempt to turn the axle side gears. This resistance is transferred to the pinion shafts that are driving the pinions. As both ends of each pinion shaft are seated in tapered ramps, and since they have some "play" at this point, this forces the shafts to slide up the ramp surfaces. This sliding movement moves both shafts in an outward direction. As each shaft moves outward, it moves its pinions in the same direction. The

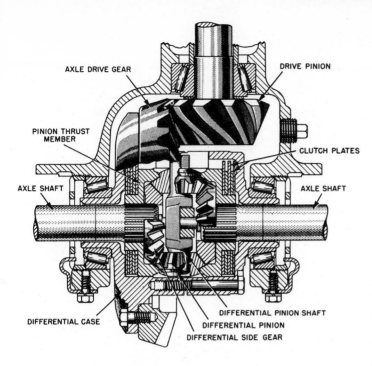

Fig. 14-13. Cross section of Sure-Grip differential. Pinion thrust members are moved against clutch units by pinion gears.

pinions press against the pinion thrust members, forcing them to lock up the clutches. This is the action when the car is traveling in a straight line. Fig. 14-14.

AXLES TURNING AT DIFFERENT SPEEDS

When the car turns a corner, the inner shaft slows down. When this happens, the pinion gears will start turning on their shafts. They will walk around the slower shaft and speed up the other shaft. This walking causes the outer shaft to rotate faster than the differential case, allowing the pinion shaft on the outer side to slide down its ramp. This releases the pressure on the outer clutches and lets the differential unit operate much like the standard model. Fig. 14-15 shows differential action when one axle is moving faster than the other. Note that the slower moving axle is receiving most of the torque since it remains clutched to the case.

This type of differential will provide better traction than the standard differential. It is particularly useful when roads are slippery and is also valuable in producing fast acceleration. A high-powered engine will often cause one wheel to spin during acceleration when using a standard differential.

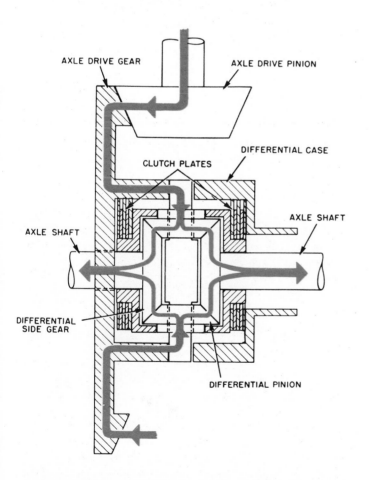

Fig. 14-14. Sure-Grip operation when vehicle is moving in straight line. Both pinion shafts are in outer positions, locking up clutch units. Differential pinions are not walking and both shafts are receiving equal torque. (Chrysler)

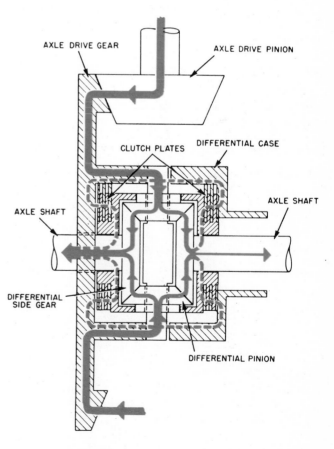

Fig. 14-15. Sure-Grip operation when vehicle is moving around corner. Left axle has slowed down, pinions walk and release pressure on outer or right-hand clutch assembly. Left-hand axle, or inner axle, is receiving greatest torque.

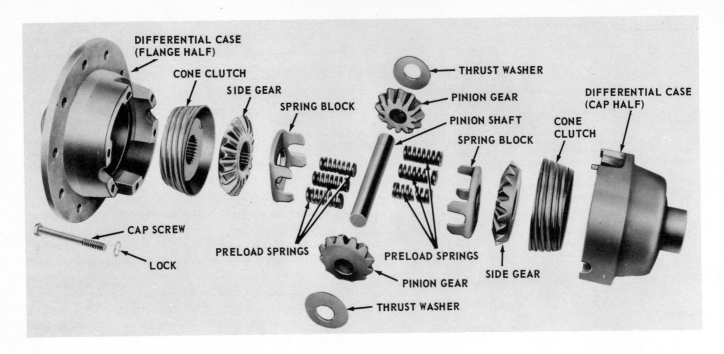

Fig. 14-16. Oldsmobile Anti-Spin traction type differential. Cone clutches are splined to axle ends. Coil spring pressure forces cones into tight engagement with case.

A somewhat different traction differential, using cone clutches under coil spring pressure, is shown in Fig. 14-16. A cross section view is pictured in Fig. 14-17.

Fig. 14-17. Cross section view of Buick Positive-Traction differential. Construction is similar to Oldsmobile's Anti-Spin Unit.

The Oldsmobile Anti-Spin differential, shown in Fig. 14-16, uses the pressure of the coil springs to force the clutch cones into tight engagement with the case. This action tends to lock the axles to the case. In order for differential action to occur, the cones must be forced to slip. If one wheel slips, the other

will still receive some driving force via the cone. The Positive-Traction differential in Fig. 14-17 functions in a similar manner.

A Chevrolet Positraction differential is illustrated in Fig. 14-18. Construction is similar to that in Figs. 14-16 and 14-17 except disc clutches are employed instead of cone clutch.

Fig. 14-18. Chevrolet Positraction differential. Disc clutches are employed instead of cone clutches. Case is one piece. 1—Fastener. 2—Differential case. 3—Side bearing. 4—Pinion shaft lock screw. 5—Ring gear. 6—Shims. 7—Clutch pack guides. 8—Clutch discs. 9—Clutch plates. 10—Side gear. 11—Spring block. 12—Thrust washer. 13—Pinion gear. 14—Pinion shaft. 15—Preload spring.
(Chevrolet)

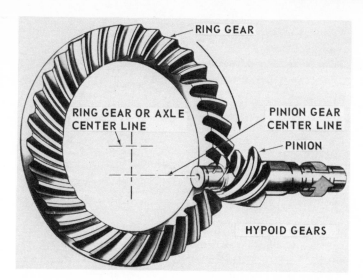

Fig. 14-19. Hypoid gearing. Pinion gear engages ring gear BELOW axle center line. (Chevrolet)

HYPOID GEARING

To facilitate lowering the propeller shaft tunnel in the floor of the car, and to allow lowering the body of the car, many ring gear pinion gears enter and drive the ring gear somewhat below the center line of the axles.

This gearing setup, using a modified spiral bevel gear, is referred to as HYPOID gearing. A special hypoid lubricant is necessary to prevent premature wear due to the sliding, wiping action that takes place between the ring and pinion gear teeth. Figs. 14-19 and 14-20.

Study the construction used in the pinion shaft assembly in Fig. 14-20. Take note of the two tapered roller bearings used to support the shaft. This type of bearing withstands both radial (forces working at right angles to the shaft) and longitudinal (lengthwise) thrust.

SPIRAL BEVEL GEARING

Another type of ring and pinion gearing uses the spiral bevel gear. This type of pinion gear meshes with the ring gear at the axle center line. The spiral tooth shape allows an overlapping tooth contact that makes for quiet operation, as well as added strength. Before one tooth rolls out of contact with another, a new tooth contact is made. This distributes the torque load over several teeth. Figs. 14-21 and 14-22.

The cross section, in Fig. 14-22, shows the path of the churning lubricant (see arrows). Notice how it is thrown up and forward where it drops down and flows back, lubricating the ring gear and pinion gear, pinion bearings, etc.

SPUR BEVEL

More antiquated, as far as differential gearing is concerned, is the spur bevel. You will notice that the teeth are straight.

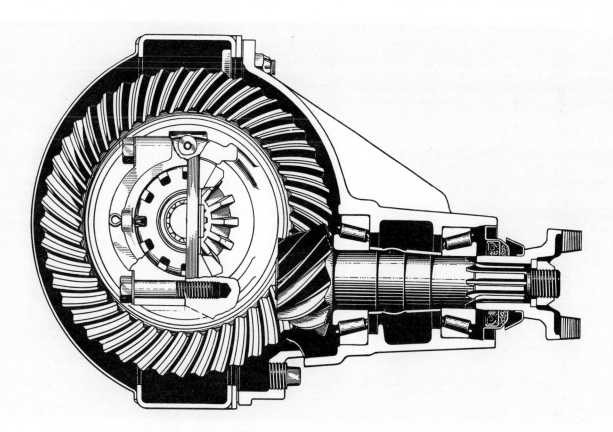

Fig. 14-20. Typical hypoid gearing — cross section. Note how pinion gear engages ring gear below axle center line.

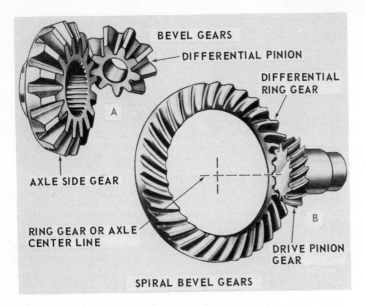

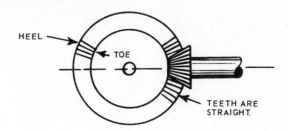

Fig. 14-23. Spur bevel gearing. Pinion on axle center line. Gears are straight cut.

Fig. 14-21. A—Bevel (spur) gears. B—Spiral bevel gears. Spiral bevel locates pinion gear in line with axle or ring gear center line. (Chevrolet)

moving the other gear), is of critical importance. Ring and pinions are always matched, and must be installed as a pair. NEVER REPLACE ONE WITHOUT THE OTHER. Fig. 14-24 illustrates tooth clearance, backlash, as well as other gear tooth nomenclature.

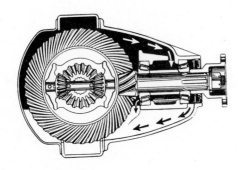

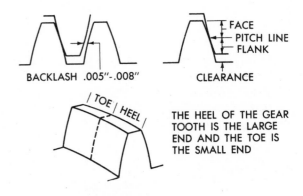

Fig. 14-24. Gear tooth nomenclature. Toe is small part of gear that faces inward. (Chevrolet)

Fig. 14-22. Cross section of differential using spiral bevel gearing. Drive pinion gear engages ring gear in line with axles. (Hillman)

This type of gear is somewhat noisy and does not offer the strength of the spiral bevel. It works well, however, for differential PINION gears. A, Fig. 14-21 and 14-23.

RING AND PINION

The tooth contact position, as well as clearance and back lash (distance one gear will move back and forth without

Correct and incorrect ring and pinion tooth contact patterns are shown in Fig. 14-25. The correct contact pattern is very important for strength, wear and quiet operation.

These patterns are brought out by coating the teeth (after cleaning) with Prussian blue or white lead compound. The gears are revolved in both directions, and the contact pattern becomes visible. The drive side is the side that contacts when the pinion is driving the ring. The coast pattern is when the ring is driving the pinion.

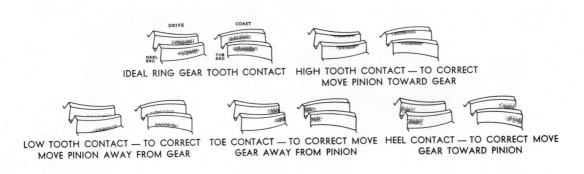

Fig. 14-25. Gear tooth contact patterns.

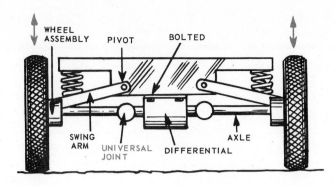

Fig. 14-26. Swing axle — rear view. Schematic of swing axle. Note two inner universal joints. Each wheel is free to move up and down by itself. Swing arms are pivoted on inner ends only. Outer ends are bolted to wheel assembly. This requires only one universal per axle. Differential housing is bolted to frame.

GEAR RATIO

A considerable reduction is effected as the ring gear has several times the number of teeth as that of the pinion. The difference is expressed as a ratio. Gear ratios vary but commonly cover a range of around 2.56-1 to nearly 4 to 1. Higher ratios (4.56-1, 5.10-1, etc.) are available for truck use.

When ratios are even (3.00-1, 4.00-1, etc.) the same teeth on the pinion would contact the same teeth on the ring every third or fourth revolution of the pinion. Any tooth damage or imperfection would then tend to be transferred to the constantly mating tooth, with the possibility of a short service life. To avoid this, most ratios are uneven (2.56-1, 2.78-1, etc.). The uneven ratio prevents frequent contact between the same teeth, providing a varying contact-wear pattern.

Lower ratios will cut down engine speed, but produce poorer acceleration. The correct ratio is determined by a careful consideration of the car weight, the type of service it will have, the engine horsepower, transmission gearing, and whether or not an overdrive is incorporated.

SWING AXLES

Some rear axles use a standard differential, but the differential housing is bolted to the frame. The rear wheels are mounted on swinging arms, and the axles are open and use a universal joint at each end. This setup allows the rear wheels to be individually sprung, and does offer some benefits over the solid axle housing. Some setups use one universal per axle. Others place the axle shafts in articulated (movable) housings. Fig. 14-26.

A partial view of the swing axle as used in the Chevrolet Corvette is shown in Fig. 14-27. Note how the differential carrier is bolted to the frame. The axle drive shaft universal joints permit the wheels to move up and down. Note transverse (across car) leaf spring.

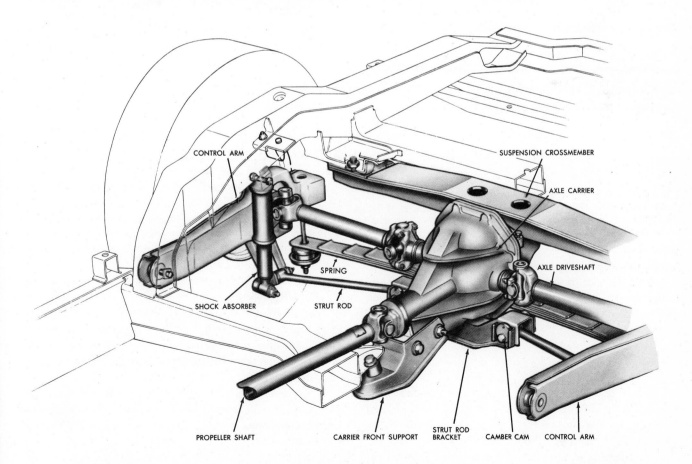

Fig. 14-27. Corvette swing axle arrangement. Differential carrier is bolted to frame. Universal joints at each end of axle shafts permit wheel up-and-down movement.

PLANETARY GEAR DIFFERENTIAL

One type of front wheel drive has used a planetary differential to reduce the space needed. Figs. 14-28 and 14-29.

The pinion gear (drive pinion) turns a ring gear. The ring gear turns an internal gear. The internal gear is meshed with the outer pinions, and the outer pinions to the inner pinions. The inner pinions engage the sun gear which is splined to the RIGHT-HAND axle. The planetary pinions rotate on shafts passed through the planet pinion carrier.

When the internal gear is revolved, it pulls the outer pinions around with it, thus driving the sun gear (right axle) and planet carrier (left axle). As long as the car is traveling in a straight line, the planet pinions do not move.

PLANETARY DIFFERENTIAL ACTION

During a right turn, the left wheel travels faster than the right. The internal gear drives the pinion carrier through the pinions. The sun gear cannot drive the right axle at the same speed as the left due to the resistance between the right tire and the road. As a result, the planet pinions walk around the sun gear, permitting the left wheel to travel faster. Fig. 14-28.

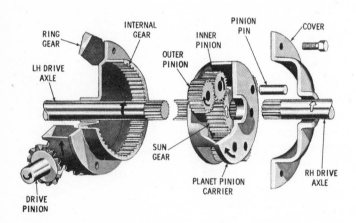

Fig. 14-28. Planetary gear type differential action during a right turn. Left axle is traveling faster than right. Planet pinions walk around sun gear. (Oldsmobile)

When the car makes a left turn, the planet pinion carrier is prevented from turning at the same speed as the sun gear, by the tire-to-road friction of the left tire. The planet pinion carrier is then forced to walk around the internal gear, allowing the right axle to travel faster than the left. Fig. 14-29.

LUBRICATION

A special heavy oil is used in rear axle housings. The viscosity will be around SAE 90. Special hypoid oils are used in hypoid assemblies. Another special oil is required for traction type differentials.

In operation, oil circulated by the ring gear is thrown over all the parts. Some installations make use of troughs to guide the returning lubricant to special spots, such as the ring and pinion contact area, and also to the pinion bearings.

The lubricant is retained by gaskets and oil seals. A drain plug is provided at the bottom of the housing, and a filler plug is located partway up the side of the housing. NEVER FILL THE HOUSING ABOVE THE LEVEL OF THIS UPPER PLUG.

Some axle outer bearings are lubricated by the housing lubricant. Others have a lubrication fitting by which wheel bearing grease may be injected at recommended intervals. NEVER USE A PRESSURE GREASE GUN ON THESE FITTINGS. GREASE MAY BE FORCED BY THE SEALS AND ENTER THE BRAKES. USE A HAND GUN AND APPLY GREASE SPARINGLY. Other types use a bearing filled with grease and sealed at the factory. Enough lubricant is provided to last the normal life of the bearing.

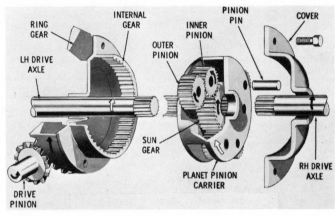

Fig. 14-29. Planetary gear differential action during a left turn. Right axle is traveling faster than left. Carrier walks around internal gear.

SUMMARY

The rear axle assembly provides a base for the springs, holds the car up, provides a foundation for the brakes, holds the rear wheels in line and drives the rear wheels while allowing them to turn at different speeds.

The axle assembly is broken into three basic units: the axles, the differential and the housing.

Axles are of three types: semifloating, three-quarter-floating, and full-floating.

The differential contains axle side gears, pinions, a pinion shaft and the case itself. The differential drives both axles while allowing their speeds to vary. A special traction differential improves the performance of the differential under adverse conditions.

The differential unit is supported by the differential carrier, and it revolves in tapered roller bearings. A large ring gear is bolted to the case. A pinion gear, supported on a shaft in the carrier, drives the ring gear.

The rear axle assembly is lubricated by a rather heavy (SAE 90 average) lubricant.

REVIEW QUESTIONS – CHAPTER 14

Like a traction differential, the ability to apply one's efforts where they will do the most good will push a person much further along the road to success. When you study, really "lean" into it. See if you can get through the following questions without getting "stuck."

1. List seven functions of the rear axle assembly.
2. For purposes of study, the rear axle assembly can be broken into three basic units or sections. Name them.
3. What are the two types of housing? Which one is more often used?
4. The outer ends of the axles run in bronze bushings. True or False?
5. Describe two methods of attaching the wheel or wheel hub to an axle.
6. Name the three different types of axles and describe them.
7. Why is a differential necessary?
8. Name the parts that make up the differential assembly.
9. Describe the action of the differential in straight-line driving.
10. Describe the action of the differential when rounding a corner.
11. What is the purpose of the differential carrier?
12. How is the differential case supported in the carrier?
13. Of what use is the special traction type of differential?
14. What parts do you find in the traction differential that you do not find in the standard model?
15. Explain the action of the traction differential when rounding a corner.
16. What is a hypoid gear setup as used in the differential?
17. Why is a hypoid used?
18. What is a spiral bevel gear?
19. What advantages does the spiral bevel gear have over the spur bevel?
20. Backlash is the amount of wear in the teeth. True or False?
21. To determine tooth contact patterns, it is necessary to coat the gear teeth with some sort of marking material such as white lead. True or False?
22. The toe part of the ring gear tooth is that part which faces the inside of the gear. True or False?
23. How do you determine the gear ratio used in the rear axle?
24. What factors limit your choice of gear ratio?
25. What type of rear axle uses universal joints and open axles?
26. A good quality engine oil is quite satisfactory as a rear axle lubricant. True or False?
27. How is the rear axle lubricant contained or kept in the housing?
28. Explain how the rear axle assembly is lubricated.
29. Outer rear wheel bearings are lubricated in three ways. Name them.

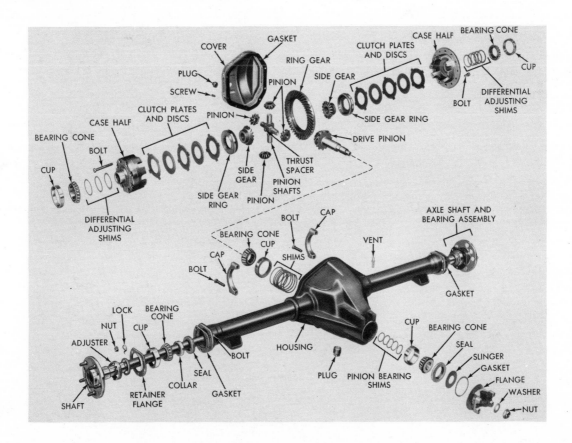

Exploded view of a complete rear axle assembly. Note use of a traction type differential.   (Plymouth)

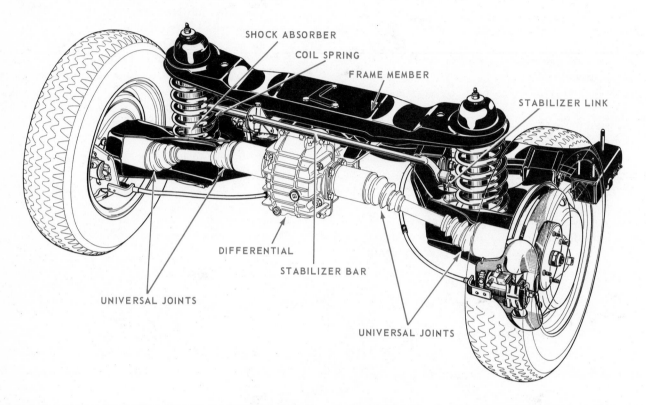

SHOCK ABSORBER

COIL SPRING

FRAME MEMBER

STABILIZER LINK

DIFFERENTIAL

STABILIZER BAR

UNIVERSAL JOINTS

UNIVERSAL JOINTS

A fully independent rear suspension. Note use of universal joints to allow wheel assemblies to move up and down. Differential is attached to the frame.   (Peugeot)

Chapter 15

FRAMES, BODIES, SPRINGS, SUSPENSION SYSTEMS

If the axles were bolted directly to the frame, every rough spot in the road would transmit a jarring force to the car. High speed and riding in comfort would be impossible.

You know that the modern auto does ride well. Its ability to negotiate rough roads and handle well at high speeds is a direct result of a properly designed suspension system.

The object of any suspension system is to allow the body of the car to travel forward with a minimum amount of up-and-down movement. It also permits the car to take turns with little roll or tendency to break frictional contact (skidding) between the tires and the road. Even though the

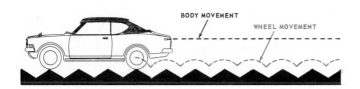

Fig. 15-1. Car body must be insulated against road shock. Black dotted line shows how car body moves forward smoothly despite rapid up and down motion of wheels.

tires and wheels must follow the road contour, the body should be influenced as little as possible. Fig. 15-1.

FRAMES

In order to provide a rigid structural foundation for the car body, and to provide a solid anchorage for the suspension system, a frame of some sort is essential. There are two types of frames in common use today: the separate steel frame to which the body is bolted at various points; the integral body and frame in which the frame and body are built as a unit and welded together.

INTEGRAL FRAME

In the integral type frame, the various body sections are used as structural strength members to help support and stiffen the entire unit. All sections are welded together. In some cases, the rigid body has reinforced points to which the suspension system is attached. Others use a partial frame, front and rear, to which the engine and suspension systems are fastened.

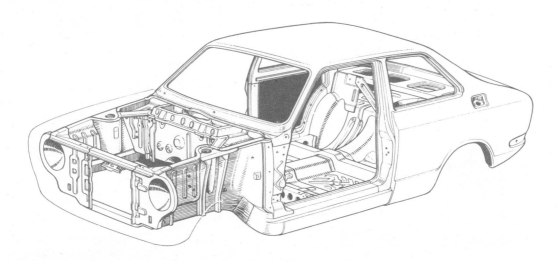

Fig. 15-2. Integral body and frame (unit) construction. By using body sections as strength members, great rigidity is secured. With this type of construction, a full length frame is not required. All sections are welded together.   (Toyota)

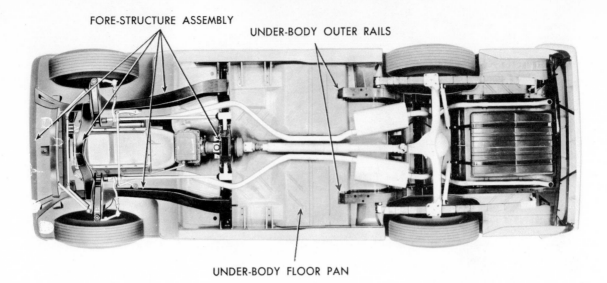

FORE-STRUCTURE ASSEMBLY

UNDER-BODY OUTER RAILS

UNDER-BODY FLOOR PAN

Fig. 15-3. Unit body construction — under view. There is no full-length frame. Notice how partial frame is used as mounting base for front suspension torsion bars, engine, rear springs, etc. (Plymouth)

An example of the unit body construction using a partial frame, is shown in Fig. 15-2.

An underside view of the unit body construction is illustrated in Fig. 15-3. Notice the short stub frame in the front end, securely bolted to the unit body. Body rails are used in the rear to form an attachment point for the rear springs.

CONVENTIONAL FRAME

The conventional type of frame is constructed as a separate unit, and the body is bolted to it at various points. Rubber insulator blocks are used between the frame and the body attachment points to minimize the transfer of bothersome vibrations.

The conventional frame must be quite rigid, since the suspension system, engine and body depend upon the frame to support and keep all units in alignment.

The frame may be constructed of steel members, cold riveted together (so that rivets will not shrink upon cooling and allow lateral movement between parts), or the frame members may be welded. The structural members are usually of the channel or box type. Fig. 15-4.

Viewed from the top, the commonly used frame would resemble those shown in Fig. 15-5.

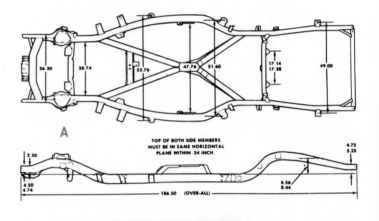

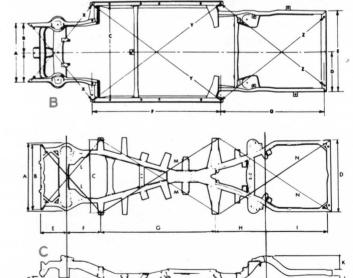

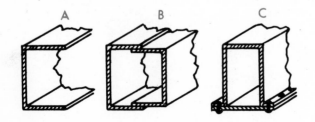

Fig. 15-4. Frame cross sections. A—Channel. B—Boxed channel. C—Boxed U-channel. B is welded. C is riveted.

Fig. 15-5. Frame shapes. A—X-frame. B—Perimeter frame. C—Tubular center X-frame. Side views of frames show front and rear kickup area. Frame is bent upward, then down to provide clearance over rear axle and for front suspension system. Current usage favors perimeter frame design.

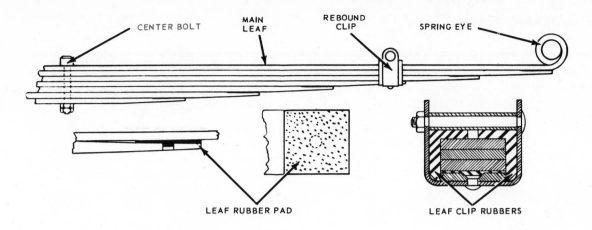

Fig. 15-6. Partial view of semielliptical leaf spring. Center bolt serves to keep leaves in line and properly spaced.   (M.G.)

SPRINGS

In many cases, a suspension system uses two or more different types of springs. There are four types of springs in general use today: leaf, coil, torsion bar and air springs.

SEMIELLIPTICAL LEAF SPRING

The semielliptical leaf spring has been used for many years. It consists of a number of flat steel springs, of varying lengths, bolted together into a single unit. The spring is fastened to the front or rear axle by means of U-bolts. The ends of the spring are shackled to the frame. The top leaf, called the main leaf, has the ends curled around to form the spring eyes. Some springs also curl the second leaf ends around the eye to provide both additional strength and safety. See Figs. 15-6, 15-8 and 15-12.

When the spring is bent downward, the leaves slide against each other. The same action takes place on the rebound (springs snapping back to their original shape). To prevent squeaks, and to ease spring action, special neoprene rubber inserts are often used between the ends of the leaves. The springs then flex these small rubber pads. Other leaf springs are covered with metal casings and are filled with grease.

Several rebound clips are placed around the spring leaves at various points. These clips prevent the leaves from separating when the spring rebounds upward. Fig. 15-6 illustrates one-half of a semielliptical spring. Rebound clips and the leaf rubber pad inserts are also shown.

Fig. 15-7 shows how the spring is fastened to the axle. Notice the use of rubber pads to eliminate spring vibration transfer. The spring center bolt passes through the spring, keeping the leaves properly spaced in a lengthwise position. The top of the center bolt fits into a depression in the mounting pad, pinning the spring firmly to the axle mounting pad when the U-bolts are tightened.

One end of the spring is mounted on a pivot shackle that will allow no fore and aft movement. The other end must be on a hinge type shackle to allow the spring to shorten and lengthen as it flexes up and down. Fig. 15-8.

Most spring shackles are of similar design but use three different materials to reduce friction.

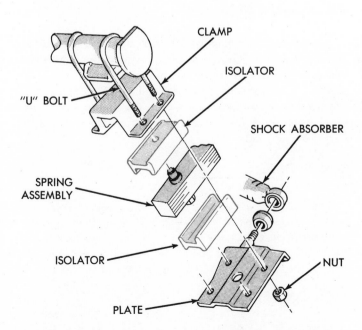

Fig. 15-7. Spring to axle housing mounting. Note use of rubber "isolator" blocks.   (Dodge)

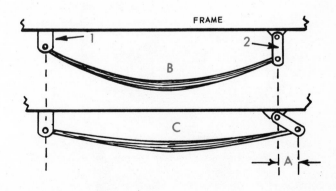

Fig. 15-8. Flexing causes a variation in spring length. 1—Stationary spring hanger. 2—Hinge type shackle. In B, normal load may produce this spring shape. In C, spring has been compressed. End 1 has merely pivoted but end 2 has moved out as spring lengthens. Difference in length is shown at A.

RUBBER—BUSHED SHACKLE

A rubber-bushed shackle uses a rubber bushing between the spring eye and the spring eye bolt. The spring and bolt do not touch. The pivoting motion is absorbed by the flexible rubber. This type requires no lubrication. Fig. 15-9.

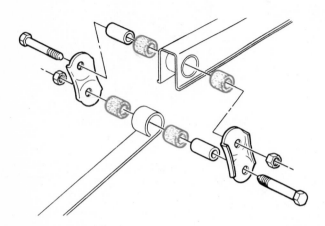

Fig. 15-9. Rubber-bushed spring shackle. Bushings are in two parts. A metal sleeve fits inside bushings, and shackle bolt passes through this sleeve. When shackle bolts are tightened, rubber bushings are somewhat compressed, causing them to expand and grip both spring eye and bushing sleeve tightly. All movement is taken within rubber bushings. (Chevrolet)

THREADED SHACKLE

The shackle bolts, in the shape of a U, are passed through the spring eye and through the spring hanger. These bolts are threaded full length. A steel bushing, threaded inside and out,

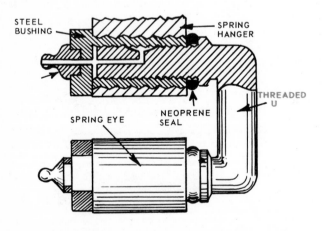

Fig. 15-10. Threaded U-shackle. Threaded sections of U pivot in steel bushings. Steel bushing is stationary in hanger and in spring eye.

is started on the bolts, then screwed in until it bottoms. Grease is inserted through a grease fitting. This type of shackle provides a wide wearing surface, prevents side motion and retains grease well. Fig. 15-10.

BRONZE—BUSHED SHACKLE

A bronze bushing is inserted between the spring eye and bolt. It must have periodic lubrication. One type of bronze bushing is illustrated in the top half of the hinge shackle, shown in Fig. 15-11.

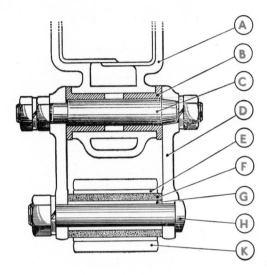

Fig. 15-11. Bronze-bushed shackle. Bronze-bushed part of shackle is shown in top half of shackle. Bottom half is rubber bushed. A—Spring or shackle hanger. B—Bronze bushing. C—Spring bolt. D—Shackle side link. E—Steel bushing. F—Rubber. G—Steel bushing. H—Spring bolt. K—Spring eye. (Austin)

A complete semielliptical spring is shown in Fig. 15-12. The left end pivots in spring hanger. Right end is supported on a hinge shackle. Rubber bushings are used in both ends.

SINGLE LEAF SPRING

A single leaf, semielliptical spring is shown in Fig. 15-13. The spring is fairly long and wide, and is quite thin even in the thicker center section. The entire spring consists of ONE leaf.

The use of leaf springs has been generally confined to the rear suspension on the modern auto. The car in Fig. 15-36 (front wheel drive) uses a dead (does not drive) rear axle with single leaf rear springs.

COIL SPRINGS

Coil springs are made of special spring rods, heated and wound in the shape of a spiral coil. They are then carefully tempered to give the proper tension. One end is secured to the frame, the other to the axle or suspension device. Pads are used on the ends to prevent vibration transfer. Fig. 15-14.

TORSION BAR SPRING

The torsion bar spring consists of a relatively long spring steel bar. One end is secured to a nonmovable mounting. The other end is free to turn.

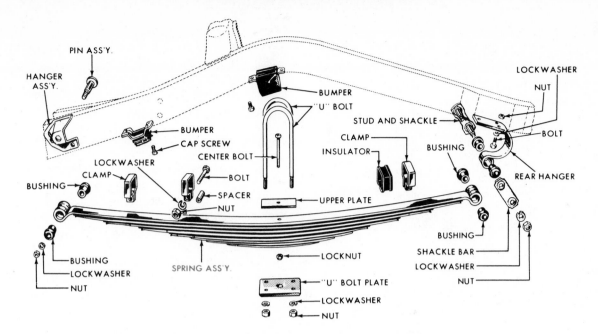

Fig. 15-12. Complete semielliptical leaf spring. Not shown is rear axle
housing to which spring is fastened via the two U-bolts.
(Lincoln)

Fig. 15-13. Single leaf spring. Long, wide and quite thin. Entire spring
is made up of one leaf.

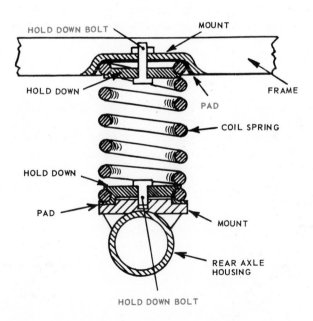

Fig. 15-14. Coil spring. Each end is bolted to mount. Rubber or fabric
pads are used at ends.

A lever arm, at right angles to the bar, is secured to the free
end. This lever arm is fastened to the suspension device. When
the wheel strikes a bump, and the lever arm is pushed upward,

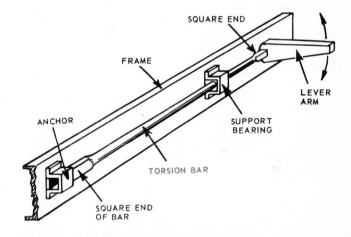

Fig. 15-15. Torsion bar spring. When lever arm is moved up or down, it
must twist long torsion bar. An anchor mount secures fixed end to
frame. Support bearing holds free end. Both anchor and free end are
square. This type of spring can be used in front or rear, or both.

it will twist the long bar. The bar will have resistance to
twisting, that has been carefully calibrated to the load forces it
is designed to bear. Fig. 15-15 illustrates a simplified form of
the torsion bar spring.

AIR SPRING

The air spring consists of a hollow steel container with a
flexible diaphragm stretched across one end. The diaphragm
extends upward part way into the container. The steel dome is
fastened to the frame. The diaphragm is secured, via a plunger,
to the axle.

The dome is filled with air, forcing the diaphragm and
plunger partially down. The spring action is imparted by the

diaphragm, flexing up and down against the compressed air inside the dome. An air compressor is used to furnish air to the dome units. Fig. 15-16 illustrates a typical air spring.

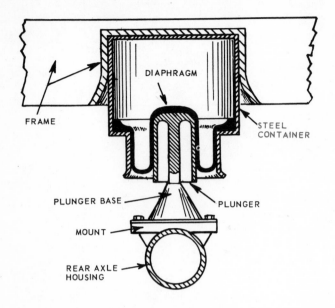

Fig. 15-16. Air spring. Proper air pressure in steel container is furnished by air compressor. Leveling valve keeps spring at proper height.

SPRING OSCILLATION

When the vehicle is traveling in a level, forward motion and the wheels strike a bump, the spring (regardless of type used) is rapidly compressed. The compressed spring will attempt to return to its normal loaded length. In so doing, it will rebound, causing the body of the car to be lifted. Since the compressed spring has stored up energy, it will rebound past its normal length. The upward movement of the car also assists in rebounding past the normal length.

The weight of the car will push the spring down, but since the car is traveling downward, the energy built up by the descending body will push the spring below its normal loaded

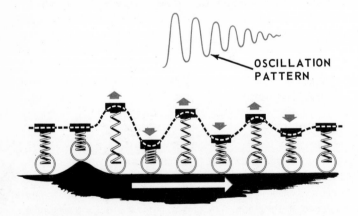

Fig. 15-17. Spring oscillation. When wheel strikes bump, spring is compressed. Even though wheel returns to level roadway, spring oscillates. Oscillation gradually diminishes and finally stops. (Ford)

height. This causes the spring to rebound again. This process, called spring oscillation, is repeated a little less each time until the car is finally still. Fig. 15-17.

This uncontrolled spring oscillation would not only cause an uncomfortable ride, but would make handling the car extremely dangerous.

SHOCK ABSORBER CONTROL ACTION

To overcome spring oscillation, a dampening device, called a SHOCK ABSORBER, is used. Many types of shock absorbers have been designed and used (vane type, opposed piston, etc.). Most popular, front and rear, is the hydraulic double direct-acting or telescoping shock. Such shocks are often referred to as airplane type shocks.

One end of the shock is mounted to the frame; the other end is attached to the axle. When the spring attempts to compress or rebound, its action is hindered by the shock absorber.

As the frame attempts to rise and fall in relation to the axle, the shock must telescope out and in. The shock's resistance to telescopic movement hinders this up-and-down motion. Instead of a long series of uncontrolled oscillations, the spring action is smoothed out, and the car quickly will return to its normal loaded level.

SHOCK CONSTRUCTION AND OPERATION

The telescopic shock absorber consists of an inner cylinder, outer cylinder, piston and piston rod and, in some case, an outer dust and rock shield. A series of valves, in the piston and at the bottom of the inner cylinder, control the movement of the hydraulic fluid within the shock.

Fig. 15-18 illustrates a typical telescoping shock absorber in three positions. The left-hand illustration shows the shock at rest with the car at normal loaded height. Notice that the inner cylinder (pressure tube) is completely filled with fluid, while the outer cylinder (reservoir tube) is only partially filled. The piston is near the midpoint of its travel in order to allow up and down movement.

The middle illustration, Fig. 15-18, shows the action of the shock during spring compression. Note that the piston and piston rod have moved down in the pressure tube, forcing the fluid to pass through the piston valve into the upper section of the pressure tube. Since the piston rod occupies space formerly filled with fluid, some fluid must be displaced through the base valve into the reservoir tube. The valves are calibrated to cause a certain resistance to the passage of fluid. Although the fluid passes through the valves quite easily when the shock is compressed slowly, any increase in compression speed causes a rapid buildup of resistance. The shock thus cushions the violence of the spring compression.

The right-hand illustration, Fig. 15-18, shows shock action during spring rebound. The piston is forced upward in the pressure tube, and the fluid trapped above moves down through the piston via a different valve. To compensate for the reduced amount of piston rod now in the pressure tube, additional fluid is pulled in from the reservoir tube through the base valve.

ROAD SPRINGS AND SHOCK ABSORBERS

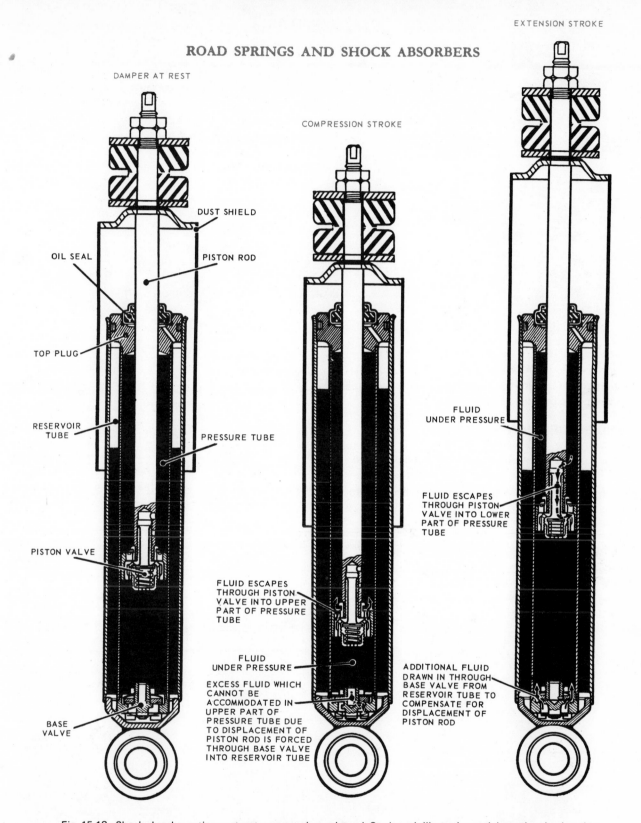

DAMPER AT REST

COMPRESSION STROKE

EXTENSION STROKE

DUST SHIELD

OIL SEAL

PISTON ROD

TOP PLUG

RESERVOIR TUBE

PRESSURE TUBE

FLUID UNDER PRESSURE

PISTON VALVE

FLUID ESCAPES THROUGH PISTON VALVE INTO UPPER PART OF PRESSURE TUBE

FLUID ESCAPES THROUGH PISTON VALVE INTO LOWER PART OF PRESSURE TUBE

FLUID UNDER PRESSURE

EXCESS FLUID WHICH CANNOT BE ACCOMMODATED IN UPPER PART OF PRESSURE TUBE DUE TO DISPLACEMENT OF PISTON ROD IS FORCED THROUGH BASE VALVE INTO RESERVOIR TUBE

ADDITIONAL FLUID DRAWN IN THROUGH BASE VALVE FROM RESERVOIR TUBE TO COMPENSATE FOR DISPLACEMENT OF PISTON ROD

BASE VALVE

Fig. 15-18. Shock absorber action — at rest, compression, rebound. Study each illustration and determine shock action shown in each.   (Standard Motor Products)

The shock action on rebound is similar to that of compression, although some shock valves are calibrated to provide more resistance on rebound than on compression. Note the rubber grommets (washers) used to fasten the shock.

Fig. 15-19 shows a typical mounting setup for a telescoping shock absorber. The top end is fastened by rubber grommets to the frame. The bottom is bolted to the lower arm of the front suspension system. The movement of the suspension

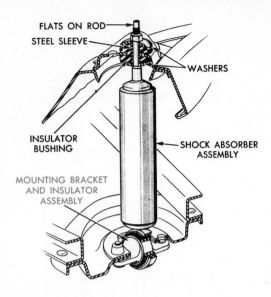

Fig. 15-19. Shock absorber mounting. (Dodge)

arm, up and down, will operate the shock.

Some shock absorbers can be set to provide a soft, normal or a firm action; others provide this feature automatically. Fig. 15-20 illustrates a type that can be set by hand.

FRONT END SUSPENSION SYSTEMS

There are two basic front end suspension systems in use today. One is the solid axle, leaf spring type; the other is the independent front wheel suspension using long and short swinging arms. There are various adaptations of these systems, but all use the same basic principle.

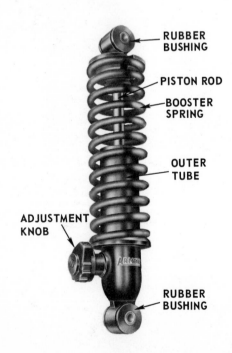

Fig. 15-20. Adjustable shock absorber. This particular shock also incorporates an overload or booster spring. When adjustment knob is turned, internal valving is altered to change hydraulic resistance to movement. (Armstrong)

SOLID AXLE

Use of the solid front suspension has been confined largely to trucks. This system uses a solid steel dead axle (does not turn with wheels) with a leaf spring at each side. The wheels swivel on each end via a pivot arrangement between the axle and the wheel spindle. Fig. 15-21.

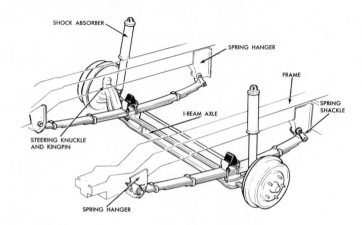

Fig. 15-21. Solid front axle suspension. Spring shackles fasten to frame. (Chevrolet)

Any up or down movement of either front wheel causes a vertical tipping effect of both wheels because they share a common axle. This effect upsets the steering geometry as well as imposes a twisting motion to the frame. Fig. 15-22. One

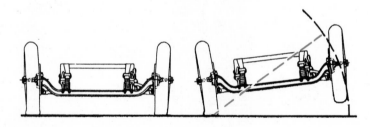

Fig. 15-22. Tipping effect of solid front axle.

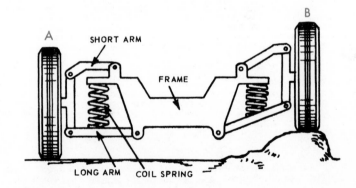

Fig. 15-23. Action of individual front wheel suspension system. Wheel B has been deflected upward. Long and short arms pivot and coil spring is compressed. Wheel A is not affected.

exception to this is the Twin-I-Beam front axle used in some Ford light trucks. In this setup, two beams are used. Each one can pivot independently. Fig. 15-33.

INDEPENDENT FRONT WHEEL SUSPENSION

With independent suspension, each front wheel is free to move up and down with a minimum (least attainable) effect on the other wheel. There is also far less twisting motion imposed on the frame.

This effect is achieved by using two arms at each side of the frame. The upper arm is considerably shorter than the lower, giving rise to the term LONG AND SHORT ARM SUSPENSION.

The inner ends of both arms are pivoted to the frame via threaded bushings (same principle as threaded U-shackle). Or, as currently in wide usage, rubber bushings. The outer ends are attached to the steering knuckle by threaded bushings or a ball joint.

When one wheel strikes a bump, it is deflected upward. This allows it to travel up or down without affecting the other wheel. Fig. 15-23.

By calculating the positioning and lengths of both upper and lower arms, it is possible for the wheel assembly to move

up or down with a minimum of tipping (change in camber angle, which will be described later). Fig. 15-24 shows inner pivot points 1 and 2, on typical long and short arm suspension. Note the different arcs transcribed by each arm.

Various examples of the threaded bushing and rubber bushing arm pivots are shown in accompanying illustrations. The threaded bushing outer arm pivots, knuckle support, steering knuckle and spindle assembly are shown in Fig. 15-25.

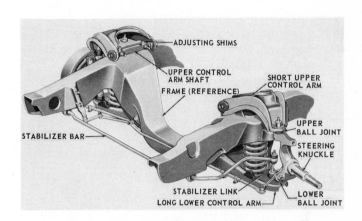

Fig. 15-26. Independent front wheel suspension system, using long and short control arms and coil springs. Note stabilizer bar.   (Buick)

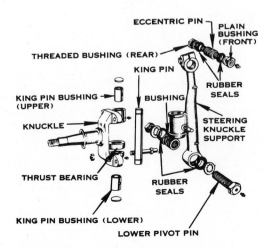

Fig. 15-24. Upper and lower arm pivot arcs of individual front wheel suspension system.

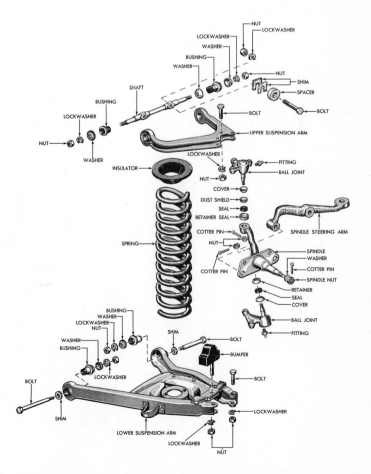

Fig. 15-25. Upper and lower control arm pivot pins and threaded bushings.

Fig. 15-27. Exploded view of long and short arm front suspension system. Study all parts, their construction and location.   (Lincoln)

INDEPENDENT FRONT SUSPENSION, USING COIL SPRING

Fig. 15-26 shows a typical independent front suspension, using rubber bushing arm pivots. The coil spring top rests in a cup-like spot against the frame. The bottom of the coil is supported by a pad on the lower control arm. The top of each telescoping shock is fastened to the frame; the bottom is attached to the lower control arm.

When the wheel strikes a bump, it is driven upward. This causes arms to pivot upward, compressing spring and shock. Note rubber bumpers used to limit arm travel and to soften the blow when limit is reached. For steering purposes, the front wheel steering knuckle pivots on ball joints.

Fig. 15-25 shows the older type of independent suspension system that utilized threaded pivot bushings on both the inner and outer arm ends. Note use of kingpin for steering pivot.

An exploded view of a current type of front suspension system (one side) is shown in Fig. 15-27. Notice that the spindle or steering knuckle pivots on BALL JOINTS. Rubber bushings are used for the control arms.

Threaded bushings and kingpins have been supplanted by rubber bushings on the control arm inner ends and by attaching the outer ends to the spindle assembly by means of ball joints. The ball joint allows the arms to pivot when they travel up and down. In addition, they allow the steering knuckle to pivot from side to side for steering, Fig. 15-28, eliminating the kingpin type pivot.

A detailed view of the ball joint steering knuckle and ball joints is shown in Figs. 15-29 and 15-30. In 15-29, notice how the knuckle and spindle assembly pivots in the ball joints.

The coil spring can be mounted on top of the upper control arm, with the upper end supported in a tall housing that is an integral part of the body and frame. Fig. 15-31.

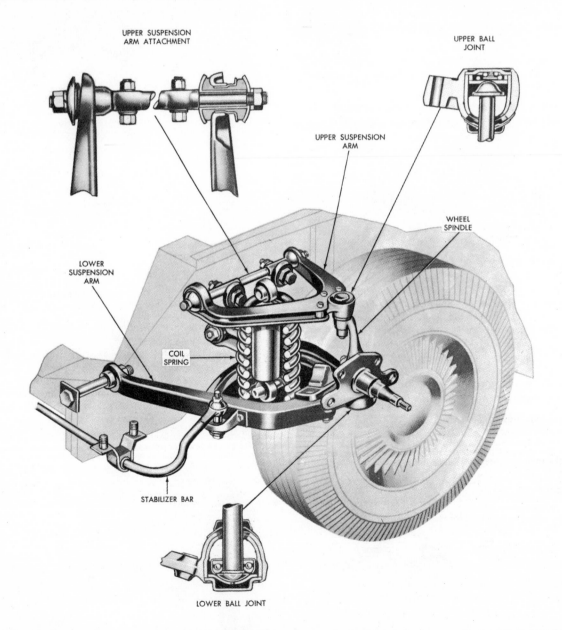

Fig. 15-28. Coil spring suspension system using ball joints on control arm outer ends. Inner ends of control arms are mounted in rubber bushings. (Lincoln)

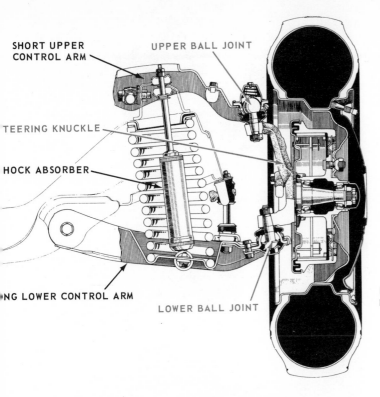

SHORT UPPER CONTROL ARM

UPPER BALL JOINT

TEERING KNUCKLE

HOCK ABSORBER

NG LOWER CONTROL ARM

LOWER BALL JOINT

Fig. 15-29. A ball joint steering knuckle. Knuckle assembly pivots on ball joints for steering action.
(Toyota)

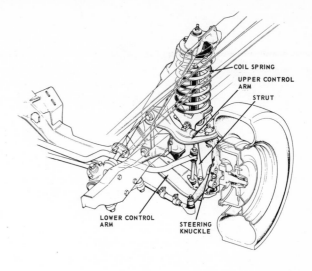

COIL SPRING

UPPER CONTROL ARM

STRUT

LOWER CONTROL ARM

STEERING KNUCKLE

Fig. 15-31. Coil spring mounted above upper control arm. Note narrow lower control arm and strut used to provide fore and aft bracing. (American Motors)

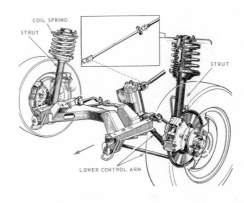

COIL SPRING

STRUT

STRUT

LOWER CONTROL ARM

Fig. 15-32. A MacPherson strut type front suspension. Wheel assembly pivots about long strut as well as moving up and down on strut. Shock absorber is built as an integral part of strut. (B.M.W.)

Another type of coil spring front suspension popular in foreign cars is the MacPherson strut type illustrated in Fig. 15-32. In this setup, the coil spring mounts on top of a heavy duty strut and pedestal assembly.

Also unusual is Ford's Twin-I-Beam front suspension on light duty trucks. It utilizes coil springs placed between the I-beam axles and the vehicle's frame. Fig. 15-33.

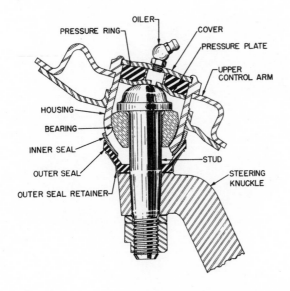

OILER

PRESSURE RING

COVER

PRESSURE PLATE

UPPER CONTROL ARM

HOUSING

BEARING

INNER SEAL

OUTER SEAL

OUTER SEAL RETAINER

STUD

STEERING KNUCKLE

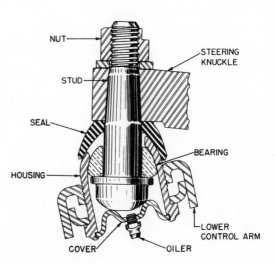

NUT

STEERING KNUCKLE

STUD

SEAL

HOUSING

BEARING

COVER

OILER

LOWER CONTROL ARM

Fig. 15-30. Cross section — upper and lower ball joints.
(Plymouth)

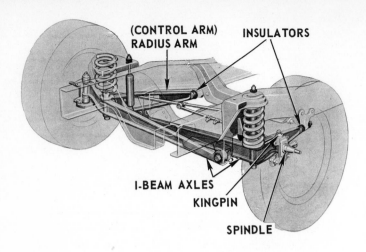

Fig. 15-33. Ford Twin-I-Beam front suspension as used in light duty (1/2, 3/4 and 1 ton) trucks. Note how one wheel can move upward without affecting other. (Ford)

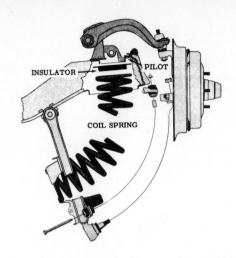

Fig. 15-34. When dismantling front suspension system, BE CAREFUL.

DANGER!

WHEN DISMANTLING A COIL SPRING FRONT SUSPENSION SYSTEM, MAKE CERTAIN THE CONTROL ARM IS SECURED TO PREVENT IT FROM SLAMMING DOWNWARD WHEN DISCONNECTED. WHEN EITHER THE INNER OR OUTER END IS DISCONNECTED, THE TERRIFIC FORCE OF THE PARTIALLY COMPRESSED COIL SPRING IS UNLEASHED, AND IT WILL PROPEL THE ARM DOWNWARD WITH LETHAL FORCE. Fig. 15-34.

TORSION BAR FRONT SUSPENSION

A torsion bar, Fig. 15-15, is located on each side of the frame in the front of the car. The lower control arm is

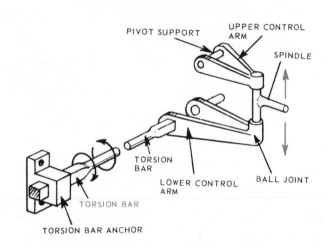

Fig. 15-35. Simplified torsion bar front suspension. When wheel spindle moves up or down, torsion bar must be twisted. Torsion bar anchor is bolted to frame. Both lower control arm support and upper control arm pivot support are fastened to frame by using support brackets.

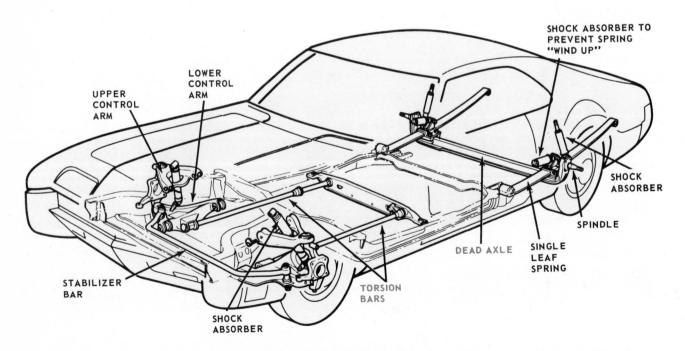

Fig. 15-36. Oldsmobile front wheel drive Toronado suspension system. Note use of torsion bars in front and a dead axle with single leaf springs in rear. (Oldsmobile)

attached to the torsion bar free end. When the wheel is driven upward, it must move the lower control arm upward, thus twisting the long spring steel bar. Fig. 15-35.

Fig. 15-36 shows a view of torsion bar suspension as used on Oldsmobile Toronado. Also see Fig. 15-3.

Fig. 15-37 illustrates one setup for hinging the lower control arm. Notice that the arm pivots on a stub shaft. The torsion bar fits into the end and does not support the arm. It merely produces spring resistance to up-and-down movement.

REAR SUSPENSION

The use of coil springs and the leaf springs dominates rear suspension systems. Fig. 15-38 illustrates a typical suspension system utilizing coil springs. The rear axle housing is mounted on springs and is attached to a set of upper and lower control arms.

The control arm pivot points are rubber bushed. One end of the arm is connected to the housing, the other to the frame. The arm arrangement allows the rear axle housing to move up and down but prevents excessive fore-and-aft and side-to-side movement.

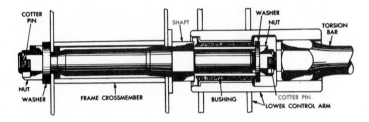

Fig. 15-37. Lower control arm pivot. Control arm rides on stub shaft. (Dodge)

The shock absorbers in Fig. 15-38 are STRADDLE mounted (tops of shocks are closer together than bottoms). In addition to controlling spring action or oscillation, straddle mounting assists in controlling side movement between frame and axle housing, and it also helps to resist tipping on turns.

There are other variations in the use of control arms. Some use a centrally located upper arm as well.

DANGER!

WHEN SUPPORTING A CAR BODY WHILE REMOVING THE REAR AXLE, WORKING ON THE SUSPENSION SYSTEM, ETC., MAKE CERTAIN THE CAR IS SECURELY BLOCKED. A CAR IS HEAVY AND CAN EASILY CRUSH A CARELESS MECHANIC — BLOCK IT SAFELY.

INDIVIDUALLY SUSPENDED REAR WHEELS

As mentioned earlier, some rear end setups bolt the differential housing to the frame, and drive the rear wheels through open (or enclosed) axles, using universal joints to allow flexibility. These assemblies may use a suspension system similar to the long and short arm front suspension. The swing axle suspension in Fig. 15-39, uses pivoted trailing arms with a traverse leaf spring. Note differential carrier is bolted to frame. Other setups utilize coil springs or torsion bars.

TYPE OF DRIVE

The type of drive, Hotchkiss or torque tube, must be considered in the selection of the proper design for the rear suspension. With the open Hotchkiss drive, some method of controlling rear axle housing twist must be incorporated. As the wheels are driven forward, the axle housing tends to

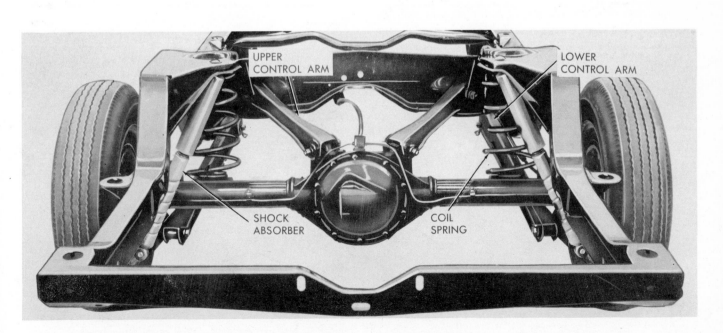

Fig. 15-38. Coil spring rear suspension system. The angle of the upper control arms, and the shock absorbers, prevents excessive side-to-side movement. Fore-and-aft movement, as well as axle housing rotation, is controlled by both the upper and lower arms. (Chevrolet)

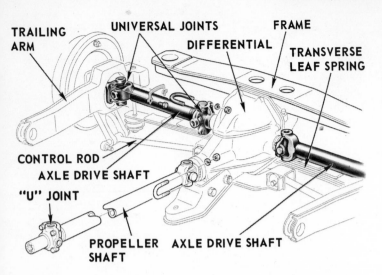

Fig. 15-39. Swing axle type of rear suspension. Note use of transverse leaf spring. U-joints on axle drive shafts allow independent wheel movement in an up-and-down direction. (Chevrolet)

revolve in the opposite direction. As a result, the housing must be held in position by some controlling device. If leaf springs are used, they will control the torque twist, and impart the drive force to the frame. If coil or air springs are used, trailing arms and other control arm designs must be incorporated since the springs themselves are incapable of handling these forces.

With the torque tube drive system, the twisting and driving forces are controlled by the torque tube. However, it is still necessary to use control arms when the coil or air spring is used, because some means must be provided to control side-to-side movement.

STABILIZER BAR

To overcome the tendency for car bodies to lean on corners, a stabilizer bar is usually incorporated in the suspension system. Made of spring steel, the stabilizer passes across the frame and is secured to the frame in two places. The outer bar ends are bent around and back. These outer ends are fastened to the lower control arms (when used in front) through short linkage arms.

Fig. 15-40 illustrates a stabilizer bar and its location on the frame. The two bushing brackets on the forward straight part of the bar fasten it to the frame. The short linkage that connects it to the control arms is shown in an exploded position.

Fig. 15-41 shows one side of a stabilizer bar. Notice how it is connected directly to the lower control arm, thus eliminating additional linkage.

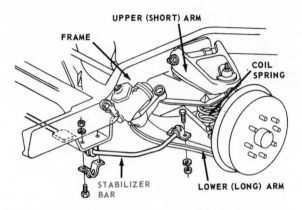

Fig. 15-41. Stabilizer bar connections. Other end (not shown) of bar is mounted in same manner. (Chevrolet)

STABILIZER ACTION

As long as the frame dips up and down in a level manner, the stabilizer bar does not affect its action. When the frame attempts to tip, one end of the bar must be bent down, and

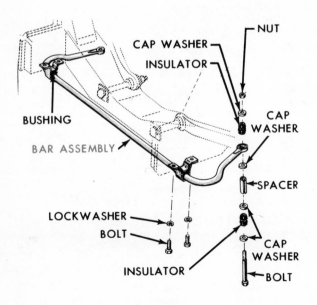

Fig. 15-40. Stabilizer bar. (Lincoln)

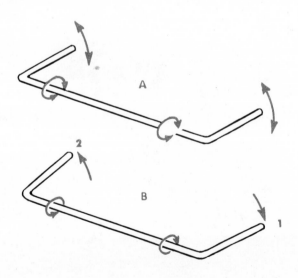

Fig. 15-42. Stabilizer bar action. In A, frame is moving up and down, but is not tipping. Bar offers no torsional resistance. In B, frame is tipping to right, pressing down on 1 and lifting up on 2. Bar must be twisted to allow this action. Resistance to twisting will minimize rolling or tipping motion of frame.

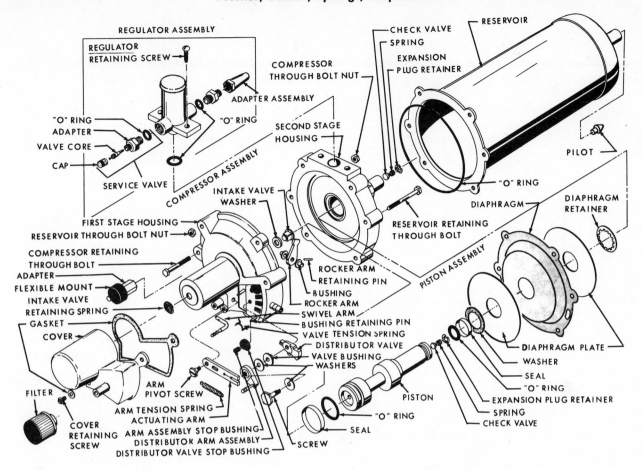

Fig. 15-43. Exploded view of vacuum-operated air compressor used to supply air to the automatic level control system. (Buick)

the other up. This imparts a twisting force to the bar, and it resists the rolling, tipping action. In so doing, it helps to keep the car level. Fig. 15-42.

SPRUNG WEIGHT

Sprung weight refers to the weight of all parts of the car that are supported by the suspension system. The frame, engine, body, etc., all make up the sprung weight.

UNSPRUNG WEIGHT

Unsprung weight refers to all parts of the car that are not supported by the suspension system, wheels, tires, steering knuckles, rear axle assembly, etc.

The ideal situation is to have the unsprung weight as small as possible in relation to the sprung weight. The unsprung weight will then have minimum effect on the riding characteristics of the car.

LUBRICATION

Where chassis grease is needed for the suspension system, grease fittings are provided. Many newer cars have joints that utilize "Teflon" (plastic material) in their construction. This, coupled with special grease and excellent seals, has extended greasing intervals to as much as 30,000 miles in some cases.

Chassis grease is a rather light and sticky type of grease that is compounded to enter fittings readily, lubricate well, adhere to the moving parts and resist water.

Where rubber bushings are used, grease is not necessary. In fact, grease should not be used since it has a deteriorating effect on the rubber.

AUTOMATIC LEVELING CONTROL

The automatic level control is designed to maintain a nearly constant rear curb height (distance from bumper, frame, etc., to ground) regardless of load changes, up to 500 lbs. over the rear axle.

The leveling installation consists of a vacuum-operated air compressor, air tank, pressure regulator valve, leveling valve (height control valve), special shock absorbers and the necessary tubing and hose.

LEVELING CONTROL NEEDS AIR PRESSURE

The air pressure for the level control is furnished by a two-stage, vacuum-operated compressor. Fig. 15-43.

A schematic cross section of the air compressor is pictured in Fig. 15-44. For the intake stroke, the sliding distributor valve (operated by linkage attached to piston) applies engine vacuum to compartment B and atmospheric pressure to

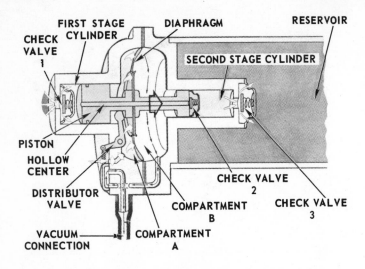

Fig. 15-44. Vacuum-operated air compressor action. As diaphragm forces piston toward reservoir, fresh air is drawn in through check valve 1. Air is compressed in second stage cylinder and forced through check valve 3, into reservoir. Reverse stroke forces air from first stage cylinder through hollow cylinder into second stage cylinder. (Pontiac)

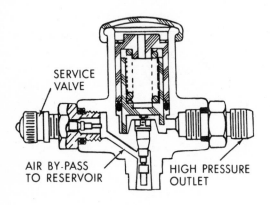

Fig. 15-45. Pressure regulator valve.

compartment A. This forces the diaphragm to move the piston toward the reservoir. This opens check valve 1, and fills the first stage with air.

When the piston reaches the end of the intake stroke, it causes the distributor valve to admit vacuum to compartment A, and atmospheric pressure to compartment B. This moves the piston back toward check valve 1, causing it to close. As the piston continues its movement, it compresses the air ahead of it. The air, being under pressure, travels through the hollow piston opening and check valve 2, allowing the air to enter the second stage cylinder. When the piston has completed its stroke, the distributor valve again reverses the vacuum to the opposite side of the diaphragm (compartment B) and the piston is forced to travel toward check valve 3. As it travels, it does two things:

1. It compresses the air in the second stage cylinder. This closes valve 2 and opens valve 3, allowing the compressed air to flow into the reservoir. At the same time, a fresh charge of air is drawn in through valve 1.

2. The compressor piston cycles (moves back and forth)

until the pressure built up in the reservoir is equal to the pressure generated by the piston in the second stage cylinder. When this point is reached, a balanced condition exists and the compressor will not operate until some air has been used by the system. The balance pressure for the unit in Fig. 15-43, would be between 150 and 275 psi, depending upon altitude. Fig. 15-44.

Air pressure from the reservoir to the system is limited to a maximum of about 125 psi by the pressure regulator valve. This protects the leveling valve and shock absorbers from damage. Fig. 15-45.

LEVELING OR HEIGHT CONTROL VALVE

A leveling valve is bolted to the vehicle frame. An overtravel lever, Fig. 15-46, is connected via linkage to the rear axle

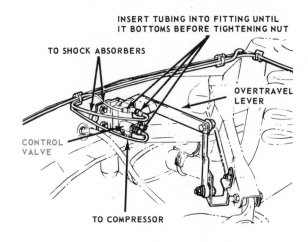

Fig. 15-46. Automatic level control valve is actuated by overtravel lever. (Pontiac)

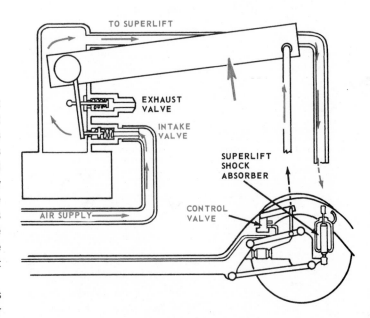

Fig. 15-47. Control valve action when a load is added to car. Car curb height is decreased, causing overtravel lever to move upward to open intake valve and pass compressed air to shock absorbers. (Buick)

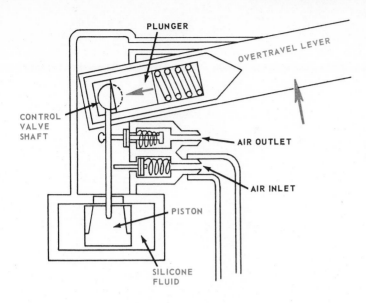

Fig. 15-48. Rapid movement of overtravel lever cannot revolve control valve shaft. In order to turn, shaft must move piston through silicone fluid. Piston will move, but slowly.

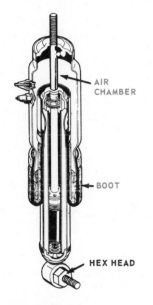

Fig. 15-49. Special Superlift shock absorber used with level control system. Note rubber boot. (Buick)

housing or to the control arm. As weight is added or removed, the distance between the frame and axle changes. This change in distance actuates the overtravel lever, causing the control valve to either admit more air to the shocks (weight added) or to exhaust air from the shocks (weight removed).

Fig. 15-47 illustrates leveling valve action when a load is added to the car. The overtravel lever is forced upward, causing the control shaft arm to open the intake valve. This admits air to the shocks. Shocks will raise car until frame to axle distance is the same as it was before the load was added. At this point, the overtravel lever will cause the valve to close.

To prevent sudden changes in frame-to-axle distance (brought about by bumps) from actuating the control valve, a time delay mechanism is built into the valve. Fig. 15-48. Note that the overtravel lever, during a rapid movement, turns around the control valve shaft without moving it. The spring loaded plunger, which normally rests flatly on the shaft moves back to allow this overtravel lever movement. The delay piston, riding in a cylinder filled with silicone fluid, retards rapid movement of the valve shaft lever. The normal delay is from 4 to 18 seconds.

SPECIAL SHOCKS

Special shock absorbers (called Superlift shocks by several makers) are basically a normal shock absorber with a rubber boot added. The boot forms a seal between the inner and outer shock sections, thus forming an air chamber. Fig. 15-49.

When air from the control valve is admitted to the shock air chamber, the shock is forced to lengthen. This action raises the car. Exhausting air from the shocks lowers the car. A minimum air pressure, around 8—15 psi, is maintained in the shock at all times to prevent boot friction.

A schematic of the entire automatic level control system is illustrated in Fig. 15-50.

SUMMARY

Car bodies must be suspended on some type of spring devices to isolate them, as much as possible, from the irregularities of the road. The body and frame unit must be rigid and strong to provide secure anchorage for the suspension

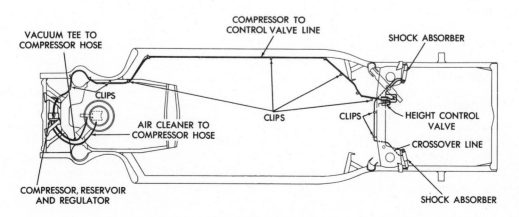

Fig. 15-50. Schematic of automatic level control system. (Cadillac)

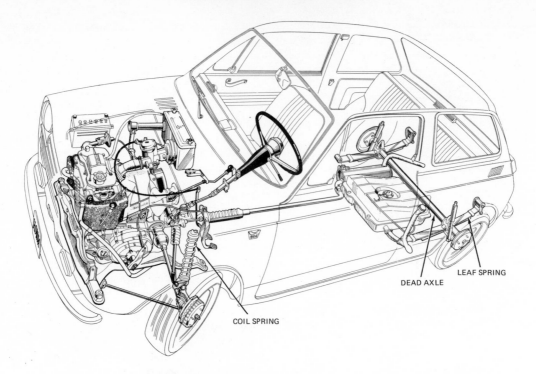

Fig. 15-51. A chassis employing independent front axle and dead axle rear suspension. (Honda)

COIL SPRING

DEAD AXLE

LEAF SPRING

system, and also to provide positive alignment and the securing of all parts.

Some cars use separate frames, while others use the integral frame and body type of construction. Where the separate frame is used, it is constructed of steel channels and cold riveted together. It has various cross members to provide mounting and bracing points.

There are four types of springs used to suspend the car: leaf spring, coil spring, torsion bar and air spring.

Modern practice uses individual wheel suspension on the front of the car. This is accomplished by mounting the wheel assembly on the ends of pivoting control arms. Any of the four springs can be used in this system.

Common practice is still utilizing the solid rear axle housing with coil, leaf or air springs. Various combinations of control arms are used, depending on the type of drive. The individually suspended swing type rear axle is finding increased application, especially among foreign imported cars.

Hydraulic shock absorbers are used to control spring oscillation. Stabilizer bars help prevent tipping or rolling on corners. Track bars keep the body and rear axle in alignment.

Fig. 15-51 illustrates a chassis employing fully independent suspension — both front and rear.

REVIEW QUESTIONS — CHAPTER 15

A thorough background of training will provide the mechanic with a lifetime "suspension system" that will do a fine job of protecting him from the rough spots to be found in the "road" of economic fluctuations. A carpenter without tools is helpless, a car without an engine is worthless, and it is obvious that an auto mechanic without training is worse than no mechanic at all. A poorly trained mechanic will spend more

of his life looking for, rather than working at, a job!

In whatever work you may choose, do your best to be the best — STUDY, PRACTICE AND WORK.

1. What is the basic job of the suspension system?
2. Describe the integral frame and body type of construction.
3. Name three types of separate frames.
4. The semielliptical spring is wound in the shape of a coil. True or False?
5. What type of spring uses spring shackles?
6. List three types of material used as bearings in spring shackles.
7. Both front and rear shackles must be of the hinge type. True or False?
8. Rebound clips are used to control spring rebound. True or False?
9. Describe the method of mounting a semielliptical spring.
10. Are ALL semielliptical springs made with more than ONE leaf?
11. Describe a coil spring.
12. How are coil springs mounted?
13. Explain how the torsion bar spring differs from the coil spring.
14. Both ends of the torsion bar are anchored to the frame. True or False?
15. Is it possible to use torsion bar springs at both front and rear?
16. What is the principle of the air spring?
17. What unit provides air for the air spring system?
18. Explain how the level control system maintains a constant rear curb height, even though weight up to 500 lbs. is added above the rear axle.
19. What is spring oscillation?

20. How does the shock absorber control spring oscillation?
21. Describe the action of the shock absorber on both compression and rebound strokes.
22. The most popular type of shock absorber in use on cars today is the _____ .
23. How are shock absorbers mounted?
24. The solid front axle is generally used on _____ .
25. What is one of the great advantages of the independent front wheel suspension system?
26. Name the basic parts used in the independent front wheel suspension system.
27. Describe the action of the independent front wheel suspension system.
28. Explain the action of the ball joints used on the outer ends of the suspension control arms.
29. List two very real dangers when working on the suspension system.
30. Why is a stabilizer bar necessary?
31. How does the stabilizer bar work?
32. Explain the difference between sprung and unsprung weight.
33. Rubber bushed shackles should never be oiled. True or False?
34. Some of the newer type of suspension joints can forego greasing for periods up to 30,000 miles. True or False?
35. If you are planning to become a mechanic, or even if you are just "interested" in learning about cars, are you doing all you can to learn?

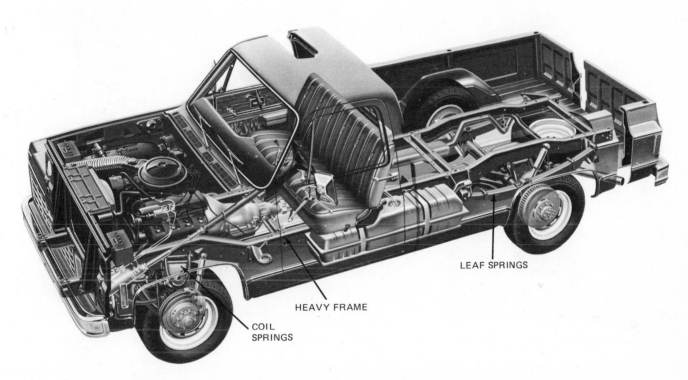

LEAF SPRINGS

HEAVY FRAME

COIL SPRINGS

Fig. 15-52. Typical pickup truck construction. This two-wheel drive, 3/4 ton pickup utilizes a heavy duty frame with coil springs in the front and leaf springs in the rear. (Chevrolet)

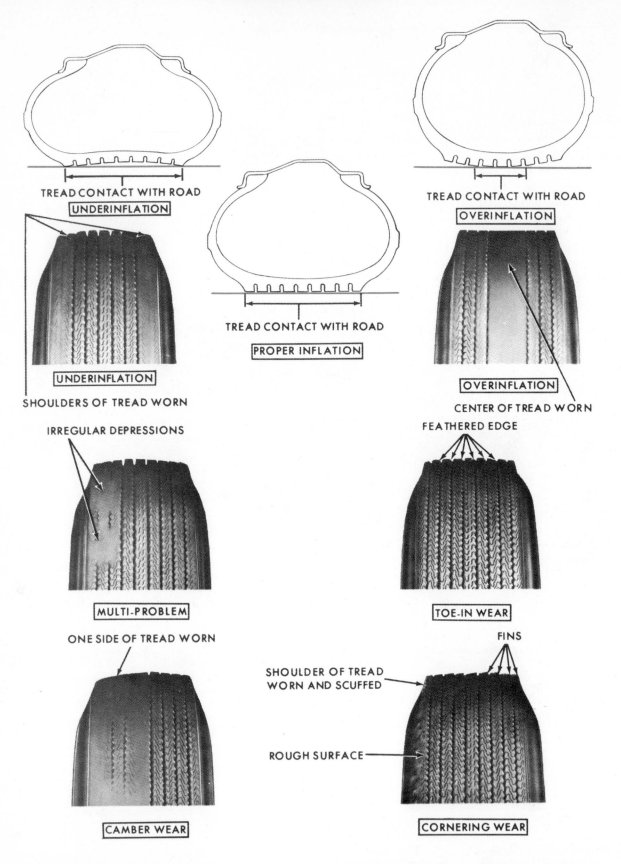

TREAD CONTACT WITH ROAD
UNDERINFLATION

UNDERINFLATION
SHOULDERS OF TREAD WORN

TREAD CONTACT WITH ROAD
PROPER INFLATION

TREAD CONTACT WITH ROAD
OVERINFLATION

OVERINFLATION
CENTER OF TREAD WORN

IRREGULAR DEPRESSIONS

MULTI-PROBLEM

FEATHERED EDGE

TOE-IN WEAR

ONE SIDE OF TREAD WORN

CAMBER WEAR

FINS
SHOULDER OF TREAD WORN AND SCUFFED

ROUGH SURFACE

CORNERING WEAR

Some common tire wear patterns. Tires must have correct air pressure and be properly aligned and balanced.   (Pontiac)

302

Chapter 16

WHEELS, HUBS, TIRES

The drop center, steel disc type wheel is in common use today. Fig. 16-1. It is made in two sections. The outer part forms the rim. The center section (often called "spider") is riveted or spot welded to the rim. The center portion has four, five or sometimes six holes used in mounting the wheel to the wheel hub.

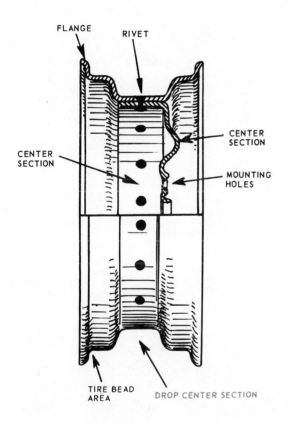

Fig. 16-1. Typical drop center wheel. Center section is riveted or spot welded to rim. Note drop center section to permit tire removal.

The center section of the rim is "dropped" to allow the tire to be removed. When one bead of the tire is placed in the dropped area, it is then possible to pull the opposite bead off the rim.

Some rims have safety ridges near the lips of the rim. In the event of a blowout, these ridges tend to keep the tire from moving into the dropped center, then coming off the wheel. Fig. 16-2.

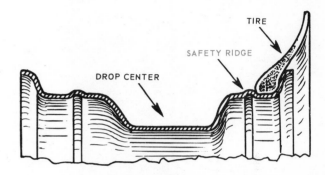

Fig. 16-2. Safety rim. Note safety ridge on each side.

The wheel must be manufactured to relatively close tolerances. When mounted, it must run within acceptable limits. Fig. 16-3.

WHEEL BOLTS TO HUB

The wheel is bolted to a hub, either by lug bolts that pass through the wheel and thread into the hub, or by studs that protrude from the hub. In the case of studs, special nuts are required.

Some cars use left-hand threads (nut or bolt turns counter-clockwise to tighten) on the driver's side, and right-hand threads (turn clockwise to tighten) on the passenger's side. Other cars use right-hand threads on both sides.

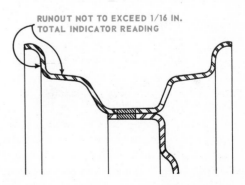

Fig. 16-3. Wheel must be accurate. Left-hand arrow indicates surface to be checked to determine side movement. Right-hand arrow indicates area checked for out-of-roundness. Wheel is spun and these areas are checked with a dial indicator that reads movement in .001 inch. (American Motors)

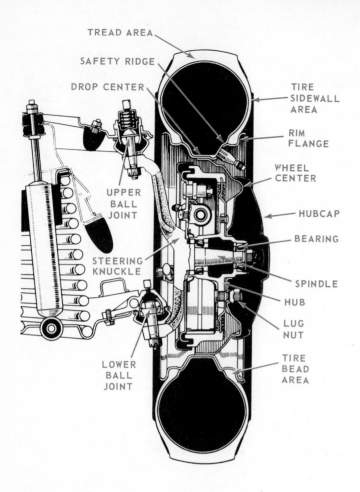

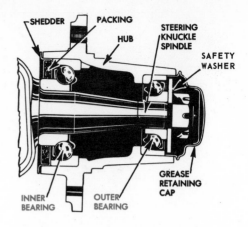

Fig. 16-5. Ball bearing front wheel hub.

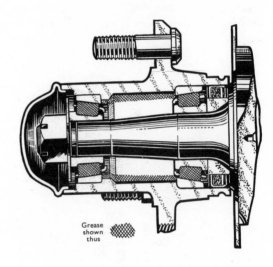

Fig. 16-4. Front wheel and tire assembly. Safety rims are used on this wheel. Notice how wheel is fastened to hub. Steering knuckle swivels on ball joints. (Toyota)

Fig. 16-6. Tapered roller bearing front wheel hub. Note that inner part of hub is NOT completely filled with grease. (Hillman)

CAUTION!

WHEN A WHEEL IS REMOVED AND REPLACED, MAKE CERTAIN THE WHEEL LUGS, OR NUTS, ARE TIGHTENED SECURELY OR THE WHEEL MAY COME OFF IN SERVICE!

A complete wheel, tire and hub assembly is shown in Fig. 16-4. Study the illustration carefully.

FRONT WHEEL HUBS

Front wheel hubs revolve about the steering knuckle spindle on either ball or tapered roller bearings. A nut on the end of the spindle draws the bearings up and removes excessive play between the bearings and hub.

A special safety washer is used between the nut and the bearing. It passes around the spindle and has a metal tab that rides in a groove in the threaded area of the spindle. When the hub and bearing spin, the rotating motion is not transferred to the nut, but is stopped by the washer since the tab prevents the washer from turning. A cotter pin is always passed through the castellated (slotted) nut and the spindle.

Study the hubs shown in Figs. 16-5, 16-6 and 16-7. Fig. 16-7 shows the spindle nut, cotter pin and safety washer.

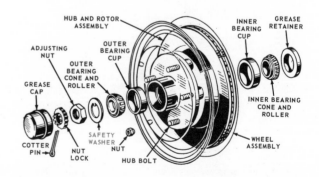

Fig. 16-7. Exploded view of a typical front wheel assembly. Safety washer, used between outer bearing and spindle nut, is found on all cars. It must always be in place. (Ford)

QUICK—CHANGE WHEELS, SPECIAL HUBS

Some sports cars feature quick-change wire wheels. The outer rim is similar to the steel disc type but, instead of a steel disc center, wire spokes are used. The wire wheel slides onto a splined hub and is retained by a single knock-off spinner nut. The wheel and tire assembly can be changed merely by striking the nut flange and spinning it off. The wheel then pulls off. To replace, merely reverse the procedure. Fig. 16-8 illustrates this type. Notice the splines on the hub and inside the wheel center section. The knock-off nut is also shown.

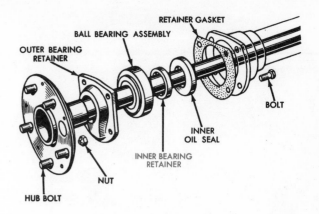

Fig. 16-9. Integral axle flange hub for rear wheel. Inner bearing retainer prevents axle from slipping through bearing and out of differential housing. (Ford)

Fig. 16-8. Quick-change wheel and hub. Single knock-off spinner nut holds wheel on hub. (Austin)

FRONT HUB LUBRICATION

Wheel hub bearings use a stringy, fibrous wheel bearing grease. Fig. 16-6 shows the location of this grease. Notice that the hollow hub is not completely filled. The bearings are thoroughly packed.

CAUTION!

ALWAYS INSTALL SAFETY WASHER AND PUT IN A NEW COTTER PIN WHEN REINSTALLING A FRONT WHEEL HUB.

REAR HUBS

One type of rear wheel hub is formed by upsetting the end of the axle itself. The wheel is then bolted directly to the axle flange. Fig. 16-9.

The other setup uses a separate hub that is secured to the tapered axle end by means of a nut and key. The key fits in a groove cut in both the hub and axle. The key is used to prevent the hub from slipping on the axle. Fig. 16-10.

TIRES AND TIRE CONSTRUCTION

Tires serve two main functions. They provide a CUSHION-ING ACTION that softens the jolts caused by road irregularities. In this respect, they are actually serving as part of the suspension system. They also provide PROPER TRACTION. Traction enables the car to drive itself forward, provides a

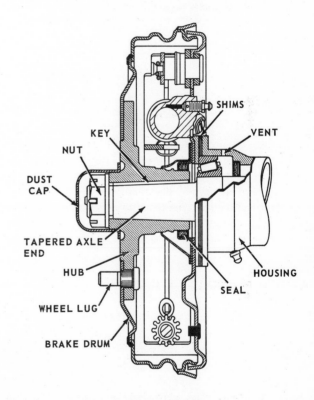

Fig. 16-10. Detachable rear axle wheel hub. Hub is kept from turning on tapered axle end by a key. (Jeep)

means of steering, and allows reasonably fast stopping. Therefore, quality tires are essential to safe vehicle operation.

Tires are constructed of various types of rubber, fabric and steel wire. The fabric casing plies (made of rayon, nylon, polyester, etc.) provide the shape and lend strength to the tire. The plies are impregnated (filled) with rubber. Fig. 16-11.

The plies are wrapped around steel bead wires, which provide strength to prevent the tire from opening up and leaving the wheel. Filler strips reinforce the bead section.

The fabric plies may be applied in a crisscross (bias) pattern. This is termed BIAS PLY construction. A cutaway of a typical bias type tire is shown in A, Fig. 16-12.

Another construction method, called BELTED BIAS, uses a

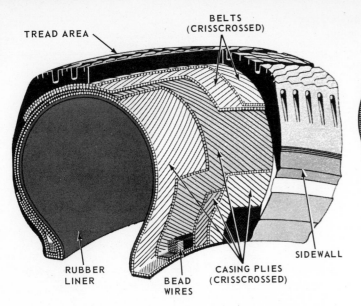

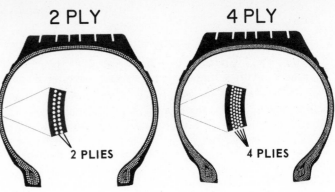

Fig. 16-13. Two-ply/four-ply rating tire has same load capacity as four-ply/four-ply rating. Note how much thicker two-ply cords are than those in four-ply.

Fig. 16-11. Typical belted bias tire construction. This specific tire uses four sidewall plies (bias arrangement) and two belts beneath tread. (Goodrich)

bias ply sidewall (two or more plies) plus several layers of tread reinforcing plies directly beneath the tread area. These plies form a continuous "belt" around the tire. See Fig. 16-11 and B, Fig. 16-12.

Another tire rapidly growing in popularity is the RADIAL PLY type. This design arranges the casing ply, or plies, so that the cords lie at right angles to the tire tread center line. Several layers or belts of reinforcing plies (fabric or steel) are placed beneath the tread section. The radial tire permits flexing with minimal tread distortion, gives longer tread life, better traction, softer ride at medium and high speeds, and has less rolling resistance. See C, Fig. 16-12.

PLY RATING AND LOAD RANGE

Ply rating is a general indicator of load-carrying ability. It does not necessarily indicate the actual number of plies used. A two-ply/four-ply rating tire has the same load-carrying ability as a tire utilizing four-ply/four-ply rating construction.

Many cars today are equipped with two-ply/four-ply rating tires. Fig. 16-13 illustrates a cross section of both a two-ply/

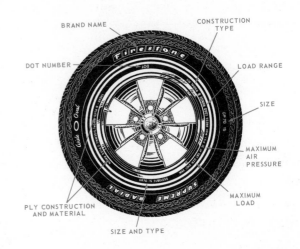

Fig. 16-14. All tires must be marked showing maximum load carrying capacity and pressure, size, load range, ply construction and material identification, manufacturer's number and construction type. (Firestone)

four-ply rating and a four-ply/four-ply rating tire. Note thicker cords in the two-ply tire.

LOAD RANGE is an indicator of a tire's load capacity as well as number of plies and as such, maximum air pressure that may be used. Load range B indicates a four-ply, C a six-ply and D an eight-ply.

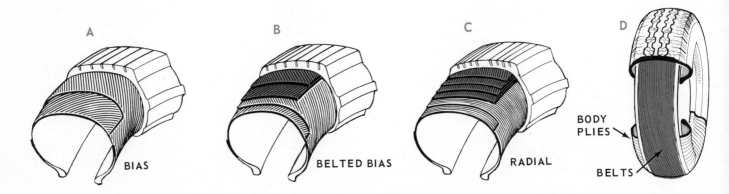

Fig. 16-12. Typical tire construction. A—BIAS PLY (sidewall plies crisscrossed, no belts). B—BELTED BIAS (sidewall plies crisscrossed, belts beneath tread area). C—RADIAL (sidewall plies parallel with each other and at right angles to tread center line, belts under tread area). D—Radial showing relationship between belts and body plies. (Pirelli)

In addition to the load range, all tires must be marked with the size, maximum load and air pressure, DOT (Department of Transportation) number, ply and material identification, construction type (radial, belted-bias, etc.). Fig. 16-14.

Tire treads and sidewalls are made of rubber. Entire tire is placed in mold and vulcanized (heated until rubber is sticky and all parts bond together). The mold imparts sidewall and tread design. Fig. 16-11 shows details of tire construction.

TUBELESS TIRES

The modern automobile tire is tubeless (no rubber inner-tube is used). The tire and rim form a leakproof unit. The tubeless tire often has a thin bonded rubber lining. The bead area has tiny sealing grooves. Tire and rim must be in good condition to hold air properly. Fig. 16-15.

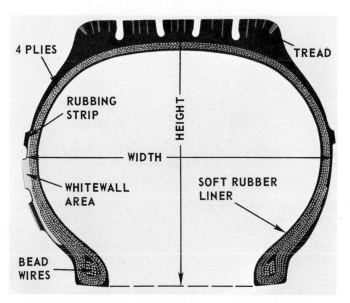

Fig. 16-15. Tubeless tire construction. Soft rubber liner prevents air leaks into plies. This particular tire is a four-ply/four-ply rating. Note width in relation to height, indicating a very low profile design.

Air is admitted to the tubeless tire by means of a valve stem placed in the rim. Fig. 16-16. Valve stem core action is shown in Fig. 16-17.

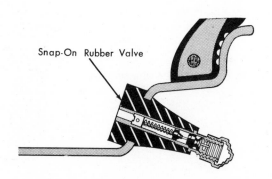

Fig. 16-16. Valve stem construction. Stem body is of rubber construction. (Cadillac)

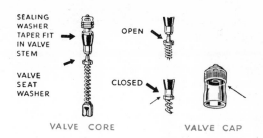

Fig. 16-17. Valve core action. Valve cap acts as second seal against air leakage as well as preventing entry of dust. (Hillman)

TIRE SIZE

For many years, tire size was related to cross-sectional width (width measured from outside of one sidewall to outside of other) and diameter across beads when mounted on rim.

The advent of the low-profile tire has made it necessary to alter tire size designations somewhat so as to indicate correct load ratings.

Super, low-profile or "wide oval" tires (extremely wide in relation to section height) use a series designation that indicates the relationship between section height and width. Section height and width is illustrated in Fig. 16-15. A "70 series" means that the average section height is 70 percent of the width. A letter prefix (D,E,F,G, etc.) denotes size range — on E being a larger section than a D, and F a larger section than on E, G larger than F, etc. Some of the current sizes in tires of this type are D70-14 (D, size section, height 70 percent of section width, and 14 in. across the beads), E70-14, H70-15, etc. Note the difference in the super, low-profile tire, as opposed to the more conventional tire. Fig. 16-18.

SPECIAL TIRES

There are various special safety, and high-speed tires available. The high-speed tire often utilizes steel wires in the

Fig. 16-18. Super, low-profile or "wide oval" type tire, shown in A, has a much greater section width in relation to height than tire shown in B. (Goodyear)

outer ply-tread area to control distortion caused by centrifugal force. One type of safety tire uses an inner two-ply nylon shield. Both inner and outer areas are filled with air. When the outer chamber is punctured, the inner chamber will support the tire for a number of miles.

Tires with various treads are available. Some are designed for normal highway driving, others for snow, mud, ice, etc.

TIRE RUNOUT

Tires must run reasonably true to prevent vibrations and unbalance. Fig. 16-19 shows one set of runout limits. Notice that tire runout is not as critical as wheel runout. As the tire is flexible, it can absorb some of its own inaccuracy.

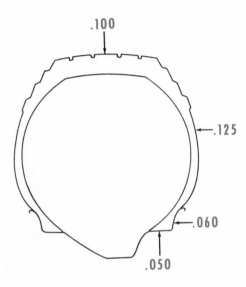

Fig. 16-19. Wheel and tire runout. (Oldsmobile)

TIRE BALANCE

To operate smoothly, it is essential that a tire and wheel be properly balanced. Tire unbalance will cause poor ride, excessive tire wear and result in steering and suspension unit wear due to the continual shaking.

The two major effects of unbalance are WHEEL TRAMP (tire and wheel hopping up and down), and WHEEL SHIMMY (shaking from side to side).

Tires must be in STATIC and DYNAMIC balance.

STATIC BALANCE

When a tire is in proper static balance, the weight mass will be evenly distributed around the axis of rotation. If the wheel is raised from the floor and spun several times and always stops in a different place, it is statically balanced.

If the wheel revolves until one side is down, it is obvious that this side is heavier than the opposite or upper side. The tire and wheel is statically unbalanced. Static unbalance is corrected by clipping lead weights to the rim edge opposite the heavy side. Fig. 16-20.

Fig. 16-20. Static balance weight. If more than one weight is necessary, it is essential that they be evenly distributed on each side of light area (indicated by small vertical line).   (Plymouth)

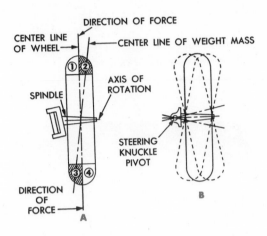

Fig. 16-21. Dynamic unbalance. A is a front view and B is a top view. (Plymouth)

DYNAMIC BALANCE

Dynamic balance is achieved when the center line of the weight mass is in the same plane as the center line of the wheel.

Study Fig. 16-21. You will see that the center line of the weight mass is not in the same plane as the center line of the wheel. When the wheel spins, centrifugal force will attempt to bring the weight mass center line into the wheel center line plane.

Weight mass, Detail 2, will attempt to move to the left. Weight mass 3 will attempt to move to the right. As the wheel revolves, the weight masses will reverse their positions: 2 will move to the right; 3 to the left. Each half turn of the wheel will reverse weight mass position. The resultant struggle, as the weight masses attempt to enter the wheel center line plane, will cause shimmy or wobble. The wobble will cause tire, suspension and steering wear. Fig. 16-21.

Dynamic unbalance is corrected by adding lead wheel weights to both sides of the rim, until weight masses 1 and 4 equal weight masses 2 and 3. When the weight mass center line is in the same plane as the wheel center line, the assembly is dynamically balanced. Fig. 16-22.

PROPER TIRE INFLATION IS IMPORTANT

Car manufacturers determine the proper tire size and pressure after extensive research. Car weight, type of suspension system and horsepower, are all taken into consideration.

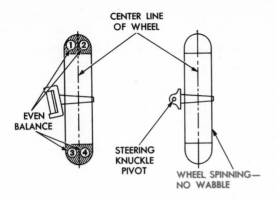

Fig. 16-22. Dynamic balance. Note absence of wheel wobble when dynamic balance is achieved.

A tire that is too small will be overloaded. A tire that is too large will add to the unsprung weight, upsetting the suspension system balance.

Proper inflation pressure is important in four ways: traction, riding comfort, fuel mileage and tire mileage. An over-inflated tire will ride hard, produce excessive wear in the tread center area and will not present the full tread for traction use.

Fig. 16-23. Inflation effect on tread contact. You can see that full contact is only possible when tire is correctly inflated. (Ford)

Under-inflation will wear the outer edges of the tread area. It will cause tire heating, produce mushy steering, and will not properly present the full tread area for traction. Fig. 16-23, shows how the tread contacts the road when under-inflated, properly inflated and over-inflated.

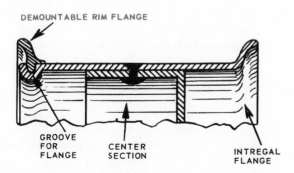

Fig. 16-24. Demountable rim flange. If not properly placed, demountable flange may fly off when inflating tire.

WHEN INFLATING A TIRE MOUNTED ON A RIM WITH A DEMOUNTABLE RIM FLANGE, (SUCH AS USED ON MANY TRUCKS), ALWAYS PLACE A CHAIN OR CAGE AROUND THE TIRE AND WHEEL ASSEMBLY. FACE THE DEMOUNTABLE RIM FLANGE DOWN OR AWAY. MAKE CERTAIN THE RIM IS SECURELY IN THE RIM GUTTER. WHEN INFLATING SUCH SETUPS, THESE RIMS HAVE BEEN KNOWN TO FLY OFF WITH LETHAL FORCE. DON'T TAKE CHANCES. Fig. 16-24.

Fig. 16-25 graphically illustrates the effect of inflation on tire wear.

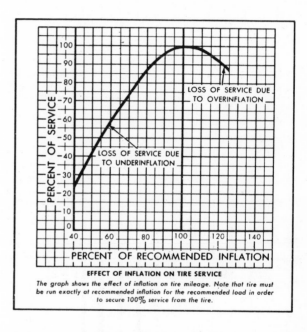

Fig. 16-25. Inflation wear graph. Notice that over-inflation is also detrimental to tire wear. (Lincoln)

HIGH SPEEDS AND OVERLOADING PRODUCE HEAT AND EXCESSIVE WEAR

The relationship between car speed, overloading and tire temperature is shown in Fig. 16-26. Note how the heat goes up faster when a given tire is subjected to a 20 percent overload as opposed to the normal load. When tires operate at 250-275 deg., they lose a great deal of strength and wear resistance.

The effect of heat on tire life is shown in the graph. Fig. 16-27. Normal tire wear is figured for 50 mph. You can see that a motorist driving at 80 mph will suffer an 80 percent loss of tread life.

TIRES SHOULD BE ROTATED

To compensate for inevitable variations in individual tire wear, it is advisable to rotate the tires about every 5,000 miles.

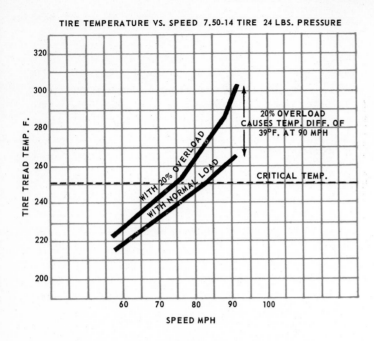

TIRE TEMPERATURE VS. SPEED 7.50-14 TIRE 24 LBS. PRESSURE

Fig. 16-26. Speed, load and temperature graph. At higher speeds, a small speed increase causes a large increase in tire temperature. Note how overload raises temperature and reduces tire life. (Rubber Mfgs. Assoc.)

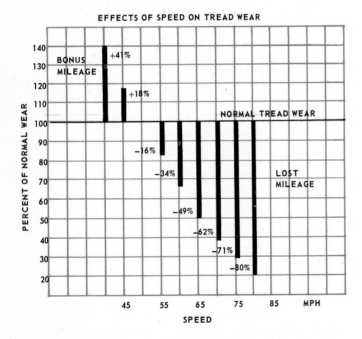

EFFECTS OF SPEED ON TREAD WEAR

Fig. 16-27. Speed and tire life graph. High speeds really "grind off" tread rubber. (Rubber Mfgs. Assoc.)

Fig. 16-28, illustrates a recommended rotation pattern for both 4 and 5 tire rotation.

SUMMARY

The drop center, steel disc type wheel is commonly used on modern cars. Wheel is stamped from steel stock, riveted together and fastened to wheel hub using four to six bolts.

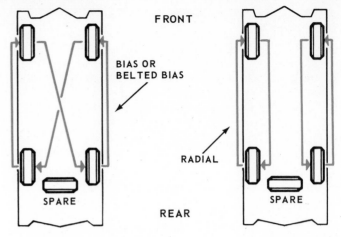

Fig. 16-28. Recommended tire rotation pattern for: Left—Four tires, bias or belted bias. Right—Four tires, radial. (Cadillac)

The front wheel hubs rotate on ball or tapered roller bearings, and are lubricated by packing the bearings with wheel bearing grease. Rear wheels are bolted to an integral or detachable hub.

A few sports cars use quick-change wheels. These require special hubs.

Tires are an important part of the suspension system. They also provide the traction necessary for driving and braking.

Rubber, fabric and steel wires are used in tire construction. Special safety-type tires are available. A wide variety of tread designs are manufactured.

Most tires are the tubeless type in which an air seal is provided by the tire and rim.

The most popular passenger car tire is of two-ply/four-ply rating construction. Ply rating is determined NOT by the number of actual plies but by load-carrying ability.

Tires and wheels must run true, and they must be balanced statically and dynamically. Special lead wheel weights are used in balancing a tire and wheel assembly.

Proper tire inflation is vital to good steering, riding quality and tire life.

High speeds are ruinous to tire life. Excessive speed promotes heating with a resultant loss of tire strength and wearing quality. Safe vehicle operation is also drastically reduced.

Tires should be rotated every 5,000 miles.

Always double check when mounting wheels on a car to make certain they are bolted on securely. If inflating truck tires — BEWARE OF THE RIM FLANGE!

REVIEW QUESTIONS — CHAPTER 16

1. Name the type of wheel most widely used today.
2. How is this wheel constructed and what material is used?
3. What is a safety rim? How does it work?
4. What safety precaution should be used when mounting wheels? (Other than supporting the car in a safe manner.)
5. What is the purpose of a wheel hub?

6. What is the major difference between a front wheel hub and a rear hub?
7. Hub bearings, when used, are either of the _____ or _____ type.
8. What different feature is offered by the quick-change hub?
9. Hubs are lubricated by using lightweight engine oil. True or False?
10. Name the three major parts of a tire.
11. Wire beads are used to give a tire better traction. True or False?
12. A four-ply rating means that a tire contains four actual layers of cord or plies. True or False?
13. How can a tire without an innertube hold air?
14. Air is injected into the tire through a valve stem and core assembly. True or False?
15. Describe two types of special tires.
16. Static unbalance causes shimmy. True or False?
17. Dynamic unbalance is when the center line of the weight mass coincides (lines up) with the center line of the wheel. True or False?
18. List four reasons why tires must be properly inflated.
19. How can a truck rim flange injure and even kill a careless mechanic?
20. How does high speed affect tire life?
21. Draw a diagram showing how tires should be rotated.
22. Tires should be rotated every _____ miles.
23. Draw a cross section of a typical tire and name the various parts.
24. Describe the difference between the construction of a conventional tire and that used in the "radial."
25. A D70-14 is a larger tire than the F70-14. True or False?

Remember — You too will go faster and further when your training is properly "balanced." Theory, practice, experimentation are all important and must not be neglected. A sound education made up of study in many fields, such as Math, English, History, etc., is essential to every student of Auto Mechanics. Learn all you can, while you can.

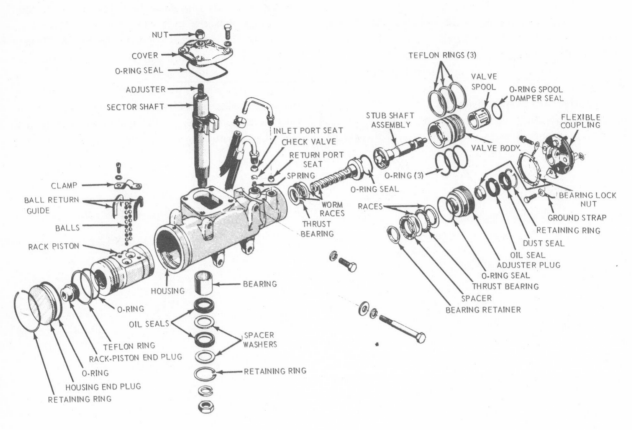

Exploded view of a power steering gear utilizing a rotary spool valve.   (Ford)

Chapter 17

FRONT WHEEL ALIGNMENT, STEERING SYSTEMS

The front wheels turn when the spindle steering knuckle units swivel, either on KINGPINS or on BALL JOINTS. Figs. 17-1 and 17-2 illustrate both the kingpin (sometimes called spindle bolt) and the ball joint. Modern practice on passenger vehicles favors the use of the ball joint.

ALIGNMENT IS IMPORTANT

It is not enough to merely place the front wheels on hubs, stand them up straight and devise a method of swiveling them from left to right. The car could be driven, but it would steer poorly. At higher speeds, it would become dangerous to

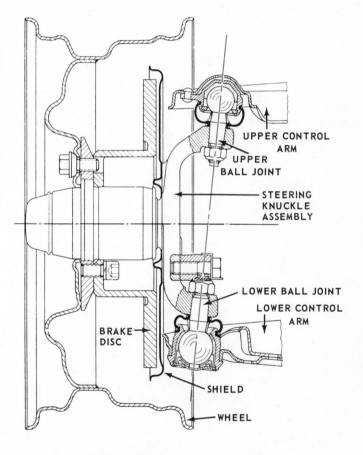

Fig. 17-2. Ball joint steering knuckle assembly. No kingpin is used in this setup. The steering knuckle swivels on two ball joints. Note center line drawn through the ball joints.   (Mercedes Benz)

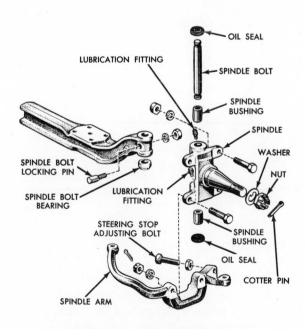

Fig. 17-1. Exploded view — spindle assembly. This unit utilizes a solid I-beam axle. Spindle assembly swivels on a kingpin (spindle bolt). Currently, this construction is limited to some trucks.   (Ford)

handle; tire life would be short. To secure easy steering, sure handling, smooth operation and good tire wear, several front wheel alignment factors must be considered.

CASTER

Everyone is familiar with the common caster used on furniture. You may have wondered why the little wheel always swivels around and trails obediently behind a chair or dresser leg. The reason is simple.

Referring to Fig. 17-3, you will see that the center line about which the caster swivels strikes the floor ahead of the caster wheel. If the caster attempts to move to either side, the

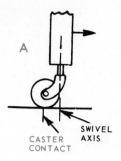

Fig. 17-3. Chair leg caster. Center line of swivel axis contacts floor ahead of caster wheel contact.

point of resistance between the caster and the floor will be to the side of the swivel center line. As the caster is encountering some resistance to forward motion, it will not move until the moving weight mass actually drags it forward. It cannot be dragged forward until all the "slack" is taken up. In this case it will not move straight forward until it is directly in back of the swivel center line.

Fig. 17-4 illustrates how resistance to forward motion forces a trailing object (caster wheel) to line up behind the moving swivel line.

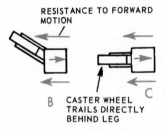

Fig. 17-4. Caster effect. Table leg in B starts to move to right. Caster wheel is off to one side and encounters a rolling resistance that forces it to line up and trail directly behind table leg in C.

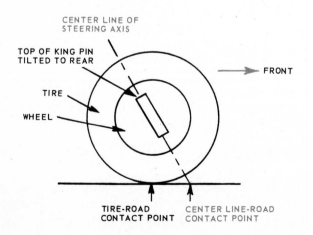

Fig. 17-5. Positive caster. Steering axis (swivel axis) strikes road in front of tire-road contact (caster contact, Fig. 17-3). Wheel will trail behind steering axis center line-road contact. (Tilt is exaggerated.)

CASTER ANGLE AS APPLIED TO STEERING SYSTEM

By tilting the top of the kingpin (or, in case of ball joint, top of steering knuckle) toward the rear of the car, the center line of the swivel, or steering point, will be in front of the point at which the tire encounters the road. As this places the point of resistance to forward motion behind the steering center line, the wheel is forced to track behind, automatically lining up the moving car. This is referred to as POSITIVE CASTER. Fig. 17-5.

Positive caster tends to force the wheels to travel in a straight ahead position. It also assists in recovery (wheels turning back to straight ahead position after making a turn). On late model cars, there is often little or no positive caster. Positive caster makes it more difficult to turn the wheels from the straight ahead position than when no caster angle is present.

Another effect of positive caster is a mild tipping effect when cornering. When making a right-hand turn, the right-hand wheel will cause the steering knuckle to raise slightly, while the left-hand wheel will allow the knuckle to lower. This creates a tipping effect, and it is obvious that raising the side of the car on the inside of the turn will have an adverse effect on its cornering ability. When the wheels are turned to the left, the left side raises and the right side drops.

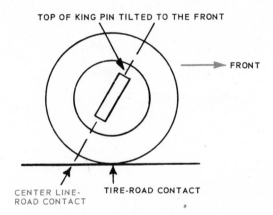

Fig. 17-6. Negative caster. Steering axis center line strikes road behind tire-road contact point. (Tilt is exaggerated.)

NEGATIVE CASTER

To ease turning, many cars employ a negative caster angle (tipping top of steering knuckle to front of car). This will ease steering and cause a mild banking effect when cornering (side of car on inside of turn will drop and outside will raise). Proper tracking is still provided by kingpin or steering knuckle inclination (to be discussed later). Fig. 17-6.

CAMBER

Positive front wheel camber is provided by tipping the tops of the wheels out. When viewing the car from the front, the top of the wheels are farther apart than the bottom. In short, the center line (C/L) of the wheel is no longer in a vertical plane (straight up and down). Fig. 17-7.

If the tops of the wheels are closer together than the bottoms, the camber is considered negative.

Positive camber places the center line of the wheel nearer the center line of the kingpin, or steering axis, where both intersect the road. This assists in reducing the tendency of the front wheels to toe-out (spread apart at front) and materially lessens road shock to the steering system. Most cars use a small amount of positive camber (less than one degree).

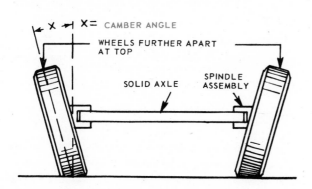

Fig. 17-7. Positive camber. When viewing car from front, tops of wheels are farther apart than bottom. (Camber angle is exaggerated.)

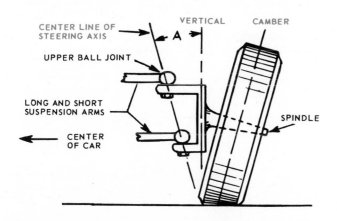

Fig. 17-8. Steering axis inclination. Front view, driver's side wheel and spindle assembly. Note how top ball joint is closer to center of car than bottom ball joint. This tips steering axis center line as shown. Steering axis inclination angle is shown at A. (Angles are exaggerated.)

KINGPIN OR STEERING AXIS INCLINATION

The top ball joints of the steering knuckles are closer together at the top than the bottom. The steering axis (view from front of car) places the center line of the steering ball joints nearer to the center line of the wheel. Fig. 17-8.

When the center line of the wheel is to the outside of the center line of the steering axis (where they intersect road), the wheels tend to toe-out. This is caused by the road-tire resistance pushing back on the spindle, causing it to swivel backward on the ball joints. Fig. 17-9.

When the center line of the wheel intersects the road at a point inside the intersection of the steering axis center line, the wheels tend to toe-in (fronts of tires are closer together than backs). Fig. 17-10.

Steering axis inclination helps to bring the center line of the wheel and the steering axis center line closer together where they intersect the road surface. This also causes suspension shocks to be transmitted to and absorbed by the heavy inner spindle and knuckle assembly. Fig. 17-11.

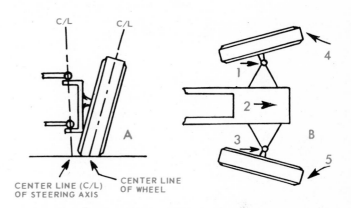

Fig. 17-9. Toe-out effect. A is front view, showing center line (C/L) of wheel striking road to outside of point where C/L of steering axis strikes or intersects road. B is a top view showing how wheels are forced to toe-out when vehicle moves forward. Arrows 1, 2 and 3 represent direction of force pushing car forward. Arrows 4 and 5 represent forces of resistance to forward motion of wheels. As resistance forces are outside of steering axis-road intersection, wheels toe-out. (Angles are exaggerated.)

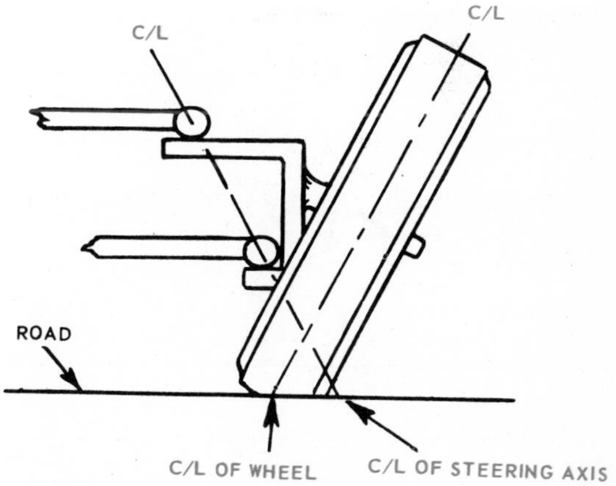

Fig. 17-10. Toe-in effect. You will note that C/L of wheel now intersects road inside of point where steering axis C/L strikes road. Resistance to forward motion will now cause wheel to swing or toe-in when the car moves forward. (Angles are greatly exaggerated.)

Steering axis inclination also helps the car wheels follow the straight ahead position. This tracking effect is caused by the fact that whenever the wheels are turned from the straight ahead position, the car is actually lifted up. Remember that the wheel spindle, when turned, is moved about the steering axis. As the axis is tilted, the spindle will move in a downward direction. Since a downward movement of the spindle is impossible (it is supported by wheel assembly), the steering knuckle is forced upward. This will raise the car a small amount. The effect is similar on both sides.

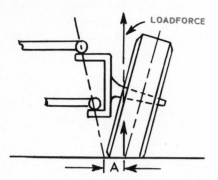

Fig. 17-11. Suspension shock transfer. Load force, transmitted in a vertical plane, will now strike heavy inner end of spindle. Camber and steering axis inclination bring this about. A shows that distance between wheel center line and steering axis center line intersection is reduced by camber and steering axis inclination. Resistance to forward motion is now closer to steering axis, reducing shock transfer to tie rods and steering system. (Angles are exaggerated.)

The weight of the car will tend to force the spindles to swivel back, returning the wheels to the straight ahead position. Fig. 17-12.

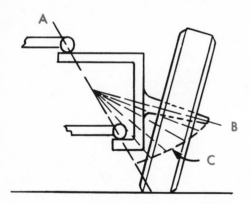

Fig. 17-12. Lifting effect of steering axis inclination. A—Steering axis C/L. B—Spindle C/L. C—Path of spindle end as wheel is turned. Note that as spindle revolves, it travels downward. Since it cannot come nearer road (wheel holds it up), steering axis area will be lifted. This action affects both wheels regardless of turning direction.

STEERING KNUCKLE ANGLE

The steering knuckle angle is determined by the arc formed between the steering axis and the center line of the spindle. Fig. 17-13 illustrates the steering knuckle angle, as well as positive camber and the steering axis. You will notice that the steering axis inclination angle, and the camber angle, are directly dependent upon each other.

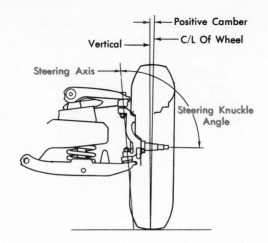

Fig. 17-13. Steering knuckle angle. If steering axis angle is changed, camber angle must also change. (Cadillac)

TOE—IN

Toe-in is accomplished by placing the front of the wheels closer together than the back, when viewed from the top.

When a wheel is cambered, the tire engages the road at an angle. Since the tire will adapt itself to the road, the rolling edge of the tire will not be at right angles to the center line of the wheel. This will cause it to roll as though it were cone-shaped, tending to roll outward while moving forward. Fig. 17-14.

To compensate for this roll out (as well as toe-out tendencies produced when steering axis intersects road inside

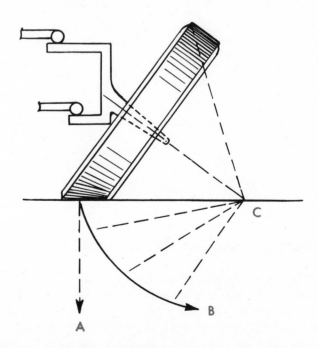

Fig. 17-14. Camber toe-out tendency. A—Straight ahead path tire should take. B—Path tire actually takes. When wheel is cambered (exaggerated), surface of tire will engage road at an angle. Tire will adapt itself to road, producing cone shape with apex at C. Tire will then roll on path B just as though it were a cone.

C/L of wheel), the wheels must be adjusted for toe-in.

This toe-in, generally 1/8 to 1/4 in., compensates for the wheel toe-out tendencies plus any wear or play in the steering linkage that would allow additional toe-out. Proper toe-in will allow the tires to move forward without a scrubbing, scraping action between tire and road.

Excessive toe-in, or toe-out, will cause rapid tire wear as shown in Fig. 17-15.

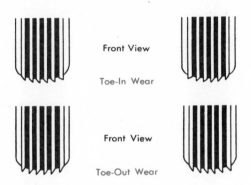

Fig. 17-15. Toe-in and toe-out wear. (Cadillac)

TOE—OUT ON TURNS

When a car turns a corner, the inner wheel must turn on a shorter radius (smaller circle) than the outer wheel. As a result, the wheels must automatically toe-out on turns to allow the inner wheel to cut more sharply. This essential action is accomplished by bending both steering arms so they angle slightly toward the center of the car. Fig. 17-16.

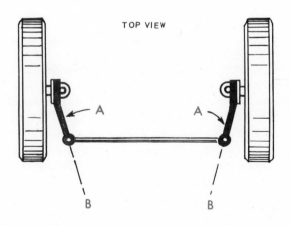

Fig. 17-16. Steering arms. Steering arms A are angled inward. If center lines B were continued, they would meet somewhere near differential.

When the wheels are turned, the steering arm on the inside of the turn swivels more sharply, due to the angle of the arm at this point. Fig. 17-17 illustrates the difference in wheel angles while turning. Note that when straight ahead, the wheels are set for toe-in.

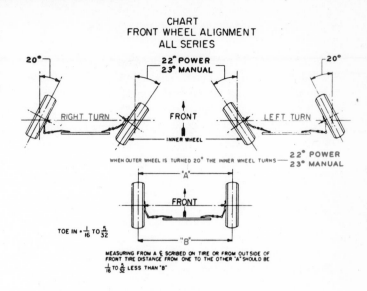

Fig. 17-17. Toe-out on turns. Note that inner wheel on a curve always turns a few degrees more sharply than outer wheel. (Buick)

SLIP ANGLES

When a car rounds a corner at high speed, the actual path taken by the tires is somewhat different than it would have taken if centrifugal force was not present.

As the wheels turn around a corner, they create a cornering force that endeavors to move the car in the exact direction the wheels are aimed. This cornering force is counteracted by centrifugal force, and the actual path the wheels take is determined by a balance between cornering and centrifugal forces. The difference in the actual path and the path with no centrifugal force is the SLIP angle.

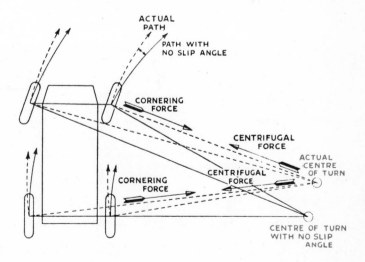

Fig. 17-18. Cornering slip angles. Actual center of turn is different from theoretical turning center with no slip angle.

The greater the slip angle, the more rapid the tire wear. Fig. 17-18 depicts slip angles.

STEERING GEOMETRY ADJUSTMENTS

Various means are provided for adjusting the different front wheel alignment angles. Eccentric bushings in the outer ends of the threaded bushing control arms enable the steering knuckle support to be tipped in or out. The knuckle support can be tipped forward or backward by turning the threaded bushing.

Shims are provided under the anchor ends of the ball joint support arms. By adding or removing shims, proper adjustment, in or out, forward or backward, is possible.

Toe-in is set by adjusting the tie rods (to be covered later).

On vehicles using a solid front axle, the kingpin inclination and the camber angles are built into the axle assembly. If these angles vary from their initial factory setting, the axle itself must be bent to reset the angles.

STEERING GEOMETRY

The term STEERING GEOMETRY is often used to describe the various angles formed by the front wheel alignment setup. In other words; caster, camber, toe-in, steering axis inclination, etc., are sometimes collectively referred to as steering geometry. The term FRONT WHEEL ALIGNMENT refers to the same thing.

STEERING SYSTEM

The steering system is designed to allow the driver to move the front wheels to the right or left, with a minimum of effort, and without excessive movement of the steering wheel.

Though the driver can move the wheels easily, road shocks are not transmitted to the driver. This absence of road shock transfer is referred to as the NONREVERSIBLE feature of steering systems.

The system may be divided into three main assemblies —

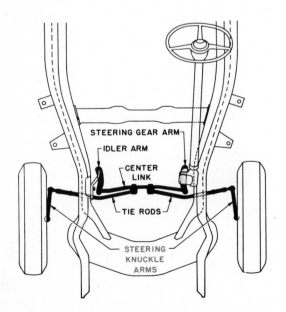

Fig. 17-19. Steering arms. Steering arms are sometimes referred to as steering knuckle arms or spindle arms.   (Chrysler)

the steering gearbox with steering wheel and connecting shaft, the spindle assembly-steering arm units, and the linkage necessary to connect the gearbox to the steering arms.

STEERING ARMS

The steering arms are either bolted to the spindle assembly or forged as an integral part. As mentioned, they are angled toward the center of the car to provide toe-out on turns. The outer ends of the steering arms have a tapered hole to which the tie rod ball sockets are attached.

When the steering arms are moved to the right or left, they force the spindle assembly to pivot on the ball joints, or on the kingpins, depending on which is used. Fig. 17-19.

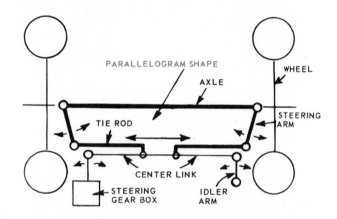

Fig. 17-20. Parallelogram linkage. Schematic top view of parallelogram steering linkage setup. Notice how axle, steering arms and tie rods form shape of a parallelogram. Small arrows show direction of motion for all parts.

STEERING LINKAGE

Linkage is necessary to connect the steering gearbox pitman arm to the steering arms.

Several variations have been used in assembling the linkage to produce the desired effect. The most commonly used arrangement today is the parallelogram type of linkage. It is so named because when viewed from the top, it assumes the shape of a parallelogram. Fig. 17-20.

Note that linkage shown in Fig. 17-19 is also of the parallelogram type.

THE LINKAGE TRAIN

PITMAN ARM

Starting at the gearbox, there is an arm called a pitman arm. This arm can be made to swing from side to side (cross steering) or from front to back (fore-and-aft steering). In the case of the parallelogram linkage train, it swings from side to side. The pitman arm is moved by the gearbox cross shaft. The pitman arm (also called steering gear arm) is shown in Figs. 17-19, 17-27, 17-29 and 17-31.

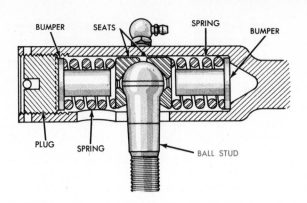

Fig. 17-21. Ball socket connection. If spring on either side of ball stud seats breaks, it cannot collapse enough to allow socket seats to spread apart and drop ball stud. Even if both springs break, seats cannot move apart enough to drop stud. Seat spreading is controlled by steel blocks inside spring. (Ford)

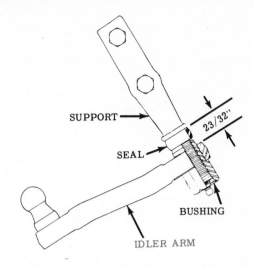

Fig. 17-23. Typical idler arm. This model uses a threaded bushing arrangement. The support bolts to the frame. (Oldsmobile)

CENTER LINK

The pitman arm is attached to a long bar, called a center link, relay rod or connecting link, Fig. 17-19, by means of a ball socket. Details of two types of ball socket connections are shown in Figs. 17-21 and 17-22.

is shown in Fig. 17-25.

Tie rods usually are of equal length and about the same length as the lower control arms. As the frame (to which center link is fastened via pitman arm and idler arm) raises and lowers, there will be minimum disturbance in toe-in, toe-out.

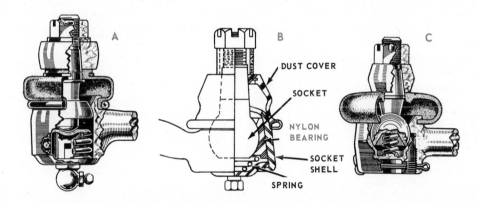

Fig. 17-22. Different types of ball socket tie rod ends. B uses a nylon bearing, while A and C employ a metal-to-metal bearing surface.

IDLER ARM

The opposite end of the center link is affixed, by means of a ball socket, to an idler arm. The idler arm pivots about a support bolted to the frame, and it is arranged so it is parallel to the pitman arm. A typical idler arm is shown in Fig. 17-23.

TIE RODS

Two tie rods, Fig. 17-19, are used to connect the steering arms to the center link. The ends are connected by ball sockets, similar to those shown in Figs. 17-21 and 17-22.

One end of each tie rod is adjustable, so tie rod length can be changed to adjust toe-in. One type of adjustable tie rod end is shown in Fig. 17-24. A typical tie rod adjustment procedure

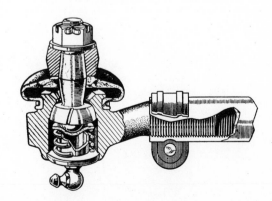

Fig. 17-24. Adjustable tie rod end. Ball socket can be screwed in or out of tie rod. When proper adjustment is procured, it is retained by tightening clamp on tie rod. This will bind tie rod threads against those of ball socket end. (Austin)

319

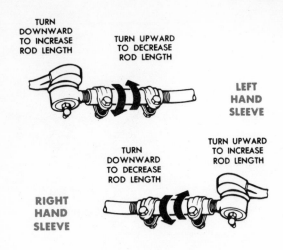

Fig. 17-25. Tie rod adjustment. This type uses a double end sleeve with two clamps. (Mercury)

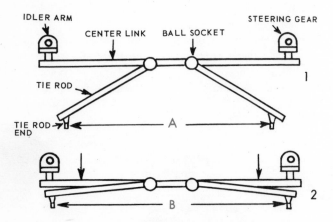

Fig. 17-26. Center link and tie rods should be nearly parallel. (1) Setup with tie rods and center link not parallel (front view). Notice distance A between tie rod ends. When body of car dips downward (2), center link is carried down with frame. Tie rod ends are fastened to steering arms and cannot move down. As center link travels down, angle between link and rods narrows. Tie rod ends must spread out (distance B), forcing steering arms out and toeing wheels in. When car is at normal curb height, link and rods should be almost parallel, as in (2). In this position, link can move up and down with a minimal effect on toe-in and toe-out.

The center link and the tie rods should be as near parallel as possible (when viewed from front). This will help eliminate toe-in and toe-out changes, as the frame raises and lowers. Fig. 17-26.

A complete linkage system (parallelogram pattern) is shown in Fig. 17-27.

An exploded view of parallelogram linkage used in another system is shown in Fig. 17-28.

A different linkage arrangement, called CENTER STEERING, is illustrated in Fig. 17-29. Notice that an intermediate steering arm replaces the parallelogram center link. A drag link rod connects the steering arm to the pitman arm. This pitman arm has fore-and-aft movement.

Another type of linkage arrangement used on a truck solid axle is shown in Fig. 17-30. Notice that the tie rod is one piece and is connected at each end to the steering or spindle arms. A drag link, working fore and aft, connects the pitman arm (sometimes referred to as steering sector shaft arm) to an extension on the spindle. When the drag link swivels this spindle, the action is transferred to the other spindle by the tie rod.

STEERING GEARBOX

The steering gearbox is designed to multiply the driver's turning torque so the front wheels may be turned easily and resist the transfer of road shock to the driver. The torque developed by the driver is multiplied through gears, then is transmitted to the wheel spindle assembly via linkage.

Present-day autos use either manual or power steering gear setups.

MANUAL OR STANDARD STEERING GEARBOX

The standard steering gearbox consists of a long steel shaft with a worm gear on one end and the steering wheel on the other. Fig. 17-31.

A pitman arm shaft passes at right angles to the worm gear. The inner end of the pitman shaft (also called cross shaft, roller shaft, sector shaft, gear shaft, etc.) engages the worm gear either through gear teeth, a round steel peg, a roller gear

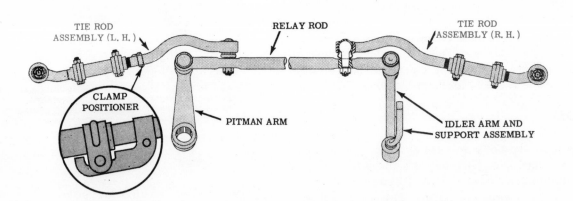

Fig. 17-27. Top view — parallelogram linkage. Each tie rod can be adjusted without changing toe-in or toe-out of opposite wheel. This allows a proper setting and leaves steering wheel in straight ahead position. (Oldsmobile)

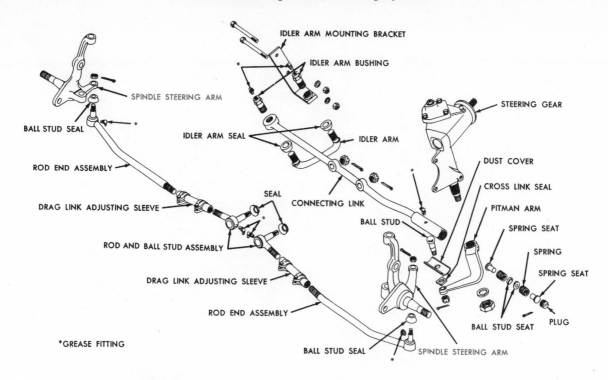

Fig. 17-28. Exploded view — parallelogram steering linkage. Steering arms in this setup are an integral part of spindle assembly. (Mercury)

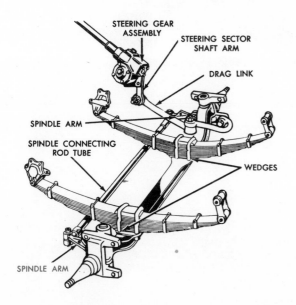

Fig. 17-29. Center steering — linkage. An intermediate steering arm (sometimes called a bell crank) is used to change fore-and-aft movement of drag link to a cross movement of tie rods.

A cast housing surrounds and holds all parts in alignment. The housing is partially filled with gear oil for lubrication. Seals prevent the escape of lubricant and keep out dust and dirt. The housing is bolted to the frame or some rigid area. The steering shaft passes up through a steering column to the steering wheel.

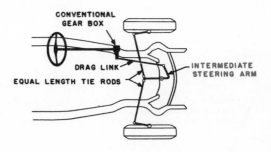

Fig. 17-30. Steering linkage. This drawing portrays a typical linkage setup on a solid axle, light truck application. Steering arm on driver's side (spindle arm) acts as a bell crank. (Ford)

or a notched arm that passes around a peg affixed to a ball nut that rides on the worm. Fig. 17-31.

The worm gear end of the steering shaft is supported by ball or roller bearings on each side of the worm. These bearings remove end and side play in the worm. Bearing adjustment is provided by shims.

The pitman arm shaft operates in bushings or needle bearings. There is an adjustment screw to control worm and pitman shaft clearance.

Fig. 17-31 shows one type of steering gearbox. The tapered peg, 10, rides in the worm grooves. When the worm is revolved, it causes the peg to follow along the worm grooves, imparting a turning motion to the pitman shaft assembly, G. Splines in the pitman arm attach it securely to the pitman shaft.

Fig. 17-32 is a cross section of the RECIRCULATING BALL WORM AND NUT GEARBOX. Instead of a taper peg riding in the worm grooves, this type uses a nut that rides up

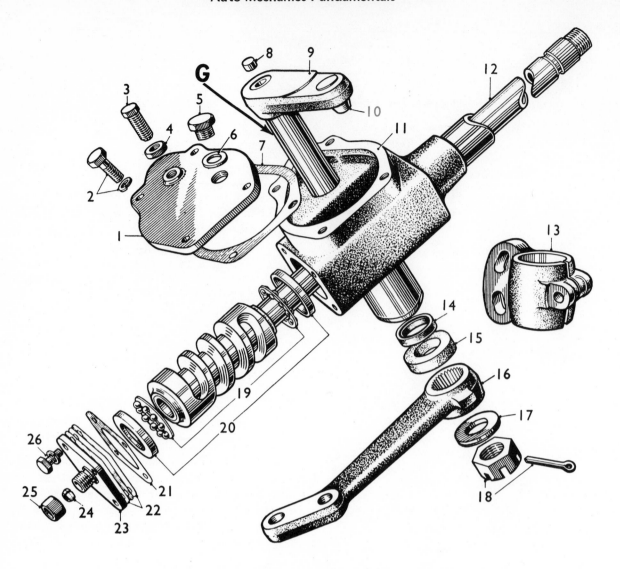

Fig. 17-31. Exploded view — worm and taper pin steering gearbox.
Taper pin (10) rides in worm groove.   (Austin Motor Co.)

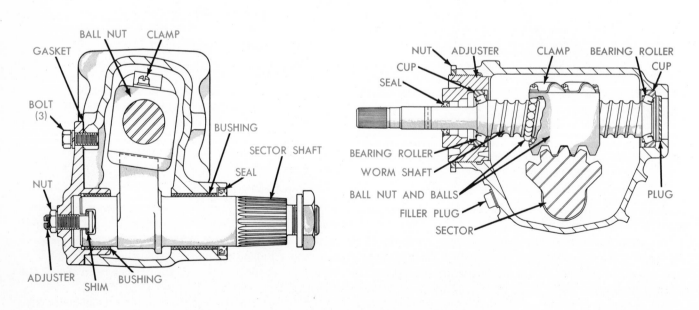

Fig. 17-32. Cross sections of a recirculating ball worm and nut steering
gearbox. Study part names, location, and function.   (Dodge)

Front Wheel Alignment, Steering Systems

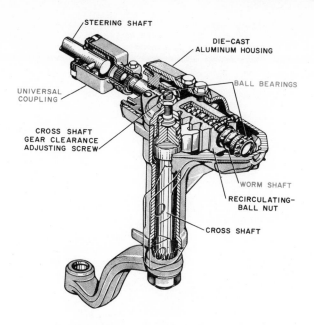

Fig. 17-33. Cutaway of a manual recirculating ball worm and nut steering gearbox. Note universal coupling. (Plymouth)

and down the worm. Ball bearings ride half in the worm and half in the nut. When the worm is turned, the ball bearings impart an axial (lengthwise of shaft) thrust to the ball nut. As the nut is prevented from turning, it will then travel up or down, depending on the worm direction.

As the ball bearings roll along and one by one, come to the end of the groove in the nut, they are forced through a return tube. Eventually, they reenter the nut groove at the opposite end, which accounts for the name, recirculating ball worm and nut.

One side of the nut has gear teeth cut into it. The pitman arm sector gear teeth engage the nut teeth. When the nut travels up or down, the pitman shaft is forced to rotate. The recirculating ball worm and nut reduces friction to a very low level, making it a very popular type of gearbox. Study Fig. 17-32. Notice the bearing and pitman shaft adjustments. Fig. 17-33 shows a cutaway view of another recirculating ball worm and nut.

Fig. 17-34 shows a different version of a recirculating ball worm and nut. Notice the ball return tube. The pitman shaft engages the ball nut by means of a slotted sector riding on the ball nut peg.

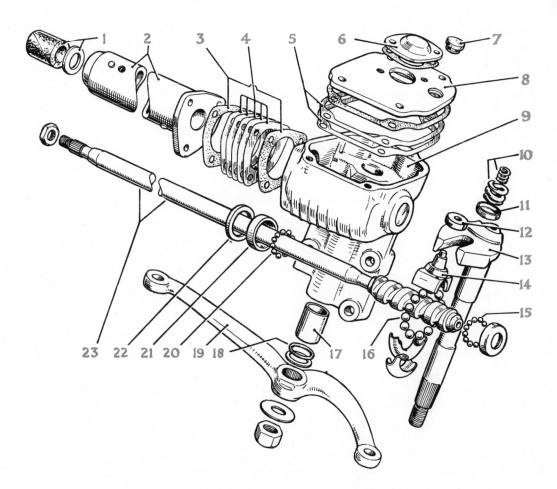

Fig. 17-34. Exploded view — recirculating ball worm and nut steering gearbox. 1—Felt bush and washer. 2—Outer column. 3—Paper gaskets (joint washers). 4—Shims. 5—Shims. 6—Spring cap. 7—Filler plug. 8—Top cover. 9—Steering box. 10—Double coil spring. 11—Damper button. 12—Guide roller. 13—Rocker shaft. 14—Nut. 15—Steel balls (lower track). 16—Steel balls (nut). 17—Rocker shaft bush. 18—Oil seal. 19—Drop arm (swing lever). 20—Steel balls (upper track). 21—Upper track. 22—Spacer. 23—Inner column and worm. (Hillman Motor Co.)

A WORM AND ROLLER gearbox is shown in Fig. 17-35. The pitman shaft engages the worm by means of a roller with gear teeth cut around it. The roller is mounted on antifriction bearings. As the roller is free to turn, the friction developed in a wiping gear action is bypassed. When the work is rotated, the roller teeth follow the worm and impart a rotary motion to the pitman shaft.

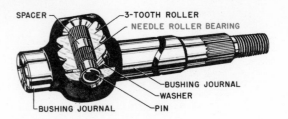

Fig. 17-36. Pitman shaft roller. Needle bearings are used to reduce friction between three-tooth roller and pin upon which it turns. (Chrysler)

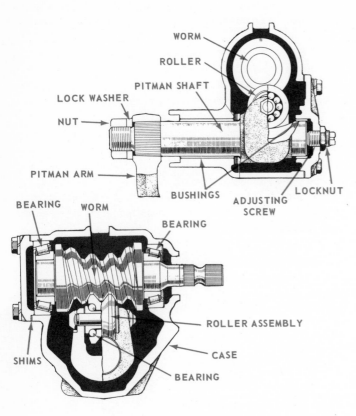

Fig. 17-35. Cross sections of a worm and roller manual gearbox. (Toyota)

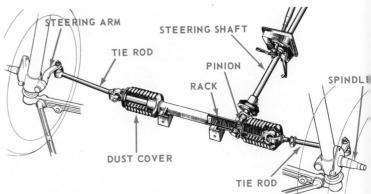

Fig. 17-37. Rack and pinion steering gearbox. Both ends of rack are connected to steering arms by means of tie rods. (Peugeot)

Fig. 17-36 illustrates a pitman shaft and roller in detail.

A RACK AND PINION steering gear is pictured in Fig. 17-37. A steering shaft from the steering wheel fastens to the pinion shaft universal joint. When the pinion is turned, it forces the rack to move, either to the right or left. The ends of the rack are connected to the steering arms by a flexible joint and a short tie rod. All gear rack movement is then transferred directly to the steering arms.

STEERING SHAFT UNIVERSAL JOINTS

Many applications make use of one or more universal joints in the steering shaft. This enables the shaft to be located at various convenient angles without moving the gearbox. It also relieves the mounting stress and deflecting (bending) forces sometimes encountered in a solid shaft.

In addition, the universal joint dampens steering shock and vibration transfer to the steering wheel. Fig. 17-38 shows a mounted gearbox with a flexible U-joint attached to the worm shaft. The steering shaft is shown in phantom (faint, indistinct) view.

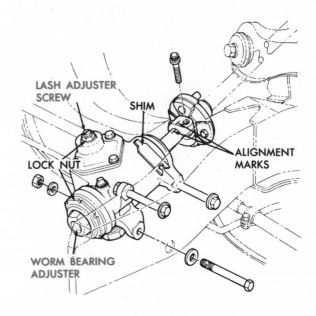

Fig. 17-38. Steering shaft flexible universal joint. Adjustments for worm bearing and cross shaft are shown. Note alignment marks at base of universal joint. (Chevrolet)

STEERING GEAR REDUCTION

The amount of gear reduction found in the various gearboxes depends on vehicle type, weight, steering characteristics (fast or slow) desired and the ultimate use of the vehicle. Gearbox ratios will vary from 15:1 to 24:1. Steering wheel

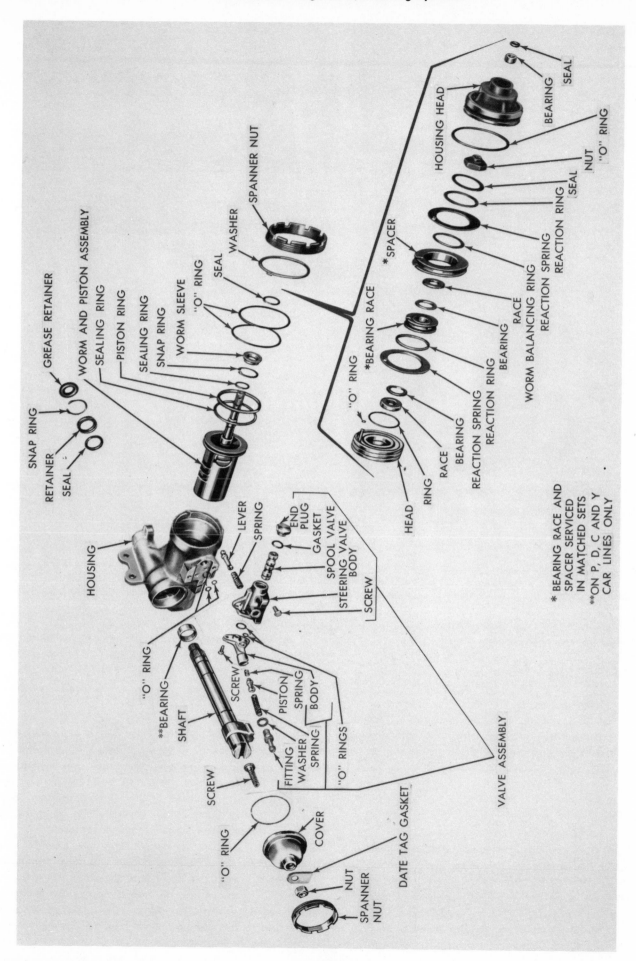

Exploded view of a power steering gear. (Plymouth)

SNAP RING
RETAINER
SEAL
GREASE RETAINER
WORM AND PISTON ASSEMBLY
SEALING RING
PISTON RING
SEALING RING
SNAP RING
WORM SLEEVE
"O" RING
SEAL
WASHER
SPANNER NUT

HOUSING
LEVER
SPRING
END PLUG
GASKET
SPOOL VALVE
STEERING VALVE BODY
SCREW

"O" RING
**BEARING
SHAFT
SCREW
SCREW
PISTON SPRING
BODY
FITTING
WASHER
SPRING
"O" RINGS
VALVE ASSEMBLY

SCREW
"O" RING
COVER
NUT
SPANNER NUT
DATE TAG GASKET

SEAL
BEARING
HOUSING HEAD
SEAL
NUT
"O" RING
SEAL
REACTION RING
REACTION SPRING
REACTION RING
RACE
WORM BALANCING RING
REACTION SPRING
BEARING
RACE
*SPACER
*BEARING RACE
"O" RING
REACTION RING
REACTION SPRING
BEARING
RACE
BEARING
HEAD
RING

* BEARING RACE AND SPACER SERVICED IN MATCHED SETS
**ON P, D, C AND Y CAR LINES ONLY

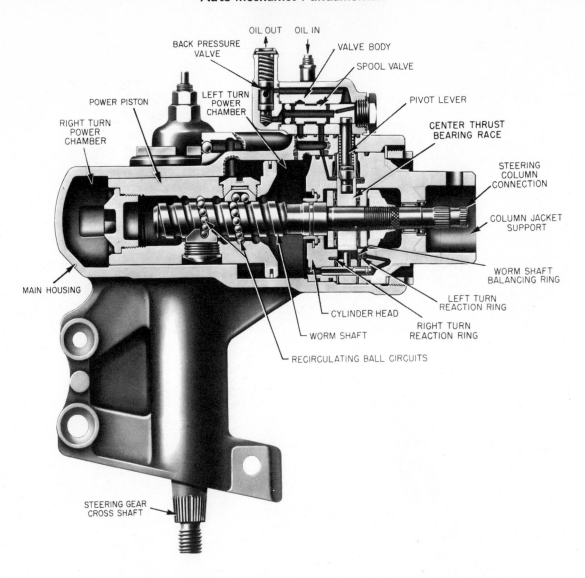

Fig. 17-39. Cross section of a power steering unit. This is self-contained, in-line type.

movement will average from 4 to 6 turns from full left to full right for standard manual gearboxes.

POWER STEERING

Power steering is designed to reduce the steering wheel turning effort by utilizing hydraulic pressure to bolster (strengthen) the normal torque developed by the steering gearbox. It should ease steering wheel manipulation and, at the same time, offer some resistance so that the driver can retain some "road feel."

ROAD FEEL

Road feel is the feeling imparted to the steering wheel by the wheels of a car in motion. This feeling is very important to the driver in sensing and predetermining vehicle steering response. A good driver can tell when the front wheels are approaching the walk (slipping sideways) point. He can tell if more or less turning effort is required to stabilize the car body on turns, in winds, etc. Complete and effortless power steering

would rob him of this valuable road feel.

To maintain road feel, power steering systems require some wheel effort, approximately 1 to 3 lbs. pull on the steering wheel rim.

TWO GENERAL TYPES OF POWER STEERING

Power steering systems are of two general types. One type controls and utilizes the hydraulic pressure directly within the steering gearbox housing. The other type uses a hydraulic cylinder and control valve attached to the linkage system. This second type uses a conventional standard gearbox. Both systems use a hydraulic pump, generally belt driven by the engine, to produce the necessary hydraulic pressure.

In both the self-contained and linkage types, a control valve is actuated by driver steering effort. This valve admits oil, under heavy pressure, to one side or the other of a hydraulic piston. The pressure the oil creates against the piston is transferred to either the pitman shaft, or to a direct connection to the steering linkage, which assist the driver in manipulating the front wheels.

SELF—CONTAINED POWER STEERING UNIT

The self-contained unit places the control valve mechanism, power piston and gears in an integral unit. Pressure developed by the unit is applied to the pitman shaft.

There are several different models of the self-contained type, but all share many basic design principles. They may be divided into two general catagories, the in-line and the offset. The in-line is in popular use today.

IN—LINE, SELF—CONTAINED POWER STEERING UNITS UTILIZING A PIVOT OPERATED, SLIDING SPOOL VALVE

Study Figs. 17-39 and 17-40. Notice that all the basic elements of the conventional manual steering gear are used. A pitman shaft (steering gear cross shaft) engages a recirculating ball nut that rides on a typical worm gear or shaft.

HOW IN—LINE UNIT WORKS

In operation, oil under pressure from the pump enters the valve body via the inlet passage. When no pressure (above 1 to 3 lbs.) is being applied to the steering wheel, the reaction unit is centered and the pivot lever is straight up and down. This keeps the spool valve in a neutral position.

Two round washer-like springs (pressure plate) hold the center thrust bearing race, Figs. 17-39 and 17-40, in a neutral position. The center thrust bearing is attached to the worm shaft, by being secured between two thrust bearings, which are in turn affixed to the worm shaft by a shoulder on the left end and a locknut on the right. The bearings revolve with the worm shaft, but the center thrust bearing remains still. The center thrust bearing can move up or down with the worm shaft, but not around with it.

The lower end of the pivot lever rests in a cutout in the center thrust bearing. When the center thrust bearing is in the neutral position, the pivot lever is straight up and holds the

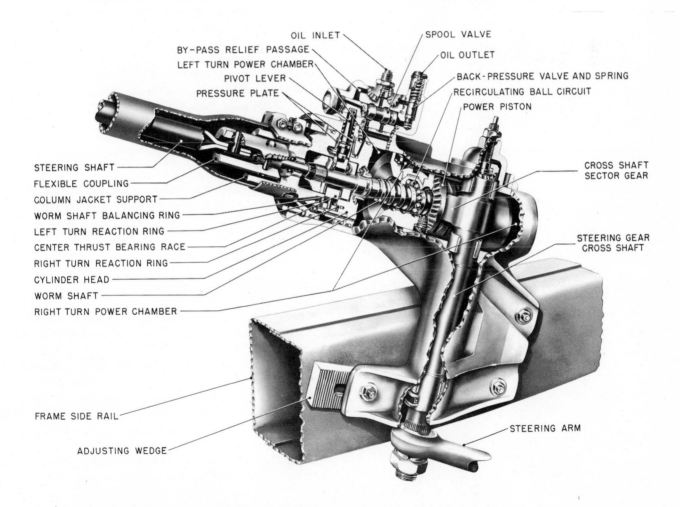

Fig. 17-40. Cutaway of a self-contained, in-line power steering unit. Note use of adjusting wedge to secure proper gearbox alignment. (Plymouth)

The basic difference is the addition of a reaction unit, a pivot lever, a valve body and the utilization of the ball nut as a power piston.

spool valve in the neutral postion. Fig. 17-41.

With the spool valve in the neutral position, oil is fed to both sides of the power piston and to both reaction rings.

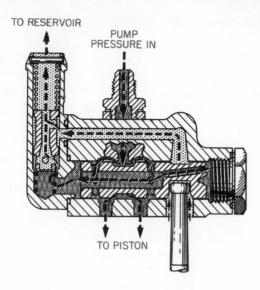

TO RESERVOIR

PUMP
PRESSURE IN

TO PISTON

Fig. 17-41. Spool valve in neutral position. Note that spool valve is centrally located and pressurized oil is passing to power piston through both piston outlets. (Plymouth)

Oil also works its way around the balls between the power piston and the worm shaft. This oil collects in the hollow formed between the left end of the worm shaft and the left end of the power piston. This pressure created on the worm shaft end is counterbalanced by pressure created on the worm shaft balancing ring, Fig. 17-39.

In the neutral position, oil pressure is equalized on all sides, and no hydraulic pressure is applied to the pitman shaft by the power piston.

LEFT TURN

When the steering wheel is turned to the left, the worm shaft attempts to impart motion to the power piston. As the

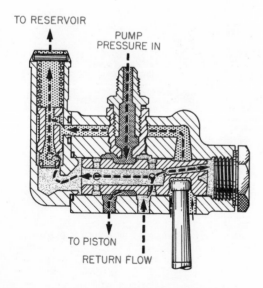

TO RESERVOIR

PUMP
PRESSURE IN

TO PISTON

RETURN FLOW

Fig. 17-42. Spool valve in left turn position. Valve has moved to left. Right piston outlet is cut off from source of pressurized oil. Return flow will now pass up through right side piston outlet.

power piston has a rack of teeth that engage the pitman arm teeth, the resistance to turning offered by the wheels holds the power piston still.

As the power piston does not want to move, the worm shaft (which has some lateral play, or looseness) will start to screw out of the power piston. As it screws out just a few thousandths of an inch, it moves the center thrust bearing race with it. The center thrust bearing race will then move the bottom of the pivot lever to the right. This causes the top of the pivot lever to move the spool valve to the left.

When the spool valve moves to the left turn position, Fig. 17-42, it opens the left turn oil passageway and oil travels to the left turn power chamber.

The right turn passageway can no longer receive pressurized oil, but has opened to allow the oil to be returned from the right turn power chamber. Fig. 17-43.

As soon as the spool valve is moved to the left, two things happen. Oil flows to the left turn power chamber and exerts a force on the power piston. Since the right turn power chamber oil is no longer pressurized, the power piston will be forced to the left. This squeezes the right turn power chamber oil out through the oil return passageways, back to the pump reservoir.

As the oil pressurizes the left turn power chamber, it also feeds to the right side reaction ring. This creates pressure that attempts to return the center thrust bearing to the neutral position. The round reaction springs also try to force it back to the neutral position.

These two forces on the center thrust bearing "fight" against the driver's turning force on the wheel and produce the essential road feel.

As soon as the driver's turning force no longer exceeds the reaction ring and spring pressure plate pressure, the center thrust bearing will be forced back to the neutral position. This pivots the spool valve to neutral and equalizes the pressure on each side of the power piston.

RIGHT TURN

A right turn will produce the same action, only the spool valve moves to the right and oil pressure is applied to the right turn power chamber.

The worm shaft is connected to the steering shaft, Fig. 17-40, by a flexible connector. This connector allows the necessary worm shaft end movement.

Maximum pump pressure for this unit is around 950 lbs. Oil leakage is prevented by neoprene rubber o-rings that ride in grooves.

Study Figs. 17-39 through 17-43 and trace out the oil flow patterns. Study the various parts and their actions.

OFFSET POWER STEERING GEAR

In the SELF-CONTAINED OFFSET type of unit, the additional force offered by the pressurized oil is not applied to the ball nut assembly (as just discussed). Instead, it is applied to the pitman shaft by a power piston rod working in a separate power cylinder. The power piston has a rack of gear

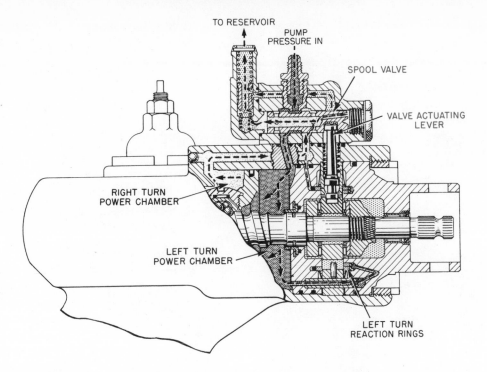

Fig. 17-43. Oil flow during left turn. Study oil flow for this turn position. A right turn is merely a reversal of this when spool valve is moved to right by valve actuating lever.

teeth cut in one side that mesh with a separate set of pitman sector teeth.

There have been several designs used for the offset type power steering gear. Fig. 17-44 depicts one type. Notice that the pitman shaft sector teeth are built on both sides. The ball nut presses on one sector, and the power piston-rack engages the other set of sector teeth.

IN—LINE UNIT WITH TORSION BAR AND ROTATING SPOOL VALVE

·A popular type of in-line, self-contained power steering unit is shown in Fig. 17-45. It employs a spool valve, but instead of sliding back and forth, the spool rotates a small amount inside the valve body. This type may be referred to as an open-center, rotary type, three-way valve. The spool valve is a very close fit in the valve body.

The steering shaft is connected to a coupling, which in turn is splined to the lower stub shaft. Fig. 17-45. The stub shaft passes through and is pinned to the spool valve. The stub shaft is pinned, at the outer end, to a torsion bar (tempered steel bar designed to twist or spring a certain amount when a turning force is applied). The inner end of the stub shaft floats (is free) on the bar.

The inner end of the torsion bar is pinned to the valve body cap. The valve body cap, in turn, is pinned to the worm shaft. A pressure line and a return line from the steering pump attach to the gear valve housing.

Oil pressure in the pressure cylinder will run around 125 psi for normal driving; about 400 psi maximum for cornering; up to 1,000 — 1,200 psi for parking. Study the construction of the unit pictured in Fig. 17-45.

ROTARY VALVE ACTION, STRAIGHT AHEAD

When the car is traveling straight ahead and no appreciable turning effort is applied to the steering wheel, the torsion bar holds the spool valve in the neutral position. In this position,

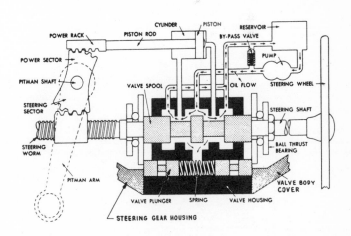

Fig. 17-44. Offset, self-contained power steering unit. In this type, power piston operates in an offset cylinder and applies force to pitman shaft through a second set of pitman shaft teeth (power sector).

the spool valve is aligned to cause oil entering the valve body from the pressure port to pass through the slots in the spool valve. From the spool valve, oil flows to the oil return port. No pressure is applied to either side of the power rack-piston. Oil flow is pictured in Fig. 17-46 and in A, Fig. 17-49.

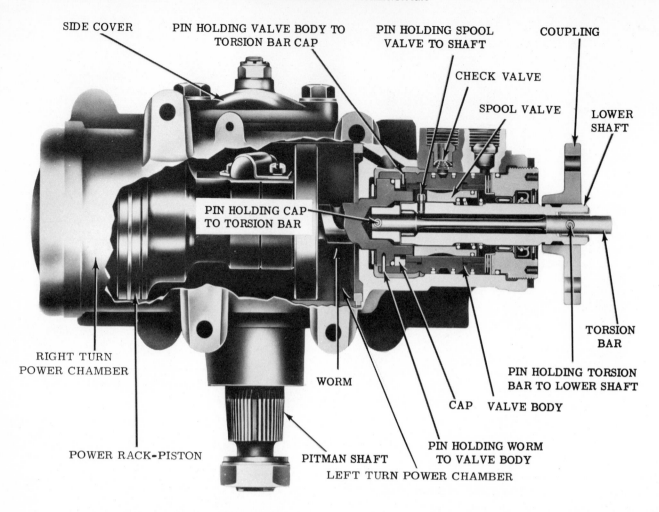

Fig. 17-45. Power steering gearbox employing a rotary valve to control oil flow to the power rack-piston. (Cadillac)

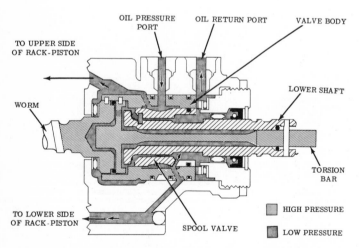

Fig. 17-46. Oil flow through rotary valve when in NEUTRAL position. Study valve construction. (Oldsmobile)

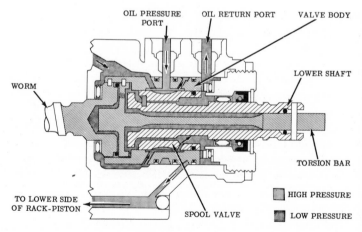

Fig. 17-47. Oil flow through rotary valve in RIGHT TURN position.

ROTARY VALVE ACTION, RIGHT TURN

When the steering wheel is moved to the right, a turning force is applied to the stub shaft and through the pin to the torsion bar. Remember that the inner end of the torsion bar is pinned to the valve body cap (valve body cap is pinned to valve body, which is pinned to worm shaft). The resistance of the front wheels to movement causes the turning effort to twist or deflect the torsion bar. As the bar twists, the stub shaft revolves enough to cause the spool valve to rotate.

As the spool valve rotates to the right, the right turn

grooves are closed off from the return grooves and open to the pressure grooves. The left turn spool grooves are opened more to the return grooves and closed to the pressure grooves. This action permits oil flow to the right turn power chamber. The oil pressure forces the power rack-piston upward, applying additional turning force to the pitman shaft sector teeth.

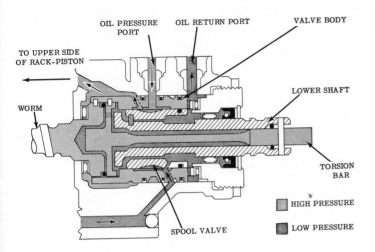

Fig. 17-48. Oil flow through rotary valve in LEFT TURN position.

On this particular unit, the initial power assist will start with about a one pound pull on the rim of the steering wheel. This turns the valve about .3 of 1 deg. Full power assist will result with about 3 1/2 lbs. pull on the wheel, rotating the spool about 4 degrees. Fig. 17-47 and B, Fig. 17-49.

ROTARY VALVE ACTION, LEFT TURN

Moving the steering wheel to the left twists the torsion bar in the opposite direction, causing the spool valve to rotate to the left. This closes off the left turn grooves from the return grooves and opens them to the pressure grooves. The right turn slots are closed to pressure and opened to return. Oil flows to the left turn power chamber and forces the power rack-piston downward. Fig. 17-48 and C, Fig. 17-49.

WHAT HAPPENS IF TORSION BAR BREAKS?

In the unlikely event the torsion bar should break, the stub shaft is connected to the worm shaft (some movement is allowed to permit torsion bar action) by two tangs that will still provide manual turning force.

ROTARY VALVE ACTION

An exploded view of a rotary valve power steering gear is shown in Fig. 17-50.

LINKAGE TYPE POWER STEERING

The linkage power steering system employs a power cylinder and control valve to provide the power assist. This system uses a conventional manual gearbox. A pitman arm actuates the power cylinder control valve. Fig. 17-51.

One end of the power cylinder is attached to the frame; the other end is connected to the steering linkage relay rod. The relay rod is attached to the control valve, which is connected

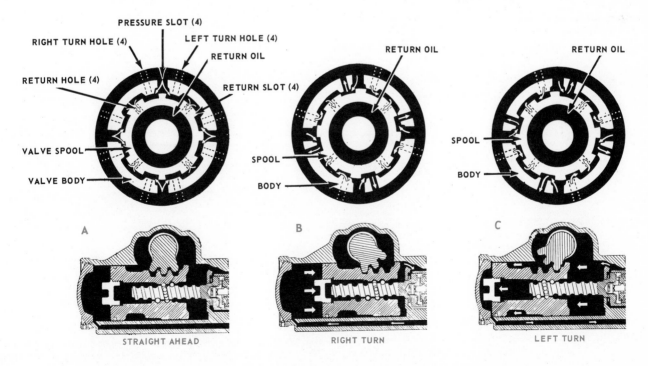

Fig. 17-49. Oil flow through rotary valve and power rack-piston nut action during straight ahead, right turn and left turn positions.

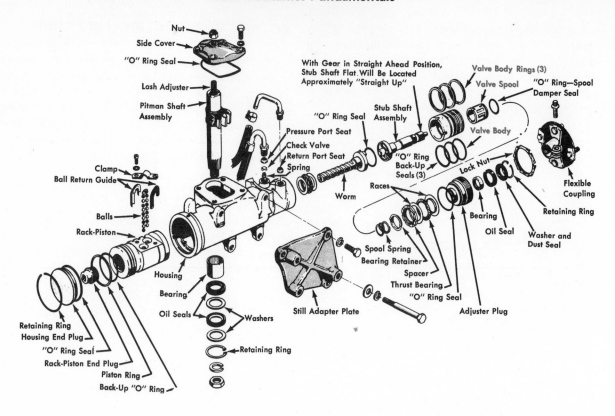

Fig. 17-50. Exploded view of a power steering gear utilizing a rotary spool valve. Study part relationship. (American Motors)

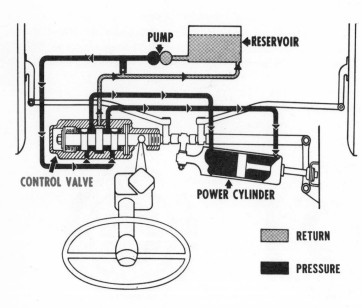

Fig. 17-51. Linkage type power steering — neutral position. Pitman arm ball actuates control valve. (Mercury)

to the power cylinder by high pressure hoses.

The end of the pitman arm is formed into a ball, which is placed in the control valve assembly in a ball socket arrangement. Pressure of the pitman shaft ball, either to the right or left, actuates the control valve. The control valve then transmits oil pressure to one side or the other of the power cylinder piston. Figs. 17-52 and 17-53.

Road feel and control valve operation are based on much the same principles as the integral type control valve. Study Figs. 17-51, 17-52 and 17-53. Follow the oil flow and valve action in neutral, right and left-hand turns.

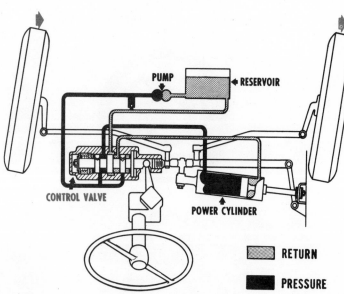

Fig. 17-52. Linkage type power steering — right turn position. Pressure from pitman arm ball has moved control valve, allowing pressurized oil to enter one side of power cylinder. Piston cannot move, so cylinder is forced away from pressurized side and exhausts oil from other side. This force is transferred to linkage to assist driver.

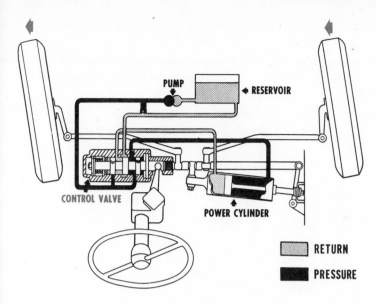

Fig. 17-53. Linkage type power steering — left turn position. Control valve has moved in other direction, reversing pressurized area in power cylinder. Power cylinder is now moving to right and will pull on linkage.

POWER CYLINDER AND CONTROL VALVE IN ONE UNIT

Another type of linkage power steering system incorporates the power cylinder and control valve in one unit. The pitman arm moves the control valve, admitting oil pressure to the power cylinder. The road feel in this system is produced by the force required to move the valve spool, which is centered by oil pressure.

Fig. 17-54 illustrates a linkage power steering system. Fig. 17-55 pictures an integral power steering system.

POWER STEERING PUMPS

A power steering pump is belt driven by the engine. It can be mounted as a separate unit or built as an integral part of the front engine plate and driven directly by the crankshaft.

The returned oil is forced into a reservoir from which the oil flows down into the pump proper.

The oil is pumped by a vane, rotor, roller vane, or a slipper vane pump. The maximum pressure (about 1,000 lbs.) is regulated by a control valve. The oil (Type-A automatic transmission fluid) provides the lubrication for the pump and steering mechanism. Fig. 17-56 illustrates a typical vane type oil pump.

Fig. 17-57 shows vane operation in another vane pump. Notice that some oil, under pressure, is admitted behind the vane ends to force them outward to contact the pump ring walls. Cutaway of a vane pump is pictured in Fig. 17-61.

Another type of power steering pump, the SLIPPER VANE, is shown in Fig. 17-58. The vanes are spring-loaded to hold them in contact with the wall.

A cutaway of a ROLLER VANE pump is shown in Fig. 17-59. A ROTOR pump is illustrated in Fig. 17-60.

POWER STEERING UNITS CAN BE OPERATED MANUALLY

In the event of pump pressure failure, power steering systems can be operated manually with somewhat more pressure than required with conventional steering gearbox.

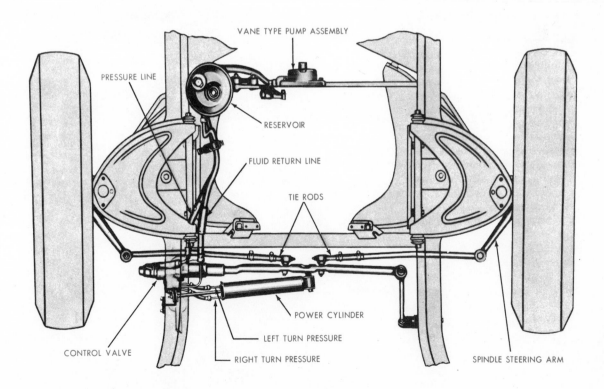

Fig. 17-54. Linkage power steering system — top view. Study arrangement of parts.   (Mercury)

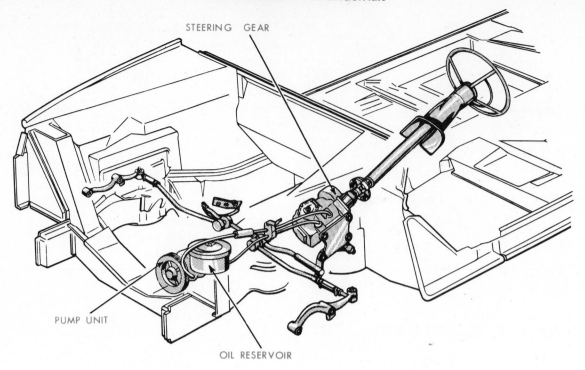

Fig. 17-55. Integral power steering system — top view. Self-contained, offset type is shown. (Lincoln)

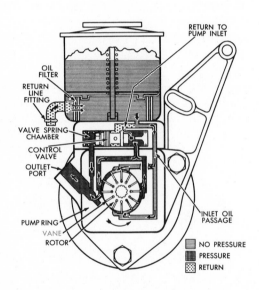

Fig. 17-56. Vane type power steering pump. Follow oil flow pattern.

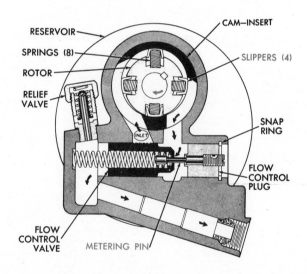

Fig. 17-58. Slipper vane power steering pump. Slippers are spring-loaded to insure wall contact. Metering pin allows heavy oil flow to gear during lower engine rpm, but reduces oil flow at higher rpm (above 3,000 rpm) to reduce fluid temperatures. (Plymouth)

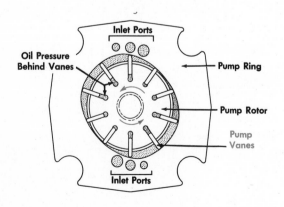

Fig. 17-57. Vane type power steering pump. (GMC)

CAUTION!

THE STEERING SYSTEM IS OF CRITICAL IMPORTANCE IN THE SAFE OPERATION OF THE CAR. IT MUST BE SERVICED AND INSPECTED ON A REGULAR BASIS.

WHENEVER REPAIRS ARE MADE, MAKE CERTAIN ALL PARTS FIT. THEY MUST BE PROPERLY ADJUSTED AND SECURE.

IF A PART IS BENT OR WORN, NEVER WELD, HEAT OR STRAIGHTEN IT. THROW IT AWAY — UNDER NO CIRCUMSTANCES CHEAT ON SAFETY!

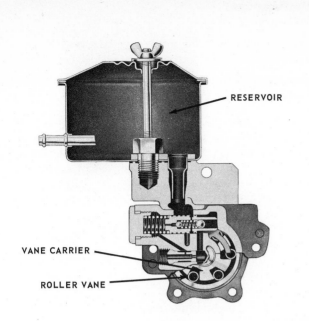

RESERVOIR

VANE CARRIER

ROLLER VANE

Fig. 17-59. Cutaway of a roller vane power steering pump. (Mercury)

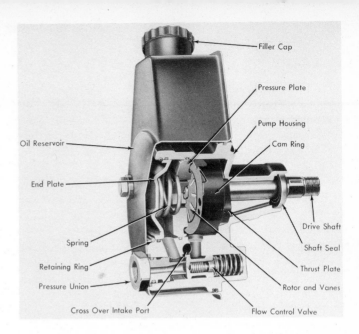

Filler Cap

Pressure Plate

Pump Housing

Cam Ring

Oil Reservoir

End Plate

Drive Shaft

Shaft Seal

Spring

Thrust Plate

Retaining Ring

Rotor and Vanes

Pressure Union

Cross Over Intake Port

Flow Control Valve

Fig. 17-61. Cutaway of a vane type power steering pump. (Cadillac)

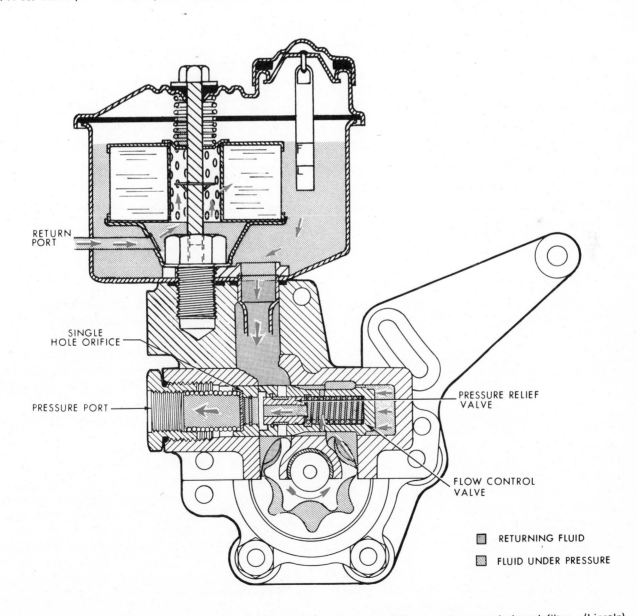

RETURN PORT

SINGLE HOLE ORIFICE

PRESSURE PORT

PRESSURE RELIEF VALVE

FLOW CONTROL VALVE

RETURNING FLUID

FLUID UNDER PRESSURE

Fig. 17-60. Rotor type power steering pump. Trace oil flow pattern and note how oil is returned to reservoir through filter. (Lincoln)

SUMMARY

To insure easy and safe handling of a car, the front wheels must be properly aligned. Several angles must be considered.

Camber is when the tops of the wheels are not the same distance apart as the bottoms. Positive camber spaces the tops farther apart than the bottoms, while negative camber spaces the tops closer together than the bottoms.

Caster is achieved by tilting the kingpin, or the steering axis, either forward or backward. When the top of the axis is tipped toward rear of car, the caster is positive. When the top is tipped toward front of car, the caster is negative.

Steering axis inclination is produced by tipping the top of the steering axis in toward the center of the car.

Toe-in is when the wheels are closer together at the front than the rear. Toe-out is when the wheels are closer together at the back than the front.

The angle formed between the steering axis and the center line of the spindle is called the steering knuckle angle.

Toe-out on turns is accomplished by angling the steering arms in toward the center of the car.

When cornering, the actual path taken by the car, and the path upon which the wheels are aimed, are different. This difference is expressed in terms of a slip angle.

The front wheels swivel from left to right on either a kingpin or a ball joint arrangement.

Steering linkage is provided to transfer the force from the pitman arm to the steering arms. The most popular arrangement for this linkage is the parallelogram. This linkage consists of a center link, tie rods and idler arm.

The various linkage parts are connected by means of ball sockets. This allows a swiveling movement without end play.

The newer type of linkage ball sockets use a special grease and excellent sealing, in combination with a special plastic bearing area, to allow extended lubrication intervals.

Cars use either a standard manual steering gearbox or a power steering unit.

Some popular types of gearing used in both types of boxes are the recirculating ball worm and nut, worm and sector, worm and taper pin and the worm and roller. The most widely used is the recirculating ball worm and nut.

There are two general types of power steering. One type is self-contained, with all hydraulic pressure applied to the pitman arm shaft. The other type uses a standard type gearbox with a power cylinder attached to the linkage system. The self-contained type is currently in widespread usage.

Both kinds of power steering systems are actuated by oil pressure. The pressurized oil is admitted to either side of the power cylinder as controlled by a valving setup.

Road feel is maintained by the use of a torsion bar, balanced oil pressure and return springs. The system uses oil pressure to assist the driver only when the pull on the steering wheel exceeds 1 to 3 lbs. The second the pressure on the wheel falls to this amount, the oil pressure is equalized on both sides of the power piston.

Oil pressure is provided by a high pressure oil pump that, generally is belt driven by the engine. The pumps can use the vane, rotor, roller vane or a slipper vane pump unit. Automatic transmission fluid or a special power steering fluid, provides lubrication for both the steering gearbox and the pump. The pump is connected to the power unit by high pressure hose.

REVIEW QUESTIONS — CHAPTER 17

If when reading, a person stumbles over a new word, and instead of looking up the pronunciation and meaning, goes merrily on his way, he can assume that his vocabulary will not improve. The same goes for the student of auto mechanics. When you see a new word, a new part, a new idea or a new theory, STOP AND FIND OUT WHAT IT MEANS, HOW IT WORKS, AND WHY IT WORKS.

1. The front wheels swivel on either a _____ or a _____ setup.
2. Why is proper wheel alignment very important?
3. What is caster?
4. Why is it used?
5. Define the meaning of camber.
6. What effect does camber have on the steering system?
7. Kingpin, or steering axis inclination, affects the steering in what way?
8. What is toe-in?
9. Of what particular use is toe-in?
10. When turning a corner, one side of the car raises while the other lowers. What is the reason for this?
11. How is toe-out on corners produced?
12. What is slip angle?
13. The term steering geometry refers to what?
14. The most popular type of steering linkage arrangement is the _____.
15. Name the linkage parts used in the above linkage arrangement.
16. Give the uses of these parts.
17. What type of steering system (self-contained or linkage) is in popular use today?
18. The _____ _____ steering gearbox reduces friction to a very low level and is therefore widely used.
19. What is a worm and sector type of gearbox?
20. Explain how a worm and roller gearbox works.
21. Of what value are U-joints as used in a steering shaft?
22. What is a self-contained power steering gearbox? How does it differ from the other type? What is the other type?
23. What is road feel and how is it maintained in the power steering systems?
24. Oil pressure for the power steering system is produced by what unit?
25. Pressures run up to _____ pounds.
26. Explain the actions involved in a typical power steering unit for straight ahead position, left turn, right turn.
27. What type of fluid is used in power steering systems?
28. Explain the action of a typical power steering pump.
29. Starting with the steering wheel, name all major parts, in order, between the steering wheel and the steering arms.
30. Name some safety factors that must be considered whenever work is done on the steering system.
31. If a power steering unit fails, what happens?
32. What is a ball socket, how is it made, and what use does it serve?

Chapter 18
AUTOMOBILE BRAKES

The modern car uses HYDRAULIC BRAKES as a stopping medium. A special fluid (hydraulic brake fluid), confined in steel tubing lines, is used to transmit both motion and pressure from the brake pedal to the wheels.

How the braking system works will be discussed, but first you must have a basic knowledge of hydraulics.

HYDRAULICS — THE SCIENCE OF LIQUID IN MOTION

A liquid, under confinement, can be used to: transmit pressure; increase or decrease pressure; transmit motion.

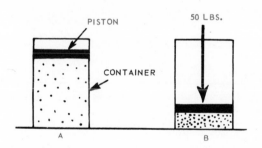

Fig. 18-1. Air is compressible. There is no pressure on piston in A. In B, pressure has forced piston down, compressing air trapped in container.

AIR IS COMPRESSIBLE

Air confined under pressure will compress, thereby reducing its volume. Fig. 18-1.

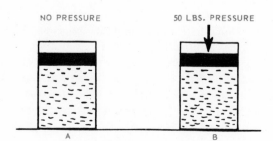

Fig. 18-2. Liquids cannot be compressed. When pressure is applied to piston in B, it does not compress liquid.

LIQUIDS CANNOT BE COMPRESSED

When a liquid is confined and placed under pressure, it cannot be compressed. A, Fig. 18-2, shows a cylinder filled with oil. A leakproof piston is placed on top of the oil. When a downward force is applied, as shown in B, the force will not compress the oil.

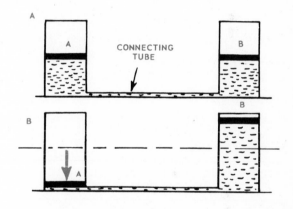

Fig. 18-3. Liquids can transmit motion. In A, both pistons are at rest. In B, piston in left cylinder has been forced down. When it traveled down, it forced liquid into other cylinder. Since other cylinder is same size, its piston will be raised equal distance other piston was lowered.

LIQUIDS CAN TRANSMIT MOTION

In Fig. 18-3, you will see that any movement of piston A will cause piston B to move a like amount.

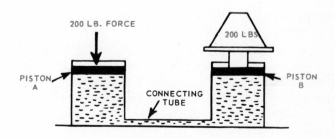

Fig. 18-4. Using a liquid to transmit force. A 200 lb. force on piston A is transmitted, via liquid, to piston B.

LIQUIDS CAN TRANSMIT FORCE

If a 200 lb. force is placed on piston A, Fig. 18-4, piston B will support 200 lbs. Both pistons are the same size.

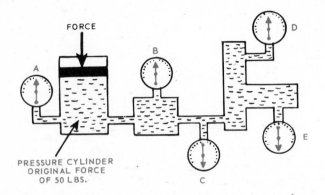

Fig. 18-5. When pressure is exerted on a confined liquid, it is transmitted undiminished. Force on piston has created a pressure of 50 lbs. psi (per square inch), upon liquid in pressure cylinder. Notice that all gages read same throughout system. Pressure is transmitted undiminished to all parts. If gage A reads 50 lbs. psi, gages B, C, D, and E will also read 50 lbs. psi.

PASCAL'S LAW

Pascal, an early French scientist, stated, "When pressure is exerted on a confined liquid, it is transmitted undiminished." Fig. 18-5 illustrates Pascal's law. Notice that the original pressure force is the same at all outlets.

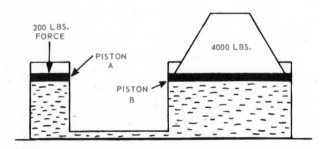

Fig. 18-6. Liquids can be used to increase force. A 200 lb. force on piston A (piston area of 1 sq. in.) is increased to 4,000 lbs. on piston B (piston area of 20 sq. in.).

LIQUIDS CAN INCREASE FORCE

When a force is applied to piston A, Fig. 18-6, it can be increased if it is transmitted to a larger piston B. If piston A has a surface area of 1 sq. in., the 200 lb. force on piston A represents a pressure of 200 lbs. per square inch (psi).

According to Pascal's law, this 200 psi force will be transmitted undiminished.

If piston B has a surface area of 20 sq. in., piston A will exert a 200 lb. force on each square inch of piston B. This would produce a mechanical advantage (MA) of twenty, and

the original 200 lb. force would be increased to 4,000 lbs.

The force may be further increased by either making piston A smaller or piston B larger.

LIQUIDS CAN DECREASE FORCE

If force is applied to piston B, the original force of 200 lbs. would have to be raised to 80,000 lbs. to produce a pressure of 4,000 lbs. on piston A. By applying the force to the large piston transmitting it to the smaller, the force increase has been reversed and the original force, as applied to the piston, will be diminished. Remember that the force involved per square inch is transmitted undiminished. Pressure producing

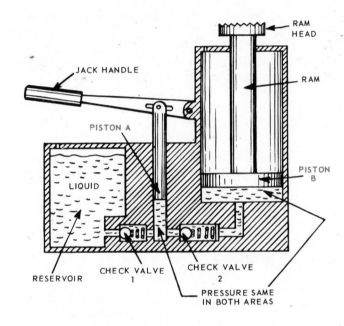

Fig. 18-7. Simple hydraulic jack. If piston B has an area 100 times greater than piston A, ram will lift 100 times more than whatever pressure is generated by piston A.

ability of this psi force can be increased or decreased by applying it to a large or to a small area. Notice in Fig. 18-7, that pressure in the pump cylinder, and pressure in the lifting cylinder, are the same yet the force applied to the pump piston has been increased many times on the head of the lift.

A HYDRAULIC JACK

Fig. 18-7 shows how a fluid, using the principle illustrated in Fig. 18-6, can be used to produce a powerful lifting force.

When the jack handle raises piston A, piston A will form a vacuum, this will draw check valve 1 open and close check valve 2. When the handle is depressed with a force that exerts 200 lbs. pressure (or any force) on piston A, check valve 1 will close, check valve 2 will open and 200 psi will be transmitted to piston B.

If piston B has a surface area one hundred times greater than the 1 sq. in. area of A, piston B will raise a weight of 20,000 lbs.

Automobile Brakes

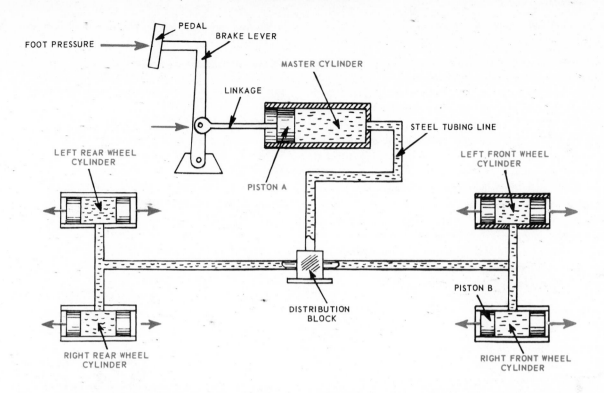

Fig. 18-8. Schematic of a simple brake system (hydraulic system). When foot pressure is applied to pedal, master cylinder piston creates pressure, thus causing wheel cylinder pistons to move outward.

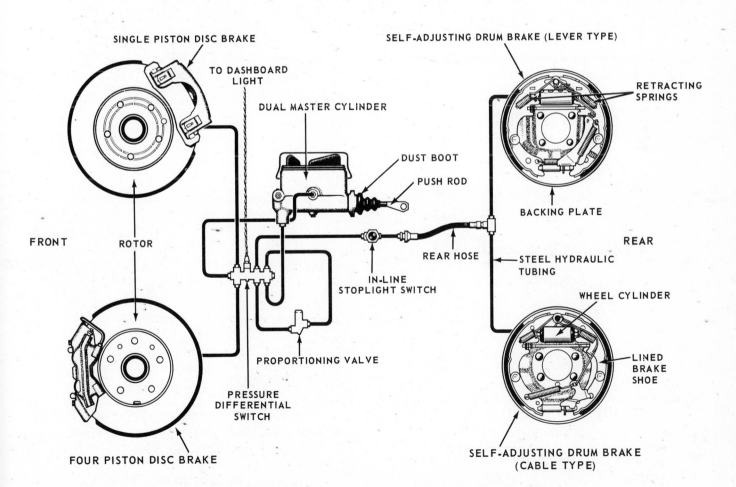

Fig. 18-9. Schematic of hydraulic brake system. Study general layout of parts and their names. Note that four different brake designs are illustrated. In practice, brake designs are paired, front or rear.   (Raybestos)

HYDRAULIC PRINCIPLES APPLIED TO CAR BRAKING SYSTEM

Basically, the car hydraulic system consists of a master cylinder (containing piston A, Fig. 18-8), steel tubing to form connecting lines, and one or two wheel cylinders (containing pistons B, Fig. 18-8) for each wheel.

When the driver depresses the brake pedal and exerts a force on the master cylinder piston, this force is transmitted, undiminished, to each wheel cylinder. The wheel cylinder pistons transfer this force (increased or decreased, depending on piston area) to the brake shoes.

When the master cylinder piston A moves, the wheel cylinder pistons will move until the brake shoes engage the revolving brake drum. Further movement is impossible. Any attempt to depress the master cylinder piston beyond this point will transmit pressure, not motion.

MASTER CYLINDER

The master cylinder is the central unit in which hydraulic pressure is developed. Pressure of the driver's foot on the brake pedal is transmitted, via various linkage arrangements, to the master cylinder piston. As the piston is forced forward in the cylinder, it pushes brake fluid ahead of it. Since the brake lines and wheel cylinders are filled with brake fluid, the piston is acting upon a solid column of fluid.

When the wheel cylinder pistons have pressed the brake shoes against the drums, fluid movement ceases and pressure rises according to the force on the master cylinder piston.

Fig. 18-9 illustrates brake pedal linkage, master cylinder, lines and wheel brace assemblies. Four systems are shown in one schematic.

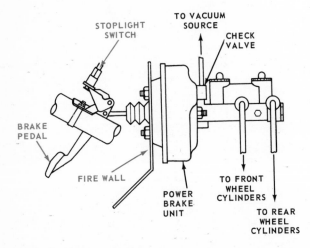

Fig. 18-10. Master cylinder located on fire wall. This location calls for a suspended brake pedal. Note stoplight switch. (Delco)

MASTER CYLINDER CONSTRUCTION

Most master cylinders are manufactured of cast iron and contain brackets with holes for mounting.

Modern practice leans heavily upon fire wall mounting (metal wall between driver and engine compartment). With the master cylinder in this location, it can be inspected and serviced easily, and it is less apt to become contaminated by water and dirt. It is operated by a suspended pedal. Fig. 18-10.

BRAKE FLUID RESERVOIR

The master cylinder generally has an integral reservoir, or reservoirs in double-piston (dual) master cylinders, for brake fluid. This will provide additional fluid, when needed, to compensate for minute leaks, lining wear requiring more fluid movement, etc. The reservoir cover is vented (has an air hole in it) to allow expansion and contraction of the fluid without forming pressure or vacuum. A rubber diaphragm contacts the fluid and follows the level up and down. The diaphragm excludes dust, moisture, etc.

A typical reservoir cover is illustrated in Fig. 18-19.

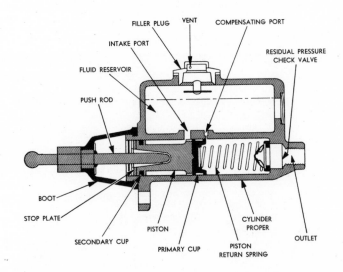

Fig. 18-11. Cross section of a typical SINGLE-PISTON master cylinder. Learn part names and relationship, one with other. Single-piston cylinder is no longer in use. (Raybestos)

CYLINDER

A very smooth wall cylinder is provided. This cylinder contains an aluminum piston. The cylinder is connected to the reservoir by two parts (holes), the compensating port and intake or breather port. Fig. 18-11.

PISTON ASSEMBLY

A close fitting aluminum piston is placed in the cylinder. The inner face of the piston presses against a rubber PRIMARY cup. This cup prevents leakage past the piston. The outer piston end has a rubber SECONDARY cup to prevent fluid from leaving the master cylinder. The inner piston head has several small bleeder ports that pass through the head to the base of the rubber primary cup.

The entire piston assembly rests in the cylinder. It is prevented from coming out by a stop plate and snap ring. Pressure is applied to the piston by means of a push rod that connects to the brake pedal linkage. Fig. 18-11.

CHECK VALVE AND PISTON SPRING

The piston is located in one end of the cylinder; a check valve is seated in the other. The check valve is held against a rubber seat by a coil spring. One end of the spring presses against the check valve; the other end is against the piston primary cup. Fig. 18-11.

The check valve is of a two-way design (illustrated in section on master cylinder operation). In operation, the piston forces fluid through the check valve into the brake lines. When the brake is released, pressure from the brake shoe return springs forces fluid back against the check valve. This pressure will force the entire check off its seat, allowing fluid to return to the cylinder. Fig. 18-17.

The check valve coil spring returns the check to its seat when the spring pressure (around 10 lbs.) is greater than the pressure of the returning fluid. This keeps pressure in the lines. This STATIC pressure will prevent the entry of air, dust, etc., into the brake system. It also assures a solid column of fluid so that movement of the master cylinder piston is immediately reflected by movement of the wheel cylinder pistons.

Study the cross section of the master cylinder illustrated in Fig. 18-11. Note the use of a rubber boot to prevent the entry of dust, water, etc., into the open end of the cylinder.

Fig. 18-12 illustrates another master cylinder. Study the part arrangement and construction.

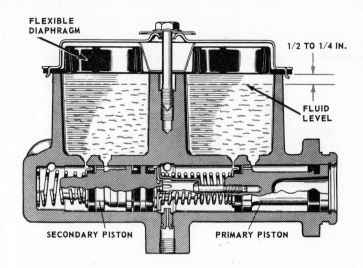

Fig. 18-13. Master cylinder fluid level. Air space is important so that reservoir fluid can expand without spewing out or creating pressure in system. Flexible diaphragm will normally ride on top of fluid. Note double-piston construction. (Bendix)

MASTER CYLINDER OPERATION, RELEASED POSITION

Fig. 18-14 shows the master cylinder in the released position. The piston is pushed back against the stop plate. There is static pressure in the brake line and wheel cylinder.

The brake shoe return springs have pulled the shoes free of the drum. The shoes, in turn, have compressed the wheel cylinder piston. The primary cup is free of the compensating port; the breather port is open to the center section of the piston. No pressure is present within the master cylinder itself.

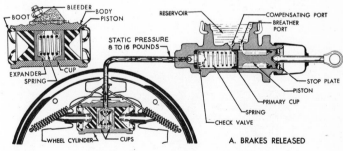

Fig. 18-14. Master cylinder in RELEASED position. Piston has returned against stop plate and only static pressure remains in brake lines. (Buick)

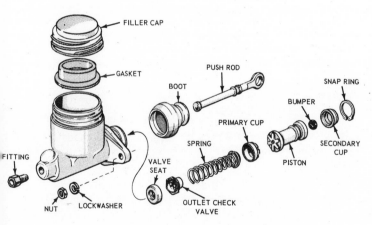

Fig. 18-12. Exploded view of single-piston master cylinder. (Ford)

MASTER CYLINDER RESERVOIR FLUID LEVEL

The master cylinder reservoir should be filled to within 1/2 to 1/4 in. of the top of the reservoir, depending upon manufacturer's recommendations.

Fig. 18-13 illustrates this, as well as the entire master cylinder. Note that a separate reservoir is used for each piston.

MASTER CYLINDER, APPLICATION

In Fig. 18-15, the brake pedal has forced the push rod toward the cylinder. This forces the piston deeper into the cylinder. As the piston moves inward, the primary cup seals off the compensating port. Reservoir fluid flows through the breather port and keeps the center of the piston filled.

With the compensating port closed, the piston traps the

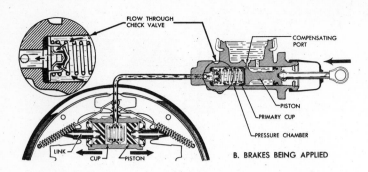

Fig. 18-15. Master cylinder in APPLIED position. Note fluid flow through check valve.

fluid ahead of it and creates pressure in the cylinder. This pressure forces the check valve inner rubber flap to open and pass fluid into the lines. As the piston continues to move inward, the wheel cylinder pistons move outward and force the shoes tightly against the spinning drum. As long as the pressure is maintained on the brake pedal, the shoes will remain pressed against the drum.

BRAKE RELEASE, START

When the driver removes his foot from the brake pedal, push rod pressure is removed from the piston. The check valve impedes the sudden return of fluid. As the piston starts to move outward in the cylinder, fluid will flow through the bleeder holes in the head of the piston. This fluid will bend the lips of the primary cup away from the cylinder wall, and fluid will flow into the cylinder ahead of the piston.

This action also allows brakes to be PUMPED up. Pumping is the repeated application of the pedal in quick movements. It is used when one full application fails to expand the shoes until they touch the drum. This condition is caused by faulty brake adjustment or worn brake shoes. Both are dangerous and

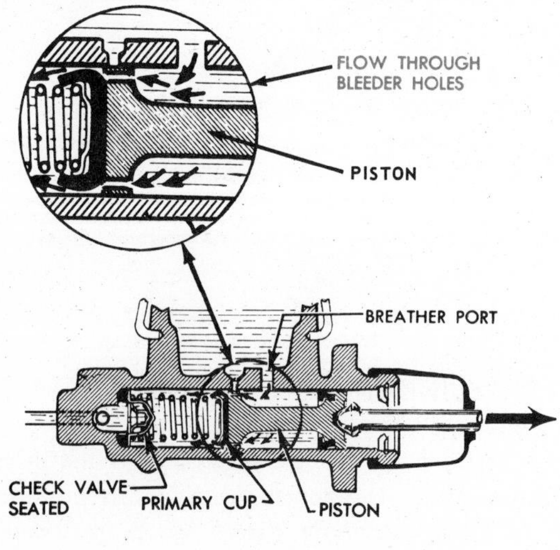

Fig. 18-16. Master cylinder at START of FAST RELEASE. Fluid is passing through bleeder holes. Rubber cup is deflected away from cylinder wall to let fluid pass.

should not be tolerated.

The flow of fluid through the bleeder holes will also prevent the possible entry of air by keeping the cylinder filled at all times. Fig. 18-16.

BRAKE RELEASE, FINISH

When pressure drops in the master cylinder, the brake shoe return springs retract the shoes. As the shoes retract, they squeeze the wheel cylinder pistons together. This, in turn, forces brake fluid flow back into the master cylinder.

The returning fluid forces the check valve rubber flap to close. The check valve is pushed off its seat, and fluid flows into the master cylinder.

As the piston returns to its full released position against the stop plate, the primary cup uncovers the compensating port and any excess fluid will flow into the reservoir.

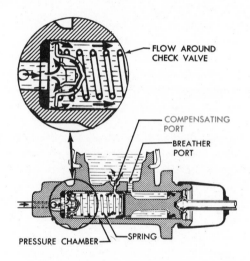

Fig. 18-17. Master cylinder at FINISH of brake RELEASE. Excess fluid is passed through compensating port. Note fluid return flow through check valve.

When the piston return spring pressure is again greater than the pressure of the returning fluid, the check valve seats, and will maintain the desired static pressure in the system. See Fig. 18-17.

DUAL (DOUBLE—PISTON) MASTER CYLINDER

For years, one real danger in the typical hydraulic brake system was the fact that a line rupture, blown wheel cylinder, etc., could cause a sudden and complete loss of braking.

To prevent this, the DUAL MASTER CYLINDER (also referred to as a TANDEM or DOUBLE-PISTON master cylinder), Figs. 18-18, 18-19 and 18-20, was developed and is now in universal use in this country.

The dual master cylinder places two pistons in a single cylinder. Each piston is served by a separate reservoir and compensating and fluid inlet ports. The pistons are kept in the released position by coil springs. Study the arrangement shown in Figs. 18-13 and 18-18.

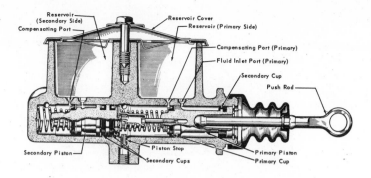

Fig. 18-18. Typical DUAL or TANDEM master cylinder. Note use of TWO pistons and separate reservoirs. This design is in common use today. (GMC)

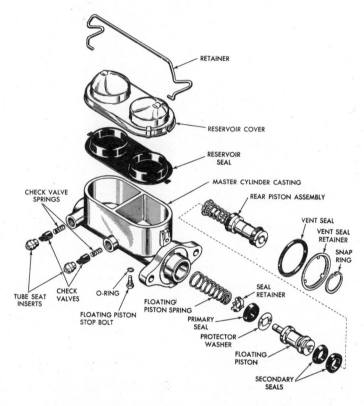

Fig. 18-19. Exploded view of a dual master cylinder. (GMC)

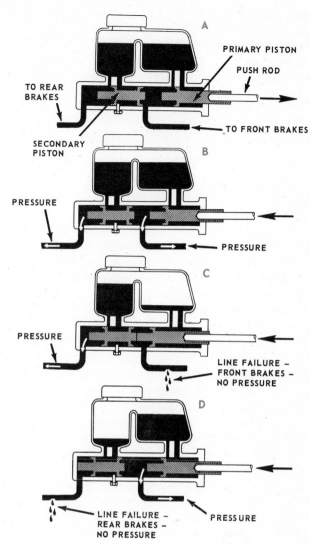

Fig. 18-20. Dual master cylinder operation. A—Released position. B—Applied position. C—Front system failure — still provides brake pressure to rear. D—Rear system failure — Still provides pressure to front. (Mercedes)

An exploded view of a similar type (dual) master cylinder is pictured in Fig. 18-19.

The secondary piston operates either the front or rear set of brakes, while the primary piston operates the other set. Both secondary and primary outlets have residual check valves (except master cylinders used with disc brakes which do not have residual valve on disc brake outlet). Operating phases are shown in Fig. 18-38.

DUAL MASTER CYLINDER OPERATION, BRAKES RELEASED

When the brake pedal is in the released position, both pistons are retracted. Each has fluid in front of it. Compensating ports are open. A, Fig. 18-20.

DUAL MASTER CYLINDER OPERATION, BRAKES APPLIED

When pedal pressure is applied to the primary piston, it moves forward and blocks off the compensating port, sealing the fluid in front of it. As it continues to move forward, it will transmit fluid pressure to the front wheel cylinders and to the base of the secondary piston. This causes the secondary piston to move foreward, block off its compensating port and apply pressure to the rear wheel cylinders. Note that pedal pressure in B, Fig. 18-20, has moved both pistons and has created pressure to both front and rear wheel cylinders.

DUAL MASTER CYLINDER OPERATION, FAILURE IN FRONT WHEEL PORTION

A failure in the front wheel portion of the system would allow fluid to escape from the system and prevent pressure buildup. Note in C, Fig. 18-20, that the primary piston moves in, forcing fluid from the burst portion of the system until the primary piston strikes the secondary piston. Additional pedal movement will cause the primary piston to physically move the secondary piston forward, transmitting normal brake pressure to the rear wheels.

DUAL MASTER CYLINDER OPERATION, FAILURE IN REAR WHEEL PORTION

If the rear brake system fails, the primary piston will force the secondary piston in until the secondary piston strikes the end of the cylinder. The primary piston will then apply normal pressure to the front system. D, Fig. 18-20.

The dual master cylinder, by providing separate systems for the front and rear brakes, will provide some braking force regardless of line failure. Although both front and rear systems could fail at the same time, such an event is highly unlikely.

PRESSURE DIFFERENTIAL WARNING SWITCH

On systems employing the dual master cylinder, failure of either front or rear systems will allow the brake pedal to travel closer to the floorboard before applying the brakes.

Either a mechanical or hydraulic switch is utilized to operate a warning light to alert the driver that the pressure in one system is low or gone. The mechanical switch is triggered by excessive brake travel. The hydraulic switch is operated by a difference in pressure (around 80 – 150 lbs.) between the front and the rear system.

One type of hydraulic pressure differential switch is shown in Fig. 18-21. A drop in pressure on either side will allow the piston assembly to move toward the side with low pressure. This will align the cutout area of the piston with the switch plunger, causing the plunger to move downward, touching the switch contact and closing the circuit to the warning light.

BRAKE LINES

The master cylinder is connected to the wheel cylinders by high quality, double thickness, steel tubing. The tubing is copper plated and lead coated to prevent rust and corrosion.

DON'T TRUST IT!

NEVER REPLACE A BRAKE LINE WITH COPPER TUBING OR INFERIOR STEEL TUBING. THAT A CAR'S BRAKES BE GOOD IS ABSOLUTELY ESSENTIAL. USE HIGH QUALITY BRAKE TUBING!

Brake tubing, where connections are made, uses a double lap flare. Fig. 18-22. Fig. 18-23 illustrates how one type of flaring tool produces a double lap flare.

VIBRATION MUST BE ELIMINATED

The brake lines must be secured, by use of clips, along their path. At no point must they be free to vibrate. Vibration will cause metal fatigue with ultimate line failure.

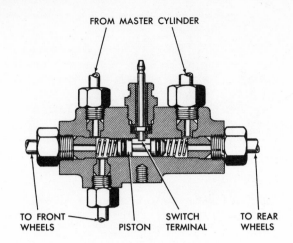

Fig. 18-21. One type of pressure differential switch used to alert driver that brake pressure in either front or rear wheel system is gone. (Chevrolet)

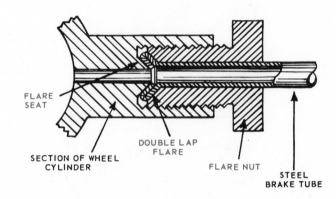

Fig. 18-22. Typical double lap flare. Notice how the flare nut draws double lap flare tightly against the flare seat, forming a pressure-tight, leakproof connection.

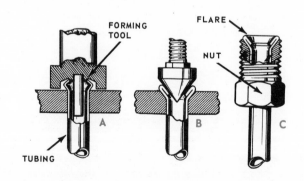

Fig. 18-23. Forming a double lap flare. A—Start of flare. B—Completion of flare. C—Completed flare with tubing nut in position. (Bendix)

FLEXIBLE HOSE

It is necessary to use a flexible, HIGH PRESSURE hose to carry the fluid to each front cylinder, and one such hose to connect the brake tubing from the rear axle to that on the body. As the wheels move up and down in relation to the

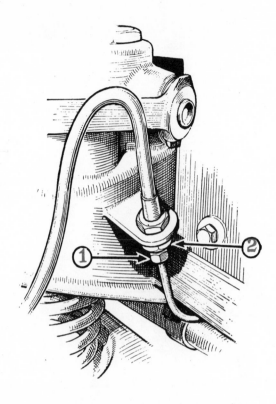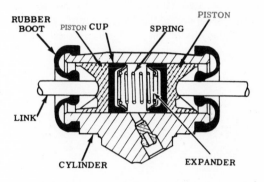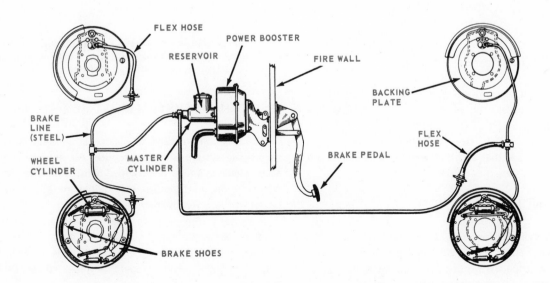

frame, this hose prevents line breakage. Fig. 18-24 shows a flex hose leading from the brake line on its way to the front wheel brake cylinder. Note that the line and hose, where connected, are securely held in a bracket.

Fig. 18-24. Flexible brake hose. To disconnect brake line, hold nut 2 while loosening nut 1. (M.G.)

Fig. 18-25 illustrates a typical hydraulic service brake system. A distributor fitting carries the line fluid to both front and rear brakes.

WHEEL CYLINDER, PURPOSE

The wheel cylinder is used to transmit the master cylinder pressure to the brake shoes and force them outward against the drum. One wheel cylinder (two, in some systems) is used at each wheel.

WHEEL CYLINDER CONSTRUCTION

The wheel cylinder assembly is of rather simple construction. It consists of a cast iron housing, two aluminum pistons (sintered iron pistons are sometimes used), two rubber cups, cup expanders (in some applications) a light-weight coil spring, two push rods and two rubber dust boots.

Fig. 18-26. Cross section of typical double-piston wheel cylinder. Stamped metal expander cups at each end of spring tend to keep lips of rubber cups in constant contact with cylinder wall.
(Oldsmobile)

The cylinder is drilled to provide an entrance for a bleeder screw (covered later) and the brake line connection. The cylinder is usually bolted to the brake backing plate (covered later). Study Figs. 18-26 and 18-27.

Fig. 18-25. Typical service brake system.

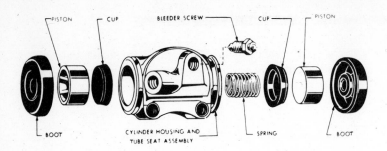

Fig. 18-27. Exploded view of typical double-piston wheel cylinder.
(Lincoln)

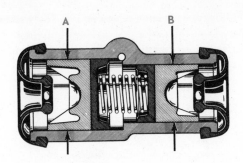

Fig. 18-29. Stepped wheel cylinder. Note difference in cylinder diameter — A being larger than B. Piston in cylinder A will exert more force than piston in cylinder B.

SINGLE—PISTON CYLINDER

In cases where two wheel cylinders are used per wheel, the cylinders are generally of the single-piston type. Since each cylinder operates one of the shoes, it is necessary to direct the pressure in only one direction. Fig. 18-28.

You will notice the cup design in the wheel cylinder is similar to the primary cup in the master cylinder. When the fluid exerts pressure against the cup, the flanged edges are pressed tightly against the cylinder. See A, Fig. 18-30.

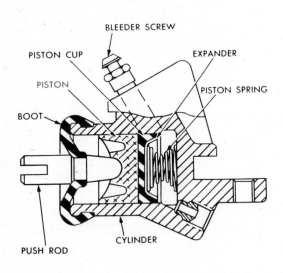

Fig. 18-28. Single-piston wheel cylinder.
(Dodge)

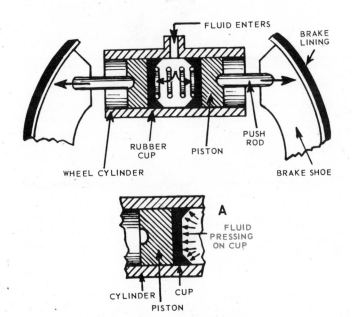

Fig. 18-30. Simplified wheel cylinder action. Fluid, under pressure, enters wheel cylinder and, by pressing on rubber cups, forces pistons and push rods outward.

STEPPED CYLINDER

When it is desired to transmit more pressure to one of the shoes than to the other, the wheel cylinder is built with a stepped cylinder. One-half of the cylinder is of one size, and the other half is another size. Recalling the section on hydraulics, you can see that each half would exert a different pressure on its brake shoe. Fig. 18-29.

WHEEL CYLINDER OPERATION

When master cylinder forces fluid into wheel cylinder, the two pistons move apart. Push rods, or links, connect each piston to a brake shoe. As pistons move outward, they force shoes against drum. Study the simplified illustration in Fig. 18-30. Also see Figs. 18-14 and 18-15.

HYDRAULIC SYSTEM MUST BE FREE OF AIR

As air is compressible, any air in a hydraulic system would render the system unfit for service. Air must be removed, so each wheel cylinder is provided with a bleeder screw. The bleeder screw is threaded into an opening leading into the center of the wheel cylinder.

Pressure is applied on the master cylinder, and the bleeder screw is loosened. This allows trapped air to escape. Then the bleeder screw is tightened before pressure is removed from the master cylinder. This process is repeated at each wheel cylinder until all air is removed.

A typical bleeder screw is shown in Fig. 18-31. When the

screw is loosened, the tapered point uncovers the bleeder hole. This permits air to move up around the point, through the cross hole and out the center passageway. When the screw is tightened, the point seals the opening.

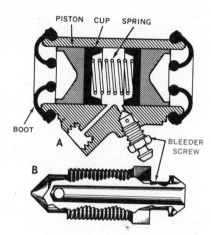

Fig. 18-31. A—Wheel cylinder showing bleeder screw. B—Cutaway of bleeder screw.

One method of bleeding brakes is shown in Fig. 18-32. One end of a rubber hose is connected to the bleeder screw. The other end is inserted in a partially filled jar of brake fluid. As the air leaves the end of the hose submerged in the fluid, air bubbles rise. Air cannot be drawn back into the system.

CAUTION!

USE HIGH QUALITY BRAKE FLUID ONLY. IT SHOULD MEET OR EXCEED SAE RECOMMENDATION J 1703 AND CONFORM TO MOTOR VEHICLE SAFETY STANDARD 116. ANY MINERAL OIL SUCH AS KEROSENE, GASOLINE, ETC., IN EVEN THE SMALLEST AMOUNTS, WILL SWELL AND DESTROY THE RUBBER CUPS IN THE SYSTEM.

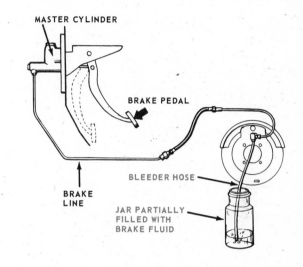

Fig. 18-32. Manual method of bleeding air from brake system. Air escaping from bleeder hose will rise to top of brake fluid.

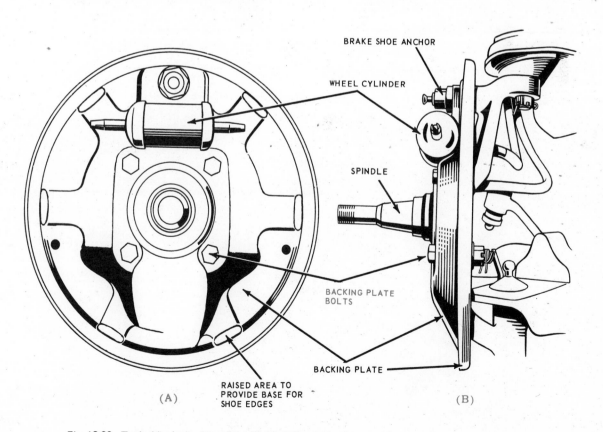

Fig. 18-33. Typical brake backing plate. Backing plate forms a foundation for brake shoes and wheel cylinder.
(Raybestos)

WHEEL BRAKE ASSEMBLY

The hydraulic system can be termed a brake actuating device. The wheel brake assemblies form the second part of the braking system.

Each wheel assembly consists of a brake backing plate, a set of brake shoes with clips, return springs, etc., and a brake drum.

BACKING PLATE

A round, stamped steel disc is bolted to the spindle steering knuckle on the front, and to the end of the axle housing in the rear. This disc, called the backing plate, is used as a foundation upon which the wheel cylinder and brake shoe assembly is fastened. The backing plate is rigid and cannot move in any direction. Fig. 18-33.

BRAKE SHOE ASSEMBLY

Most wheel brake assemblies use two brake shoes. These shoes are of stamped steel and have brake linings either riveted or bonded (glued) to the outer surface.

Although of similar design, there are some minor differences in the shapes of the shoes. All utilize a T-shape cross section. The web is used to give the shoe rigidity. When forced

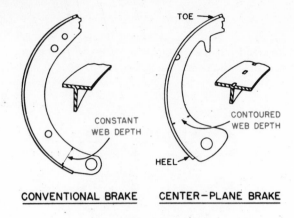

Fig. 18-34. Two brake shoe designs. (Dodge)

against the drum, it will exert braking pressure over the full lining width and length. Fig. 18-34. The ends of the brake shoes may be free-floating, or they may have one end fastened to an anchor.

Brake return, or retracting, springs are used to pull the shoes together when hydraulic pressure is released. Small spring clips of various designs are used to keep the shoes against the backing plate to insure shoe alignment and prevent rattle. Study the arrangement of parts in Fig. 18-35.

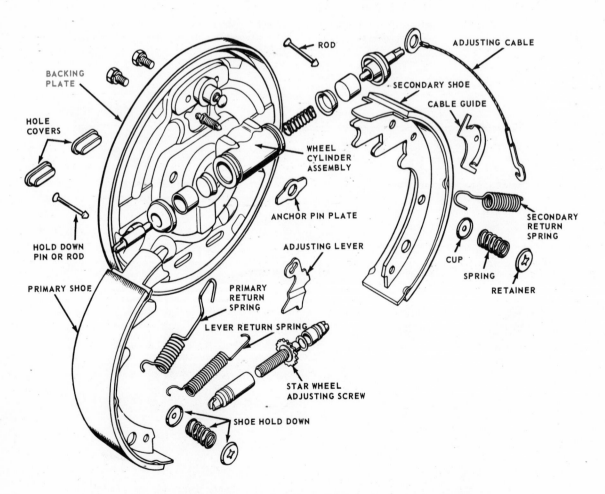

Fig. 18-35. Backing plate and brake shoe assembly — front wheel. (Dodge)

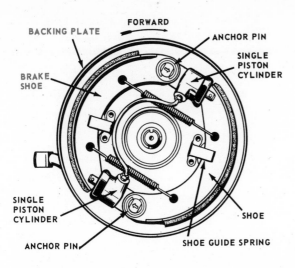

Fig. 18-36. Backing plate and shoe assembly in place. Note this setup uses two single-piston wheel cylinders. Eccentric anchor pins adjust to center shoes with drum and also hold wheel cylinders in place. (Bendix)

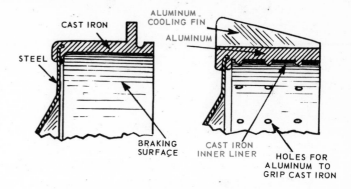

Fig. 18-37. Sections of typical brake drums. Note cast iron inner liner when aluminum is used to facilitate cooling.

A backing plate and shoe assembly is shown on a front wheel spindle steering knuckle in Fig. 18-36. This is the two cylinder, single-piston type. Each shoe pivots on one end; the wheel cylinders shove the other end outward.

PRIMARY SHOE

The PRIMARY (forward or leading) brake shoe is the shoe that faces toward the front of the car. It often has different lining than the other shoe.

SECONDARY SHOE

The SECONDARY (reverse or trailing) brake shoe is the shoe that faces the rear of the car.

BRAKE LINING

Brake lining is made of asbestos impregnated with special compounds to bind the asbestos fibers together. Some linings are woven of asbestos threads and fine copper wire. The most widely used lining is molded (asbestos fibers ground up and pressed into shape). Various lining compositions are used to produce specific braking characteristics. For heavy duty braking, some manufacturers offer special metallic lining that is very resistant to brake fade.

BRAKE DRUM

The brake drum bolts to the wheel hub between the hub and the wheel. It completely surrounds the brake shoe assembly and comes very close to the backing plate so that water and dust will be excluded.

The center section is of stamped steel construction, while the outer braking rim is of cast iron. An aluminum outer rim is sometimes used to aid in cooling. When aluminum is used, a cast iron braking surface is fused to the aluminum rim.

The outer rim is cast to the steel center section. Cooling fins are often cast into the rim. Fig. 18-37. The braking area of the drum must be smooth, round and parallel to the shoe surface.

When the wheels turn, the drum revolves around the stationary brake shoes. Some drums have a stiff spring stretched around them, on the outside rim, which helps to reduce drum vibration.

A complete wheel brake assembly is pictured in Fig. 18-38. Study this drawing carefully. Note how the brake drum meets the backing plate (dust shield). The plate edge covers the drum for a short distance to provide a good seal.

HOW BRAKE ASSEMBLY WORKS

When master cylinder operation creates pressure in the wheel cylinder, the wheel cylinder pistons and links shove the brake shoes outward into contact with the revolving brake drum. As the shoes cannot revolve, they will stop the drum and wheel.

The kinetic energy (energy stored up in a moving object) in the moving car is converted into heat by the brakes. A fast moving car has tremendous energy. To bring it to a stop will produce a great amount of heat. This heat is given off, mostly to the surrounding air, by the brake drums. Some will pass to the shoes and to the backing plate.

BRAKE FADE

When brake lining becomes overheated, it loses a great deal of its frictional properties. Even hard pedal pressure will not produce fast, even stops.

Some cars, designed for fast and rigorous driving, use special metallic linings that have great resistance to fade.

BRAKE SHOE ARRANGEMENT

There are many arrangements used in mounting brake shoes. Several popular ones are illustrated in this chapter. Some utilize a servo action (one shoe helps to apply other), as well as a self-energizing action (using frictional force to increase shoe to drum pressure).

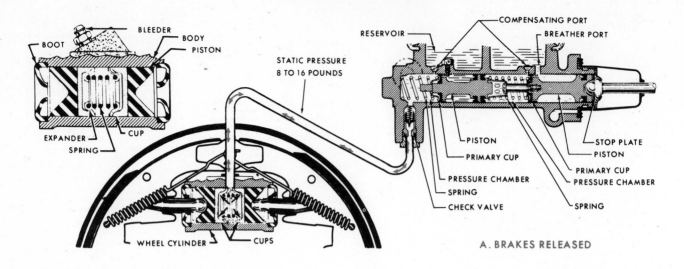

A. BRAKES RELEASED

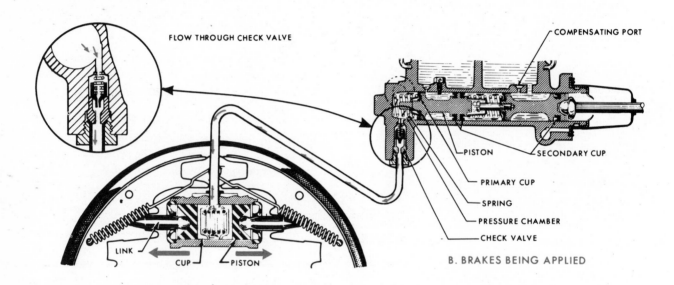

B. BRAKES BEING APPLIED

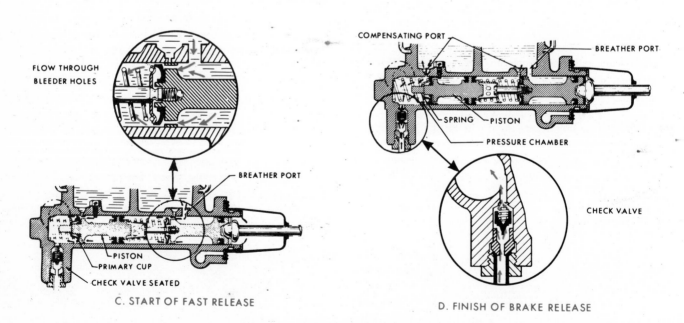

C. START OF FAST RELEASE

D. FINISH OF BRAKE RELEASE

Fig. 18-38. Dual master cylinder. Cylinder operation is shown in released (A), applied (B), start of fast release (C). Study action of brake master cylinder in each operational phase. (Buick)

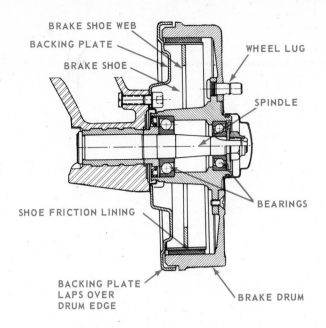

Fig. 18-39. Cross section of complete wheel brake assembly. Note how shoe friction lining engages drum. Backing plate laps over drum outer edge to help prevent entry of water, dirt, etc. (Peugeot)

SERVO ACTION, SELF—ENERGIZING BRAKES

Servo and self-energizing action is produced by hooking the heel of the primary shoe to the toe of the secondary shoe. When the wheel cylinder forces the top ends of the shoes out against the drum, the forward shoe will tend to stick to the revolving brake drum and be carried around with it. As the primary shoe attempts to revolve, it will jam the secondary shoe against the single anchor pin. This stops both shoes and produces a binding effect that actually helps the brake shoes apply themselves. This servo and self-energizing action reduces the amount of pedal pressure necessary. Study Fig. 18-40.

Note, in Fig. 18-40, how the primary shoe attempts to

rotate in the direction of the drum. Since the adjusting screw connects it to the toe of the secondary shoe, the heel of the secondary shoe is jammed against the anchor pin. The arrows illustrate the braking force direction. When the car is traveling in reverse, the secondary shoe applies the primary shoe.

The primary shoe on self-energizing brakes generally has a smaller amount of lining and, usually, it is of a different composition. The secondary shoe does more of the braking; less is needed on the primary shoe.

A servo action, self-energizing brake assembly is shown in Fig. 18-41. Note that the primary shoe has less lining. This is a

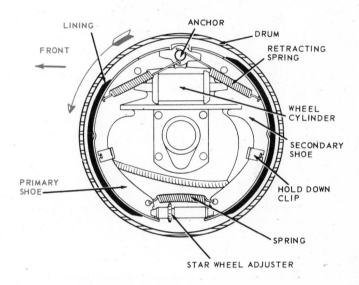

Fig. 18-41. Servo action, self-energizing brake. Primary shoe (left) has a short lining when used in this setup. (Wagner Electric Corp.)

rear wheel brake. The star wheel at the bottom is used to adjust the shoes to keep them close to the drum. Figs. 18-41 and 18-42.

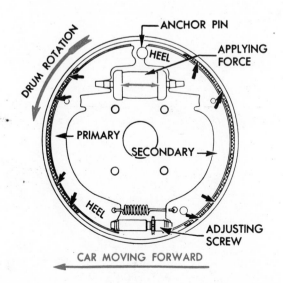

Fig. 18-40. Servo action, self-energizing brake. Primary shoe starts to move with drum and in so doing, applies secondary shoe. (Buick)

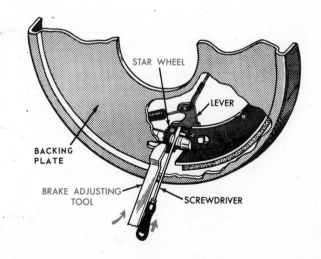

Fig. 18-42. Star wheel shoe adjuster. Special brake adjusting tool engages star wheel teeth to make initial brake adjustment. By moving handle one way or the other, adjuster can be shortened or lengthened to adjust shoe-to-drum clearance. Screwdriver is being used to keep automatic adjuster lever out of engagement with star wheel. (Plymouth)

DOUBLE ANCHOR BRAKE

Notice that each shoe is anchored at the bottom. Fig. 18-43. The primary shoe is self-energizing; the rear shoe is not. Spring clips at the outer ends hold the shoes against the backing plate.

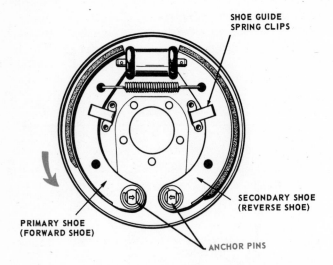

Fig. 18-43. Double anchor brake. Each shoe is fastened to a separate anchor. Shoe-to-drum alignment is secured by rotating eccentric shaped anchor pins.

SINGLE ANCHOR, SELF—CENTERING BRAKE

The brake shown in Fig. 18-44 has no servo action, but the primary shoe is self-energizing. It is self-centering in that the shoe anchor does not fix the location of shoes in any specific relationship to the drum. The shoes can move up or down until centered. The brakes in Fig. 18-41 are also self-centering.

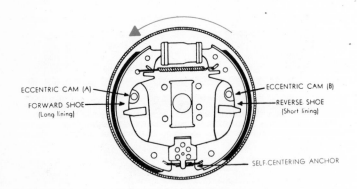

Fig. 18-44. Single anchor, self-centering brake. Shoe ends, where they butt against anchor, are free to move up or down and automatically align with drum. (Raybestos)

DOUBLE ANCHOR, DOUBLE CYLINDER BRAKE

In the brake assembly in Fig. 18-45, the free end of each shoe is actuated toward the direction of rotation, and both

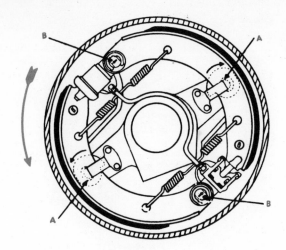

Fig. 18-45. Double anchor, double cylinder brake. Apply pistons move each shoe in such a way as to make them self-energizing. (Raybestos)

shoes are self-energizing. Two single-piston cylinders are used.

Brake adjustment is provided by eccentrics. A in Fig. 18-45. The anchors, B, are also eccentrics so that the shoes may be centered in relationship to the drum.

SELF—ADJUSTING BRAKES

Modern brake systems utilize a device that automatically adjusts the drum to lining clearance as the linings wear down. Manual adjustment is provided by turning the star wheel with a special adjusting lever.

The adjustment systems are of three general types: cable, link and lever. Some systems operate only when braking in reverse; others, when braking in a forward direction, and still another operates when braking in either direction.

A special adjustment lever is attached to one of the brake shoes in such a manner that it is free to pivot back and forth. One end engages the teeth in the star wheel, while the other is attached to a link or a cable. Figs. 18-46 and 18-47. The cable, or the link, is attached either to the anchor (adjustment will

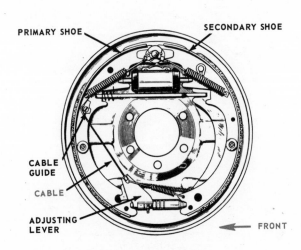

Fig. 18-46. Cable type automatic brake adjuster. This specific setup works when braking in both directions.

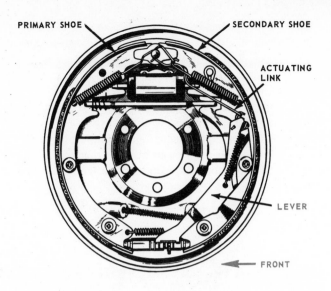

Fig. 18-47. Lever type automatic brake adjuster. This unit functions only when braking in reverse.

The amount of lever movement is very small, and normally the cable will not draw the lever up far enough to engage a new tooth on the star wheel. When sufficient lining wear has taken place, however, the out and around movement will be enough to cause the cable to draw the lever up to the next tooth on the star wheel. When the brakes are released, the return spring will draw the lever down, forcing it to rotate the star wheel one tooth.

This adjustment action will take place during the entire life of the lining. Fig. 18-46 depicts a typical cable operated adjustment system.

A lever type adjuster is shown in Fig. 18-47. Since the actuating link is attached to the anchor, it will operate only in one direction. On this unit, where the lever is pivoted to the secondary shoe, it functions only during reverse braking.

The link type adjuster is very similar to the cable type. The utilization of links (wires) instead of the cable is the basic difference. Fig. 18-48.

BRAKING RATIO

When a car is stopped, there is transfer of weight to the front of the car (notice how front end dips down during quick stops). This weight transfer forces the front tires more tightly against the road while it relieves the rear tires somewhat.

To compensate for this effect, and to also take advantage of it, the front brakes usually are designed to produce more stopping power than the rear. An average ratio would be: Front brakes — 55 to 60 percent. Rear brakes — 45 to 40 percent of the stopping force.

DISC BRAKE

The DISC brake differs from drum brakes in the use of a heavy disc or rotor. The disc is bolted to and revolves with the wheel hub. The disc, Figs. 18-49 and 18-50, may be solid or slotted for improved cooling.

A brake CALIPER, housing two or four hydraulic pistons (equal number on each side of disc) is placed around the disc. The caliper is bolted to the spindle. Brake friction pads are so arranged that when hydraulic pressure is built up behind the pistons, the pads will be forced against the disc, providing braking effort. A typical disc brake is pictured in Fig. 18-49.

Another type of caliper uses a single piston on one side only. This caliper is free to slide sideways to allow brake pads to align with disc.

CALIPER CONSTRUCTION AND OPERATION

The heavy, rigid brake caliper, of one or two piece construction, contains one, two (one each side), three or four (two each side) hydraulic cylinders. The caliper is bolted securely to the spindle.

Pistons are fitted to the caliper cylinders with the outer ends resting against friction pads. Rubber boots exclude the entry of dirt and moisture.

One design allows the brake pads to drag VERY LIGHTLY against the rotating disc at all times. Another type operates

take place when braking in one direction only) or to the other brake shoe (will function when braking in either direction). A spring holds the adjustment lever in a downward position.

Study the construction of the cable type adjuster in Fig. 18-46. This particular unit passes the actuating cable up over the anchor and attaches it to the secondary shoe, causing it to function when braking in either direction.

When the brakes are applied, in a forward direction for example, the shoes move out till they contact the rotating drum. They will then move around with the drum until the secondary shoe heel engages the anchor. The small out and around movement tightens the actuating cable, and it pulls upward on the adjustment lever a small amount. When the brakes are released, the shoes are retracted, causing the actuating cable to become slack and the return spring pulls the adjustment lever back down. This action is repeated during each brake application.

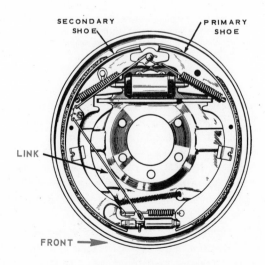

Fig. 18-48. Link type automatic brake adjuster. This unit functions during reverse braking.

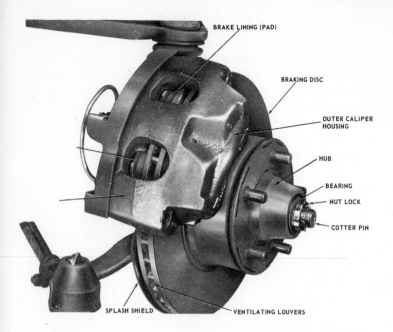

Fig. 18-49. Typical front wheel disc brake. Disc brake has excellent cooling characteristics, making it highly resistant to brake "fade." (Plymouth)

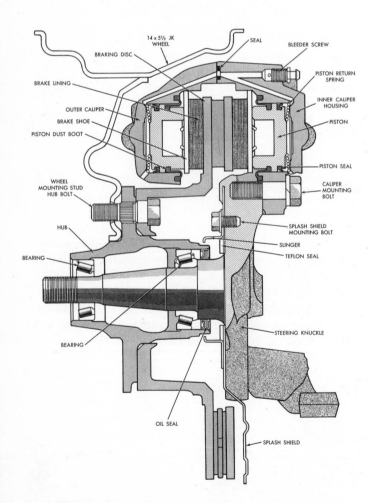

Fig. 18-50. Cross section of disc brake, showing caliper construction. On this unit, piston return spring causes piston to bear against friction pad (brake lining), causing lining to rub lightly against disc even when brake is released. Disc is made of cast iron.

with only a minimal pad to disc clearance (about .005 in.).

A hydraulic line from the master cylinder leads to one side of the caliper. The other side will receive pressure through a crossover (external) line or bypassing the fluid internally through the casting as shown in Fig. 18-50.

The brake lining, or friction pads, are self-centering and press on each side of the disc with equal pressure. Study the caliper construction illustrated in Fig. 18-50.

An exploded view, showing caliper construction, is pictured in Fig. 18-51. Note that the friction pad or lining is affixed to a metal brake shoe.

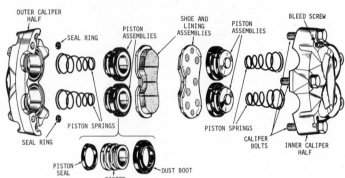

Fig. 18-51. Two-piece disc brake caliper, using four apply pistons. (Bendix)

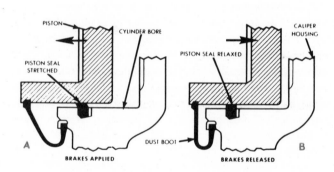

Fig. 18-52. Piston seal pulls piston back, providing pad-to-disc clearance. A—Brake applied. Piston moves outward and stretches seal. B—Brakes released. Seal rolls back to original shape, drawing piston back with it.

The caliper piston shown in Fig. 18-52 operates against a snug seal ring snapped into a groove in the cylinder wall. When the brake is applied, the piston moves outward. In so doing, it stretches the seal to one side. A in Fig. 18-52. When brake pressure is released, the seal returns to its normal position, B. This seal "roll" action pulls the piston back around .005 in., providing a small amount of pad (lining)-to-disc clearance. As the lining pads wear, the piston moves out through the seal, automatically keeping the proper pad to disc clearance.

A cutaway view of a caliper, using the seals to withdraw the pistons upon brake release, is pictured in Fig. 18-53.

Current practice in this country, when disc brakes are employed, is to place discs on the front with conventional

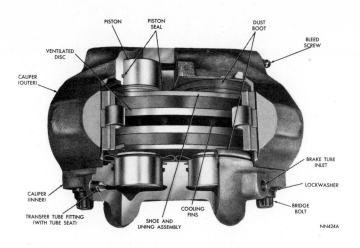

Fig. 18-53. Cutaway of four-piston caliper. Note ventilated disc and piston seals. (Plymouth)

drum brakes on the rear. However, some models are offered with disc brakes on all four wheels.

METERING VALVE

In order to insure that the back drum brakes apply at about the same time as the front discs, and that the drum brakes do an equal amount of work during light stopping, a metering valve is placed in the line to the disc brakes.

The metering valve prevents brake fluid movement to the discs until a specified psi has built up in the system (75 – 125 lbs.). By the time system pressure reaches this point, the force of the heavy drum brake shoe retracting springs has been overcome, and the shoes are starting to contact the drum. The metering valve then opens and allows fluid pressure to apply the disc brakes. A typical metering valve is illustrated in Fig. 18-54. Note that it is part of a combination valve that also incorporates a brake warning switch and a rear wheel

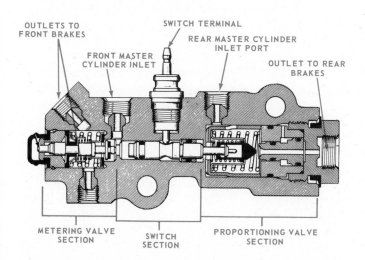

Fig. 18-54. This combination valve incorporates a metering valve (left portion), a pressure differential switch (center portion) and a proportioning valve (right portion). Metering valve blocks fluid pressure to front disc brakes until a certain pressure has been reached. This assures that rear drum brakes are applied. (Buick)

proportioning valve. No residual pressure is maintained in the front disc system.

PROPORTIONING VALVE

Another valve, called a proportioning valve, can be placed in the line to the rear wheels. This valve limits the amount of pressure to the rear drum brakes to prevent rear wheel lockup during rapid stops. The amount of pressure to the rear wheels is directly proportional to the amount of pressure developed within the master cylinder. Fig. 18-55 illustrates a system using metering and proportioning valves. See Fig. 18-54 also.

Disc brakes are highly efficient and resist brake fade (loss of lining-to-disc friction due to overheating) to a high degree. This is due to the disc rotating in the open air and to the smaller friction area. Despite the fact that the disc is exposed to water and dirt, the disc brakes work well and without grabbing, during wet or dry operation.

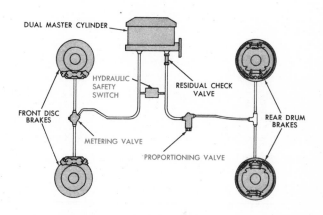

Fig. 18-55. Disc brake system shown employs a proportioning valve, a metering valve and a safety switch (pressure differential switch). (Ammco)

Due to the relatively small friction area, disc brakes require somewhat higher hydraulic pressure than conventional drum brakes. This condition is amplified by the lack of either a servo or self-energizing action found in some drum brakes.

EMERGENCY OR PARKING BRAKES

Two types of emergency brakes are in general use. One type uses a mechanical linkage system to operate the brake shoes (usually the rear). The other type uses a brake drum fastened to the transmission output shaft to apply stopping force through the drive line and axles. The transmission emergency brake is no longer in general use on automobiles.

BRAKE SHOE EMERGENCY OR PARKING BRAKE

A foot pedal or a dash mounted handle is connected through linkage and cables, to both rear wheel brake assemblies. When the emergency brake is placed in the ON position, the cables pull a lever in each rear brake assembly.

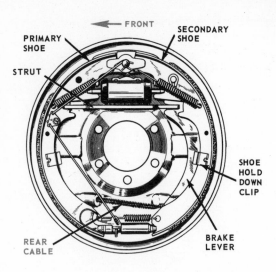

Fig. 18-56. Rear wheel emergency brake assembly. Rear cable pulls on brake lever, thus causing strut to spread and apply brake shoes. (Bendix)

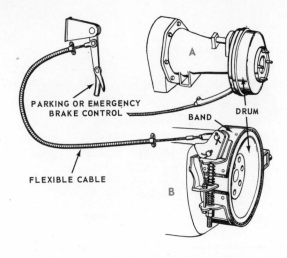

Fig. 18-57. Two types of drive line emergency brakes. "A" shows internal type that uses two brake shoes operating against inside of drum. "B" is external type that draws a friction-lined band against outside of drum to prevent drive line from turning. (Bendix)

The top of the lever is bolted to one shoe, and one end of a strut (steel bar) is notched into the lever below the top bolt. The other end of the strut is notched into the opposite shoe. The emergency brake cable is attached to the lower end of the lever. Fig. 18-56.

When the cable pulls the bottom of the lever toward the left, the lever forces the strut to the left until the strut presses the shoe against the drum. At this point, continued movement of the lever causes the lever to pivot on the strut, moving the top of the lever, and the shoe to which it is bolted, to the right until that shoe engages the drum. Further lever movement will now apply force to both shoes.

DRIVE LINE EMERGENCY BRAKE

The drive line emergency uses a brake drum mounted on the end of the transmission output shaft. Two stationary brake shoes are mounted inside. The emergency linkage cable operates a lever that applies these shoes much as the rear wheel shoes are applied.

When the drum is stopped, it holds the drive line stationary, and the car (when both rear wheels are on the ground) cannot roll. A in Fig. 18-57.

EMERGENCIES WILL ARISE

ADJUSTMENT OF THE EMERGENCY OR PARKING BRAKE IS OFTEN NEGLECTED. KEEP THE EMERGENCY BRAKE PROPERLY ADJUSTED, ALL LINKAGE FREE AND LUBRICATED. IT WILL BE EAISER TO USE WHEN IT IS REALLY NEEDED!

BRAKE LIGHTS

Brake lights are operated either by a hydraulically operated switch placed somewhere in the brake line, or by a mechanically operated switch actuated by the brake pedal.

POWER BRAKES

Many cars are equipped with power brakes. Such brakes are designed to reduce the amount of pedal pressure necessary to stop the car. Another feature is that pedal travel (distance from release position to full brake application) can be shortened. This makes it possible to have the pedal approximately the same distance from the floorboard as the accelerator pedal. The driver, to apply the brakes, merely tips his toe from the accelerator to the brake. The power brake unit is an additional part that operates in conjunction with the regular brake system already discussed.

HOW POWER BRAKE IS CONSTRUCTED

The power brake unit is built to take advantage of the vacuum produced in the car engine. In very elementary terms, it is a closed cylinder with a piston inside. One side of the piston is connected to the master cylinder piston. The other side is connected to the brake pedal. A vacuum control valve is placed between the brake and the piston.

The vacuum control valve admits vacuum to one side of the piston, while the normal atmospheric pressure is allowed to exist on the other. This valve can also allow atmospheric pressure to reach both sides of the piston. Fig. 18-58.

HOW THE POWER BRAKE WORKS

When the driver presses the brake pedal down, the vacuum control valve closes off the atmospheric pressure to the brake cylinder side of the piston. Further movement of the brake pedal opens a vacuum inlet passage to this same side. As there is atmospheric pressure on one side of the piston and a partial vacuum on the other, the piston will be forced toward the vacuum side. Since the piston is connected to the master cylinder piston, it will apply pressure to the brake system.

The power brake has three stages of operation:

1. Brakes released (released position).

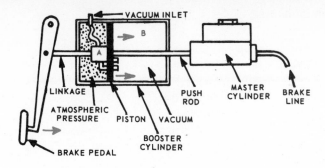

Fig. 18-58. Schematic of a simplified power brake. Engine vacuum is admitted through vacuum inlet. Control valve A can apply either vacuum or atmospheric pressure to B section of booster cylinder. As shown, brake pedal is being applied. Pedal linkage has caused control valve A to cut off atmospheric pressure and to apply vacuum to B side. Atmospheric pressure is forcing piston and push rod to right, building pressure in master cylinder.

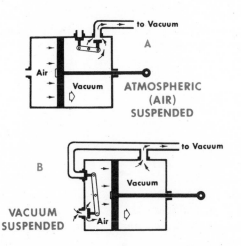

Fig. 18-59. Atmospheric and vacuum suspended power boosters. (Wagner)

Fig. 18-60. Typical power booster. 1—Master cylinder. 2—Vacuum check valve. 3—Grommet. 4—Diaphragm. 5—Diaphragm plate. 6—Rear shell. 7—Diaphragm spring. 8—Reaction plate. 9—Valve plunger. 10—Bearing seal. 11—Poppet valve. 12—Poppet valve spring. 13—Poppet retainer. 14—Dust boot. 15—Valve operating rod. 16—Filter and silencers. 17—Valve rod and plunger return spring. 18—Mounting stud. 19—Air valve lock plate. 20—Diaphragm. 21—Front shell. 22—Push rod seal. 23—Cylinder-to-shell seal. 24—Hydraulic push rod. 25—Adjusting screw. (GMC)

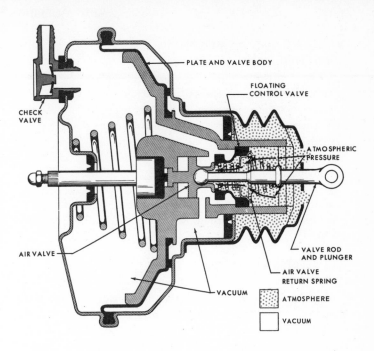

Fig. 18-61. Power booster in RELEASED POSITION. Vacuum exists on both sides of diaphragm (vacuum suspended), allowing spring to move diaphragm plate to right. No pressure is applied to master cylinder. (Buick)

2. Applying brakes (applied position).
3. Holding constant apply pressure (holding position).

TYPICAL POWER BOOSTER

Study the sectional view in Fig. 18-60. This unit is the same as the one pictured in Figs. 18-61, 18-62 and 18-63. Use it for part name and identification when studying Figs. 18-61, 18-62 and 18-63.

Fig. 18-62. Power booster in APPLIED POSITION. Poppet valve has closed vacuum port. Valve plunger has moved to left, opening atmospheric port. Atmospheric pressure is forcing diaphragm plate toward vacuum side. (Pontiac)

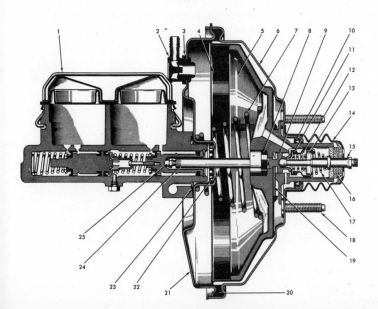

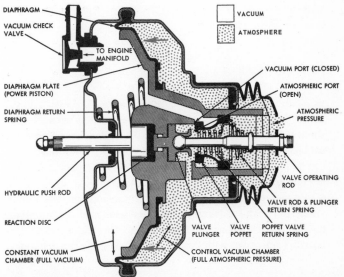

VACUUM AND ATMOSPHERIC SUSPENDED BOOSTER

In reference to vacuum and atmospheric (air) suspension, the vacuum suspended power booster has a vacuum existing on both sides of the diaphragm or piston when the booster is in the released position. When the booster is applied, atmospheric pressure is admitted to one side to cause the necessary movement of the piston. B in Fig. 18-59.

An atmospheric suspended booster has atmospheric pressure on each side of the piston in the released position. To cause booster action, vacuum is admitted to one side. A, Fig. 18-59.

RELEASED POSITION

With engine running and brakes fully released, the valve operating rod and valve plunger are moved to the right by the rod and plunger return spring. Fig. 18-61. This presses the right end of the valve plunger against the face of the poppet valve, closing off the atmospheric port. The vacuum port is open and a vacuum is formed on both sides of the diaphragm plate (side to left of diaphragm is subjected to constant vacuum regardless of valve plunger position).

With a vacuum on both sides, the diaphragm return spring moves the diaphragm plate to the right, releasing all pressure on the master cylinder. Fig. 18-61.

APPLIED POSITION

When the driver depresses the brake pedal, the valve operating rod is moved to the left. Fig. 18-62. This causes the valve plunger to move to the left. As the plunger moves, it will compress the valve return spring and move the poppet valve into contact with the vacuum port seat in the valve housing. This closes the vacuum port leading to the right side of the diaphragm.

Further application of the brake pedal will cause the valve rod to force the valve plunger away from the poppet, OPENING the atmospheric port. As the port is opened, atmospheric air will rush into the area to the right of the diaphragm (control vacuum area) and force the diaphragm plate to the left. This, of course, applies pressure to the master cylinder push rod. (Note how rubber reaction disc is compressed.)

BRAKE FEEL

As the diaphragm plate builds up hydraulic pressure in the brake system, a reaction pressure is transmitted back through the master cylinder hydraulic push rod. The compressed rubber reaction disc transmits this reaction pressure to both the diaphragm plate and the valve plunger. The pressure on the plunger valve tends to force it to the right (without moving diaphragm plate), causing the valve to close off the atmospheric port. The movement of the valve plunger is transmitted to the valve operating rod and on to the brake pedal. This provides the driver with "brake feel," so that he can accurately

determine the amount of braking effort being exerted.

This reaction force is directly proportional to pressure created within the master cylinder. Some power boosters use a reaction plate and levers, diaphragms, etc., instead of the rubber disc. Study the reaction disc in Fig. 18-61. Note that it is not compressed, and that a portion engages the valve plunger. Compare this with the compressed reaction disc in Fig. 18-62.

RUNOUT POINT

The vacuum runout is that point where a full vacuum has built up all the pressure possible on the power piston or diaphragm plate. If the driver desires more pressure than a full vacuum can produce, he can exert additional pressure on the pedal. This will hold the vacuum port wide open and transmit all of the driver's effort directly to the master cylinder piston.

Generally, wheel lockup (brake shoes will seize brake drum so tightly, tires will stop revolving and will slide) will occur before the vacuum runout point is reached.

HOLDING POSITION

As long as the driver increases pedal pressure, the valve plunger will move to the left. When the driver exerts the desired amount of pressure, plunger movement will cease. The reaction force, transmitted to the reaction disc, will force the valve plunger (without moving diaphragm plate) a slight amount to the right. This shuts off the atmospheric port. The vacuum port also remains shut.

When this balanced condition occurs, the right side of the diaphragm plate will be subjected to a specific and unchanging amount of atmospheric pressure. This maintains a constant pressure in the brake system.

If additional pedal force is applied, the atmospheric valve is reopened, causing the diaphragm plate to exert additional pressure on the master cylinder. It will maintain this pressure until pedal and reaction forces are again balanced.

If the pedal is released, the atmospheric port is closed and the vacuum port opened and both sides of the diaphragm plate are subjected to a vacuum. As this removes any force from the plate, the diaphragm return spring will push the plate and valve assembly to the right of the booster cylinder.

Fig. 18-63 shows a power booster in the holding position. Both vacuum and atmospheric ports are closed. Note how the reaction disc has moved the valve plunger a little to the right (in relation to diaphragm plate), as compared with the applied position in Fig. 18-62. This action closes the atmospheric port. Fig. 18-63.

OTHER BOOSTER TYPES

There are many variations in booster design, however, the basic principles are much the same.

The heavy-duty tandem booster, shown in Fig. 18-64, uses two diaphragm plates to increase booster-to-master cylinder pressure.

A reaction plate and lever setup, to provide "brake feel," is

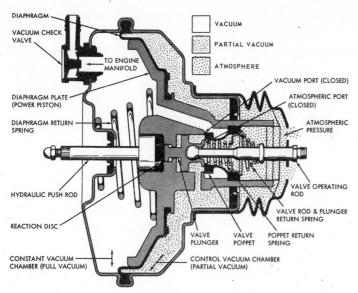

Fig. 18-63. Power booster in HOLDING POSITION. Both vacuum and atmospheric ports are closed. This will retain degree of braking desired. Note how reaction disc has forced valve plunger enough to right to close atmospheric port.

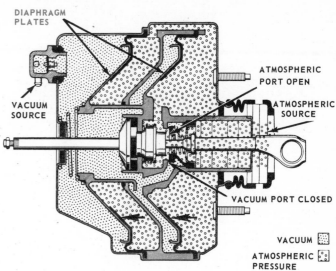

Fig. 18-64. Tandem (double) diaphragm power booster. Note use of two diaphragm plates. Booster is shown in applied position.   (Ford)

illustrated in Fig. 18-65. Counterforce from the master cylinder push rod exerts a force on the reaction plate. The plate transfers this force to the levers, causing them to tip to the right and move the air valve (valve plunger) to the right to

close off the atmospheric port. This movement is transmitted to the brake pedal through the push rod.

PISTON TYPE BOOSTER

Boosters discussed so far have been of the diaphragm type and are in popular use today. Some boosters have used a

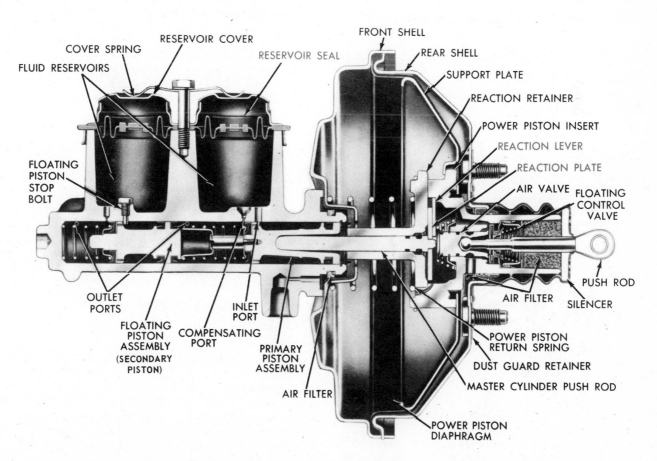

Fig. 18-65. Power booster employing reaction plate and levers to create "brake feel." Note fluid reservoir seals.   (Cadillac)

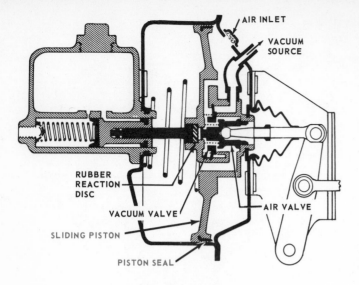

Fig. 18-66. Power booster using a sliding piston instead of a diaphragm and plate. Note piston seal. (Wagner)

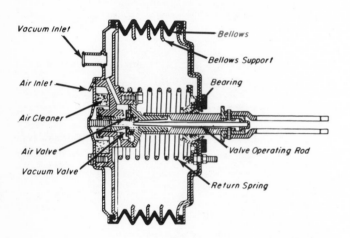

Fig. 18-67. Bellows type vacuum booster. When vacuum is applied to bellows, bellows shortens in a lengthwise direction to apply necessary pressure. (Bendix)

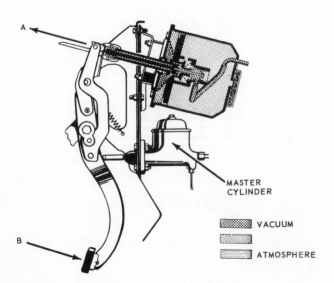

Fig. 18-68. Power booster attached to brake pedal. This setup applies booster pressure to brake pedal instead of directly to master cylinder. End result is similar — reduced foot pressure. (Plymouth)

sliding piston instead of a diaphragm and plate. A sealing ring is used around the piston to provide an airtight seal. This arrangement is pictured in Fig. 18-66.

BELLOWS TYPE BOOSTER

The bellows booster (it gets its name from rubber bellows construction) works on the same principle of vacuum and atmospheric pressure, but no cylinder and piston are used.

When vacuum is admitted to the bellows, the entire unit draws together. In Fig. 18-67, the right side is bolted to the fire wall. When the unit draws together, the left end moves toward the fire wall. This particular bellows booster applies pressure to the brake pedal above its pivot point.

PISTON TYPE BOOSTER ATTACHED TO BRAKE PEDAL

Fig. 18-68 illustrates a piston booster attached to a brake pedal. The booster is in the applied position, forcing the brake pedal out at A and in at B to actuate the master cylinder.

EXPLODED VIEW OF A DIAPHRAGM BOOSTER

An exploded view of a typical diaphragm power booster (similar to one shown in Figs. 18-61, 18-62 and 18-63) is pictured in Fig. 18-69. Study the part names and location.

VACUUM RESERVOIR TANK

As long as the engine is running, vacuum is available for the booster. The second the engine stops, the vacuum is gone. To provide several brake applications without engine vacuum, a vacuum reservoir may be incorporated in the booster system.

A vacuum check valve keeps a constant vacuum in the reservoir and in the booster up to the valve assembly. If the engine stops, the check valve retains the vacuum. One or two brake applications can be made from the vacuum existing in the reservoir. Fig. 18-70.

BOOSTER FAILURE

In the event the booster or the vacuum fails, the brakes may still be applied by foot pressure alone. It will require somewhat more pressure than when the booster is working, but the system will still function.

COMPLETE BRAKE SYSTEM, ANTI—SKID SYSTEM

Study Fig. 18-71, showing a complete brake system. Note the booster system, the emergency brake setup, lines and wheel assemblies.

Fig. 18-72 is a schematic of a typical two wheel (rear wheels) anti-skid braking system. Other optional systems are also available. The anti-skid system permits rapid straight line stops by preventing rear wheel lockup. Some systems affect all four wheels.

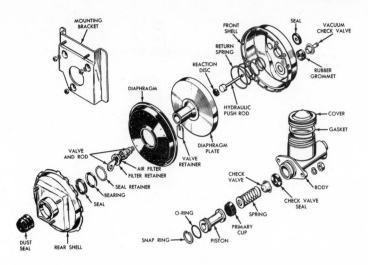

Fig. 18-69. Exploded view of a typical diaphragm type power booster. (Ford)

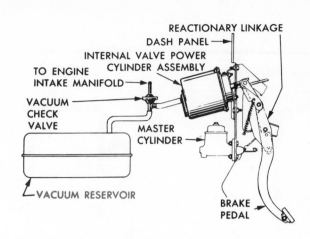

Fig. 18-70. Vacuum reservoir tank. Reservoir allows one or two brake applications after engine vacuum is gone. (Plymouth)

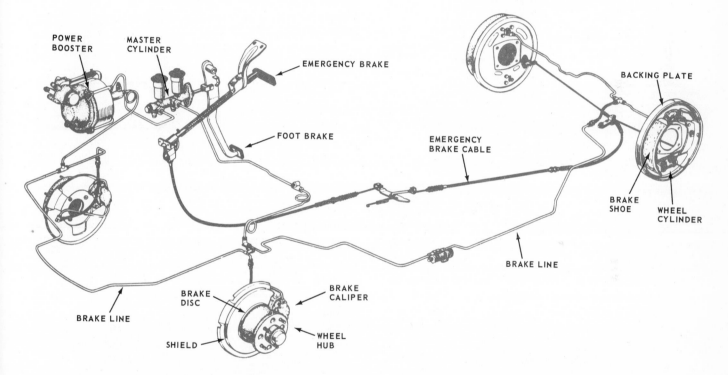

Fig. 18-71. Complete braking system. Movement of brake pedal causes pistons in master cylinder to force fluid through tubing to each of rear wheel cylinders and front brake calipers. This applies shoes in rear and disc friction pads in front. Emergency brake applies rear wheel shoes only. Note that in THIS SPECIFIC installation, brake booster is a separate unit and functions to increase front wheel disc brake line pressure only. (Toyota)

SUMMARY

Liquids, under confinement, can be used to transmit motion and to increase or decrease pressure. Air is compressible. Liquids are not. Pascal's law states, "When pressure is exerted on a confined liquid, it is transmitted undiminished."

The brake system can be divided into two principal parts, hydraulic system and wheel brake assemblies.

When the brake pedal is depressed, the master cylinder piston compresses brake fluid in the master cylinder. This causes fluid to move in the brake lines. This fluid movement will expand the wheel cylinders, causing them to apply the brake shoes to the brake drums. Both movement and pressure are used in the hydraulic system.

Cars now employ double-piston or dual master cylinders. This provides a separate braking force for front and back wheels.

Brake lines are made of double thickness, plated steel. No other material is satisfactory. Flexible hoses are used in both the front and in the rear.

There are two basic types of wheel cylinders: single-piston type and double-piston type. Wheel cylinders can also be of

HYDRAULIC CIRCUIT
VACUUM CIRCUIT
ELECTRICAL CIRCUIT

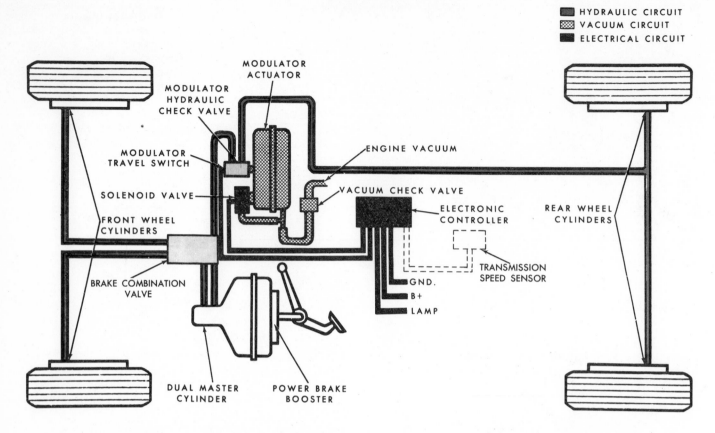

MODULATOR
ACTUATOR

MODULATOR
HYDRAULIC
CHECK VALVE

ENGINE VACUUM

MODULATOR
TRAVEL SWITCH

VACUUM CHECK VALVE

SOLENOID VALVE

ELECTRONIC
CONTROLLER

REAR WHEEL
CYLINDERS

FRONT WHEEL
CYLINDERS

TRANSMISSION
SPEED SENSOR

GND.

B+

LAMP

BRAKE COMBINATION
VALVE

DUAL MASTER
CYLINDER

POWER BRAKE
BOOSTER

Fig. 18-72. Schematic showing one type of anti-skid (rear wheels only) braking system. When rear wheel rpm starts to slow down below usual rpm for a given road speed, transmission sensor (on rear wheels on some models) sends an electrical signal to electronic controller, causing modulator to cut off pressure to rear wheel brakes. This releases brakes and prevents lockup. As soon as wheel rpm is what it should be in proportion to car road speed, sensor causes modulator to restore full braking pressure. This ON-OFF braking cycle takes place in about 1/3 of a second. Front wheel brake pressure remains constant. By cycling rear brake pressure, rapid straight-line stops are possible. (Cadillac)

stepped construction.

All air must be expelled from a hydraulic system. The removal of this air is termed BLEEDING.

Never use any fluid other than high quality brake fluid in the system. Anything else is likely to ruin the rubber cups and make the brakes inoperative.

The wheel brake assembly consists of a backing plate, brake shoes and springs, and a brake drum. The backing plate is attached in a rigid manner, and the brake shoes are attached to the backing plate. The shoes are arranged so they may be expanded by the wheel cylinders. As the shoes expand, they will stop the brake drum and wheel.

There are two brake shoes per wheel. There is one primary shoe and one secondary shoe. Some shoes are arranged so that servo action assists in applying the brakes.

Brake shoe servo action is when one shoe helps apply the other. Self-energizing shoes are arranged so shoe-to-drum friction helps apply the shoe.

Some shoes are self-adjusting, others must be adjusted periodically to compensate for shoe lining wear.

The front brakes generally provide somewhat more braking force than the rear brakes. This is to compensate for the transfer of weight to the front during stops.

The disc brake provides high resistance to both brake fade and grabbing. The use of the disc necessitates incorporating a

proportioning and a metering valve into the system. The metering valve prevents disc application until a definite line pressure is reached while the proportioning valve controls pressure to the rear wheels.

There are two types of emergency or parking brakes in general use. One operates the regular service brake shoes, generally in the rear. The other uses a drive line brake.

Power brakes are used on many cars today. The power brake is essentially the regular brake system with a vacuum or power booster added, to make braking easier for the driver.

There are three basic types of power boosters: piston, diaphragm and bellows types. The diaphragm type is in common use today. They may be either vacuum or atmospheric suspended. All work in essentially the same manner. Atmospheric pressure can be admitted to both sides of the piston. Engine vacuum can, upon application of the brake pedal, be applied to one side of the piston. With atmospheric pressure on one side and a vacuum on the other, the piston is forced to move in the direction of the vacuum. As it moves, it applies force either to the master cylinder piston or to the brake pedal itself.

A valve setup is used to control the admission of atmospheric pressure and the application of the vacuum. The booster works in three stages: released, being applied, holding.

The power booster is designed to provide a braking feel for

the driver. This is essential in controlling the amount of brake pressure.

The vacuum runout point for a booster is when a full vacuum is applied and the booster has exerted all possible pressure on the system.

A vacuum reservoir is provided so that one or two brake applications can be made after the engine stops. Some systems provide a small electric motor driven vacuum pump to produce sufficient vacuum for brake application. This works only when the engine is off.

OF ALL THE PARTS OR SYSTEMS OF THE AUTOMOBILE, THE BRAKES RATE AS ONE OF THE MOST IMPORTANT FROM A SAFETY STANDPOINT. THEY MUST BE SERVICED AND INSPECTED ON A REGULAR BASIS. ANY DEFECTIVE PART MUST BE REPLACED AT ONCE.

REVIEW QUESTIONS — CHAPTER 18

When disassembling, inspecting, repairing, then reassembling automotive components, always study each part. Notice how it is made, of what, how it performs its particular job, and its relationship to the entire unit. Only in this way will you develop the keen insight into part design and automotive engineering principles that are a MUST for mechanics that aspire to be the best.

1. Air is not compressible. True or False?
2. Liquids can transmit both _____ and _____ .
3. What is Pascal's law?
4. Explain the workings of a simple hydraulic jack.
5. If a 100 lb. force is applied to a piston with 1 sq. in. of area, how much pressure would a piston with an area of 10 sq. in. exert, if both of these pistons rested with an airtight seal on a column of oil? Assume that both cylinders or columns are connected via a pipe.
6. What is the purpose of a master cylinder?
7. Explain how a master cylinder works in the released position, in the applied position, quick release, and final release positions.
8. The master cylinder reservoir must be filled to within about _____ of the filler plug.
9. The reservoir plug must be vented. True or False?
10. Why must a reservoir be provided?
11. Of what use is the master cylinder check valve?
12. Brake lines must be made of _____ .
13. Brake lines, where connections are made, must have a _____ flare.
14. Name the parts of a wheel cylinder.
15. Wheel cylinders are generally fastened to the brake drum. True or False?
16. Of what use is flexible high pressure brake hose?
17. Name two types of wheel cylinders.
18. Why are some wheel cylinders of stepped construction?
19. Bleeder screws are furnished in the brake lines. True or False?
20. How does the shape of the rubber pistons in the master cylinder and wheel cylinders help them to seal?
21. A small amount of air in the brake system is permissible. True or False?
22. Kerosene is a satisfactory substitute for brake fluid in an emergency. True or False?
23. Of what use is the backing plate?
24. Give the names of the two brake shoes used in each wheel assembly.
25. Why are brake shoe retracting springs used?
26. Brake lining is made of _____ .
27. What part of the wheel assembly dissipates the greatest amount of heat?
28. Explain what happens in the wheel assembly when the master cylinder builds up pressure in the brake lines.
29. What is brake fade?
30. Explain servo action. What is meant by self-energizing?
31. What unit is used to keep brake shoes in close contact with the drum?
32. What are two types of emergency brakes?
33. How does a power brake work?
34. Explain the action of the power booster in the released, applied and hold positions.
35. How is brake feel provided in a power booster?
36. Name three types of power boosters.
37. Of what use is a vacuum reservoir tank?
38. Of what value is the double-piston or dual master cylinder?
39. A proportioning valve is used with disc brake systems. True or False?
40. The metering valve in disc (front wheels only) brake systems:
    a. Controls pressure to rear wheels.
    b. Controls pressure to front wheels.
    c. Prevents both front and rear systems from simultaneous failure.
    d. Prevents disc brake application until line pressure reaches a certain point.
41. What does vacuum suspended mean when referring to power booster? Atmospheric suspended?
42. What type of power booster is in popular use today:
    a. Bellows.
    b. Diaphragm.
    c. Piston.
    d. Bellows-piston.

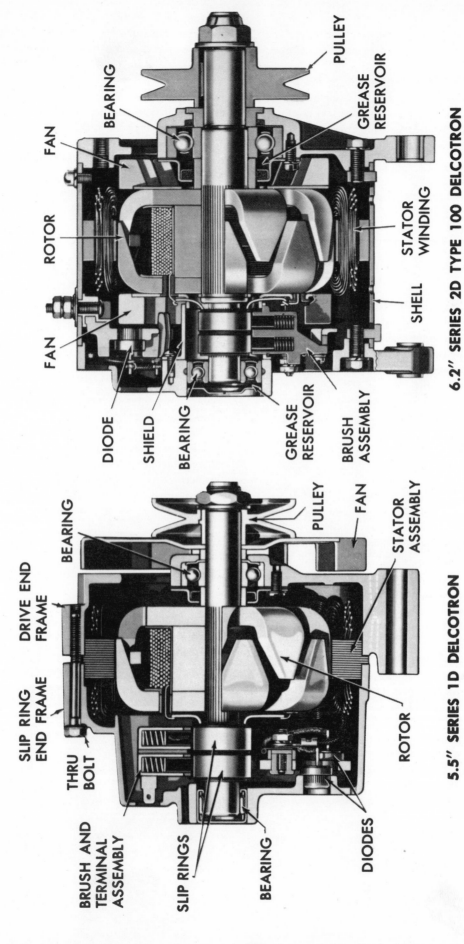

PULLEY

BEARING

GREASE RESERVOIR

FAN

ROTOR

STATOR WINDING

SHELL

FAN

DIODE

SHIELD

BEARING

GREASE RESERVOIR

BRUSH ASSEMBLY

6.2" SERIES 2D TYPE 100 DELCOTRON

BEARING

DRIVE END FRAME

PULLEY

FAN

SLIP RING END FRAME

STATOR ASSEMBLY

THRU BOLT

BRUSH AND TERMINAL ASSEMBLY

SLIP RINGS

BEARING

ROTOR

DIODES

5.5" SERIES 1D DELCOTRON

Left, normal duty; right, heavy duty alternator.   (Delcotron)

Chapter 19

ELECTRICAL SYSTEM FUNDAMENTALS

The field of electricity covers a wide and complicated range. Much is known about electricity and much remains to be discovered. A thorough coverage of electricity is not within the scope of this text on auto mechanics. However, a basic understanding of electrical fundamentals and their application to the auto, is a MUST for all students of auto mechanics.

A BAFFLING FORCE

Men have learned to control electricity, predict its actions and harness it in many ways. No person has literally seen it

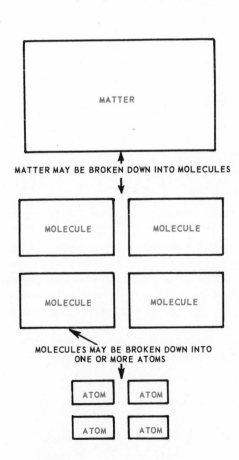

MATTER MAY BE BROKEN DOWN INTO MOLECULES

MOLECULES MAY BE BROKEN DOWN INTO ONE OR MORE ATOMS

Fig. 19-1. Breakdown of matter.

flow through a wire. For many years no one was even sure of the direction it moved or what actually did move.

THE ELECTRON THEORY

The accepted theory regarding electricity is termed the ELECTRON THEORY. This theory has explained many of the strange behavior patterns as well as opening the way to many new discoveries, not only in electricity, but in other fields as well.

The electron theory is based on the assumption that electrical phenomenon (fact which may be observed) is produced by a flow of electrons from one area to another. It also states that the flow of electrons is caused by having an excess number in one area, and a shortage in the other. The area with the surplus electrons is referred to as NEGATIVE and the area in short supply, the POSITIVE.

MATTER

Anything that takes up space and has mass is considered matter. Matter can be a solid, a liquid or a gas.

MOLECULE

A molecule is the smallest portion that matter can be broken down to, and still retain all the properties of the original matter. The molecule is made up of atoms.

ATOMS

There are around one hundred different types of atoms. Various combinations of these atoms will produce different molecules of matter.

A molecule of pure water is made up of one atom of oxygen and two atoms of hydrogen. Both oxygen and hydrogen are gases.

All matter is made up of either different atoms, or the same type of atoms. Fig. 19-1 illustrates a progressive breakdown of matter.

STRUCTURE OF THE ATOM

The atom itself is made up of Neutrons, Protons and Electrons.

The Electron is a negatively (−) charged particle. The Proton is a positively (+) charged particle. The Neutron is a neutral particle with no charge.

Protons and neutrons form the center, or nucleus, of the atom. The electrons rotate rapidly around this nucleus. As unlike charges attract each other, the electrons are held in their orbital path. They will not draw in any closer, because the centrifugal force built up by their whirling flight counterbalances the mutual attraction force. Fig. 19-2.

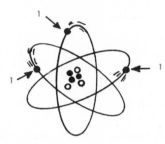

Fig. 19-2. Structure of atom. Electrons, 1, are whirling about nucleus. Solid dots in nucleus represent neutrons. White dots are protons.

There can be many different atoms, each with a different number of protons, neutrons and electrons. Every atom has the same number of electrons as it has protons. Protons and neutrons are alike in any type of atom. The number of neutrons, protons and electrons, and their arrangement, produces different types of matter.

A simple atom is that of hydrogen. The hydrogen atom consists of a single neutron and a single proton. A more complex atom is that of copper. This atom consists of 34 neutrons, 29 protons and 29 electrons. Fig. 19-3.

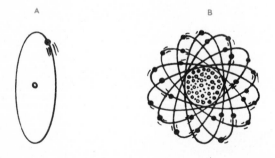

Fig. 19-3. Two different atoms. A—Simple hydrogen atom consisting of one proton and one electron. B—More complex copper atom has 34 neutrons, 29 protons and 29 electrons.

FREE ELECTRONS

You will notice in Fig. 19-4 that the electrons move in different size orbits. The inner orbits contain what is referred

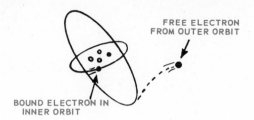

Fig. 19-4. Free and bound electrons. Free electron in outer orbit has been knocked from its path.

to as BOUND electrons. The outer orbits contain the FREE electrons.

The bound electrons are so named because it is difficult to force them from their orbits. The outer free electrons can be more readily moved from their orbits.

POSITIVE AND NEGATIVE CHARGE ATOMS

If some of the free electrons are moved out of their orbit, the atom will have more protons than electrons. This will give the atom a positive charge. If the electrons are returned, until they equal the number of protons, the atom will become electrically neutral. If a surplus of electrons are present, the atom becomes negatively charged.

A negatively charged body will repel electrons, while the positively charged body will attract electrons. Fig. 19-5.

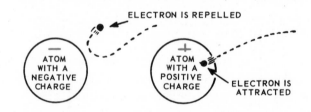

Fig. 19-5. Negative and positive charge atoms. Atom with negative charge repels electrons. Positive charged atom attracts electrons.

CURRENT

An electrical current is a mass movement of electrons through a conductor (wire). This may be likened to water flowing through a pipe. The amount of electrons passing any given point in the circuit in one second determines the amount of current flowing. The amount of current flow is measured in AMPERES.

Before a current will flow, there must be a surplus of electrons at one end and a scarcity at the other end of the conductor.

CURRENT FLOW

Fig. 19-6 illustrates a schematic of a simple circuit utilizing a battery as the source. Note that the electron flow is from negative to positive.

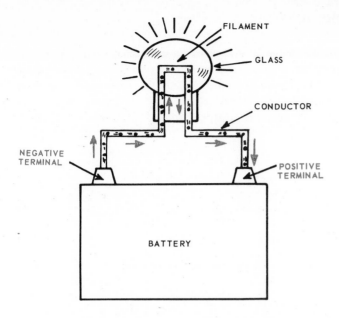

Fig. 19-6. Current flow. Electrons move from negative post through conductor, on to positive post.

The battery, through chemical action (covered later), has built up a surplus of electrons at the negative post and a lack of electrons at the positive post. As soon as the copper wire (conductor) is connected to both posts, current (electrons) will flow from the negative to the positive post.

On their way from the negative post, the electrons will pass through the light bulb. As the filament in the bulb creates a resistance to the flow of electrons, it will heat up until it is white hot and glows. After passing through the filament, the electrons pass on to the positive post.

When an electron leaves the negative source, this single electron does not speed through the conductor. It strikes an adjacent electron and knocks it from its orbit. This electron strikes its neighbor and knocks another electron from orbit. As one electron knocks another out of orbit, it replaces the one it knocked out.

To help illustrate this action, imagine placing enough corks in a pipe to fill it from end to end. If you push one cork

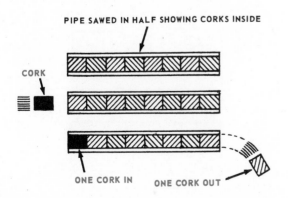

Fig. 19-7. Cork (electron) in a pipe (conductor). Even though cork has just entered pipe, it causes a chain of pressure that results in a different cork being forced out opposite end.

(electron) into one end of the pipe (conductor), one cork (electron) will pop out the other. As soon as one cork presses on the column of corks, its movement is transmitted through the column. Even though it enters only a small distance, its influence is felt at the other end almost at once. Fig. 19-7.

VOLTAGE (Electromotive Force — EMF)

As mentioned, current flow is produced by having a difference in electrical potential, or pressure, at each end of a conductor (one end with surplus of electrons and other lacking electrons).

This pressure differential (also known as electromotive force — EMF) causes the current to flow. To keep the current flowing, it is necessary to maintain the electrical pressure or, as it is commonly called, VOLTAGE.

In the automotive electrical system, voltage is produced either by a battery (electrochemical action) or by a generator (electromagnetic action).

The amount of pressure or voltage is measured in VOLTS. The higher the voltage, the more current flow it can create. Cars commonly use a 12 volt system.

The battery in A, Fig. 19-8, is dead. Both negative and

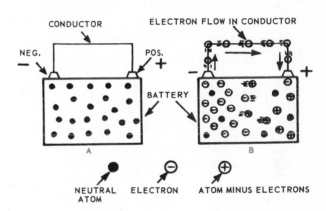

Fig. 19-8. Battery voltage. A—Battery discharged, no voltage, no current flow. B—Battery charged, normal voltage, electrons massed on negative side. Current (electrons) flow through conductor.

positive sides have a balanced electron conditon. There is no voltage or pressure difference.

In B, Fig. 19-8, the battery is charged. The negative side has great numbers of surplus electrons, while the positive side has a deficiency. When a wire is connected across the terminals, the voltage will cause current to flow from negative to positive.

RESISTANCE

When current flows through a conductor, it meets resistance. A relatively poor conductor offers greater resistance than a good conductor.

As the electrons attempt to flow, they bump into other atoms in the conductor. This causes a heating effect. As the conductor heats up, the movement of the atoms becomes

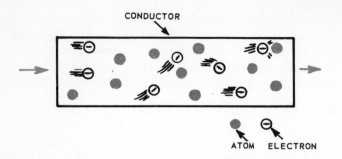

Fig. 19-9. Resistance in a conductor. As electrons attempt to move through conductor, they strike atoms. This deflects them and in effect resists their efforts to pass through conductor.

more agitated. This further increases the number of collisions. If current flow is increased, the conductor will become hotter and hotter until it will literally burn up.

This heating action is one of the great uses for electricity. Such things as lights, electric burners, heaters, etc., make use of resistance found in conductors. In order to function, the type of conductor and amount of voltage and current flow must be carefully balanced, so the wire will reach a specific desired heat.

A large conductor, of a given material, will offer less resistance than a smaller one. The longer the conductor, the greater its resistance. The amount of resistance in a conductor is measured in OHMS. Fig. 19-9.

TEST GAGES

To measure the rate of current flow in amperes, a gage called an AMMETER is used. The ammeter is connected in series with the circuit. Fig. 19-10.

Voltage is determined by using a VOLTMETER. The voltmeter is connected across the circuit. Fig. 19-11.

Resistance in ohms is calculated by using an OHMMETER. The ohmmeter test leads are connected across the resistance being measured. Fig. 19-12.

MATERIALS USED FOR CONDUCTORS

Because it has a great number of free electrons, and is relatively inexpensive, copper is widely used as a conductor. There are other materials, such as silver, that are better conductors than copper. Since they are expensive, their use is limited to certain units in an automotive system.

ELECTRICITY MUST HAVE A PATH

In order for electrons to flow, they must have a path or conductor. If more than one path is available, current will flow in the one with the least resistance. As you learned in the chapter on the Ignition System, air itself will act as a conductor when voltage is high enough.

To make certain current flows where desired, the conductors in the car's electrical system must be insulated.

INSULATORS

Any material in which the atoms have more than four electrons in their outer orbit will make a good insulator. Since electrons tend to be bound, and resist being knocked from their orbits, current cannot flow.

Glass, ceramics, plastics, rubber, fiber, etc., are excellent

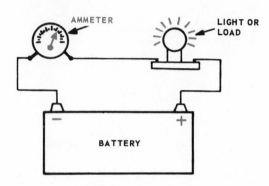

Fig. 19-10. Using an ammeter to check current flow. Ammeter is connected in series with circuit.

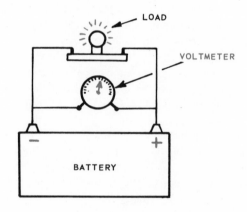

Fig. 19-11. Using a voltmeter to check circuit voltage. Voltmeter is connected in parallel with circuit.

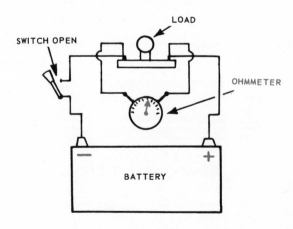

Fig. 19-12. Using an ohmmeter to check load resistance. Note that circuit is open to prevent damaging ohmmeter. Ohmmeter has its own source of current. Ohmmeter circuit is hooked directly to the load.

insulators. All are used in automotive electrical systems. The wiring itself uses a rubber or plastic coating to prevent the escape of electrons to adjacent metal parts or to other wires. Spark plugs use ceramic insulators. Switches, relays, etc., use fiber, rubber, glass, etc., for specific points where insulation is required.

CIRCUITS

A circuit consists of a power source, a unit to be operated and the necessary wiring to provide a path for electron flow. The circuit must be complete in order to function.

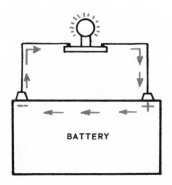

Fig. 19-13. A simple circuit. Current flows from negative to positive through the unit.

SIMPLE CIRCUIT

A very basic circuit, Fig. 19-13, consists of a power source (battery), a single unit to be operated (bulb) and connecting wires. The current flows from the negative terminal through the conductor, through the unit, and then on through the second conductor to the positive terminal of the battery. It then passes from the positive terminal to the negative terminal through the inside of the battery.

SERIES CIRCUIT

A series circuit consists of two or more resistance units that are so wired that the current has but one path to follow. To flow, and complete the circuit, the current must pass through each resistance unit.

As each resistance unit impedes the rate of current flow (amperes), the total amount of resistance offered by all units will determine the amperage flow. The more units added to a specific circuit, the more the resistance, with a resultant lowering of current flow.

If amperage flow is checked at several points in a series circuit, all readings will be the same.

VOLTAGE DROP

As the available voltage is lowered by each resistance unit, the voltage will drop as the units occur. The voltage may be up

to a full rated 12 volts at the battery, but will become progressively lower farther out on the circuit. Study Fig. 19-14. Notice that the current must pass through one unit to get to another, that the ampere flow is the same throughout, that the voltage is dropping, and that if any one resistance unit is broken or burned out (wire severed), the current cannot flow.

Any type of circuit will have a certain amount of normal voltage drop. Even the conductor itself somewhat impedes electron flow. When voltage drop becomes excessive, poor unit operation will result. Voltage drop is controlled by using the proper size wires, ample source voltage, proper insulation and clean connections.

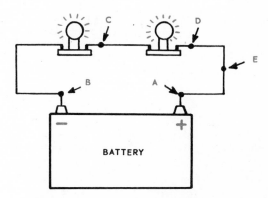

Fig. 19-14. A series circuit. If ammeter were inserted in circuit at B, C, D and E, all readings would be same. If voltmeter were installed across or between A and B, it would show full battery voltage. If connected between A and C, it would show a voltage drop (less voltage). If connected between A and D, it would show still more drop.

PARALLEL CIRCUIT

The parallel circuit consists of two or more resistance units connected in such a way as to allow current flow to pass through all units at once. This eliminates the necessity of the current going through one before reaching the other. Also, if one unit is burned out, the others will still function.

If you were to measure the total resistance in a parallel circuit, you would find it to be less than any one of the unit resistances. You will recall that in the series circuit, the current has to pass through the resistance units, one after the other. This gave a resistance to current movement that was equal to the sum of all the units. In a parallel circuit, the current has several units that it can pass through, actually reducing the total resistance in relation to any one unit resistance value.

EQUAL VOLTAGE

Equal voltage is applied to all resistance units in a parallel circuit. The amount of current flow through each unit will vary according to its individual resistance, but the total current flowing will equal the sum of the current flowing through all the units. Fig. 19-15.

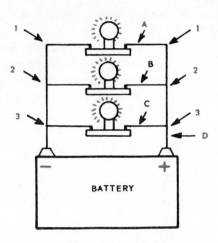

Fig. 19-15. A parallel circuit. Voltage across 1—1, 2—2, and 3—3 would be equal. Current flow in amperes at A, B and C would depend upon unit. Current flow at D would equal sum of current flowing through A, B and C. Total resistance of A, B and C would be less than any single one.

SERIES — PARALLEL CIRCUIT

A series-parallel circuit is essentially one in which the series and the parallel circuits are combined. It requires the connecting of three or more resistance units in a combination series and parallel hookup.

To determine the resistance of this circuit, the total resistance of the parallel circuit is added to that of the series portion. Current flow will be determined by the total resistance.

Voltage drop will be the drop across the parallel circuit, plus that of any series resistance unit. Fig. 19-16.

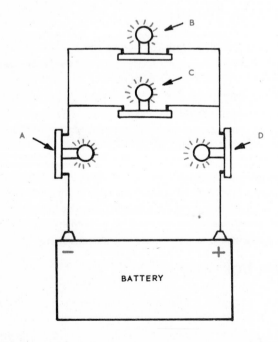

Fig. 19-16. Series — parallel circuit. Units A and D are in series. Units B and C are in parallel.

CAR BODY ACTS AS ONE WIRE

A circuit requires one wire from the source to the unit, and one wire from the unit back to the source.

In an automobile where the body, engine, etc., are made of steel, it is not necessary to have a two-wire circuit. The steel structure of the car can act as one of the wires. One of the battery posts (negative post in this country) is connected (grounded) to the steel framework of the vehicle. The other post or terminal (positive) is connected to the various electrical units with an insulated wire. Fig. 19-17 illustrates the frame being used in place of a wire.

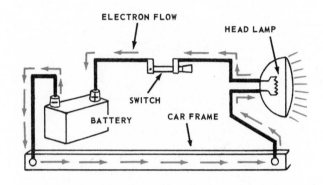

Fig. 19-17. Using car frame as one wire in a circuit. Battery and electrical unit are grounded to the frame. Electron flow is from the negative (—) battery terminal to the frame, from the frame to the head lamp and back to the battery through the closed switch.

GROUND SYMBOL

When looking at wiring diagrams, where the circuit is connected to the steel structure of the vehicle, you will see a ground symbol. This symbol indicates that the circuit is assumed, and is not actually drawn. Fig. 19-18.

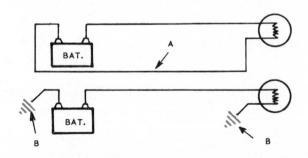

Fig. 19-18. Ground symbol. In top circuit, wire A, (indicating steel structure of car) has been drawn in. In bottom circuit, ground symbols B have eliminated drawing wire A portion of circuit.

OHM'S LAW

Ohm's law can be expressed in three different ways:

1. Amperes equals volts divided by ohms ($I = \dfrac{E}{R}$).

2. Volts equals amperes multiplied by ohms (E = IR).

3. Ohms equals volts divided by amperes ($R = \frac{E}{I}$).

Ohm's law, in its three forms, can be used to find any one unknown factor when the other two are known. A handy form in which to remember Ohm's law in its three applications is illustrated in Fig. 19-19.

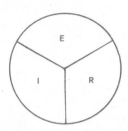

Fig. 19-19. Handy form of Ohm's law. Try three combinations using this form.

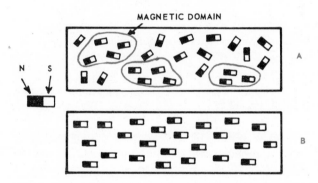

Fig. 19-20. Magnetic domains. In A, magnetic domains are small, and they are surrounded by atoms that are not aligned. In B, metal is magnetized and domains enlarged and line up. Combined force creates a strong magnetic field. Note that N and S poles are aligned.

MAGNETISM

A great deal remains to be learned about the strange force termed MAGNETISM. It is probably best understood by observing some of its effects and powers.

One theory regarding magnetism states that electrons rotating about the nucleus contain the essentials of magnetism. These electrons not only rotate about the nucleus, but also spin on their own axes (electron spins). Even though each electron creates a feeble magnetic field, the electrons are orbiting in so many directions that one field tends to cancel out the other.

DOMAINS

In an unmagnetized material, there are many minute sections in which the atoms line up to produce a magnetic field. As these magnetic domains are scattered through the material in all directions, one domain tends to cancel out the force of the next one.

To magnetize material (such as steel), it is necessary to orient or line up these domains so that most of them point in the same direction. This is done by placing the steel in a strong magnetic field produced by a current passing through a coil of wire (covered later).

The domains will grow in size as they move into alignment. This growth involves lining up neighboring atoms so the magentic effect is increased. Fig. 19-20.

LINES OF FORCE

The first thing to remember about the term MAGNETIC LINES OF FORCE is that it is a term used to describe the invisible force involved.

If a simple bar magnet is placed under a sheet of glass, and fine iron filings are sprinkled on the surface of the glass, the filings will line up in accordance with the so-called lines of force. Fig. 19-21.

The lines of force leave the north (N) pole of the magnet, and enter the south (S) pole. They then pass from the south to the north pole through the body of the magnet.

Notice the effect produced on the iron filings when a horseshoe magnet is placed under the glass. The glass should be tapped to allow the filings to line up. Fig. 19-22.

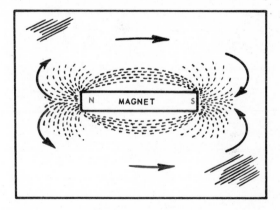

Fig. 19-21. Lines of force around a bar magnet. Lines of force leave N pole and travel along magnet, then enter S pole.

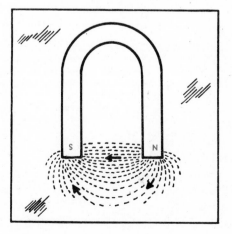

Fig. 19-22. Lines of force around a horseshoe magnet.

LINES OF FORCE DO NOT CROSS EACH OTHER

You will note that these fine lines of force do not cross. The lines tend to be drawn in as nearly a straight line as possible, and at the same time they push each other apart.

UNLIKE POLES ATTRACT, LIKE POLES REPEL

If two bar magnets are shoved together with unlike poles on the approaching ends, they will snap together. If the approaching ends have similar poles, they will repel each other. If forced together, they will snap apart when released.

By studying Fig. 19-23, you will see that the magnetic force lines are leaving the north pole. Remember that these lines tend to draw into as straight a line as possible, yet they will not cross as one line repels the other. As the two magnets near each other, the force lines leaving the N pole of magnet number one, instead of swinging up and back to their own south pole, leap across the gap to the other closer south pole of magnet number two. The lines leaving the N pole of magnet number two, will now swing along both magnets to reach the south pole of magnet number one. As the lines stretch as tightly as possible, the magnets cling tightly to each other. They have, in effect, become one long bar magnet. Fig. 19-23.

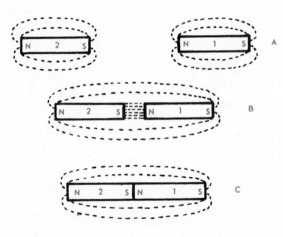

Fig. 19-23. Unlike poles attract. In A, both magnets are apart and lines of force assume their normal state. In B, magnets approach each other causing force lines from N pole of magnet No. 1 to enter S pole of magnet No. 2. This draws them together, as in C, and the two magnets act as one bar magnet.

When like poles near each other, the magnetic lines entering each pole will repel each other and will refuse to cross. This causes each magnet to push away from the other, until the force fields are far enough apart to become too weak to cause further movement.

Magnetic lines of force will follow metal more readily than air. If a metal bar is placed across the ends of a horseshoe magnet, the force lines follow the bar from N to S pole.

ELECTRICAL SYSTEM, BATTERY

An electric current can be produced chemically by means of a battery.

The typical 12 volt battery is constructed of six elements. The element itself is made up of two groups of plates. One group forms the Positive group; the other, the Negative group. The plates (number can vary with battery capacity but will not affect open circuit voltage) are lined up with space between, then welded to a post strap. Both groups are sandwiched together so that a negative plate adjoins a positive plate — neg., pos., neg., pos., etc.

Insulating separators of nonconducting microporous (tiny pores) rubber, cellulose fiber or some other type of material are placed between each set of plates. The separators have vertical grooved ridges to allow the electrolyte (mixture of sulphuric acid and distilled water) to move about freely. The separators prevent the plates from touching each other and short-circuiting.

POSITIVE PLATE

The positive plate grid (wire-like, skeletal framework of the plate, made of lead and antimony) is filled with lead peroxide. The lead peroxide is in sponge form, which allows the electrolyte to penetrate the plate.

NEGATIVE PLATE

The negative plate grid is filled with porous lead, plus expanders to prevent the lead from returning to an inactive solid state. Fig. 19-24 illustrates an element consisting of positive and negative plates, with separators, being sandwiched together.

The six elements are assembled in a hard rubber or plastic battery box or case. The case has partitions to divide it into six compartments or CELLS. Fig. 19-25.

Fig. 19-26 shows the elements in place in the battery box. After installation, each element was connected to its neighbor by a cell connector lead strap. Note side terminal connectors.

Another battery is shown in Fig. 19-27. Note how the elements are kept away from the floor of the container. This allows room for shedding plate material to deposit in the form of sediment. If this sediment touches the plates, it will cause a short circuit.

ELECTROLYTE

The battery is then filled (slightly over top of plates) with an electrolyte solution made up of sulphuric acid and distilled water. The water makes up about 64 percent (by weight) of the electrolyte.

HOW A BATTERY FUNCTIONS

The battery does not actually "store" electricity. When charging, it converts electricity into chemical energy. Chemical action of electrolyte, working on active material in the plates,

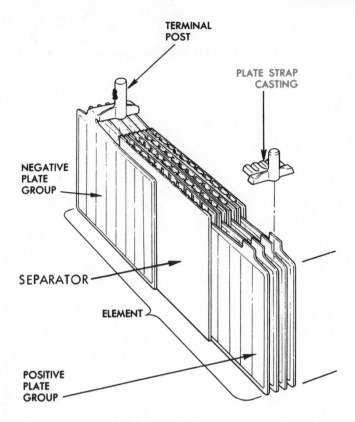

Fig. 19-24. Battery element. Each group is welded to plate strap casting. (Delco—Remy)

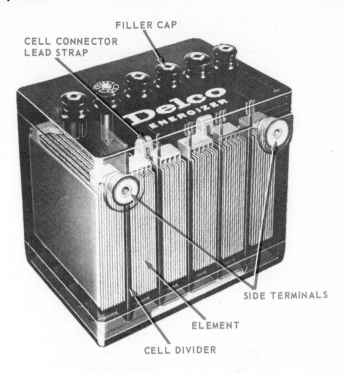

Fig. 19-26. A 12V battery is shown, with elements in place, cell connector straps attached to elements and cover installed. Note side terminals. (Delco)

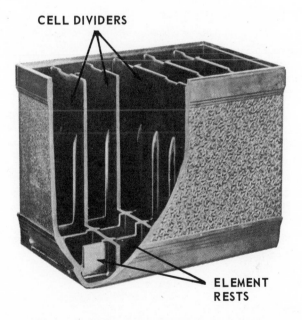

Fig. 19-25. Battery box or container. Note cell dividers and raised bottom (element rests).

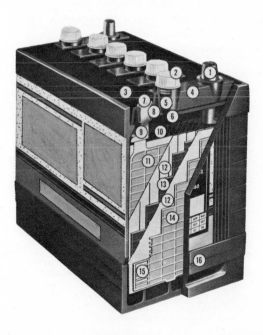

Fig. 19-27. Section view of complete 12V battery. Learn names of all parts. 1—Terminal post. 2—Vent plug. 3—Sealing compound. 4—Cell cover. 5—Filling tube. 6—Electrolyte level mark. 7—Inter-cell connector. 8—Lead insert. 9—Plate strap. 10—Separator protector. 11—Negative plate. 12—Separator. 13—Positive plate. 14—Negative plate with active material removed to show: 15—Plate grid. 16—Container. (Exide)

causes a transfer of electrons from positive plates to negative plates. When battery is placed in a closed (completed) circuit, surplus of electrons at negative post flows to positive post. An electric current is produced. The battery is now converting chemical energy into electricity.

As the current flows, the battery starts discharging. The

sulphate in the electrolyte combines with the plate materials, forming lead sulphate. As the current continues flowing (constantly growing weaker), the electrolyte becomes weaker and weaker. Finally the battery is completely discharged.

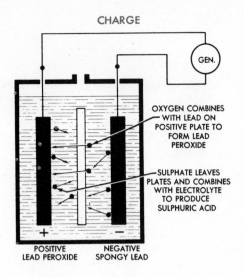

Fig. 19-28. Battery charge. (Ford)

BATTERY CHEMICAL ACTION REVERSES

By passing an electric current back into the battery (with a battery charger or car generator) in a reverse direction, the chemical action is reversed. The sulphate leaves the plates and returns to the electrolyte in the form of sulphuric acid, and the plates revert to lead peroxide and sponge lead. After charging for a sufficient time, the battery will become fully charged. Figs. 19-28 and 19-29.

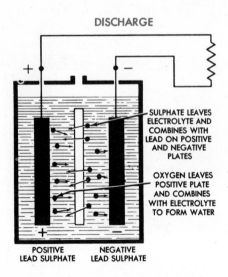

Fig. 19-29. Battery discharge. (Ford)

SPECIFIC GRAVITY OF THE ELECTROLYTE

When the battery is fully charged, the specific gravity (relative weight of a given volume of a certain material as compared with an equal volume of water) of the battery is around 1.280. This is determined by drawing the electrolyte up into a HYDROMETER. Fig. 19-30.

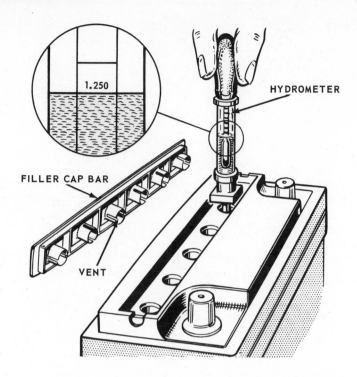

Fig. 19-30. Hydrometer. Electrolyte is being drawn up into glass barrel of hydrometer. (Plymouth)

The depth at which the float rests in the electrolyte determines its specific gravity. The higher the float, the higher the charge. The float is marked, and the mark at the top of the electrolyte is read. Fig. 19-31 shows how to read a hydrometer; also how various stages of discharge affect the specific gravity.

TEMPERATURE CORRECTION

When reading a hydrometer, remember that the temperature of the electrolyte affects the specific gravity. Many hydrometers have built-in thermometers with temperature correction scales. Fig. 19-32.

TEMPERATURE INFLUENCES BATTERY EFFICIENCY

The chart in Fig. 19-33 illustrates how temperature affects the efficiency of a battery.

BATTERIES CAN FREEZE

The electrolyte in batteries, if discharged too much, can freeze and the battery will burst. The freezing point, depending on specific gravity, is illustrated in Fig. 19-34.

DISCHARGE RATE

A battery without being charged, will not last indefinitely. Fig. 19-35 shows the typical current load for modern cars. The generator must produce sufficient current output to cover all possible electrical needs.

IDENTIFYING BATTERY TERMINALS

The positive post of the battery, which is usually marked with a POS, or +, is somewhat larger than the negative post. Its top is sometimes painted red. The negative post is the smaller of the two, and is often painted green. It is marked with a NEG, or −.

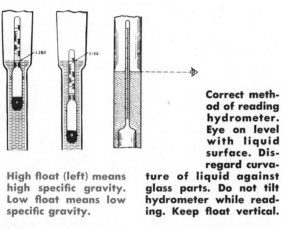

High float (left) means high specific gravity. Low float means low specific gravity.

Correct method of reading hydrometer. Eye on level with liquid surface. Disregard curvature of liquid against glass parts. Do not tilt hydrometer while reading. Keep float vertical.

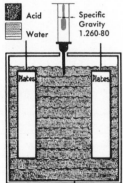

Acid

Water

Specific Gravity 1.260-80

Plates   Plates

FULLY CHARGED

Acid in water gives electrolyte specific gravity of 1.270.

Specific Gravity below 1.230-50

GOING DOWN

As battery discharges, acid begins to lodge in plates. Specific gravity drops.

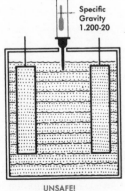

Specific Gravity 1.200-20

UNSAFE!

Battery half discharged. More acid in plates, less in electrolyte. Starting failure in sight if battery is allowed to remain in car.

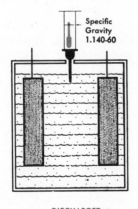

Specific Gravity 1.140-60

DISCHARGED

Acid almost entirely in plates, leaving weak electrolyte behind. Specific gravity lower, almost that of water.

Fig. 19-31. Specific gravity as related to battery state of charge. As battery charge falls off, float sinks lower and lower into electrolyte.

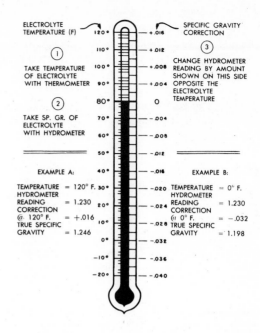

ELECTROLYTE TEMPERATURE (F)

SPECIFIC GRAVITY CORRECTION

① TAKE TEMPERATURE OF ELECTROLYTE WITH THERMOMETER

③ CHANGE HYDROMETER READING BY AMOUNT SHOWN ON THIS SIDE OPPOSITE THE ELECTROLYTE TEMPERATURE

② TAKE SP. GR. OF ELECTROLYTE WITH HYDROMETER

EXAMPLE A:

TEMPERATURE = 120° F.
HYDROMETER
READING = 1.230
CORRECTION
@ 120° F. = +.016
TRUE SPECIFIC
GRAVITY = 1.246

EXAMPLE B:

TEMPERATURE = 0° F.
HYDROMETER
READING = 1.230
CORRECTION
@ 0° F. = −.032
TRUE SPECIFIC
GRAVITY = 1.198

Fig. 19-32. Temperature correction scale. (Buick)

BATTERY EFFICIENCY AT VARIOUS TEMPERATURES	
Temperature	Efficiency of a Fully Charged Battery
80° F.	100%
50° F.	82%
30° F.	64%
20° F.	58%
10° F.	50%
0° F.	40%
−10° F.	33%

Fig. 19-33. Temperature affects battery efficiency. It is obvious why a fully charged battery is a MUST for cold weather starting. (Ford)

Specific Gravity	Freezing Point Degrees F.	Specific Gravity	Freezing Point Degrees F.
1.100	+18	1.220	−31
1.120	+13	1.240	−50
1.140	+ 8	1.260	−75
1.160	+ 1	1.280	−92
1.180	− 6	1.300	−95
1.200	−17		

Fig. 19-34. Specific gravity affects freezing point. (GMC)

BATTERY CAPACITY RATINGS

There are several standards set up around battery performance. Perhaps one of the most used is the 20-Hour Rating in Ampere-Hours.

The 20-hour rating is used to determine the lighting ability of a battery. The test is conducted as follows:

A new, fully charged battery, at a temperature of 80 deg., is discharged at a rate equivalent to one-twentieth of its published 20-hour capacity in ampere-hours. It is discharged at this rate until the terminal voltage drops to 10.5V (5.5V for a 6 volt battery). The hours of discharge required to drop the terminal voltage to 5.25V, multiplied by the discharge rate in amperes, determines the ampere-hour capacity of the battery and its 20-hour rating.

TYPICAL CURRENT LOAD OF MODERN CARS

	12 Volt System	6 Volt System
Head Lamps	8.0 Amps	13.5 Amps
Ignition	3.0 Amps	3.3 Amps
Instrument Lights ..	1.5 Amps	1.2 Amps
Tail Lights	1.0 Amps	1.5 Amps
License Light	0.4 Amps	0.6 Amps
Heater	5.0 Amps	8.7 Amps
Radio	1.8 Amps	5.5 Amps
	20.7 Amps	34.3 Amps
Summer Starting	100 - 225 Amps*	125 - 300 Amps*
Winter Starting	225 - 500 Amps*	300 - 700 Amps*

Fig. 19-35. Typical current load for various electrical units. Load will vary somewhat depending upon type of ignition system, radio, etc.

If a 120-ampere-hour battery (advertised rating) is to be tested, it will be discharged at one-twentieth of its advertised rating, or, in this case, 6 amps. To pass the test, this battery should withstand 20 hours of 6 amp. discharge before terminal voltage drops below that stated above. Fig. 19-36.

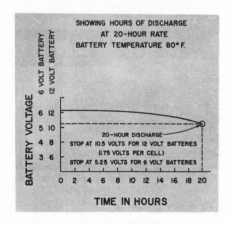

Fig. 19-36. Battery manufacturers 20-hour rating discharge.

There are several other tests, such as the cold zero deg. cranking test, life cycle test, etc.

BATTERY CAPACITY

Plate size, number of plates and amount of acid in the electrolyte determine battery capacity. The heavy amperage draw starting capacity of a battery depends to a great extent on the total area.

BATTERY VOLTAGE

Battery voltage is determined by the number of cells (compartments containing elements). Each cell has an open circuit voltage of 2.1V. Three such cells produce a 6V battery; six cells, a 12V battery.

CAUTION!

A battery will occasionally need water added to the electrolyte. Do not overfill. Always use DISTILLED water. Some tap waters contain harmful chemicals that will reduce the life and efficiency of a battery. (Freezer and refrigerator frost, when melted, makes a good source of distilled water.)

DRY—CHARGED BATTERY

The dry-charged battery is popular because it is shipped with the plates charged and the battery dry. If this battery is kept in a cool, dry spot, it will last a long time without the necessity of trickle charging (charging constantly at a very low rate).

When the battery is delivered to the customer, the electrolyte is added and, in this way, the buyer gets a NEW battery even though it has been stored for some time.

DANGER — ACID BURNS!

BATTERY ELECTROLYTE IS A STRONG ACID. IT CAN BE HIGHLY INJURIOUS TO THE EYES, SKIN AND CLOTHING. IF ACCIDENTALLY SPLASHED WITH ACID, FLUSH AFFECTED AREA WITH LARGE AMOUNTS OF WATER. FOLLOW IMMEDIATELY BY BATHING AREA IN A MIXTURE OF BAKING SODA AND WATER. SODA, BEING AN ALKALI, WILL NEUTRALIZE THE ACID. WHEN WORKING WITH BATTERIES, BE CAREFUL!

DANGER — BATTERIES CAN EXPLODE!

Never bring a spark or open flame near a battery. When charging, the battery gives off hydrogen and oxygen gas. This gas, which is highly explosive, fills the space above the plates and even the atmosphere immediately surrounding the battery. When a battery explodes, it can throw lethal quantities of acid in all directions. BE CAREFUL!

JOB OF THE BATTERY

The main purpose of the battery is to supply current for the ignition system and starter motor UNTIL the car is running. When the engine starts, the generator supplies the electrical needs of the car and recharges the battery.

ELECTRICAL SYSTEM, GENERATOR

The generator is an electromagnetic device driven by the engine. It is used to produce electricity to meet the needs of the electrical system. It also keeps the battery charged.

GENERATOR USES MAGNETISM TO PRODUCE ELECTRICITY

When a wire is passed through a magnetic field, voltage is generated in the wire. If the wire is part of a closed circuit, current will flow. Fig. 19-37.

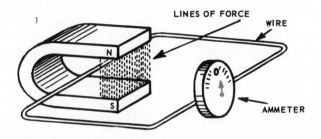

Fig. 19-37. Inducing voltage in a wire. In 1, wire is stationary. No current is flowing. In 2, wire is being drawn swiftly through magnetic lines of force (in direction A). Wire, as it passes through field, will have voltage induced in it. As it makes a complete circuit, current flows (see needle on ammeter).

If the wire is passed through the magnetic field in the opposite direction, the current flow in the wire will be reversed.

The amount of current produced in the wire will depend on two factors:

1. Strength of magnetic field.
2. Speed at which wire cuts or moves through field.

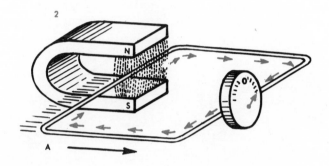

Fig. 19-38. Wire cutting through a magnetic field. Wire is passing through field in direction A. Note how lines of force are bent as wire moves through field. Two arrows around wire indicated direction in which magnetic field moves around wire. Using left-hand rule for conductors (Fig. 19-39), you will see that any current flow in wire would be directly away from you.

As the wire cuts the field, the lines of force are distorted or bent around the leading side. You will notice in Fig. 19-38 that the arrow around the wire indicates the direction in which the magnetic field moves around the wire.

LEFT—HAND RULE FOR CONDUCTORS

By placing your left hand, with the fingers wrapped around the wire in the direction of the magnetic field, your thumb will point in the direction the current is moving.

By the same token, if you know the direction the current is moving, grasp the wire with the left-hand thumb pointing in the direction of current flow. Your fingers will point in the direction the magnetic field is moving. Fig. 19-39.

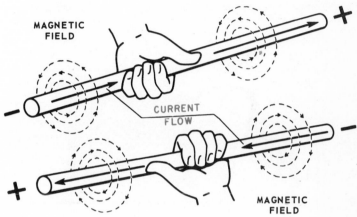

Fig. 19-39. Left-hand rule for conductors. Thumb points in direction of current flow and fingers wrap around conductor in direction of magnetic field.

ELECTRICITY PRODUCES MAGNETISM

If a current carrying coil is placed around a bar of steel, the bar will become magnetized. The more turns of wire and the stronger the current, the more powerful the magnet.

This principle is used in the generator and in other items of equipment such as relays, voltage regulators, starter solenoids, etc. Fig. 19-40.

LEFT—HAND RULE FOR COILS

You can determine either polarity (which end is N, which end is S) by grasping the coil so the fingers of the left hand wrap around the coil in the direction the current is moving. The thumb will then point to the north pole. If the pole is known, grasp the wire with the thumb pointing in the direction of the N pole, and the fingers will point in the direction the current is flowing. Fig. 19-41.

ELECTROMAGNET CORE

By placing a soft iron core within the coil, the magnetic force lines are concentrated and strengthened. As there is less

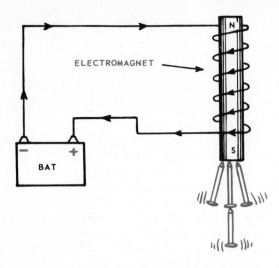

Fig. 19-40. An electromagnet. When current is passed through coil surrounding iron bar, bar becomes magnet. Notice how bar attracts and holds nails while current is flowing.

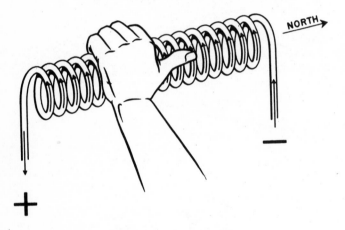

Fig. 19-41. Left-hand rule for coils. (Autolite)

magnetic reluctance (resistance) in the core than in the air, the force lines will follow the core. How the generator makes use of this principle will be explained.

ALTERNATING CURRENT GENERATOR

If a looped wire is rotated between two magnets (pole shoes), it will have voltage induced in it. If the wire is part of a closed circuit, current will flow. If each end of the wire loop is connected to a round ring (slip ring), the loop can rotate in the field and at the same time, remain in contact with the remainder of the circuit. Each end of the rest of the circuit is attached to a carbon brush (block of carbon) that rubs against the slip ring. Fig. 19-42 shows the action of an alternating current (AC) generator. Notice that a voltmeter is placed across the circuit.

In A, Fig. 19-42, the loop is cutting through the field in a counterclockwise direction. This means that the top part of the right-hand side of the loop is the leading edge, while the bottom side is the leading edge on the other half. This causes

the magnetic lines to wrap around the loop in one direction in one half, and the other direction in the other half. This creates a current flow in the direction of the arrows. Notice that the voltmeter has swung to the right, indicating that voltage is now being generated.

In B, Fig. 19-42, the loop has continued turning until it is out of the magnetic force field. As the loop is not cutting any lines, it creates no voltage, and no current flows.

When the loop again passes through the field, the part of the loop moving up through the force lines will be moving down through the force lines, and vice versa. As this reverses the direction the force lines wrap around the loop, the voltage again rises, but the direction of current flow is reversed. This is termed alternating current and is similar to that used in your home.

You will notice in C, Fig. 19-42, that the voltage starts at zero, rises to a peak, falls to zero, reverses directions, rises to a peak and falls again. This all takes place in one complete revolution of the loop.

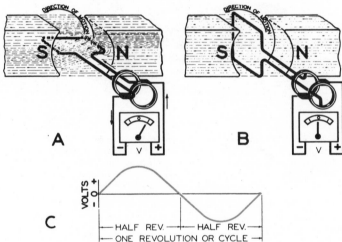

Fig. 19-42. Alternating current generator. This shows only one loop. Actual generator uses many loops. (Autolite)

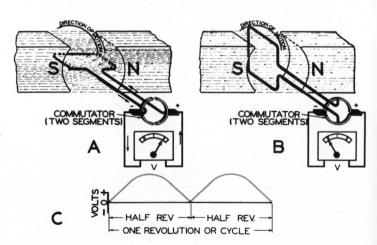

Fig. 19-43. Direct current generator. Note use of commutator segments instead of slip rings.

DIRECT CURRENT GENERATOR

The direct current (DC) generator operates on the same principle as the AC generator. The main difference is that the current moves in one direction only.

This is accomplished by splitting a single ring (commutator) and connecting the circuit brushes to it in such a manner that when the loop has passed out of the field and is ready to reverse, the brushes connect with opposite halves of the ring.

The DC generator shown in Fig. 19-43 starts out at zero voltage, rises to a peak as the loops cut through the strongest part of the field, drops to zero and before it can reverse, starts rising to a peak again with the current traveling in the same direction.

GENERATOR NEEDS MORE LOOPS

With only one loop passing through the field, the voltage produced is erratic. You noticed in Figs. 19-42 and 19-43, that the voltage rose and fell. By adding more loops, the voltage would not fall back to zero because one loop will be leaving the field while another is entering.

As more loops are added, the voltage fluctuations will smooth out. A higher voltage can be produced with more loops. Fig. 19-44 illustrates the effect of adding another loop. Notice that the commutator is now split into four pieces (segments). The brushes contact the commutator segments at exactly the right time to prevent current reversal. Note the graphic illustration of the resultant voltage and the voltage variation.

POLE SHOES

A car generator does not use permanent magnets. Instead, it incorporates two pole shoes. Many turns of wire are wound

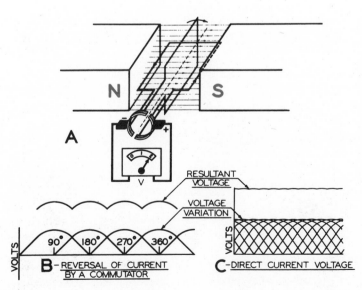

Fig. 19-44. Effect of adding a loop to DC generator. Commutator now contains four segments. Note how voltage pulsation is smoothed out by use of extra loop. (Autolite)

around the pole shoes. When current passes through these windings, the pole shoes become electromagnets. The more current that flows, the stronger the field. These coils are called FIELD COILS or FIELD WINDINGS.

The two field windings are connected together in series and wound so that one pole shoe becomes the north pole and the other the south pole. Fig. 19-45 shows the two pole shoes, field frame and field winding. Only one turn of wire is shown for the field winding; actually, there are many turns.

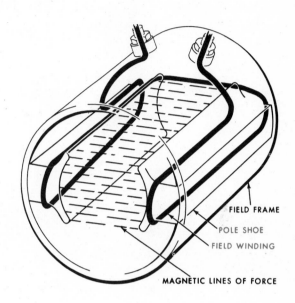

Fig. 19-45. Field windings and pole shoes. In actual generator, field windings consist of many turns of wire. (Lincoln)

ARMATURE

The loops (armature windings) that are to rotate in the magnetic field, are wound on a special laminated (many pieces) holder called an armature.

The armature is supported in bearings set in end plates. Generally a ball bearing is used at the pulley end and a bronze bushing at the other end. The armature is turned by placing a pulley on one end and driving it with a V-belt from the crankshaft pulley. Figs. 19-46, 19-47 and 19-48.

COMMUTATOR

The segments of the commutator are fastened to the armature, and the ends of the armature windings are soldered to the segments. The armature segments are insulated from the armature and from each other. The mica insulation between the segments is undercut (cut below level of commutator) to prevent excessive wear of the carbon brushes. See Figs. 19-46, 19-47 and 19-48.

BRUSHES

Two carbon brushes, in spring-loaded holders, are attached to the end plate. Some generators mount the holders in the

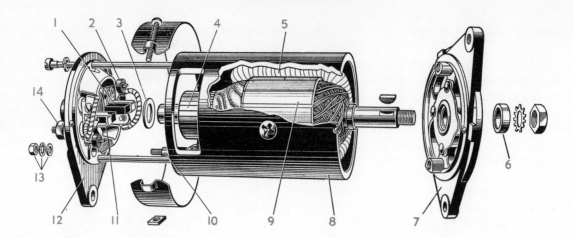

Fig. 19-46. Exploded view of typical DC generator. Learn names and location of all parts. 1—Brush. 2—Brush springs. 3—Thrust washer. 4—Commutator. 5—Field coil. 6—Spacer. 7—Driving end bracket. 8—Field frame. 9—Armature. 10—Field terminal. 11—Brush holder. 12—Commutator end bracket. 13—Field terminal nuts and washer. 14—Terminal.   (Austin—Healey)

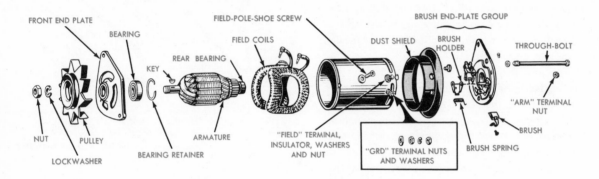

Fig. 19-47. Exploded view of a typical DC generator.   (Ford)

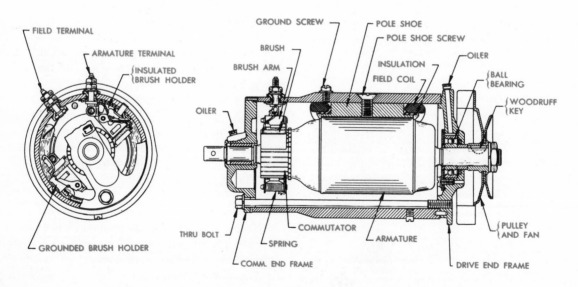

Fig. 19-48. Cross section of typical DC generator.   (American)

frame. The brush held in an insulated holder picks up the voltage developed by the armature windings. This insulated brush is connected to the armature (ARM.) terminal of the generator by a short wire.

The other brush holder is grounded to the frame, completing the circuit. The spring pressure on both holders causes the brushes to remain in firm contact with the commutator at all times.

GENERATOR CIRCUITS

Two types of DC generator circuits are in common use. Both types attach the insulated brush to the ARM. terminal. The main difference is in the grounding of the field windings. One type grounds the field externally, A, Fig. 19-49. The other is grounded internally, B.

HOW GENERATOR WORKS

When the generator armature first begins to spin, there is a weak residual magnetic field in the soft iron pole shoes. As the armature spins, it begins to build voltage. Some of this voltage is impressed on the field windings through the generator regulator. This impressed voltage builds up a stronger winding current, increasing the strength of the magnetic field. The increased field, in turn, produces more voltage in the armature. This builds more current in the field windings, with a resultant higher armature voltage. When the voltage has reached an acceptable peak, further increase is prevented by limiting the amount of voltage impressed on the field windings. A schematic view of a simple generator is shown in Fig. 19-50.

GENERATOR COOLING AND LUBRICATION

Most generators use small fans on the pulley ends for cooling. Such a fan draws a stream of cooling air through the generator. Fig. 19-51.

Lubrication is provided by applying light engine oil to oil holes. The holes lead to a felt wick that administers the oil to the bearings. Avoid over oiling, as the excess oil will be thrown onto the commutator.

GENERATOR REGULATOR

It is necessary to control both the voltage and the current produced by the generator. This is done by incorporating a voltage and current regulator in the generator circuit.

Excessive generator output will damage the battery, burn out the lights and damage other units in the electrical system.

The generator described in the last few paragraphs is termed a shunt generator. As such, it has no internal method of controlling its output. Therefore, it must be controlled by a regulator.

REGULATOR CONSTRUCTION AND OPERATION

The commonly used regulator has three separate control units. One unit controls the voltage (voltage regulator), another controls the current output (current limiter), and the third connects and disconnects the generator from the battery circuit (circuit breaker or cutout relay).

CIRCUIT BREAKER OR CUTOUT RELAY

The circuit breaker or cutout relay is a simple magnetic make-and-break switch. It is used to connect the generator to

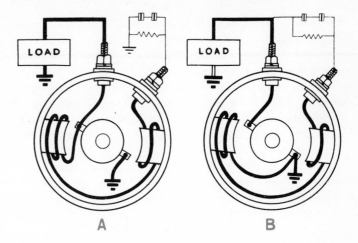

Fig. 19-49. Generator circuits. Circuit A is most widely used of two. A grounds field externally in regulator. B runs field current through regulator to field coils, then grounds circuit inside generator. (Autolite)

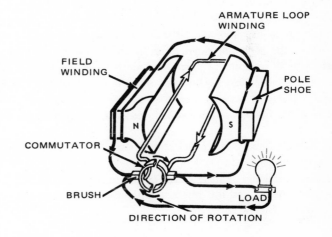

Fig. 19-50. Simple generator. (GMC)

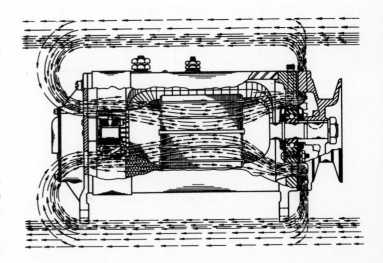

Fig. 19-51. Air cooling generator. Pulley fan draws air through generator body. (Autolite)

the battery circuit when generator voltage builds up to the desired value. It will disconnect the generator when the generator slows down or stops. This prevents the battery from discharging through the generator.

A soft iron core is magnetized to pull down a hinged armature. When the armature is pulled down, it closes a set of contact points and completes the circuit. When the core field is broken, the spring pulls the armature up, breaking the contacts and opening the circuit.

A fine wire from the generator circuit is given many turns around an iron core (shunt winding) and is then grounded. A heavy wire from the generator is given a few turns around the coil and is connected to the upper contact point. The other contact point is connected to the battery circuit.

When the generator first starts turning, the cutout armature keeps the contact points open. As generator voltage builds up, an ever increasing magnetic field is created by the current flowing through the fine windings around the core. When cut-in voltage is reached, the shunt magnetic field is strong enough to pull the armature down and close the contacts. This connects the generator to the battery circuit and current flows from generator to battery. This also causes the heavy windings to add to the magnetic pull of the fine winding.

When the generator slows down or stops, battery voltage will exceed generator voltage, and current will flow from battery to generator. This reversal of current direction in the large winding reverses its magnetic field, which works against that of the fine winding. This reduces magnetic pull on the armature, and the armature spring pulls the contacts open, breaking the circuit. This prevents the battery from discharging through the generator.

When generator voltage builds up again, the armature will be drawn down, and the generator will again be connected to the circuit. Fig. 19-52.

VOLTAGE REGULATOR CONSTRUCTION

The external end of the field coil circuit is connected to one of the voltage control unit contact points. The other point (regulator armature point) leads directly to ground. A spring keeps the armature point against the other point, completing the field coil circuit.

A shunt circuit going to ground through a resistance unit is placed just ahead of the points. When the points are closed, the field circuit takes the easy route to ground through the points. When the points are open, the field circuit must pass through the resistance to get to ground.

An iron core, similar to that of the circuit breaker, is placed directly beneath the armature. A coil of fine wire (one end connected to ground, other end to generator armature circuit in circuit breaker) is placed around the core. Fig. 19-53.

VOLTAGE REGULATOR ACTION

When the generator is operating (if battery is low or a number of electrical units are operating), the generator voltage may stay below that for which the control is set. In this event, the flow of current from the generator through the core shunt

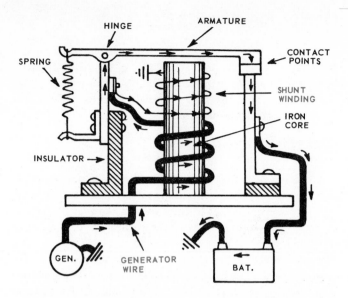

Fig. 19-52. Schematic of a cutout relay. Current passing through both heavy generator wire coil and shunt coil will magnetize iron core, causing it to keep armature pulled down. This allows current from generator to pass through battery. Follow arrows.

coil will be too weak to pull the armature down, and the generator field will continue to go to ground through the points. This will complete the field circuit, with no resistance, and the generator will continue to build up voltage.

If some of the electrical load is removed or if the battery becomes charged, the generator voltage will increase until it reaches the unit setting. At this point, current flow through the core shunt coil will be high enough to pull the armature down and separate the points.

When the points separate, the field circuit must flow through the field resistance unit. This cuts down the current flow to the generator field coils and weakens the generator magnetic field. This immediately cuts down generator output. When output drops, the current flow through the shunt

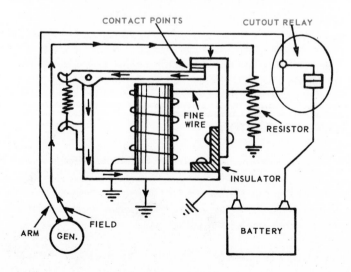

Fig. 19-53. Voltage regulator unit. Trace movement of current in field circuit. Note that voltage has not yet reached a point high enough to cause electromagnet to pull contact points apart.

winding drops, and the armature closes the points. When the points close, field resistance is removed and output goes up.

This cycle is repeated over and over, and the points actually have a vibrating action. They open and close about 50 to 200 times per second. This action then controls or limits the maximum voltage that the generator can produce.

The voltage regulator limits maximum voltage. The condition of the battery and the electrical load actually control the amount of voltage produced. The voltage regulator only comes into play when the generator attempts to exceed predetermined voltage limits. Fig. 19-54 illustrates voltage regulator action. The top illustration shows the points closed when generator voltage is low. When it reaches regulator limits, the points open and the field circuit passes through the resistor. Study each illustration and trace the flow of current in each case.

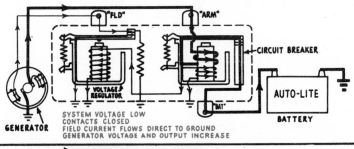

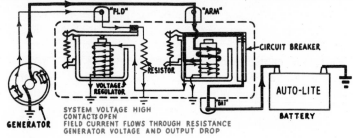

Fig. 19-54. Voltage regulator action. Top view shows regulator action when generator voltage is below that for which regulator is set. Bottom view shows regulator action when generator voltage attempts to exceed proper setting. (Autolite)

CURRENT REGULATOR OR LIMITER

Even though generator voltage is controlled, it is possible (when battery is low or a heavy load is on) for generator current to run extremely high. This would overheat the generator and it would soon be ruined. The current, as well as the voltage, must be limited.

CURRENT REGULATOR CONSTRUCTION

A current regulator is built in much the same fashion as the voltage regulator. Instead of a shunt coil of fine wire, a lesser number of turns of heavy wire is wrapped around the core. This heavy wire is connected in series with the generator armature circuit. In operation, the field circuit passes through the current limiter points on its way to the voltage regulator points. See thin black line in Fig. 19-55.

CURRENT REGULATOR ACTION

For purposes of explanation, imagine the generator speed as fairly slow, voltage low and little current flow. Both the current regulator points and the voltage regulator points are closed, and the field circuit is completed without passing through any resistances.

As generator speed increases, current output goes up. If the battery is low and generator voltage remains low, the voltage regulator points remain closed.

Current flow increases up to the setting of the unit. At this time, current flow through the heavy coil windings will cause the core to draw the armature down, opening the current regulator points. In order to be complete, the field circuit must pass through a resistor. This lowers current output, points close, output increase, points open, output down, points close, and so on. The points will vibrate open and closed, fast enough to limit the current to a smooth maximum.

When current flow is low and voltage reaches the voltage regulator setting, current regulator points remain closed but voltage regulator points open and close, limiting the voltage.

Either the voltage regulator or the current regulator can function, but they cannot function at the same time. In order for the field circuit to reach ground without passing through a resistance, both sets of points must be closed.

Study Fig. 19-55 and trace the field circuit through both current and voltage regulator units. Follow the armature circuit through the current regulator coil to the cutout relay and on to the battery. Note the field resistances. One of the resistance units becomes part of the field circuit when either the voltage regulator or current regulator points are open. The other resistance unit is connected to the cutout unit.

When either point set opens, there is a reduction in field current. This causes an induced voltage (similar to action in ignition coil) in the field coils. The resistances, in addition to increasing field circuit resistance, also help prevent arcing at the points. Study Figs. 19-55 and 19-56.

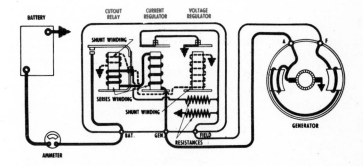

Fig. 19-55. Three unit generator regulator. Trace out all circuits until you understand function of each.

DOUBLE CONTACT VOLTAGE REGULATOR

Another type of regulator uses a double set of contact points on the voltage regulator unit. When the bottom

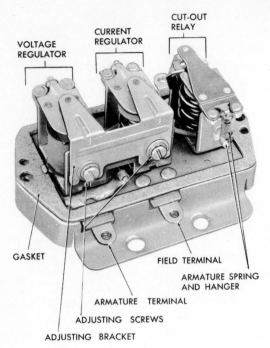

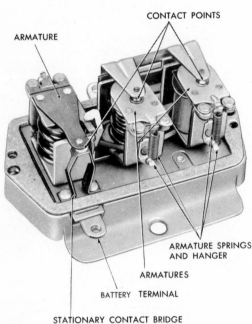

Fig. 19-56. Typical regulator. Regulator cover is not shown. (Plymouth)

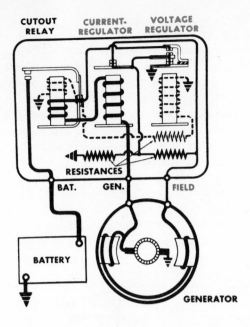

Fig. 19-57. Double contact voltage regulator. Regulator is shown in position it would assume when generator voltage is below that of unit setting. Field circuit is going directly to ground.

TEMPERATURE COMPENSATION

By using bimetallic hinge strips on the circuit breaker (cutout relay) and current regulator, it is possible to allow increased current flow during cold operation. As the system warms up, contact armature resistance changes and current flow is lowered. One type uses a special magnetic shunt that lowers the magnetic pull on the voltage regulator armature when the unit is cold. This allows the voltage to increase until enough magnetic pull is developed to separate the points. When the unit is hot, most of the magnetic field acts on the contact armature, and voltage is reduced.

Regulator coil magnetic efficiency varies from cold to hot. This must be controlled by temperature compensating devices. Fig. 19-58 shows how one type of regulator is temperature compensated.

AIR GAP

The term REGULATOR AIR GAP refers to distance between contact armature and top of iron core that is magnetized to draw the armature down.

GENERATOR POLARITY

POLARITY can be defined as the directional flow of current leaving the generator. When the generator is stopped, a certain amount of residual magnetism remains in the pole shoes. The polarity of the shoes at this time, depends on the direction of current flow in the field coil windings prior to stopping.

If, during generator testing, current is accidentally caused to flow in an opposite to normal direction, the pole shoes will change polarity. N becomes S, and S becomes N. When the

contacts are touching, the field circuit is completed without passing through a resistance unit.

As voltage builds up, the armature moves down and neither upper nor lower contacts are touching. This inserts a series resistance unit into the field circuit.

As generator speed and field current build up beyond the control of the series resistance, the armature is pulled down, bringing the upper contacts together. This short circuits the field, bringing the voltage within controlled limits. A double contact gives longer service with higher generator output loads than a single contact type. It also allows higher field currents with lower generator speeds. Fig. 19-57.

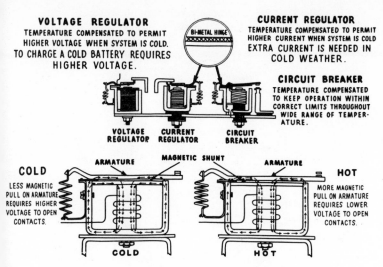

VOLTAGE REGULATOR
TEMPERATURE COMPENSATED TO PERMIT HIGHER VOLTAGE WHEN SYSTEM IS COLD. TO CHARGE A COLD BATTERY REQUIRES HIGHER VOLTAGE.

BI-METAL HINGE

CURRENT REGULATOR
TEMPERATURE COMPENSATED TO PERMIT HIGHER CURRENT WHEN SYSTEM IS COLD EXTRA CURRENT IS NEEDED IN COLD WEATHER.

CIRCUIT BREAKER
TEMPERATURE COMPENSATED TO KEEP OPERATION WITHIN CORRECT LIMITS THROUGHOUT WIDE RANGE OF TEMPERATURE.

VOLTAGE REGULATOR   CURRENT REGULATOR   CIRCUIT BREAKER

COLD
LESS MAGNETIC PULL ON ARMATURE REQUIRES HIGHER VOLTAGE TO OPEN CONTACTS.

ARMATURE   MAGNETIC SHUNT   ARMATURE

HOT
MORE MAGNETIC PULL ON ARMATURE REQUIRES LOWER VOLTAGE TO OPEN CONTACTS.

COLD   HOT

Fig. 19-58. Regulator temperature compensation.   (Autolite)

generator is installed and the engine started, the reversed polarity will cause current to flow in the wrong direction.

The reversed current flow will damage the cutout relay points by doubling system voltage at the points (generator and battery will be in series). The operating coil magnetic strength will be insufficient to hold the cutout points closed. As a result, they will vibrate and burn. In time, they will actually weld together, and the battery will discharge through the generator. A dead battery, as well as possible damage to the generator, will result. To prevent this, the generator should be polarized before starting the engine.

GENERATOR POLARIZING

When a generator has been disconnected and reinstalled, it must be polarized prior to starting the engine.

When the generator grounds the field circuit externally, A, Fig. 19-49, the generator is polarized by connecting a jumper wire across the Gen. and Bat. terminals of the regulator. Hold one end of the wire against the generator terminal, and scratch the other end across the battery terminal. This one flash causes

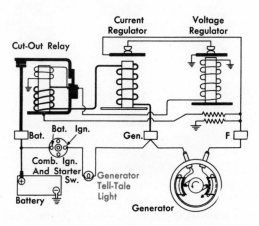

Fig. 19-59. Generator charge indicator light.
(Cadillac)

current to flow through the field coil in the proper direction, insuring correct polarity.

To polarize the internally grounded field circuit, B, Fig. 19-49, DISCONNECT the field lead from the regulator and momentarily scratch it on the battery terminal of the regulator.

Remember this rule for polarizing generators — ALWAYS PASS CURRENT THROUGH THE FIELD COILS IN A DIRECTION THAT WILL CAUSE THE GROUND SIDE OF THE COILS TO BE CONNECTED TO THE GROUND SIDE OF THE BATTERY.

Remember, always polarize a generator after installation!

OTHER TYPES

There are other types of DC generators such as the split field, third brush, etc. Since their uses are limited, they will not be discussed in this text.

A TEAM

The generating system is made up of the battery, generator and regulator. All are dependent on each other. In order for the generating system to function properly, all parts must be in proper working order. The action of any one part is governed by the action of the other two parts.

AMMETER

An ammeter is installed in the generating system to measure the flow of electricity (in amperes) to or from the battery. The ammeter will let the operator know if the battery is being charged, and the rate of charge. The ammeter is connected into the circuit in series. Fig. 19-55 illustrates the ammeter in a typical system.

GENERATOR INDICATOR LIGHT

An indicator light located on the instrument panel may be used instead of an ammeter. When the generator is charging, the light is off. When the generator is not charging, the light will come on. The light does not tell how much the generator is charging. Fig. 19-59 shows the wiring circuit for a typical generator indicator light.

ALTERNATING CURRENT (AC) GENERATOR

The AC generator (also called an alternator) is widely used on cars today.

The modern car has made heavy demands on the electrical system. This, coupled with the fact that much driving is of the stop-and-go city type, has made it difficult to maintain the battery in a fully charged state. A DC generator is not very efficient of the lower traffic and idling speeds.

An AC generator provides a better output at lower engine speeds and, in the heavier models, provides a very high output at cruising speeds. Fig. 19-60 is a graphic comparison between the DC and the AC generator.

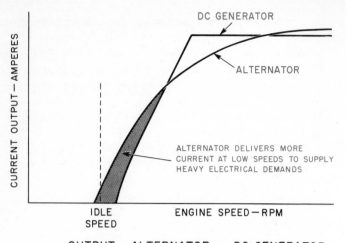

Fig. 19-60. Alternator — generator comparison. Note low speed superiority of alternator. (Plymouth)

SEVERAL BASIC DIFFERENCES

The alternator delivers alternating current (AC), while the DC generator delivers direct current (DC).

The DC generator spins the armature (output) coils inside of two or more stationary field coils. The AC generator spins the field coil inside a set of stationary stator windings (output).

The AC generator field is excited (current passed through it) by connecting it to a battery. The current draw is relatively light, around 2 amps. It varies with generator size.

ALTERNATOR CONSTRUCTION

As mentioned, the alternator spins a field coil with alternating N and S poles inside the stator windings. A round iron core is placed on a shaft, and the field winding is placed around the core. Two iron pole pieces are slid on the shaft. They cover the field winding and are arranged so the fingers are interspaced. The fingers on one pole piece all form N poles. The fingers on the other form S poles. As the fingers are interspaced, they form alternate N—S—N—S poles. Fig. 19-61 shows the complete rotor assembly in exploded form.

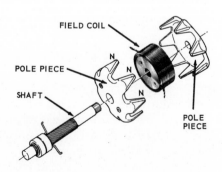

Fig. 19-61. AC rotor assembly. Pole pieces and field winding assembly are pressed on shaft. (Delco—Remy)

Fig. 19-62 shows a simple rotor turning inside a single conductor (stator winding). As the rotating rotor magnetic fields cut through the conductor, voltage is generated and current flows in the complete circuit. As the alternating N and S pole fields pass around the conductor, the current will alternate; first one direction and then the other.

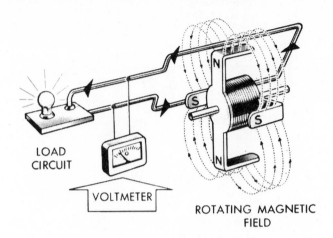

Fig. 19-62. Schematic of simple AC generator. Magnetic field cuts through stator windings. (Delco—Remy)

A small motorcycle alternator stator and rotor are shown in Fig. 19-63. In this installation, the rotor is fastened to the engine crankshaft and whirls inside the stator assembly. Notice the stator is made up of laminated iron rings. The stator reduces field reluctance, thereby intensifying the magnetic field. Fig. 19-63.

Direct current from the battery is fed through the field coil by using brushes rubbing against slip rings. One end of the field coil is fastened to the insulated brush, while the other end is attached to the grounded brush.

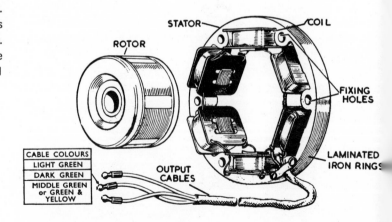

Fig. 19-63. Small motorcycle alternator. (Joe Lucas Ltd.)

Figs. 19-64 and 19-65 illustrate an AC generator using slip rings. As the pole fields pass through the conductor in Fig. 19-64, voltage is imparted in the conductor, and current flows

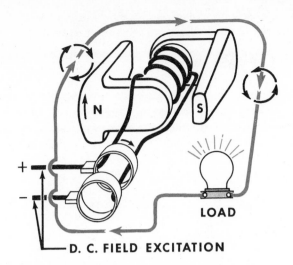

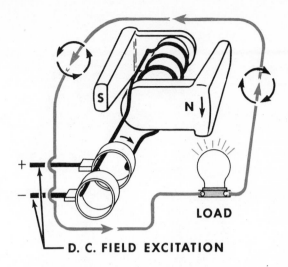

Fig. 19-64. Schematic of a simple single-phase AC generator. Brushes contact slip rings. (Delco—Remy)

Fig. 19-65. Schematic of simple single-phase AC generator. You will notice that pole pieces have reversed position they assumed in Fig. 19-64. This causes reversal of current in load line or conductor.

in one direction. In Fig. 19-65, the rotor turns 180 deg., and the N pole and S pole are at opposite positions. This causes current to flow in the opposite direction.

An exploded view of one type of alternator is shown in Fig. 19-66. Note the end plates or covers containing bearings that support the rotor shaft. The stator and stator windings are supported by the covers.

A cross-sectional view of a typical alternator is given in Fig. 19-67. Study the location of the various components. The use of DIODES will be explained shortly.

An assembled alternator is illustrated in Fig. 19-68. The alternator is mounted in a manner similar to that of the DC generator.

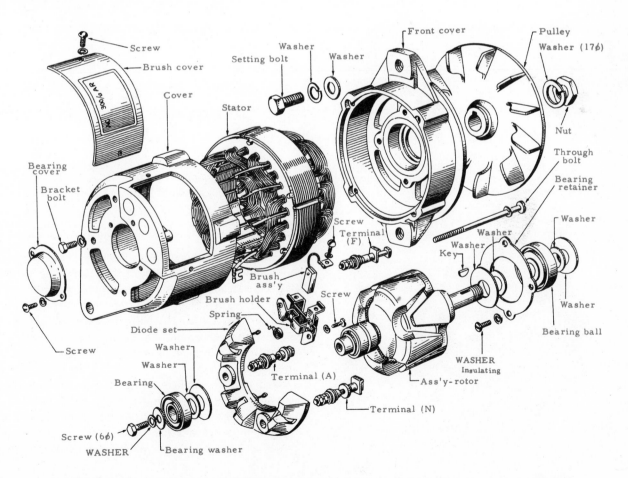

Fig. 19-66. Exploded view of one type of alternator (AC generator). Learn names and location of parts. (Datsun)

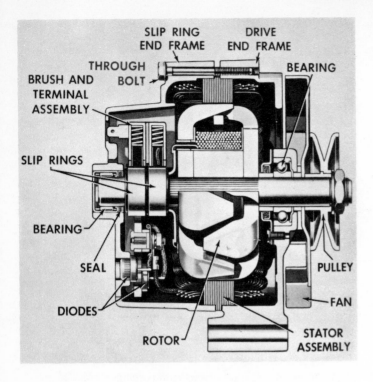

Fig. 19-67. Cross section of typical alternator or AC generator. (Cadillac)

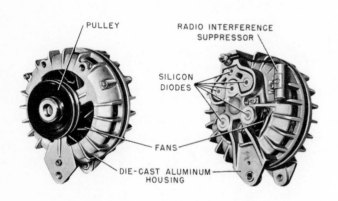

Fig. 19-68. Assembled alternator. Fans and cooling ribs keep unit at a workable temperature. (Plymouth)

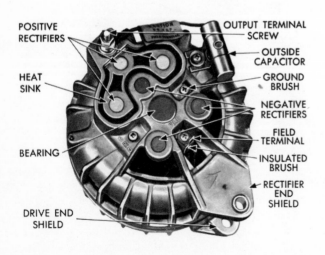

Fig. 19-69. Alternator — end view. (Plymouth)

An end view of an alternator is shown in Fig. 19-69. Note the location of the field, ground and output terminals.

THREE—PHASE AC

The stator windings are made up of three separate windings. This produces what is known as three-phase AC. When only one winding is used, single-phase current results. A curve illustrating single-phase AC current flow is shown in Fig. 19-70.

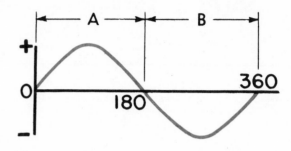

Fig. 19-70. Single-phase AC. A represents positive half of wave and B negative half. (Delco—Remy)

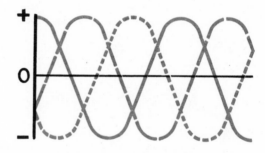

Fig. 19-71. Three-phase AC. Phase outputs overlap but do not coincide. (Delco—Remy)

Fig. 19-71 illustrates three-phase AC. Notice that the distance between high points in the output is shorter.

DIODES

The electrical system of the car is designed to utilize DC, so the AC flow from the alternator must be rectified (changed) to DC. This is accomplished by using an outside rectifier or, as is generally done, by passing the AC flow into silicon diodes.

Diodes have a peculiar ability to allow current to flow readily in one direction. When the current reverses and attempts to flow through in the other direction, current flow is stopped.

The diodes or rectifiers are located in one end plate or shield. In the alternator illustrated in Fig. 19-69, three negative polarity rectifiers are pressed into the aluminum end plate. These are on the ground side. Three positive polarity rectifiers are pressed into a heat sink which is insulated from the end plate.

The arrangement of the rectifiers will allow current to flow

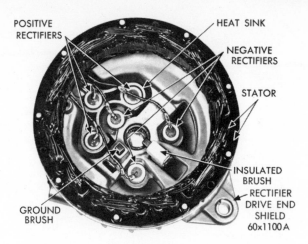

Fig. 19-72. Rectifier wiring arrangement. (Plymouth)

from the alternator to the battery, but will not allow current flow from the battery to the alternator. This eliminates the need for a generator cutout relay.

The wiring arrangement of the rectifiers is shown in Fig. 19-72. A cross-sectional view of a diode is given in Fig. 19-73. Fig. 19-74 pictures a graphic representation of rectified current flow.

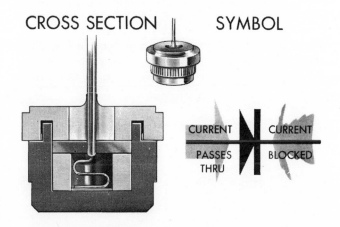

Fig. 19-73. Cross section of a diode. Current is free to pass one way only. Diode, like a transistor, is a semi-conductor (conducts electricity under certain conditions, acts as an insulator under others). (Delco—Remy)

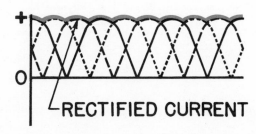

Fig. 19-74. Rectified current flow. This represents three-phase AC current depicted in Fig. 19-71, after it has been rectified. (Delco—Remy)

ALTERNATOR OPERATION

When the ignition key is turned on, the battery is connected to the No. 4 terminal of the regulator, Fig. 19-75. Current flow is through both the indicator lamp (ignition terminal of ignition switch) and through the resistor (accessory terminal of ignition switch). Combined current flow is about 1 amp.

From the No. 4 regulator terminal, the current path is through the lower voltage regulator points to the regulator F terminal, then through the alternator field to ground. This current flow excites the field, causing the pole pieces to become electromagnets, and making the "no charge" indicator lamp light.

When the engine is started, it will drive the alternator rotor. As the whirling rotor pole force fields cut through the stator windings, an AC voltage is generated.

This initial voltage is impressed from one phase of the stator to the R terminal on the alternator or generator. The R terminal is connected to the No. 2 regulator terminal and from there to the field relay coil winding. Current flow through the field relay coil windings causes the relay points to close. This action connects the No. 3 regulator terminal to the lower regulator points. Since the No. 3 terminal is connected to the battery (through starter solenoid), field current flow will be directly from the battery.

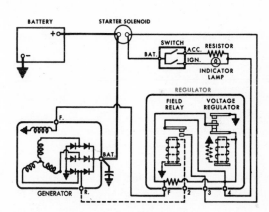

Fig. 19-75. Typical alternator wiring diagram. Note use of a two unit regulator. Study diagram to determine how field relay operates the "no charge" indicator lamp. (Pontiac)

At the same time, closure of the relay points impresses full battery voltage on BOTH SIDES of the No. 4 regulator terminal, stopping current flow through the indicator lamp and resistor. This causes the indicator lamp to go out.

In this regulator, the field relay acts as an indicator lamp relay, as well as completing the circuit between the battery and alternator field. Fig. 19-75.

A typical two unit AC generator regulator is illustrated in Fig. 19-76.

Note that the voltage regulator shown in Fig. 19-75 has a double set of contact points. When the alternator is inoperative, a spring holds the bottom contacts together. At lower

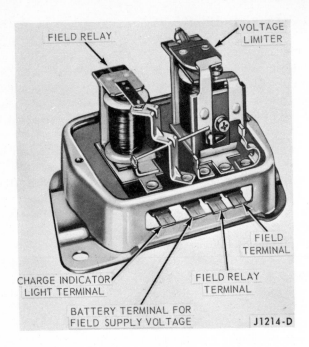

Fig. 19-76. Two unit voltage regulator used in AC generator circuit. (Ford)

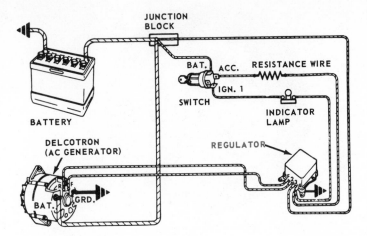

Fig. 19-77. Wiring diagram for Delcotron (AC generator) charging system using a two unit regulator. (Buick)

generator speeds, or when battery or accessory current demand is quite high, the lower contact points remain closed. This permits full current flow (about 2 amps) to the field windings.

As the alternator rpm increases, the voltage at the battery terminal also increases. This causes a heavier current flow through the voltage regulator shunt winding, creating enough magnetic pull to draw the lower points apart. This forces the field circuit to pass through a resistance unit, cutting field current to around 3/4 amp. This, in turn, lowers the value of the pole piece magnetic field, causing a fall off in alternator voltage.

With the lowering of voltage, the current through the voltage regulator shunt winding falls off. This reduces the magnetic pull and allows the spring to once again close the lower points. This opening and closing action is very rapid and will maintain an even voltage setting unless alternator speed becomes too high.

When alternator speed becomes high enough, the 3/4 amp. field current (lower points open) will permit the voltage to rise above that desired. The rise in voltage increases the current flow through the voltage regulator shunt windings, intensifying the magnetic pull on the point armature to the point where the upper contact points are drawn together. This grounds the field circuit, stopping ALL current flow in the alternator field windings. Voltage immediately falls off and the upper points open. Voltage increase, and the points close. The vibrating, opening and closing action will limit maximum voltage.

When engine speed is reduced, the lower set of points will limit the voltage satisfactorily. Study the voltage regulator circuit in Fig. 19-75.

Fig. 19-77 shows a wiring diagram of the Delco AC generator (called a Delcotron) charging circuit. This setup uses a two unit regulator.

SINGLE UNIT REGULATOR

When an ammeter is incorporated into the charging circuit (instead of an indicator lamp), a single unit regulator will suffice.

Since the diode rectifiers prevent current flow back into the alternator, no cutout relay is required. The alternator current (not voltage) output is self-limiting. Therefore no current regulator is needed. This makes it possible to utilize the single unit regulator (contaning only a voltage regulator).

The single unit regulator circuit is pictured in Fig. 19-78. Operation of the voltage regulator is similar to that described in the two unit regulator. Study the circuit. Note the diode rectifiers.

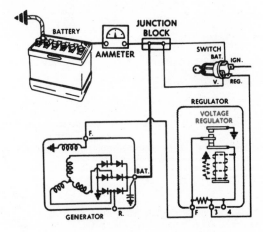

Fig. 19-78. Single unit voltage regulator. Note use of ammeter, eliminating need for a field relay.

TRANSISTOR VOLTAGE REGULATOR

The transistor, or solid state (no moving parts) voltage regulator, shown in Fig. 19-79, contains transistors, resistors, diodes, a capacitor and a thermistor.

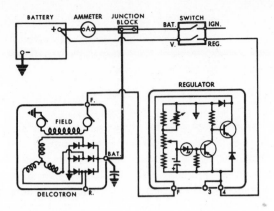

Fig. 19-79. Charging circuit employing a transistor regulator.

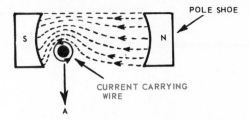

Fig. 19-80. Current carrying conductor in magnetic field. Field around wire (see small arrows) will oppose that of pole field. This will cause wire to be forced out in direction of arrow A.

The Zener diode is nonconducting until a specific voltage is reached. At this point, the diode "breaks down" (becomes conductive) and transmits a flow of electricity. This current flow, acting through the transistors, stops the flow of field current to the AC generator. This causes the voltage output to fall below diode "break down" voltage, at which time the diode once again becomes nonconductive. This, in turn, stops the current flow to the transistors and allows resumption of field current flow. The cycle is repeated as fast as 2,000 times per second. The thermistor compensates for changes in temperature.

A charging circuit employing a transistor regulator, is shown in Fig. 19-79. Some regulators are built as part of the alternator.

TRANSISTORIZED VOLTAGE REGULATOR

The transistorized (not transistor) regulator utilizes a single transistor working in conjuction with a vibrating contact point. The transistor makes possible less current flow across the contact points, prolonging point service life.

STARTER SYSTEM

In order to start the engine, it must be cranked by a starting motor. The starting motor is mounted on the clutch or flywheel housing and operates a small gear that can be meshed with a large ring gear on the flywheel.

To energize the starter motor, an electrical circuit is completed by the drive. When the starter motor armature begins to turn, the starter gear moves out and engages the ring gear. This spins the flywheel and starts the engine. When the engine starts, the driver breaks the starter circuit, and the starter gear moves out of mesh with the ring gear.

STARTER MOTOR PRINCIPLES

If a current carrying wire is placed in a magnetic field, the field set up around the wire by the moving current will oppose that of the magnetic field. If the wire is free to move, it will move out of the field. This fact forms the basis for an electric motor. Fig. 19-80.

ARMATURE LOOP

If the conductor is bent in the form of a loop, each side of the loop will set up a magnetic field. Since the current is traveling in opposite directions, the two sides will have opposing fields. Fig. 19-81.

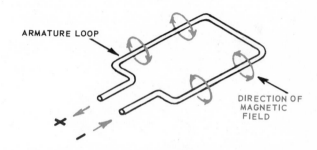

Fig. 19-81. Armature loop. Note direction of magnetic field on each side of loop.

When the armature loop is placed in a magnetic field, both sides of the loop will oppose the magnetic field, but in opposite directions. This will cause the loop to rotate out of the field. Fig. 19-82.

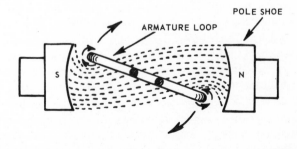

Fig. 19-82. Armature loop will rotate. Each side of loop has a field moving in opposite directions. These fields oppose pole shoe field causing loop to rotate.

COMMUTATOR SEGMENTS

In order to pass an electrical current into the revolving loop, the loop ends must be fastened to commutator segments

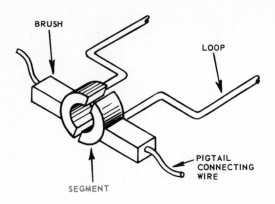

Fig. 19-83. Commutator segments. In actual starter motor, many loops and segments are used. Each segment is insulated from shaft and other segments.

(much like DC generator). Copper brushes rub against the turning segments. Fig. 19-83.

One loop would not give the starter motor sufficient power, so many loops are used. The ends of each loop are connected to a copper segment. Fig. 19-84.

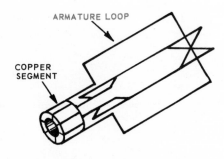

Fig. 19-84. Many loops are required. Addition of loops will increase power or torque of starter motor.

The loops are formed over and insulated from a laminated iron core. The ends of all loops are connected to copper commutator segments. The segments are insulated from the shaft and from each other. The core is supported on a steel shaft which, in turn, is supported in bronze bushings.

When current is fed into two of the four brushes, it flows through all the loops and out the other brushes. This creates a magnetic field around each loop. As the armature turns, the loop will move to a position where the current flow reverses. If this were not done, the magnetic field around the loops would push them up out of one side of the field. And when they entered the other side of the pole field, they would be repelled. This constant reversal of current flow is accomplished by the commutator segments moving under the brushes. Two brushes are insulated and two are grounded.

FIELD FRAME AND POLE SHOES

Field pole shoes, four or more, are placed inside a steel cylinder (field frame). Two or more of the shoes are wound with heavy copper strips. Then, when current is passed through the coils, the pole shoes become powerful magnets. The field frame provides a low reluctance magnetic path and further strengthens the magnetic field.

STARTER ELECTRICAL CIRCUITS

The copper loops and field windings are heavy enough to carry a large amount of current with minimum resistance. Since they draw heavy amounts of current, they must not be operated on a continuous basis for longer than 30 seconds. After cranking for 30 seconds, wait a couple of minutes to let the starter motor dissipate some of its heat. A starter heats quickly, so prolonged use can cause serious damage.

Various wiring arrangements are used in starter construction. Detail A, Fig. 19-85, illustrates a four pole, two field winding connected in parallel but in series with the armature. By wiring the field coil in parallel, resistance is lowered, and the field coils produce stronger magnetic fields. The other two poles have no windings but serve to strengthen the field.

Detail B, Fig. 19-85, shows a four pole, four winding setup. Windings are parallel and in series with the armature.

Detail C, Fig. 19-85, pictures a four pole, three winding setup. Two of the windings are in series with themselves and the armature. One winding does not pass through the armature, but goes directly to the ground. This SHUNT WINDING aids with additional starting torque. However, as starter speed increases, the shunt still draws a heavy current and tends to keep starter speed within acceptable limits.

Detail D shows a four pole, four winding setup. Three windings are in series, one is a shunt winding.

Detail E is a four pole, four winding setup with all windings in series.

There are other setups such as the six pole, six windings, etc. The ones illustrated in Fig. 19-85 are most commonly used.

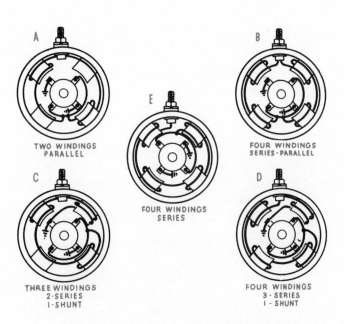

Fig. 19-85. Starter circuits. (Autolite)

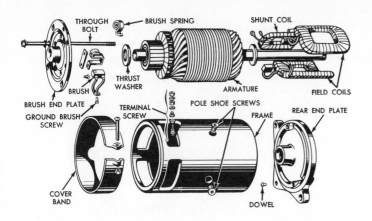

Fig. 19-86. Exploded view of typical starter motor. Learn names and location of all parts. (Ford)

STARTER MOTOR CONSTRUCTION

Fig. 19-86 illustrates, in exploded form, the various parts of a typical starter motor employing a Bendix type drive. This particular starter has a four pole, three winding circuit. One of the windings is a shunt winding.

Fig. 19-87 shows a cutaway view of a typical starter using an overrunning clutch starter drive. Study construction. Note how the armature shaft is supported at both ends by a bronze bushing.

FLYWHEEL RING GEAR

In order to provide a means for the starting motor to turn the flywheel, a large ring gear is either welded to, or heated and shrunk onto, the flywheel. (Metal ring is heated to expand and while hot is fitted on flywheel. When cool, metal contracts and ring grasps flywheel securely.) Fig. 19-88.

STARTER PINION GEAR

A small gear, called a starter pinion gear, is attached to the starter armature shaft. The pinion gear is much smaller than the ring gear and usually turns 15 to 20 times in order to crank the flywheel one revolution.

PINION ENGAGEMENT

It is necessary to mesh the pinion gear for cranking, and when the engine has started, to demesh (disengage) the pinion.

Fig. 19-87. Cutaway of starter motor using an overrunning clutch starter drive. Note shift lever and solenoid. 1—Shift lever. 2—Lever pivot. 3—Solenoid plunger. 4—Steel washer. 5—Rubber washer. 6—Solenoid windings. 7—Contact plate. 8—Terminal. 9—Field lead wire. 10—Fastener. 11—Gasket. 12—Shims. 13—Snap ring. 14—Bushing. 15—End frame. 16—Adjusting washer. 17—Brush holder. 18—Brush. 19—Brush spring. 20—Commutator. 21—Armature. 22—Pole shoe. 23—Frame. 24—Field winding. 25—Drive end frame. 26—Roller bearing. 27—Pinion. 28—Stop ring. 29—Snap ring. 30—Bushing. (Volvo)

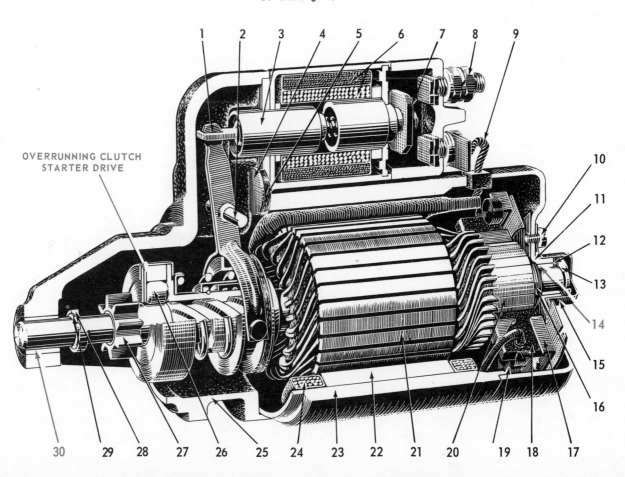

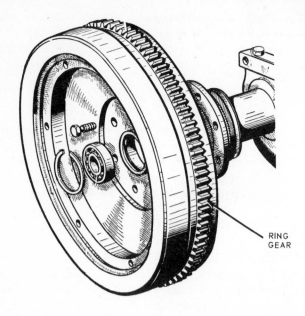

Fig. 19-88. Flywheel ring gear. Ring gear can be cut as an integral part or can be shrunk on. (Citroen)

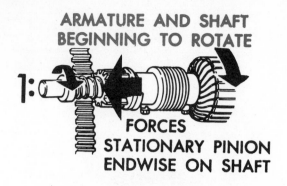

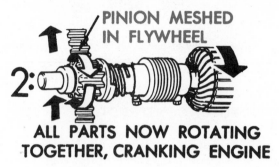

Fig. 19-89. Bendix starter drive action. (Delco—Remy)

There are two basic types of pinion engagement devices, and several designs of each basic type. The self-engaging or inertia type is quite popular. The mechanically engaged type is also used.

SELF—ENGAGING STARTER DRIVE, BENDIX TYPE

With the Bendix self-engaging pinion drive, the armature shaft spins and the drive pinion stands still while the threaded sleeve spins inside the pinion. As the sleeve spins, the pinion slides out and meshes with the ring gear. As soon as the pinion reaches its stop, the turning sleeve causes the pinion to turn with it to crank the engine.

The pinion sleeve fits loosely on the armature shaft and is connected to a large spring that is, in turn, fastened to the armature shaft. The spring absorbs the sudden shock of pinion engagement.

When the engine starts, the pinion will spin faster than the armature turns the pinion sleeve. This will reverse its travel on the threaded pinion, and it will spin backward, free of the ring gear. A small coil spring will prevent it from working back into mesh.

Fig. 19-89 illustrates the action of the Bendix type pinion engaging device. In detail 1, the armature shaft and pinion sleeve begin to spin. The stationary pinion is sliding to the left. In detail 2, the sliding action is stopped, and the pinion begins to turn and crank the engine. When the engine starts, detail 3, the pinion is spinning faster than the threaded pinion sleeve. This spins the pinion back out of mesh with the ring gear.

The various parts of a Bendix starter drive are shown in Fig. 19-90. The drive head is attached to the armature shaft.

A Bendix drive, in place on a starter motor, is pictured in Fig. 19-91. Notice the threaded pinion sleeve and pinion. The pinion is weighted on one side, so it will tend to resist being turned until it has slid into mesh with the ring gear.

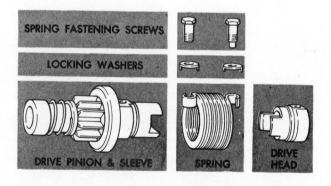

Fig. 19-90. Bendix starter drive construction. (Delco—Remy)

BENDIX FOLO—THRU

The Bendix Folo-Thru drive has a feature not found on the drive just discussed.

When the pinion slides into mesh with the ring gear, a small, spring-loaded detent pin drops into a notch in the threaded

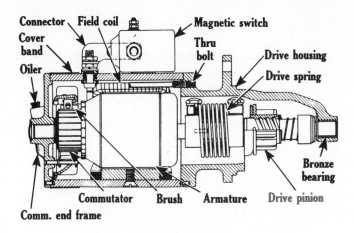

Fig. 19-91. Starter motor with Bendix drive.

pinion sleeve. This locks the Bendix in the engaged position and prevents it from being demeshed until the engine speed reaches about 400 rpm. When the engine reaches this speed, the pinion will be spinning fast enough so that centrifugal force will throw the detent pin outward. This action unlocks the pinion, and it will spin out of mesh.

There is a ratchet type clutch drive, detail A, Fig. 19-92, that will allow the pinion to travel faster than the armature shaft. It is the Bendix Folo-Thru drive. The advantage of this type of drive is that it prevents the pinion from being demeshed until the engine is actually running. With the first type of Bendix drive, if the engine reaches 200 or 300 rpm, then stalls, there could be a problem. The operator may engage the starter again just as the engine stops and it will kick back. If the starter engages the ring gear at this time, serious damage may result to the starter. The Bendix Folo-Thru drive eliminates this possibility.

OVERRUNNING CLUTCH, MECHANICALLY ENGAGED

The overrunning clutch starter drive is engaged with the ring gear by manually operated linkage or an electric solenoid that actuates the shift linkage.

In operation, the pinion is turned by the overrunning clutch. When the pinion is slid into mesh with the ring, the shifting mechanism completes the starter circuit and the one-way clutch drives the pinion.

When the engine starts, the ring will spin the pinion faster than the starter armature is turning. The pinion is free to run faster because the overrunning clutch engages the pinion only when the armature is driving it.

Fig. 19-93 illustrates a typical overrunning clutch starter

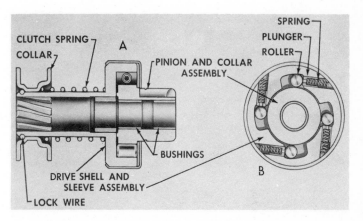

Fig. 19-93. Overrunning clutch starter drive. Drive rollers lock pinion and collar assembly to drive shell and sleeve assembly, when armature shaft is driving shell and sleeve assembly faster than pinion. (Chevrolet)

drive. Detail A is a cross section view. Note the pinion, drive shell and sleeve assembly, coil spring and shift collar. The coil spring comes into use when the pinion teeth butt against the ring teeth. The actuating lever, or shift lever, will continue to move the shift collar, compressing the spring. When the lever has moved far enough to close the starter switch, the pinion will turn and the compressed spring will force it into mesh with the ring gear.

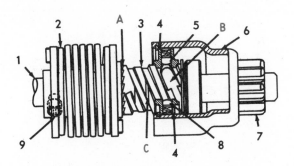

Fig. 19-92. Bendix Folo-Thru drive. Detent pin 5, drops into groove B, and keeps pinion engaged until engine speed is sufficient to throw pin 5 out of groove. 1—Armature shaft. 2—Drive spring. 3—Screw shaft. 4—Control nut. 5—Detent spring and pin. 6—Pinion barrel. 7—Pinion. 8—Anti-drift pin and spring. 9—Setscrew. A—Ratchet. B—Screw shaft detent notch. C—Screw shaft anti-drift slope.

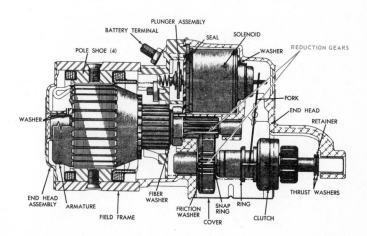

Fig. 19-94. Overrunning clutch starter drive. This starter drive is engaged by an electric solenoid. This starter increases cranking torque by using reduction gears. (Dodge)

B in Fig. 19-93, is an end view showing the spring loaded clutch rollers. When the drive sleeve turns faster than the pinion and collar assembly, the rollers are wedged against the pinion, causing it to turn. When pinion speed exceeds that of the sleeve, the rollers retract and the pinion is free of the shell and sleeve assembly.

Fig. 19-94 shows the overrunning clutch drive in place on a starter. Note the solenoid used to operate the shift lever.

STARTER SOLENOID

If an iron core is placed inside a coil of wire, and electric current is passed through the coil, the iron core will be drawn into the coil. Such a unit is called an electric solenoid. A solenoid has many applications and can be used where a push or pull motion is desired.

Fig. 19-95 shows a simple solenoid, and how it can be used to pull on a starter shift lever.

You will note that the solenoid in Fig. 19-95, not only pulled the starter shift lever until the pinion meshed, but it also closed the starter switch after the meshing was almost complete.

STARTER SOLENOID IN OPERATION

Fig. 19-96 shows the starter solenoid in three stages of operation. The disengaged position finds the pinion demeshed. The solenoid plunger is released and no current is flowing.

The partially engaged stage shows current from the starter switch flowing through both the pull-in and the hold-in coils.

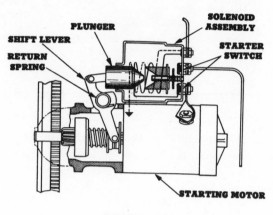

DISENGAGED

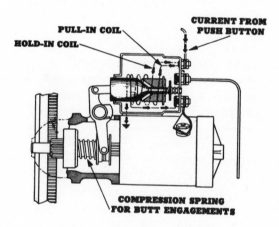

PINION PARTIALLY ENGAGED

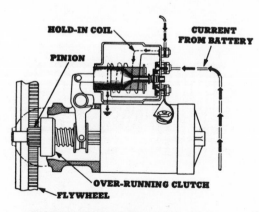

PINION FULLY ENGAGED AND STARTING MOTOR CRANKING

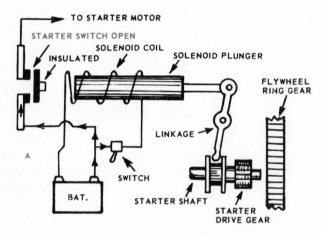

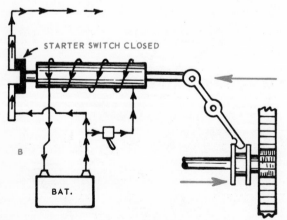

Fig. 19-95. Solenoid used to engage starter drive. In A, the solenoid plunger has moved to right, opening starter switch and retracting starter drive gear. Dash switch is off. In B, dash switch is on, energizing solenoid. Solenoid has drawn plunger into coil. Plunger has engaged drive gear and closed starter motor switch. Starter will now crank engine.

Fig. 19-96. Electric solenoid operation. Note operation in each stage. (Chevrolet)

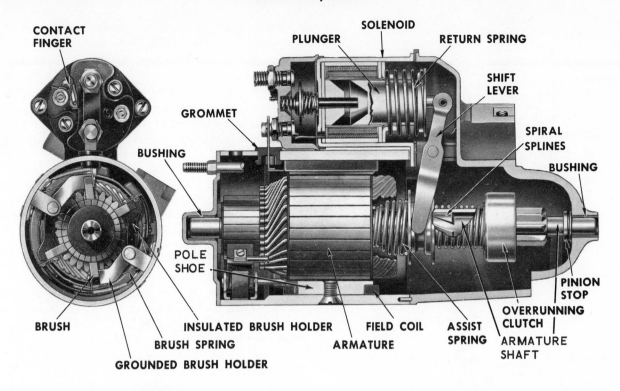

CONTACT FINGER

PLUNGER   SOLENOID   RETURN SPRING

SHIFT LEVER

GROMMET

SPIRAL SPLINES

BUSHING

BUSHING

POLE SHOE

PINION STOP

BRUSH   INSULATED BRUSH HOLDER   FIELD COIL   ASSIST SPRING   OVERRUNNING CLUTCH

BRUSH SPRING   ARMATURE   ARMATURE SHAFT

GROUNDED BRUSH HOLDER

Fig. 19-97. Sectional view of starter with solenoid operated drive mechanism.

Both coils are drawing the plunger inward, causing it to pull the shift lever. The pinion is just starting to engage.

The cranking stage shows the plunger pulled into the coil. The pinion is fully engaged. The hold-in coil is operating, but the pull-in winding has cut out. The plunger has operated the starting switch and current is flowing from the battery through the switch and to the starter. Cranking will begin.

When the engine has started, the hold-in coil will cut out and the plunger will move out, retracting the pinion and opening the starter switch. Fig. 19-96.

Starters with self-engaging Bendix type drives generally use a solenoid to open and close the starter switch.

A starter motor employing a solenoid and overrunning clutch drive is shown in Fig. 19-97. Study the solenoid construction.

A somewhat different method of actuating the starter drive is pictured in Fig. 19-98. In this setup, a movable pole shoe is connected to the starter drive shift lever.

When the starter is in the OFF position, one of the field coils is grounded through a set of contacts. When the starter is first actuated, a very heavy flow of current through this field coil pulls on the movable pole shoe. The shoe, in turn, engages the starter drive.

When fully engaged, the field coil contacts open and normal starter operation takes place. A holding coil keeps the sliding pole shoe in the engaged position. When current to the starter is interrupted, the pole shoe moves back and disengages the starter drive.

STARTER ACTUATING SWITCHES

In addition to the solenoid operated switch that directs battery current to the starter, it is necessary to furnish current to the solenoid. This is done in several ways. On most cars, turning the key to the fullest travel closes a switch to the solenoid. On older cars, a plunger switch on the dash is used to complete the solenoid circuit. Switches have been used that were operated by the clutch pedal, accelerator pedal, etc.

On cars with automatic transmissions, a safety switch is used. To actuate the starter, the drive selector must be placed in Neutral or Park position to complete the circuit through the switch. This prevents the car from being started when the transmission is in gear.

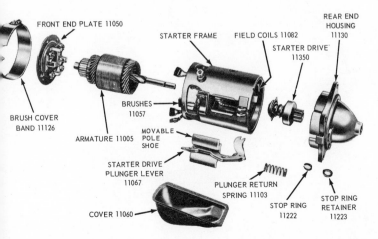

FRONT END PLATE 11050

STARTER FRAME   FIELD COILS 11082

REAR END HOUSING 11130

STARTER DRIVE 11350

BRUSH COVER BAND 11126

BRUSHES 11057

ARMATURE 11005

MOVABLE POLE SHOE

STARTER DRIVE PLUNGER LEVER 11067

PLUNGER RETURN SPRING 11103

STOP RING 11222

STOP RING RETAINER 11223

COVER 11060

Fig. 19-98. Starter design using one movable pole shoe as a means of actuating the starter drive.
(Ford)

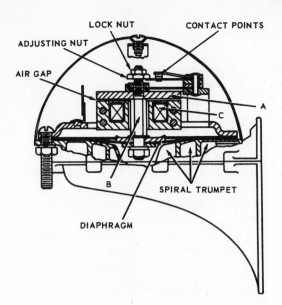

Fig. 19-99. Cross section of typical electric horn. (GMC)

HORNS, LIGHTS, RELAYS AND ELECTRICAL CIRCUITS

The electric horn makes use of an electromagnet, vibrator points, and a diaphragm. Referring to Fig. 19-99, you will see that the electromagnet coil, C, will attract metal washer, A, whenever the contact points are closed. Current flows through the contact points energizing the electromagnet. The electromagnet draws the washer down and, since the washer is attached to rod, B, the diaphragm (thin metal disc) will be bent downward.

When washer, A, moves downward, it causes the points to open and break the circuit. The electromagnet releases the washer and it moves upward, flexing the diaphragm in the other direction. As it moves upward, the points close and the washer is again drawn downward. This cycle is repeated very fast. These small up-and-down pulsations of the diaphragm cause the sound. Sound is magnified by passing it through a long trumpet or a spiral trumpet such as horn in Fig. 19-99.

RELAY

A relay is an electromagnetic device used to make and break a circuit. The generator cutout relay is one application of this device.

Remember that the farther current is forced to travel, the lower its voltage becomes. If you were to have the horn circuit switch on the steering wheel, it would be necessary to run a wire from the battery, up the steering column, through the switch, down the column and out to the horn. Another possibility would be a wire from the battery to the horn, and from the horn to the switch. The switch in this case would be a ground switch in order to complete the circuit. At any rate, the wire would be unnecessarily long.

By using the relay between the battery and the horn, it is possible to run a wire from the battery to the relay and from the relay to the horn. When the driver actuates the relay, it will complete the circuit and the horn will blow. As this is the shortest possible route, voltage drop will be low.

Fig. 19-100 shows a typical horn circuit using a relay. This particular circuit is inoperative until ignition key is turned on.

When the ignition key is on, voltage is available for the electromagnetic coil, B. When the driver depresses the horn button, the electromagnetic coil circuit is completed and

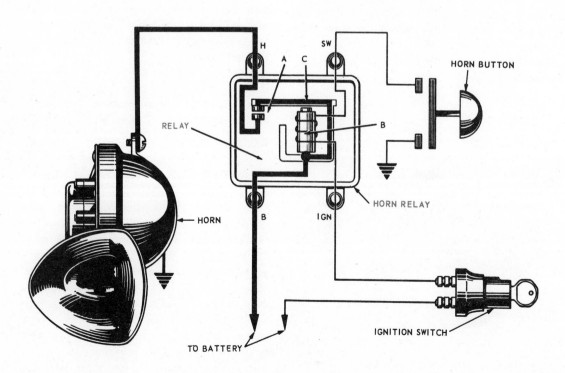

Fig. 19-100. Horn circuit using a relay. (Dodge)

current will flow.

The current flow energizes electromagnet, B. This draws armature, C, down, closing switch points, A. Current will flow through the heavy wire to the relay, across the closed points, on to the horn and to ground. As soon as the horn button is released, the electromagnet is deenergized, the armature pops up, breaking the points, and the horn stops blowing.

By using the relay, it is possible to use a short run of heavy wire to avoid voltage drop. It is also possible to actuate the relay from a distance. Headlight relays, when used, are constructed in a similar fashion. Fig. 19-100.

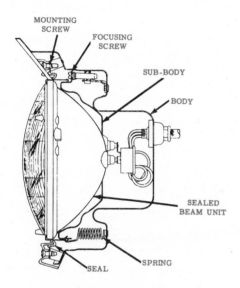

Fig. 19-101. Head lamp assembly.
(Oldsmobile)

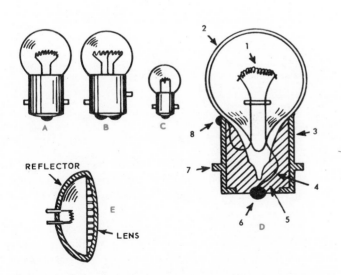

Fig. 19-102. Various types of light bulbs. A—Single contact. B—Double contact. C—Midget type. D—Cross section of a typical bulb. 1—Filament. 2—Glass. 3—Brass base. 4—Filament lead. 5—Glass-like cement. 6—Contact. 7—Holding pins. 8—Filament ground lead soldered to brass base. Current travels up from contact through filament, then to ground through brass base. E—Cross section of sealed beam head lamp. Unit is one piece. Reflector is silvered to provide good reflective properties. Lens controls and directs light.

HEADLIGHTS

The modern headlight is of sealed beam construction. The filament, reflector and lens are fused into one airtight unit. Prongs that fit into a socket connector make removal easy.

Fig. 19-101 shows a typical head lamp assembly. Note the focusing screw used to adjust headlight aim.

BULBS

Light bulbs used throughout the car are either of the single or double contact type.

Some bulbs have two filaments. This allows the bulb to be used for two separate lighting jobs. The taillight bulb is an example of this type. It is used for both a taillight and stoplight.

Bulbs must be of the proper candlepower (measurement of light intensity) and voltage.

Various bulbs are shown in Fig. 19-102. Note that the sealed beam is actually a large bulb.

HEAD LAMP SWITCH

The head lamp switch closes the head lamp circuit; operates the parking lights, taillights, instrument and other dash lights, etc. It may be of the push-pull or the turn type. Some cars use a toggle (up-and-down motion) switch. Fig. 19-103 illustrates one type.

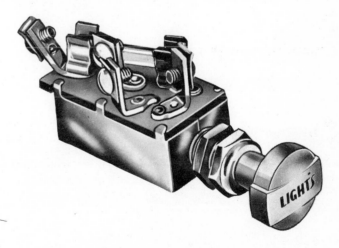

Fig. 19-103. Head lamp switch. Head lamp switches vary a great deal from car to car.
(Cole-Hersee)

DIMMER SWITCH

A dimmer switch is located in the head lamp circuit. It is a simple switch that will direct current to one of the two terminals but never to both. One terminal leads to the high beam filament; the other, to the low beam. The high beam terminal also feeds current to the high beam indicator lamp on the dash. Find the dimmer switch in Figs. 19-107 and 19-108.

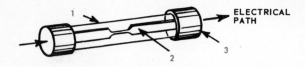

Fig. 19-104. Fuse. 1—Glass. 2—Conductor strip. 3—Metal cap.

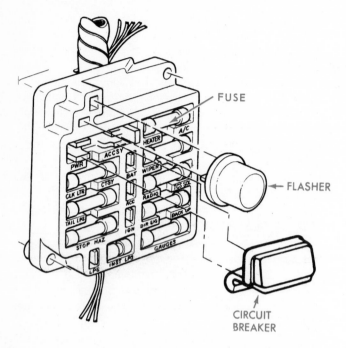

Fig. 19-105. Typical fuse block or panel. (Chevrolet)

FUSES AND CIRCUIT BREAKERS

It is necessary to protect electrical circuits against short circuits (part of circuit accidentally going to ground) that could burn up the wiring. The electrical units must be protected against overloads also.

Fuses and circuit breakers are used to secure adequate protection. When current flow exceeds that for which the circuit is designed, the fuse will "blow" and break the circuit. The circuit breaker points will open and close, constantly breaking the flow of current.

FUSES

A fuse consists of a glass tube with metal end caps. A soft conductor carries the current from one cap to the other. This conductor is designed to carry a specific maximum load. When amperage becomes excessive, the conductor will melt. Fig. 19-104.

A typical fuse block or panel is pictured in Fig. 19-105. Notice how the fuses fit into spring clips.

FUSIBLE LINK (WIRE)

A circuit may be protected by the use of a special wire, called a FUSIBLE LINK, that acts as a fuse. The wire is covered with HYPALON insulation.

When the fusible link or wire is subjected to an overload, it will begin to heat, causing the Hypalon insulation to puff up and smoke. If the overload is continued, the Hypalon will finally rupture and the fusible link will burn out, breaking the circuit.

CIRCUIT BREAKER

A circuit breaker feeds the current through a bimetallic strip, a set of points and on through the remainder of the circuit.

When amperage exceeds that for which the breaker is built, the bimetallic strip heats and bends to separate the points. On cooling, it will close and reestablish the circuit. If the overload condition still exists, it will heat and reopen. The circuit breaker will open and close quite rapidly.

This gives it an advantage over a fuse when used on a head lamp circuit. With the fuse, an overload will burn out the fuse and the headlights will go out, and stay out until the fuse is replaced. This could cause an accident. The circuit breaker will cause the lights to flicker on and off rapidly but will still produce enough light to allow the driver to pull safely off the road. Fig. 19-106.

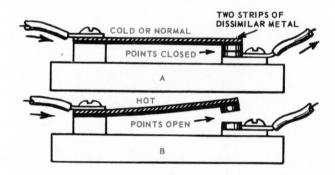

Fig. 19-106. Circuit breaker. A—Breaker is carrying a normal load. Contact points closed. Circuit complete. B—Breaker overloaded, bimetallic strip heats up. Strip curves upward separating points. Circuit broken.

MANY CIRCUITS

The modern auto has a large number of electrical circuits. A typical wiring diagram showing the ignition, generating, starting, head lamp and horn circuits is pictured in Fig. 19-107.

A more complete electrical diagram is illustrated in Fig. 19-108. The electrical circuits, with the addition of many accessory items, are becoming so complex that wiring diagrams are often broken down by sections: starting, charging, body wiring, instrument panel, etc. Study the circuits shown in Figs. 19-107 and 19-108. Can you trace the various circuits?

GAGES

Gages, including oil pressure, temperature, gasoline and ammeter, were discussed in the chapters covering systems

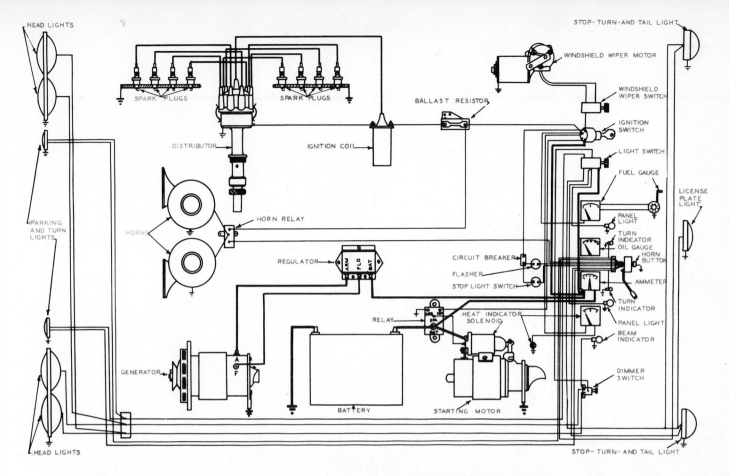

Fig. 19-107. Typical electrical system diagram. The modern car uses such a multitude of electrical components that it is very difficult to show everything on one drawing. As a result, a number of diagrams, each showing one or more systems, are usually required.

using these gages. The wiring circuits for these gages are shown in Fig. 19-107. You will note that the oil pressure gage is not wired. It is of the direct pressure Bourdon tube type and, as such, uses no electricity.

CAUTION!

When installing electrical components, make certain they are properly fused. Wires must be insulated, kept from hot areas and must be secured to prevent chafing. When passing through metal parts, install grommets (rubber insulating washers) to protect the wires. Terminals should be clean and tight.

When working on the electrical system, disconnect the battery to prevent accidental short circuiting.

Use great care when repairing head lamp circuits. If the lights suddenly go out, it could be tragic.

Never replace a fuse with a fuse of higher capacity. Never replace a fuse with a bolt, wire, etc. The fuse of correct capacity must be in the circuit.

SUMMARY

ELECTRICAL FUNDAMENTALS

The electron theory states that electricity is caused by a flow of electrons through a conductor. There is a shortage of electrons at one end of the conductor and a surplus at the other. Whenever this condition is present, current will flow.

An atom is made of neutrons, protons and electrons. The electron is a negative charge of electricity, while the proton is positively charged. The neutron is neutral, electrically.

The free electrons in an atom are those that are relatively easy to "shake" out of orbit. The free electrons are those that cause the current flow.

The pressure that causes current to flow is VOLTAGE. The amount of current flowing is measured in AMPERES. Every conductor will have some RESISTANCE to the movement of electrons.

Insulators are made of materials containing atoms with very few free electrons. The bound electrons resist current flow.

In order for electricity to flow, it must have a circuit or path that is complete. Three circuits are: Series, Parallel and Series-Parallel.

The lining up or orienting of the magnetic domains in a material is what causes the material to become magnetized.

A magnet is surrounded by lines of force. They leave the magnet at the north pole, and enter the south pole. Like poles repel; unlike poles attract.

BATTERY

A battery is an electrochemical device. It contains positive and negative plates that are connected in such a way as to produce three or six groups or cells. The plates are covered with electrolyte.

A charged battery will amass (gather a great quantity) a surplus of electrons at the negative post. When the battery is

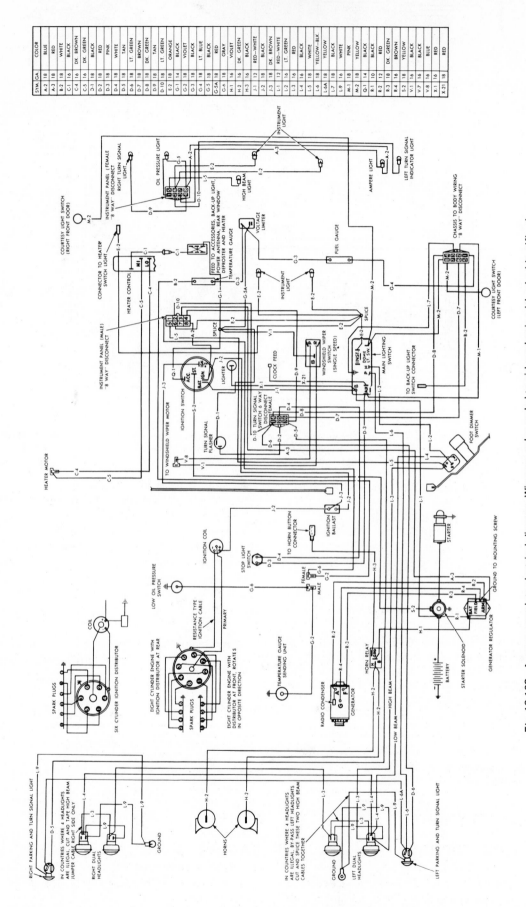

Fig. 19-108. A more complete electrical diagram. Wires are color coded to aid mechanic with wire identification. (Plymouth)

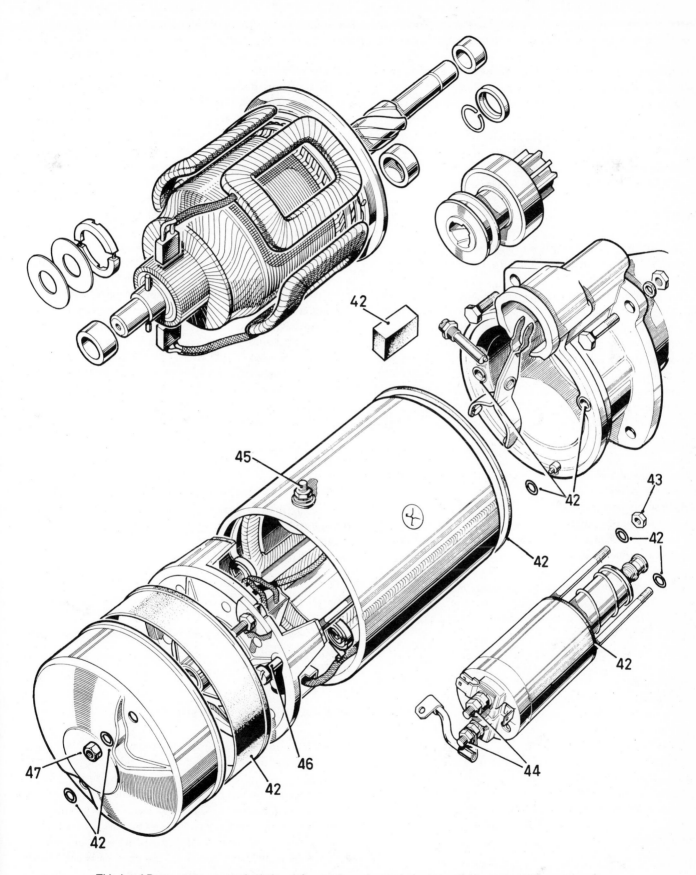

This Land-Rover starter motor is designed for rough service. It is waterproofed by means of special seals where housings connect and solenoid attaches. 42—Seals. 43—Solenoid attaching nuts. 44—Solenoid terminals. 45—Starter field terminal. 46—Through bolt. 47—Commutator cover attaching nut.

placed in a completed circuit, the electrons will flow from the negative post, through the circuit, and on to the positive post.

The battery must be kept charged by passing electricity through it in a reverse direction to that of battery current flow.

The battery is checked for charge by using a hydrometer to determine the specific gravity of the electrolyte.

Batteries can explode and should never be exposed to sparks or open flame.

The battery is used to supply current for starting the car. Upon starting, the generator will recharge the battery.

Car batteries are of the 12V type.

Be careful of battery electrolyte — it is dangerous!

GENERATOR

The primary job of the generator is to supply the necessary current for the electrical needs of the car and to keep the battery charged.

The generator functions by spinning an armature in a strong magnetic field produced by field coils. As the armature loops spin in the magnetic field, current is induced in the loops. A commutator at one end of the armature is contacted by carbon brushes. The insulated brush picks up the voltage generated by the whirling armature and conducts it to the rest of the electrical system. Both alternating current and direct current generators are in use.

GENERATOR REGULATOR

It is necessary to control the maximum voltage and current output developed by the generator. This is done by using the generator regulator.

The regulator contains three units: a cutout relay to connect or disconnect the generator from the system; a voltage control; a current control.

When either current or voltage output attempts to exceed that for which the generator is designed, an electromagnet breaks a set of contact points, and inserts a resistance in the generator field circuit. With a resistance in the field, voltage and current will lower.

The voltage regulator and the current regulator both operate points in a vibrating manner. As the points vibrate, they produce a smooth, controlled generator output. Either unit can work, but they will not function together.

The battery, generator and regulator work as a team, and the function of any one unit depends upon the action of the other two.

Battery condition and electrical load determine the current and voltage requirements. The generator produces the required voltage and current, and the regulator controls the generator output on a level consistent with system needs.

STARTER

The starter is used to crank the engine for starting purposes.

The starter motor utilizes a current carrying series of armature loops placed in a strong magnetic field produced by field coils. As the armature loops are repelled by the magnetic field, the armature is forced to spin. By arranging a sufficient number of loops and connecting them to commutator bars or segments, current in the loops keeps reversing so the repelling force will remain constant. This allows the armature to continue spinning.

The starter uses large copper conductors in both the field and armature circuits. The field and armature are connected in series via four brushes. Since the field strength is very strong, the starter motor will produce enough torque to crank the engine.

Starters draw heavy current loads and therefore heat quite rapidly. They must not be operated continuously for longer than thirty seconds without a brief rest in which to dissipate the heat.

A starter can be energized by closing the starter switch either manually or by using an electric solenoid.

The starter pinion that engages the flywheel ring gear is engaged in two ways. One method uses the Bendix self-engaging type of drive. This type spins a pinion sleeve that causes the pinion to slide out on the sleeve until it engages the ring. The Bendix Folo-Thru is one type of self-engaging drive.

The manual engagement type moves the starter pinion into mesh by means of a shift lever. The lever can be actuated either by a solenoid or mechanically by the drive. When this type starter drive is used, the overrunning clutch is incorporated so that when the engine starts, the pinion will be free to spin until such time as it is drawn out of mesh.

RELAY

A relay is an electromagnetic unit designed to make and break a circuit. The relay can be a distance from the operator and will still function with only a ground wire and switch.

Some relays, such as those used in regulators, are not controlled by the operator.

HEADLIGHTS, FUSES

The headlight circuit is of vital importance. It must be properly maintained.

Fuses are used to protect the car wiring circuits. They must be of the proper size.

MECHANIC IS RESPONSIBLE

Obviously, when a car leaves the shop, the mechanic stays behind. However, the mechanic's care, skill, knowledge and reputation leave with each car entrusted to his hands.

If anything goes wrong, and you were the mechanic on the job, you are responsible. The car owner's comfort, peace of mind and, indeed, his life may depend on you. This is a heavy responsibility and one to accept with a great deal of seriousness. Work carefully, think, use good parts and proper procedures, keep parts clean, avoid the use of slipshod methods and, above all, double-check your work. Remember: A real mechanic is not satisfied with anything but the BEST!

REVIEW QUESTIONS — CHAPTER 19

1. What is the electron theory?
2. What is a molecule?
3. Describe the makeup of an atom.
4. About how many kinds of atoms are there?
5. The proton is a negative charge of electricity. True or False?
6. Neutrons orbit around the electrons. True or False?
7. What is a bound electron? A free electron?
8. Explain how electrons flow through a conductor.
9. What is a current of electricity?
10. Current flows from negative to positive. True or False?
11. What is voltage?
12. Define resistance.
13. The flow of electricity is measured in _____ .
14. Name the instrument used to check each of the following:
    1. Voltage.
    2. Amperage.
    3. Resistance.
15. What material is widely used as a conductor in the car electrical system?
16. Name and describe three different types of circuits.
17. List several materials that are used for insulation.
18. Draw the symbol for ground.
19. Of what use is Ohm's law?
20. Explain how a metal bar can be magnetized.
21. What are lines of force?
22. Unlike poles attract; like poles repel. True or False?
23. Magnetic lines of force leave the south pole and enter the north pole. True or False?
24. Describe briefly the construction of a typical automotive battery.
25. The battery contains _____ and _____ plates.
26. A 12V battery will contain three elements. True or False?
27. The battery electrolyte contains_____ and _____ .
28. How is a battery charged?
29. What is a hydrometer and what is its use?
30. A battery will have a _____ and a _____ post.
31. How are the battery posts identified?
32. A battery should, when water is needed, be filled with _____ water.
33. What is a dry-charged battery?
34. Battery acid or electrolyte will burn the skin and clothing. True or False?
35. A battery can explode if a spark or flame is brought near. True or False?
36. What is the main job of the battery?
37. The generator is an _____ device.
38. Explain how a generator works.
39. Explain the left-hand rule for conductors; for coils.
40. Name the major parts of the generator.
41. Of what use is the commutator?
42. Generator brushes are made of _____ .
43. What are the major differences between the DC and the AC generator?
44. What are the two types of generator circuits in common use on a DC generator?
45. How is generator output controlled?
46. Most generators are oil cooled. True or False?
47. Of what use is the generator regulator?
48. Describe the action of the regulator when controlling voltage, current, and connecting or disconnecting the generator from the battery circuit.
49. The voltage regulator and current regulator can both function at the same time. True or False?
50. What is the double contract voltage regulator? How is it different from the single contact?
51. Does temperature affect the voltage regulator? If so, is there anything done to correct it?
52. What is polarity?
53. Why is it important to polarize a generator after installing?
54. What will happen if the generator polarity is accidentally reversed?
55. How is a generator polarized? There are two methods, depending on the type of circuit. Describe both.
56. How does an ammeter work?
57. Describe the action of the AC generator.
58. How is the AC changed over to DC?
59. What is the main task of the generator?
60. Explain the operation of the starter motor.
61. Name several different starter internal circuits.
62. Of what use is the shunt circuit in the starter motor?
63. Starter motors can be used continuously up to ten minutes before stopping. True or False?
64. Name two types of starter drive mechanisms.
65. Describe the construction and operation of the drive units named in answering question 64.
66. What is a solenoid and what are some of its uses?
67. The electric horn operates by flexing a strong spring. True or False?
68. Draw a simple relay.
69. What is a bulb filament?
70. The dimmer switch is connected in series with the head lamp circuit. True or False?
71. What is the difference between a circuit breaker and a fuse?
72. Draw the following:
    1. Starter circuit.
    2. Generator circuit.
    3. Head lamp circuit.
73. What is a fusible link?
74. When an ammeter is used instead of an indicator light in an AC charging circuit, a field relay must be incorporated in the regulator. True or False?

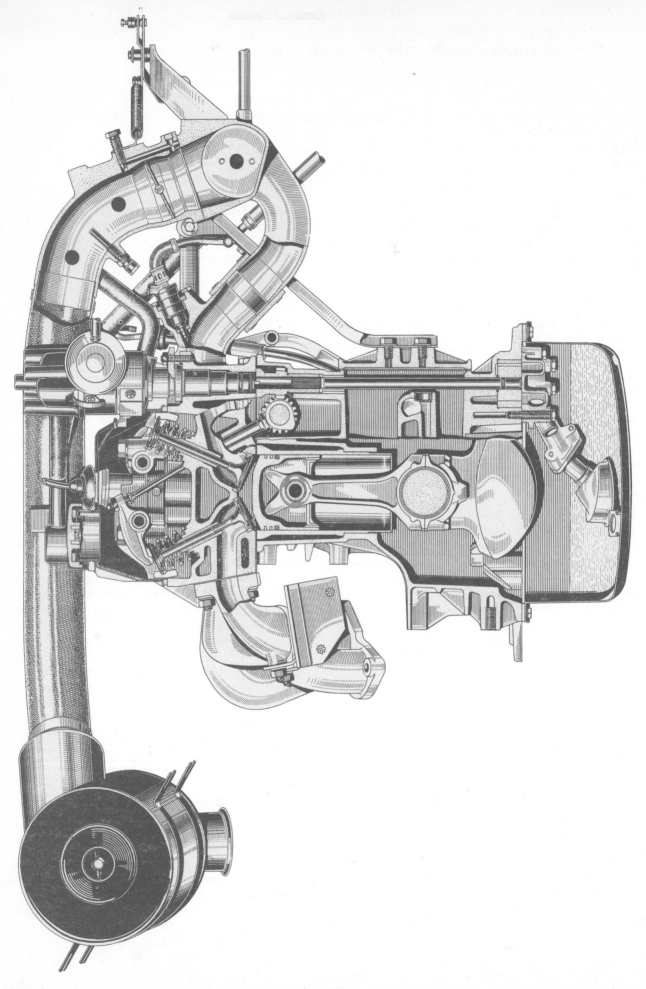

Industry illustration. Cross-sections shows details of Renault engine.

Chapter 20

AUTOMOBILE
AIR CONDITIONING

Automobile air conditioning makes driving far more comfortable in hot weather. It is a method whereby air entering the car is cleaned, dehumidified (excess moisture removed), brought to and kept at a selected temperature.

PRINCIPLES OF REFRIGERATION

The automotive air conditioning system makes use of the principles of refrigeration. Fig. 20-1. To provide a better insight into the functions of the system, it is wise to study some of the basic principles involved.

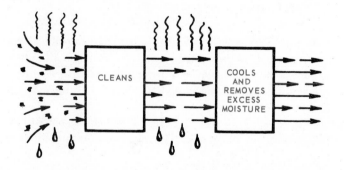

Fig. 20-1. Schematic showing airstream being cleaned, cooled and dehumidified.

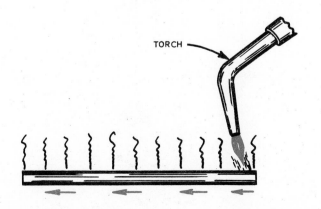

Fig. 20-2. Transfer of heat by conduction. Heat imparted by torch is transferred through rod by conduction.

WHAT IS COOLING?

The term COOLING may be defined as a process of removing heat from an object.

Technically speaking, there is no such thing as cold. When we speak of something being cold, we are really saying that the object is not as hot as something else. COLD actually is a reference to a certain portion of the temperature range. The normal use of the word cold is generally determined by a comparison to the human sense of comfort.

EFFECTS OF HEAT

Heat can be defined as a form of energy that can increase, decrease or maintain a specific temperature. Some of the more commonly observed effects of heat would be the expansion that takes place when a substance is heated or raised to a higher temperature. The added heat creates a more violent agitation of the molecules, increasing the length of their travel with a resultant increase in physical size.

Water, when heated sufficiently, vaporizes. When heat is lowered enough, water becomes a solid. Wood will reach the ignition point if heat is added, causing it to burn. Water freezing and wood burning are actual changes of state produced by heat.

TRANSFERENCE OF HEAT

Heat may be transferred from one object to another in three ways:
1. Conduction. 2. Convection. 3. Radiation.

CONDUCTION

Conduction is accomplished by actual physical contact between the two bodies. The agitated action of the molecules of the hotter body transfer their actions to that of the cooler body. If a copper wire is held in a flame, the heat of the flame is rapidly conducted through the wire. It travels from molecule to molecule. Some substances transfer heat much more rapidly than others. Fig. 10-2.

CONVECTION

Heat can be transferred from one object to another by the air around them. This is one form of convection. As the air heated by the hotter body rises, it comes in contact with the cooler body, raising its temperature.

In a room, the air circulation will make a full circle. Even though hot air rises, circulation as it cools will finally bring it to the floor level and heat all objects in the room. Fig. 20-3.

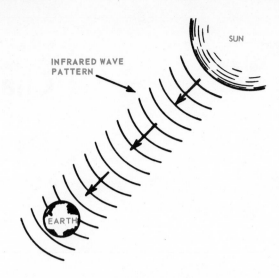

Fig. 20-4. Transfer of heat by radiation. Sun heats earth by transmitting infrared rays. Conduction and convection cannot help since there is no physical contact or air.

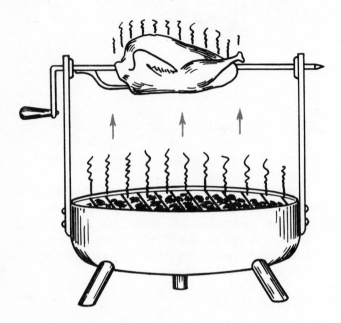

Fig. 20-3. Transfer of heat by convection and radiation. Air above hot coals is heated, rises and transfers heat to bird on spit. A great deal of heat is also transferred to bird by radiation.

RADIATION

Even in the absence of air or physical contact, heat can be transferred through space by infrared rays. (Infrared rays are not visible to the human eye.) Energy is emitted from molecules and atoms because of internal changes.

When heat is transferred through conduction and convection, the process is relatively slow. Heat transferred by radiation travels at the speed of light.

To get an idea of how heat is transmitted by radiation, assume a body containing heat gives off infrared rays. These rays travel through space and, even though the rays may not contain heat, they have the ability to cause molecules of other substances to increase their speed, and inpart heat into the other object. Fig. 20-4.

HEAT TRANSFER IS FROM HOT TO COLD

By studying convection, conduction and radiation, it can be seen that when two bodies are placed in reasonable proximity to each other, the hotter body will give up heat to the colder body. Eventually the hot body will cool, and the cool body will heat, until they reach an equal temperature level.

BTU (British Thermal Unit)

To provide a means of measuring the amount of heat transfer from one object to another, a unit of measurement called the British thermal unit (Btu) was devised. By definition, when one pound of water is heated one degree Fahrenheit, one Btu of heat has been transferred to the water.

LATENT (HIDDEN) HEAT

If heat is imparted to ice, the ice will start to melt. As it is melting, it will not grow warmer even if a great amount of heat is applied to it. This heat that disappears into the ice with no temperature rise, is called LATENT HEAT. It is there, but it cannot be detected.

When a pound of water is heated, it will absorb one Btu for each one degree of temperature rise. It will reach the boiling

°F.	# Pressure	°F.	# Pressure
—40	11.0*	50	46.7
—35	8.3*	55	52.0
—30	5.5*	60	57.7
—25	2.3*	65	63.7
--20	0.6	70	70.1
—15	2.4	75	76.9
—10	4.5	80	84.1
— 5	6.8	85	91.7
— 0	9.2	90	99.6
5	11.8	95	108.1
10	14.7	100	116.9
15	17.7	105	126.2
20	21.1	110	136.0
25	24.6	115	146.5
30	28.5	120	157.1
32	30.1	125	167.5
35	32.6	130	179.0
40	37.0	140	204.5
45	41.7	150	232.0
*Inches of Vacuum			

Fig. 20-5. Chart illustrating relationship between temperature and pressure for Refrigerant—12. (American Motors)

Air Conditioning

point at 212 deg. F. (atmospheric pressure at sea level). Heat applied at this point, even a great amount, will not cause a further increase in temperature. Once again heat is absorbed but is not apparent.

The amount of heat being absorbed to change the water into vapor is great, but if one were to measure the temperature of the vapor, it too would register 212 deg. F. The heat then, is "hidden" in the water vapor.

When this vapor comes in contact with cooler air, latent heat will transfer to the cooler air and vapor will condense (change back to a liquid). This particular law is known as the LATENT HEAT OF VAPORIZATION LAW.

This latent heat of vaporization (heat transfer involved in changing a liquid to a vapor and vapor back to a liquid) is the basic principle of the refrigeration process used in air conditioning.

PRESSURE AFFECTS BOTH VAPORIZATION AND CONDENSATION TEMPERATURES

Water boils at sea level (unconfined) at 212 deg. F. If water at 20,000 feet in the air is heated, it will boil at a lower temperature. The pressure on the water at 20,000 feet is less, so the boiling point will be lower. When pressure is raised above normal atmospheric pressure, the boiling point will also rise. The condensation point is likewise affected.

The refrigeration process makes use of this pressure affect on vaporization and condensation. The liquid used in the refrigeration system is generally referred to as REFRIGERANT-12. It is also called by trade names such as Freon-12, Isotron-12, Prestone-12, etc.

REFRIGERANT—12

Refrigerant-12 will boil or vaporize at minus 21.7 deg. F. when subjected to zero pounds of pressure (sea level). If the pressure is raised enough, the boiling point can be moved up well beyond the temperature encountered on the hottest day. This low boiling point, plus the fact that it can be passed through the system endlessly without loss of efficiency, makes Refrigerant-12 an ideal refrigeration medium. Fig. 20-5 illustrates the relationship between temperature and pressure.

HOW REFRIGERATION SYSTEM WORKS

The basic refrigeration system used to air condition automobiles consists of a compressor, condenser, receiver, expansion valve, evaporator and, in many instances, a suction throttling valve. Mufflers may also be incorporated. These various units are connected together by high pressure lines. Fig. 20-6.

In order to understand just how a refrigeration system functions, start tracing the flow of refrigerant through the system. Keep in mind that the principle of refrigeration lies in the fact that by controlling the pressure on the refrigerant, we can make it boil, or condense, and thus absorb or give off heat. Start at the receiver.

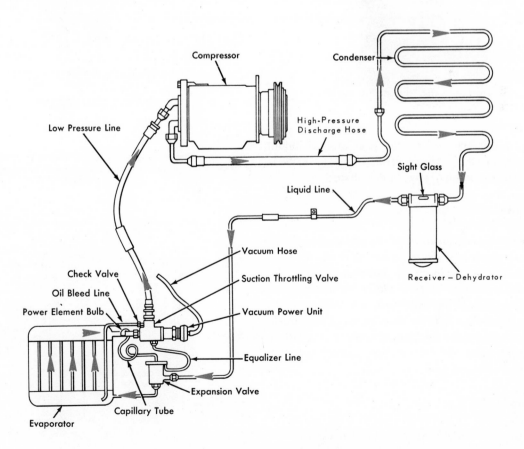

Fig. 20-6. Schematic of refrigeration system. Arrows indicate direction of refrigerant flow. (Cadillac)

RECEIVER—DEHYDRATOR

The receiver is a metal tank that collects the liquefied refrigerant as it leaves the condenser. It strains the refrigerant and removes moisture by means of a special desiccant (drying agent).

The receiver may also contain a safety valve. In the event of fire or other overheating, the valve will release the refrigerant charge when pressure in the system exceeds around 500 lbs. Fig. 20-7.

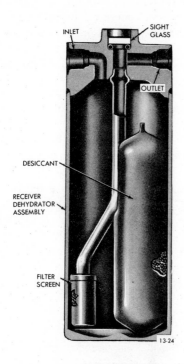

Fig. 20-7. Receiver—dehydrator. Note strainer and desiccant. (Buick)

From the receiver-dehydrator, the high pressure liquid refrigerant is piped to the expansion valve. A SIGHT GLASS is located in the line (or in top of receiver-dehydrator).

SIGHT GLASS

A sight glass serves two functions. It provides a visual means of making an inspection of the refrigerant as it moves to the expansion valve. The presence of bubbles or foam (at temperature above 70 deg. F.), may indicate the system is low on refrigerant and must be charged (filled). Fig. 20-6.

Some sight glasses also indicate the presence of moisture in the refrigerant. If excess moisture is present, a moisture sensitive element turns pink. If moisture content is within limits, the element remains blue.

EXPANSION VALVE

The Refrigerant-12 arriving at the expansion valve is in a liquid state and is under heavy pressure (100 to 250 psi).

The expansion valve serves two functions. It meters refrigerant into the evaporator coils, and also reduces the pressure. It must administer the correct amount of refrigerant for all conditions, neither too much nor too little. A typical expansion valve is shown in Fig. 20-8.

Notice that refrigerant from the receiver enters through a fine screen in the high pressure liquid inlet. Further movement of the refrigerant is controlled by the spring-loaded valve.

The action of this valve is controlled by three forces:
1. Valve spring tends to close valve opening.
2. Suction manifold (area at end of evaporator on compressor side) pressure, working through equalizer line, tends to close valve opening.

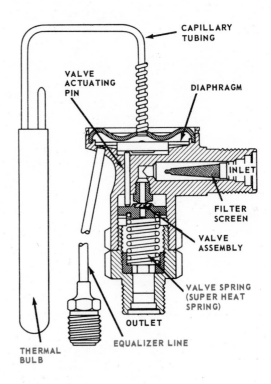

Fig. 20-8. Typical expansion valve. This valve is controlled by three forces. 1—Valve spring pressure. 2—Equalizer line pressure. 3—Thermal bulb pressure. (GMC)

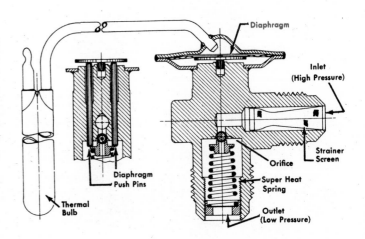

Fig. 20-9. A somewhat different type of expansion valve. This valve does not utilize an equalizer line.

3. Pressure of thermal bulb tends to open valve.

You will note that a diaphragm is located in the top of the expansion valve. This diaphragm contacts the valve actuating pins. When the valve pins move down, the valve opens. When the pins move up, the valve closes.

If the compressor is started and the system activated, the high pressure liquid refrigerant flows in as far as the needle valve.

As the suction manifold is still warm, considerable pressure is built up in the thermal bulb. This pressure is transmitted to the diaphragm via the capillary tube. The bulb, capillary tube and area above the diaphragm are filled with both liquid and vapor refrigerant. Some systems fill this with carbon dioxide.

The thermal bulb pressure on the diaphragm will overcome the pressure of both the valve spring and equlizer tube pressure, causing the diaphragm to flex downward. This downward movement is transmitted to the valve through the actuating pins.

When the valve opens, low pressure refrigerant (liquid) enters the evaporator coils, and warm air passing the coils gives up its heat to the colder refrigerant (around 7 deg.). Refrigerant passing through the evaporator coils begins to boil and, before it leaves the evaporator, completely vaporizes.

Refrigerant vapor leaving the evaporator passes through the suction manifold on its way to the compressor. Even though it is boiling, it is still quite cold. This cools the thermal bulb and reduces its pressure on the diaphragm.

As the refrigerant passes into the evaporator, suction tube pressure is increased. This pressure is applied to the bottom of the diaphragm (through equalizer line) and attempts to move the diaphragm upward.

When the downward pressure of the thermal bulb is nearly equalized by the upward pressure of the suction manifold, the valve spring will close the valve.

When the valve is closed, and refrigerant no longer enters the evaporator, the manifold pressure drops, and its temperature rises. The drop in pressure reduces the upward equalizer pressure on the diaphragm. At the same time, the thermal bulb pressure is increased by the rise in suction manifold temperature. This causes the diaphragm to flex downward, opening the valve and admitting more refrigerant to the evaporator.

By the balance between these three forces, the cooling activity of the evaporator is closely controlled. (A slightly different type of expansion valve is shown in Fig. 20-9.) The expansion valve then strains the liquid refrigerant, reduces the pressure and meters the proper amount of low pressure refrigerant into the evaporator coils in such a way that a careful balance is maintained between evaporator cooling efficiency and heat load.

THE EVAPORATOR

The evaporator consists of a long tube passed back and forth through numerous cooling fins. The fins are firmly attached to the tube.

An airstream produced by an electric blower, is passed through the cooling fins and around the coils. This air is warmer than the refrigerant and gives up its heat to the fins,

coils and then to the refrigerant itself.

As the low pressure refrigerant moves through the evaporator, heat given up by the air passing through the evaporator causes the refrigerant to begin to boil. By the time the refrigerant has passed through the evaporator, it has become a vapor.

Like the latent heat of vaporization found in boiling water vapor, the refrigerant has a great deal of latent heat in its vapor. As the heat is absorbed by the boiling refrigerant, the coils and fins turn cold and in turn cool the air passing over them. This cool air is blown into the interior of the car.

The evaporator also dehumidifies (removes moisture) the air passing through the fins. Excess moisture that may be in the air will condense, as it strikes the cold evaporator fins. This condensation is drained off in the form of water.

Evaporator construction is shown in Fig. 20-10. Note that the expansion valve and the suction throttling valve are connected.

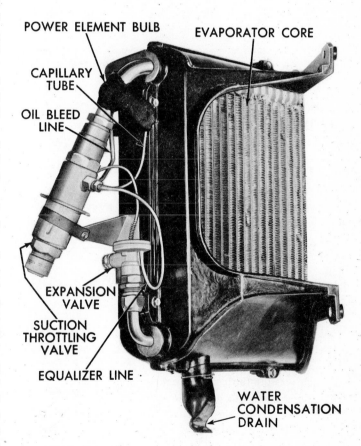

Fig. 20-10. Typical evaporator construction. Note that this evaporator is constructed along lines of a cooling system radiator. It is made of aluminum. (Cadillac)

EVAPORATOR TEMPERATURE IS IMPORTANT

The normal evaporator coil temperature varies from 33 deg. F. to 60 deg. F. If the temperature is permitted to drop to 32 deg. F. or below, condensation that forms on the coils will freeze. Such icing will destroy cooling efficiency.

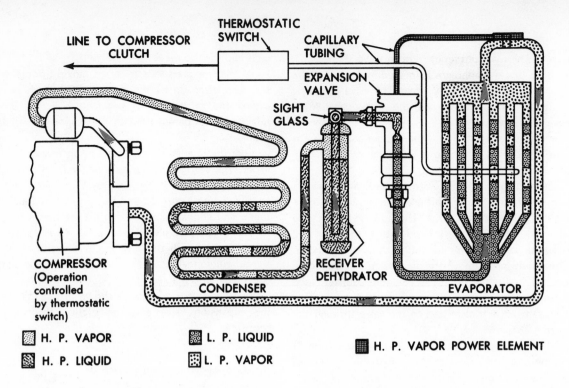

Fig. 20-11. Capillary tubing is contact with evaporator core controls a thermostatic switch that energizes compressor magnetic clutch. This prevents icing (freezing) of evaporator coils.   (Chevrolet)

EVAPORATOR TEMPERATURE CONTROL, THERMOSTATIC SWITCH

Evaporator coil temperature may be controlled by using a thermostatic switch to disconnect the compressor magnetic clutch, when evaporator temperature drops to a critical point. The sensing bulb that operates the switch is placed in contact with the evaporator core. When coil temperature reaches an acceptable level, the switch cuts in the magnetic clutch to once again drive the compressor. Fig. 20-11.

EVAPORATOR TEMPERATURE CONTROL, HOT GAS BYPASS VALVE

At one time the hot gas bypass valve was used to control evaporator icing. The hot gas bypass valve, located on the outlet side of the evaporator, meters a controlled amount of hot, high pressure refrigerant (from compressor back into the evaporator outlet. This hot gas joins the vaporized refrigerant leaving the evaporator and is pulled into the compressor. The refrigerant from the hot gas bypass raises evaporator pressure. The pressure increase raises the boiling point of the refrigerant with a resultant loss in cooling efficiency. This in turn raises the evaporator temperature.

The compressor runs constantly (when AC system is turned on) in systems using the hot gas bypass valve. A typical system with hot gas bypass valve is pictured in Fig. 20-12.

EVAPORATOR TEMPERATURE CONTROL, SUCTION THROTTLING VALVE (STV)

The suction throttling valve has replaced the hot gas bypass valve as a means of controlling evaporator temperature. It is

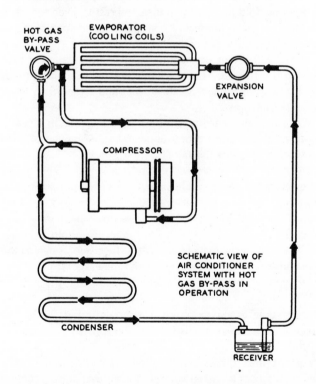

Fig. 20-12. Hot gas bypass valve. Note how bypass valve feeds hot, high pressure refrigerant back into evaporator outlet.
(Buick)

installed at the evaporator outlet. Fig. 20-13.

The evaporator outlet pressure acts on the STV piston and, through small bleeder holes, against a diaphragm. Opposing this force is the valve spring and atmospheric pressure on the other side of the diaphragm.

When the evaporator pressure rises to a predetermined level

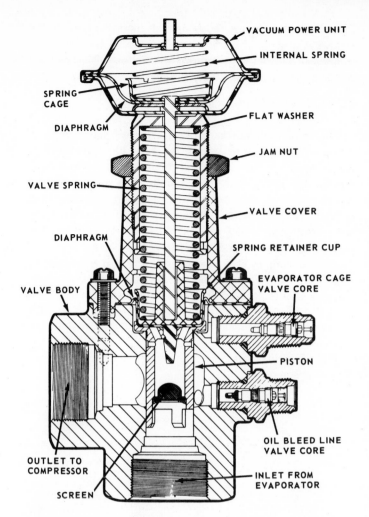

Fig. 20-13. Typical suction throttling valve (STV). This valve balances evaporator pressure very accurately at either 30 psi or at 33.5 psi, controlling evaporator temperature. When engine vacuum is removed from vacuum power unit, small internal spring adds 3.5 psi pressure to that exerted by valve spring. This will increase controlled pressure to 33.5 psi and will protect against evaporator icing at high elevations. (Harrison)

(usually around 30 to 33 psi), the piston will be moved in an opening direction and pass the excess pressure into the compressor inlet line. When pressure drops, the piston will close enough to restrict vapor flow to the compressor until pressure builds up to the desired level.

The STV will keep the evaporator pressure balanced very accurately and closely control the evaporator temperature. In Fig. 20-13, note the vacuum power unit. By applying engine vacuum to this unit, valve spring pressure to piston is reduced, allowing the piston to open under less pressure. This lowers evaporator pressure for maximum cooling. See Figs. 20-5 and 20-13.

POA SUCTION THROTTLING VALVE

The POA (Pilot Operated Absolute) suction throttling valve represents an improvement over the suction throttling valve just discussed. It controls evaporator pressure at the desired minimum level regardless of compressor speed, evaporator load or changes in altitude.

The POA valve, Fig. 20-14, contains a bronze vacuum bellows. The sealed bellows contains almost a perfect vacuum. Since the interior of the bellows has absolutely no pressure, it is referred to as the PILOT OPERATED ABSOLUTE valve.

Note that in Fig. 20-14, the system is off and pressure is equal on both sides of piston (around 70 psi, ambient air temperature 70-80 deg. F.) allowing spring A to force the piston closed. Since pressure surrounding the bellows is over 28.5 psi, the bellows has contracted. This opens the needle valve and exposes the bellows chamber and head of piston to compressor pressure (lower than inlet or evaporator pressure when system is operating).

When the system is activated, Fig. 20-15, compressor suction starts reducing pressure on the outlet side of the valve. Since the needle valve is still open, pressure will also lower around the bellows and top of the piston. Evaporator pressure will force the piston toward the lower pressure area around the piston head, opening the piston and allowing refrigerant vapor to pass through the valve.

When the pressure surrounding the bellows has lowered to 28.5 psi, the bellows expands and the needle valve closes. Fig. 20-16. Evaporator pressure, entering through an orifice in the piston head, begins to increase pressure on the piston. The piston is still open, valve still closed.

When pressure on both sides of the piston becomes nearly

Fig. 20-14. Schematic of a typical PILOT OPERATED ABSOLUTE SUCTION THROTTLING VALVE (POA). System is OFF, pressure is equal on both sides of piston, so spring A has forced piston closed. (Pontiac)

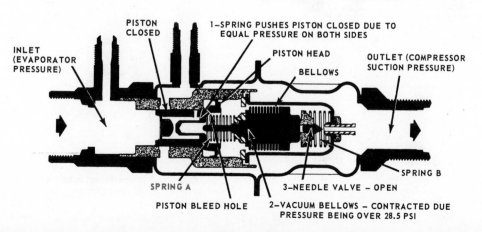

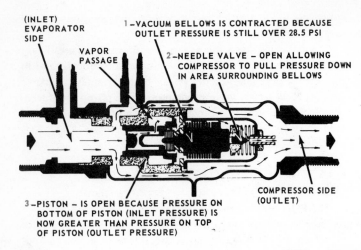

(INLET) EVAPORATOR SIDE

VAPOR PASSAGE

1 — VACUUM BELLOWS IS CONTRACTED BECAUSE OUTLET PRESSURE IS STILL OVER 28.5 PSI

2 — NEEDLE VALVE — OPEN ALLOWING COMPRESSOR TO PULL PRESSURE DOWN IN AREA SURROUNDING BELLOWS

COMPRESSOR SIDE (OUTLET)

3 — PISTON — IS OPEN BECAUSE PRESSURE ON BOTTOM OF PISTON (INLET PRESSURE) IS NOW GREATER THAN PRESSURE ON TOP OF PISTON (OUTLET PRESSURE)

Fig. 20-15. POA valve. System activated. Compressor lowering pressure on outlet side, around bellows and on head of piston. Pressure differential has overcome spring tension and forced piston to open. Needle valve is still open.

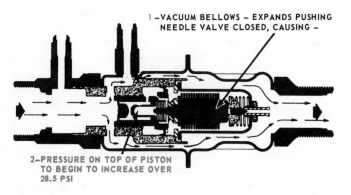

1 — VACUUM BELLOWS — EXPANDS PUSHING NEEDLE VALVE CLOSED, CAUSING —

2 — PRESSURE ON TOP OF PISTON TO BEGIN TO INCREASE OVER 28.5 PSI

Fig. 20-16. POA valve. Outlet pressure around bellows has reduced to 28.5 psi, causing bellows to expand and close needle valve.

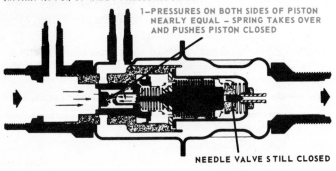

THE PRESSURE SURROUNDING BELLOWS AND ON TOP OF PISTON HAS NOW INCREASED SUFFICIENTLY OVER 28.5 PSI TO BECOME NEARLY EQUAL (WITHIN 1.3 PSI) OF INLET PRESSURE. SINCE —

1 — PRESSURES ON BOTH SIDES OF PISTON NEARLY EQUAL — SPRING TAKES OVER AND PUSHES PISTON CLOSED

NEEDLE VALVE STILL CLOSED

Fig. 20-17. POA valve. Increase in pressure on top of piston has permitted spring to close piston. Pressure in bellows area is still not high enough to contract bellows and open valve.

equal, spring A will force the piston closed. The valve remains closed until pressure increases to the point that causes the bellows to contract. Fig. 20-17 shows the POA valve with the piston closed.

When pressure in the bellows area is increased enough, the bellows will contract and the valve will open. This will apply compressor vacuum to the bellows and piston head. The piston will be drawn open, Fig. 20-15, and the cycle repeated.

COMPRESSOR

The compressor can be of a typical two cylinder, reciprocating piston type or of axial design. Figs. 20-18 and 20-19. It is belt driven by the engine. The compressor draws in the low pressure refrigerant from the evaporator and compresses it to around 100-250 psi, depending on compressor speed and ambient air temperature. Note that the inlet side of the compressor is generally referred to as the LOW SIDE. The outlet side is referred to as the HIGH SIDE.

When a compressor compresses refrigerant, it also raises the refrigerant temperature. The temperature rise from compression is sufficient to bring the temperature of the compressed refrigerant to above that of the surrounding air. This compressed and heated refrigerant is then forced into the condenser.

Fig. 20-18 illustrates one type of two cylinder, reciprocating piston compressor.

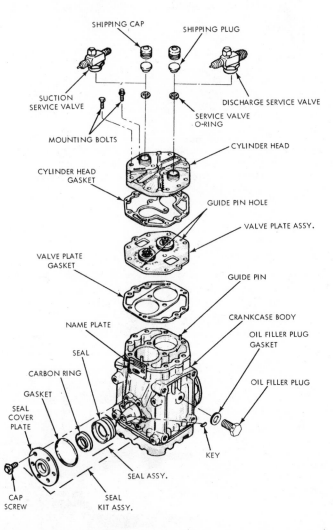

SHIPPING CAP

SHIPPING PLUG

SUCTION SERVICE VALVE

DISCHARGE SERVICE VALVE

SERVICE VALVE O-RING

MOUNTING BOLTS

CYLINDER HEAD

CYLINDER HEAD GASKET

GUIDE PIN HOLE

VALVE PLATE ASSY.

VALVE PLATE GASKET

GUIDE PIN

NAME PLATE

CRANKCASE BODY

OIL FILLER PLUG GASKET

SEAL

CARBON RING

OIL FILLER PLUG

GASKET

SEAL COVER PLATE

KEY

CAP SCREW

SEAL ASSY.

SEAL KIT ASSY.

Fig. 20-18. Typical two cylinder, reciprocating piston, air conditioning compressor. (Ford)

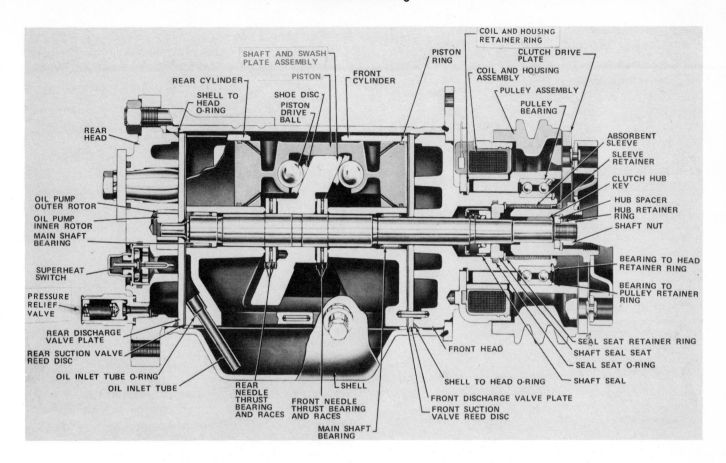

Fig. 20-19. Axial type air conditioning compressor. As Swash plate turns, it moves double-ended pistons back and forth. Note magnetic clutch drive pulley.   (Buick)

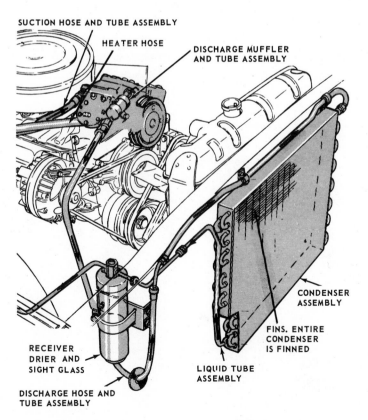

Fig. 20-20. Typical condenser. (Fins are not all drawn in.) Unit mounts in front of radiator.   (Plymouth)

Fig. 20-19 shows a cross-sectioned view of the axial type of compressor. This type uses three, double-ended pistons activated by a wabble or "swash" plate. The design effectively produces a six cylinder compressor.

CONDENSER

A condenser is constructed much like an evaporator. It is mounted in front of the radiator so that an ample stream of air is forced through the cooling fins. Both the fan and forward motion of the car cause air to flow through the condenser.

As the hot high pressure Refrigerant-12 gas or vapor passes into the condenser, its temperature is higher than that of the cooling air passing through the condenser. As a result, the refrigerant gives up its heat to the airstream. The compression heat and the latent heat of vaporization both escape.

As the refrigerant passes through the coils, it cools and condenses back into a high-pressure liquid. As it condenses, it flows into the receiver-dehydrator.

The refrigerant must leave the receiver-dehydrator and enter the expansion valve as a liquid. If refrigerant gas is passed to the expansion valve, very little cooling will take place.

A typical condenser is shown in Fig. 20-20. Note inlet line from compressor and outlet line to receiver-dehydrator.

Individual units that make up one car air conditioning refrigeration system are shown in Fig. 20-21. This system does not use a suction throttling valve.

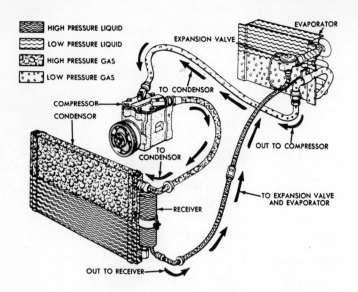

Fig. 20-21. Air conditioning refrigeration system. Refrigerant flow is from receiver-dehydrator to expansion valve, to evaporator, to compressor, to condenser, to receiver-dehydrator. (Ford)

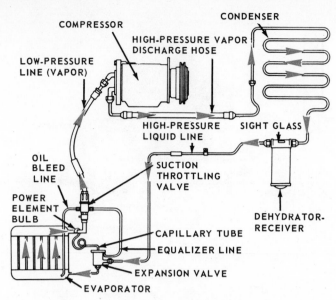

Fig. 20-23. Schematic of an air conditioning system, similar to that shown in Fig. 20-22. Arrows indicate direction of refrigerant flow. (Cadillac)

Another system, incorporating a muffler and POA suction throttling valve, is pictured in Fig. 20-22. A schematic of a similar system, is illustrated in Fig. 20-23.

MAGNETIC CLUTCH

The compressor pulley turns whenever the engine is running. The compressor does not run until the pulley is engaged to the compressor driving shaft.

This engagement is accomplished through the use of a magnetic clutch. Fig. 20-24 pictures one type of magnetic clutch. When the system is turned on, current is passed to the electromagnet coil, causing it to attract the armature plate.

Through the strong magnetic pull, the armature plate is drawn firmly against the side of the revolving pulley. Effectively locked together, the revolving pulley drives the armature plate which, in turn, drives the compressor.

When the current is stopped, the magnetic pull is broken and flat springs pull the armature plate away from the pulley surface. Note in Fig. 20-24, that the magnetic coil does not revolve. Its magnetic pull is transmitted through the pulley to the armature. The armature plate and hub assembly are fastened to the compressor drive shaft. The magnetic clutch pulley, when not driving the compressor, spins on double-row ball bearings.

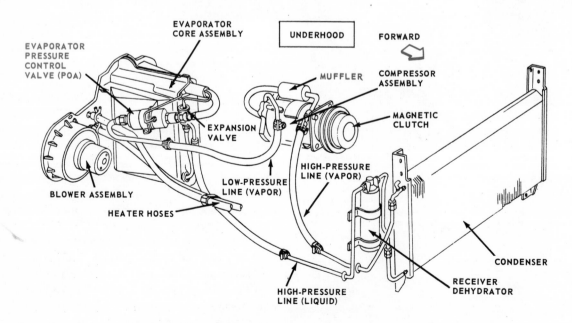

Fig. 20-22. This air conditioning system employs a POA evaporator pressure control valve (suction throttling valve) to control coil temperature. Note use of a muffler to quiet pumping noises. Sometimes a muffler is used on inlet or suction line as well. (GMC)

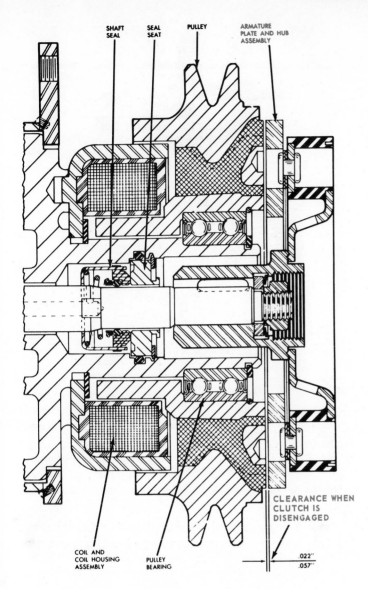

Fig. 20-24. Magnetic clutch. Pulley revolves whenever engine is running. When current is passed through electromagnet coil, magnetic pull draws armature plate against pulley, driving compressor. (Pontiac)

AIR CONDITIONER AIRFLOW

Many air conditioning systems incorporate a hot water heating core in the setup. By means of a control door, the air can be made to travel through either the heater core or the evaporator. Control of the door may be by manually operated dash controls or, as is becoming increasingly popular, control by automatic means.

Some systems pass all air through the evaporator core first. If heating is necessary, heated air is blended in to produce the required outlet air temperature. This method has the advantage of dehumidifying the air, preventing window fogging during damp weather conditions.

Most systems are arranged so that three air settings are possible:
1. Air within car can be recirculated.
2. All air entering car can be drawn from outside.
3. Some outside air can be mixed with inside air that is recirculating.

A schematic cross section illustrating a typical airflow pattern is pictured in Fig. 20-25. Note that all incoming air, recirculated or outside, is first passed through the evaporator core.

Another airflow pattern is shown in Fig. 20-26. Note how all air is first passed through the evaporator, then blended.

MUFFLER

A muffler is often installed on the discharge side (high pressure) of the compressor. The muffler reduces pumping noise and line vibrations. It also smoothes out the surge action produced by the compressor. Note muffler in Fig. 20-22.

REFRIGERANT OIL

A specific amount (about 10 ounces) of special refrigeration system oil is placed in the system to provide lubrication for the compressor. Since Refrigerant-12 has great affinity

Fig. 20-25. Schematic showing typical airflow pattern. All air first passes through evaporator core, where it is cleaned, cooled and dehumidified. Then, for full cooling, all air bypasses around heater core. For full heat, all air passes through heater core. For blending to desired temperature, some air passes through heater core and some bypasses it. (Chevrolet)

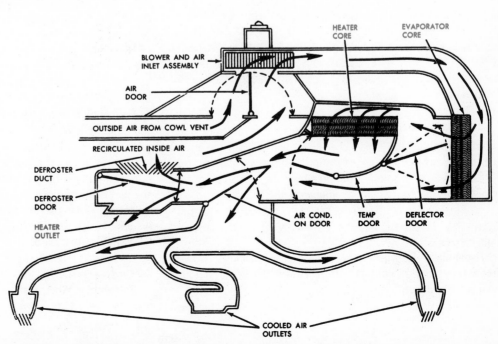

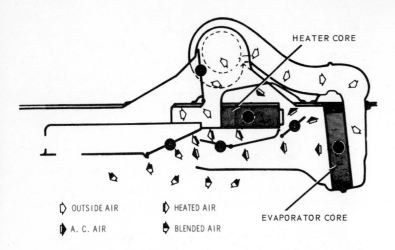

HEATER CORE

EVAPORATOR CORE

◇ OUTSIDE AIR   ◈ HEATED AIR

◆ A. C. AIR   ◈ BLENDED AIR

Fig. 20-26. Schematic showing how outside air is first cooled, then blended to proper temperature by passing some air through heater core and mixing it with cooled air. (Oldsmobile)

(attraction) for the oil, a certain amount of oil circulates throughout the system. Refrigeration oil is highly refined and must be ABSOLUTELY FREE of moisture.

DISCHARGE AND SUCTION SERVICE VALVES

All automobile air conditioning systems provide discharge and suction service valves. These valves permit the system to be evacuated (emptied), as well as charged (filled) and checked with pressure gages.

Fig. 20-27 pictures one type of service valve. Sketch 1 shows the valve in a position for evacuating. Number 2 is the operating position. Number 3 is the position used for charging.

Some systems employ a spring-loaded valve, called a Schrader Valve, instead of the discharge and suction service valve illustrated in Fig. 20-27. Connections can be made quickly and easily.

DANGER!

REFRIGERANT-12 (DICHLORODIFLUOROMETHANE), DESPITE THE FACT THAT IT IS COLORLESS, NONEXPLOSIVE AND NONFLAMMABLE, MUST BE HANDLED WITH EXTREME CARE. STUDY THE FOLLOWING SAFETY PRECAUTIONS THAT A MECHANIC MUST ALWAYS OBSERVE!

1. KEEP SERVICE AREA WELL VENTILATED.

Remember: Refrigerant-12, at atmospheric pressure and room temperature, is heavier than air. It will displace the air in the room and can cause suffocation. Ample ventilation also helps to prevent poisoning from breathing the fumes caused by allowing refrigerant to contact an open flame.

2. ALWAYS WEAR PROTECTIVE GOGGLES WHEN YOU ARE WORKING ON OR NEAR AN AIR CONDITIONING SYSTEM.

Remember: Refrigerant-12 vaporizes so quickly, it will freeze anything it contacts. If it enters the eyes, serious damage may occur.

3. KEEP STERILE MINERAL OIL AND A WEAK BORIC

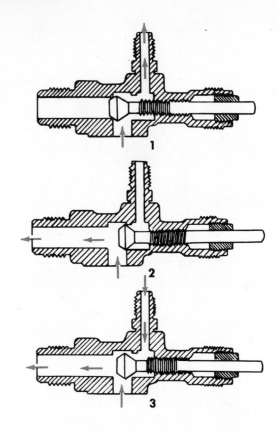

Fig. 20-27. Typical service valve. 1—Evacuating position. 2—Position for operating. 3—Charging position. Arrows indicate refrigerant flow in the various positions. (American Motors)

ACID SOLUTION ON HAND.

Remember: If, by some chance, refrigerant enters the eyes, DO NOT PANIC! Splash large amounts of cold water into the eyes as a means of raising the temperature. Do not rub the eyes. Apply several drops of sterile (clean) mineral oil to the eye. The oil will absorb the refrigerant and help flush it from the eyes. This may be followed by a liberal application of boric acid solution. ALWAYS SEEK THE SERVICES OF A DOCTOR AT ONCE — EVEN IF THE PAIN SEEMS TO HAVE DISAPPEARED.

4. KEEP REFRIGERANT AWAY FROM SKIN.

Remember: If refrigerant contacts the skin, treat in the manner recommended for the eyes.

5. AVOID DISCHARGING REFRIGERANT DIRECTLY INTO SERVICE AREA.

Remember: Refrigerant will vaporize at room temperature and since it is heavier than air, it will settle down. If enough refrigerant is discharged into an area without excellent ventilation, it can displace the air and cause suffocation. When refrigerant is discharged into the service area, it may contact an open flame and produce poisonous phosgene gas.

Always discharge the system into the service area exhaust system. If no such facility is present, discharge the system OUTSIDE of the building.

6. NEVER DISCHARGE REFRIGERANT INTO AN AREA CONTAINING AN OPEN FLAME.

Remember: When Refrigerant-12 comes in contact with an open flame, it produces a poisonous gas (phosgene). In

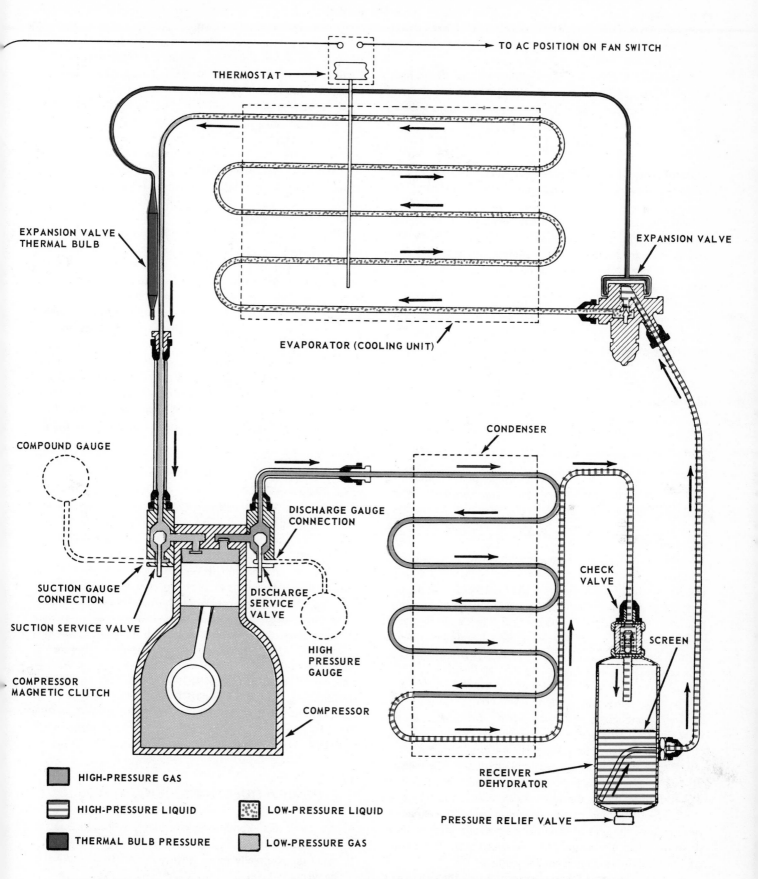

TO AC POSITION ON FAN SWITCH

THERMOSTAT

EXPANSION VALVE
THERMAL BULB

EXPANSION VALVE

EVAPORATOR (COOLING UNIT)

COMPOUND GAUGE

CONDENSER

DISCHARGE GAUGE
CONNECTION

SUCTION GAUGE
CONNECTION

DISCHARGE
SERVICE
VALVE

SUCTION SERVICE VALVE

HIGH
PRESSURE
GAUGE

CHECK
VALVE

SCREEN

COMPRESSOR
MAGNETIC CLUTCH

COMPRESSOR

RECEIVER
DEHYDRATOR

PRESSURE RELIEF VALVE

HIGH-PRESSURE GAS

HIGH-PRESSURE LIQUID

LOW-PRESSURE LIQUID

THERMAL BULB PRESSURE

LOW-PRESSURE GAS

Fig. 20-28. Study this schematic carefully. Note various units, refrigerant flow direction (arrows) and refrigerant state (liquid, gas, low pressure, high pressure, etc.) in all sections of system. Note that this system uses a thermostatic control to shut off compressor, to control evaporator coil icing instead of suction throttling valve employed by some other systems.

addition to being dangerous to humans, it will tarnish bright metal surfaces.

7. NEVER STEAM CLEAN, WELD, SOLDER, BAKE BODY FINISHES OR, IN ANY WAY, SUBJECT THE AIR CONDITIONING SYSTEM TO EXCESS HEAT.

Remember: Refrigerant when closed to the atmosphere, will build up high pressures with heat. These pressures may easily burst the system or blow out the safety plug, admitting the entire refrigerant charge into the atmosphere.

8. HANDLE REFRIGERANT—12 DRUMS (OR SMALL CANS) WITH CARE.

Remember: If exposed to excess heat (even direct rays from the sun), the drum may blow the safety plug. The protective screw cap should always be over the valve to prevent physical damage. Drums have been known to explode (even when equipped with a safety plug). If tank warming during servicing is needed, use nothing but warm water or warm wet rags. Never use a torch, gas stove, steam cleaner, etc., to heat the tank. Never heat drum above 125 deg. F.

9. NEVER FILL A TANK COMPLETELY.

Remember: When filling a small tank from a larger one, never completely fill the tank. Allow ample space for refrigerant expansion due to heating. A full tank can be very dangerous.

SUMMARY

Cooling is a process of removing heat from a given object.

Heat is a form of energy that can affect a temperature change in any object.

Heat produces various effects. Expansion, contraction, vaporization, burning, freezing, etc., are all effects of the removal or addition of heat.

Heat transfer is affected by convection, conduction and radiation.

Heat always transfers from a hot object to a colder object.

Latent (hidden) heat is the heat that is absorbed during a change of state. When water is changed from a liquid to a vapor, a great deal of heat can be applied without affecting a change in temperature. This latent heat is hidden in the vapor.

Refrigerant-12 is the substance used in refrigeration systems. It boils at minus 21.7 deg. F. at zero pressure (sea level). It is colorless, tasteless and heavier than air in either the liquid or vapor state. It must be handled with great care.

The automotive refrigeration system consists basically of a receiver, sight glass, expansion valve, evaporator, suction throttling valve, compressor, muffler and condenser.

With the system operating, high pressure refrigerant, in a liquid state, collects in the receiver-dehydrator. The receiver-dehydrator also strains impurities and removes moisture from the refrigerant. From the receiver-dehydrator, the refrigerant moves through a line to the expansion valve. The expansion valve reduces the pressure and meters a certain amount of low pressure liquid refrigerant into the evaporator. The amount admitted will depend upon cooling load.

As the low pressure liquid refrigerant enters the evaporator, it is warmed by air passing over the coils. As it moves through the evaporator, it begins to boil. More and more heat is absorbed from the passing air. When refrigerant reaches the evaporator outlet, it is completely vaporized. The vapor is laden with hidden or latent heat.

From the evaporator, the refrigerant vapor is forced through (in some systems) a suction throttling valve. The suction throttling valve will maintain a regulated evaporator outlet pressure to control evaporator temperature.

After leaving the suction throttling valve, the refrigerant vapor is drawn into the compressor. The compressor raises the pressure of the vapor and, at the same time, effects a rapid rise in temperature. The vapor on its way to the condenser from the compressor, is under 100-250 psi and is hotter than the ambient (surrounding) air temperature. It still retains all the latent heat of vaporization that it picked up on its way through the evaporator.

When the hot high pressure refrigerant gas or vapor enters the condenser, it begins to give up heat to the airstream moving over the condenser fins. By the time it passes through the condenser, it has lost sufficient heat to return to the liquid state. Still under heavy pressure from the compressor, the liquid refrigerant travels from the condenser into the receiver-dehydrator. From the receiver-dehydrator, it starts another cycle through the system.

This continual cycling of liquid refrigerant to the evaporator, where it vaporizes and is drawn out, compressed and condensed, produces a steady cooling effect on the air passing through the evaporator core. In this way, heat is taken from the air entering the car and is discharged into the atmosphere at the condenser.

The system cools, dehumidifies and filters air passing into the car. Follow the path of refrigerant through the system. Figs. 20-28 and 20-29.

Refrigerant-12 is dangerous. Observe all safety precautions when handling drums, or when working on or near the system.

A good mechanic is NEVER satisfied with his present level of skill and knowledge. He CONSTANTLY seeks to improve both his level of skill and his background of information. This requires a good deal of study, listening and observing.

Take advantage of special automotive classes, read quality automotive magazines and journals, maintain an up-to-date library of top-notch books in your field. Above all, never miss an opportunity to observe, question or listen to, other men in your field. Remember: WHEN YOU THINK YOU KNOW IT ALL — YOU STOP GROWING. IT WILL NOT BE LONG UNTIL THE JOB AT HAND STANDS LIKE A GIANT IN RELATION TO YOUR ABILITY TO PERFORM IT. WHEN THIS HAPPENS, YOUR USEFULLNESS AS A MECHANIC IS GONE.

REVIEW QUESTIONS – CHAPTER 20

1. A good air conditioning system will _____, _____ and _____ the air entering the car.
2. What do you really mean when you refer to something as being cold?
3. Name several readily observed effects produced by the addition or removal of heat.

4. What is heat?
5. What is meant by Btu?
6. How is the law of latent heat of vaporization utilized in the refrigeration system?
7. _____ affects both the vaporization and condensation temperature of refrigerant.
8. Heat is transferred from one object to another by _____, _____ and _____ .
9. Heat is transferred from a hot object to the colder one. True or False?
10. What is Refrigerant-12?
11. Of what use is the receiver in the system?
12. The sight glass makes it possible to visually inspect the refrigerant for the presence of _____ or _____ .
13. The expansion valve performs two important functions. Name them.
14. Explain the function of the evaporator.
15. Why is it necessary to have a compressor in the system?

16. The condenser is used to change the refrigerant from a liquid into a vapor. True or False?
17. Describe the use and action of the magnetic clutch used on the compressor.
18. Name two methods of preventing the evaporator coils from icing up.
19. Of what use are the discharge and suction service valves?
20. List nine safety precautions to be observed when working on or near the air conditioning system.
21. Starting at the receiver-dehydrator, explain just what happens as the refrigerant makes a complete circuit through the system. Describe the areas in which the refrigerant is a vapor and the areas in which it is a liquid.
22. The suction throttling valve is located:
   a. At the inlet to the evaporator.
   b. At the outlet of the evaporator.
   c. At the inlet to the condenser.
   d. At the outlet of the compressor.

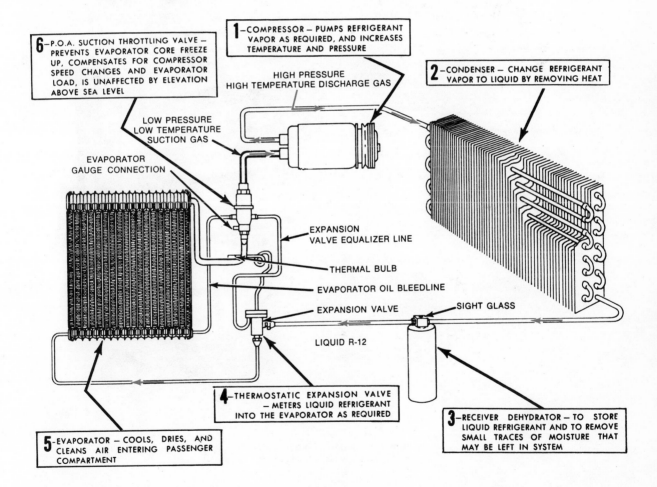

6—P.O.A. SUCTION THROTTLING VALVE — PREVENTS EVAPORATOR CORE FREEZE UP, COMPENSATES FOR COMPRESSOR SPEED CHANGES AND EVAPORATOR LOAD, IS UNAFFECTED BY ELEVATION ABOVE SEA LEVEL

1—COMPRESSOR — PUMPS REFRIGERANT VAPOR AS REQUIRED, AND INCREASES TEMPERATURE AND PRESSURE

2—CONDENSER — CHANGE REFRIGERANT VAPOR TO LIQUID BY REMOVING HEAT

HIGH PRESSURE HIGH TEMPERATURE DISCHARGE GAS

LOW PRESSURE LOW TEMPERATURE SUCTION GAS

EVAPORATOR GAUGE CONNECTION

EXPANSION VALVE EQUALIZER LINE

THERMAL BULB

EVAPORATOR OIL BLEEDLINE

EXPANSION VALVE

SIGHT GLASS

LIQUID R-12

4—THERMOSTATIC EXPANSION VALVE — METERS LIQUID REFRIGERANT INTO THE EVAPORATOR AS REQUIRED

3—RECEIVER DEHYDRATOR — TO STORE LIQUID REFRIGERANT AND TO REMOVE SMALL TRACES OF MOISTURE THAT MAY BE LEFT IN SYSTEM

5—EVAPORATOR — COOLS, DRIES, AND CLEANS AIR ENTERING PASSENGER COMPARTMENT

Fig. 20-29. Schematic shows a typical refrigerant cycle. Trace flow through various units by following "RED" arrows.
(Chevrolet)

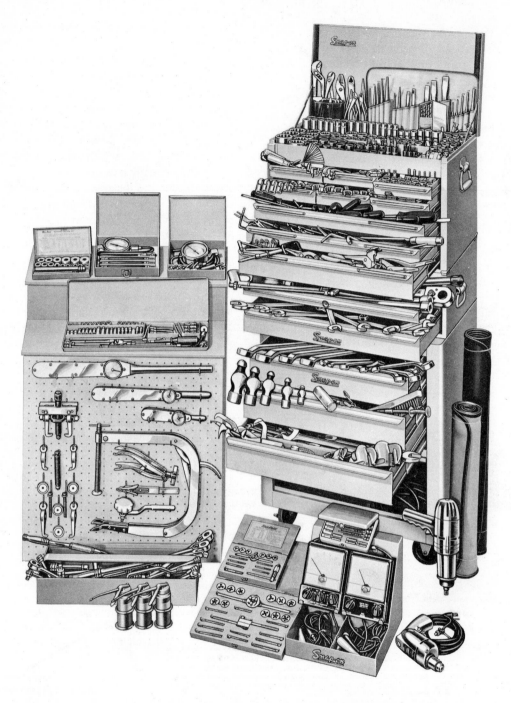

A set of quality hand tools is a must for the mechanic. How many of these tools can you identify? (Snap-On Tools)

Chapter 21

TOOL IDENTIFICATION, TOOL USE

Today's mechanic must be familiar with and understand the use of a large number of tools. Proper tool selection will improve both the quality and speed of any repair operation. Many repair jobs, in fact, would be exceedingly difficult to perform without the right tools for the job.

A good mechanic owns a wide selection of quality tools. He constantly strives to add to his collection and takes great pride in keeping his tools clean and orderly.

WHAT TO LOOK FOR IN TOOLS

Some mechanics prefer one brand of tools; some, another. You will find, however, that mechanics agree on several important features that will be found in quality tools.

TOOL MATERIAL

The better tools are made of high strength alloy steel and, as such, can be made without a great deal of bulk. These tools are light and easy to use in tight quarters. Heavy, "fat" tools are useless on many jobs.

The quality alloy also gives the tool great strength. Its working areas will stand abuse, and useful life will be greatly extended. Quality material makes it possible for the manufacturer to give a good guarantee on the tool.

TOOL CONSTRUCTION

Quality tools receive superior heat treating. Their openings and working surfaces are held to closer tolerances. Sharp edges are removed and the tools are carefully polished. This imparts a finish that is easy to wipe clean and a tool that is more comfortable to use.

Tools worthy of a good mechanic will be slim, efficient and properly designed for the job at hand. They will be strong, easy to clean and a pleasure to use. A good guarantee plus efficient service and parts replacement will be offered. This can be an important feature; otherwise usable tools often have to be discarded due to the loss or failure of some minor part that is no longer available.

REMEMBER: In buying tools, you usually get just what you pay for. Good tools are a good investment.

USE OF TOOLS

Use tools designed for the job at hand. Keep your tools orderly and clean in a good, roll-type cabinet, tool chest and "tote" tray (a small tray that may be carried to the job to keep a few selected tools close at hand). This will assist you in keeping tools in good shape and readily available.

Any tool subjected to rust should be cleaned and lightly oiled. Cutting tools, such as files and chisels, should be separated to preserve the cutting edges. Delicate measuring tools should be placed in protective containers. Keep heavy tools by themselves, and arrange tools so those most often used are handy.

Repeat — KEEP YOUR TOOLS CLEAN. You cannot hope to successfully assemble a fine piece of machinery with grubby tools. The slightest bit of dirt or abrasive, that finds its way into moving parts can create havoc when the unit is placed in operation. A mechanic's tools are a good indication of his worth. Dirty, beat up and jumbled tools reveal that the mechanic has a lot to learn. His workmanship is likely to be just like his tools.

TYPES OF TOOLS

This chapter does not attempt to cover all the tools used in the automotive trade. Large garages often utilize hundreds of special tools designed for specific jobs, models and units. Basic tools commonly used will be discussed.

HAMMERS

Mechanics find great use for the ball peen, plastic-tipped and brass-tipped hammers. The ball peen is used for general striking work and is available in weights ranging from a couple of ounces to several pounds. Brass and plastic-tipped hammers are used where there is danger of the steel ball peen marring the surface. Fig. 21-1.

CAUTION!

BE CAREFUL WHEN USING A HAMMER. DO NOT SWING THE HAMMER IN A DIRECTION THAT WOULD

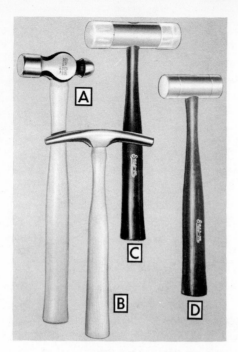

Fig. 21-1. Several types of hammers. A—Ball peen. B—Upholstery. C—Plastic-tipped. D—Brass-tipped. (Snap-on Tools)

ALLOW IT TO STRIKE SOMEONE IF IT SLIPS FROM YOUR GRASP. KEEP THE HAMMER HANDLE TIGHT IN THE HAMMER HEAD, AND KEEP THE HANDLE CLEAN AND DRY.

CHISELS

Several sizes and types of chisels are essential for cutting bolts, rivets, etc. When chiseling, grasp the chisel firmly and strike it squarely. Attempt to keep the fingertips around the chisel body to prevent it from flying from your grasp. On the other hand, avoid grasping it too tightly because a poorly aimed blow may cause serious injury to the hand.

Keep the cutting edge of the chisel sharp, and the striking surface chamfered (edges properly tapered) to reduce the danger of mushroom particles flying about. Fig. 21-2. Wear goggles when using a chisel.

The flat cold chisel, illustrated in Fig. 21-2, is used for general cutting. Special chisels such as the cape, half-round and diamond point are used when their shape fits a definite need.

PUNCHES

A starting punch is a punch that tapers to a flat tip. It is used in starting to punch out pins, and to drive out rivets after the heads have been cut off.

Once the pin has been started, the starting punch (because of its taper) can no longer be used. A drift punch, which has the same diameter for most of its length, is used to complete the job.

A pin punch is similar to a drift punch but is smaller in diameter. Fig. 21-3.

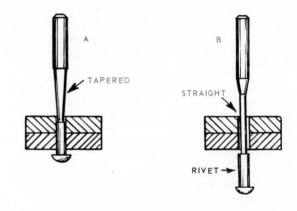

Fig. 21-3. Starting and drift punches. Starting punch A is used to start rivet from hole. Drift punch B is used to drive rivet from hole.

An aligning punch, which has a long, gradual taper, is useful to shift parts and bring corresponding holes into alignment. Fig. 21-4.

A center punch is used to mark the material before drilling. It leaves a small, V-shaped hole that assists in aligning the drill bit. A center punch may also be used to mark parts so they

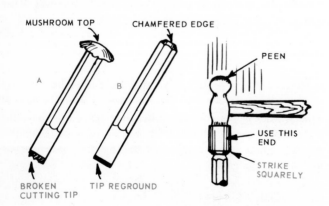

Fig. 21-2. Keep chisels in good condition. Chisel in A is in a dangerous condition. B is same chisel after grinding.

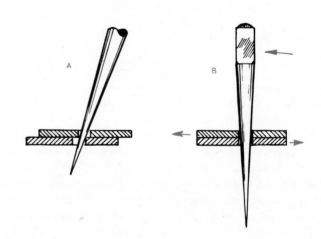

Fig. 21-4. Aligning punch. This punch is shoved through holes in A, then pulled upright and shoved deeper into holes in B. This lines up the two holes.

Tools

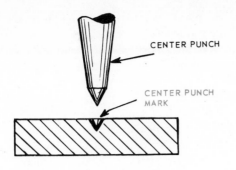

Fig. 21-5. Center punch is used for marking parts and starting drills.

will be assembled in the correct positions. Fig. 21-5.

Be careful when sharpening chisels and punches. Keep the edges at the proper angle and avoid turning them blue (overheating) on the grinder. Overheating will draw (remove) the temper and render the tip soft and useless. Grind slowly and keep quenching (dipping) the tool in water. Fig. 21-6 illustrates several commonly used chisels and punches. Learn

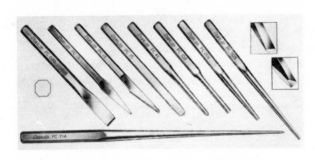

Fig. 21-6. Chisel and punch assortment.

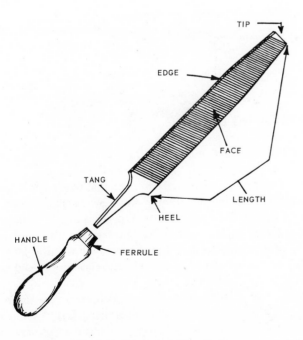

Fig. 21-7. Typical mill file. Single-cut.

their names. WEAR GOGGLES WHEN GRINDING OR CHIPPING WITH A CHISEL.

FILES

Commonly used files are the flat mill, half-round, round, square and triangular. All come in different sizes and with fine-to-coarse cutting edges. Some files have one or more cutting edges. Some files have one or more safe (surface with no cutting edges) sides.

A file should have the handle firmly affixed to the tang. The handle gives a firm grip, and it eliminates the danger of the sharp tang piercing the hand. A typical mill file is shown in Fig. 21-7.

FILE CUT AND SHAPE

When a file has a single series of cutting edges that are parallel to each other, the file is referred to as a single-cut file. A file with two sets of cutting edges that cross at an angle is called a double-cut file. Fig. 21-8.

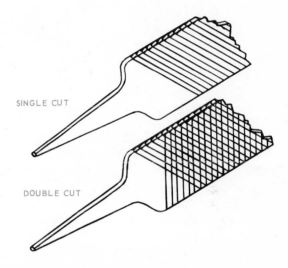

Fig. 21-8. Single and double-cut files.

A file, from rough to smooth, depends on the number and size of the cuts. Fig. 21-9 illustrates several classifications of cut for both single and double-cut files.

Common file shapes are illustrated in Fig. 21-10. The thin, flat, point file also shown in Fig. 21-10 is used to file distributor and regulator points.

USING A FILE

When filing, hold the file with both hands. On the forward cut, bear down with pressure sufficient to produce good cutting. On the return stroke, raise the file to avoid battering the cutting edges. One hand is on the handle, while the other hand grasps the tip of the file.

Control the file to prevent it from rocking. Practice is essential. Remember that a file is NOT a crude tool. In the

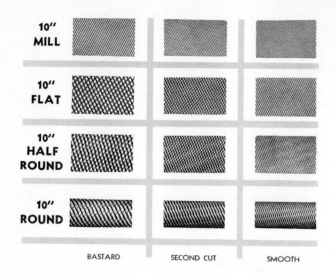

Fig. 21-9. File cuts. Three different file cuts: bastard, second-cut and smooth. (Simonds File Co.)

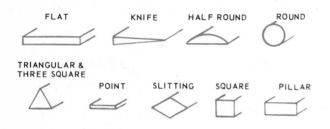

Fig. 21-10. Nine common file shapes, handy for the mechanic.

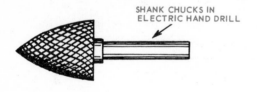

Fig. 21-11. Rotary file.

hands of an expert, a file becomes a precision tool.

Choose a file that is correct for the job at hand. Use a coarse cut file on soft material to prevent clogging. A slight drag on the return stroke, when filing soft material, will help keep the cutting teeth clear of filings.

Chalking (rubbing file with chalk) will help remove oil from the file and keep it from clogging. A file card (short bristle wire hand brush) is helpful in clearing the file. An occasional tap of the handle will also help.

Keep the file clean and free from oil and grease. A handle should always be used. Secure it by hammering the handle, with file in place, on a solid object. Do not hammer the file, it may break.

When you finished filing, place the file in a dry spot, protected from contact with other files and tools. A good file is a fine tool. Treat it with respect, and you will do more and better work.

There are many files for special purposes, such as those used in body and fender repair. Swiss needle files are miniature files that come in almost all shapes and are very handy for delicate operations.

ROTARY FILES

A selection of rotary files is often found in the mechanic's toolbox. These files are designed to be chucked in an electric hand drill. They are useful in working in "blind" holes, etc., where a normal file is useless. Fig. 21-11.

DRILLS

A twist drill is used by mechanics for drilling holes. Twist drills are available in fractional inch sizes (1/4, 1/2, etc.), letter sizes (A, B to Z) and number sizes (wire gage, numbers 1 to 80).

For general auto shop use, a set of fractional size drills (29 drills from 1/16 to 1/2), plus a set of number drills (totaling 60, 1 to 60), will usually suffice. A number 1 drill has a diameter of .228 in. The number 60 drill has a diameter of .040 in.

Twist drills with a straight shank are used in hand drills and portable electric drills. For heavy duty power drills, the taper shank drill is often used. Fig. 21-12.

Drills are commonly furnished in carbon steel and high speed steel. Carbon drills will require more frequent sharpening, and they will not last nearly as long as high speed steel drills.

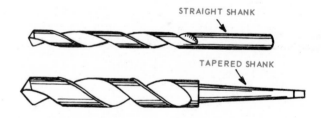

Fig. 21-12. Drill shanks.

SHARPENING DRILLS

Different cutting angles are used to make drills more efficient in various metals. Lip clearance angles also vary.

When sharpening a drill, keep in mind that both cutting lips must be the same length and angle. Avoid overheating the drill. If a carbon drill turns blue from overheating, it will be worthless until the blue area of the drill is ground off.

The general purpose drill should have the lips cut at 59 to 60 deg. to the center line of the drill. Lip clearance should be around 12 deg. Fig. 21-13.

Proper drill grinding must be learned by practice. Use an old drill to practive on. A simple drill grinding gage is pictured in Fig. 21-14. This gage will help you maintain proper lip length and angle.

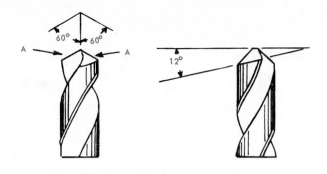

Fig. 21-13. Drill lip angles. Lengths at A must be equal. Angles should be the same on both sides.

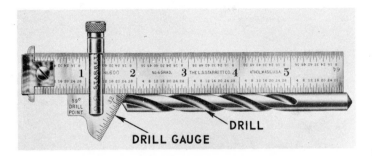

Fig. 21-14. Using a drill gage. Gage checks angle and length of tip. (Starrett)

A drill can be sharpened on either the side or face of the grinder. Start the cut at the leading edge of the lip and finish at the heel. Keep the shank of the drill slightly lower than the tip. The cut is taken with a slight rocking, pivoting motion.

Without starting the grinder, place a new drill against the stone. Move it through the sharpening motion and see how well the stone follows the lip angles. Try your skill on an old drill, then carefully compare it with a new one. Keep at it until you have mastered this skill. If your drills are correctly sharpened, they will cut readily without grabbing. Both lips will produce equal amounts of metal chips. Fig. 21-15.

USING DRILLS

Always securely fasten the piece to be drilled. Chuck the drill tightly and center punch the spot to be drilled. After the drill has started, apply a small quantity of cutting oil. The oil will make cutting easier, faster and will lengthen drill life. When drilling thin body metal or cast iron, oil is not required.

Hold the drill at the proper angle and apply enough pressure to achieve good cutting. When the drill is ready to break through the work, let up on the pressure to avoid grabbing. On thin metal, hold the work down, as well as keep it from turning. Thin work attempts to "climb" up the drill.

SAFETY HINTS

Large portable electric drills develop considerable torque. Make certain you have a firm grip before engaging the drill.

Never install or remove a drill in a chuck without

unplugging the cord. If it is accidentally started when you are grasping the drill or the chuck key, you could be injured.

Keep loose clothing (sleeves, pant legs, ties, etc.) away from the drill.

Always secure the work firmly. If the drill grabs and the work is loose, the work can swing around with a vicious cutting force.

Wear goggles when grinding drills.

Never use a power tool that is not properly grounded, and do not use power tools when standing on a wet surface.

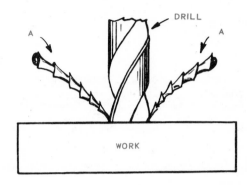

Fig. 21-15. Drill cutting properly. Chips, A, of equal size indicates both lips are cutting.

Several twist drills are illustrated in Fig. 21-16.

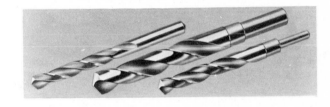

Fig. 21-16. Twist drills.

Portable electric drills are pictured in Fig. 21-17. Note the different handle setups.

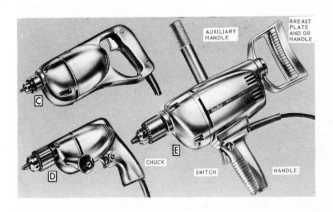

Fig. 21-17. Portable electric drills. Note different handle arrangements.

REAMERS

Reamers are used to enlarge or shape holes. They produce holes that are more accurate in size and smoother than those produced by drills. The reamer should not be used to make deep cuts in metal. A cut of a few thousandths of an inch at a time is all that should be attempted.

Some reamers are adjustable in size, while others are of a fixed size. Tapered reamers are handy for removing burrs, assisting in tap starting, etc.

Use cutting oil when reaming. Always turn a reamer clockwise, both when entering and when leaving the hole.

A few reamer types are shown in Figs. 21-18 and 21-19.

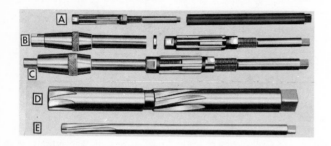

Fig. 21-18. Reamers. A—Adjustable. B, C—Adjustable with a pilot to help align reamer. D—Kingpin reamer. E—Valve guide reamer. (Snap-on Tools)

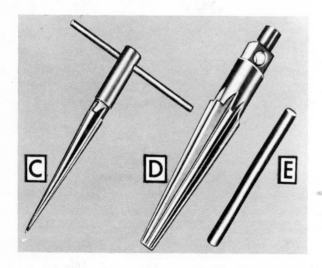

Fig. 21-19. Tapered reamers. (Snap-on Tools)

IMPACT WRENCH

A power-operated (electric or air) impact wrench is a must for fast work. Parts can be removed and replaced in a fraction of the time required with hand wrenches.

Although the impact wrench in no way replaces hand-operated wrenches, it can be used often and should be included in the tool collection. Various sizes are available. Fig. 21-20 pictures an air-operated impact wrench.

Fig. 21-20. An impact (power) wrench is a must for every mechanic. This wrench is air-operated. (Skil)

HACKSAWS

The hand hacksaw is a much used tool. It is excellent for cutting bolts, tubing, etc.

Hacksaw blades are generally furnished with 14, 18, 24 or 32 teeth per inch. The 14-tooth blade is handy for cutting fairly thick articles. The 18-tooth blade handles work of medium thickness. The 24-tooth blade is best used on heavy sheet metal, medium tubing, brass, copper, etc. For thin sheet metal and thinwall tubing, use a 32-tooth blade. Fig. 21-21 illustrates typical work for each type of blade.

Blades are made of various materials. Some blades are hardened across the full blade width; others have only the

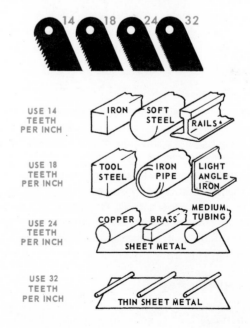

Fig. 21-21. The type of work to be cut determines the best blade tooth per inch selection. These are typical. (Crescent Tools)

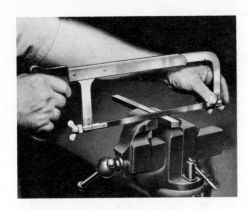

Fig. 21-22. When using a hacksaw, blade must be tight, teeth must face away from handle. Hold work securely and cut at 40–50 strokes per minute. (Starrett)

operated by an electric drill is also handy for drilling large holes in thin sheet metal. Fig. 21-23.

A conventional hacksaw, as well as two special, tight-quarter saws, are pictured in Fig. 21-24.

Fig. 21-24. Hacksaw types. A, B, C—Standard type hacksaw frames. D—Thin junior type for close quarters. E—Jab type will hold broken blade sections. It is handy in restricted areas. F—Hole saw used to drill holes in thin sheet metal.

tooth edge hardened.

When cutting tough alloy steels, select one of the better quality blades, such as high speed (tungsten) steel. These will outwear several of the cheaper blades.

USING THE HACKSAW

When hacksawing, keep the blade tight in the saw frame. Make certain the teeth cutting edges face away from the handle. Fig. 21-22. See that the work is held securely to prevent slippage that would result in a broken blade.

Grasp the hacksaw frame firmly, one hand on the grip and the other on the front. Apply pressure enough on the forward stroke to insure cutting, then, raise the blade slightly on the return stroke. Avoid starting the cut on a sharp edge; this may chip the saw teeth.

Try to saw at a speed that will produce 40 to 50 strokes per minute (forward strokes). A little oil will help.

SPECIAL HACKSAWS

The auto mechanic will sometimes find use for a special hacksaw that allows cutting in restricted quarters. A hole saw,

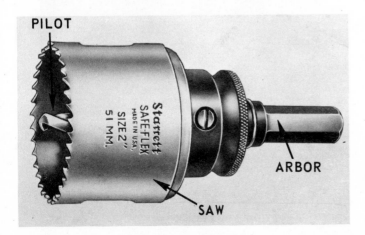

Fig. 21-23. Hole saws are handy for cutting openings in thin metal, as well as for making large diameter holes.

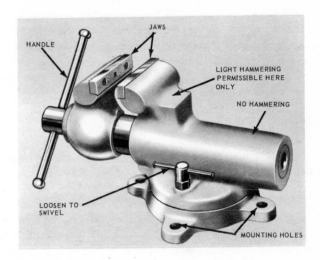

Fig. 21-25. Typical bench vise.

VISE

A vise is a work-holding device. A typical bench vise is shown in Fig. 21-25. If the work is easily marred, place special copper covers over the serrated steel jaws.

Keep your vise clean. Oil all working parts. Do not use a vise for an anvil. Do not hammer the handle to tighten. Fig. 21-25.

TAPS

Taps are used for cutting internal threads. Fig. 21-26. For general garage use, a set of taper, plug, bottom and machine screw taps should be provided. Each type of tap will have to be provided in both National Fine and National Coarse threads, as well as in the various sizes, 1/4, 5/16, etc.

The taper tap is used to tap completely through the hole. Notice that it has a long gradual taper that allows the tap to start easily.

The plug tap is used to tap threads partway through.

The bottoming tap is used to cut threads all the way to the

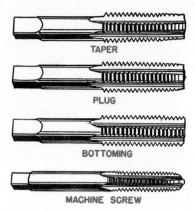

Fig. 21-26. Four kinds of taps.

TAP DRILL SIZES
Recommended for
AMERICAN NATIONAL SCREW THREAD PITCHES

COARSE STANDARD THREAD (N. C.) Formerly U. S. Standard Thread				SPECIAL THREAD (N. S.)			
Sizes	Threads Per Inch	Outside Diameter of Screw	Tap Drill Sizes	Decimal Equivalent of Drill	Sizes	Threads Per Inch	Outside Diameter of Screw	Tap Drill Sizes	Decimal Equivalent of Drill

Sizes	Threads Per Inch	Outside Diameter of Screw	Tap Drill Sizes	Decimal Equivalent of Drill	Sizes	Threads Per Inch	Outside Diameter of Screw	Tap Drill Sizes	Decimal Equivalent of Drill
1	64	.073	53	0.0595	1	56	.0730	54	0.0550
2	56	.086	50	0.0700	4	32	.1120	45	0.0820
3	48	.099	47	0.0785	4	36	.1120	44	0.0860
4	40	.112	43	0.0890	6	36	.1380	34	0.1110
5	40	.125	38	0.1015	8	40	.1640	28	0.1405
6	32	.138	36	0.1065	10	30	.1900	22	0.1570
8	32	.164	29	0.1360	12	32	.2160	13	0.1850
10	24	.190	25	0.1495	14	20	.2420	10	0.1935
12	24	.216	16	0.1770	14	24	.2420	7	0.2010
1/4	20	.250	7	0.2010	1/16	64	.0625	3/64	0.0469
5/16	18	.3125	F	0.2570	3/32	48	.0938	49	0.0730
3/8	16	.375	5/16	0.3125	1/8	40	.1250	38	0.1015
7/16	14	.4375	U	0.3680	5/32	32	.1563	1/8	0.1250
1/2	13	.500	27/64	0.4219	5/32	36	.1563	30	0.1285
9/16	12	.5625	31/64	0.4843	3/16	24	.1875	26	0.1470
5/8	11	.625	17/32	0.5312	3/16	32	.1875	22	0.1570
3/4	10	.750	21/32	0.6562	7/32	24	.2188	16	0.1770
7/8	9	.875	49/64	0.7656	7/32	32	.2188	12	0.1890
1	8	1.000	7/8	0.875	1/4	24	.250	4	0.2090
1-1/8	7	1.125	63/64	0.9843	1/4	27	.250	3	0.2130
1-1/4	7	1.250	1-7/64	1.1093	1/4	32	.250	7/32	0.2187
					5/16	20	.3125	17/64	0.2656
FINE STANDARD THREAD (N. F.) Formerly S.A.E. Thread					5/16	27	.3125	J	0.2770
Sizes	Threads Per Inch	Outside Diameter of Screw	Tap Drill Sizes	Decimal Equivalent of Drill	5/16	32	.3125	9/32	0.2812
0	80	.060	3/64	0.0469	3/8	24	.375	21/64	0.3281
1	72	.073	53	0.0595	3/8	27	.375	R	0.3390
2	64	.086	50	0.0700	7/16	24	.4375	X	0.3970
3	56	.099	45	0.0820	7/16	27	.4375	Y	0.4040
4	48	.112	42	0.0935	1/2	12	.500	27/64	0.4219
5	44	.125	37	0.1040	1/2	24	.500	29/64	0.4531
6	40	.138	33	0.1130	1/2	27	.500	15/32	0.4687
8	36	.164	29	0.1360	9/16	27	.5625	17/32	0.5312
10	32	.190	21	0.1590	5/8	12	.625	35/64	0.5469
12	28	.216	14	0.1820	5/8	27	.625	19/32	0.5937
1/4	28	.250	3	0.2130	11/16	11	.6875	19/32	0.5937
5/16	24	.3125	I	0.2720	11/16	16	.6875	5/8	0.6250
3/8	24	.375	Q	0.3320	3/4	12	.750	43/64	0.6719
7/16	20	.4375	25/64	0.3906	3/4	27	.750	23/32	0.7187
1/2	20	.500	29/64	0.4531	7/8	12	.875	51/64	0.7969
9/16	18	.5625	0.5062	0.5062	7/8	18	.875	53/64	0.8281
5/8	18	.625	0.5687	0.5687	7/8	27	.875	27/32	0.8437
3/4	16	.750	11/16	0.6875	1	12	1.000	59/64	0.9219
7/8	14	.875	0.8020	0.8020	1	27	1.000	31/32	0.9687
1	14	1.000	0.9274	0.9274					
1-1/8	12	1.125	1-3/64	1.0468					
1-1/4	12	1.250	1-11/64	1.1718					

Fig. 21-27. Tap drill size chart.   (South Bend Lathe)

bottom of a blind hole. The plug tap should precede the bottoming tap, since the bottoming tap will not start well.

The plug tap is the most widely used. It will work satisfactorily, except in the case of running threads completely to the bottom of a blind hole.

The machine screw tap handles the small diameter, fine thread jobs. Fig. 21-26.

USING THE TAP

After determining the diameter, and number of threads per inch of the screw or stud that will enter the tapped hole, use a tap drill size chart to find what size hole to drill.

For example, suppose you find that you want a threaded hole for a 3/8 stud with coarse (National Coarse) threads. Referring to the chart in Fig. 21-27, you will find the 3/8 N.C. will have 16 threads per inch. Looking directly across from the 3/8 size, you will find that the tap drill size is listed as 5/16. This means that for a 3/8 N.C. stud, you must drill a hole 5/16 in diameter.

If the hole is to be tapped partway through, use a 3/8 x 16 N.C. plug tap. Place the tap in a tap handle, and carefully start the tap in the hole. Place some tap lubricant on the tap. After threading the tap in one or two turns, back the tap up about a quarter to one-half turn to break the chip. Repeat this process as your tapping continues.

Be careful the hole does not clog with chips. It may be necessary to withdraw the tap and remove the chips. Taps are quite brittle. Use them with care, and make certain you use the proper size tap drill.

What size hole would you drill to tap threads for a 1/2 N.F. bolt? (Answer 29/64.) Figs. 21-27 and 21-28.

DIES

Dies are used to cut external threads. A die of the correct size is placed in a diestock (handle) and turned. Use lubricant, back up every one or two turns, and keep free of chips.

Often dies are adjustable in size, so you can enlarge or reduce (slightly) the outside diameter of a threaded area.

Both taps and dies should be cleaned, lightly oiled and placed in a protective box for storage.

Fig. 21-28. Tapping operation must be performed carefully. Keep tap square with hole, use lubrication when required and keep backing tap to break chip. Use proper size tap drill.

SPECIAL TAPS AND DIES

The mechanic will also find use for a few special taps and dies as illustrated in Fig. 21-29. One form of rethreading tool

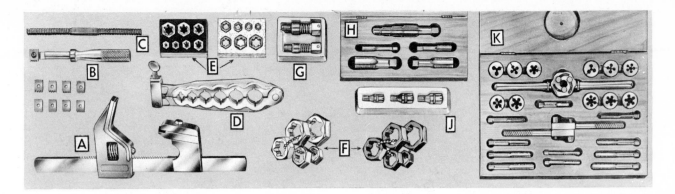

Fig. 21-29. Taps and dies.

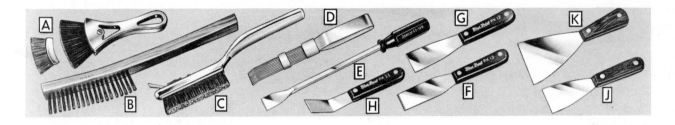

Fig. 21-30. Cleaning tools. A—Sparkproof bristle brush. B—Wire scratch brush. C—Wire brush and scraper combined. D—Flexible fingers carbon scraper. E—Rigid carbon scraper. F, G, H—Scraper blades. J, K—Putty knives.

is shown in A. It is placed on the thread and turned. B shows an internal thread chaser used to clean up dirty or damaged internal threads. A thread restorer, C, is handy for quickly reconditioning external threads. The axle rethreader, D, is placed around the good thread area, clamped shut, and is then turned back over the damaged area.

Nut or rethreading dies, E and F, can be turned on a damaged thread. An ordinary box wrench can be used to turn them. G and H show spark plug hole taps. These are very handy to clean up damaged or carboned plug hole threads. A combination tap and die set for tube flare fittings is illustrated in J. A combination tap and die set is pictured in K.

CLEANING TOOLS

Various cleaning tools are pictured in Fig. 21-30. These different brushes and scrapers provide valuable assistance in the thorough cleaning of parts.

Carbon cleaning brushes that may be operated in a 1/4 in. electric drill are shown in Fig. 21-31. A mandrel, D, is also shown. It is used to hold a small wire wheel.

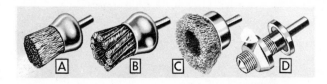

Fig. 21-31. Power carbon cleaning brushes. D, is a mandrel that may be chucked in the electric hand drill to operate a wire wheel.

A wire wheel that may be used on a portable electric drill or mounted on a grinder mandrel is shown in Fig. 21-32.

Fig. 21-32. Power wire wheel.

CAUTION!

ALWAYS WEAR GOGGLES WHEN USING POWER GRINDERS OR WIRE WHEELS.

PLIERS

A mechanic needs a good variety of pliers. Study the pliers shown in Figs. 21-33 through 21-37. Learn the names and uses of each pliers.

Never use pliers in place of a wrench. Use pliers designed for the job. Avoid cutting hardened parts with your lineman,

Fig. 21-33. Needle nose pliers.

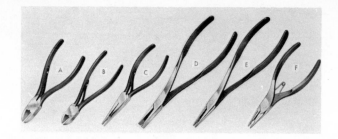

Fig. 21-35. Assorted pliers. A, B—Long nose diagonal cutters. C—Needle nose. D—Duck bill. E—Short needle nose. F—Cutter with long reach cutting jaws.

Fig. 21-34. Assorted pliers.

diagonal or other cutting pliers. Needle nose pliers are delicate; use them carefully.

SCREWDRIVERS

Screwdrivers of various lengths and types are required. The standard, Phillips, Clutch-type and Reed and Prince tips will

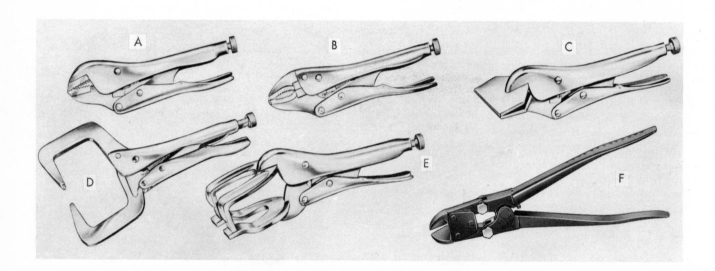

Fig. 21-36. Above. Vise-grip pliers. A, B—Standard vise-grip. C—Vise-grip bending tool. D—Vise-grip C-clamp. E—Vise-grip welding pliers. F—High leverage cutter for wire and small bolts.   Fig. 21-37. Below. Special purpose pliers. A—Hose clamp pliers. B—Lock ring. C—Lock washer. D and E—Brake spring pliers.

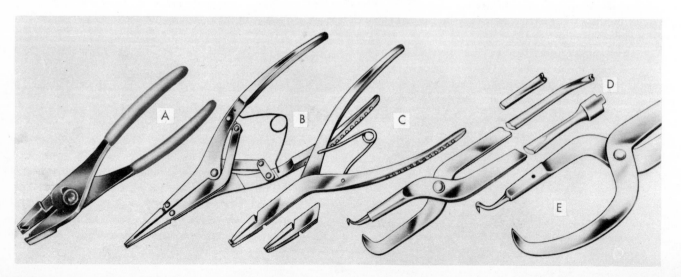

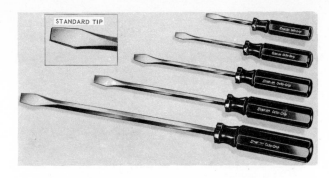

Fig. 21-38. Standard tip screwdrivers.

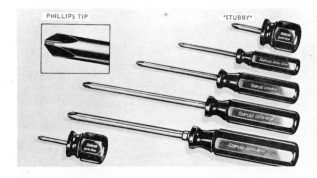

Fig. 21-39. Phillips tip screwdrivers.

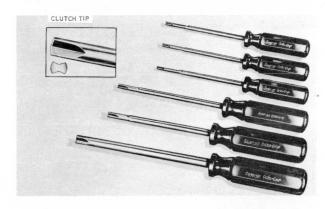

Fig. 21-40. Clutch tip screwdrivers.

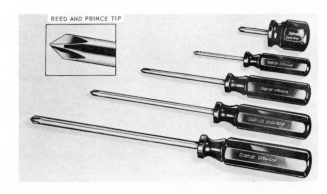

Fig. 21-41. Reed and Prince tip screwdrivers.

cover the various types of jobs ordinarily encountered when servicing cars.

It is wise to have in your tool kit several large and heavy types that will stand some prying and hammering.

The standard tip screwdriver is pictured in Fig. 21-38. It can have a round or square shank.

A number of Phillips tip screwdrivers are shown in Fig. 21-39. Note the stubby screwdriver. It is essential when working in close quarters.

The clutch tip or, as it is sometimes called, the "butterfly," is illustrated in Fig. 21-40.

A set of Reed and Prince tip screwdrivers is shown in Fig. 21-41. At first glance, it looks like a Phillips tip. Close study will show the two tips are of different design.

OTHER TYPES

An assortment of offset screwdrivers makes it easy to remove and drive screws in difficult areas. Fig. 21-42.

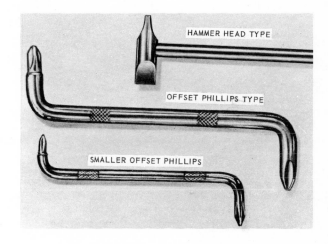

Fig. 21-42. Offset screwdrivers.

The long, thin shank, electrical-type screwdriver is handy for working on small intricate assemblies. Another essential screwdriver is the type designed to hold screws while they are being started.

USING SCREWDRIVERS

As with all tools, use a screwdriver that is designed for the job. Unless the screwdriver is designed for the purpose, do not use a hammer or pry with it.

If the screwdriver tip becomes worn or damaged, grind it slowly to avoid overheating the metal. Attempt to retain the original shape. On the standard tip, do not grind a sharp taper. This will cause the screwdriver to ride up out of the slot.

CAUTION!

WHEN HOLDING SMALL UNITS IN THE HAND, DO NOT SHOVE DOWN ON THE SCREWDRIVER; IT MAY

SLIP AND PIERCE YOUR HAND. IF WORKING ON ELECTRICAL EQUIPMENT, SHUT OFF THE CURRENT. WHERE IT IS IMPOSSIBLE TO SHUT OFF THE CURRENT, USE AN INSULATED (FULL LENGTH) SCREWDRIVER AND KEEP YOUR HANDS FREE OF ANY WIRES.

Fig. 21-44. Open end wrench.

BOX END WRENCHES

The box end wrench is an excellent tool because it grips the nut on all sides. This reduces the chances of slipping with resultant damage to the nut and possibly the hand.

A box wrench is designed with either a 6 or 12-point opening. For general use, the 12-point works well. It allows the wrench to be removed and replaced without moving the handle over such a long swing. For stubborn jobs, damaged nuts, or when there is danger of collapsing the nut, the 6-point will do a better job. It grips the nut across each flat, reducing slippage to a minimum.

The box end wrench is available in either a double offset or a 15 deg. angle offset. Sets in standard size openings, as well as standard lengths, will handle most jobs. Sets are available in midget length for small work in cramped areas. Each end on the box end is a different size.

CAREFUL!

IN USING ALL WRENCHES, PULL, DO NOT PUSH. IF YOU MUST PUSH ON THE WRENCH, PUSH WITH THE PALM OF THE HAND WITH THE FINGERS OUTSPREAD TO AVOID SMASHING THEM IF THE WRENCH SLIPS.

are subject to slipping under a heavy pull. There are places, however, where they must be used. Fig. 21-44 illustrates an open end wrench for adjusting valve tappets; it is long with a thin profile. The standard open end wrench is somewhat shorter and of a huskier design.

The open end wrench has the head set at an angle. When the wrench has traveled as far as allowable, it may be flipped over and placed on the nut in an arc of 30 deg. Each end has a different size opening. Open ends are also available with the heads offset more than 15 deg.

COMBINATION BOX AND OPEN END WRENCH

The combination wrench has a box end head on one end, and an open end on the other. Generally, the open end is offset 15 deg. from the wrench center line. Both ends are of the same size.

The combination wrench makes a very convenient tool since the box end can be used both for breaking loose and final tightening. The open end is for faster fastener removal or installation. Fig. 21-45.

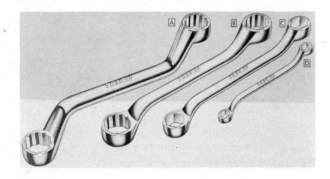

Fig. 21-43. Box end wrenches. A—Double offset, 12-point. B—Standard offset, 12-point. C and D—Standard offset, 6—point. These are short length box end wrenches.

Fig. 21-43 shows several box end wrenches. Detail A has an extra deep offset for additional clearance. B is a typical double offset. C and D are double offset with 6-point openings. Can you see why the 6-point will grasp the nut or cap screw more securely? These are short length box ends. The standard length is of the same design, but longer.

OPEN END WRENCHES

Open end wrenches are handy but not as dependable as box end wrenches. They grasp the nut on only two of its flats, and

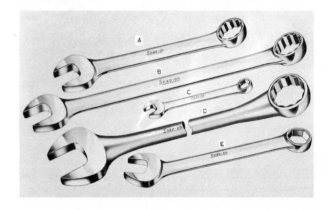

Fig. 21-45. Combination box and open end wrenches. A—Short length. B—Long length. C—Midget length. D—Large nut size. E—Double offset, both ends set off at a 15 deg. angle from center line of wrench. (Snap-on Tools)

SOCKET WRENCHES

A socket wrench is very convenient and, in most instances, is faster than the other wrenches.

Sockets are available with 6-point, 12-point and double-square openings. The 1/4 in., 3/8 in., 1/2 in. and 3/4 in. drives will cover a wide range. The drive size refers to the size of the square hole into which the socket handle fits. The larger the drive, the heavier and bulkier the socket.

The 1/4 in. drive is for small work in difficult areas. The 3/8 in. drive will handle a lot of general work where the torque (tightness) requirements are not too high. The 1/2 in. drive is for all-around service. The 3/4 in. drive is for large work with high torque settings.

Sockets are furnished in standard length; also deep sockets for work such as spark plugs requiring a longer than ordinary reach. Swivel sockets are good for angle work. Fig. 21-46.

Sockets should be kept clean, including the inside, and stored according to size.

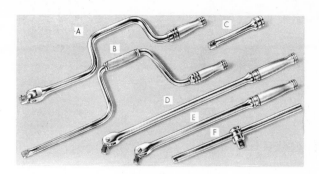

Fig. 21-47. Socket handles. A—Swivel head speed handle. B—Standard speed handle. C—Extension bar. D—Long flex handle. E—Short flex handle. F—Sliding T-handle.

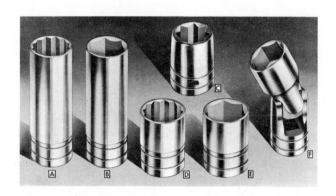

Fig. 21-46. Sockets. A—Deep, 12-point. B—Deep, 6-point. C—Double-square. D—Standard, 12-point. E—Standard, 6-point. F—Swivel.

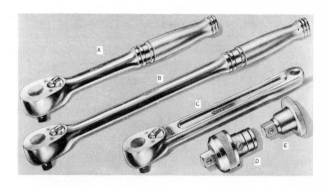

Fig. 21-48. Socket ratchet handles. A—Short ratchet. B—Long ratchet. C—Ratchet. D—Ratchet adapter. E—Ratchet spinner.

SOCKET HANDLES

Various handles are available for sockets. The speed handle is used for fast operation because it can be turned rapidly.

Flex handles in varying lengths allow the socket to be turned with great force, and at odd angles.

Extension bars of different lengths allow the mechanic to lengthen the socket setup to reach difficult areas.

The sliding T-handle varies the handle length. Fig. 21-47.

The ratchet handle is most versatile. It allows the user to either tighten or remove a nut without removing the socket. On the backstroke, the handle "ratchets." Fig. 21-48.

A ratchet that may be used with the speed handle, flex handle, etc., as well as an adopter that allows the mechanic to spin the socket with his fingers until too tight, is also shown in Fig. 21-48.

Fig. 21-49. Socket attachments. A—Weatherhead fitting. B—Handle size adapter. C—Drag link socket. D—Pan screwdriver. E—Crowfoot.

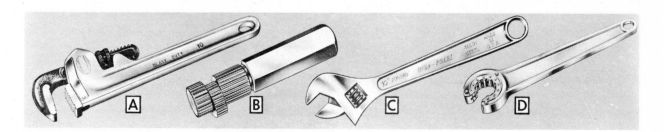

Fig. 21-50. Wrenches. A—Pipe wrench. B—Inside pipe wrench. C—Adjustable wrench. D—Cam-lock wrench.

SOCKET ATTACHMENTS

Many socket attachments are available. Such items as screwdriver heads, drag link sockets, pan screw sockets, crow foot attachment, Allen wrench heads, etc., all combine to make a socket set a fine tool. A few attachments are shown in Fig. 21-49.

PIPE WRENCHES

Pipe wrenches, both inside and outside, are useful for grasping large or irregular objects. They are strong and exert a powerful holding force. A and B, Fig. 21-50.

CRESCENT OR ADJUSTABLE WRENCH

The Crescent wrench is handy in that it can be adjusted for size. However, it tends to slip, so it is a poor wrench to use for most jobs when other tools are available. When used, adjust the jaws firmly. Also, make certain the wrench is placed so the pull on the handle is toward the bottom side. This relieves heavy pressure on the adjustable jaw. See C, Fig. 21-50.

OTHER USEFUL WRENCHES

The flexible head socket wrench, A in Fig. 21-51, as well as the ratcheting box end, B, are useful additions to the mechanic's toolbox.

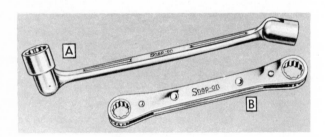

Fig. 21-51. Flex head socket and ratchet box end wrenches. A—Flex head socket. B—Ratchet box.

Cylinder head wrenches are used for the removal of cylinder head nuts and cap screws. Note the various shapes. Fig. 21-52.

A stud remover, obviously, is used to remove various studs. Two types are shown in Fig. 21-53.

Allen and fluted wrenches are often used in tightening or removing setscrews, cap screws, etc. They are strong and provide a good grip. Fig. 21-54.

TORQUE WRENCHES

The torque wrench is a MUST for all mechanics. It will enable you to tighten bolts, nuts, etc., to exact torque (tightness) specifications supplied by the manufacturer.

It is of utmost importance that bolts be pulled up to proper

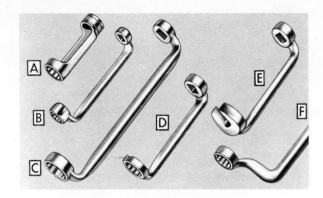

Fig. 21-52. Cylinder head wrenches.

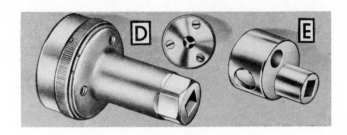

Fig. 21-53. Stud removers. D—Three jaw stud puller. E—Wedge type stud puller. (Snap-on Tools)

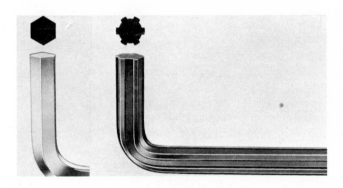

Fig. 21-54. Allen and fluted wrenches. Hex Allen wrench (left) is more popular type.

specifications. Improper and varying torque on units or assemblies will cause distortion. When certain torque is specified (heads, main and rod bearings, etc.), use an accurate torque wrench. Fig. 21-55.

Torque wrenches are available in both foot-pound and inch-pound types. The inch-pound wrench is used for delicate, low torque applications.

PULLERS

Auto mechanics will find use for a wide variety of pullers. They are used for pulling gears, bearings, hubs, etc. An assortment of pullers is shown in Fig. 21-56.

Another excellent puller is the slide hammer type. The

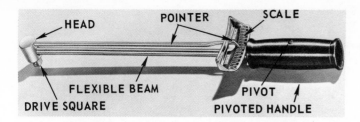

Fig. 21-55. One popular design torque wrench. Torque wrenches insure correct fastener tension.

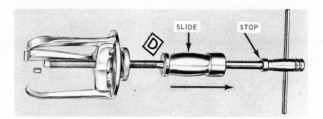

Fig. 21-57. Slide hammer puller.

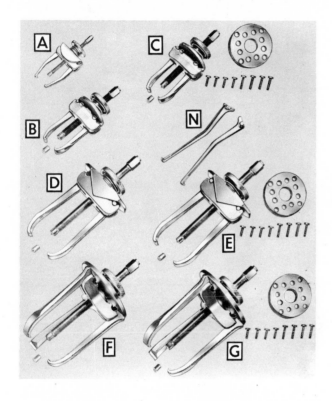

Fig. 21-56. Pullers. Pullers are available in various styles and sizes.

puller jaws grasp the work, and the weighted slide is banged against the stop. This type of puller is fast and efficient on many types of jobs. Fig. 21-57.

SPECIAL PURPOSE TOOLS

Other tools shown in accompanying illustrations are considered special purpose tools. Fig. 21-58 pictures brake service tools. Fig. 21-59 features engine service tools. Some tools for regulator and distributor work are shown in Fig. 21-60.

Useful tire and wheel tools are illustrated in Fig. 21-61. Body tools are shown in Fig. 21-62. Some very handy tools that enable the mechanic to see in difficult areas, as well as remove loose parts where the hand will not reach, are pictured in Fig. 21-63.

A tubing cutter and a tube flaring tool are shown in Fig. 21-64. All tube fittings should be removed and replaced with special tubing flare nut wrenches. They are designed to provide a maximum amount of contact. Slide the open point over the tube and then down on the nut. Fig. 21-65. Miscellaneous hand tools are illustrated in Fig. 21-66.

SOLDERING EQUIPMENT

For soldering all automotive wiring, use ROSIN CORE solder only. Acid core solder leaves a residue that will cause

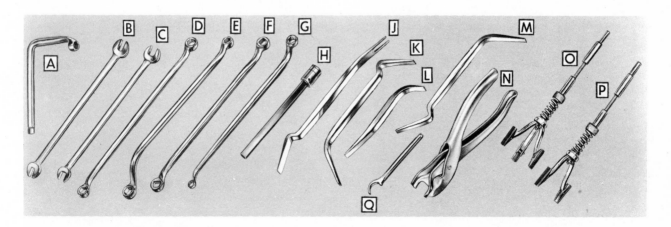

Fig. 21-58. Brake service specialty tools. A—Right angle bleeder wrench. B—Bendix eccentric cam wrench. C—Ford eccentric cam wrench. D, E—Offset-head bleeder wrenches. F—Taunus brake wrench. G—Compact car bleeder wrench. H—Datsun brake adjusting tool. J—Bendix brake adjusting tool. K—Pontiac brake adjusting tool. L—Chevrolet brake adjusting tool. M—Star wheel adjusting tool. N—C-washer pliers. O—Three arm hone for brake cylinders. P—Two arm hone for brake cylinders. Q—Parking brake tool.

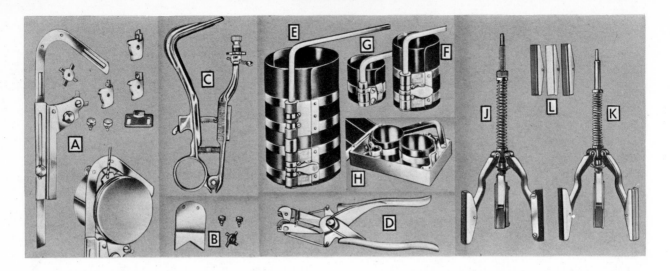

Fig. 21-59. Piston and cylinder service tools. A—Ring groove cleaner and cutter. B—Ring groove cleaner attachments. C—Ring groove cleaner. D—Piston ring spreader. E, F, G, H—Ring compressors. J, K—Cylinder wall deglazers. L—Deglazer stones.   (Snap-on Tools)

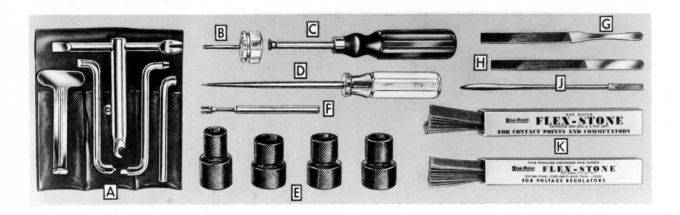

Fig. 21-60. Regulator and distributor tools. A—Voltage regulator tools. B—Hex tool for distributor point adjustment. C—Distributor cleaning brush. D—Neon spark tester. E—Distributor point rubbing block surfacers. F—Distributor spring tension tool. G—Distributor point file. H—Regulator point file. J—Point riffler file. K—Flex stones for cleaning points.   (Snap-on Tools)

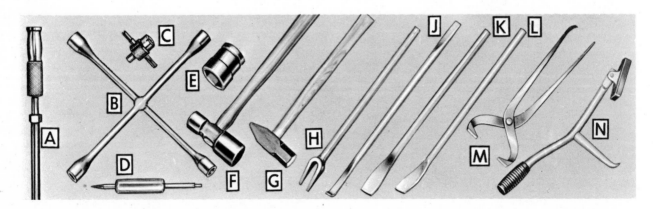

Fig. 21-61. Tire and wheel tools. A—Tire bead remover. B—Wheel lug wrench. C, D—Valve core tools. E—Impact socket. F—Rubber-tipped steel hammer. G—Cross peen hammer. H—Tie rod separator. J—Split rim tool. K, L—Tire removing tools. M—Grease cap tool. N—Hub cap tool.

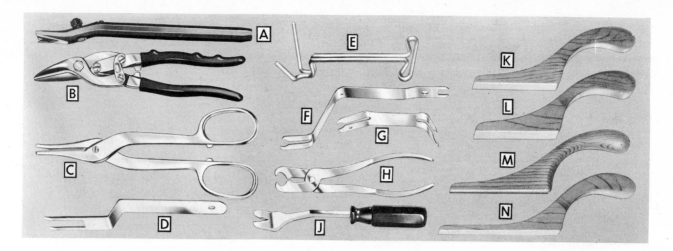

Fig. 21-62. Body tools. A—Body panel cutter. B—Metal cutting shears. C—Tin snips. D, E, F, G—Door handle tools. H—Door handle pliers. J—Door panel remover. K, L, M, N—Body solder paddles.

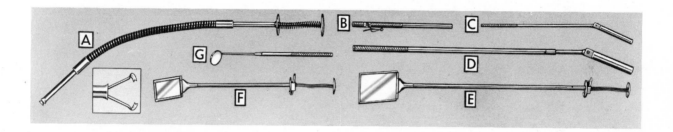

Fig. 21-63. Probing tools. Various mirrors and pickup tools for probing into areas that are otherwise inaccessible.

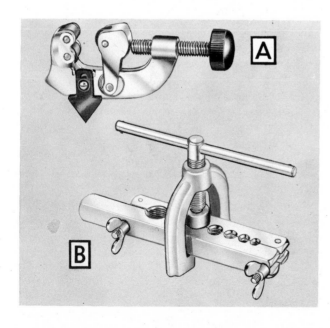

Fig. 21-64. Tubing tools. A—Tubing cutter. B—Tubing flaring tool.

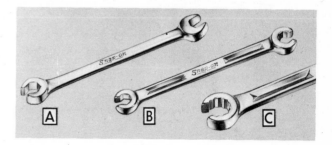

Fig. 21-65. Tubing wrenches. A, B—Six-point. C—Twelve-point.

the solder bonds properly. Cool and wipe clean. Tape soldered wiring with plastic tape. Irons are shown in Fig. 21-67.

MEASURING TOOLS

A mechanic is often called on to make PRECISION measurements. To do this, he must have, and understand the use of: outside micrometer, inside micrometer, dial gages, calipers, dividers, depth gages, combination square.

MICROMETER

Both inside and outside micrometers are used to make readings accurate to a fraction of a thousandth of an inch. Fig.

corrosion in electrical units, and should never be used for wiring. Acid core is satisfactory for radiator work.

All parts to be soldered should be clean and bright. Heat the parts with the soldering iron and add solder. Make certain

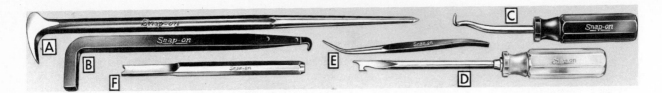

Fig. 21-66. Other mechanics hand tools. A—Pry bar. B—Cotter pin remover. C, D, E—Cotter pin removers. F—Bushing cutter.

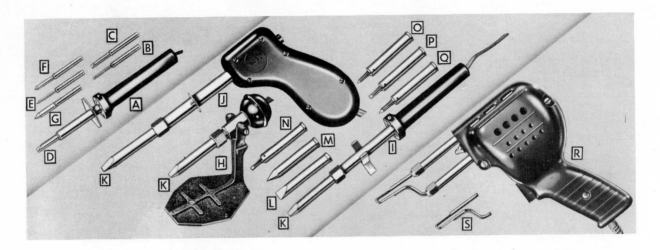

Fig. 21-67. Soldering irons. Several types of irons and tips, all electric.

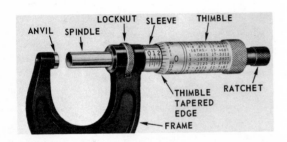

Fig. 21-68. Typical outside micrometer. Learn names of parts. (Starrett)

21-68. These precision tools must be kept immaculate. When not in use, keep them wiped down with an oily rag. When stored, do not allow the spindle end to contact the anvil. Keep them in protective containers.

In use, an outside micrometer is closed around the work until the micrometer may be pulled across the highest point of the work with a slight drag. Never force the micrometer. Both micrometer and work must be clean. Practice on something of a known diameter until you develop the proper "feel."

READING THE MICROMETER

Micrometers are made so that every turn of the thimble will move the spindle .025 in. You will notice that the sleeve is marked with a series of lines. Each of these lines represents .025. Every fourth one of these .025 markings is marked 1,2,3,4,5,6,7,8,9. These numbers on the sleeve indicate .100,

.200, .300, etc., (one hundred thousandths, etc.). The micrometer sleeve then is marked out for one inch in .025 (twenty-five thousandths inch) markings. They will read from .000 to 1.000.

The tapered end of the thimble has twenty-five lines marked around it. They will read 0,5,10,15 and 20. In that one turn of the thimble moves the thimble edge exactly .025, or one mark on the sleeve, the distance between marks is determined by reading the thimble line that is even with the long line drawn the length of the sleeve markings.

Look at the markings on the micrometer (mike) in Fig. 21-69. How many numbers are visible on the sleeve? There are three. This 3 indicates that the mike is open at least .300 (three hundred thousandths). You can see that the thimble edge is actually past the 3 but not to the 4. By a careful study you will see that the thimble edge has moved exactly two additional marks past the 3. This means that the edge is lined up two marks past the 3. As each mark represents .025, it is obvious that the edge is actually stopped at .300 plus .050 or

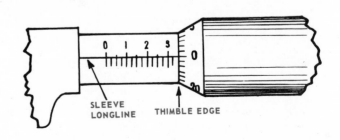

Fig. 21-69. Micrometer reads — 0.350.

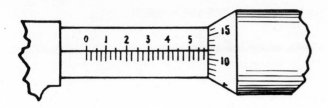

Fig. 21-70. Micrometer reads — 0.587.

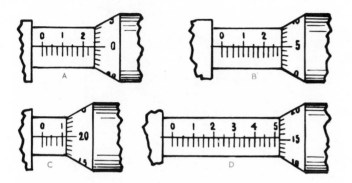

Fig. 21-71. Reading the "mike." A—0.250. B—0.280.
C—0.120. D—0.515-1/2, or 0.5155.

.350 (three hundred and fifty thousandths of an inch). In that the thimble edge 0 marking is in line with the line through the sleeve markings, the micrometer is set exactly on .350. The reading then, if this were a one inch mike (reads from 0-1 inch), would be .350.

In Fig. 21-70, the micrometer has been opened to a wider measurement. You will see that the thimble edge is no longer exactly on a sleeve marking but is somewhere in between.

How many numbers are visible on the sleeve? There are 5 or .500 (five hundred thousandths). The thimble edge has moved three marks or .075 past the .500 mark. This makes a total of .575. The thimble edge has moved past the third mark. In that the fourth mark is not visible, we know it is somewhere between the third and fourth mark.

By examining the thimble edge marks, you will see that the twelfth mark is aligned with the sleeve long line. This means that the thimble edge has moved twelve spindle marks past the third sleeve mark. In that each thimble mark is worth .001 (one thousandth of an inch), the thimble has actually moved .012 (twelve thousandths of an inch) past the third sleeve mark. Your reading then would be .500 (largest sleeve number visible) PLUS .075 (three sleeve marks past sleeve number) PLUS .012 (twelve thimble marks past the third sleeve mark), making a total reading of 0.587 (five hundred and eighty-seven thousandths of an inch).

Study the readings shown in Fig. 21-71. Compare your answers with those shown. Make your reading in FOUR steps.

1. Read the largest sleeve number that is visible — each one is worth .100.
2. Count the number of full sleeve marks past this number. Each one is worth .025.
3. Count the number of thimble marks past this last sleeve number. Each one is worth .001. If the thimble marks

are not quite in line with the sleeve long line, estimate the fraction of a mark.
4. Add the readings in steps 1, 2 and 3.

The total is the correct micrometer reading.

Micrometers are available in one inch, two inch, three inch, etc., sizes. The range of the mike would actually be zero to one inch, one inch to two inches, two inches to three inches, etc. For example, if you are using a two inch to three inch mike, and your total reading is .094, the actual diameter of the object being measured would be 2.094. If you were using a zero to one inch mike, the diameter would be 0.094.

The inside micrometer (reads the same way) is illustrated in Fig. 21-72. It is being used to measure the diameter of a cylinder.

INSIDE AND OUTSIDE CALIPERS

Calipers are useful tools for rough measurements. Fig. 21-73 illustrates a pair of outside calipers being used to measure the diameter of a shaft.

Fig. 21-74 shows how the inside calipers can be used to measure the size of a hole.

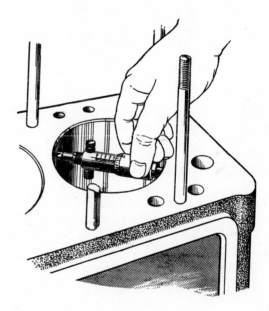

Fig. 21-72. Using an inside micrometer to check cylinder bore size.
(Austin—Healey)

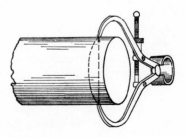

Fig. 21-73. Measuring diameter of shaft, using an outside caliper.
(South Bend Lathe)

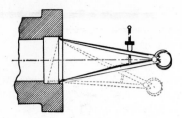

Fig. 21-74. Using inside caliper to measure diameter of counterbore.

To determine the reading, hold the calipers on an accurate steel rule.

DIVIDERS

Dividers are made somewhat like calipers, but have straight shanks and pointed ends. They are handy for marking circles, surface measurements, etc. Fig. 21-75 shows dividers being used to find the center of a steel shaft.

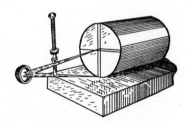

Fig. 21-75. Dividers being used to find center of shaft. (South Bend Lathe)

DIAL GAGE OR INDICATOR

A dial indicator is used to read movement in thousandths of an inch. Common uses are checking end play in shafts, backlash in gear teeth, valve lift, shaft runout, taper in cylinders, etc.

Fig. 21-76 shows how a dial indicator is used to check a ring gear for runout. The dial button is placed against the work until the needle revolves about one-half turn. The dial face is set to zero, and the gear revolved in the V-block stand. Any runout (wobble) of the ring gear will cause the dial needle to move. This movement is read in one thousandth of an inch.

FEELER GAGES

Feeler gages are thin strips of specially hardened and ground steel. The thickness of each strip is marked in thousandths of an inch. They are used to check clearances between two parts: valve clearance, piston ring end gap, etc. Fig. 21-77.

Space limitations will not permit covering the multitude of tools available for the mechanic. The student would be wise to procure catalogs from tool manufacturers. Study the catalogs to learn tool identification and practice.

REMEMBER:

GOOD TOOLS make your work easier, faster and more efficient. Use them properly and give them good care. If treated with intelligence, they will give you many years of satisfying use.

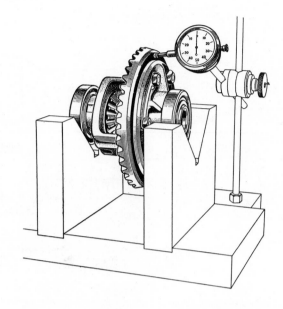

Fig. 21-76. Checking ring gear runout with a dial indicator. (Austin—Healey)

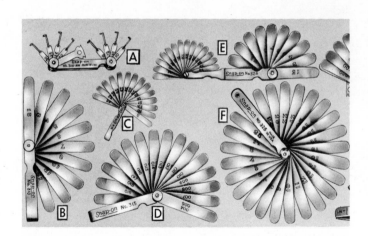

Fig. 21-77. Feeler gages. A—Wire gap gages. B, C, D, E, F—Various feeler gages.

REVIEW QUESTIONS — CHAPTER 21

1. What material is commonly used in the construction of quality tools?
2. Name three things to look for when buying tools.
3. Give a few general rules for the use of tools.
4. Name three types of hammers useful to the mechanic.
5. Name several types of chisels the mechanic should have.

6. Of what use are punches?
7. What is the difference between starting, drift and aligning punches?
8. A file with two sets of cutting edges at angles to each other is referred to as a _____ file.
9. Give some safety rules for the use of the hammer, chisels and files.
10. Give two methods that are helpful in keeping the file clean.
11. Drills made of carbon steel will last longer than those made of high speed steel. True or False?
12. For general drilling in steel, what lip angles would you recommend?
13. Describe the procedure to use when sharpening drills.
14. Give some safety precautions to observe when using a portable hand drill.
15. Hacksaw teeth should point away from the handle. True or False?
16. Reamers are used for drilling large holes. True or False?
17. A husky vise will also be useful as an anvil. True or False?
18. If you wish to cut internal threads in a hole, you would use a tap. True or False?
19. If you wish to cut threads on a bolt, you would use a tap. True or False?
20. Describe some useful cleaning tools.
21. When grinding, always wear _____.
22. Name five types of pliers.
23. List five types of screwdrivers.
24. A box end wrench is one of the best to use. True or False?
25. Open end wrenches grip a nut on all but one flat. True or False?
26. Combination box end and open end wrenches have a different size opening on each end. True or False?
27. A spark plug would require a _____ socket.
28. A quarter inch drive set of sockets would be handy for removing head bolts. True or False?
29. Name four types of socket handles.
30. A pipe wrench may be used for gripping a crankshaft journal. True or False?
31. For tightening head bolts, connecting rod bolts, etc., always use a _____ wrench.
32. Describe an Allen wrench.
33. Gears are usually removed by using a _____.
34. See how many other hand tools you can name.
35. Describe how to read a micrometer.
36. Give some general rules for the use and care of tools.

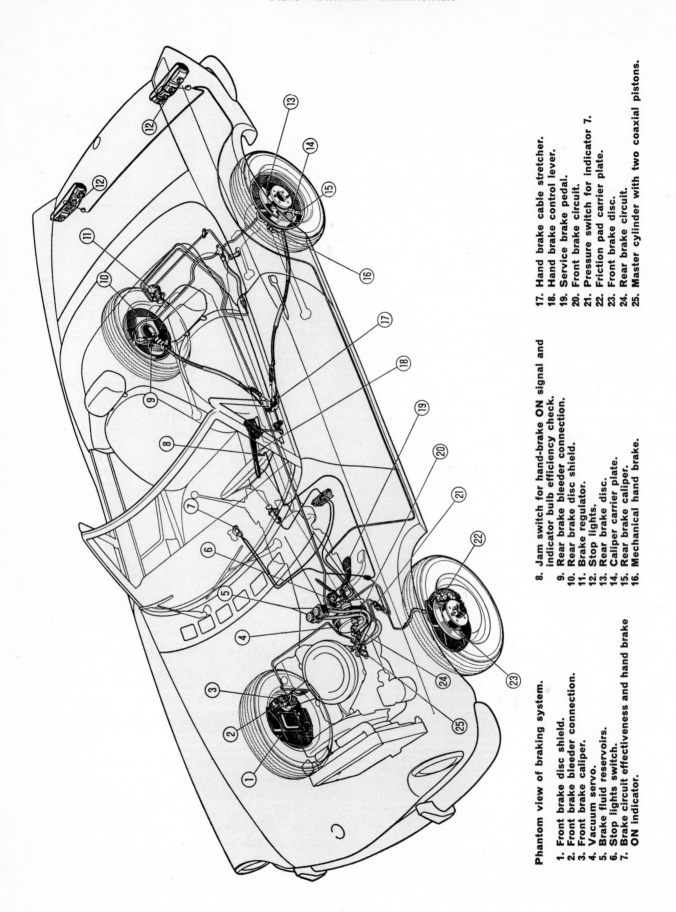

Phantom view of braking system.

1. Front brake disc shield.
2. Front brake bleeder connection.
3. Front brake caliper.
4. Vacuum servo.
5. Brake fluid reservoirs.
6. Stop lights switch.
7. Brake circuit effectiveness and hand brake ON indicator.
8. Jam switch for hand-brake ON signal and indicator bulb efficiency check.
9. Rear brake bleeder connection.
10. Rear brake disc shield.
11. Brake regulator.
12. Stop lights.
13. Rear brake disc.
14. Caliper carrier plate.
15. Rear brake caliper.
16. Mechanical hand brake.
17. Hand brake cable stretcher.
18. Hand brake control lever.
19. Service brake pedal.
20. Front brake circuit.
21. Pressure switch for indicator 7.
22. Friction pad carrier plate.
23. Front brake disc.
24. Rear brake circuit.
25. Master cylinder with two coaxial pistons.

Many career opportunities exist in the automotive field. A brake specialist works with the entire brake system as shown above. (Fiat)

Chapter 22

CAREER OPPORTUNITIES IN AUTOMOTIVE FIELDS

A VAST FIELD

Our country has been aptly referred to as a "nation on wheels." In fact, the very essence of our economic life and growth is dependent, in a great part, upon the continued improvement and advancement of the automobile.

The automotive field has become so large that today it employs about one out of every seven workers in our nation.

TYPES OF CAREERS AVAILABLE

This chapter will concern itself with careers directly connected with automotive maintenance and repair as found in the modern garage or service center.

CLEANING

In cleaning work, you will be called upon to steam clean the engine and underbody portion of the car. You will wash, and often wax, the exterior finish. Vacuuming, window and upholstery cleaning and installation of seat covers will undoubtedly be delegated to you. Part of your day will be spent in the delivery and moving of cars, plus an occasional assist to the mechanics, parts men or other personnel.

Your work in the cleaning department will provide you with an opportunity to demonstrate your ability as a hard-working and responsible man. There will be plenty of chances to observe the work involved in other areas and, coupled with additional study, you can prepare yourself for advancement. When ready, many men move from the cleaning to the lubrication department.

LUBRICATION

The lubrication work area involves lubricating the working parts of the car, checking oil levels in the engine, transmission, differential, etc. It includes inspection of battery, brake master cylinder, radiator and a general inspection of parts critical to safe operation, such as the steering system, springs, shock absorbers, brake lines, parking brake linkage, tires, etc.

Work in this area will provide you with an opportunity to

learn a great deal about the various systems on the car. It could pave the way to advancement toward more complicated mechanical service.

MECHANIC, LIGHT REPAIR

Part of the work as a light repair mechanic consists of checking out new cars before delivery. You will go over the entire car to make certain that all systems are functioning and in proper adjustment.

You will probably be responsible for making the normal checkup offered the car buyer after a certain mileage has been reached. At this time you would normally go over the entire car, torque the head bolts, check the timing and valve clearance, inspect the brakes, steering, etc. You would discuss the car's performance with the owner and check up on any complaints he may have regarding its operation.

Training in this job, even though it generally involves minor repair and adjustment, is invaluable if you are interested in the more complicated heavy repair work or, for that matter, in some specialty such as tune-up, brakes, etc.

MECHANIC, HEAVY REPAIR

In the heavy repair department, you will be called on to service, dismantle, check, repair, reassemble and check engines and units such as transmissions and differentials. This job will require a great deal of study, practice and experience.

At the start, you probably will work under a qualified and experienced man in this department. Your success in this area will largely depend upon your aptitude (natural ability), interest, ambition and a sincere desire to learn.

Once having mastered the work involved, you will be a valuable man in any garage and will have placed yourself in a good position for advancement.

SPECIALIZATION

Today the automobile is steadily incorporating new devices and special features. It is becoming more and more difficult for any one man to attempt the mastery of the entire car.

A mechanic using tune-up equipment.   (Sun Electric Corp.)

When the volume of work will permit it, specialization in one certain area will provide customers with faster and more efficient service in that the repairman handling the job will be a specialist in that particular area.

The specialist will have advanced and concentrated training and experience in his speciality. As he works full time in the one area, it follows that he will become highly proficient.

TUNE—UP SPECIALIST

The tune-up specialist will handle jobs involving ignition timing and ignition system service, carburetor cleaning, adjusting and checking. Generator and starter work, as well as electrical wiring, headlights, etc., will be in his field.

Tune-up work requires a high level of competency. It is a good paying job and will provide an excellent chance for further advancement.

Mechanic using torque wrench to provide correct tension.
(Sturtevant Co.)

DIAGNOSTIC SPECIALIST

A relatively new type of service center, specializing in diagnostic work, is becoming very popular. Each car moves through various inspection areas. Each area is manned by diagnostic specialists who check a specific system and indicate their findings on a check list. When the car has cleared all areas or stations, the customer is given a summary of recommended services and usually an estimated cost.

The idea behind the diagnostic center is to offer a fast, complete, accurate and unbiased inspection service in which the customer is made aware of any repair or maintenance work that his car may require. The center may or may not have repair facilities. In any event, the customer is free to take his car to a shop of his own choice.

Mechanic making notes relative to brake condition.
(Bendix Corp.)

Work in this type of shop requires a high degree of skill within a given specialty. The mechanic may be assigned to a single station or may be asked to work in more than one.

PARTS SPECIALIST

Most garages and service centers of any size find it essential to maintain a rather extensive stock of parts.

The ordering, cataloging, storage and distribution of these parts requires the services of a parts specialist. The parts man must possess a thorough knowledge of all aspects regarding this operation. He must maintain and use numerous parts catalogs. He will be required to advise mechanics regarding parts modification, parts interchangeability, etc. The parts specialist is a very important part of the overall garage operation.

Checking wheel and drum assembly for static and dynamic balance.
(Bear Mfg. Co.)

BRAKE SPECIALIST

This work involves such work as drum turning, lining installation and truing, shoe adjustment, master and wheel cylinder repair, bleeding, line replacement. Special training in brake trouble diagnosis and power brake units is required.

TRANSMISSION SPECIALIST

This man will handle all transmission work, both standard and automatic. He will possess special skills in testing, diagnosis, disassembly, checking, repair and reassembly of

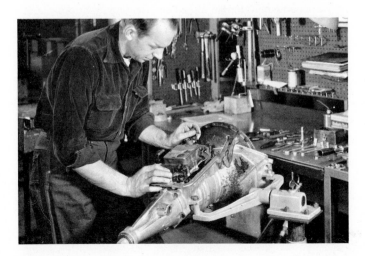

Transmission rebuilding specialist replacing transmission control valve body.   (Spring Div., Borg-Warner Corp.)

transmissions. He must have a thorough background of practice as well as special training in this field.

FRONT END SPECIALIST

Wheel balancing and alignment, as well as work on the steering gear box, steering linkage and spindles, falls to the front end specialist. He will also check and repair the springs and control arms. In short, it is his duty to see that the car steers and handles easily and safely, and that the tires run smoothly and wear properly.

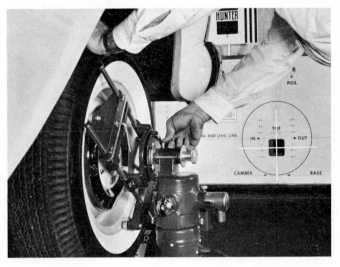

Front end specialist checking toe-in. This particular front end alignment setup uses both a flashing light and a buzzer to indicate when predetermined toe setting is as specified.   (Hunter Engineering Co.)

ELECTRICAL SPECIALIST

The modern car makes wide use of electrical units. The electrical specialist will handle work on such items as radios, electric seats, window-operating devices, convertible top power units, instruments, generators, starters, etc. This job requires an extensive knowledge of electricity as well as the repair and adjustment of all electrical units.

AIR CONDITIONING SPECIALIST

This job involves checking, adjusting and repairing automotive air conditioning systems. Air conditioning is rapidly becoming more popular and will produce many jobs for men with special training in this area.

BODY AND FENDER SPECIALIST

Body and fender work, although not thought of in the general sense of mechanical repair, is an important part of garage work. This is a specialty that requires a great deal of training. As a result, it pays quite well.

The body and fender man will repair damage to the car's metal structure as well as to the upholstering and inside trim.

He will install glass, repair locks, handles, etc. This job requires proficiency in cutting, welding, metal bumping, filling, priming, painting, etc.

Within the field of body and fender work, painting is rapidly becoming a specialty of its own.

SUB—SPECIALISTS

In some of the larger garages, men specialize within a speciality. In a specialized area, the work can become involved enough to provide full time work on specific units within the specialty.

SUPERVISORY POSITIONS

Most men, upon entering a particular field, look forward to advancement in pay and position. This is as it should be, both in the garage and service station fields, as in many other fields.

Advancement is dependent on your knowledge and skill, as well as a demonstration of your ability to cope with the problems connected with the position.

SHOP FOREMAN

The shop foreman is in charge of the mechanics in the repair department. He is held directly responsible for the work turned out by his mechanics.

The foreman must be highly competent mechanic, familiar with and able to perform, all jobs that enter his shop.

In larger garages, the foreman's time will be spent supervising his mechanics. He will check on their work, give suggestions, and in general, see that the shop runs smoothly and efficiently. In a small garage, the foreman may spend part of his day doing repair work himself, with the remainder of the time devoted to supervision.

SERVICE MANAGER

The service manager holds a very responsible position. He is in charge of the overall garage service operation. He must see that customers get prompt, efficient, and fair-priced service. It is his responsibility to see that the customer is pleased, that the mechanics and others in the service departments are satisfied and are doing good work.

The service manager's job usually entails handling of employee training programs. He cooperates closely with factory representatives to see that the latest and best service techniques are employed.

He must have an insight into the job itself. He must possess leadership ability, a good personality, training, knowledge and the ability to work with others. Being a service manager is not an easy job, but certainly a worthwhile ambition.

OVERALL JOB PICTURE

Other jobs found in the modern garage, in addition to those discussed, are shown on the dealer organization chart.

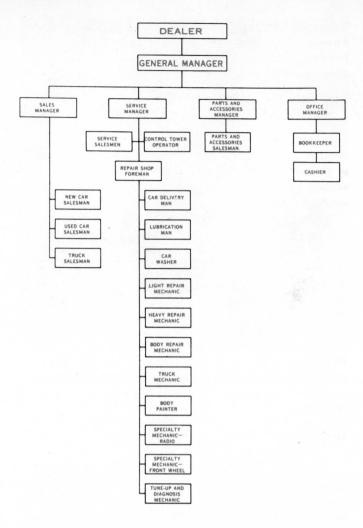

This chart will give you an idea of how a modern automobile dealership is organized.

SMALL GARAGE MECHANIC

The mechanic in the small garage will often be called upon to meet customers, diagnose problems and prepare cost estimates and billings. He must have training in all aspects of mechanics embraced by the work done in his garage.

WORKING CONDITIONS

The garage of today is a far cry from the crowded, dark, cold and poorly ventilated garage of yesterday. The modern garage is well lighted, roomy and provides a pleasing place in which to work.

Many modern garages feature lunchrooms, showers, lockers, etc., for their employees. The addition of heavy power equipment has removed much of the sheer labor involved, and also provides easier and more comfortable access to the various parts of the car.

Most mechanics work inside, but when a job incorporates roadside repair, the mechanic may be faced with varying weather conditions.

Many mechanics, especially those in urban areas and in large garages, belong to one of several unions. Among the unions are the International Association of Machinists; the International Brotherhood of Teamsters, Chauffeurs, Warehousemen and Helpers of America; the United Automobile, Aircraft and Agricultural Implement Workers of America.

SALARY

It is difficult to say what salary you can expect. Much depends on type of work performed, location (geographical) of the job, employer, prevailing business conditions, and job supply and demand. All of these exert a definite influence on mechanics' salaries.

Suffice it to say that most mechanics earn a good living and that their salaries are in line with other types of technical jobs.

Many auto mechanics receive a certain percentage of the customer labor charge. As the rates for various jobs are fixed, the skilled mechanic will be able to do more jobs in one day than the unskilled mechanic, and as a result, will earn more money.

When starting to learn the trade under an apprentice program, the student or apprentice mechanic will earn around 55 percent of the journeyman mechanic's income. The rate is gradually increased until the apprentice will earn up to 90 percent of the journeyman rate in his last six months of apprenticeship.

HOURS

Working hours per week may vary from 40 to 48 hours. In some instance, mechanics exceed these hours. Overtime salary is generally paid for all hours above those set by the union. Holidays and vacations with pay are becoming increasingly available to mechanics.

AVAILABILITY OF EMPLOYMENT

The increasing use of cars, plus the addition of more complex items on the car, has provided a lucrative field for qualified mechanics. If you really learn the trade and are a hardworking, conscientious person, you should be able to readily find employment.

Many small, self-owned shops have been started by mechanics. Many service stations offer light-duty repair, and employment possibilities.

On the average, most mechanics work in shops employing one to five men. However, some of the larger garages employ over one hundred men.

About a third of all mechanics work in new and used car dealer garages. Another third works in repair shops that specialize in a multitude of automotive repairs such as body and fender work, tune-up, brake, front end, radiator work, etc. Other mechanics work for large fleet owners such as the utility companies, dairies, produce companies and federal, state and county and city organizations.

If you are sincerely interested in cars, enjoy working on and studying about cars and are in normal health, the chances are good that you would like this work and do well in it.

HOW DOES A PERSON BECOME AN AUTO MECHANIC?

A person develops into an auto mechanic by study, observation, instruction and experience.

High Schools, Trade Schools, The Armed Forces and Apprentice Programs all offer excellent opportunities to acquire the training necessary to gain employment in this field.

Several sound reasons for finishing high school, as well as recommended courses, will be discussed first.

Remember: Many high schools offer fine training programs in Auto Mechanics. Take advantage of them.

STAY IN HIGH SCHOOL

Above all else, GRADUATE FROM HIGH SCHOOL. High school dropouts are considered poor risks by employers and for good reasons.

Facing facts, we know that the vast majority of students that drop out of school do so for one of several reasons. Some of these reasons for dropping out are:

1. The student lacks the intelligence to compete in school.
2. The student is just plain lazy.
3. The student is bored and feels that school will do nothing for him.
4. The student wants to start earning money so that he may feel grown up and independent.
5. The student cannot get along with others and rebels against authority.

The actual reason for dropping out may be any one or a combination of those mentioned. At any rate, the prospective employer will usually have high school graduates applying for the job. If high school graduates are available, why should he take a chance on a dropout who did not have what it takes to finish high school.

We know that there are some very fine, intelligent and hard-working students that drop out of school, but the employer is not equipped, nor even inclined, to ferret out these people. Actual practice has shown that on the average, it is much safer to hire a high school graduate in preference to a dropout.

National figures on salaries and employment show that the average dropout is unemployed a great deal of the time. When he is able to find work, it is usually in some low-pay job. The world today is moving forward at an ever-increasing speed, and there will be little chance for the unskilled and uneducated dropout. STAY IN SCHOOL!

WHAT TO TAKE WHILE IN SCHOOL

Contrary to what many young men in school think, English, Social Studies, Math and Science are extremely valuable to anyone planning to enter one of the trades.

ENGLISH

The ability to express yourself, both orally and in writing, is important, since you will have many opportunities to meet the public and also to make written reports. The impression you make through your speaking and writing will have a definite bearing on your worth to any employer as well as on your chances for advancement. WORK in your English courses — it will pay off.

MATH

You often hear the statement, "What good does math do — I'm going to be a mechanic?" It does a great deal of good. Mechanics make many precision measurements, customer billings, cost estimates, etc. To fully understand automotive theory requires a good grasp of mathematics. Do not forget that there are many times outside the shop when you will need a reasonable skill in math. Business Math, Algebra, Geometry and Trigonometry will all be useful tools for your mental toolbox.

SCIENCE

Chemistry and physics will be invaluable aids in understanding the construction and operation of the automobile. The car embodies the use of chemistry and physics in hundreds of applications.

SOCIAL STUDIES

A thorough knowledge of our world, past and present, its people, its problems, its aims, etc., is essential if you hope to make an intelligent contribution to society in general. It will set you apart as a person "in the know" and will allow you to discuss countless things in an intelligent fashion. Your contributions to your work, your country, and your chances of advancement will be greater with study in this area.

English and Social Studies, as well as some Math and Science, are often required courses. Make certain you take and get all you can from them!

OTHER IMPORTANT COURSES

As elective courses, take such things as Mechanical Drawing, Blueprint Reading, Art, Electricity, Machine Shop, Welding, Fundamentals of Auto Mechanics (theory), Auto Mechanics (shop experience), and if offered by your school, On the Job Training in this field.

USE YOUR COUNSELOR AND INSTRUCTORS

This is a big order. To make certain you are on the right track and are getting what you need, see your school counselor. Most high schools, through their guidance office, will provide you with valuable assistance in determining your assets and liabilities, your vocation, the courses necessary, as well as providing you with information on furthering your training after graduation from high school. Talk with your shop instructors, as they can be of great help.

AFTER GRADUATION?

In the event that you desire additional training after graduation, the following areas will prove valuable.

TRADE SCHOOL

There are numerous trade schools across the nation. Many offer excellent courses aimed at teaching the auto mechanics trade. Before selecting any certain school, it is well to run a careful check on the school itself. You should determine several things:

1. Is the school a reputable institution?
2. Does it offer the courses you need?
3. Does it have top quality instructors and plenty of late model equipment?
4. Does it have a job placement agency?
5. How far along the road to complete training will it take you?
6. What are the costs?
7. Do they offer aid in securing a part-time job if you must work to help pay expenses?

To help you check on these items, contact the school itself, the Better Business Bureau in the city in which the school is located, your High School Guidance Office, the Labor Union to which the mechanics in the area of the school belong, and the Vocational Department of the State Department of Education.

A thorough investigation of the school of your choice will be well worth your time. DO NOT FAIL TO MAKE IT!

ARMED FORCES

If you have decided to join one of the Armed Forces, it may be possible for you to receive training in the field of auto mechanics. The Services have top quality schools. If you have the opportunity to receive the training they offer, plus the experience your enlistment may give you, you will be well on your way to becoming a good mechanic.

If you desire to enter apprenticeship training upon your discharge, most apprentice programs give credit for previous experience and training. How much credit is given will depend on the individual employer or the Joint Apprenticeship Committee involved.

For information regarding training in the Armed Forces, see your local recruiting agent and your High School Guidance Office.

APPRENTICE PROGRAM

Many men in the field of auto mechanics learn the trade by enrolling in an apprentice program.

The apprentice program is designed to teach a specific trade. This is done by placing a man on the job and seeing that he is exposed to, and instructed in, all facets of the trade. The program will generally include some off the job training in physics, math and other related subjects. These classes will be either in a vocational school, or as is sometimes done, taught at the place of business. Correspondence courses are sometimes required.

As the apprentice gradually develops skills, he will be exposed to the more technical aspects of the job. In a good apprentice program, the apprentice will learn the trade thoroughly and, upon completion of his apprenticeship, will be a highly skilled and valued employee.

HOW TO GET IN THE APPRENTICE PROGRAM

If you are interested in becoming an apprentice, you may contact an employer in the trade you wish to enter. He may be interested in the program and willing to hire you. The local labor union representing the trade of your choice may be able to help you. Many cities have local joint apprenticeship committees. You may also contact your local state employment service office.

QUALIFICATIONS ESSENTIAL FOR APPRENTICESHIP

You should have a sincere interest in the trade, be willing to study, observe, take instruction and WORK! You should have average or above ability to work with your hands in addition to your head. Although high school dropouts are sometimes placed, many programs prefer high school graduates.

Learning a trade will equip you with valuable skills and place you in demand. The national apprenticeship program is a fine way in which to master your trade.

ADDITIONAL INFORMATION

If you are interested in the field of auto mechanics, and think you would like to learn the trade, ask your high school guidance office (they will be glad to help you even if you have dropped out or graduated) for all the information they have available. In addition, you write to the following sources for very informative booklets that can aid you in your decision:

U. S. Department of Labor
Bureau of Apprenticeship and Training
Washington, D. C. 20006

Motor Vehicle Manufacturers Association
320 New Center Building
Detroit, Michigan 48202

National Automobile Dealers Association
2000 K Street, N.W.
Washington, D.C. 20006

The following car manufacturers, as well as others, may be able to provide you with brochures, booklets, etc. All offer training programs and are glad to explain the program and how to enroll. If you desire booklets, phamplets, etc., only, address the Manager, Educational Publications Department, at the company address. For training program information, address:

AMERICAN MOTORS CORPORATION

College Graduates:
    Training Supervisor
    American Motors Corporation
    14250 Plymouth Road
    Detroit, Michigan 48232

High School, Vocational School Graduates:
    Personnel Department
    American Motors Corporation
    14250 Plymouth Road
    Detroit, Michigan 48232

CHRYSLER CORPORATION

College Graduates:
    College Recruitment Services
    Personnel Department
    Chrysler Corporation
    341 Massachusetts Avenue
    Detroit, Michigan 48231

High School, Vocational School Graduates:
    Manager
    Personnel Department
    Chrysler Corporation
    341 Massachusetts Avenue
    Detroit, Michigan 48231

FORD MOTOR COMPANY

Placement and College Recruitment
Ford Motor Company
The American Road
Dearborn, Michigan 48121

GENERAL MOTORS CORPORATION

Salaried Placement Personnel
General Motors Corporation
General Motors Building
Detroit, Michigan 48202

JOB CORPS

Job Corps is a training program for men and women between 16 and 21. Many trades, including auto mechanics, are taught.

The training centers provide essential materials such as medical and dental care, books, etc.

For information on joining the Job Corps, see your local State Employment Service or write to: U.S. Department of Labor, Manpower Administration, Job Corps, Washington, D.C. 20210.

PLAN YOUR CAREER EARLY

Don't "float" through high school wondering what you want to do. Make every effort to set your sights on a goal: college, technical school, armed forces, apprenticeship, etc.

Your guidance office will be able to help you determine your real talents and can give you helpful information and guidance in choosing a career.

Start your high school career by taking subjects that will allow you to select a college, commercial, vocational, etc., course in your Junior or Senior year. By doing this, you will not sell yourself short if you change your plans. You will then have a background of studies that will enable you to go any one of several ways.

DON'T BECOME A LOST ART

There is little room in this fast moving age for unskilled labor. Unless you learn a trade, you may find yourself out of work most of the time and when you do get employment you may be poorly paid and perhaps not in work of your choice.

SEE YOUR AUTO SHOP INSTRUCTOR

If you have been thinking about a career in the field of auto mechanics, make certain to discuss your ideas and plans with your auto mechanics instructor. He is vitally interested in this field and will be in a position to answer your questions.

REMEMBER THESE WORDS

THINK, PLAN, STUDY, WORK, LEARN, PRACTICE, QUESTION. These are good words and are most significant to anyone entering a trade. If you decide to be a mechanic, decide to be the BEST. If you do, these words will become very familiar to you.

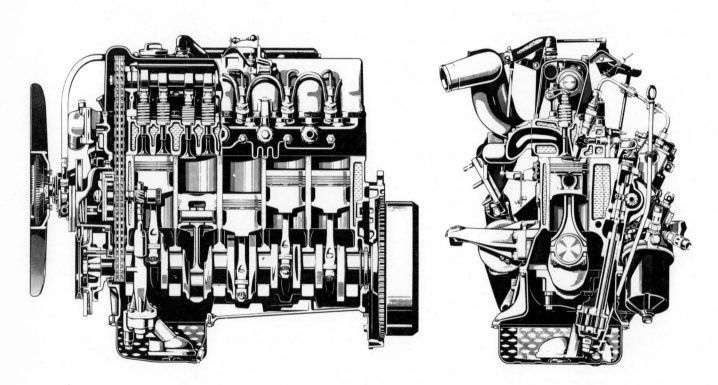

A recently developed FIVE cylinder diesel engine for automotive use.   (Mercedes-Benz)

METRIC TABLES

SOME COMMON ABBREVIATIONS			
ENGLISH		**METRIC**	
UNIT	ABBREVIATION	UNIT	ABBREVIATION
inch	in	kilometer	km
feet	ft	hectometer	hm
yard	yd	decameter	dkm
mile	mi	meter	m
grain	gr	decimeter	dm
ounce	oz	centimeter	cm
pound	lb	millimeter	mm
teaspoon	tsp	cubic centimeter	c^3
tablespoon	tbsp	kilogram	kg
fluid ounce	fl oz	hectogram	hg
cup	c	dekagram	dkg
pint	pt	gram	g
quart	qt	decigram	dg
gallon	gal	centigram	cg
cubic inch	in^3	milligram	mg
cubic foot	ft3	kiloliter	kl
cubic yard	yd^3	hectoliter	hl
square inch	in^2	decaliter	dkl
square foot	ft^2	liter	l
square yard	yd^2	centiliter	cl
square mile	mi^2	milliliter	ml
Fahrenheit	°F	decastere	dks
barrel	bbl	square kilometer	km^2
fluid dram	fl dr	hectare	ha
board foot	bd ft	are	a
rod	rd	centare	ca
dram	dr	tonne	t
bushel	bu	Celsius	°C

Auto Mechanics Fundamentals

MEASURING SYSTEMS

ENGLISH	METRIC

LENGTH

ENGLISH	METRIC
12 inches = 1 foot	1 kilometer = 1000 meters
36 inches = 1 yard	1 hectometer = 100 meters
3 feet = 1 yard	1 decameter = 10 meters
5,280 feet = 1 mile	1 meter = 1 meter
16.5 feet = 1 rod	1 decimeter = 0.1 meter
320 rods = 1 mile	1 centimeter = 0.01 meter
6 feet = 1 fathom	1 millimeter = 0.001 meter

WEIGHT

ENGLISH	METRIC
27.34 grains = 1 dram	1 tonne = 1,000,000 grams
438 grains = 1 ounce	1 kilogram = 1000 grams
16 drams = 1 ounce	1 hectogram = 100 grams
16 ounces = 1 pound	1 dekagram = 10 grams
2000 pounds = 1 short ton	1 gram = 1 gram
2240 pounds = 1 long ton	1 decigram = 0.1 gram
25 pounds = 1 quarter	1 centigram = 0.01 gram
4 quarters = 1 cwt	1 milligram = 0.001 gram

VOLUME

ENGLISH	METRIC
8 ounces = 1 cup	1 hectoliter = 100 liters
16 ounces = 1 pint	1 decaliter = 10 liters
32 ounces = 1 quart	1 liter = 1 liter
2 cups = 1 pint	1 deciliter = 0.1 liter
2 pints = 1 quart	1 centiliter = 0.01 liter
4 quarts = 1 gallon	1 milliliter = 0.001 liter
8 pints = 1 gallon	1000 milliliter = 1 liter

AREA

ENGLISH	METRIC
144 sq. inches = 1 sq. foot	100 sq. millimeters = 1 sq. centimeter
9 sq. feet = 1 sq. yard	100 sq. centimeters = 1 sq. decimeter
43,560 sq. ft. = 160 sq. rods	100 sq. decimeters = 1 sq. meter
160 sq. rods = 1 acre	10,000 sq. meters = 1 hectare
640 acres = 1 sq. mile	

TEMPERATURE

FAHRENHEIT		CELSIUS (centigrade)
32 degrees F	Water freezes	0 degrees C
68 degrees F	Reasonable room temperature	20 degrees C
98.6 degrees F	Normal body temperature	37 degrees C
173 degrees F	Alcohol boils	78.34 degrees C
212 degrees F	Water boils	100 degrees C

Metric Tables

USEFUL CONVERSIONS

WHEN YOU KNOW:	MULTIPLY BY:	TO FIND:

TORQUE		
Pound - inch	0.11298	newton-meters (N-m)
Pound - foot	1.3558	newton-meters
LIGHT		
Foot candles	1.0764	lumens/meters^2 (lm/m^2)
FUEL PERFORMANCE		
Miles/gallon	0.4251	kilometers/liter (km/1)
SPEED		
Miles/hour	1.6093	kilometers/hr (km/h)
FORCE		
kilogram	9.807	newtons (n)
ounce	0.278	newtons
pound	4.448	newtons
POWER		
Horsepower	0.746	kilowatts (kw)
PRESSURE OR STRESS		
Inches of water	0.2491	kilopascals (kPa)
Pounds/sq. in.	6.895	kilopascals
ENERGY OR WORK		
BTU	1055.0	joules (J)
Foot - pound	1.3558	joules
Kilowatt-hour	3600000.0	joules (J = one W's)

CONVERSION TABLE
METRIC TO ENGLISH

WHEN YOU KNOW: ⇩	MULTIPLY BY: * = Exact		TO FIND: ⇩
	VERY ACCURATE	APPROXIMATE	
LENGTH			
millimeters	0.0393701	0.04	inches
centimeters	0.3937008	0.4	inches
meters	3.280840	3.3	feet
meters	1.093613	1.1	yards
kilometers	0.621371	0.6	miles
WEIGHT			
grains	0.00228571	0.0023	ounces
grams	0.03527396	0.035	ounces
kilograms	2.204623	2.2	pounds
tonnes	1.1023113	1.1	short tons
VOLUME			
milliliters		0.2	teaspoons
milliliters	0.06667	0.067	tablespoon
milliliters	0.03381402	0.03	fluid ounces
liters	61.02374	61.024	cubic inches
liters	2.113376	2.1	pints
liters	1.056688	1.06	quarts
liters	0.26417205	0.26	gallons
liters	0.03531467	0.35	cubic feet
cubic meters	61023.74	61023.7	cubic inches
cubic meters	35.31467	35.0	cubic feet
cubic meters	1.3079506	1.3	cubic yards
cubic meters	264.17205	264.0	gallons
AREA			
square centimeters	0.1550003	0.16	square inches
square centimeters	0.00107639	0.001	square feet
square meters	10.76391	10.8	square feet
square meters	1.195990	1.2	square yards
square kilometers		0.4	square miles
hectares	2.471054	2.5	acres
TEMPERATURE			
Celsius	*9/5 (then add 32)		Fahrenheit

Metric Tables

CONVERSION TABLE
ENGLISH TO METRIC

WHEN YOU KNOW:	MULTIPLY BY: * = Exact		TO FIND:
	VERY ACCURATE	APPROXIMATE	
LENGTH			
inches	* 25.4		millimeters
inches	* 2.54		centimeters
feet	* 0.3048		meters
feet	* 30.48		centimeters
yards	* 0.9144	0.9	meters
miles	* 1.609344	1.6	kilometers
WEIGHT			
grains	15.43236	15.4	grams
ounces	* 28.349523125	28.0	grams
ounces	* 0.028349523125	.028	kilograms
pounds	* 0.45359237	0.45	kilograms
short ton	* 0.90718474	0.9	tonnes
VOLUME			
teaspoon		5.0	milliliters
tablespoon		15.0	milliliters
fluid ounces	29.57353	30.0	milliliters
cups		0.24	liters
pints	* 0.473176473	0.47	liters
quarts	* 0.946352946	0.95	liters
gallons	* 3.785411784	3.8	liters
cubic inches	* 0.016387064	0.02	liters
cubic feet	* 0.028316846592	0.03	cubic meters
cubic yards	* 0.764554857984	0.76	cubic meters
AREA			
square inches	* 6.4516	6.5	square centimeter
square feet	* 0.09290304	0.09	square meters
square yards	* 0.83612736	0.8	square meters
square miles		2.6	square kilometer
acres	* 0.40468564224	0.4	hectares
TEMPERATURE			
Fahrenheit	* 5/9 (after subtracting 32)		Celsius

DIMENSIONAL AND TEMPERATURE CONVERSION CHART

INCHES			DECIMALS	MILLI-METRES	INCHES TO MILLIMETRES		MILLIMETRES TO INCHES		FAHRENHEIT & CENTIGRADE			
					Inches	m.ms.	m.ms.	Inches	°F	°C	°C	°F
		1/64	.015625	.3969	.0001	.00254	0.001	.000039	-20	-28.9	-30	-22
	1/32		.03125	.7937	.0002	.00508	0.002	.000079	-15	-26.1	-28	-18.4
		3/64	.046875	1.1906	.0003	.00762	0.003	.000118	-10	-23.3	-26	-14.8
1/16			.0625	1.5875	.0004	.01016	0.004	.000157	-5	-20.6	-24	-11.2
		5/64	.078125	1.9844	.0005	.01270	0.005	.000197	0	-17.8	-22	-7.6
	3/32		.09375	2.3812	.0006	.01524	0.006	.000236	1	-17.2	-20	-4
		7/64	.109375	2.7781	.0007	.01778	0.007	.000276	2	-16.7	-18	-0.4
1/8			.125	3.1750	.0008	.02032	0.008	.000315	3	-16.1	-16	3.2
		9/64	.140625	3.5719	.0009	.02286	0.009	.000354	4	-15.6	-14	6.8
	5/32		.15625	3.9687	.001	.0254	0.01	.00039	5	-15.0	-12	10.4
		11/64	.171875	4.3656	.002	.0508	0.02	.00079	10	-12.2	-10	14
3/16			.1875	4.7625	.003	.0762	0.03	.00118	15	-9.4	-8	17.6
		13/64	.203125	5.1594	.004	.1016	0.04	.00157	20	-6.7	-6	21.2
	7/32		.21875	5.5562	.005	.1270	0.05	.00197	25	-3.9	-4	24.8
		15/64	.234375	5.9531	.006	.1524	0.06	.00236	30	-1.1	-2	28.4
1/4			.25	6.3500	.007	.1778	0.07	.00276	35	1.7	0	32
		17/64	.265625	6.7469	.008	.2032	0.08	.00315	40	4.4	2	35.6
	9/32		.28125	7.1437	.009	.2286	0.09	.00354	45	7.2	4	39.2
		19/64	.296875	7.5406	.01	.254	0.1	.00394	50	10.0	6	42.8
5/16			.3125	7.9375	.02	.508	0.2	.00787	55	12.8	8	46.4
		21/64	.328125	8.3344	.03	.762	0.3	.01181	60	15.6	10	50
	11/32		.34375	8.7312	.04	1.016	0.4	.01575	65	18.3	12	53.6
		23/64	.359375	9.1281	.05	1.270	0.5	.01969	70	21.1	14	57.2
3/8			.375	9.5250	.06	1.524	0.6	.02362	75	23.9	16	60.8
		25/64	.390625	9.9219	.07	1.778	0.7	.02756	80	26.7	18	64.4
	13/32		.40625	10.3187	.08	2.032	0.8	.03150	85	29.4	20	68
		27/64	.421875	10.7156	.09	2.286	0.9	.03543	90	32.2	22	71.6
7/16			.4375	11.1125	.1	2.54	1	.03937	95	35.0	24	75.2
		29/64	.453125	11.5094	.2	5.08	2	.07874	100	37.8	26	78.8
	15/32		.46875	11.9062	.3	7.62	3	.11811	105	40.6	28	82.4
		31/64	.484375	12.3031	.4	10.16	4	.15748	110	43.3	30	86
1/2			.5	12.7000	.5	12.70	5	.19685	115	46.1	32	89.6
		33/64	.515625	13.0969	.6	15.24	6	.23622	120	48.9	34	93.2
	17/32		.53125	13.4937	.7	17.78	7	.27559	125	51.7	36	96.8
		35/64	.546875	13.8906	.8	20.32	8	.31496	130	54.4	38	100.4
9/16			.5625	14.2875	.9	22.86	9	.35433	135	57.2	40	104
		37/64	.578125	14.6844	1	25.4	10	.39370	140	60.0	42	107.6
	19/32		.59375	15.0812	2	50.8	11	.43307	145	62.8	44	112.2
		39/64	.609375	15.4781	3	76.2	12	.47244	150	65.6	46	114.8
5/8			.625	15.8750	4	101.6	13	.51181	155	68.3	48	118.4
		41/64	.640625	16.2719	5	127.0	14	.55118	160	71.1	50	122
	21/32		.65625	16.6687	6	152.4	15	.59055	165	73.9	52	125.6
		43/64	.671875	17.0656	7	177.8	16	.62992	170	76.7	54	129.2
11/16			.6875	17.4625	8	203.2	17	.66929	175	79.4	56	132.8
		45/64	.703125	17.8594	9	228.6	18	.70866	180	82.2	58	136.4
	23/32		.71875	18.2562	10	254.0	19	.74803	185	85.0	60	140
		47/64	.734375	18.6531	11	279.4	20	.78740	190	87.8	62	143.6
3/4			.75	19.0500	12	304.8	21	.82677	195	90.6	64	147.2
		49/64	.765625	19.4469	13	330.2	22	.86614	200	93.3	66	150.8
	25/32		.78125	19.8437	14	355.6	23	.90551	205	96.1	68	154.4
		51/64	.796875	20.2406	15	381.0	24	.94488	210	98.9	70	158
13/16			.8125	20.6375	16	406.4	25	.98425	212	100.0	75	167
		53/64	.828125	21.0344	17	431.8	26	1.02362	215	101.7	80	176
	27/32		.84375	21.4312	18	457.2	27	1.06299	220	104.4	85	185
		55/64	.859375	21.8281	19	482.6	28	1.10236	225	107.2	90	194
7/8			.875	22.2250	20	508.0	29	1.14173	230	110.0	95	203
		57/64	.890625	22.6219	21	533.4	30	1.18110	235	112.8	100	212
	29/32		.90625	23.0187	22	558.8	31	1.22047	240	115.6	105	221
		59/64	.921875	23.4156	23	584.2	32	1.25984	245	118.3	110	230
15/16			.9375	23.8125	24	609.6	33	1.29921	250	121.1	115	239
		61/64	.953125	24.2094	25	635.0	34	1.33858	255	123.9	120	248
	31/32		.96875	24.6062	26	660.4	35	1.37795	260	126.6	125	257
		63/64	.984375	25.0031	27	690.6	36	1.41732	265	129.4	130	266

CAPACITY CONVERSION U.S. GALLONS TO LITERS

Gallons	0	1	2	3	4	5
	Liters	Liters	Liters	Liters	Liters	Liters
0	00.0000	3.7853	7.5707	11.3560	15.1413	18.9267
10	37.8533	41.6387	45.4240	49.2098	52.9947	56.7800
20	75.7066	79.4920	83.2773	87.0626	90.8480	94.6333
30	113.5600	117.3453	121.1306	124.9160	128.7013	132.4866
40	151.4133	155.1986	158.9840	162.7693	166.5546	170.3400

Metric Tables

MILLIMETER CONVERSION CHART

mm. Ins.	15 = .5905	30 = 1.1811	45 = 1.7716	60 = 2.3622	75 = 2.9527	90 = 3.5433	105 = 4.1338	120 = 4.7244
.25 = .0098	15.25 = .6004	30.25 = 1.1909	45.25 = 1.7815	60.25 = 2.3720	75.25 = 2.9626	90.25 = 3.5531	105.25 = 4.1437	120.25 = 4.7342
.50 = .0197	15.50 = .6102	30.50 = 1.2008	45.50 = 1.7913	60.50 = 2.3819	75.50 = 2.9724	90.50 = 3.5630	105.50 = 4.1535	120.50 = 4.7441
.75 = .0295	15.75 = .6201	30.75 = 1.2106	45.75 = 1.8012	60.75 = 2.3917	75.75 = 2.9823	90.75 = 3.5728	105.75 = 4.1634	120.75 = 4.7539
1 = .0394	16 = .6299	31 = 1.2205	46 = 1.8110	61 = 2.4016	76 = 2.9921	91 = 3.5827	106 = 4.1732	121 = 4.7638
1.25 = .0492	16.25 = .6398	31.25 = 1.2303	46.25 = 1.8209	61.25 = 2.4114	76.25 = 3.0020	91.25 = 3.5925	106.25 = 4.1831	121.25 = 4.7736
1.50 = .0591	16.50 = .6496	31.50 = 1.2402	46.50 = 1.8307	61.50 = 2.4213	76.50 = 3.0118	91.50 = 3.6024	106.50 = 4.1929	121.50 = 4.7885
1.75 = .0689	16.75 = .6594	31.75 = 1.2500	46.75 = 1.8405	61.75 = 2.4311	76.75 = 3.0216	91.75 = 3.6122	106.75 = 4.2027	121.75 = 4.7933
2 = .0787	17 = .6693	32 = 1.2598	47 = 1.8504	62 = 2.4409	77 = 3.0315	92 = 3.6220	107 = 4.2126	122 = 4.8031
2.25 = .0886	17.25 = .6791	32.25 = 1.2697	47.25 = 1.8602	62.25 = 2.4508	77.25 = 3.0413	92.25 = 3.6319	107.25 = 4.2224	122.25 = 4.8130
2.50 = .0984	17.50 = .6890	32.50 = 1.2795	47.50 = 1.8701	62.50 = 2.4606	77.50 = 3.0512	92.50 = 3.6417	107.50 = 4.2323	122.50 = 4.8228
2.75 = .1083	17.75 = .6988	32.75 = 1.2894	47.75 = 1.8799	62.75 = 2.4705	77.75 = 3.0610	92.75 = 3.6516	107.75 = 4.2421	122.75 = 4.8327
3 = .1181	18 = .7087	33 = 1.2992	48 = 1.8898	63 = 2.4803	78 = 3.0709	93 = 3.6614	108 = 4.2520	123 = 4.8425
3.25 = .1280	18.25 = .7185	33.25 = 1.3091	48.25 = 1.8996	63.25 = 2.4901	78.25 = 3.0807	93.25 = 3.6713	108.25 = 4.2618	123.25 = 4.8524
3.50 = .1378	18.50 = .7283	33.50 = 1.3189	48.50 = 1.9094	63.50 = 2.5000	78.50 = 3.0905	93.50 = 3.6811	108.50 = 4.2716	123.50 = 4.8622
3.75 = .1476	18.75 = .7382	33.75 = 1.3287	48.75 = 1.9193	63.75 = 2.5098	78.75 = 3.1004	93.75 = 3.6909	108.75 = 4.2815	123.75 = 4.8720
4 = .1575	19 = .7480	34 = 1.3386	49 = 1.9291	64 = 2.5197	79 = 3.1102	94 = 3.7008	109 = 4.2913	124 = 4.8819
4.25 = .1673	19.25 = .7579	34.25 = 1.3484	49.25 = 1.9390	64.25 = 2.5295	79.25 = 3.1201	94.25 = 3.7106	109.25 = 4.3012	124.25 = 4.8917
4.50 = .1772	19.50 = .7677	34.50 = 1.3583	49.50 = 1.9488	64.50 = 2.5394	79.50 = 3.1299	94.50 = 3.7205	109.50 = 4.3110	124.50 = 4.9016
4.75 = .1870	19.75 = .7776	34.75 = 1.3681	49.75 = 1.9587	64.75 = 2.5492	79.75 = 3.1398	94.75 = 3.7303	109.75 = 4.3209	124.75 = 4.9114
5 = .1968	20 = .7874	35 = 1.3779	50 = 1.9685	65 = 2.5590	80 = 3.1496	95 = 3.7401	110 = 4.3307	125 = 4.9212
5.25 = .2067	20.25 = .7972	35.25 = 1.3878	50.25 = 1.9783	65.25 = 2.5689	80.25 = 3.1594	95.25 = 3.7500	110.25 = 4.3405	125.25 = 4.9311
5.50 = .2165	20.50 = .8071	35.50 = 1.3976	50.50 = 1.9882	65.50 = 2.5787	80.50 = 3.1693	95.50 = 3.7598	110.50 = 4.3504	125.50 = 4.9409
5.75 = .2264	20.75 = .8169	35.75 = 1.4075	50.75 = 1.9980	65.75 = 2.5886	80.75 = 3.1791	95.75 = 3.7697	110.75 = 4.3602	125.75 = 4.9508
6 = .2362	21 = .8268	36 = 1.4173	51 = 2.0079	66 = 2.5984	81 = 3.1890	96 = 3.7795	111 = 4.3701	126 = 4.9606
6.25 = .2461	21.25 = .8366	36.25 = 1.4272	51.25 = 2.0177	66.25 = 2.6083	81.25 = 3.1988	96.25 = 3.7894	111.25 = 4.3799	126.25 = 4.9705
6.50 = .2559	21.50 = .8465	36.50 = 1.4370	51.50 = 2.0276	66.50 = 2.6181	81.50 = 3.2087	96.50 = 3.7992	111.50 = 4.3898	126.50 = 4.9803
6.75 = .2657	21.75 = .8563	36.75 = 1.4468	51.75 = 2.0374	66.75 = 2.6279	81.75 = 3.2185	96.75 = 3.8090	111.75 = 4.3996	126.75 = 4.9901
7 = .2756	22 = .8661	37 = 1.4567	52 = 2.0472	67 = 2.6378	82 = 3.2283	97 = 3.8189	112 = 4.4094	127 = 5.0000
7.25 = .2854	22.25 = .8760	37.25 = 1.4665	52.25 = 2.0571	67.25 = 2.6476	82.25 = 3.2382	97.25 = 3.8287	112.25 = 4.4193	
7.50 = .2953	22.50 = .8858	37.50 = 1.4764	52.50 = 2.0669	67.50 = 2.6575	82.50 = 3.2480	97.50 = 3.8386	112.50 = 4.4291	
7.75 = .3051	22.75 = .8957	37.75 = 1.4862	52.75 = 2.0768	67.75 = 2.6673	82.75 = 3.2579	97.75 = 3.8484	112.75 = 4.4390	
8 = .3150	23 = .9055	38 = 1.4961	53 = 2.0866	68 = 2.6772	83 = 3.2677	98 = 3.8583	113 = 4.4488	
8.25 = .3248	23.25 = .9153	38.25 = 1.5059	53.25 = 2.0965	68.25 = 2.6870	83.25 = 3.2776	98.25 = 3.8681	113.25 = 4.4587	
8.50 = .3346	23.50 = .9252	38.50 = 1.5157	53.50 = 2.1063	68.50 = 2.6968	83.50 = 3.2874	98.50 = 3.8779	113.50 = 4.4685	
8.75 = .3445	23.75 = .9350	38.75 = 1.5256	53.75 = 2.1161	68.75 = 2.7067	83.75 = 3.2972	98.75 = 3.8878	113.75 = 4.4783	
9 = .3543	24 = .9449	39 = 1.5354	54 = 2.1260	69 = 2.7165	84 = 3.3071	99 = 3.8976	114 = 4.4882	
9.25 = .3642	24.25 = .9547	39.25 = 1.5453	54.25 = 2.1358	69.25 = 2.7264	84.25 = 3.3169	99.25 = 3.9075	114.25 = 4.4980	
9.50 = .3740	24.50 = .9646	39.50 = 1.5551	54.50 = 2.1457	69.50 = 2.7362	84.50 = 3.3268	99.50 = 3.9173	114.50 = 4.5079	
9.75 = .3839	24.75 = .9744	39.75 = 1.5650	54.75 = 2.1555	69.75 = 2.7461	84.75 = 3.3366	99.75 = 3.9272	114.75 = 4.5177	
10 = .3937	25 = .9842	40 = 1.5748	55 = 2.1653	70 = 2.7559	85 = 3.3464	100 = 3.9370	115 = 4.5275	
10.25 = .4035	25.25 = .9941	40.25 = 1.5846	55.25 = 2.1752	70.25 = 2.7657	85.25 = 3.3563	100.25 = 3.9468	115.25 = 4.5374	
10.50 = .4134	25.50 = 1.0039	40.50 = 1.5945	55.50 = 2.1850	70.50 = 2.7756	85.50 = 3.3661	100.50 = 3.9567	115.50 = 4.5472	
10.75 = .4232	25.75 = 1.0138	40.75 = 1.6043	55.75 = 2.1949	70.75 = 2.7854	85.75 = 3.3760	100.75 = 3.9665	115.75 = 4.5571	
11 = .4331	26 = 1.0236	41 = 1.6142	56 = 2.2047	71 = 2.7953	86 = 3.3858	101 = 3.9764	116 = 4.5669	
11.25 = .4429	26.25 = 1.0335	41.25 = 1.6240	56.25 = 2.2146	71.25 = 2.8051	86.25 = 3.3957	101.25 = 3.9862	116.25 = 4.5768	
11.50 = .4528	26.50 = 1.0433	41.50 = 1.6339	56.50 = 2.2244	71.50 = 2.8150	86.50 = 3.4055	101.50 = 3.9961	116.50 = 4.5866	
11.75 = .4626	26.75 = 1.0531	41.75 = 1.6437	56.75 = 2.2342	71.75 = 2.8248	86.75 = 3.4153	101.75 = 4.0059	116.75 = 4.5964	
12 = .4724	27 = 1.0630	42 = 1.6535	57 = 2.2441	72 = 2.8346	87 = 3.4252	102 = 4.0157	117 = 4.6063	
12.25 = .4823	27.25 = 1.0728	42.25 = 1.6634	57.25 = 2.2539	72.25 = 2.8445	87.25 = 3.4350	102.25 = 4.0256	117.25 = 4.6161	
12.50 = .4921	27.50 = 1.0827	42.50 = 1.6732	57.50 = 2.2638	72.50 = 2.8543	87.50 = 3.4449	102.50 = 4.0354	117.50 = 4.6260	
12.75 = .5020	27.75 = 1.0925	42.75 = 1.6831	57.75 = 2.2736	72.75 = 2.8642	87.75 = 3.4547	102.75 = 4.0453	117.75 = 4.6358	
13 = .5118	28 = 1.1024	43 = 1.6929	58 = 2.2835	73 = 2.8740	88 = 3.4646	103 = 4.0551	118 = 4.6457	
13.25 = .5217	28.25 = 1.1122	43.25 = 1.7028	58.25 = 2.2933	73.25 = 2.8839	88.25 = 3.4744	103.25 = 4.0650	118.25 = 4.6555	
13.50 = .5315	28.50 = 1.1220	43.50 = 1.7126	58.50 = 2.3031	73.50 = 2.8937	88.50 = 3.4842	103.50 = 4.0748	118.50 = 4.6653	
13.75 = .5413	28.75 = 1.1319	43.75 = 1.7224	58.75 = 2.3130	73.75 = 2.9035	88.75 = 3.4941	103.75 = 4.0846	118.75 = 4.6752	
14 = .5512	29 = 1.1417	44 = 1.7323	59 = 2.3228	74 = 2.9134	89 = 3.5039	104 = 4.0945	119 = 4.6850	
14.25 = .5610	29.25 = 1.1516	44.25 = 1.7421	59.25 = 2.3327	74.25 = 2.9232	89.25 = 3.5138	104.25 = 4.1043	119.25 = 4.6949	
14.50 = .5709	29.50 = 1.1614	44.50 = 1.7520	59.50 = 2.3425	74.50 = 2.9331	89.50 = 3.5236	104.50 = 4.1142	119.50 = 4.7047	
14.75 = .5807	29.75 = 1.1713	44.75 = 1.7618	59.75 = 2.3524	74.74 = 2.9429	89.75 = 3.5335	104.75 = 4.1240	119.75 = 4.7146	

DICTIONARY OF
AUTOMOTIVE TERMS

AAA: American Automobile Association.

ABDC: After Bottom Dead Center.

ABSOLUTE ZERO: A state in which no heat is present. Believed to be -459.7 deg. F. or -273.16 deg. C.

A BONE: MODEL "A" Ford.

AC: Alternating current.

ACCELERATOR: Floor pedal used to control, through linkage, throttle valve in carubretor.

ACCELERATOR PUMP: Small pump, located in carburetor, that sprays additional gasoline into air stream during acceleration.

ACCUMULATOR PISTON (Automatic Transmission): Unit designed to assist the servo to apply brake band quickly, yet smoothly.

ACETYLENE: Gas commonly used in welding or cutting operations.

ACKERMAN PRINCIPLE: Bending outer ends of steering arms slightly inward so that when car is making a turn, inside wheel will turn more sharply than outer wheel. This principle produces toe-out on turns.

ADDITIVE: Solution, powder, etc., added to gasoline, oil, grease, etc., in an endeavor to improve characteristics of original product.

ADVANCE (Ignition timing): To set ignition timing so a spark occurs earlier or more degrees before TDC.

AEA: Automotive Electric Association.

AERA: Automotive Engine Rebuilders Association.

AHRA: American Hot Rod Association.

AIR: Air Injection Reactor system of reducing objectionable exhaust emissions.

AIR CLEANER: Device used to remove dust, abrasive, etc., from air being drawn into an engine, compressor, power brake, etc.

AIR COOLED: An object cooled by passing a stream of air over its surface.

AIR FOIL: Device, similar to a stubby wing, mounted onto a racing car or dragster to provide high speed stability. The air foil is mounted in a horizontal position.

AIR-FUEL RATIO: Ratio (by weight or by volume) between air and gasoline that makes up engine fuel mixture.

AIR GAP (Regulator): Distance between contact armature and iron core that when magnetized, draws armature down.

AIR GAP (Spark Plugs): Distance between center and side electrodes.

AIR HORN (Carburetor): Top portion of air passageway through carburetor.

AIR HORN (Warning): Warning horn operated by compressed air.

AIR POLLUTION: Contamination of earth's atmosphere by various natural and man-made pollutants such as smoke, gases, dust, etc.

AIR SPRING: Container and plunger separated by air under pressure. When container and plunger attempt to squeeze together, air compresses and produces a spring effect. Air spring has been used on some suspension systems.

ALIGN: To bring various parts of unit into correct positions in respect to each other or to a predetermined location.

ALLOY: Mixture of two or more materials.

ALNICO MAGNET: Magnet using (Al) aluminum, (Ni) nickel, and (Co) cobalt in its construction.

ALTERNATOR: Device similar to generator but which produces AC current. The AC must be rectified before reaching the car's electrical system.

ALTERNATING CURRENT (AC): Electric current that first flows one way in circuit and then other. Type used in homes.

AMA: Automobile Manufacturers Association.

AMBIENT TEMPERATURE: Temperature of air surrounding an object.

AMMETER: Instrument used to measure rate of current flow. (In amperes.)

AMPERE: Unit of measurement used in expressing rate of current flow in a circuit.

AMPERE HOUR CAPACITY: Measurement of storage battery ability to deliver specified current over specified length of time.

ANNEAL: To remove hardness from metal by heating, usually to a red color, then allowing it to cool slowly. Unlike steel, copper is annealed by heating, and then plunging it into cold water.

ANODE: In an electrical circuit — the positive pole.

ANTIBACKFIRE VALVE: Valve used in air injection reaction (exhaust emission control) system to prevent backfiring during period immediately following sudden deceleration.

ANTIFREEZE: Chemical added to cooling system to prevent coolant from freezing in cold weather.

ANTIFRICTION BEARING: Bearing containing rollers or balls plus an inner and outer race. Bearing is designed to roll, thus minimizing friction.

ANTIPERCOLATOR: Device for venting vapors from main discharge tube, or well, of a carburetor.

API: American Petroleum Institute.

APRA: Automotive Parts Rebuilders Association.

ARC OR ELECTRIC WELDING: Welding by using electric current to melt both metal to be welded and welding rod or electrode that is being added.

ARCING: Electricity leaping the gap between two electrodes.

ARMATURE (Relay, regulator, horn, etc.): The movable part of the unit.

ARMATURE (Starter or Generator): The portion that revolves between the pole shoes, made up of wire windings on an iron core.

ASBESTOS: Heat resistant and nonburning fibrous mineral widely used for brake shoes, clutch linings, etc.

ASLE: American Society of Lubrication Engineers.

ASME: American Society of Mechanical Engineers.

ASSE: American Society of Safety Engineers.

ASTM: American Society for Testing Materials.

ATA: American Trucking Association.

ATDC: After Top Dead Center.

ATMOSPHERIC PRESSURE: Pressure exerted by atmosphere on all things exposed to it. Around fifteen pounds per square inch at sea level. (14.7).

ATOM: Tiny particle of matter made up of electrons, protons and neutrons. Atoms or combinations of atoms make up molecules. The electrons orbit around the center or nucleus made up of the protons and neutrons.

AUTOMATIC CHOKE: A carburetor choke device that automatically positions itself in accordance with carburetor needs.

AWL (Tire): Sharp pointed steel tool used to probe cuts, nail holes, etc., in tires.

AXIAL: Direction parallel to shaft or bearing hole.

AXLE (Full-floating): Axle used to drive rear wheels. It does not hold them on nor support them.

AXLE (Semi or one-quarter floating): Axle used to drive wheels, hold them on, and support them.

AXLE (Three-quarter floating): Axle used to drive rear wheels as well as hold them on. It does not support them.

AXLE END GEARS: Two gears, one per axle, that are splined to the inner ends of drive axles. They mesh with and are driven by "spider" gears.

AXLE RATIO: Relationship or ratio between the number of times the propeller shaft must revolve to turn the axle drive shafts one turn.

BACKFIRE (Intake system): Burning of fuel mixture in intake manifold. May be caused by faulty timing, crossed plug wires, leaky intake valve, etc.

BACKFIRE (Exhaust system): Passage of unburned fuel mixture into exhaust system where it is ignited and causes an explosion (backfire).

BACKLASH: Amount of "play" between two parts. In case of gears, it refers to how much one gear can be moved back and forth without moving gear into which it is meshed.

BACK PRESSURE: Refers to resistance to flow of exhaust gases through exhaust system.

BAFFLE: Obstruction used to slow down or divert the flow of gases, liquids, sound, etc.

BALANCE (Tire): See Static Balance and Dynamic Balance.

BALL BEARING: (Antifriction): Bearing consisting of an inner and outer hardened steel race separated by a series of hardened steel balls.

BALL JOINT: Flexible joint utilizing ball and socket type of construction. Used in steering linkage setups, steering knuckle pivot supports, etc.

BALL JOINT STEERING KNUCKLE: Steering knuckle that pivots on ball joints instead of on a kingpin.

BALL JOINT ROCKER ARMS: Rocker arms that instead of being mounted on shaft, are mounted upon ball-shaped device on end of stud.

BALLAST RESISTOR: Resistor constructed of special type wire, properties of which tend to increase or decrease voltage in direct proportion to heat of wire.

BASE CIRCLE: As applied to camshaft - - lowest spot on cam. Area of cam directly opposite lobe.

BATTERY: Electrochemical device for producing electricity.

BATTERY CHARGING: Process of renewing battery by passing electric current through battery in reverse direction.

BBDC: Before Bottom Dead Center.

BCI: Battery Council International.

BDC: Bottom Dead Center.

BEAD (Tire): Steel wire reenforced portion of tire that engages the wheel rim.

BEARING: Area of unit in which contacting surface of a revolving part rests.

BEARING CLEARANCE: Amount of space left

between shaft and bearing surface. This space is for lubricating oil to enter.

BELL HOUSING (Clutch housing): Metal covering around flywheel and clutch, or torque converter assembly.

BENDIX TYPE STARTER DRIVE: A self-engaging starter drive gear. Gear moves into engagement when starter starts spinning and automatically disengages when starter stops.

BEVEL GEAR: Gear in which teeth are cut in a cone shape, as found in axle end gears.

BEVEL SPUR GEAR: Gear in which teeth are cut in a cone shape. Teeth are aligned with cone center line, as found in some differential gears.

BEZEL: Crimped edge of metal that secures glass face to an instrument.

BHP: Brake horsepower. Measurement of actual power produced by engine.

BINDERS: Car brakes.

BLEEDING: Removing air, pressure, fluid, etc., from a closed system, as in air conditioning.

BLEEDING THE BRAKES: Refers to removal of air from hydraulic system. Bleeder screws are loosened at each wheel cylinder, (one at a time) and brake fluid is forced from master cylinder through lines until all air is expelled.

BLOCK: Part of engine containing cylinders.

BLOW-BY: Refers to escape of exhaust gases past piston rings.

BLOWER: Supercharger.

BLUEPRINTING (Engine): Dismantling engine and reassembling it to EXACT specifications.

BODY PUTTY: Material designed to smooth on dented body areas. Upon hardening, putty is dressed down and area painted.

BOILING POINT: Exact temperature at which a liquid begins to boil.

BONDED BRAKE LINING: Brake lining that is attached to brake shoe by adhesive.

BONNET: British term for car hood.

BOOSTER: Device incorporated in car system (such as brakes and steering), to increase pressure output or decrease amount of effort required to operate or both.

BOOT: British term for trunk.

BORE: May refer to cylinder itself or to diameter of the cylinder.

BORE DIAMETER: Diameter of cylinders.

BORING: Renewing cylinders by cutting them out to a specified size. Boring bar is used to make cut.

BORING BAR (Cylinder): Machine used to cut engine cylinders to specific size. As used in garages, to cut worn cylinders to a new diameter.

BOTTLED GAS: LPG (Liquefied Petroleum Gas) gas compressed into strong metal tanks. Gas, when confined in tank, under pressure, is in liquid form.

BOUND ELECTRONS: Electrons in inner orbits around nucleus of atom. They are difficult to move out of orbit.

BOURDON TUBE: Circular, hollow piece of metal used in some instruments. Pressure on hollow section causes it to attempt to straighten. Free end then moves needle on gage face.

BOX: Transmission.

BOXED ROD: Connecting rod in which I-beam section has been stiffened by welding plates on each side of the rod.

BRAKE ANCHOR: Steel stud upon which one end of brake shoes is either attached to or rests against. Anchor is firmly affixed to backing plate.

BRAKE ANTI-ROLL DEVICE: Unit installed in brake system to hold brake line pressure when car is stopped on upgrade. When car is stopped on upgrade and brake pedal released, anti-roll device will keep brakes applied until either clutch is released or, as on some models, accelerator is depressed.

BRAKE BACKING PLATE: Rigid steel plate upon which brake shoes are attached. Braking force applied to shoes is absorbed by backing plate.

BRAKE BAND: Band, faced with brake lining, that

encircles a brake drum. Used on several parking brake installations.

BRAKE BLEEDING: See Bleeding the Brakes.

BRAKE CYLINDER: See Wheel Cylinder.

BRAKE — DISC TYPE: Braking system that instead of using conventional brake drum with internal brake shoes, uses steel disc with caliper type lining application. When brakes are applied, section of lining on each side of spinning disc is forced against disc thus imparting braking force. This type of brake is very resistant to brake fade.

BRAKE DRUM: Cast iron or aluminum housing, bolted to wheel, that rotates around brake shoes. When shoes are expanded, they rub against machined inner surface of brake drum and exert braking effect upon wheel.

BRAKE DRUM LATHE: Machine to refinish inside of a brake drum.

BRAKE FADE: Reduction in braking force due to loss of friction between brake shoes and drum. Caused by heat buildup.

BRAKE FEEL: Discernible, to driver, relationship between the amount of brake pedal pressure and the actual braking force being exerted. Special device is incorporated in power brake installations to give driver this feel.

BRAKE FLUID: Special fluid used in hydraulic brake systems. Never use anything else in place of regular fluid.

BRAKE FLUSHING: Cleaning brake system by flushing with alcohol or brake fluid. Done to remove water, dirt, or any other contaminant. Flushing fluid is placed in master cylinder and forced through lines and wheel cylinders where it exists at cylinder bleed screws.

BRAKE HORSEPOWER (bhp): Measurement of actual useable horsepower delivered at crankshaft. Commonly computed using an engine on a chassis dynamometer.

BREAK-IN: Period of operation between installation of new or rebuilt parts and time in which parts are worn to the correct fit. Driving at reduced and varying speed for a specified mileage to permit parts to wear to the correct fit.

BRAKE LINING: Friction material fastened to brake shoes. Brake lining is pressed against rotating brake drum thus stopping car.

BRAKE — PARKING: Brake used to hold car in position while parked. One type applies rear brake shoes by mechanical means and other type applies brake band to brake drum installed in drive train.

BRAKES — POWER: Conventional hydraulic brake system that utilizes engine vacuum to operate vacuum power piston. Power piston applies pressure to brake pedal, or in some cases, directly to master cylinder piston. This reduces amount of pedal pressure that driver must exert to stop the car.

BRAKE SHOE GRINDER: Grinder used to grind brake shoe lining so it will be square to and concentric with brake drum.

BRAKE SHOE HEEL: End of brake shoe adjacent to anchor bolt or pin.

BRAKE SHOE TOE: Free end of shoe, not attached to or resting against an anchor pin.

BRAKE SHOES: Part of brake system, located at wheels, upon which brake lining is attached. When wheel cylinders are actuated by hydraulic pressure they force brake shoes apart and bring lining into contact with drum.

BRAZE: To join two pieces of metal together by heating edges to be joined and then melting drops of brass or bronze on area. Unlike welding, this operation is similar to soldering, only a higher melting point material is used.

BREAKER (Tire): Rubber or fabric (or both) strip placed under tread to provide additional protection for main tire carcass.

BREAKER ARM: Movable arm upon which one of breaker points is affixed.

BREAKER POINTS (Ignition): Pair of movable points that are opened and closed to break and

make the primary circuit.

BREATHER PIPE: Pipe opening into interior of engine. Used to assist ventilation. Pipe usually extends downward to a point just below engine so passing air stream will form a partial vacuum thus assisting in venting engine.

BROACH: Bringing metal surface to desired shape by forcing multiple-edged cutting tool across surface.

BRUSH: Pieces of carbon, or copper, that rub against the commutator on generator and starter motor.

B & S GAUGE (Brown and Sharpe): Standard measure of wire diameter.

BTDC: Before Top Dead Center.

BTU (British thermal unit): Measurement of the amount of heat required to raise temperature of one pound of water, one degree Fahrenheit.

BUDC: Before Upper Dead Center. Same as BTDC.

BURNISH: To bring a surface to a high shine by rubbing with hard, smooth object.

BUSHING: Bearing for shaft, spring shackle, piston pin, etc., of one piece construction which may be removed from part.

BUTANE: Petroleum gas that is liquid, when under pressure. Often used as engine fuel in trucks.

BUTTERFLY VALVE: Valve in carburetor that is so named due to its resemblance to insect of same name.

BYPASS FILTER: Oil filter that constantly filters PORTION of oil flowing through engine.

BYPASS VALVE: Valve that can open and allow fluid to pass through in other than its normal channel.

CALIBRATE: As applied to test instruments-- adjusting dial needle to correct zero or load setting.

CALIPERS (inside and outside): Adjustable measuring tool placed around, or within, an object and adjusted until it just contacts. It is then withdrawn and distance measured between contacting points.

CALORIE (Gram): A unit of heat. Amount of heat required to raise the temperature of one gram of water 1 deg. centigrade.

CALORIFIC VALUE: Measurement of the heating value of fuel.

CALORIMETER: Measuring instrument used to determine amount of heat produced when a substance is burned; also friction and chemical change heat production.

CAM: Offset portion of shaft that will, when shaft turns, impart motion to another part such as valve lifters.

CAM ANGLE or DWELL (Ignition): Number of degrees breaker cam rotates from time breaker points close until they open again.

CAM GROUND: Piston ground slightly egg-shaped. When heated, it becomes round.

CAMBER: Tipping top wheel center line outward produces positive camber. Tipping wheel center line inward at top produces negative camber. When camber is positive, tops of tires are further apart than bottom.

CAMSHAFT: Shaft with cam lobes (bumps) used to operate valves.

CAMSHAFT GEAR: Gear that is used to drive camshaft.

CANDLE POWER: Measurement of light producing ability of light bulb.

CAP: Cleaner Air Package System of reducing amount of unburned hydrocarbons in automobile exhaust.

CAPACITANCE: Property of condenser that permits it to receive and retain an electrical charge.

CAPACITOR: See Condenser.

CARBON: Used to describe hard, or soft, black deposits found in combustion chamber, on plugs, under rings, on and under valve heads, etc.

CARBONIZE: Building up of carbon on objects such as spark plugs, pistons, heads, etc.

CARBON MONOXIDE: Deadly, colorless, odorless, and tasteless gas found in engine exhaust. Formed by incomplete burning of hydrocarbons.

CARBURETOR: Device used to mix gasoline and air

in correct proportions.

CARBURETOR ADAPTER: Adapter used to fit or place one type of carburetor on an intake manifold that may not be originally designed for it. Also used to adapt four-barrel carbs to two-barrel manifolds.

CARBURETOR CIRCUITS: Series of passageways and units designed to perform a specific function — idle circuit, full power circuit, etc.

CARBURETOR ICING: Formation of ice on throttle plate or valve. As fuel nozzles feed fuel into air horn it turns to a vapor. This robs heat from air and when weather conditions are just right (fairly cold and quite humid) ice may form.

CARBURIZING FLAME: Welding torch flame in which there is an excess of acetylene.

CARDAN JOINT: Type of universal joint.

CARRIER BEARINGS: Bearings upon which differential case is mounted.

CASE-HARDENED: Piece of steel that has had outer surface hardened while inner portion remains relatively soft.

CASTER: Tipping top of kingpin either forward or toward the rear of car. When tipped forward it is termed negative caster. When tipped toward rear it is termed positive caster.

CASTING: Pouring metal into a mold to form an object.

CASTLE or CASTELLATED NUT: Nut having series of slots cut into one end, into which cotter pin may be passed to secure nut.

CATHODE: In electric circuit — the negative pole.

CCS: Controlled Combustion System of reducing unburned hydrocarbon emission from engine exhaust.

CEC: Combination Emission Control.

CELL (Battery): Individual (separate) compartments in battery which contain positive and negative plates suspended in electrolyte. Six-volt battery has three cells, twelve-volt battery six cells.

CELL CONNECTOR: Lead strap or connection between battery cell groups.

CENTER LINE: Imaginary line drawn lengthwise through center of an object.

CENTER OF GRAVITY: Point in object, if through which an imaginary pivot line were drawn, would leave object in balance. In car, the closer the weight to the ground, the lower the center of gravity.

CENTER STEERING LINKAGE: Steering system utilizing two tie rods connected to steering arms and to central idler arm. Idler arm is operated by drag link that connects idler arm to pitman arm.

CENTIGRADE: Thermometer on which boiling point of water is 100 deg. and freezing point is 0 deg.

CENTRIFUGAL ADVANCE (Distributor): Unit designed to advance and retard ignition timing through action of centrifugal force.

CENTRIFUGAL CLUTCH: Clutch that utilizes centrifugal force to expand a friction device on driving shaft until it is locked to a drum on driven shaft.

CENTRIFUGAL FORCE: Force which tends to keep moving objects traveling in straight line. When moving car is forced to make a turn, centrifugal force attempts to keep it moving in straight line. If car is turning at too high a speed, centrifugal force will be greater than frictional force between tires and road and the car will slide off the road.

CERAMIC FILTER: Filtering device utilizing porous ceramic as filtering agent.

CETANE NUMBER: Measurement of diesel fuel performance characteristics.

CFM: Cubic feet per minute. A measure of air flow.

CHAMFER: To bevel (or a bevel on) edge of an object.

CHANGE OF STATE: Condition in which substance changes from a solid to a liquid, a liquid to a gas, a liquid to a solid, or a gas to a liquid.

CHANNELED: Car body lowered down around frame.

CHARGE (Battery): Passing electric current through battery to restore it to active (charged) state.

CHASE: To repair damaged threads.

CHASSIS: Generally, chassis refers to frame, engine, front and rear axles, springs, steering system and gas tank. In short, everything but body and fenders.

CHASSIS DYNAMOMETER: See Dynamometer.

CHECK VALVE: Valve that opens to permit passage of fluid or air in one direction and closes to prevent passage in opposite direction.

CHILLED IRON: Cast iron possessing hardened outer skin.

CHOKE: Butterfly valve located in carburetor used to enrichen mixture for starting engine when cold.

CHOKE STOVE: Heating compartment in or on exhaust manifold from which hot air is drawn to automatic choke device.

CHOP: Lowering height of some area of car — roof, hood, etc.

CHOPPED WHEEL: Lightened flywheel.

CHRISTMAS TREE: Device, using series of lights, to start cars on timed 1/4 mile drag run.

CID: Cubic Inch Displacement.

CIRCUIT (Electrical): Source of electricity (battery), resistance unit (headlight, etc.) and wires that form path for flow of electricity from source through unit and back to source.

CIRCUIT BREAKER (Lighting system): Protective device that will make and break flow of current when current draw becomes excessive. Unlike fuse, it does not blow out but vibrates on and off thus giving driver some light to stop by.

CLEARANCE: Given amount of space between two parts — between piston and cylinder, bearing and journal, etc.

CLOCKWISE: Rotation to right as that of clock hands.

CLUSTER or COUNTER GEAR: Cluster of gears that are all cut on one long gear blank. Cluster gears ride in bottom of transmission. Cluster provides a connection between transmission input shaft and output shaft.

CLUTCH: Device used to connect or disconnect flow of power from one unit to another.

CLUTCH DIAPHRAGM SPRING: Round dish-shaped piece of flat spring steel. Used to force pressure plate against clutch disc in some clutches.

CLUTCH DISC: Part of clutch assembly splined to transmission clutch or input shaft. Faced with friction material. When clutch is engaged, disc is squeezed between flywheel and clutch pressure plate.

CLUTCH EXPLOSION: Clutches have literally flown apart (exploded) when subjected to high rpm. Scatter shield is used on competition cars to protect driver and spectators from flying parts in event clutch explodes.

CLUTCH HOUSING or BELL HOUSING: Cast iron or aluminum housing that surrounds flywheel and clutch mechanism.

CLUTCH PEDAL FREE TRAVEL: Specified distance clutch pedal may be depressed before throw-out bearing actually contacts clutch release fingers.

CLUTCH PILOT BEARING: Small bronze bushing, or in some cases ball bearing, placed in end of crankshaft or in center of flywheel depending on car, used to support outboard end of transmission input shaft.

CLUTCH PRESSURE PLATE: Part of a clutch assembly that through spring pressure, squeezes clutch disc against flywheel thereby transmitting driving force through the assembly. To disengage clutch, pressure plate is drawn away from flywheel via linkage.

CLUTCH SEMI-CENTRIFUGAL RELEASE FINGERS: Clutch release fingers that have a weight attached to them so that at high rpm release fingers place additional pressure on clutch pressure plate.

CLUTCH THROW-OUT FORK: Device or fork that straddles throw-out bearing and used to force throw-out bearing against clutch release fingers.

CO: Symbol for carbon monoxide.

COEFFICIENT OF FRICTION: Measurement of amount of friction developed between two objects in physical contact when one object is drawn across the other. If a book were placed on a table and measuring scale used to pull the book, amount of weight or pull registered on scale would be the coefficient of friction.

COIL (Ignition): Unit used to step up battery voltage to point necessary to fire spark plugs.

COIL SPRING: Section of spring steel rod wound in spiral pattern or shape. Widely used in both front and rear suspension systems.

COLD: Little or no perceptible heat.

COLLAPSED (Piston): Piston whose skirt diameter has been reduced due to heat and forces imposed upon it during service in engine.

COMBUSTION: Process involved during burning.

COMBUSTION CHAMBER: Area above piston with piston on TDC. Head of piston, cylinder and head form the chamber.

COMBUSTION CHAMBER VOLUME: Volume of combustion chamber (space above piston with piston on TDC) measured in cc (cubic centimeters).

COMMUTATOR: Series of copper bars connected to armature windings. Bars are insulated from each other and from armature. Brushes, (as in generator or starter) rub against whirling commutator.

COMPENSATING PORT: Small hole in brake master cylinder to permit fluid to return to reservoir.

COMPENSATOR VALVE (Automatic Transmission): Valve designed to increase pressure on brake band during heavy acceleration.

COMPOUND: Two or more ingredients mixed together.

COMPRESSION: Applying pressure to a spring, or any springy substance, thus causing it to reduce its length in direction of compressing force. Applying pressure to gas, thus causing reduction in volume.

COMPRESSION CHECK: Testing compression in all cylinders at cranking speed. All plugs are removed, compression gage placed in one plug hole, throttle cracked wide open and engine cranked until gage no longer climbs. Compression check is a fine way in which to determine condition of valves, rings and cylinders.

COMPRESSION GAGE: Gage used to test compression in cylinders.

COMPRESSION RATIO: Relationship between cylinder volume (clearance volume) when piston is on TDC and cylinder volume when piston is on BDC.

COMPRESSION STROKE: Portion of piston's movement devoted to compressing the fuel mixture trapped in engine's cylinder.

CONCENTRIC: Two or more circles so placed as to share common center.

CONDENSE: Turning vapor back into liquid.

CONDENSER (Ignition): Unit installed between breaker points and coil to prevent arcing at breaker points. Condenser has ability to absorb and retain surges of electricity.

CONDENSER (Refrigeration): Unit in air conditioning system that cools hot compressed refrigerant and turns it from vapor into liquid.

CONDENSATION: Moisture, from air, deposited on a cool surface.

CONDUCTION: Transfer of heat from one object to another by having objects in physical contact.

CONDUCTOR: Material forming path for flow of current.

CONE CLUTCH: Clutch utilizing cone-shaped member that is forced into a cone-shaped depression in flywheel, or other driving unit, thus locking two together. Although no longer used on cars, cone clutch finds some applications in small riding tractors, heavy power mowers, etc.

CONNECTING ROD: Connecting link between piston and crankshaft.

CONSTANT MESH GEARS: Gears that are always in mesh with each other — driving or not.

CONSTANT VELOCITY UNIVERSAL JOINT: Universal joint so designed as to effect smooth transfer of torque from driven shaft to driving shaft without any fluctuations in speed of driven shaft.

CONTACT POINTS also called BREAKER POINTS: Two removable points or areas that when pressed together, complete circuit. These points are usually made of tungsten, platinum or silver.

CONTRACTION (Thermal): Reduction in size of object when cooled.

CONVECTION: Transfer of heat from one object to another when hotter object heats surrounding air and air in turn heats other object.

COOLANT: Liquid in cooling system.

CORE: When referring to casting — sand unit placed inside mold so that when metal is poured, core will leave a hollow shape.

CORONA (Electrical): Luminous discharge of electricity visible near surface of an electrical conductor under high voltage.

CORRODE: Removal of surface material from object by chemical action.

COUNTERBALANCE: Weight attached to some moving part so part will be in balance.

COUNTERBORE: Enlarging hole to certain depth.

COUNTERCLOCKWISE: Rotation to the left as opposed to that of clock hands.

COUNTERSHAFT: Intermediate shaft that receives motion from one shaft and transfers it to another. It may be fixed (gears turn on it) or it may be free to revolve.

COUNTERSINK: To make a counterbore so that head of a screw may set flush, or below the surface.

COUPLING: Connecting device used between two objects so motion of one will be imparted to other.

COUPLING POINT: This refers to point at which both pump and turbine in torque converter are traveling at same speed. The drive is almost direct at this point.

COWL: Part of car body between engine firewall and front of dash panel.

CRANKCASE: Part of engine that surrounds crankshaft. Not to be confused with the pan which is a thin steel cover that is bolted to crankcase.

CRANKCASE DILUTION: Accumulation of unburned gasoline in crankcase. Excessively rich fuel mixture or poor combustion will allow certain amount of gasoline to pass down between pistons and cylinder walls.

CRANKSHAFT: Shaft running length of engine. Portions of shaft are offset to form throws to which connecting rods are attached. Crankshaft is supported by main bearings.

CRANKSHAFT GEAR: Gear mounted on front of crankshaft. Used to drive camshaft gear.

CROSS SHAFT (Steering): Shaft in steering gearbox that engages steering shaft worm. Cross shaft is splined to pitman arm.

CRUDE OIL: Petroleum in its raw or unrefined state. It forms the basis of gasoline, engine oil, diesel oil, kerosene, etc.

CUBES: Cubic inches, or cubic inch displacement of an engine.

CU. IN. (C.I.): Cubic inch.

CUNO FILTER: Filter made up of a series of fine discs or plates pressed together in a manner that leaves very minute space between discs. Liquid is forced through these openings to produce straining action.

CURRENT: Movement of free electrons through conductor.

CUTOUT (Exhaust): Form of bypass valve, located in exhaust line, used to divert the flow of exhaust from one pipe to another. Often used to bypass muffler into straight pipe.

CUTOUT (Regulator): Device to connect or disconnect generator from battery circuit. When generator is charging, cutout makes circuit. When generator stops, cutout breaks circuit. Also referred to as cutout relay, and circuit breaker.

CYCLE: Reoccurring period during which series of events take place in definite order.

CYLINDER: Hole, or holes, in cylinder block that contain pistons.

CYLINDER BLOCK: See Block.

CYLINDER HEAD: Metal section bolted on top of block. Used to cover tops of cylinders. In many cases cylinder head contains the valves. Also forms

part of combustion chamber.

CYLINDER HONE: Tool that uses an abrasive to smooth out and bring to exact measurements such as engine cylinders, wheel cylinders, bushings, etc.

CYLINDER LINER: See Cylinder Sleeve.

CYLINDER SLEEVE: Replaceable cylinder. It is made of a pipe-like section that is either pressed or pushed into the block.

DASHBOARD: Part of body containing driving instruments, switches, etc.

DASHPOT: Unit utilizing cylinder and piston, or cylinder and diaphragm, with small vent hole, to retard or slow down movement of some part.

DC (Electrical): Direct Current.

DC (Piston position): Dead Center. Piston at extreme top or bottom of its stroke.

DEAD AXLE: Axle that does not rotate but merely forms base upon which to attach wheels.

DEAD CENTER (Engine): Point at which piston reaches its uppermost or downmost position in cylinder. Rod crank journal would be at 12 o'clock UDC or 6 o'clock LDC.

DE DION (De Dion): Rear axle setup in which driving wheels are attached to curved dead axle attached to frame by a central pivot. Differential unit is bolted to frame and is connected to the driving wheels by drive axles utilizing universal joints.

DEFLECTION RATE (Springs): Measurement of force, in lbs., required to compress leaf spring a distance of one inch.

DEGLAZER: Abrasive tool used to remove glaze from cylinder walls so a new set of rings will seat.

DEGREE (Circle): 1/360 part of a circle.

DEGREE WHEEL: Wheel-like unit attached to engine crankshaft. Used to time valves to a high degree of accuracy.

DEMAGNETIZE: Removing residual magnetism from an object.

DEPOLARIZE: Removal of residual magnetism thereby destroying or removing the magnetic poles.

DESICCANT: Material, such as silica-gel, placed within a container to absorb and retain moisture.

DETENT BALL AND SPRING: Spring loaded ball that snaps into a groove or notch to hold some sliding object in position.

DETERGENT: Chemical added to engine oil to improve its characteristics (sludge control, non-foaming, etc.).

DETONATION: Fuel charge firing or burning too violently, almost exploding.

DEUCE: Hot rod built around a 1932 Ford coupe body.

DIAGNOSIS: Process of analyzing certain symptoms, readings, etc., in order to determine underlying reason for trouble at hand.

DIAL GAGE OR INDICATOR: Precision micrometer type instrument that indicates reading via needle moving across dial face.

DIAPHRAGM: Flexible cloth-rubber sheet stretched across an area thereby separating two different compartments.

DIE (Forming): One of a matched pair of hardened steel blocks that are used to form metal into a desired shape.

DIE (Thread): Tool for cutting threads.

DIE CASTING: Formation of an object by forcing molten metal, plastic, etc., into a die.

DIESEL ENGINE: Internal combustion engine that uses diesel oil for fuel. True diesel does not use an ignition system but injects diesel oil into cylinders when piston has compressed air so tightly that it is hot enough to ignite diesel fuel without spark.

DIESELING: Condition in which engine continues to run after ignition key is turned off. Also called "running on."

DIFFERENTIAL: Unit that will drive both rear axles at same time but will allow them to turn at different speeds when negotiating turns.

DIFFERENTIAL CASE: Steel unit to which the ring gear is attached. Case drives spider gears and forms

an inner bearing surface for axle and gears.

DIG OUT: To accelerate at top power.

DIODE: Unit having ability to pass electric current readily in one direction but resisting current flow in the other.

DIPSTICK: Metal rod that passes into oil sump. Used to determine quantity of oil in engine.

DIRECT CURRENT (DC): Electric current that flows steadily in one direction only.

DIRECT DRIVE: Such as high gear when crankshaft and drive shaft revolve at same speed.

DIRECTIONAL STABILITY (Steering): Ability of car to move forward in straight line with minimum of driver control. Car with good directional stability will not be unduly affected by side wind, road irregularities, etc.

DISCHARGE (Battery): Drawing electric current from battery.

DISC WHEEL: Wheel constructed of stamped steel.

DISPLACEMENT: Total volume of air displaced by piston traveling from BDC to TDC.

DISTILLATION: Heating a liquid and then catching and condensing the vapors given off by heating process.

DISTRIBUTION TUBES (Cooling System): Tubes used in engine cooling area to guide and direct flow of coolant to vital areas.

DISTRIBUTOR (Ignition): Unit designed to make and break the ignition primary circuit and to distribute resultant high voltage to proper cylinder at correct time.

DISTRIBUTOR CAP (Ignition): Insulated cap containing central terminal with series (one per cylinder) of terminals that are evenly spaced in circular pattern around central terminal. Secondary voltage travels to central terminal where it is then channeled to one of outer terminals by the rotor.

DOHC: Refers to an engine with double (two) overhead camshaft.

DOUBLE FLARE: End of tubing, especially brake tubing, has a flare so made that flare area utilizes two wall thicknesses. This makes a much stronger joint and from safety standpoint, it is a must.

DOWEL PIN: Steel pin, passed through or partly through, two parts to provide proper alignment.

DOWNDRAFT CARBURETOR: A carburetor in which air passes downward through carburetor into intake manifold.

DOWNSHIFT: Shifting to lower gear.

DRAG: To accelerate a car from standing start, over course one-fourth mile in length.

Also used by some drivers when referring to challenging another driver to an acceleration race.

DRAG LINK: A steel rod connecting pitman arm to one of steering knuckles. On some installations drag link connects pitman arm to a center idler arm.

DRAG WHEEL: Special steering wheel used on some dragsters. Often consists of cross-bar spoke and portion of rim on each end.

DRAGSTER: Car especially built for drag racing.

DRAW (Electrical): Amount of electrical current required to operate electrical device.

DRAW (Forming): To form (such as wire) by pulling wire stock through series of hardened dies.

DRAW (Temper): Process of removing hardness from a piece of metal.

DRAW-FILING: Filing by passing file, at right angles, up and down the length of work.

DRIER (Receiver-Drier): Tank, containing desiccant, inserted in air conditioning system to absorb and retain moisture.

DRILL: Tool used to bore holes.

DRILL PRESS: Nonportable machine used for drilling.

DRIVE-FIT: Fit between two parts when they must be literally driven together.

DRIVE LINE: Propeller shaft, universal joints, etc., connecting transmission output shaft to axle pinion gear shaft.

DRIVE OR PROPELLER SHAFT SAFETY STRAP: A metal strap or straps, surrounding drive shaft to prevent shaft from falling to ground in event of a universal joint or shaft failure.

DRIVE SHAFT: Shaft connecting transmission output shaft to differential pinion shaft.

DROP CENTER RIM: Center section of rim being lower than two outer edges. This allows bead of tire to be pushed into low area on one side while the other side is pulled over and off the flange.

DROP FORGED: Part that has been formed by heating steel blank red hot and pounding it into shape with a powerful drop hammer.

DROPPED AXLE: Front axle altered so as to lower the frame of car. Consists of bending axle downward at outer ends. (Solid front axle.)

DRY CELL or DRY BATTERY: Battery (like flashlight battery) that uses no liquid electrolyte.

DRY CHARGED BATTERY: Battery with plates charged but lacking electrolyte. When ready to be placed in service, electrolyte is added.

DRY SLEEVE: Cylinder sleeve application in which sleeve is supported in block metal over its entire length. Coolant does not touch sleeve itself.

DUAL BRAKES: Tandem or dual master cylinder to provide separate brake system for both front and rear of car.

DUAL BREAKER POINTS (Ignition): Distributor using two sets of breaker points to increase cam angle so that at high engine speeds, sufficient spark will be produced to fire plugs.

DUALS: Two sets of exhaust pipes and mufflers — one for each bank of cylinders.

DUNE BUGGY: Off-road vehicle set up to run in sand.

DWELL: See Cam Angle.

DYNAMIC BALANCE: When center line of weight mass of a revolving object is in same plane as center line of object, that object would be in dynamic balance. For example, weight mass of the tire must be in the same plane as center line of wheel.

DYNAMO: Another word for generator.

DYNAMOMETER: Machine used to measure engine horsepower output. Engine dynamometer measures horsepower at crankshaft and chassis dynamometer measures horsepower output at wheels.

EARTH (Electrical): British term for ground.

EARTH WIRE: British term for ground wire.

ECCENTRIC (Off center): Two circles, one within the other, neither sharing the same center. A protrusion on a shaft that rubs against or is connected to another part.

ECONOMIZER VALVE: Fuel flow control device within carburetor.

EEC: Evaporative Emission Control.

EGR: Exhaust Gas Recirculation.

ELECTROCHEMICAL: Chemical (battery) production of electricity.

ELECTRODE (Spark plug): Center rod passing through insulator forms one electrode. The rod welded to shell forms another. They are referred to as center and side electrodes.

ELECTRODE (Welding): Metal rod used in arc welding.

ELECTROLYTE: Sulphuric acid and water solution in battery.

ELECTROMAGNET: Magnet produced by placing coil of wire around steel or iron bar. When current flows through coil, bar becomes magnetized and will remain so as long as current continues to flow.

ELECTROMAGNETIC: Magnetic (generator) production of electricity.

ELECTRON: Negatively charged particle that makes up part of the atom.

ELECTROPLATE: Process of depositing gold, silver, chrome, nickel, etc., upon an object by placing object in special solution and then passing an electric current through solution. Object forms one terminal, special electrode the other. Direct current is used.

ELEMENT (Battery): Group of plates. Three elements for a six volt and six elements for the twelve volt battery. The elements are connected in series.

ELLIOT TYPE AXLE: Solid bar front axle on which ends span or straddle steering knuckle.

EMF: Electromotive force. (Voltage.)

EN-BLOCK: One piece — such as an engine cylinder block cast in one piece.

ENDPLAY: Amount of axial (lengthwise) movement between two parts.

ENERGY (Physics): Capacity for doing work.

ENGINE ADAPTER: Unit that allows a different engine to be installed in a car — and still bolt up to original transmission.

ENGINE (Auto): Device that converts heat energy into useful mechanical motion.

ENGINE DISPLACEMENT: Volume of space through which head of piston moves in full length of its stroke — multiplied by number of cylinders in engine. Result is given in cubic inches.

EP LUBRICANT (Extreme Pressure): Lubricant compounded to withstand very heavy loads imposed on gear teeth.

ESC: Electronic Spark Control.

ET (Elapsed Time): Length of time it takes a dragster to complete one-fourth mile run.

ETHYL GASOLINE: Gasoline to which Ethyl fluid has been added to improve gasoline's resistance to knocking. Slows down burning rate thereby creating a smooth pressure curve that will allow the gasoline to be used in high compression engines.

ETHYLENE GLYCOL: Chemical solution added to cooling system to protect against freezing.

EVAPORATOR: Unit in air conditioning system used to transform refrigerant from a liquid to a gas. It is at this point that cooling takes place.

EXCITE: To pass an electric current through a unit such as field coils in generator.

EXHAUST CUTOUT: Y-shaped device placed in exhaust pipe ahead of muffler. Driver may channel exhaust through muffler or out other leg of the Y where exhaust passes out without going through the muffler.

EXHAUST GAS ANALYZER: Instrument used to check exhaust gases to determine combustion efficiency.

EXHAUST MANIFOLD: Connecting pipes between exhaust ports and exhaust pipe.

EXHAUST PIPE: Pipe connecting exhaust manifold to muffler.

EXHAUST STROKE: Portion of piston's movement devoted to expelling burned gases from cylinder.

EXHAUST TUNING: Cutting exhaust pipe to length that provides maximum efficiency.

EXHAUST VALVE (ENGINE): Valve through which burned fuel charge passes on its way from cylinder to exhaust manifold.

°F: Temperature measurement in degrees Fahrenheit.

FAHRENHEIT: Thermometer on which boiling point of water is 212 deg. and freezing point is 32 deg. above zero.

FARAD: Unit of capacitance; capacitance of condenser retaining one coulomb of charge with one volt difference of potential.

FEELER GAGE: Thin strip of hardened steel, ground to an exact thickness, used to check clearances between parts.

FENDER SKIRT: Plate designed to cover portion of rear fender wheel opening.

FERROUS METAL: Metal containing iron or steel.

F-HEAD ENGINE: Engine having one valve in the head and the other in the block.

FIBER GLASS: Mixture of glass fibers and resin that when curved (hardened) produces a very light and strong material. Used to build boats, car bodies, repair damaged areas, etc.

FIELD: Area covered or filled with a magnetic force.

FIELD COIL: Insulated wire wrapped around an iron or steel core. When current flows through wire, strong magnetic force field is built up.

FILAMENT: Fine wire inside light bulb that heats to incandescence when current passes through it. The filament produces the light.

FILLET: Rounding joint between two parts connected at an angle.

FILTER: Device designed to remove foreign substances from air, oil, gasoline, water, etc.

FINAL DRIVE RATIO: Overall gear reduction (includes transmission, overdrive, auxiliary transmission, etc., gear ratio as well as rear axle ratio) at rear wheels.

FINISHING STONE (Hone): Fine stone used for final finishing during honing.

FIREWALL: Metal partition between driver's compartment and engine compartment.

FIRING ORDER: Order in which cylinders must be fired — 1, 5, 3, 6, 2, 4, etc.

FIT: Contact area between two parts.

FLARING TOOL: Tool used to form flare connections on tubing.

FLASH POINT: The point in the temperature range at which a given oil will ignite and flash into flame.

FLAT CRANK: Crankshaft having one of the bearing journals out-of-round.

FLAT HEAD: Engine with all the valves in block.

FLAT SPOT: Refers to a spot experienced during an acceleration period where the engine seems to "fall on its face" for a second or so and will then begin to pull again.

FLOAT BOWL: The part of the carburetor that acts as a reservoir for gasoline and in which the float is placed.

FLOAT LEVEL: Height of fuel in carburetor float bowl. Also refers to specific float setting that will produce correct fuel level.

FLOODING: Condition where fuel mixture is overly rich or an excessive amount has reached cylinders. Starting will be difficult and sometimes impossible until condition is corrected.

FLUID COUPLING: Unit that transfers engine torque to transmission input shaft through use of two vaned units (called a torus) operating very close together in a bath of oil. Engine drives one torus causing it to throw oil outward and into other torus which then begins to turn the transmission input shaft. A fluid coupling cannot increase torque above that produced by crankshaft. (Engine torque.)

FLUTE: Groove in cutting tool that forms a passageway for exit of chips removed during the cutting process.

FLUX (Magnetic): Lines of magnetic force moving through magnetic field.

FLUX (Soldering, brazing): Ingredient placed on metal being soldered or brazed, to remove and prevent formation of surface oxidization which would make soldering or brazing difficult.

FLYWHEEL: Relatively large wheel that is attached to crankshaft to smooth out firing impulses. It provides inertia to keep crankshaft turning smoothly during periods when no power is being applied. It also forms a base for starter ring gear and in many instances, for clutch assembly.

FLYWHEEL RING GEAR: Gear on outer circumference of flywheel. Starter drive gear engages ring gear and cranks engine.

FOOT POUND: Measurement of work involved in lifting one pound one foot.

FOOT POUND (Tightening): One pound pull one foot from center of an object.

FORCE: Pressure (pull, push, etc.) acting upon body that tends to change state of motion, or rest, of the body.

FORCE-FIT: Same as drive fit.

FORGE: To force piece of hot metal into desired shape by hammering.

FOUR BANGER, SIX BANGER, ETC.: Four cylinder, six cylinder engine, etc.

FOUR-ON-THE-FLOOR: Four-speed manual transmission with floor mounted shift.

FOUR-STROKE CYCLE ENGINE: Engine requiring two complete revolutions of crankshaft to fire each piston once.

FOUR-WHEEL DRIVE: Vehicle, such as Jeep, in which front wheels, as well as rear, may be driven.

FREE ELECTRONS: Electrons in outer orbits

around nucleus of atom. They can be moved out of orbit comparatively easy.

FREEWHEEL: Usually refers to action of car on downgrade when overdrive over-running clutch is slipping with resultant loss of engine braking. This condition will only occur after overdrive unit is engaged but before balk ring has activated planetary gearset.

FREEZING: When two parts that are rubbing together heat up and force lubricant out of area, they will gall and finally freeze or stick together.

FREON-12: Gas used as cooling medium in air conditioning and refrigeration systems.

FRICTION: Resistance to movement between any two objects when placed in contact with each other. Friction is not constant but depends on type of surface, pressure holding two objects together, etc.

FRICTION BEARING: Bearing made of babbitt, bronze, etc. There are no moving parts and shaft that rests in bearing merely rubs against friction material in bearing.

FUEL: Combustible substance that is burned within (internal) or without (external) an engine so as to impart motion to pistons, vanes, etc.

FUEL BURNER or FUELER: Competition car with an engine set up to burn alcohol, nitro, etc. mixture instead of standard pump gasoline.

FUEL INJECTION: Fuel system that uses no carburetor but sprays fuel either directly into cylinders or into intake manifold just ahead of cylinders.

FUEL MIXTURE: Mixture of gasoline and air. An average mixture, by weight, would contain 16 parts of air to one part of gasoline.

FUEL PUMP: Vacuum device, operated either mechanically or electrically, that is used to draw gasoline from tank and force it into carburetor.

FULCRUM: Support on which a lever pivots in raising an object.

FULL-FLOATING AXLE: Rear drive axle that does not hold wheel on nor does it hold wheel in line or support any weight. It merely drives wheel. Used primarily on trucks.

FULL-FLOW OIL FILTER: Oil filter that filters ALL of oil passing through engine — before it reaches the bearings.

FULL HOUSE: Engine that is fully modified and equipped for all-out performance.

FULL TIME FOUR-WHEEL DRIVE: Setup in which all four wheels are driven — all the time — off road or on. Addition of a third differential, located at transfer case, permits front and rear wheels to operate at different speeds.

FUNNY CAR: Car equipped with a powerful engine, used for drag racing. Usually has special body (such as fiber glass) mounted on special lightweight frame and suspension system.

FUSE: Protective device that will break flow of current when current draw exceeds capacity of fuse.

FUSION: Two metals reaching the melting point and flowing or welding themselves together.

GAL: Gallon.

GALVONOMETER: Instrument used to measure pressure, amount of, and direction of an electric current.

GAS: A nonsolid material. It can be compressed. When heated, it will expand and when cooled, it will contract. (Such as air.)

GAS BURNER or GASSER: Competition car with engine set up to operate on standard pump gasoline instead of an alcohol, nitro, etc., mixture.

GASKET: Material placed between two parts to insure proper sealing.

GASOLINE: Hydrocarbon fuel used in the internal combustion engine.

GASSING: Small hydrogen bubbles rising to top of battery electrolyte during battery charging.

GEAR: Circular object, usually flat edged or cone-shaped, upon which a series of teeth have been cut. These are meshed with teeth of another gear and

when one turns, it also drives the other.

GEAR RATIO: Relationship between number of turns made by driving gear to complete one full turn of driven gear. If driving gear turns four times to turn driven gear once, gear ratio would be 4 to 1.

GENERATOR: Electromagnetic device for producing electricity.

GLASS: Term used for the material "fiber glass."

GLASS PACK MUFFLER: Straight through (no baffles) muffler utilizing fiber glass packing around perforated pipe to deaden exhaust sound.

GLAZE: Highly smooth, glassy finish on cylinder walls.

GLAZE BREAKER or DEGLAZER: Abrasive tool used to remove glaze from cylinder walls prior to installation of new piston rings.

GOVERNOR: Device designed to automatically control speed or position of some part.

GPM: Gallons Per Minute.

GRID: Lead screen or plate to which battery plate active material is affixed.

GRIND: To remove metal from an object by means of revolving abrasive wheel, disc or belt.

GROUND (Battery): Terminal of battery connected to metal framework of car. In this country, NEGATIVE terminal is grounded.

GROWLER: Instrument used in testing starter and generator armature.

GUDGEON PIN: British term for piston or wrist pin.

GUM (Fuel system): Oxidized portions of fuel that form deposits in fuel system or engine parts.

GUT: To strip the interior of car. May also refer to removing internal baffles from muffler.

GUTTED MUFFLER: Muffler with no silencing baffles. Makes a very loud sound.

GVW: Gross Vehicle Weight. Total weight of vehicle including vehicle passengers, load, etc. Used as indicator of how heavy vehicle can be loaded (GVW minus vehicle curb weight = payload).

HALF-MOON KEY: Driving key serving same purpose as regular key but it is shaped somewhat like a half circle.

HARMONIC BALANCER: See Vibration Damper.

HC: Symbol for hydrocarbon.

HEADERS: Special exhaust manifolds that replace stock manifold. Designed with smooth flowing lines to prevent back pressure caused by sharp bends, rough castings, etc.

HEAT CROSSOVER (V-8 engine): Passage from one exhaust manifold up, over and under carburetor and on to other manifold. Crossover provides heat to carburetor during engine warmup.

HEAT ENGINE: Engine operated by heat energy released from burning fuel.

HEAT EXCHANGER: Device, such as radiator, either used to cool or heat by transferring heat from one object to another.

HEAT RANGE (Spark plugs): Refers to operating temperature of given style plug. Plugs are made to operate at different temperatures depending upon thickness and length of porcelain insulator as measured from sealing ring down to tip.

HEAT RISER: Area, surrounding portion of the intake manifold, through which exhaust gases can pass to heat fuel mixture during warmup.

HEAT SINK: Device used to prevent overheating of electrical device by absorbing heat and transferring it to atmosphere.

HEAT TREATMENT (Metal): Application of controlled heat to metal object in order to alter its characteristics (toughness, hardness, etc.).

HEEL (Brake): End of brake shoe which rests against anchor pin.

HEEL (Gear Tooth): Wide end of tapered gear tooth such as found in differential gears.

Hg: Abbreviation for the word MERCURY. Vacuum is measured in inches of mercury.

HELICAL: Spiraling shape such as that made by a coil spring.

HELICAL GEAR: Gear that has teeth cut at an angle to center line of gear.

HEMI: Engine using hemispherical-shaped (half of globe) combustion chambers.

HEMISPHERICAL COMBUSTION CHAMBER: A round, dome-shaped combustion chamber that is considered by many to be one of the finest shapes ever developed. Hemispherical-shape lends itself to use of large valves for improved breathing and suffers somewhat less heat loss than other shapes.

HERRINGBONE GEARS: Two helical gears operating together and so placed that angle of the teeth form a "V" shape.

HIGH COMPRESSION HEADS: Cylinder head with smaller combustion chamber area thereby raising compression. Head can be custom built or can be a stock head milled (cut) down.

HIGH LIFT ROCKER ARMS: Custom rocker arms designed so that standard lift of push rod will depress or open valve somewhat more than stock lifter.

HIGH-RISE MANIFOLD: Intake manifold designed to mount carburetor or carburetors, considerably higher above engine than is done in standard manifold, done to improve angle at which fuel is delivered.

HIGH TENSION: High voltage from ignition coil. May also indicate secondary wire from the coil to distributor and wires from distributor to plugs.

HONE: To remove metal with fine grit abrasive stone to precise tolerances.

HOOD PINS: Pins designed to hold hood closed.

HOPPING UP: Increasing engine performance through various modifications.

HORIZONTAL-OPPOSED ENGINE: Engine possessing two banks of cylinders that are placed flat or 180 deg. apart.

HORSEPOWER: Measurement of engine's ability to perform work. One horsepower is defined as ability to lift 33,000 pounds one foot in one minute. To find horsepower, total rate of work in foot pounds accomplished is divided by 33,000. If a machine was lifting 100 pounds 660 feet per minute, its total rate of work would be 66,000 foot pounds.

Dividing this by 33,000 foot pounds (1 horsepower) you find that the machine is rated as 2 horsepower (hp).

HORSEPOWER (Brake): See Brake Horsepower.

HORSEPOWER (Gross): Maximum horsepower developed by engine without a fan, air cleaner, alternator, exhaust system, etc.

HORSEPOWER (Net): Maximum horsepower developed by engine equipped with fan, air conditioning, air cleaner, exhaust system, and all other systems and items normally present when engine is installed in car.

HORSEPOWER — WEIGHT FACTOR: Relationship between total weight of car and horsepower available. By dividing weight by horsepower, number of pounds to be moved by one horsepower is determined. This factor has a great effect on acceleration, gas mileage and all around performance.

HOTCHKISS DRIVE: Method of connecting transmission output shaft to differential pinion by using open drive shafts. Driving force of rear wheels is transmitted to frame through rear springs or through link arms connecting rear axle housing to frame.

HOT ROD: Car that has been modified to produce high performance, (extra power, better traction, superior gearing, better suspension, etc.).

HOT SHOT BATTERY: A dry cell battery generally of six volts.

HOT SPOT: Localized area in which temperature is considerably higher than surrounding area.

HOT WIRE: Wiring around key switch so as to start car without key.

Wire connected to battery or to some part of electrical system in which a direct connection to battery is present. A current-carrying wire.

Hp: Horsepower.

HUB (Wheel): Unit to which wheel is bolted.

HYATT ROLLER BEARING (antifriction): Similar to conventional roller bearing except that rollers are hollow and are split in a spiral fashion from end to end.

HYDRAULIC: Refers to fluids in motion. Hydraulics is science of fluid in motion.

HYDRAULIC BRAKES: Brakes that are operated by hydraulic pressure. Master cylinder provides operating pressure that is transmitted via steel tubing to wheel cylinders that in turn apply brake shoes to brake drums.

HYDRAULIC LIFTER: Valve lifter that utilizes hydraulic pressure from engine's oiling system to keep it in constant contact with both camshaft and valve stem. They automatically adjust to any variation in valve stem length.

HYDRAULICS: The science of liquid in motion.

HYDROCARBON: A mixture of hydrogen and carbon.

HYDROCARBON—UNBURNED: Hydrocarbons that were not burned during the normal engine combustion process. Unburned hydrocarbons make up about 0.1 percent of engine exhaust emission.

HYDROCARBONS: Combination of hydrogen and carbon atoms. All petroleum based fuels (gasoline, kerosene, etc.) consist of hydrocarbons.

HYDROMETER: Float device for determining specific gravity of electrolyte in a battery. This will determine the state of charge.

HYPOID GEARING: System of gearing wherein pinion gear meshes with ring gear below center line of ring gear. This allows a somewhat lower drive line thus reducing hump in the floor of car. For this reason hypoid gearing is used in differential on many cars.

ICEI: Internal Combustion Engine Institute.

ICING: Formation of ice (under certain atmospheric conditions) on throttle plate, air horn walls, etc., caused by lowering of fuel mixture temperature as it passes through air horn.

ID: Inside diameter.

IDLE: Indicates engine operating at its normal slow speed with throttle closed.

IDLE VALVE OR IDLE NEEDLE: Needle used to control amount of fuel mixture reaching cylinders during idling. It, or they, may be adjusted by turning the exposed heads.

IGNITION: Lighting or igniting fuel charge by means of a spark (gas engine) or by heat of compression (diesel engine).

IGNITION SYSTEM: Portion of car electrical system, designed to produce a spark within cylinders to ignite fuel charge. Consists basically of battery, key switch, resistor, coil, distributor, points, condenser, spark plugs and necessary wiring.

I-HEAD ENGINE: Engine having both valves in the head.

IHP: Indicated Horsepower.

IMI: Ignition Manufacturers' Institute.

IMPACT WRENCH: An air or electrical driven wrench that tightens or loosens nuts, cap screws, etc., with series of sharp, rapid blows.

IMPELLER: Wheel-like device upon which fins are attached. It is whirled to pump water, move and slightly compress air, etc.

IMPULSE COUPLING (Magneto): Device that speeds up rotating magnet to increase voltage output at cranking speeds.

IN.: Inch.

INCLUDED ANGLE (Steering): Angle formed by center lines drawn through steering axis (kingpin inclination) and center of wheel (camber angle) as viewed from front of car. Combines both steering axis and camber angles.

INDEPENDENT SUSPENSION: A suspension system that allows each wheel to move up and down without undue influence on other wheels.

INDICATED HORSEPOWER (ihp): Measure of power developed by burning fuel within cylinders.

INDUCTION: Imparting of electricity into one object, not connected, to another by the influence of magnetic fields.

INERTIA: Force which tends to keep stationary

object from being moved, and tends to keep moving object in motion.

INHIBITOR: Substance added to oil, water, gas, etc., to prevent action such as foaming, rusting, etc.

INJECTOR (Carburetion): Refers to pump system (used in fuel injection system) that squirts or injects measured amount of gasoline into intake manifold in vicinity of intake valve. In diesel engine, fuel is injected directly into cylinder.

IN-LINE ENGINE: Engine in which all cylinders are arranged in straight row.

INPUT SHAFT: Shaft delivering power into mechanism. Shaft from clutch into transmission is transmission input shaft.

INSERT BEARING: Removable, precision made bearing which insures specified clearance between bearing and shaft.

INSULATOR (Electrical): Unit made of material that will not (readily) conduct electricity.

INTAKE MANIFOLD: Connecting tubes between base of carburetor and port openings to intake valves.

INTAKE STROKE: Portion of piston's movement devoted to drawing fuel mixture into engine cylinder.

INTAKE VALVE (Engine): Valve through which fuel mixture is admitted to cylinder.

INTEGRAL: Part of. (The cam lobe is an integral part of camshaft.)

INTERMEDIATE GEAR: Any gear in auto transmission between 1st and high.

INTERMITTENT: Not constant but occurring at intervals.

INTERNAL COMBUSTION ENGINE: Engine that burns fuel within itself as means of developing power.

ION: Electrically charged atom or molecule produced by electrical field, high temperature, etc.

IONIZE (Air): To convert wholly or partly, into ions. This causes air to become a conductor of electricity.

JERRY CAN: Five gallon container used by many off-road fans to carry extra fuel, water, etc.

JET: Small hole or orifice used to control flow of gasoline in various parts of carburetor.

JOURNAL: Part of shaft prepared to accept a bearing. (Con rod, main bearing.)

JUICE BRAKES: Hydraulic brakes.

KEY: Parallel-sided piece inserted into groove cut part way into each of two parts, which prevents slippage between two parts.

KEYWAY: Slot cut in shaft, pulley hub, wheel hub, etc. Square key is placed in slot and engages a similar keyway in mating piece. Key prevents slippage between two parts.

KICKDOWN SWITCH: Electrical switch that will cause transmission, or overdrive unit, to shift down to lower gear. Often used to secure fast acceleration.

KILL SWITCH: Special switch designed to shut off ignition in case of emergency.

KILOMETER: Metric measurement equivalent to 5/8 of mile.

KINGPIN: Hardened steel pin that is passed through the steering knuckle and axle end. The steering knuckle pivots about the kingpin.

KINGPIN or STEERING AXIS INCLINATION: Tipping the tops of the kingpins inward towards each other. This places the center line of steering axis nearer center line of tire-road contact area.

KNOCKING (Bearing): Noise created by part movement in a loose or worn bearing.

KNOCKING (Fuel): Condition, accompanied by audible noise, that occurs when gasoline in cylinders burns too quickly. Also referred to as detonation.

KNURL: To roughen surface of piece of metal by pressing series of cross-hatched lines into the surface and thereby raising area between these lines.

LACQUER (Paint): Fast drying automotive body paint.

LAMINATED: Something made up of many layers.

LAND: Metal separating a series of grooves.

LANDS (Ring): Piston metal between ring grooves.

LAP: One complete trip around race track or route laid out for racing.

LAP or LAPPING: To fit two surfaces together by coating them with abrasive and then rubbing them together.

LATENT HEAT: Amount of heat (Btu's) beyond boiling or melting point, required to change liquid to a gas, or a solid to a liquid.

LATENT HEAT OF EVAPORATION: Amount of heat (Btu's) required to change a liquid to a vapor state without elevating vapor temperature above that of the liquid.

LB: Pound.

LEAD BURNING: Connecting two pieces of lead by melting edges together.

LEAF SPRING: Suspension spring made up of several pieces of flat spring steel. Varying numbers of leaves (individual pieces) are used depending on intended use. One car uses single leaf in each rear spring.

LETTER DRILLS: Series of drills in which each drill size is designated by letter of alphabet — A, B, C, etc.

L-HEAD ENGINE: Engine having both valves in block and on same side of cylinder.

LIGHTENED VALVES: Valves in which all possible metal has been ground away to reduce weight. This will allow higher rpm without valve float.

LIMITED-SLIP DIFFERENTIAL: Differential unit designed to provide superior traction by transferring driving torque, when one wheel is spinning, to wheel that is not slipping.

LINKAGE: Movable bars or links connecting one unit to another.

LIQUID TRACTION: Special liquid applied to tires of drag racers to provide superior traction.

LIQUID WITHDRAWAL (LPG): Drawing LPG from bottom of tank to insure delivery of liquid LPG. Withdrawal from top of the tank will deliver LPG in the gaseous state.

LITER: Metric measurement of capacity — equivalent to 2.11 pints. Five liters equals 1.32 gallon.

LIVE AXLE: Axle upon which wheels are firmly affixed. Axle drives the wheels.

LIVE WIRE: See Hot Wire.

LOAD RANGE (Tire): Letter system (A, B, C, etc.) used to indicate specific tire load and inflation limit.

LOG MANIFOLD: Special intake manifold generally designed to accept four or more carburetors. Each side has bases for carburetors set on a pipe-like log area.

LONG and SHORT ARM SUSPENSION: Suspension system utilizing upper and lower control arm. Upper arm is shorter than lower. This is done so as to allow wheel to deflect in a vertical direction with a minimum change in camber.

LONGITUDINAL LEAF SPRING: Leaf spring mounted so it is parallel to length of car.

LOUVER: Ventilation slots such as sometimes found in hood of automobile.

LOW BRAKE PEDAL: Condition where brake pedal approaches too close to floorboard before actuating the brakes.

LOW LEAD FUEL: Gasoline containing not much more than 0.5 grams of tetraethyl lead per gallon.

LOW PIVOT SWING AXLE: Rear axle setup that attaches differential housing to frame via a pivot mount. Conventional type of housing and axle extend from differential to one wheel. The other side of differential is connected to other driving wheel by a housing and axle that is pivoted at a point in line with differential to frame pivot point.

LPG: Liquefied petroleum gas.

LUBRICANT: Any material, usually of a petroleum nature such as grease, oil, etc., that is placed between two moving parts in an effort to reduce friction.

LUBRICATION: Reducing friction between two

parts by coating them with oil, grease, etc.

LUG (Engine): To cause engine to labor by failing to shift to a lower gear when necessary.

MAG: Magneto.

MAGNAFLUX: Special chemical process, used to check parts for cracks.

MAGNET (Permanent): Piece of magnetized steel that will attract all ferrous material. Permanent magnet does not need electricity to function and will retain its magnetism over a period of years.

MAGNETIC FIELD: Area encompassed by magnetic lines of force surrounding either a bar magnet or electromagnet.

MAGNETO: Engine driven unit that generates high voltage to fire spark plugs. It needs no outside source of power such as battery.

MAGS or MAG WHEEL: Lightweight, sporty wheels made of magnesium. Term mag is often applied to aluminum and aluminum and steel combination wheels.

MAIN BEARING SUPPORTS: Steel plate installed over main bearing caps to increase their strength for racing purposes.

MANDREL: Round shaft used to mount stone, cutter, saw, etc.

MANIFOLD: Pipe or number of pipes connecting series of holes or outlets to common opening. See Exhaust and Intake Manifold.

MANIFOLD HEAT CONTROL, VALVE: Valve placed in exhaust manifold, or in exhaust pipe, that deflects certain amount of hot gas around base of carburetor to aid in warmup.

MANOMETER: Instrument to measure pressure (vacuum).

MASTER CYLINDER: Part of hydraulic brake system in which pressure is generated.

MECHANICAL BRAKES: Service brakes that are actuated by mechanical linkage connecting brakes to brake pedal.

MECHANICAL EFFICIENCY: Engine's rating as to how much potential horsepower is wasted through friction within moving parts of engine.

MEGOHM: 1,000,000 ohms.

MEMA: Motor and Equipment Manufacturers' Association.

MEP: Mean Effective Pressure. Pressure of burning fuel (average) on power stroke subtracted by average pressure on other three strokes. Pressure is in pounds per square inch.

MESH: To engage teeth of one gear with those of another.

METAL FATIGUE: Crystallizing of metal due to vibration, twisting, bending, etc. Unit will eventually break. Bending a piece of wire back and forth to break it is a good example of metal fatigue.

METERING ROD: Movable rod used to vary opening area through carburetor jet.

METRIC SIZE: Units made to metric system measurements.

MICROFARAD: 1/1,000,000 farad.

MICROMETER (Inside and outside): Precision measuring tool that will give readings accurate to within fraction of one thousandth of an inch.

MIKE: Either refers to micrometer or to using micrometer to measure an object.

MILL: Often used to refer to engine.
    To remove metal through use of rotating toothed cutter.

MILLIMETER: Metric measurement equivalent to .039370 of an inch.

MILLING MACHINE: Machine that uses variety of rotating cutter wheels to cut splines, gears, keyways, etc.

MISFIRE: Fuel charge in one or more engine cylinders which fails to fire or ignite at proper time.

MODULATOR (Transmission): Pressure control or adjusting valve used in hydraulic system of automatic transmission.

MOLD: Hollow unit into which molten metal is poured to form a casting.

MOLECULE: Smallest portion that matter may be

divided into and still retain all properties of original matter.

MONOBLOCK: All cylinders cast as one unit.

MOTOR: Electrically driven power unit (electric motor). Term is often incorrectly applied to internal combustion engine.

MOTOR (Generator): Attaching generator to battery in such a way it revolves like an electric motor.

MPH: Miles per hour.

MUFFLER: Unit through which exhaust gases are passed to quiet sounds of running engine.

MULTIPLE DISC CLUTCH: Clutch utilizing several clutch discs in its construction.

MULTI-VISCOSITY OILS: Oils meeting S.A.E. requirements for both low temperature requirements of light oil and high temperature, requirements of heavy oil. Example: (S.A.E. 10W — 30).

NADA: National Automobile Dealers' Association.

NASCAR: Letters denoting National Association for Stock Car Auto Racing.

NBFU: National Board of Fire Underwriters.

NEEDLE BEARING (Antifriction): Roller type bearing in which rollers have very narrow diameter in relation to their length.

NEEDLE VALVE: Valve with long, thin, tapered point that operates in small hole or jet. Hole size is changed by moving needle valve in or out.

NEGATIVE TERMINAL: Terminal (such as on battery) from which current flows on its path to positive terminal.

NEUTRON: Neutral charge particle forming part of an atom.

NEWTON'S LAW: For every action there is an equal, an opposite reaction.

NHRA: National Hot Rod Association.

NHTSA: National Highway Traffic Safety Administration.

NITROGEN OXIDES: In combustion process, nitrogen from air combines with oxygen to form nitrogen oxides.

NLGI: National Lubricating Grease Institute.

NONFERROUS METALS: All metals containing no iron — (except in very minute quantities).

NORTH POLE (Magnet): Magnetic pole from which lines of force emanate; travel is from north to south pole.

NOx: See Oxides of Nitrogen.

NOZZLE: Opening through which fuel mixture is directed into carburetor air stream.

NSC: National Safety Council.

NUMBER DRILLS: Series of drills in which each size is designated by number (0-80).

OCTANE RATING: Rating that indicates a specific gasoline's ability to resist detonation.

OD: Outside diameter.

ODOMETER: Device used to measure and register number of miles traveled by car.

OEM: Original Equipment Manufacturer.

OFF-ROAD VEHICLE: Vehicle designed to operate in rough country (hills, sand, mud, etc.) without benefit of regular roads.

OHM: Unit of measurement used to indicate amount of resistance to flow of electricity in a given circuit.

OHMMETER: Instrument used to measure amount of resistance in given unit or circuit. (In ohms.)

OIL BATH AIR CLEANER: Air cleaner that utilizes a pool of oil to insure removal of impurities from air entering carburetor.

OIL BURNER: Engine that consumes an excessive quantity of oil.

OIL — COMBINATION SPLASH and PRESSURE SYSTEM: Engine oiling system that uses both pressure and splash oiling to accomplish proper lubrication.

OIL FILTER: Device used to strain oil in engine thus removing abrasive particles.

OIL — FULL PRESSURE SYSTEM: Engine oiling system that forces oil, under pressure, to moving parts of engine.

OIL GALLERY: Pipe or drilled passageway in engine used to carry engine oil from one area to another.

OIL — ML (Motor Light): Engine oil designed for light duty service under favorable conditions.

OIL — MM (Motor Medium): Engine oil designed for moderate duty service with occasional high speeds.

OIL — MS (Motor Severe): Engine oil designed for high speed, heavy duty operation. Also for a great deal of stop and go driving.

OIL PUMP: Device used to force oil, under pressure to various parts of the engine, it is driven by gear on camshaft.

OIL PUMPING: Condition wherein an excessive quantity of oil passes piston rings and is consumed in combustion chamber.

OIL SEAL: Device used to prevent oil leakage past certain area.

OIL SLINGER: Device attached to revolving shaft so any oil passing that point will be thrown outward where it will return to point of origin.

OIL — SPLASH SYSTEM: Engine oiling system that depends on connecting rods to dip into oil troughs and splash oil to all moving parts.

OPEN CIRCUIT: Circuit in which a wire is broken or disconnected.

OSCILLATING ACTION: Swinging action such as that in pendulum of a clock.

OSCILLOSCOPE: Testing unit which projects visual reproduction of the ignition system spark action onto screen of cathode-ray tube.

OTTO CYCLE: Four-stroke cycle consisting of intake, compression, firing and exhaust strokes.

OUTPUT SHAFT: Shaft delivering power from within mechanism. Shaft leaving transmission, attached to propeller shaft, is transmission output shaft.

OVERDRIVE: Unit utilizing planetary gearset so actuated as to turn drive shaft about one-third faster than transmission output shaft.

OVERHEAD CAMSHAFT: Camshaft mounted above the head, driven by long timing chain.

OVERHEAD VALVES: Valves located in head.

OVERRUNNING CLUTCH: Clutch mechanism that will drive in one direction only. If driving torque is removed or reversed, clutch slips.

OVERRUNNING CLUTCH STARTER DRIVE: Starter drive that is mechanically engaged. When engine starts, overrunning clutch operates until drive is mechanically disengaged.

OVERSQUARE ENGINE: Engine in which bore diameter is larger than length of stroke.

OVERSTEER: Tendency for car, when negotiating a corner, to turn more sharply than driver intends.

OXIDES OF NITROGEN (NOx): Undesirable exhaust emission, especially prevalent when combustion chamber flame temperatures are high.

OXIDIZE (Metal): Action where surface of object is combined with oxygen in air to produce rust, scale, etc.

OXIDIZING FLAME: Welding torch flame in which an excess of oxygen exists. Free or unburned oxygen tends to burn molten metal.

OXYGEN: Gas, used in welding, made up of colorless, tasteless, odorless, gaseous element oxygen found in atmosphere.

PAN: Thin stamped cover bolted to the bottom of crankcase, It forms a sump for engine oil and keeps dirt, etc., from entering engine.

PANCAKE ENGINE: Engine in which cylinders are on a horizontal plane. This reduces overall height and enables them to be used in spots where vertical height is restricted.

PAPER AIR CLEANER: Air cleaner that makes use of special paper through which air to carburetor is drawn.

PARALLEL CIRCUIT: Electrical circuit with two or more resistance units so wired as to permit current to flow through both units at same time. Unlike series circuit, current in parallel circuit does not have to pass through one unit to reach the other.

PARALLELOGRAM STEERING LINKAGE: Steering system utilizing two short tie rods connected to steering arms and to a long center link. The link is supported on one end on an idler arm and the other end is attached directly to pitman arm. Arrangement forms a parallelogram shape.

PARKING BRAKE: Hand operated brake which prevents vehicle movement while parked by locking rear wheels, or transmission output shaft.

PARTICULATES (Lead): Tiny particles of lead found in engine exhaust emissions when leaded fuel is used.

PASCAL'S LAW: "When pressure is exerted on confined liquid, it is transmitted undiminished."

PAWL: Stud or pin that can be moved or pivoted into engagement with teeth cut on another part — such as parking pawl on automatic transmission that can be slid into contact with teeth on another part to lock rear wheels.

PAYLOAD: Amount of weight that may be carried by vehicle. Computed by subtracting vehicle curb weight from GVW.

PCV (Positive Crankcase Ventilation): System which prevents crankcase vapors from being discharged directly into atmosphere.

PEEL or BURN RUBBER: Rear wheels slipping on highway during acceleration.

PEEN: To flatten out end of a rivet, etc., by pounding with round end of a hammer.

PENETRATING OIL: Special oil used to free rusted parts so they can be removed.

PERIPHERY: Outside edge or circumference.

PERMANENT MAGNET: Magnet capable of retaining its magnetic properties over very long period of time.

PETROL: Gasoline.

PETROLEUM: Raw material from which gasoline, kerosene, lube oils, etc., are made. Consists of hydrogen and carbon.

PHILLIPS HEAD SCREW: Screw having a fairly deep cross slot instead of single slot as used in conventional screws.

PHOSPHOR-BRONZE: Bearing material composed of tin, lead and copper.

PHOTOCHEMICAL: Relates to branch of chemistry where radiant energy (sunlight) produces various chemical changes.

PHOTOCHEMICAL SMOG: Fog-like condition produced by sunlight acting upon hydrocarbon and carbon monoxide exhaust emissions in atmosphere.

PIEZOELECTRIC IGNITION: System of ignition that employs use of small section of ceramic-like material. When this material is compressed, even a very tiny amount, it emits a high voltage that will fire plugs. This system does not need a coil, points, or condenser.

PILOT SHAFT: Dummy shaft that is placed in a mechanism as a means of aligning parts. It is then removed and regular shaft installed.

PINGING: Metallic rattling sound produced by the engine during heavy acceleration when ignition timing is too far advanced for grade of fuel being burned.

PINION CARRIER: Part of rear axle assembly that supports and contains pinion gear shaft.

PINION (Gear): Small gear either driven by or driving, a larger gear.

PIPES: Exhaust system pipes.

PISTON: Round plug, open at one end, that slides up and down in cylinder. It is attached to connecting rod and when fuel charge is fired, will transfer force of explosion to connecting rod then to crankshaft.

PISTON BOSS: Built-up area around piston pin hole.

PISTON COLLAPSE: Reduction in diameter of piston skirt caused by heat and constant impact stresses.

PISTON DISPLACEMENT: Amount (volume) of air displaced by piston when moved through full length of its stroke.

PISTON HEAD: Portion of piston above top ring.

PISTON LANDS: Portion of piston between ring grooves.

PISTON PIN or WRIST PIN: Steel pin that is passed through piston. Used as base upon which to fasten upper end of connecting rod. It is round and is

usually hollow.

PISTON RING: Split ring installed in a groove in piston. Ring contacts sides of ring groove and also rubs against cylinder wall thus sealing space between piston and wall.

PISTON RING (Compression): Ring designed to seal burning fuel charge above piston. Generally there are two compression rings per piston and they are located in two top ring grooves.

PISTON RING (Oil Control): Piston ring designed to scrape oil from cylinder wall. Ring is of such design as to allow oil to pass through ring and then through holes or slots in groove. In this way oil is returned to pan. There are many shapes and special designs used on oil control rings.

PISTON RING END GAP: Distance left between ends of the ring when installed in cylinder.

PISTON RING EXPANDER: See Ring Expander.

PISTON RING GROOVE: Slots or grooves cut in piston head to receive piston rings.

PISTON RING SIDE CLEARANCE: Space between sides of ring and ring lands.

PISTON SKIRT: Portion of piston below rings. (Some engines have an oil ring in skirt area.)

PISTON SKIRT EXPANDER: Spring device placed inside piston skirt to produce an outward pressure which increases diameter of skirt.

PISTON SKIRT EXPANDING: Enlarging diameter of piston skirt by inserting an expander, by knurling outer skirt surface, or by peening inside of piston.

PITMAN ARM: Short lever arm splined to steering gear cross shaft. Pitman arm transmits steering force from cross shaft to steering linkage system.

PITS: Area at a race track for fueling, tire changing, making mechanical repairs, etc.

PIT STOP: A stop at the pits by racer, for fuel, tires, repairs, etc.

PIVOT: Pin or shaft about which a part moves.

PLANET CARRIER: Part of a planetary gearset upon which planet gears are affixed. Planet gears are free to turn on hardened pins set into carrier.

PLANET GEARS: Gears in planetary gearset that are in mesh with both ring and sun gear. Referred to as planet gears in that they orbit or move around central or sun gear.

PLANETARY GEARSET: Gearing unit consisting of ring gear with internal teeth, sun or central pinion gear with external teeth, and series of planet gears that are meshed with both the ring and the sun gear.

PLATES (Battery): Thin sections of lead peroxide or porous lead. There are two kinds of plates — positive and negative. The plates are arranged in groups, in an alternate fashion, called elements. They are completely submerged in the electrolyte.

PLATINUM: Precious metal sometimes used in the construction of breaker points. It conducts well and is highly resistant to burning.

PLAY: Movement between two parts.

PLEXIGLAS: Trade name for an acrylic plastic, made by the Rhom and Haas Co.

PLIES (Tire): Layers of rubber impregnated fabric that make up carcass or body of tire.

PLUG GAPPING: Adjusting side electrode on spark plug to provide proper air gap between it and the center electrode.

PLY RATING (Tires): Indication of tire strength (load carrying capacity). Does not necessarily indicate actual number of plies. Two-ply four-ply rating tire would have load capacity of a four-ply tire of same size but would have only two actual plies.

POLARITY (Battery Terminals): Indicates if the battery terminal (either one) is positive or negative (plus or minus) (+ or —).

POLARITY (Generator): Indicates if pole shoes are so magnetized as to make current flow in a direction compatible with direction of flow as set by battery.

POLARITY (Magnet): Indicates if end of a magnet is north or south pole (N or S).

POLARIZING (Generator): Process of sending quick

surge of current through field windings of generator in direction that will cause pole shoes to assume correct polarity. This will insure that the generator will cause current to flow in same direction as normal.

POLE (MAGNET): One end, either north or south, of a magnet.

POLE SHOES: Metal pieces about which field coil windings are placed. When current passes through windings, pole shoes become powerful magnets. Example: pole shoes in a generator or starter motor.

PONY CAR: Small, sporty car along the lines of the Mustang, Firebird, Camaro, etc.

POPPET VALVE: Valve used to open and close valve port entrances to engine cylinders.

PORCELAIN (Spark Plug): Material used to insulate center electrode of spark plug. It is hard and resistant to damage by heat.

POROSITY: Small air or gas pockets, or voids, in metal.

PORT: Openings in engine cylinder blocks for exhaust and intake valves and water connections.

To smooth out, align and somewhat enlarge intake passageway to the valves.

POSITIVE TERMINAL: Terminal (such as on battery), to which current flows,

POST (Battery): Round, tapered lead posts protruding above top of battery to which battery cables are attached.

POT: Carburetor.

POTENTIAL: An indication of amount of available energy.

POUR POINT: Lowest temperature at which fluid will flow under specified conditions.

POWER STEERING: Steering system utilizing hydraulic pressure to increase the driver's turning effort. Pressure is utilized either in gearbox itself or in hydraulic cylinder attached to steering linkage.

POWER or FIRING STROKE: Portion of piston's movement devoted to transmitting power of burning fuel mixture to crankshaft.

PPM (Parts-per-million): Term used in determining extent of pollution existing in given sample of air.

PRECISION INSERT BEARING: Very accurately made replaceable type of bearing. It consists of an upper and lower shell. The shells are made of steel to which a friction type bearing material has been bonded. Connecting rod and main bearings are generally of precision insert type.

PREHEATING: Application of some heat prior to later application of more heat. Cast iron is preheated to avoid cracking when welding process is started. A coil (ignition) is preheated prior to testing.

PREHEATING (Metal): Process of raising temperature of metal to specific level before starting subsequent operations such as welding, brazing, etc.

PREIGNITION: Fuel charge being ignited before proper time.

PRELOADING: Adjusting antifriction bearing so it is under mild pressure. This prevents bearing looseness under a driving stress.

PRESS-FIT: Condition of fit (contact) between two parts that requires pressure to force parts together. Also referred to as drive or force fit.

PRESSURE BLEEDER: Device that forces brake fluid, under pressure, into master cylinder so that by opening bleeder screws at wheel cylinders, all air will be removed from brake system.

PRESSURE CAP: Special cap for radiator. It holds a predetermined amount of pressure on water in cooling system. This enables water to run hotter without boiling.

PRESSURE RELIEF VALVE: Valve designed to open at specific pressure. This will prevent pressures in system from exceeding certain limits.

PRIMARY CIRCUIT (Ignition System): Low voltage (6 or 12) part of ignition system.

PRIMARY, FORWARD, or LEADING BRAKE SHOE: Brake shoe installed facing front of car. It will be a self-energizing shoe.

PRIMARY WINDING (Coil): Low voltage (6 or 12 volt) winding in ignition coil. The primary winding is heavy wire; secondary winding uses fine wire.

PRIMARY WIRES: Wiring which serves low voltage part of ignition system. Wiring from battery to switch, resistor, coil, distributor points.

PRINTED CIRCUIT: Electrical circuit made by connecting units with electrically conductive lines printed on a panel. This eliminates actual wire and task of connecting it.

PROGRESSIVE LINKAGE: Carburetor linkage designed to open throttle valves of multiple carburetors. It opens one to start and when certain opening point is reached, it will start to open others.

PRONY BRAKE: Device utilizing friction brake to measure horsepower output of engine.

PROPANE (LPG): Petroleum product, similar to and often mixed with butane, useful as engine fuel. May be referred to as LPG.

PROPELLER SHAFT: Shaft connecting transmission output shaft to differential pinion shaft.

PROTON: Positive charge particle, part of atom.

PSI: Pounds per square inch.

PULL IT DOWN (Engine): Term often used in reference to dismantling and overhauling an engine.

PULSATION DAMPER: Device to smooth out fuel pulsations or surges from pump to carburetor.

PUMPING THE GAS PEDAL: Forcing accelerator up and down in an endeavor to provide extra gasoline to cylinders. This is often cause of flooding.

PURGE: Removing impurities from system. See Bleeding.

PUSH ROD: Rod that connects valve lifter to rocker arm. Used on valve-in-head installations.

PYLON: Marker for controlling traffic.

QUADRANT (Gearshift): Gearshift selector indicator marked PRNDL.

QUADRA-TRAC: See Full Time Four-Wheel Drive.

QUENCHED (Flame): Flame front in combustion chamber being extinguished as it contacts colder cylinder walls. This sharply elevates hydrocarbon emissions.

QUENCHING: Dipping heated object into water, oil or other substance, to quickly reduce temperature.

QUICKSILVER: Metal mercury. Often used in thermometers.

RACE (Bearing): Inner or outer ring that provides a contact surface for balls or rollers in bearing.

RACE CAMSHAFT: Camshaft, other than stock, designed to improve performance by altering cam profile. Provides increased lift, faster opening and closing, earlier opening and later closing, etc. Race camshafts are available as semi-race or street grind, three-fourths race or full race. Grinds in between these general categories are also available.

RACING SLICK: Type of tire used in "drag racing" as well as some "stock car" applications. Tread surface of tire is completely smooth, for maximum rubber contact with track surface.

RACK AND PINION GEARBOX (Steering): Steering gear utilizing pinion gear on end of steering shaft. Pinion engages long rack (bar with teeth along one edge). Rack is connected to steering arms via rods.

RADIAL (Direction): Line at right angles (perpendicular) to shaft, cylinder, bearing, etc., center line.

RADIAL ENGINE: Engine possessing various numbers of cylinders so arranged that they form circle around crankshaft center line.

RADIATION: Transfer of heat from one object to another when hotter object sends out invisible rays or waves that upon striking colder object, cause it to vibrate and thus heat.

RADIUS RODS: Rods attached to axle and pivoted on frame. Used to keep axle at right angles to frame and yet permit an up and down motion.

RAIL: Dragster built around a relatively long pipe frame. The only body panels used are around the driver's cockpit area.

RAKED: Ground clearance, at front or rear of car, reduced or increased, giving tilted appearance.

RAM AIR: Air "scooped" up by an opening due to vehicle forward motion.

RAM INDUCTION: Using forward momentum of car to scoop air and force it into carburetor via a suitable passageway.

RAM INTAKE MANIFOLD: Intake manifold that has very long passageways that at certain speeds aid entrance of fuel mixture into cylinders.

RATED HORSEPOWER (Engine): Indication of horsepower load that may safely be placed upon engine for prolonged periods of time. This would be somewhat less than the engine maximum horsepower.

RATIO: Fixed relationship between things in number, quantity or degree. For example, if fuel mixture contains one part of gas for fifteen parts of air, ratio would be 15 to 1.

REACTOR: See Stator.

REAM: To enlarge or smooth hole by using round cutting tool with fluted edges.

REAR AXLE (Banjo type): Rear axle housing from which differential unit may be removed while housing remains in place on car. Housing is solid from side to side.

REAR AXLE HOUSING (Split type): Rear axle housing made up of several pieces and bolted together. Housing must be split apart to remove differential.

RECEIVER-DRIER: See Drier.

RECIPROCATING ACTION: Back-and-forth movement such as action of pistons.

RECIRCULATING BALL WORM AND NUT: Very popular type of steering gear. It utilizes series of ball bearings that feed through and around and back through grooves in worm and nut.

RECTIFIER: Device used to change AC (alternating current) into DC (direct current).

RED LINE: Top recommended engine rpm. If a tachometer is used, it will have a mark (Red line) indicating maximum rpm.

REDUCING FLAME: Welding flame in which there is an excess of acetylene.

REFRIGERANT: Liquid used in refrigeration systems to remove heat from evaporator coils and carry it to condenser.

REFRIGERANT-12: Name applied to refrigerant generally used in automotive air conditioning systems.

REGULATOR (Electrical): Device used to control generator voltage and current output.

REGULATOR (Gas or Liquid): Device to reduce and control pressure.

RELAY: Magnetically operated switch used to make and break flow of current in circuit. Also called "cutout, and circuit breaker."

RELIEVE: Removing, by grinding, small lip of metal between valve seat area and cylinder — and removing any other metal deemed necessary to improve flow of fuel mixture into cylinders. Porting is generally done at same time.

RESISTANCE (Electrical): Measure of conductors ability to retard flow of electricity.

RESISTOR: Device placed in circuit to lower voltage. It will also decrease flow of current.

RESISTOR SPARK PLUG: Spark plug containing resistor designed to shorten both capacitive and inductive phases of spark. This will suppress radio interference and lengthen electrode life.

RESONATOR: Small muffler-like device that is placed into exhaust system near end of tail pipe. Used to provide additional silencing of exhaust.

RETARD (Ignition timing): To set the ignition timing so that spark occurs later or less degrees before TDC.

REVERSE-ELLIOT TYPE AXLE: Solid bar front axle on which steering knuckles span or straddle axle ends.

REVERSE FLUSH: Cleaning cooling system by pumping a powerful cleaning agent through system in a direction opposite to that of normal flow.

REVERSE IDLER GEAR: Gear used in transmission to produce a reverse rotation of transmission output shaft.

RICARDO PRINCIPLE: Arrangement in which portion of combustion chamber came in very close contact with piston head. Other portion, off to one side, contained more space. As the piston neared TDC on compression stroke, fuel mixture was squeezed tightly between piston and head thus causing mixture to squirt outward into larger area in very turbulent manner. This produced a superior mixture and allowed compression ratios to be raised without detonation.

RIDING THE CLUTCH: Riding the clutch refers to driver resting his foot on clutch pedal while car is being driven.

RING (Chrome): Ring on which the outer edge has a thin layer of chrome plate.

RING (Pinned): Steel pin, set into piston, is placed in space between ends of ring. Ring is thus kept from moving around in groove.

RING GAP: Distance between ends of piston ring when installed in cylinder.

RING EXPANDER: Spring device placed under rings to hold them snugly against cylinder wall.

RING GEAR: Large gear attached to differential carrier or to outer gear in planetary gear setup.

RING GROOVES: Grooves cut into piston to accept rings.

RING JOB: Reconditioning cylinders and installing new rings.

RING RIDGE: Portion of cylinder above top limit of ring travel. In a worn cylinder, this area is of smaller diameter than remainder of cylinder and will leave ledge or ridge that must be removed.

RIVET: Metal pin used to hold two objects together. One end of the pin has head and other end must be set or peened over.

RMA: Rubber Manufacturer's Association.

ROAD FEEL: Feeling imparted to steering wheel by wheels of car in motion. This feeling can be very important in sensing and predetermining vehicle steering response.

ROCKER ARM: Arm used to direct upward motion of push rod into a downward or opening motion of valve stem. Used in overhead valve installations.

ROCKER ARM SHAFT: Shaft upon which rocker arms are mounted.

ROCKER PANEL: Section of car body between front and rear fenders and beneath doors.

ROCKWELL HARDNESS: Measurement of the degree of hardness of given substance.

ROD: Refers to a car, driving a car hard, or to a connecting rod.

RODDING THE RADIATOR: Top and sometimes, the bottom tank of the radiator is removed. The core is then cleaned by passing a cleaning rod down through tubes. This is done when radiators are quite clogged with rust, scale and various mineral deposits.

ROLL BAR: Heavy steel bar that goes from one side of frame, up and around in back of the driver, and back down to the other side of frame. It is used to protect driver in the event his car rolls over.

ROLLER BEARING: Bearing utilizing a series of straight, cupped or tapered rollers engaging an inner and outer ring or race.

ROLLER CLUTCH: Clutch, utilizing series of rollers placed in ramps, that will provide drive power in one direction but will slip or freewheel in the other direction.

ROLLER TAPPETS or LIFTERS: Valve lifters that have roller placed on end contacting camshaft. This is done to reduce friction between lobe and lifter. They are generally used when special camshafts and high tension valve springs have been installed.

ROLLING RADIUS: Distance from road surface to center of wheel with vehicle moving under normal load. Rolling radius is dependent on tire size.

ROTARY ENGINE: Piston engine in which the crankshaft is fixed (stationary) and in which cylinders rotate around crankshaft.

ROTARY ENGINE (Wankel): Internal combustion engine which is not of a reciprocating (piston) engine design. Central rotor turns in one direction only and yet effectively produces required intake, compression, firing and exhaust strokes.

ROTARY FLOW (Torque Converter): Movement of oil as it is carried around by pump and turbine. Rotary motion is not caused by oil passing through pump, to turbine, to stator, etc., as is case with vortex flow. Rotary flow is at right angles to center line of converter whereas vortex flow is parallel (more or less depending on ratio between speeds of pump and turbine).

ROTARY MOTION: Continual motion in circular direction such as performed by crankshaft.

ROTOR (Distributor): Cap-like unit placed on end of distributor shaft. It is in constant contact with distributor cap central terminal and as it turns, it will conduct secondary voltage to one of the outer terminals.

ROUGHING STONE (Hone): Coarse stone used for quick removal of material during honing.

RPM: Revolutions per minute.

RUNNING-FIT: Fit in which sufficient clearance has been provided to enable parts to turn freely and to receive lubrication.

RUNNING ON: See Dieseling.

SAE: Society of Automotive Engineers.

SAE or RATED HORSEPOWER: A simple formula of long standing is used to determine what is commonly referred to as the SAE or Rated Horsepower. The formula is:

$$\frac{\text{Bore Diameter}^2 \times \text{Number of Cylinders}}{2.5}$$

This formula is used primarily for licensing purposes and is not too accurate a means of determining actual brake horsepower.

SAFETY FACTOR: Providing strength beyond that needed, as an extra margin of insurance against part failure.

SAFETY HUBS: Device installed on the rear axle to prevent wheels leaving car in event of a broken axle.

SAFETY RIM: Rim having two safety ridges, one on each lip, to prevent tire beads from entering drop center area in event of a blowout. This feature keeps tire on rim.

SAFETY VALVE: Valve designed to open and relieve pressure within a container when container pressure exceeds predetermined level.

SAND BLAST: Cleaning by the use of sand propelled at high speeds in an air blast.

SAYBOLT VISCOMETER: Instrument used to determine fluidity or viscosity (resistance to flow) of an oil.

SCALE (Cooling System): Accumulation of rust and minerals within cooling system.

SCATTER SHIELD: Steel or nylon guard placed around bell or clutch housing to protect driver and spectator from flying parts in event of part failure at high rpm. Such a shield is often placed around transmissions and differential units.

SCAVENGING: Referring to a cleaning or blowing out action in reference to the exhaust gas.

SCHRADER VALVE: Valve, similar to spring loaded valve used in tire stem, used in car air conditioning system service valves.

SCORE: Scratch or groove on finished surface.

SCREW EXTRACTOR: Device used to remove broken bolts, screws, etc., from holes.

SCS: Speed Control Switch. (Speed sensitive spark advance control.)

SEALED BEAM HEADLIGHT: Headlight lamp in which lens, reflector and filament are fused together to form single unit.

SEALED BEARING: Bearing that has been lubricated at factory and then sealed, it cannot be lubricated during service.

SEAT: Surface upon which another part rests or seats. Example: Valve seat is matched surface upon which valve face rests.

SEAT (Rings): Minor wearing of piston ring surface

during initial use. Rings then fit or seat properly against the cylinder wall.

SECONDARY CIRCUIT (Ignition System): High voltage part of ignition system.

SECONDARY, REVERSE, or TRAILING BRAKE SHOE: Brake shoe that is installed facing rear of car.

SECONDARY WIRES: High voltage wire from coil to distributor tower and from tower to spark plugs.

SECTION MODULUS: Relative structural strength measurement of member (such as frame rail) that is determined by cross-sectional area and member shape.

SECTION WIDTH (Tire): Overall width minus height of any lettering or pattern extending outward from sidewalls.

SEDIMENT: Accumulation of matter which settles to bottom of a liquid.

SEIZE: See Freezing.

SELF-ENERGIZING: Brake shoe (sometimes both shoes) that when applied develops wedging action that actually assists or boosts braking force applied by wheel cylinder.

SEMA: Specialty Equipment Manufacturer's Association.

SEMI-ELLIPTICAL SPRING: Spring, such as commonly used on truck rear axles, consisting of one main leaf and number of progressively shorter leaf springs.

SEMI-FLOATING AXLE: Type of axle commonly used in modern car. Outer end turns wheel and supports weight of car; inner end which is splined, "floats" in differential gear.

SEPARATORS (Battery): Wood, rubber or plastic sheets inserted between positive and negative plates to prevent contact.

SERIES CIRCUIT: Circuit with two or more resistance units so wired that current must pass through one unit before reaching other.

SERIES-PARALLEL CIRCUIT: Circuit of three or more resistance units in which a series and a parallel circuit are combined.

SERVO (Transmission): Oil operated device used to push or pull another part — such as tightening the transmission brake bands.

SERVO ACTION: Brakes so constructed as to have one end of primary shoe bearing against end of secondary shoe. When brakes are applied, primary shoe attempts to move in the direction of the rotating drum and in so doing applies force to the secondary shoe. This action, called servo action, makes less brake pedal pressure necessary and is widely used in brake construction.

SHACKLE: Device used to attach ends of a leaf spring to frame.

SHAVE: Removal of some chrome or decorative part.

SHAVE (Engine): Removal of metal from contact surface of cylinder head or block.

SHIFT FORKS: Devices that straddle slots cut in sliding gears. Fork is used to move gear back and forth on shaft.

SHIFT POINT: Point, either in engine rpm or road speed, at which transmission should be shifted to next gear.

SHIFT RAILS: Sliding rods upon which shift forks are attached. Used for shifting the transmission (manual).

SHOCK ABSORBER: Oil filled device used to control spring oscillation in suspension system.

SHORT BLOCK: Engine block complete with crankshaft and piston assemblies.

SHROUD: Metal enclosure around fan, engine, etc., to guide and facilitate flow of air.

SHIM: Thin spacer installed between two units to increase distance between them.

SHIMMY: Front wheels shaking from side to side.

SHORT or SHORT CIRCUIT: Refers to some "hot" portion of the electrical system that has become grounded. (Wire touching a ground and providing a completed circuit to the battery.)

SHRINK-FIT: Fit between two parts which is so

tight, outer or encircling piece must be expanded by heating so it will fit over inner piece. In cooling, outer part shrinks and grasps inner part securely.

SHUNT: An alternate or bypass portion of an electrical circuit.

SHUNT WINDING: Wire coil forming an alternate or bypass circuit through which current may flow.

SIDE-DRAFT CARBURETOR: Carburetor in which air passes through carburetor into intake manifold in a horizontal plane.

SILENCER: Muffler.

SILVER SOLDER: Similar to brazing except that special silver solder metal is used.

SINGLE-BARREL, DOUBLE-BARREL and FOUR-BARREL CARBURETORS: Number of throttle openings or barrels from the carburetor to the intake manifold.

SINTERED BRONZE: Tiny particles of bronze pressed tightly together so that they form a solid piece. The piece is highly porous and is often used for filtering purposes.

SKID PLATE: Stout metal plate or plates attached to underside of vehicle to protect oil pan, transmission, fuel tank, etc., from damage caused by "grounding out" on rocks, etc.

SKINS: Tires.

SKIRTS: Cover for the rear fender cutout.

SKIVING (Tire): Cutting out tread injury on bevel.

SLANT ENGINE: In-line engine in which cylinder block has been tilted from vertical plane.

SLICKS: Very wide tire, without tread pattern, designed to provide maximum amount of traction.

SLIDING-FIT: See Running-Fit.

SLIDING GEAR: Transmission gear splined to the shaft. It may be moved back and forth for shifting purposes.

SLIP ANGLE: Difference in actual path taken by a car making a turn and path it would have taken if it had followed exactly as wheels were pointed.

SLIP JOINT: Joint that will transfer driving torque from one shaft to another while allowing longitudinal movement between two shafts.

SLINGSHOT: Form of dragster using rather long thin frame with a very light front axle and wheel assembly.

SLUDGE: Black, mushy deposits throughout interior of the engine. Caused from mixture of dust, oil, and water being whipped together by moving parts.

SMOG: Fog made darker and heavier by chemical fumes and smoke.

SNAP RING: Split ring snapped into a groove in a shaft or in a groove in a hole. It is used to hold bearings, thrust washers, gears, etc., in place.

SNUBBER: Device used to limit travel of some part.

SODIUM VALVE: Valve in which stem has been partially filled with metallic sodium to speed up transfer of heat from valve head, to stem and then to guide and block.

SOHC: Engine with single overhead camshaft.

SOLDERING: Joining two pieces of metal together with lead-tin mixture. Both pieces of metal must be heated to insure proper adhesion of melted solder.

SOLENOID: Electrically operated magnetic device used to operate some unit. Movable iron core is placed inside of coil. When current flows through coil, core will attempt to center itself in coil. In so doing, core will exert considerable force on anything it is connected to.

SOLVENT: Liquid used to dissolve or thin other material. Examples: Alcohol thins shellac; gasoline dissolves grease.

SOUPING: Hopping up or increasing engine performance through various modifications.

SPARK: Bridging or jumping of a gap between two electrodes by current of electricity.

SPARK ADVANCE: Causing spark plug to fire earlier by altering position of distributor breaker points in relation to distributor shaft.

SPARK ARRESTOR: Device used to prevent sparks (burning particles of carbon) from being discharged from exhaust pipe. Usually used on off-road equipment to prevent forest fires.

SPARK GAP: Space between center and side electrode tips on a spark plug.

SPARK KNOCK: See Preignition.

SPARK PLUG: Device containing two electrodes across which electricity jumps to produce a spark to fire fuel charge.

SPECIFIC GRAVITY: Relative weight of a given volume of specific material as compared to weight of an equal volume of water.

SPEEDOMETER: Instrument used to determine forward speed of an auto in miles per hour.

SPIDER GEARS: Small gears mounted on shaft pinned to differential case. They mesh with, and drive, the axle end gears.

SPINDLE (Wheel): Machined shaft upon which inside races of front wheel bearings rest. Spindle is an integral part of steering knuckle.

SPIRAL BEVEL GEAR: Ring and pinion setup widely used in automobile differentials. Teeth of both ring and pinion are tapered and are cut on a spiral so that they are at an angle to center line of pinion shaft.

SPLINE: Metal, land, remaining between two grooves. Used to connect parts.

SPLINED JOINT: Joint between two parts in which each part has a series of splines cut along contact area. The splines on each part slide into grooves between splines on other part.

SPLIT MANIFOLD: Exhaust manifold that has a baffle placed near its center. An exhaust pipe leads out of each half.

SPONGY PEDAL: When there is air in brake lines, or shoes that are not properly centered in drums, brake pedal will have a springy or spongy feeling when brakes are applied. Pedal normally will feel hard when applied.

SPOOL BALANCE VALVE (Automatic Transmission): Hydraulic valve that balances incoming oil pressure against spring control pressure to produce a steady pressure to some control unit.

SPOOL VALVE: Hydraulic control valve shaped somewhat like spool upon which thread is wound.

SPORTS CAR: Term commonly used to describe a relative small, low slung, car with a high performance engine.

SPOT WELD: Fastening parts together by fusing, at various spots. Heavy surge of electricity is passed through the parts held in firm contact by electrodes.

SPRAG CLUTCH: Clutch that will allow rotation in one direction but that will lock up and prevent any movement in the other direction.

SPRING (Main Leaf): Long leaf on which ends are turned to form an "eye" to receive shackle.

SPRING BOOSTER: Device used to "beef" up sagged springs or to increase the load capacity of standard springs.

SPRING CAPACITY AT GROUND: Total vehicle weight (sprung and unsprung) that will be carried by spring bent or deflected to its maximum normal loaded position.

SPRING CAPACITY AT PAD: Total vehicle sprung weight that will be carried by spring bent or deflected to its normal fully loaded position.

SPRING LOADED: Device held in place, or under pressure from a spring or springs.

SPRING STEEL: Heat treated steel having the ability to stand a great amount of deflection and yet return to its original shape or position.

SPRING WINDUP: Curved shape assumed by rear leaf springs during acceleration or braking.

SPROCKET: Toothed wheel used to drive chain.

SPRUNG WEIGHT: Weight of all parts of car that are supported by suspension system.

SPUR GEAR: Gear on which teeth are cut parallel to shaft.

SPURT or SQUIRT HOLE: Small hole in connecting rod big end that indexes (aligns) with oil hole in crank journal. When holes index, oil spurts out to lubricate cylinder walls.

SQUARE ENGINE: Engine in which bore diameter and stroke are of equal dimensions.

SQ. FT.: Square Foot.

SQ. IN.: Square Inch.

STABILIZER BAR: Transverse mounted spring steel bar that controls and minimizes body lean or tipping on corners.

STAMPING: Sheet metal part formed by pressing between metal dies.

STATIC BALANCE: When a tire, flywheel, crankshaft, etc., has an absolutely even distribution of weight mass around axle of rotation, it will be in static balance. For example, if front wheel is jacked up and tire, regardless of where it is placed, always slowly turns and stops with the same spot down, it would not be in static balance. If, however, wheel remains in any position in which it is placed, it would be in static balance. (Bearings must be free, no brake drag, etc.)

STATIC ELECTRICITY: Electricity generated by friction between two objects. It will remain in one object until discharged.

STATIC PRESSURE (Brakes): Certain amount of pressure that always exists in brake lines — even with brake pedal released. Static pressure is maintained by a check valve.

STATIC RADIUS: Distance from road surface to center of wheel with vehicle normally loaded, at rest.

STATOR: Small hub, upon which series of vanes are affixed in radial position, that is so placed that oil leaving torque converter turbine strikes stator vanes and is redirected into pump at an angle conducive to high efficiency. Stator makes torque mutliplication possible. Torque multiplication is highest at stall when the engine speed is at its highest and the turbine is standing still.

STEEL PACK MUFFLER: Straight-through (no baffled) muffler utilizing metal shavings surrounding a perforated pipe. Quiets exhaust sound.

STEERING ARMS: Arms, either bolted to, or forged as an integral part of steering knuckles. They transmit steering force from tie rods to knuckles, thus causing wheels to pivot.

STEERING AXIS INCLINATION: See Kingpin Inclination.

STEERING GEAR: Gears, mounted on lower end of steering column, used to multiply driver turning force.

STEERING GEOMETRY: Term sometimes used to describe various angles assumed by components making up front wheel turning arrangement, camber, caster, toe-in, etc.

Also used to describe related angles assumed by front wheels when car is negotiating a curve.

STEERING KNUCKLE: Inner portion of spindle affixed to and pivots on either a kingpin or on upper and lower ball joints.

STEERING KNUCKLE ANGLE: Angle formed between steering axis and center line of spindle. This angle is sometimes referred to as Included Angle.

STETHOSCOPE: Device (such as used by doctors) to detect and locate abnormal engine noises. Very handy tool for troubleshooter.

STICK SHIFT: Transmission that is shifted manually through use of various forms of linkage. Often refers to upright gearshift stick that protrudes through floor.

Either floor or steering column mounted manual shift device for transmission.

STOCK CAR: Car as built by factory.

STORMER: Hot car that really moves out.

STOVEBOLT: Generally refers to Chevrolet (GMC) 6-cylinder, in-line, valve-in-head (push rod operated) engine.

STREET ROD: Slightly modified rod that will give good day-to-day performance on the streets.

STRESS: To apply force to an object. Force or pressure an object is subjected to.

STRIP: Area used for drag racing.

Removing tires and wheels, battery, hubcaps and other items of value as done by thieves.

STRIPING TOOL: Tool used to apply paint in long narrow lines.

STROBOSCOPE: See Timing Light.

STROKE: Distance piston moves when traveling from TDC to BDC.

STROKED CRANKSHAFT: Crankshaft, either special new one or stock crank reworked, that has con rod throws offset so that length of stroke is increased.

STROKER: Engine using crankshaft that has been stroked.

STUD: Metal rod with threads on both ends.

STUD PULLER: Tool used to install or remove studs.

SUCTION: See Vacuum.

SUCTION THROTTLING VALVE: Valve placed between air conditioning evaporator and compressor which controls evaporator pressure to provide maximum cooling without icing evaporator core.

SUMP: Part of oil pan that contains oil.

SUN GEAR: Center gear around which planet gears revolve.

SUPER CAR: Car with high horsepower engine that will provide fast acceleration and high speed.

SUPERCHARGER: Unit designed to force air, under pressure, into cylinders. Can be mounted between carburetor and cylinders or between carburetor and atmosphere.

SUPER STOCK: Factory car (stock) with engine, suspension, running gear, etc., modified to increase horsepower and overall performance.

SWEATING: Joining two pieces of metal together by placing solder between them and then clamping them tightly together while heat, sufficient to melt the solder, is applied.

SWING AXLE: Independent rear suspension system in which each driving wheel can move up or down independently of other, differential unit is bolted to frame and various forms of linkage are used upon which to mount wheels. Drive axles, utilizing one or more universal joints, connect differential to drive wheels.

SYNCHROMESH TRANSMISSION: Transmission using device (synchromesh) that synchronizes speeds of gears that are being shifted together. This prevents "gear grinding." Some transmissions use synchromesh on all shifts, while others synchronize second and high gearshifts.

SYNCHRONIZE: To bring about a timing that will cause two or more events to occur simultaneously; plug firing when the piston is in correct position, speed of two shafts being the same, valve opening when piston is in correct position, etc.

TACHOMETER: Device used to indicate speed of engine in rpm.

TAIL PIPE: Exhaust piping running from muffler to rear of car.

TAP: To cut threads in a hole, or can be used to indicate fluted tool used to cut threads.

TAP AND DIE SET: Set of taps and dies for internal and external threading — usually covers a range of the most popular sizes.

TAPERED ROLLER BEARING (Antifriction): Bearing utilizing series of tapered, hardened steel rollers operating between an outer and inner hardened steel race.

TAPPET: Screw used to adjust clearance between valve stem and lifter or rocker arm.

TAPPET NOISE: Noise caused by lash or clearance between valve stem and rocker arm or between valve stem and valve lifter.

TCS: Transmission Controlled Spark.

TDC: Top Dead Center.

TEFLON: Plastic with excellent self-lubricating (slippery) bearing properties.

TEMPER: To effect a change in physical structure of piece of steel through use of heat and cold.

TENSION: Pulling or stretching stress applied to an object.

TERMINAL: Connecting point in electric circuit.

When referring to battery, it would indicate two battery posts.

T-FORD or T-BONE: Model-T Ford car.

T-HEAD ENGINE: Engine having intake valve on one side of cylinder and exhaust on other.

THERMAL EFFICIENCY: Percentage of heat developed in burning fuel charge that is actually used to develop power determines thermal efficiency. Efficiency will vary according to engine design, use, etc. If an engine utilizes great deal of heat to produce power, its thermal efficiency would be high.

THERMOSTAT: Temperature sensitive device used in cooling system to control flow of coolant in relation to temperature.

THIRD BRUSH (Generator): Generator in which a third, movable brush is used to control current output.

THREE-QUARTER RACE CAMSHAFT: Description of custom camshaft indicating type of lobe grinding which, in turn, dictates type of use. Other grinds are one-quarter race, full-race, street-grind, etc.

THROTTLE VALVE: Valve in carburetor. It is used to control amount of fuel mixture that reaches cylinders.

THROW: Offset portion of crankshaft designed to accept connecting rod.

THROWING A ROD: When an engine has thrown a connecting rod from crankshaft. Major damage is usually incurred.

THRUST BEARING: Bearing designed so as to resist side pressure.

THRUST WASHER: Bronze or hardened steel washer placed between two moving parts. The washer prevents longitudinal movement and provides a bearing surface for thrust surfaces of parts.

TIE ROD: Rod, or rods, connecting steering arms together. When tie rod is moved, wheels pivot.

TIG: Gas tungsten arc welding (Tungsten Inert Gas).

TIMING CHAIN: Drive chain that operates camshaft by engaging sprockets on camshaft and crankshaft.

TIMING GEARS: Both the gear attached to the camshaft and the gear on the crankshaft. They provide a means of driving the camshaft.

TIMING LIGHT: Stroboscopic unit that is connected to secondary circuit to produce flashes of light in unison with firing of specific spark plug. By directing these flashes of light on whirling timing marks, marks appear to stand still. By adjusting distributor, timing marks may be properly aligned, thus setting timing.

TIMING MARKS (Ignition): Marks, usually located on vibration damper, used to synchronize ignition system so plugs will fire at precise time.

TIMING MARKS (Valves): One tooth on either the camshaft or crankshaft gear will be marked with an indentation or some other mark. Another mark will be found on other gear between two of teeth. Two gears must be meshed so that marked tooth meshes with marked spot on other gear.

TINNING: Coating piece of metal with a very thin layer of solder.

TIRE BALANCE: In that tires turn at relatively high speeds, they must be carefully balanced both for static balance and for dynamic balance.

TIRE BEAD: Portion of tire that bears against rim flange. Bead has a number of turns of steel wire in it to provide great strength.

TIRE CASING: Main body of tire exclusive of tread.

TIRE PLIES: Layers of nylon, rayon, etc., cloth used to form casing. Most car tires are two ply with a four ply rating. Two ply indicates two layers of cloth or plies.

TIRE ROTATION: Moving front tires to rear and rear to front to equalize any wear irregularities.

TIRE SIDEWALL: Portion of tire between tread and bead.

TIRE TREAD: Part of tire that contacts road.

TOE-IN: Having front of wheels closer together than the back (front wheels). Difference in measurement across front of wheels and the back will give amount of toe-in.

TOE-OUT: Having front of wheels further apart than the back.

TOE-OUT OF TURNS: When car negotiates a curve, inner wheel turns more sharply and while wheels remain in this position, a condition of toe-out exists.

TOGGLE SWITCH: Switch actuated by flipping a small lever either up and down or from side to side.

TOLERANCE: Amount of variation permitted from an exact size or measurement. Actual amount from smallest acceptable dimension to largest acceptable dimension.

TOOTH HEEL (Differential Ring Gear): Wider outside end of tooth.

TOOTH TOE (Differential Ring Gear): Narrower inside end of tooth.

TOP OFF: Fill a container to full capacity.

TORQUE: Turning or twisting force such as force imparted on drive line by engine.

TORQUE CONVERTER: Unit, quite similar to fluid coupling, that transfers engine torque to transmission input shaft. Unlike fluid coupling, torque converter can multiply engine torque. This is accomplished by installing one or more stators between torus members. In torque converter driving torus is referred to as "pump" and driven torus as "turbine."

TORQUE (Gross): Maximum engine torque developed by engine without fan, air cleaner, alternator, exhaust system, etc.

TORQUE (Net): Maximum torque developed by engine equipped with fan, air cleaner, exhaust system, and all other systems or units normally present when engine is installed in car.

TORQUE MULTIPLICATION (Automatic Transmission): Increasing engine torque through the use of a torque converter.

TORQUE TUBE DRIVE: Method of connecting transmission output shaft to differential pinion shaft by using an enclosed drive shaft. Drive shaft is enclosed in torque tube that is bolted to rear axle housing on one end and is pivoted through a ball joint to rear of transmission on other. Driving force of rear wheels is transferred to frame through torque tube.

TORQUE WRENCH: Wrench used to draw nuts, cap screws, etc., up to specified tension by measuring torque (turning force) being applied.

TORSIONAL VIBRATION: Twisting and untwisting action developed in shaft. It is caused either by intermittent applications of power or load.

TORSION BAR: Long spring steel rod attached in such a way that one end is anchored while other is free to twist. If an arm is attached, at right angles, to free end, any movement of arm will cause rod or bar to twist. Bar's resistance to twisting provides a spring action. Torsion bar replaces both coil and leaf springs in some suspension systems.

TORSION BAR SUSPENSION: Suspension system that makes use of torsion bars in place of leaf or coil spring.

TRACK: Distance between front wheels or distance between rear wheels. They are not always the same.

TRACTION BAR: Articulated bar or link attached to both frame and rear axle housing to prevent spring windup (with resultant wheel hop) during heavy acceleration or braking.

TRACTION DIFFERENTIAL: See Limited-Slip Differential.

TRAMP: Hopping motion of front wheels.

TRANSAXLE: Drive setup in which transmission and differential are combined into a single unit.

TRANSFER CASE: Gearbox, driven by transmission, that will provide driving force to both front and rear propeller shafts on four-wheel drive vehicle.

TRANSFORMER: Electrical device used to increase or decrease voltage. Car ignition coil transforms voltage from 12 volts to upward of 20,000 volts.

TRANSISTOR IGNITION: Form of ignition system utilizing transistors and a special coil. Conventional distributor and point setup is used. With transistor unit, voltage remains constant, thus permitting high engine rpm without resultant engine "miss." Point life is greatly extended as transistor system passes a very small amount of current through points.

TRANSMISSION: Device that uses gearing or torque conversion to effect a change in ratio between engine rpm and driving wheel rpm. When engine rpm goes up in relation to wheel rpm, more torque but less speed is produced. Reduction in engine rpm in relation to wheel rpm produces a higher road speed but delivers less torque to driving wheels.

TRANSMISSION ADAPTER: A unit that allows a different make or year transmission to be bolted up to original engine.

TRANSMISSION (Automatic): Transmission that automatically effects gear changes to meet varying road and load conditions. Gear changing is done through series of oil operated clutches and bands.

TRANSMISSION (Standard or Conventional): Transmission that must be shifted manually to effect a change in gearing.

TRANSVERSE LEAF SPRING: Leaf spring mounted so it is at right angles to length of car.

TRAPS: Area over which car is raced for timing purposes.

TREAD: Distance between two front or two rear wheels.

TREAD (Tire): Portion of tire which contacts roadway.

TREAD WIDTH (Tire): Distance between outside edges of tread as measured across tread surface.

TRIP ODOMETER: Auxiliary odometer that may be reset to zero at option of driver. Used for keeping track of mileage on trips up to one thousand miles.

TROUBLESHOOTING: Diagnosing engine, transmission, etc., problems by various tests and observations.

TRS: Transmission Regulated Spark.

TUBE CUTTER: Tool used to cut tubing by passing a sharp wheel around and around tube.

TUNE-UP: Process of checking, repairing and adjusting carburetor, spark plugs, points, belts, timing, etc., in order to obtain maximum performance from engine.

TURBINE: Wheel upon which series of angled vanes are affixed so moving column of air or liquid will impart a turning motion to wheel.

TURBINE ENGINE: Engine that utilizes burning gases to spin a turbine, or series of turbines, as a means of propelling the car.

TURBOCHARGER: Exhaust powered supercharger.

TURBULENCE: Violent, broken movement or agitation of a fluid or gas.

TURNING RADIUS: Diameter of circle transcribed by outer front wheel when making a full turn.

TV ROD: Throttle valve rod that extends from foot throttle linkage to throttle valve in automatic transmission.

TVS: Thermostatic Vacuum Switch.

TWIST DRILL: Metal cutting drill with spiral flutes (grooves) to permit exit of chips while cutting.

TWO-STROKE CYCLE ENGINE: Engine requiring one complete revolution of crankshaft to fire each piston once.

UNDERCOATING: Soft deadening material sprayed on underside of car, under hood, trunk lid, etc.

UNDER-SQUARE ENGINE: Engine in which bore diameter is smaller than length of stroke.

UNDERSTEER: Tendency for car, when negotiating a corner, to turn less sharply than driver intends.

UNIT BODY: Car body in which body itself acts as frame.

UNIVERSAL JOINT: Flexible joint that will permit changes in driving angle between driving and driven shaft.

UNSPRUNG WEIGHT: All parts of car not supported by suspension system, wheels, tires, etc.

UPDRAFT CARBURETOR: Carburetor in which the air passes upward through the carburetor into the intake manifold.

UPSET: Widening of diameter through pounding.

UPSHIFT: Shifting to a higher gear.

VACUUM: Enclosed area in which air pressure is below that of surrounding atmospheric pressure.

VACUUM ADVANCE (Distributor): Unit designed to advance and retard ignition timing through action of engine vacuum working on a diaphragm.

VACUUM BOOSTER: Small diaphragm vacuum pump, generally in combination with fuel pump, that is used to bolster engine vacuum during acceleration so vacuum operated devices will continue to operate.

VACUUM GAGE: Gage used to determine amount of vacuum existing in a chamber.

VACUUM PUMP: Diaphragm type of pump used to produce vacuum.

VACUUM RUNOUT POINT: Point reached when vacuum brake power piston has built up all the braking force it is capable of with vacuum available.

VACUUM TANK: Tank in which vacuum exists. Generally used to provide vacuum to power brake installation in event engine vacuum cannot be obtained. Tank will supply several brake applications before vacuum is exhausted.

VALVE: Device used to either open or close an opening. There are many different types.

VALVE CLEARANCE (Engine): Space between end of valve stem and actuating mechanism (rocker arm, lifter, etc.).

VALVE DURATION: Length of time, measured in degrees of engine crankshaft rotation, that valve remains open.

VALVE FACE: Outer lower edge of valve head. The face contacts the valve seat when the valve is closed.

VALVE FLOAT: Condition where valves in engine are forced back open before they have had a chance to seat. Brought about (usually) by extremely high rpm.

VALVE GRINDING: Renewing valve face area by grinding on special grinding machine.

VALVE GUIDE: Hole through which stem of poppet valve passes. It is designed to keep valve in proper alignment. Some guides are pressed into place and others are merely drilled in block or in head metal.

VALVE HEAD (Engine): Portion of valve above stem.

VALVE-IN-HEAD ENGINE: Engine in which both intake and exhaust valves are mounted in the cylinder head and are driven by pushrods or by an overhead camshaft.

VALVE KEEPER or VALVE KEY or VALVE RETAINER: Small unit that snaps into a groove in end of valve stem. It is designed to secure valve spring, valve spring retaining washer and valve stem together. Some are of a split design, some of a horseshoe shape, etc.

VALVE LASH: Valve tappet clearance or total clearance in the valve operating train with cam follower on camshaft base circle.

VALVE LIFT: Distance a valve moves from full closed to full open position.

VALVE LIFTER or CAM FOLLOWER: Unit that contacts end of valve stem and camshaft. Follower rides on camshaft and when cam lobes move it upward, it opens valve.

VALVE MARGIN: Width of edge of valve head between top of valve and edge of face. Too narrow a margin results to preignition and valve damage through overheating.

VALVE OIL SEAL: Neoprene rubber ring placed in groove in valve stem to prevent excess oil entering area between stem and guide. There are other types of these seals.

VALVE OVERLAP: Certain period in which both intake and exhaust valve are partially open. (Intake is starting to open while exhaust is not yet closed.)

VALVE PORT: Opening, through head or block, from intake or exhaust manifold to valve seat.

VALVE ROTATOR: Unit that is placed on end of valve stem so that when valve is opened and closed, the valve will rotate a small amount with each opening and closing. This gives longer valve life.

VALVE SEAT: Area onto which face of poppet seats when closed. Two common angles for this seat are forty-five and thirty degrees.

VALVE SEAT GRINDING: Renewing valve seat area by grinding with a stone mounted upon a special mandrel.

VALVE SEAT INSERT: Hardened steel valve seat that may be removed and replaced.

VALVE SPRING: Coil spring used to keep valves closed.

VALVE STEM (Engine): Portion of valve below head. The stem rides in the guide.

VALVE TAPPET: Adjusting screw to obtain specified clearance at end of valve stem (tappet clearance). Screw may be in top of lifter, in rocker arm, or in the case of ball joint rocker arm, nut on mounting stud acts in place of a tappet screw.

VALVE TIMING: Adjusting position of camshaft to crankshaft so that valves will open and close at the proper time.

VALVE TRAIN: Various parts making up valve and its operating mechanism.

VALVE UMBRELLA: Washer-like unit that is placed over end of the valve stem to prevent the entry of excess oil between the stem and the guide. Used in valve-in-head installations.

VANE: Thin plate affixed to rotatable unit to either throw off air or liquid, or to receive thrust imparted by moving air or liquid striking the vane. In the first case it would be acting as a pump and in the second case as a turbine.

VAPORIZATION: Breaking gasoline into fine particles and mixing it with incoming air.

VAPOR LOCK: Boiling or vaporizing of the fuel in the lines from excess heat. Boiling will interfere with movement of the fuel and will in some cases, completely stop the flow.

VAPOR SEPARATOR: A device used on cars equipped with air conditioning to prevent vapor lock by feeding vapors back to the gas tank via a separate line.

VARIABLE PITCH STATOR: Stator that has vanes that may be adjusted to various angles depending on load conditions. Vane adjustment will increase or decrease efficiency of stator.

VARNISH: Deposit on interior of engine caused by engine oil breaking down under prolonged heat and use. Certain portions of oil deposit themselves in hard coatings of varnish.

VENTURI: That part of a tube, channel, pipe, etc., so tapered as to form a smaller or constricted area. Liquid, or a gas, moving through this constricted area will speed up and as it passes narrowest point, a partial vacuum will be formed. Taper facing flow of air is much steeper than taper facing away from flow of air. Venturi principle is used in carburetor.

VIBRATION DAMPER: Round weighted device attached to front of crankshaft to minimize torsional vibration.

VISCOSIMETER: Device used to determine viscosity of a given sample of oil. Oil is heated to specific temperature and then allowed to flow through set orifice. Length of time required for certain amount to flow determines oil's viscosity.

VISCOSITY: Measure of oil's ability to pour. (Thick, thin.)

VISCOSITY INDEX: Measure of oil's ability to resist changes in viscosity when heated.

VOLATILE: Easily evaporated.

VOLATILITY: Property of gasoline, alcohol, etc., to evaporate quickly and at relatively low temperatures.

VOLT: Unit of electrical pressure or force that will move a current of one ampere through a resistance of one ohm.

VOLTAGE: Difference in electrical potential between one end of a circuit and the other. Also called EMF (electromotive force). Voltage causes current to flow.

VOLTAGE DROP: Lowering of voltage due to excess length of wire, undersize wire, etc.

VOLTAGE REGULATOR: See Regulator — Voltage.

VOLTMETER: Instrument used to measure voltage in given circuit. (In volts.)

VOLUME: Measurement, in cubic inches, cubic feet, etc., of amount of space within a certain object or area.

VOLUMETRIC EFFICIENCY: Comparison between actual volume of fuel mixture drawn in on intake stroke and what would be drawn in if cylinder were to be completely filled.

VORTEX: Mass of whirling liquid or gas.

VORTEX FLOW (Torque Converter): Whirling motion of oil as it moves around and around from pump, through turbine, through stator and back into pump and so on.

VULCANIZATION: Process of heating compounded rubber to alter its characteristics — making it tough, resilient, etc.

WANDERING (Steering): Condition in which front wheels tend to steer one way and then another.

WANKEL ENGINE: Rotary combustion engine that utilizes one or more three-sided rotors mounted on drive shaft operating in specially shaped chambers. Rotor turns constantly in one direction yet produces an intake, compression, firing and exhaust stroke.

WATER JACKET: Area around cylinders and valves that is left hollow so that water may be admitted for cooling.

WEDGE: Engine using wedge-shaped combustion chamber.

WEDGE COMBUSTION CHAMBER: Combustion chamber utilizing wedge shape. It is quite efficient and lends itself to mass production and as a result is widely used.

WEIGHT (Curb): Weight of vehicle (no passengers) with all systems (fuel, cooling, lubrication) filled.

WEIGHT (Shipping): Basic vehicle weight including all standard items but without fuel or coolant.

WEIGHT (Sprung): See Sprung Weight.

WEIGHT DISTRIBUTION: Percentage of total vehicle weight as carried by each axle (front and rear).

WELD: To join two pieces of metal together by raising area to be joined to point hot enough for two sections to melt and flow together. Additional metal is usually added by melting small drops from end of metal rod while welding is in progress.

WET SLEEVE: Cylinder sleeve application in which water in cooling system contacts a major portion of sleeve itself.

WHEEL ALIGNER: Device used to check camber, caster, toe-in, etc.

WHEEL BALANCER: Machine used to check wheel and tire assembly for static and dynamic balance.

WHEELBASE: Distance between center of front wheels and center of rear wheels.

WHEEL CYLINDER: Part of hydraulic brake system that receives pressure from master cylinder and in turn applies brake shoes to drums.

WHEEL HOP: Hopping action of rear wheels during heavy acceleration.

WHEELIE BARS: Short arms attached to rear of a drag racer to prevent front end from rising too far off ground during heavy acceleration. Arms are usually of spring material and have small wheels attached to ends that contact ground.

WHEEL LUG or LUG BOLT: Bolts used to fasten wheel to hub.

WIDE TREADS, WIDE OVAL, etc.: Wide tires. Tire height, bead to tread surface is about 70 percent of tire width across outside of carcass.

WINDING THE ENGINE: Running engine at top rpm.

WINDSCREEN: British term for windshield.

WIRING DIAGRAM: Drawing showing various electrical units and wiring arrangement necessary for them to function properly.

WISHBONE: Radius rod setup used in many older Ford cars to keep axle square with frame.

WITNESS MARKS: Punch marks used to position or

locate some part in its proper spot.

WORM GEAR: Coarse, spiral shaped gear cut on shaft. Used to engage with and drive another gear or portion of a gear. As used in steering gearbox, it often engages cross shaft via a roller or by a tapered pin.

WORM AND ROLLER: Type of steering gear utilizing a worm gear on steering shaft. A roller on one end of cross shaft engages worm.

WORM AND SECTOR: Type of steering gear utilizing worm gear engaging sector (a portion of a gear) on cross shaft.

WORM AND TAPER PIN: Type of steering gear utilizing worm gear on steering shaft. End of cross shaft engages worm via taper pin.

WRIST PIN: See Piston Pin.

YIELD STRENGTH (Elastic Limit): Maximum force (in pounds per square inch) that can be sustained by given member and have that member return to its original position, length, shape, etc., when force or pressure is removed.

ACKNOWLEDGMENTS

The production of a book of this nature would not be possible without the cooperation of the Automotive Industry. In preparing the manuscript for AUTO MECHANICS FUNDAMENTALS, the industry has been most cooperative. The author acknowledges the cooperation of these companies with great appreciation:

Accurate Products, Inc., AC Spark Plug Div. of General Motors Corp., Aeroquip Corp., Air Lift Co., Air Reduction, Albertson and Co., Alemite Div. of Stewart-Warner, Alfa Romeo Cars, Allen Test Products, Alondra, Inc., Aluminum Co. of America, A.L.C. Co., American Brake Shoe Co., American Bosch Arma Corp., American Hammered Automotive Replacement Div., American Iron and Steel Institute, American Manufacturers Assn., American Motors Corp., American Optical Co., American Standards Assn., Inc., Ammco Tools, Inc., Anti-Friction Bearing Manufacturers Assn., Inc., AP Parts Corp., Armstrong Patents Co., Ltd., Armstrong Tool Co., Arnolt Corp., Audi, Automotive Electric Assn., Automotive Service Industry Assn., Baldwin, J. A., Mfg. Co., Barbee Co., Inc., Battery Council International, Bear Mfg. Co., Belden Mfg. Co., Bendix Automotive Service Div. of Bendix Corp., Bethlehem Steel Co., Big Four Industries, Inc., Binks Mfg. Co., Black and Decker Mfg. Co., Blackhawk Mfg. Co., Bonney Forge and Tool Works, Borg & Beck, Borg Warner Corp., Bosch, Robert, Corp., Branick Mfg. Co., Inc., Breeze Corp., Inc., Bremen Bearing Co., British Leyland Motors, Inc., British Motor Corp. — Hambro, Inc., Brown and Sharpe, Indus. Prod. Div., Cadillac Div. of General Motors Corp., Carter Div. of ACF Industries, Inc., Cedar Rapids Eng. Co., Champion Pneumatic Machinery Co., Champion Spark Plug Co., Chevrolet Div. of General Motors Corp., Chicago Rawhide Mfg. Co., Chrysler Plymouth Div. of Chrysler Corp., Citroen Cars Corp., Cleveland Graphite, Bronze Div. of Clevite Corp., Clevite Service Div. of Clevite Corp., Cole-Hersee Co., Colt Industries, Continental Motors Corp., Cooper Tire and Rubber Co., Cornell, William Co., Corning, Cox Instrument, Cummins Engine Co., Inc., Dana Corp., Datson, Deere & Co., Delco-Remy Div. of General Motors Corp., DeVilbiss Co., Dodge Div. of Chrysler Corp., Dole Valve Co., Dow Corning Corp., Dual Drive, Inc., Duff-Norton, Dura-Bond Engine Parts Co., Eaton Corp., Echlin Mfg. Co., Edelmann, E., and Co., E. I. du Pont de Nemours and Co., Inc., EIS Automotive Div., ESB Brands, Inc., Ethyl Corp., Eutectic Welding Alloys Corp., Exxon Company USA, Fafnir Bearing Co., FAG Bearing, Ltd., Federal-Mogul Corp., Fel-Pro, Inc., Ferrari Cars, Fiat, Firestone Tire and Rubber Co., Fiske Brothers Refining Co., FMC Corp., Ford Div. of Ford Motor Co., Gates Rubber Co., Gatke Corp., General Electric, Girling Ltd., Globe Hoist Co., GMC Truck and Coach Div. of General Motors Corp., Goodall Mfg. Co., Goodrich Co., Goodyear Tire and Rubber Co., Gould Inc., Gray Co., Inc., Graymills Corp., Grey-Rock Div. of Raybestos-Manhattan, Inc., Guide Lamp Div. of General Motors, Gulf Oil Corp., Gunite Foundries Div. of Kelsey-Hayes Co., Gunk Chemical Div. of Radiator Specialty Co., Halibrand Eng. Corp., Harrison Radiator Div. of General Motors, Hastings Mfg. Co., Hein-Werner Corp., H. K. Porter, Inc., Homestead Industries, Inc., Honda, Hub City Iron Co., Huck Mfg. Co., Hunter Eng. Co., Hydramatic Division of General Motors, Ideal Corp.,

Ignition Manufacturers Inst., Imperial Eastman Corp., Inland Mfg. Co., International Harvester Co., Iskenderian Racing Cams, Jaguar Cars, Ltd., Johnson Bronze Co., Johns-Manville, Kal-Equip. Co., K-D Mfg. Co., Kelly-Springfield Tire Co., Kelsey-Hayes Co., Kent Moore Org., Kester Solder Co., Kleer-Flo Co., K. O. Lee Co., Land-Rover, Leece-Neville Co., Lenroc Co., Lincoln Electric Co., Lincoln Eng. Co., Lincoln-Mercury Div. of Ford Motor Co., Lucas, Joseph, Ltd., Lufkin Rule Co., Mack Trucks, Inc., MacMillan Petroleum Corp., Magnaflux Corp., Mansfield Tire & Rubber Co., Marquette Corp., Martin Senour Paints, Marvel-Schebler Products Div. of Borg-Warner Corp., Maserati, Mazda, McCord Corp., Mercedes-Benz, Merit Industries, Inc., Midland-Ross Corp., Mobil Oil Corp., Monroe Auto Equipment Co., Moog Industries, Inc., Morton-Norwich Products, Inc., Motorola Automotive Products, Inc., Motor Wheel Corp., Muskegon Piston Ring Co., N. A. P. A. Micro Test, National Board of Fire Underwriters, Nice Ball Bearing Co., Nicholson File Co., Nissan, Nugier, F. A., Co., Oakite Products, Inc., Oldsmobile Div. of General Motors Corp., Owatonna Tool Co., Packard Electric, P and G Mfg. Co., Paxton Products, Pennsylvania Refining Co., Perfect Circle Corp., Permatex Co., Inc., Peugeot, Inc., Pontiac Div. of General Motors Corp., Porsche, Porter, H. K., Inc., Prestolite Co., Proto Tool Co., Purolator Products, Inc., Questor, Raybestos Div. of Raybestos-Manhattan, Inc., Rinck-McIlwaine, Inc., Renault, Rochester Division of General Motors, Rockford Clutch Div. of Borg-Warner Corp., Rootes Motors, Inc., Rottler Boring Bar Co., Rubber Manufacturers Assn., Saginaw Steering Gear, Salisbury Corp., Schrader Div. of Scovill Mfg. Co., Inc., Sealed Power Corp., Shell Oil Co., Sherwin-Williams Co., SKF Industries, Inc., Skil, Slep Electronics, Snap-on Tools Corp., Society of Automotive Eng., Inc., Solex Ltd., Sornberger Equip. Sales, South Bend Lathe, Inc., Spicer, Standard Motor Products, Standard Oil Co. of Calif., Standard-Thomson Corp., Stant Mfg. Co., Inc., Star Machine and Tool Co., Starrett, L. S., Co., Stemco Mfg. Co., Storm-Vulcan, Inc., Straza Industries, Sturtevant, P. A., Co., Subaru, Sun Electric Corp., Sunnen Products Co., Testing Systems, Inc., Texaco, Inc., Thompson Products Replacement Div. of Thompson-Ramo-Wooldridge, Inc., Thor Power Tool Co., 3-M Company, Timken Roller Bearing Co., Toyota, Traction Master Co., Trucut (Frank Wood and Co.), Union Carbide Corp., Uniroyal, Inc., United Parts Div. of Echlin Mfg. Co., U.S. Cleaner Corp., United Tool Processes Corp., Utica-Herbrand Div. of Kelsey-Hayes Co., Valvoline Oil Co., Van Norman Machine Co., Vellumoid Co., Victor Mfg. And Gasket Co., Volkswagen of America, Inc., Wagner Electric Corp., Walbro, Walker Mfg. Co., Warner-Gear-Warner Motive, Weatherhead Co., Weaver Mfg. Div. of Dura Corp., White Engine Co., Williams, J. H., and Co., Wilton Corp., Wix Corp., World Bestos Div. of the Firestone Tire and Rubber Co., Wudel Mfg. Co., Young Radiator Co.

INDEX